# Selected Titles in This Series

62 **János Kollár, Robert Lazarsfeld, and David R. Morrison, Editors,** Algebraic geometry—Santa Cruz 1995 (University of California, Santa Cruz, July 1995)

61 **T. N. Bailey and A. W. Knapp, Editors,** Representation theory and automorphic forms (International Centre for Mathematical Sciences, Edinburgh, Scotland, March 1996)

60 **David Jerison, I. M. Singer, and Daniel W. Stroock, Editors,** The legacy of Norbert Wiener: A centennial symposium (Massachusetts Institute of Technology, Cambridge, October 1994)

59 **William Arveson, Thomas Branson, and Irving Segal, Editors,** Quantization, nonlinear partial differential equations, and operator algebra (Massachusetts Institute of Technology, Cambridge, June 1994)

58 **Bill Jacob and Alex Rosenberg, Editors,** $K$-theory and algebraic geometry: Connections with quadratic forms and division algebras (University of California, Santa Barbara, July 1992)

57 **Michael C. Cranston and Mark A. Pinsky, Editors,** Stochastic analysis (Cornell University, Ithaca, July 1993)

56 **William J. Haboush and Brian J. Parshall, Editors,** Algebraic groups and their generalizations (Pennsylvania State University, University Park, July 1991)

55 **Uwe Jannsen, Steven L. Kleiman, and Jean-Pierre Serre, Editors,** Motives (University of Washington, Seattle, July/August 1991)

54 **Robert Greene and S. T. Yau, Editors,** Differential geometry (University of California, Los Angeles, July 1990)

53 **James A. Carlson, C. Herbert Clemens, and David R. Morrison, Editors,** Complex geometry and Lie theory (Sundance, Utah, May 1989)

52 **Eric Bedford, John P. D'Angelo, Robert E. Greene, and Steven G. Krantz, Editors,** Several complex variables and complex geometry (University of California, Santa Cruz, July 1989)

51 **William B. Arveson and Ronald G. Douglas, Editors,** Operator theory/operator algebras and applications (University of New Hampshire, July 1988)

50 **James Glimm, John Impagliazzo, and Isadore Singer, Editors,** The legacy of John von Neumann (Hofstra University, Hempstead, New York, May/June 1988)

49 **Robert C. Gunning and Leon Ehrenpreis, Editors,** Theta functions – Bowdoin 1987 (Bowdoin College, Brunswick, Maine, July 1987)

48 **R. O. Wells, Jr., Editor,** The mathematical heritage of Hermann Weyl (Duke University, Durham, May 1987)

47 **Paul Fong, Editor,** The Arcata conference on representations of finite groups (Humboldt State University, Arcata, California, July 1986)

46 **Spencer J. Bloch, Editor,** Algebraic geometry – Bowdoin 1985 (Bowdoin College, Brunswick, Maine, July 1985)

45 **Felix E. Browder, Editor,** Nonlinear functional analysis and its applications (University of California, Berkeley, July 1983)

44 **William K. Allard and Frederick J. Almgren, Jr., Editors,** Geometric measure theory and the calculus of variations (Humboldt State University, Arcata, California, July/August 1984)

43 **François Trèves, Editor,** Pseudodifferential operators and applications (University of Notre Dame, Notre Dame, Indiana, April 1984)

42 **Anil Nerode and Richard A. Shore, Editors,** Recursion theory (Cornell University, Ithaca, New York, June/July 1982)

41 **Yum-Tong Siu, Editor,** Complex analysis of several variables (Madison, Wisconsin, April 1982)

40 **Peter Orlik, Editor,** Singularities (Humboldt State University, Arcata, California, July/August 1981)

39 **Felix E. Browder, Editor,** The mathematical heritage of Henri Poincaré (Indiana University, Bloomington, April 1980)

(*Continued in the back of this publication*)

# Algebraic Geometry
# Santa Cruz 1995

Proceedings of Symposia in
PURE MATHEMATICS

Volume 62, Part 2

# Algebraic Geometry
# Santa Cruz 1995

Summer Research Institute on
Algebraic Geometry
July 9—29, 1995
University of California, Santa Cruz

János Kollár
Robert Lazarsfeld
David R. Morrison
Editors

**American Mathematical Society**
Providence, Rhode Island

PROCEEDINGS OF A SUMMER RESEARCH INSTITUTE ON ALGEBRAIC GEOMETRY HELD AT THE UNIVERSITY OF CALIFORNIA, SANTA CRUZ JULY 9–29, 1995

with support from the
National Science Foundation, Grant DMS 9414979.

---

Any opinions, findings, and conclusions or recommendations expressed in this material are those of the authors and do not necessarily reflect the views of the National Science Foundation.

---

1991 *Mathematics Subject Classification.* Primary 14–XX.

---

**Library of Congress Cataloging-in-Publication Data**

Summer Research Institute on Algebraic Geometry (1995 : University of California, Santa Cruz)
Algebraic geometry—Santa Cruz 1995 : Summer Research Institute on Algebraic Geometry, July 9–29, 1995, University of California, Santa Cruz / János Kollár, Robert Lazarsfeld, David R. Morrison, editors.
p. cm. — (Proceedings of symposia in pure mathematics, ISSN 0082-0717 ; v. 62)
Includes bibliographical references.
ISBN 0-8218-0493-6 (set).—ISBN 0-8218-0894-X (pt. 1).—ISBN 0-8218-0895-8 (pt. 2)
1. Geometry, Algebraic—Congresses. I. Kollár, János. II. Lazarsfeld, R. (Robert) III. Morrison, David R., 1955– . IV. Title. V. Series.
QA564.S86 1995 pt. 2
516.3′5—dc21

97-24634
CIP

---



10 9 8 7 6 5 4 3 2 1 02 01 00 99 98 97

# Contents

## Part 1

## VI. Fundamental Groups and Non-Abelian Hodge Theory



## VII. Complex Geometry



## VIII. Toric Geometry

# Preface

The 1995 AMS Summer Research Institute on Algebraic Geometry was held at the University of California, Santa Cruz, from July 9 through July 29, 1995. The organizing committee consisted of Henri Gillet, János Kollár, Robert Lazarsfeld (chairman), Robert MacPherson, David Morrison and Yum-Tong Siu. The present volumes contain write-ups of all the plenary lecture series with the exception of E. Witten's course on duality in supersymmetric gauge theories[1] and A. J. de Jong's lectures on alterations of singularities[2]. They also contain a number of survey articles growing out of seminar talks.

One of the main thrusts of the Institute was to provide an introduction to some of the exciting new areas and directions that have recently opened up in geometrically oriented algebraic geometry. At the same time, the conference provided a forum for the dissemination of important developments in the more classical branches of the field. A summary of the complete scientific program appears on the following pages. We hope that the papers in these Proceedings will give a sense of the breadth and vitality of algebraic geometry as it emerged at Santa Cruz. However the editors are well aware that, in spite of their best efforts, important areas remain under-represented in these volumes.

Finally, on behalf of the whole organizing committee, we would like to thank Chris Harkness and the AMS for their work in the day-to-day organization of the Institute, David Cox and David Perkinson for organizing the graduate student seminar, and the seminar organizers and speakers for their contributions to the scientific program. But we'd particularly like to express our gratitude to all the participants, whose good humor and enthusiasm ensured the success of the Institute.

JÁNOS KOLLÁR, ROBERT LAZARSFELD, DAVID R. MORRISON

[1] A much expanded version of Witten's lecture series was delivered at the Institute for Advanced Study during 1996-97, and is being written up for publication.

[2] de Jong's lectures reported on A. J. de Jong, Smoothness, semi-stability and alterations. Inst. Hautes Etudes Sci. Publ. Math. No. 83 (1996), 51–93.

# Scientific Program of the Institute

## FIRST WEEK

### Plenary Lecture Series and Mini-Courses

C. LeBrun, An introduction to twistors
C. Simpson, Hodge theory and fundamental groups
E. Witten, Duality in supersymmetric guage theories

### Seminar and Contributed Lectures

V. Alexeev, On the automorphisms of surfaces of log-general type
J. Amorós, On nonfibered Kähler groups
D. Arapura, Solvable Kähler groups
P. Aspinwall, Quantum algebraic geometry and K3 surfaces
L. Barbieri-Viale, Bloch's conjecture revisited
V. Batyrev, Mirror symmetry and toric geometry
T. Bauer, Higher order embeddings of abelian varieties
F. Catanese, Images of canonical surfaces in three-space and nodal surfaces
A. Corti, Some ideas and problems in the Mori program and the Sarkisov program
O. Debarre, Subvarieties with minimal class on Jacobians
A. Dickenstein, Residues in toric varieties
I. Dolgachev, Mirror symmetry for K3 surfaces
H. Esnault, Higher Chern classes of flat bundles in the Deligne bundle are torsion on a complex projective manifold (after A. Reznikov)
R. Friedman, Algebraic surfaces and Seiberg–Witten invariants
M. Green, What comes after the Abel–Jacobi map?
M. Gross, Deforming Calabi–Yau threefolds
Z. Guan, Classification of almost homogeneous spaces
R. Hain, Hodge theory and mapping class group
R. Hartshorne, Classification of space curves and Zeuthen's problem
Y. Hu, Relative GIT and universal moduli spaces
K. Hulek, Moduli of abelian and K3 surfaces
A. Hwang, Examples of Kähler metrics with constant scalar curvature
E. Izadi, Cubic hypersurfaces
H. Kaji, On the homogeneous projective vaarieties with degenerate secants
A. Kaplan, Hodge cycles at infinity
Y. Karphishpan, Riemann reciprocity in higher dimensions

S. Katz, Mirror symmetry and curves of higher genus
L. Katzarkov, On the Shafarevich maps
K. Konno, The slope of fibered surfaces
H. Lange, Quadrics containing a Prym-canonical curve
J. Li, Donaldson invariants and virtual moduli cycle
N. Mohan Kumar, Deformation of rank two vector bundles on $\mathbb{P}^3$
D. Morrison, Black holes and Calabi–Yau manifolds
D. Morrison, Evidence for mirror symmetry from toric enumerative geometry
T. Napier, Covering spaces of families of curves
K. O'Grady, Moduli of vector bundles on surfaces: some global questions
J. Rabin, Supercurves and their Jacobians
M. Ramachandran, Subvarieties, local systems and harmonic maps
M. Reid, Freedom and very ampleness
Y. Ruan, Higher genus pseudo-holomorphic curves
T. Szemberg, Primitive line bundles and higher order embeddings
M. Teicher, Braid groups and algebraic surfaces
D. Toledo, Projective varieties with non-residually finite fundamental groups
J. Wahl, Cohomology of the square of an ideal sheaf and geometry of canonical curves
P.H.M. Wilson, Symplectic classes on Calabi–Yau threefolds

# SECOND WEEK

## Plenary Lecture Series and Mini-Courses

J.-P. Demailly, Hyperbolicity of projective algebraic varieties and jet bundles
W. Fulton, Enumerative geometry via quantum cohomology: an introduction to the work of Kontsevich and Manin
J. Kollár, Higher dimensional geometry and its applications

## Seminar and Contributed Lectures

M. Andreatta, Extremal contractions from a 4-fold
C. Bartocci, Fourier–Mukai transform and moduli spaces of stable heaves on $K3$ surfaces
A. Beauville, Recent progress in moduli spaces of vector bundles on curves
A. Bertram, Quantum Schubert calculus
I. Biswas, Hodge cycles on moduli spaces of vector bundles on curves
R. Brussee, The canonical and the Seiberg–Witten classes of a Kähler manifold
L. Caporaso, Rational curves on rational ruled surfaces
J. Cheah, Hilbert function of the Hilbert scheme of points
C. Ciliberto, On the bicanonical map for surfaces of general type
E. Colombo, Families of abelian varieties on a complete smooth curve
G. Dethloff, Iitaka–Severi conjecture for complex 3-folds
R. Donagi, Seiberg–Witten integrable systems
L. Ein, Recent work on linear series
H. Esnault, An algebraic invariant of flat bundles
R. Hain, $C - C^{-}$

A. Kabanov, The second cohomology of the moduli space of algebraic curves with symplectic coefficients
Y. Kachi, Flips from certain fibered 4-folds
N. Katz, Informal report on rigid local systems
Y. Kawamata, On Fujita's freeness conjecture
S. Kleiman, Gorenstein algebras, symmetric matrices, self-linked ideals, and symbolic powers
S. Kovács, Smooth families over rational and elliptic curves
H. Kurke, Framed vector bundles and framed local jumps
J. M. Landsberg, Differential-geometric characterizations of complete intersections
R. Lazarsfeld, Local positivity of ample line bundles
E. Looijenga, Cohomology of moduli spaces of curves
J. McKernan, Rational curves on quasi-projective surfaces
Y. Miyaoka, Vector fields on Calabi–Yau manifolds in characteristic $p$
R. Morelli, Counting lattice points in polytopes
S. Mori, Quotients by groupoids
S. Müller-Stach, Higher Chow groups of surfaces
I. Nieto, The singular $H_{2,2}$-invariant quartic surfaces in $\mathbb{P}_3$
M. Ohno, On degenerate secant varieties which have the smallest contact loci with their tangent spaces
R. Pandharipande, The canonical class of $\overline{\mathcal{M}}_{0,n}(\mathbb{P}^r, d)$
T. Pantev, Motivic local systems on curves
U. Persson, Geography and moduli of surfaces: new insights
M. Pikaart, On the stable cohomology of $M_g$ (and $\overline{M}_g$)
S. Popescu, Calabi–Yau 3-folds and moduli of abelian surfaces
E. Previato, Poncelet's theorem in space
M. S. Ravi, Some applications of algebraic geometry to systems and control theory
M. Reid, Some flips
A. Rudakov, Classification of stable vector bundles on Del Pezzo surfaces
B. Runge, On complex Shimura varieties
F.-O. Schreyer, Surfaces in $\mathbb{P}^4$ with applications
C. Schuhmann, Mapping threefolds onto three-quadrics
K. Smith, Fujita conjecture in positive characteristic
E. Szabó, Automorphisms of varieties of general type
M. Thaddeus, Birational approximation of moduli spaces
B. Totaro, The cohomology of configuration spaces, and Fulton–MacPherson's compactification
I.-H. Tsai, Varieties dominated by compact Hermitian symmetric spaces
E. Viehweg, Moduli of polarized manifolds and normal varieties with weak singularities
V. Vinnikov, Commuting nonselfadjoint operators and algebraic curves
G. Xu, On the complement of a generic curve in the projective plane
A. Yekutieli, Adeles, Chern classes and the residue complex
D.-Q. Zhang, Hypersurfaces on Fano 3-folds with log terminal singularities
S. Zucker, Hodge theory and spectres on locally symmetric varieties

# THIRD WEEK

## Plenary Lecture Series and Mini-Courses

S. Bloch, Motives and motivic cohomology
J. de Jong, Dominating varieties by smooth varieties
C. Soulé, Hermitian vector bundles on arithmetic varieties

## Seminar and Contributed Lectures

D. Abramovich, Uniformity of points on curves
J.-B. Bost, Canonical polygons of hermitian vector bundles and transcendence methods
R.-O. Buchweitz, Hodge modules of weighted homogeneous complete intersections
J. Burgos Gil, Arithmetic $K$-theory and Beilinson regulator
M.-C. Chang, Asymptotics on Hilbert schemes and moduli spaces
W. Cherry, Non-Archimedean analytic curves in abelian varieties
A. Collino, Torelli for the infinitesimal invariant of $C - C^{-}$ and $K$-theory
J.-L. Colliot-Thélène, Zero-cycles on varieties over $p$-adic fields
M. Cook, Using generic initial ideals to lower the bound of degree of surfaces not of general type in $\mathbb{P}^4$
H. D'Souza, A complete classification theorem for a class of 3-folds
H. Gillet, Motives and the weight filtration on $\mathbb{Z}$-cohomology
M. Grinberg, Nearby cycles in linear algebra
S. Gusein-Zade, The Arf-invariant and the Arnold invariants of plane curves
B. Hassett, Uniformity of points on surfaces of general type
H. Hauser, Initial ideals and resolution of surface singularities
R. Hoobler, Étale cohomology of geometric local rings is Galois cohomology
E. Hrushovski, Model theory and the Mordell–Lang conjecture
T. Iarrobino, Interpolation problem at points of $\mathbb{P}^n$ and the $s$-secant variety of the Veronese embedding
S. Ishii, The canonical modifications by weighted blowing ups
P. Jaworski, On the topological triviality along the moduli of deformations of quasi-homogeneous singularities
M. Kapranov, Hecke operators on moduli space of vector bundles on an algebraic surface
N. Katz, What we do and don't know about curves over finite fields
S. Katz, Schubert: a Maple package for intersection theory
S. Kimura, Finite dimensionality of Chow motives and Bloch's conjecture
R. Kobayashi, Value distribution theory and algebraic geometry
D. McLaughlin, Reciprocity
G. Megyesi, Singularities of configurations of curves and a generalization of the Miyaoka–Yau inequality
B. Moonen, Linearity properties of Shimura varieties
A. Moriwaki, Bogomolov conjecture over function fields
M. Nakamaye, A new proof of Dyson's lemma
J. Nekovář, Syntomic cohomology, $p$-adic regulators and Chow groups
A. Némethi, The mixed Hodge structure of an ICIS
Y. Nisnevich, A generalization of Quillen's work on the Gersten conjecture and algebraic cycles onto singular varieties

M. Nori, Torsion and divisibility of algebraic cycles
K. Pardue, Free resolutions and deformation classes of graded modules
T. Peternell, The role of the tangent bundle in the classification theory of algebraic varieties
R. Pink, Compactification of the moduli space of Drinfeld modules
M. Rapoport, P-adic period domains
W. Raskind, Higher Abel–Jacobi maps and filtrations on Chow groups
C. Schoen, On the $\ell$-adic Abel–Jacobi map
N. Shepherd-Barron, Icosahedral Galois representations, mod 2 and 5
V. Shokurov, A projectivity criterion
M. Singer, Computational questions in differential Galois theory
Y.-T. Siu, Hyperbolicity of complements of plane curves
F. Sottile, Enumerative geometry of the real Grassmannian
V. Srinivas, Torsion zero cycles on singular varieties
J. Stevens, Degenerations of elliptic curves and cusp singularities
M. Stillman, MACAULAY – The new generation
B. Sturmfels, Defining equations of toric varieties
P. Vojta, Roth's theorem with moving targets
C. Weibel, Roitman's theorem for singular complex surfaces
S. Zhang, Equidistributions for torsion points and small points
S. Zhang, Heights of Heegner cycles and derivatives of L-series

# V. Quantum Cohomology and Connections with Physics

Proceedings of Symposia in Pure Mathematics
Volume 62.2, 1997

# Seiberg-Witten Integrable Systems

Ron Y. Donagi

## Contents



---

1991 Mathematics Subject Classification. Primary: 81T13, 81T60, 58F07.
Secondary: 14D20, 14H40.
Work partially supported by NSF grant DMS95-03249, and by grants from the Univesity of Pennsylvania Research Foundation and The Harmon Duncombe Foundation.

## Introduction

While many mathematicians are aware of the monopole equations [**W**] and of the impact they have had in topology, probably not very many are as familiar with the works [**SW1, SW2**] of Seiberg and Witten on which the monopole equations are based. The appearance of these works was a major breakthrough in Quantum Field Theory (QFT), and has to a large extent revolutionized and transformed the subject. The actual achievement in [**SW1**] was an *exact* solution to a problem in QFT (pure $N=2$ supersymmetric Yang-Mills in four dimensions with gauge group $SU(2)$) which could traditionally be solved only approximately, by computing the first few terms in a perturbative expansion. The solution is close to algebraic geometry both in its details (involving families of elliptic curves) and in its general approach, which is to use global properties (monodromy, symmetries) of a physical theory to uniquely characterize a geometric model of the theory. This work opened the doors to solutions, along similar lines, of many specific quantum theories, some of which are reviewed here. It also led directly to the ongoing revolution in string theory involving string dualities and the appearance of higher dimensional objects (branes) on more or less equal footing with the strings themselves.

The purpose of these notes is to try to make this exciting and rapidly evolving chapter of high energy physics accessible to mathematicians, particularly to algebraic geometers. I have had to make some choices: I decided to focus on the original work and its further development within QFT, thus omitting the spectacular (but well documented) applications to topology and strings. Even so, it was impossible to render with mathematical rigor the enormous amount of background material needed from the physics. I opted instead, in §1, for the telling of one uniform but very imprecise story of the physics. My hope is that it will afford, to a mathematical reader with no prior knowledge of the physics, at least a general sense of the physical theory, of how the different parts interact, and of what some of the basic concepts mean. The remaining, mathematical, sections of these notes discuss what is now known (and what is still missing) about Seiberg-Witten type solutions of four-dimensional $N = 2$ supersymmetric Yang-Mills theories in algebro-geometric terms, with the emphasis on an interpretation via integrable systems.

Section 1 starts with Maxwell's classical description of electromagnetism. We arrive at our central object of study, the duality in super Yang-Mills theory, after discussing separately the three directions in which Maxwell's theory needs to be extended: nonabelian gauge groups, quantization, and supersymmetry. We are then ready for an overview of the Seiberg-Witten solution, [**SW1**], emphasizing its algebro-geometric elements. The section concludes with a discussion of the other theories which could be exactly solvable by algebro-geometric techniques. These theories occupy us in the remainder of these notes.

Algebraically integrable systems are quickly reviewed in §2. The upshot is that all the structures arising in the Seiberg-Witten physics of §1 are naturally realized here, so algebraically integrable systems are the natural context for describing the physics (cf. [**DW**], [**F**]). We also clarify the role of the Seiberg-Witten differential, which occurs naturally in any integrable system of *spectral* type. One important feature of this differential, the linear dependence of its residues on the mass parameters, turns out to be part of a very general phenomenon. We state and outline a proof of an apparently new result, which is a complex analogue of the theorem of Duistermaat-Heckman [**DH**] on variation of the symplectic structure in a family of symplectic quotients. The complex case is somewhat subtler than the compact version due to the presence of non-semisimple (e.g. nilpotent) orbits.

Section 3 is mostly an attempt to rewrite the solution [**DW**] of the SU($n$) theory with adjoint matter, with an emphasis on the math rather than the physics. We review the work of Markman [**Mn**] and others on the integrable systems of meromorphic Higgs bundles; among these systems we pick the particular subsystem we need. The system (i.e. its spectral curves) is described explicitly, then compared to the elliptic soliton description [**T, TV**]. A major part of [**DW**] conisted of "tests" showing that this system accurately reproduces physical properties of the theory; we describe some of these "tests" here.

In the final section we discuss the other major class of integrable systems which is known to solve the pure $N = 2$ SYM theories: the periodic Toda lattices [**MW1**]. In particular, we compare the various forms assumed by the spectral curves. We then move to a common generalization of both systems, the elliptic Calogero-Moser systems, which were proposed in [**Mc**] as a solution of the theory with adjoint matter and arbitrary gauge group. This seems quite promising, but it still lacks a satisfactory algebro-geometric description. We also mention briefly some other theories where at least the family of spectral curves is known, and conclude by mentioning some recent ideas in string theory which exhibit the QFT integrable systems discussed here as limits of stringy systems.

It is a pleasure to acknowledge pleasant and useful conversations, about some of the topics covered here, with D. Freed, L. Jeffrey, A. Kirillov, D. Lowe, N. Nekrasov, R. Plesser, N. Seiberg, S. Singer, R. Sjammar, J. Weitsman, and E. Witten, as well as helpful comments on the manuscript by D. Freed, A. Grassi, A. Ksir, N. Nekrasov, M. Verbitsky and E. Witten.

## §1. Some Physics Background

### Electromagnetism

The electric field $E$ and the magnetic field $B$ in $\mathbb{R}^3$ are governed by Maxwell's equations, of T-shirt fame. Think of $E$ and $B$, respectively, as a 1-form and a 2-form on $\mathbb{R}^3$. Both are time-dependent, so we combine them into the electromagnetic field $F := E \wedge dt + B$, a 2-form on $\mathbb{R}^4$. Maxwell's equations in vacuo are:

$$dF = 0 \qquad\qquad d * F = 0.$$

Here $*$ is the Hodge star operator on forms on $\mathbb{R}^4$, determined by the Minkowski metric $g = dx^2 + dy^2 + dz^2 - c^2dt^2$. (The speed of light $c$ is usually set to 1 by adjusting units.)

The same equations make sense for an unknown 2-form $F$ on any (pseudo) Riemannian manifold $(X, g)$. They are clearly invariant under the group of global isometries of $(X, g)$. When $\dim X = 4$, the equations are also invariant under the duality $F \mapsto *F$. When $X = \mathbb{R}^4$ with the Minkowski metric, this duality interchanges the electric and magnetic fields. The group of isometries in this case is the Poincare group, an extension of the Lorentz group $SO(3,1)$ of rotations by the group $(\mathbb{R}^4, +)$ of translations.

Geometrically, we fix a $U(1)$-bundle $L$ on $X$ and consider a connection $A$ on $L$ as a new variable. The first equation $dF = 0$ is interpreted as saying that (locally) $F = dA$ is the curvature. The remaining equation then is $d * dA = 0$. The group of isometries of the triple $(X, g, L)$ acts on the connections and preserves the equation. This is an extension of the group of those isometries of $(X, g)$ which preserve $L$ by the *gauge group* of $C^\infty$ maps from $X$ to $U(1)$. The duality is still there, but is now hidden.

Now impose an external field $j$. (In $\mathbb{R}^4$, $j$ is a 3-form whose $dx \wedge dy \wedge dz$ component is the charge density, while the other 3 components give the current.) The equations become

$$dF = 0, \qquad d * F = j,$$

and duality is lost. Geometrically, once we think of $F$ as curvature, the first equation becomes a tautology (the Bianchi identity), while the second is an extra condition which may or may not hold. Physically, we explain this asymmetry as due to the absence of magnetic charges ("monopoles").

Dirac proposed that some magnetic charges can be introduced by changing the topology, or allowing singularities. For example, a monopole at the origin of $\mathbb{R}^3$ is represented by a $U(1)$-bundle on $(\mathbb{R}^3_{x,y,z} \setminus 0) \times \mathbb{R}_t$. Such bundles are characterized topologically by their Chern class $c_1 \in \mathbb{Z}$, and this $c_1$ is interpreted as the magnetic charge at the origin. Indeed, the "magnetic charge" at the origin is just the total magnetic flux across a small sphere $S^2$ around the origin in $\mathbb{R}^3$; this is the integral over $S^2$ of $B$, which equals the integral of $F$, which is $c_1$. We can extend $F$ to $\mathbb{R}^4$ as a distribution, but now $dF$ is the $\delta$-function $c_1.\delta_{x,y,z}$ instead of $dF = 0$. This is not yet sufficient to restore the symmetry to Maxwell's equations: the electric charge is arbitrary, but the magnetic charge is allowed only in discrete "quanta". This does raise the possibility, though, that complete symmetry could be restored in a quantum theory, where, as in the real world, electric charge also comes in discrete units. Indeed, Dirac discovered that in quantum mechanics, where an electrically charged particle is represented by a wave function satisfying Schrodinger's equation, the electric charge $e$ of one particle and the magnetic charge $g$ of another (in appropriate units) satisfy $eg \in \mathbb{Z}$ . This has the curious interpretation that all electric charges are automatically quantized as soon as one magnetic monopole exists somewhere in the universe.

In [**MO**], Montonen and Olive conjectured that in certain supersymmetric Yang-Mills quantum field theories the electric-magnetic duality is indeed restored. This involves an extension of the basic picture in three separate directions: replacing the "abelian" Maxwell equations by the non-abelian Yang-Mills; quantizing the classical theory; and adding supersymmetry to the Poincaré invariance.

**Yang-Mills theory**

Yang-Mills theory is quite familiar to mathematicians. It involves replacing the structure group $U(1)$ by an arbitrary reductive group $G$. We thus have a connection $A$ on some $G$-bundle $V$ over our fourfold $X$, the curvature is now the $\mathrm{ad}(V)$-valued two-form $dA + \frac{1}{2}[A, A]$, and the equations are as before,$dF = 0$, $d * F = 0$, except that the covariant derivative $d = d_A$ on $\mathrm{ad}(V)$ now replaces the exterior derivative $d$ on the trivial bundle $\mathrm{ad}(L)$.

These equations can be, and usually are, considered as variational equations for the Lagrangian

$$L(A) := \int_X \mathrm{Tr}(F \wedge *F),$$

where the trace Tr can be any non-degenerate invariant pairing on the Lie algebra $\mathfrak{g}$ of $G$. This Lagrangian is considered as a functional on an appropriate space of connections $A$ (if $X$ is not compact, we must impose some decay condition at infinity for the integral to converge). The equation $dF = 0$ is again the Bianchi identity, while $d * F = 0$ is the equation of the critical locus of $L$. Either from the Lagrangian description or directly from the equations, it is clear that the symmetry group includes global isometries of $X$ which lift to $V$ ("the Poincare group") as well as local diffeomorphisms of the bundle ("the gauge group") and, in dimension 4, the duality $F \mapsto *F$. In the original model of Yang and Mills, the group $G$ was SU(2), while in QCD it is SU(3). For any "realistic" model of all of particle physics, $G$ would need to contain at least $U(1) \times \mathrm{SU}(2) \times \mathrm{SU}(3)$.

**Quantization**

Quantization is an art form which, when applied to classical physical theories, yields predictions of subatomic behavior which are in spectacular agreement with experiments. The quantization of the classical mechanics of particles leads to quantum mechanics, which can be described, in a Hamiltonian approach, via operators and rays in Hilbert space, or in a Lagrangian formulation, via path integrals. Quantization of continuous fields leads to QFT: quantum field theory. It has a Hamiltonian formulation, where the Hilbert space is replaced by the "larger" Fock space, as well as a Lagrangian formulation, based on averaging the action (= exp of the Lagrangian) over "all possible histories" via path integrals. These are generally ill-defined integrals over infinite-dimensional spaces. Nevertheless, there exist powerful tools for expanding them perturbatively, as power series in some "small" parameters such as Planck's constant, around a point corresponding to a "free" limit where the integral becomes Gaussian and can be assigned a value. (Actually, in modern QFT the "small parameters" used are often taken to be dimensionless coupling constants rather than Planck's constant.) The coefficients of the expansion are finite dimensional integrals encoded combinatorially as Feynman diagrams. These power series typically diverge and need to be "regularized" and "renormalized". These processes are somewhat akin to the way we make the Weierstrass $\wp$-function converge, by subtracting a correction term from each summand in an infinite series: a first attempt to write down a doubly periodic meromorphic function in terms of an infinite series of functions leads, unfortunately, to an everywhere divergent series; nevertheless we can obtain a series which does converge, and indeed to a doubly periodic meromorphic function, by subtracting an appropriate constant from each term. A physicist might say that we "cancelled the infinity" of the original series by subtracting "an infinite constant". In any case, it is some of these renormalized values which are confirmed to amazing accuracy by experiment, providing physicists with unshakable confidence in the validity of the technique.

For a physicist, a "theory", either classical or quantum, is usually specified by its Lagrangian $L$, a functional given by integration over the underlying space $X$ of a Lagrangian density $\mathcal{L}$, which is a local expression involving the values and derivatives of a given collection of fields. The classical theory considers extrema of $L$, while the quantum theory averages the action over all possible fields. The largest weight is thus still assigned to the extrema of $L$, but all "quantum fluctuations" around the extrema are now included. The Lagrangian density typically consists of a quadratic part, plus higher order terms which in a perturbative approach are interpreted as small perturbations. In a theory consisting of scalar fields $\varphi_i$, the quadratic term might be $\sum_i(|d\varphi_i|^2 + m_i^2\varphi_i^2)$, where $m_i$ are the masses. A perturbation might include terms like $g_I\varphi^I$, where $I = (i_1, \ldots, i_k)$ is a multiindex, and $g_I$ is a (supposedly small) *coupling constant* giving the strength of an interaction involving the $k$ fields $\varphi_i, i \in I$. (Other types of theories might involve also some fields which are vectors or spinors with respect to the Poincare group.) A typical quantity to be computed is the n-point function

$$\langle x_1, \ldots, x_n \rangle := \int \varphi(x_1) \ldots \varphi(x_n) \exp iL(\varphi) D\varphi,$$

where integration is over the infinite-dimensional space of all $\varphi$'s, with respect to a (non-existent) measure $D\varphi$, and $L(\varphi) := \int_X \mathcal{L}(\varphi)$.

A formal approach to this integral is to interpret it as a power series in the coupling constants. Terms of the expansion are described by *Feynman diagrams.* The types of edges in a diagram correspond to the types of fields in the theory. The vertex types correspond to the multiindices $I$ with non-zero $g_I$. A $\varphi^4$ term thus corresponds to a vertex where four $\varphi$ edges meet (this may represent two incoming particles which collide and scatter away), while a $\varphi\psi\bar{\psi}$ vertex can represent an electron $\psi$ annihilating a positron $\bar{\psi}$ to produce a photon $\varphi$. The *Feynman rules* assign to each diagram a $4\ell$ dimensional integral, where $\ell$ is the number of loops, i.e. first Betti number, of the diagram, and 4 is the dimension of space-time: the integration is over all ways of assigning a momentum (a 4-vector) to each oriented edge, subject to Kirchhoff's law of null-sum at each vertex. The integrand for each diagram is the product over all edges of a basic function (*the propagator*) of the momentum assigned to that edge, times a factor involving the coupling constant $g_I$ at each vertex of type $I$, and an overall numerical factor accounting for graph automorphisms. The n-point function, or rather its Fourier tranform, is then expanded as the sum of all possible diagram integrals with $n$ external legs.

One problem is that some of these Feynman integrals may be divergent. Of course, even if they converge individually, their sum may not. So the individual integrals need to be *regularized*, and the entire theory needs to be *renormalized.* A theory is said to be *superrenormalizable* if all divergences in it disappear after a finite number of its divergent diagrams are made finite by adjusting some of the parameters. In a *renormalizable* theory there are infinitely many divergent diagrams, but they can all be renormalized by iteratively adjusting the values of just a finite collection of parameters. There is an easily verifiable necessary condition for renormalizability: each field $\varphi$ carries a certain dimensionality $[\varphi]$. (This is measured in powers of the mass; time and distance are converted to mass-inverse by setting the speed of light $c$ and Planck's constant $h$ to 1.) Now $L$, which occurs as an exponent, must have no dimensionality (it is a pure number, independent of units), so each term in $\mathcal{L}$ must have dimensionality 4. This determines the dimensionality of each $g_I$. The obvious necessary condition for renormalizability is that $[g_I] \geq 0$ for all $I$: otherwise, the insertion of a vertex of type $I$ in an already divergent diagram will give a new diagram with divergence which is worse.

Similarly, a necessary condition for superrenormalizability is that $[g_I] > 0$ for all $I$, since otherwise insertions into one divergent diagram would yield others with equally bad divergence. These conditions serve to point out the very few candidates for renormalizable theories. Actually proving renormalizability in each case tends to be much harder.

For regularization, we introduce an auxiliary variable $\Lambda$ so that the integral can be interpreted as (the "limit" of) a function of $\Lambda$ going to infinity with $\Lambda$. For example, $\Lambda$ can be a momentum cutoff, meaning that the integral is carried out only over the ball of momenta $p$ satisfying $|p| \leq \Lambda$. (Or better, use some approximation of unity by smooth compactly supported cutoff functions.) Another popular variation is dimensional regularization: one writes down the expression for a given integral in a $d$-dimensional space, and observes that the answer is an analytic function in $d$ for sufficiently small $d$, typically acquiring a singularity at the relevant dimension $d = 4$. The analogue of $\Lambda$ in this version is $\exp \frac{1}{4-d}$.

The modern approach to renormalization is based on the *renormalization group flow*. Very crudely, one might illustrate this process as follows. The integral for each diagram $D$ translates into a function $f_D(m, g, p, \Lambda)$ of the masses, coupling constants, external momenta and cutoff (or whatever else we used for $\Lambda$). If the number of divergent diagrams is finite (and $\leq$ the number of parameters in $L$) then the simultaneous level sets of the corresponding $f_D$ will be (or will contain) a family of curves on which $\Lambda$ is unbounded. So we may hope to find a flow parametrized by $\Lambda$ along these curves:

$$m = m(\Lambda), g = g(\Lambda), p = p(\Lambda).$$

In a general renormalizable theory there may be infinitely many $f_D$, but each may be modified (by a quantity which is bounded as a function of $\Lambda$) so there is still a flow tangent to all of them. Instead of attempting to fix the parameters at their "bare" values and taking $\Lambda$ to infinity, renormalization is accomplished by flowing along this renormalization group flow. The effective, physically measurable value of the parameters thus varies with the scale $\Lambda$. Similar phenomena are rather common in statistical physics. For example, the electric charge of a particle in a dielectric appears to be reduced, or *screened*, with distance, as a result of the alignment of the dielectric's charges around it. In QFT this occurs even in vacuum, because of the quantum fluctuations.

A famous series of experiments at the Stanford Linear Accelerator showed that, when bombarded by very high energy electrons, a proton behaves as if it is made of three separate subparticles (quarks) which move freely within it. Yet at low energies (or equivalently, as viewed on a larger distance scale), the quarks are confined within the proton. A major success of QFT was its interpretation of these results in terms of the varying coupling constant $g$ for the strong force. The relevant theory (QCD: quantum chromodynamics) is *asymptotically free*, which means that $g \to 0$ as $\Lambda \to \infty$, and $g \to \infty$ as $\Lambda \to 0$. Thus at high energy scale $\Lambda$, the quarks behave as if $g = 0$, i.e. as in the free theory, while at large distance $\Lambda \to 0$ the force becomes stronger and is presumably sufficient to confine the quarks. (This last point is not rigorously proved; Seiberg and Witten obtained a model for confinement in the supersymmetric analogue.)

An often computable quantity which describes the behavior of a coupling constant is its *beta function*, $\beta := \Lambda \partial g / \partial \Lambda$. An asymptotically free theory corresponds to $\beta < 0$ for small values of $g$. In the boundary case where $\beta$ is identically 0, the theory is scale-invariant.

An additional complication is involved in the quantization of theories which are (that is, their Lagrangian is) gauge-invariant. The problem with gauge-invariant theories stems from the degeneracy of their Lagrangian. In classical gauge theory, the classical solutions (extrema of the action) are not isolated, but come in entire gauge-group orbits. The equivalence of the Lagrangian and Hamiltonian descriptions breaks down: this equivalence is based on a Legendre transform, which makes sense only near isolated points. In the quantized theory, the path integral needs to be taken not over all paths, but over gauge-equivalence classes. This is seen already for the free (unperturbed) theory: the value of a Gaussian integral is given by the inverse determinant of the operator in the exponent; degeneracy (or gauge invariance) implies that this determinant vanishes, and needs to be replaced by a determinant on an appropriate transversal slice or quotient. For quantized Yang-Mills, or gauge, theories, a good reference is [**FS**]. The reward for handling these additional complications is a collection of theories which are renormalizable and sometimes asymptotically free: the beta-function of the basic theory (corresponding to Maxwell's in vacuo) is negative. The coupling to matter increases beta, but it remains negative when the number of quarks is small enough.

**Supersymmetry**

Supermathematics is the habit of adding the prefix "super" to ordinary, commutative objects, to denote their sign-commutative generalization. Thus, a super space is a locally ringed topological space whose structure sheaf $\mathcal{O} = \mathcal{O}_0 \oplus \mathcal{O}_1$ is $\mathbb{Z}_2$ graded, $\mathcal{O}_0$ is central, and multiplication $\mathcal{O}_1 \times \mathcal{O}_1 \to \mathcal{O}_0$ is skew symmetric. Affine super space $\mathbb{R}^{m,n}$ is the topologcal space $\mathbb{R}^m$ with the sheaf $C^\infty(\mathbb{R}^m) \otimes \Lambda^\cdot(\mathbb{R}^n)$ and the $\mathbb{Z}_2$ grading coming from the $\mathbb{Z}$-grading of the exterior algebra. An $(m, n)$-dimensional super manifold is a super space which is locally isomorphic to $\mathbb{R}^{m,n}$ (The isomorphisms are, of course, isomorphisms of $\mathbb{Z}_2$-graded algebras, and need not preserve any $\mathbb{Z}$-grading. A typical automorphism of $\mathbb{R}^{1,2}$, with even coordinate $x$ and odd coordinates $\theta_1$, $\theta_2$, may send $x$ to $x + \theta_1\theta_2$. So the identification of $\mathcal{O}$ as $C^\infty$ tensor exterior algebras is not preserved under coordinate change.) All of calculus extends to super manifolds. The only non-obvious point is that the transformation law for differential forms involves the determinant of the transformation in the even variables, but the inverse determinant of the transformation in the odd variables, and a hybrid – the *Berezinian* – for a general transformation mixing the parities. The resulting operation of *integration* looks, in the odd directions, more like usual *differentiation*, i.e., in local coordinates $x_1, \dots, x_m, \theta_1, \dots, \theta_n$ it reads off the coefficient of the top odd part, $\Pi_{i=1}^n d\theta_i$.

The infinitesimal symmetries of a super manifold are described by a super Lie algebra: a $\mathbb{Z}_2$-graded algebra $\mathfrak{g} = \mathfrak{g}_0 \oplus \mathfrak{g}_1$ with a [ , ] operation which is sign-antisymmetric and satisfies the only reasonable version of the Jacobi identity:

$$\sum_{i=1}^{3} (-1)^{x_{i-1}x_{i+1}} [X_{i-1}, [X_i, X_{i+1}]] = 0,$$

where the $X_i$ $(i \in \mathbb{Z}_3)$ are in $\mathfrak{g}_{x_i}$. In particular $\mathfrak{g}_0$ is an ordinary Lie algebra, and $\mathfrak{g}_1$ is a $\mathfrak{g}_0$-module with a $\mathfrak{g}_0$-valued symmetric bilinear form [ , ] with respect to which

$\mathfrak{g}_0$ acts as derivations:

$$[X, [\theta_1, \theta_2]] = [[X, \theta_1], \theta_2] + [\theta_1, [X, \theta_2]],$$

for $X \in \mathfrak{g}_0$, $\theta_i \in \mathfrak{g}_1$. The basic example is the Poincare super algebra $\mathbf{p} = \mathbf{p}_0 + \mathbf{p}_1$, whose even part $\mathbf{p}_0$ is the usual Poincare algebra, and whose odd part $\mathbf{p}_1$ is the spinor module $S$ of the Lorentz group SO(3,1). (The Poincare group $P_0$ acts, via its quotient SO(3,1), on $\mathbf{p}_1$; this induces the infinitesimal action of $\mathbf{p}_0$.) The pairing $\gamma: \mathbf{p}_1 \times \mathbf{p}_1 \to \mathbf{p}_0$ is Clifford multiplication, with image the translation subgroup of $\mathbf{p}_0$. A physicist would write everything in coordinates, so $\gamma$ becomes the collection of Pauli or Dirac matrices (depending on whether real or complex forms are used). The $N$-extended super Poincaré algebra has the same $\mathbf{p}_0$, but $\mathbf{p}_1$ is replaced by $S^{\oplus N}$, the sum of $N$ orthogonal copies of the spinors.

A supersymmetric ("SUSY") space is the super analogue of a homogeneous space for the super Poincaré algebra or its $N$-extended version. An affine example is $(V, \mathcal{O})$, where $V$ is a vector space with non-degenerate quadratic form (of signature (3,1), in our case), and $\mathcal{O}$ is given in terms of the spinor module $S = S(V)$ as $C^\infty(V) \otimes \Lambda^\cdot(S^{\oplus N})$. For more details on SUSY spaces, we refer to [**Be**], where they are defined to be super manifolds modelled on the above affine example.

An $N = 1$ SUSY theory is a theory (i.e. a Lagrangian involving certain fields) with an infinitesimal action of the super Poincaré algebra. Likewise, a theory "with $N$ supersymmetries" has an action of the $N$-extended Poincaré super algebra. These theories can often, though not always, be described in terms of a super Poincaré invariant Lagrangian involving superfields on a SUSY space. The ordinary Lagrangian, on ordinary space, is recovered by Berezin integration along all the odd directions. The classic reference on supersymmetry is [**WB**]. A possibly more friendly introduction is [**Fr**], and the most geometrical version is [**Be**]. The first two references include a study of the supersymmetric version of Yang-Mills theory.

In physics, there are two types of particles: bosons, which combine with each other freely, and fermions, whose combinations are restricted by Pauli's exclusion principle. Mathematically, the distinction is roughly this: consider a collection of $n$ particles with distinct physical properties. If $H_i$ is the quantum Hilbert space of states of the i$^{\text{th}}$ particle, then the space of states of the entire collection is $\otimes_{i=1}^n H_i$. Now if instead we have $n$ identical particles, there are fewer distinct states of the ensemble: for bosons we get $\text{Sym}^n H$, while for fermions, $\Lambda^n H$. In four dimensional theories, bosonic particles ordinarily live in representations of the Poincare group with integer spin, while fermions have half-integer (that is, non-integer) spin. (This is the "spin-statistics" theorem.) Most "matter" particles (electrons, protons, quarks) are fermions; photons are bosons.

At a somewhat superficial level, the motivation for introducing supersymmetry is to be able to treat the two types of particles uniformly. A complementary and deeper reason is based on the no-go theorem of Coleman and Mandula (cf. [**Fr**], [**WB**]). This says that the most general algebra of symmetries of a class of quantum theories satisfying some rather reasonable non-degeneracy assumptions (one of these is non-freeness, another is the existence of massive particles) is of the form $\mathbf{p}_0 \oplus \mathfrak{g}$, where $\mathbf{p}_0$ is the Poincaré algebra of space-time isometries, and $\mathfrak{g}$ is the gauge algebra. The point is that this is a direct sum (of algebras), so there can be no non-trivial mixing of the two symmetries. Since symmetries are equivalent to conserved quantities, this implies for instance that any conserved quantity ("current") other than energy-momentum and angular momentum (which come from $\mathbf{p}_0$) must transform as a scalar under $\mathbf{p}_0$. This is unsatisfactory, since some free theories (e.g. of one real scalar field plus one real vector field, [**Fr**]; such theories are not covered by the Coleman-Mandula theorem) do admit conserved vectors and tensors, which should survive under small perturbation.

The tasks of mixing the two particle types and of avoiding the Coleman-Mandula restriction can fortunately be accomplished simultaneously through the introduction of super Lie algebras: even elements preserve particle types, odd elements exchange them. The even part of the algebra obeys Coleman-Mandula, but there is room for odd symmetries and the corresponding odd, or spinorial, conserved quantities. A theorem of Haag, Sohnius and Lopuszanski classifies the possible super Lie algebras of symmetries of physically interesting theories: there are the $N$-extended super Poincare algebras, as well as some central extensions and variants including an (even) gauge algebra $\mathfrak{g}$ acting on the odd sector.

**N=2 Super Yang-Mills**

The setting for Montonen-Olive duality [**MO**], for the work of Seiberg-Witten [**SW1, SW2**], and for the development discussed in later sections, is $N=2$ supersymmetric Yang-Mills theory (SYM) in four-dimensional space. The Lagrangian for this theory is usually written in terms of super fields on an affine SUSY space [**WB**], making the super Poincare invariance evident. When written out explicitly in terms of its component fields, the Lagrangian is quite complicated. It starts with a purely bosonic term analogous to the usual Yang-Mills Lagrangian $\int \mathrm{Tr}(F \wedge * F)$. Additional terms, involving additional fields, are required for the $N=2$ supersymmetry. The fields involved are grouped into $N=1$ multiplets (representations of the $N=1$ super Poincare algebra), which in turn combine into $N=2$ multiplets.

Such a theory depends, of course, on the choice of a compact gauge group $G$. For a given $G$ there is the pure $N=2$ theory, analogous to Maxwell's equations in vacuo, containing only those fields and terms in the Lagrangian required for supersymmetry. There are also various theories in which some combination of additional fermionic ("matter") particles is thrown in. We will return to these shortly.

Each SYM theory has a moduli space $B$ of vacuum states (= eigenstates of the Hamiltonian or energy operator corresponding to the lowest eigenvalue). This is of course a consequence of the degeneracy, or of the gauge invariance, of the Lagrangian: a non-degenerate Lagrangian should correspond to a unique vacuum, and an ordinary degeneracy is obtained when two eigenvalues happen to coincide, so there are two (or more) independent vacua. The situation in gauge theory is that there is a continuum $B$ of independent vacua. The complex dimension of $B$ is the rank $r$ of $G$, and we can in fact describe a coordinate system $(u_i)$, $i=1,\ldots,r$, on it. First, the classical analogue: the classical potential is

$$V(\varphi) = \frac{1}{g^2}\mathrm{Tr}[\varphi, \varphi^\dagger]^2,$$

where $g$ is a coupling constant (a parameter in the Lagrangian), and $\varphi$ is a Higgs field – a complex field transforming as a scalar under Poincare, and in the adjoint representation of $G$. Since $[\varphi, \varphi^\dagger]$ is self-adjoint, the minima occur precisely when $\varphi$ commutes with $\varphi^\dagger$. This locus is $G$-invariant: it is the union of the $G$-orbits of elements in the Cartan subalgebra $\mathbf{t}$ of the complexified Lie algebra $\mathfrak{g}_{\mathbb{C}}$. The quotient by $G$ is thus $B = \mathbf{t}/W$, where $W$ is the Weyl group. For example, when

$G = \mathrm{SU}(n)$ we have $\mathbf{t} \approx \mathbb{C}^{n-1}$, and the quotient is again $\mathbb{C}^{n-1}$, with coordinates $u_2 = \operatorname{Tr} \varphi^2, \ldots, u_n = \operatorname{Tr} \varphi^n$. (The reader will no doubt recognize this as a (real) symplectic reduction: the action of $G$ on $\mathfrak{g}_{\mathbb{C}}$ is Hamiltonian with respect to the natural translation-invariant symplectic structure induced on the real vector space underlying $\mathfrak{g}_{\mathbb{C}}$ by the imaginary part of the Killing form. The moment map sends $\varphi$ to $\operatorname{Tr}[\varphi, \varphi^{\dagger}]$, and the symplectic quotient is $\mathfrak{g}_{\mathbb{C}}//G = B = \mathbf{t}/W$. We will run into the complex analogue of symplectic quotients in the discussion of linearity in the next section.)

Now the constraints imposed by $N = 2$ supersymmetry (i.e. the difficulty of writing down a supersymmetric Lagrangian) imply that the quantum moduli space remains the same $B$, as an abstract algebraic variety, so it still has the coordinates $u_i$. But we will see that just about all other features of $B$ change in going from the classical to the quantum: for example, this applies to the Kähler metric on it, the locus of singularities (of the metric), the flat structure, and the relation of the $u$ coordinates to the natural fields in the theory.

A central quantity in any SYM theory is the electric charge $a$, which is the vector of eigenvalues of the Higgs field $\varphi$. It lives in the Cartan, $\mathbf{t}$. For SU(2) this is a complex number, as is $u$. We can take $\varphi$ to have eigenvalues $\pm \frac{a}{2}$, and then $u = \operatorname{Tr} \varphi^2 = \frac{a^2}{2}$, in the classical theory. In the quantum theory, this relationship still holds approximately as $u \to \infty$, by the asymptotic freedom of the theory. But for finite $u$ we will see that the relationship is very different.

A crucial property of these theories is that their low-energy (read: real-world observable) behavior can be described in terms of only a finite number of parameters – the coordinates $u_i$ on the quantum moduli space $B$. This description is via a "low energy effective Lagrangian" $\mathfrak{L}_{\text{eff}}$, obtained from the full Lagrangian through a process of averaging over all heavy degrees of freedom. So all measurable quantities become (locally) functions on $B$. In particular, this applies to the components $a_i = a_i(u)$ of the electric charge. Supersymmetry implies that these functions are holomorphic, and we will use them as alternative local coordinates on $B$.

While the existence of $\mathfrak{L}_{\text{eff}}$ is guaranteed by the general theory, its actual computation can be extremely difficult. The supersymmetry allows $\mathfrak{L}_{\text{eff}}$ to be expressed locally in terms of a single holomorphic function $\mathfrak{F} = \mathfrak{F}(a)$, the prepotential. (This is due to Seiberg, cf. [**SW1**], (2.7) and (2.11). Note that we need $\mathfrak{F}$ as a function of $a$, not of $u$.) All other quantities in the theory can then be expressed in terms of $\mathfrak{F}$ and the coordinates $a$. This includes the vector

$$a^D = \frac{d\mathfrak{F}}{da},$$

whose physical significance will appear shortly; the matrix

$$\tau(a) = \frac{d^2\mathfrak{F}}{da^2} = \frac{da^D}{da},$$

which is a complexified coupling constant (for SU(2), the low-energy effective values of the real coupling constant $g$ (encountered in the potential $V(\varphi)$) and of a phase angle $\theta$ are given by $\tau = \frac{\theta}{2\pi} + \frac{4\pi i}{g^2}$. In particular, $\mathfrak{F}(a)$ must be such that $\operatorname{Im}(\tau) > 0$); the Kähler metric

$$ds^2 = \operatorname{Im}(\tau \, da \, d\bar{a}) = \operatorname{Im}(da^D \, d\bar{a})$$

which was mentioned above; and its Kähler potential

$$K = K(a, \bar{a}) = \mathrm{Im}\left(\sum_i a_i^D \bar{a}_i\right) = \mathrm{Im}\left(\sum_i \frac{d\mathcal{F}}{da_i}\bar{a}_i\right).$$

The effective Lagrangian itself is given as a sum of kinetic and potential parts, each a Berezin integral over an $N = 1$ SUSY affine space of dimensions (4,4) and (4,2), respectively:

$$\mathcal{L}_{\text{eff}} = \frac{1}{4\pi}\int K(\Phi, \bar{\Phi})d^4\eta + \frac{1}{8\pi}\mathrm{Im}\int \tau(\Phi)|W|^2 d^2\eta.$$

Here $\Phi$ and $W$ are the two $N = 1$ multiplets which together make up the $N = 2$ multiplet containing the Higgs field $\varphi$. ($\varphi$ and a fermion $\psi$ are the fields in the "chiral multiplet" $\Phi$. The "vector multiplet" $W$ contains another fermion $\lambda$ and the gauge field $A$. It can be thought of as the field strength, $N = 1$ super analogue of the curvature $F$ in ordinary $YM$.) The integrands involve an analytic function ($\tau$) or the real part of one ($K$); this function is to be applied formally to the super fields $\Phi$, $W$ and their conjugates. The integration over the four (or two) odd space directions $\eta$ is interpreted via Berezin's recipe as a derivative, or as the reading off of the relevant coefficients. (The second term is actually written as an integral over a (4,2)-dimensional "chiral" quotient of the full (4,4)-dimensional SUSY space, in which two of the odd directions have been reduced.

What is missing is an explicit expression for $\mathcal{F}$. Traditionally, $\mathcal{F}$ was written as an infinite sum involving its classical ("tree level") value, the one-loop correction which is logarithmic in $a$ (higher loop contributions were known to vanish), and an infinite sum of mysterious instanton corrections involving negative powers of $a$. The breakthrough came when Seiberg and Witten succeeded, in [**SW1**], in computing $\mathcal{F}$ (and hence $\mathcal{L}_{\text{eff}}$, $ds^2$, etc.) exactly, for the pure $N = 2$ theory with $G = \mathrm{SU}(2)$. Their solution is based on the global properties of the theory.

**Duality**

The original conjecture of Montonen and Olive [**MO**] was that the low energy theory should be invariant under a duality exchanging electric and magnetic charges, and also exchanging the coupling constant $g$ with its inverse. In other words: electric charge in the strongly coupled theory (large $g$) behaves like magnetic charge at weak coupling, and vice versa. Seiberg and Witten found that a modified version of this duality holds in pure $N = 2$ SYM for SU(2). They applied the super version of the Hodge $*$ operator to the Lagrangian $\mathcal{L}_{\text{eff}}$, added some variables (Lagrange multipliers) and integrated out others, and ended up with a new Lagrangian $\mathcal{L}_{\text{eff}}^D$ which is equivalent to the original, and has the same form, but in terms of a new quantity $a^D := \frac{d\mathcal{F}}{da}$:

$$\mathcal{L}_{\text{eff}}^D(a^D) = \mathcal{L}_{\text{eff}}(a).$$

The same transformation sends $a^D$ to $-a = -\frac{d\mathcal{F}}{da^D}$. (This must be so, since the first term in $\mathcal{L}_{\text{eff}}$ changes sign when $a$, $a^D$ are simply exchanged.) The *physical meaning* of $a^D$ is therefore that of the dual, or magnetic, charge. As to the coupling constants, the defining relations

$$\tau = \frac{da^D}{da}, \quad \tau^D = \frac{d(-a)}{da^D}$$

together with the chain rule, imply that the actual duality transformation is

$$\tau^D = \tau^D(a^D) = -\tau(a)^{-1}.$$

This specializes to the Montonen-Olive transformation

$$g^D = g^{-1}$$

when $\theta = 0$, but not otherwise.

There is another, more elementary, transformation which preserves the Lagrangian: the $\theta$ angle is defined only modulo $2\pi\mathbb{Z}$, so $\tau$ is defined modulo $\mathbb{Z}$. There is therefore a transformation which fixes $a$ and sends $\tau \mapsto \tau + 1$. Since $\tau = \frac{da^D}{da}$, this requires that $a^D$ goes to $a^D + a$.

The two transformations can be represented by the matrices

$$S = \begin{pmatrix} 0 & 1 \\ -1 & 0 \end{pmatrix} \qquad T = \begin{pmatrix} 1 & 1 \\ 0 & 1 \end{pmatrix},$$

acting linearly on the 2-vector $\begin{pmatrix} a^D \\ a \end{pmatrix}$ and fractional-linearly on $\tau$. Since $S$ and $T$ generate $\mathrm{SL}(2,\mathbb{Z})$,we see that the theory has an $\mathrm{SL}(2,\mathbb{Z})$ action. (For other groups, $\mathrm{SL}(2,\mathbb{Z})$ is replaced by a subgroup of finite index in $\mathrm{Sp}(2r,\mathbb{Z})$.)

**The Seiberg-Witten solution**

The foregoing suggests that what lives intrinsically over a generic $u \in B$ is not the electric charge $a(u)$ but the unimodular lattice $\mathbb{Z}a(u) + \mathbb{Z}a^D(u)$ of all charges (electric, magnetic, and "dyonic", or mixed). As $u$ varies, we get a $\mathbb{Z}^2$ local system $V$ over $B$ minus some singular locus, together with a homomorphism to (the trivial bundle with fiber) $\mathbb{C}$. Dually, this homomorphism is equivalent to a holomorphic section $\begin{pmatrix} a^D \\ a \end{pmatrix}$ of the vector bundle $V_{\mathbb{C}}$ (i.e. of the sheaf $V \otimes \mathcal{O}_{B\smallsetminus \mathrm{sing}}$). The plan is to identify the local system (variation of Hodge structure, really) $V$, and then to fix the section $\begin{pmatrix} a^D \\ a \end{pmatrix}$, which in turn will determine the prepotential $\mathcal{F}$ (up to constant).

Seiberg and Witten identify $V$ by finding its monodromy. Near $u = \infty$ we know that $u \sim \frac{1}{2}a^2$, so the monodromy around $\infty$ sends $a \mapsto -a$. The known non-trivial logarithmic (one-loop) term in the perturbative expansion of $\mathcal{F}$ then fixes the shift in $a^D$: the monodromy is

$$M_\infty = \begin{pmatrix} -1 & 2 \\ 0 & -1 \end{pmatrix}.$$

In particular, the local system is non trivial, so $\tau$ is a non-constant holomorphic function on the universal cover of $B\smallsetminus\mathrm{sing}$, with values in $\mathbb{H}$. The singular locus must therefore contain at least two points (in addition to $\infty$). From a physical argument, Seiberg and Witten deduce that the system is as simple as possible: exactly three singularities, which can be relocated to $u = \pm 1$ and $\infty$ by an automorphism. Duality instructs that the finite monodromies should be conjugate to $-M_\infty$. Seiberg and Witten label things so that $a^D$ vanishes at $u = 1$, yielding the monodromy

$$M_1 = \begin{pmatrix} 1 & 0 \\ -2 & 1 \end{pmatrix}$$

which then determines $M_{-1}$.

The monodromies generate the level-2 congruence subgroup $\Gamma(2) \subset \mathrm{SL}(2,\mathbb{Z})$, and the local system itself can now be identified with the fiber cohomology of the universal level-2 elliptic curve

$$E_u\colon y^2 = (x+1)(x-1)(x-u)$$

over the $u$-line $\mathbb{C} \setminus \{\pm 1\}$.

The complexification $V_{\mathbb{C}}$ has a global trivialization in terms of the holomorphic 1-form $\lambda_1 = \frac{dx}{y}$ and the residueless meromorphic form $\lambda_2 = \frac{xdx}{y}$. We can also fix an integral homology basis, say $\gamma =$ loop around the branch points $0, 1$, and $\gamma^D =$ loop around $1$, $u$. To determine $a(u)$ and $a^D(u)$, we use the condition that $\mathrm{Im}(\tau) > 0$, where $\tau(u) := \frac{da^D/du}{da/du}$. One geometric solution is to take $\tau(u)$ as the period

$$\tau_u := \frac{\oint_{\gamma^D} \lambda_1}{\oint_{\gamma} \lambda_1}$$

of the elliptic curve $E_u$; rigidity implies that this is in fact the only solution. We then find that, up to a multiplicative constant, $a$ and $a^D$ are the periods over $\gamma$, $\gamma^D$ of the meromorphic 1-form

$$\lambda = \frac{y\,dx}{x^2-1} = \frac{(x-u)dx}{y} = \lambda_2 - u\lambda_1.$$

**Adding matter**

In a quantum field theory, a matter particle is represented by a fermion field $\psi$ which, in the case of a gauge theory, needs to be in some representation of $G$. It is added (or: coupled) to the theory by adding to the YM Lagrangian a term corresponding to the Lagrangian of the free particle (this includes a parameter, the particle mass $m$), and another term corresponding to its interaction with the YM field. In SYM the field needs to be in a representation $\rho$ of $G$, tensored with an $N=2$ multiplet, to preserve the SUSY. (The relevant multiplet, which is built from the fermion $\psi$, is called a *hypermultiplet*; we have already encountered the other basic type, the $N=2$ *vector multiplet*, which contained the scalar Higgs field $\varphi$.) If the number of particles added is not too large, the theory remains asymptotically free ($\beta < 0$) or becomes scale invariant ($\beta = 0$), and in either case retains many properties of the pure SYM theory. There still is an $r$-dimensional moduli space $B$. At $u \in B$ there are $r$ electric charges $a_i(u)$ and $r$ dual, magnetic, charges, $a_i^D(u)$, each charge living in the Cartan $\mathbf{t} \approx \mathbb{C}^r$ as before. Duality is modified somewhat: instead of $\mathrm{Sp}(2r,\mathbb{Z})$ transformations acting on $\binom{a^D}{a}$, one gets affine symplectic transformations of the form

$$\begin{pmatrix} a^D \\ a \end{pmatrix} \longmapsto R\begin{pmatrix} a^D \\ a \end{pmatrix} + \sum_{i=1}^{N_f} m_i \begin{pmatrix} n_i^D \\ n_i \end{pmatrix}$$

where $R \in \mathrm{Sp}(2r,\mathbb{Z})$, the $m_i$ are the masses of the $N_f$ particles added, and $n_i$, $n_i^D$ are integral $r \times r$ matrices. ("$N_f$" stands for number of *flavors*. The number of *colors* $N_c$ is the dimension of the representation space of $\rho$. The terminology comes from quantum chromodynamics (QCD), a gauge theory with $G = \mathrm{SU}(3)$.)

It is reasonable to guess that this structure can be implemented by a family of $r$-dimensional abelian varieties $A_u$ over $u \in B \setminus (\mathrm{sing})$: The $r \times 2r$ period matrix of

$A_u$ in some integral homology basis $\gamma^D$, $\gamma$ is $\left(\frac{da^D(u)}{du}, \frac{da(u)}{du}\right)$. (The polarization on the complex torus corresponding to the lattice generated by columns of this matrix is given by the Dirac quantization condition.) On the total space of the family there is a meromorphic 1-form $\lambda$, the *Seiberg-Witten differential.* There are $N_f$ divisors $D_i$ along which $\lambda$ acquires a pole with constant residue $\frac{m_i}{2\pi i}$. The charges $a^D$, $a$ can then be recovered as the periods of $\lambda$ over $\gamma^D$, $\gamma$. These have precisely the ambiguity needed: monodromy of the cycles $\gamma^D$, $\gamma$ acts as $\mathrm{Sp}(2r, \mathbb{Z})$, while moving such a cycle across the divisor $D_i$ changes the period by an integer multiple of $m_i$. The masses $m_i$ acquire a cohomological interpretation: the 2-form $\sigma = d\lambda$ is holomorphic (in fact, it is a holomorphic symplectic form) on the total space, and its de Rham cohomology class can be expressed in terms of the fundamental classes of the $D_i$:

$$[\sigma] = \sum m_i[D_i].$$

In particular, as we vary the masses, we obtain a family of algebraic symplectic manifolds in which the symplectic class varies *linearly* in the parameters.

In the case of scale-invariant theories ($\beta = 0$), the theory and therefore also its geometric model depend on one parameter in addition to the masses. This is $\tau_{c\ell}$, the classical limit ($u \to \infty$) of the quantum coupling constant $\tau = \tau(u)$. In asymptotically free theories ($\beta < 0$) we have

$$\lim_{u \to \infty} \tau(u) = \infty,$$

and instead of $\tau$ the theory depends on a scale parameter $\Lambda$. In the [**SW1**] solution, this $\Lambda$ is set to 1 for convenience.

In [**SW2**] Seiberg and Witten show that the above geometric model indeed leads to an exact solution of the various SU(2) SYM theories with matter. For each of these, the solution requires a family parametrized by the vector $m = (m_i)$ of masses (and by the additional parameter $\tau$ or $\Lambda$, which we omit from notation), of elliptic surfaces

$$X_m \to B \approx \mathbb{C}$$

with meromorphic 1-forms $\lambda_m$ whose residues equal $\frac{m}{2\pi i}$. The solution satisfies numerous consistency checks (the way one theory is known to flow to another in the limit where some parameter, such as a mass, goes to 0 or $\infty$, is reflected in corresponding degenerations of one geometric model to another) and has all the internal symmetry of each of the physical theories.

The number of particles which can be added to a SYM theory is determined as follows. The Killing form on the reductive Lie algebra $\mathfrak{g}$ determines an element $c_2$ in the center of the universal enveloping algebra $U(\mathfrak{g})$. By Schur's lemma, $c_2$ acts on the total space of each irreducible representation $\rho$ as multiplication by some scalar $c_2(\rho)$. (This $c_2$ is closely related to the second chern class of a bundle: Let $V$, $V'$ be two vector bundles on a surface associated via representations $\rho$, $\rho'$ to the same $G$-bundle. Then

$$\frac{c_2(\rho')}{c_2(\rho)} = \frac{c_2(V')}{c_2(V)},$$

where $c_2$ on the right denotes Chern classes. This can be used to give another definition of $c_2$ of a representation.) In any case, the $\beta$ function of the SYM theory with particles in representations $\rho_i$, $i = 1, \ldots, N_f$, is

$$\beta = c_2(\text{ad}) - \sum_i c_2(\rho_i)$$

where ad is the adjoint representation. For the classical algebras we can adjust the scaling of the Killing form so that $c_2(F) = 1$ where $F$ is the fundamental representation. With this normalization, $c_2(\text{ad}) = 2h^\vee_{\mathfrak{g}}$, where $h^\vee_{\mathfrak{g}}$ is the dual Coxeter number. For $SU(n)$ we have $h^\vee_{\mathfrak{g}} = n$, so the SU(2) theory, for example, remains asymptotically free with$N_f = 0, 1, 2$ or 3 particles in the fundamental representation $F$, and becomes scale invariant for $N_f = 4$ fundamental particles or for one particle in the adjoint representation. These are precisely the cases studied and solved in [**SW2**].

It is natural to ask whether the four dimensional, $N{=}2$ SYM theory with arbitrary gauge groups $G$ and collection of particles $\rho_i$ with $\beta \le 0$ can be similarly solved in geometric terms via families of abelian varieties and meromorphic one-forms. This question will occupy us in the remainder of this work.

## §2. Why integrable systems

### Algebraically integrable systems

An algebraically integrable system consists of a complex algebraic manifold $X$, an everywhere non-degenerate, closed holomorphic $(2,0)$-form $\sigma$ on $X$, and a morphism $\pi : X \to B$ whose general fiber is a (polarized) abelian subvariety $X_b := \pi^{-1}(b)$ which is Lagrangian with respect to $\sigma$. In practice, the object which is naturally given is often an affine open $X_0 \subset X$, the fibers being affine subsets of abelian varieties. This is a straightforward algebraic analogue of the (real, $C^\infty$) notion of integrable system in classical mechanics. As there, any function $H$ ("Hamiltonian") on the base $B$ determines a 1-form $d\pi^* H$ on $X$, and when contracted with the symplectic form $\sigma$ it gives a vector field $v_H$ on $X$, tangent to the fibers of $\pi$. The pullbacks $\pi^* H_1, \pi^* H_2$ of two functions on $B$ Poisson-commute, and in particular the corresponding vector fields commute.

Locally on the base of an algebraically integrable system there is a natural holomorphic flat structure, depending on a discrete choice. The fiber homology $H_1(X_b, \mathbb{Z})$ has a non-degenerate skew-symmetric pairing (induced by the polarization). The choice involved is that of a Lagrangian subspace $V^*_{\mathbb{Z}}$ for this pairing. This is often achieved by specifying a symplectic basis $\alpha_1, \ldots, \alpha_g, \beta_1, \ldots, \beta_g$ of $H_1(X_b, \mathbb{Z})$, and taking the subspace $V^*_{\mathbb{Z}}$ spanned by the $\alpha$ cycles. This choice needs to be made for one $b \in B$, and (by Gauss-Manin) is propagated to nearby fibers. Integration over these $\alpha$-cycles gives a local trivialization of the relative cotangent bundle $\pi_* \omega_X$, and the symplectic form $\sigma$ converts this to a flat structure on (the tangent bundle of) $B$, modelled on $V := Hom(V^*_{\mathbb{Z}}, \mathbb{C})$.

Locally on the flat base $B$, we can ask what kind of data is required to specify an algebraically integrable system inducing the given flat structure. We think of the family of Abelian varieties, plus choice of $\alpha$ and $\beta$ cycles, as specified by its period map

$$p : B \to \mathbb{H}_g$$

to the Siegel half space $\mathbb{H}_g$ of $g \times g$ symmetric complex matrices with positive imaginary part. The total space $\pi : X \to B$ is then retrieved as pullback of the universal abelian variety $\mathfrak{X}_g \to \mathbb{H}_g$. The question then is to understand the condition on $p$ for there to be a symplectic form on $X$ for which $\pi$ is a Lagrangian fibration. The fixing of the flat structure is achieved by thinking of $\mathbb{H}_g$ as open subset of $Sym^2V^*$, where $V$ is the tangent space to $B$ (at any $b \in B$). The derivative of the period map is

$$dp_b \in Hom(V, Sym^2V^*) = V^* \otimes Sym^2V^*.$$

A straightforward computation done in [**DM2**] shows that the condition is that $dp_b$ lives in the subspace $\mathrm{Sym}^3V^*$, for all $b \in B$. So an integrable system with base $B$ is specified by a field of cubics on the tangent bundle of $B$.

This field of cubics is subject, of course, to a strong integrability condition. In flat local coordinates $z_i$ on $B$, let $(p_{ij}(z))$ be the period matrix and $(c_{ijk}(z))$ the field of cubics. Then equality of mixed partials

$$\frac{\partial p_{ij}}{\partial z_k} = c_{ijk} = \frac{\partial p_{ik}}{\partial z_j}$$

implies the existence of functions $w_i$ (locally) on $B$ such that

$$p_{ij} = \frac{\partial w_i}{\partial z_j}.$$

These again have symmetric partials,

$$\frac{\partial w_i}{\partial z_j} = p_{ij} = \frac{\partial w_j}{\partial z_i},$$

so there is a single function $\mathcal{F}$ on $B$, called the *prepotential*, satisfying:

$$w_i = \frac{\partial \mathcal{F}}{\partial z_i}$$
$$p_{ij} = \frac{\partial^2 \mathcal{F}}{\partial z_i \partial z_j}$$
$$c_{ijk} = \frac{\partial^3 \mathcal{F}}{\partial z_i \partial z_j \partial z_k}$$

The prepotential determines also a Kahler potential on $B$

$$K := Im(\sum_i w_i \bar{z}_i) = Im(\sum_i \frac{\partial \mathcal{F}}{\partial z_i} \bar{z}_i)$$

and its associated Kahler form

$$\frac{i}{2}\partial\bar{\partial}K = \frac{i}{2}\sum_{i,j} Im(p_{ij}) dz_i \wedge d\bar{z}_j.$$

Equivalently, this Kahler (1,1)-form on $B$ is obtained by integration over the $g$-dimensional fibers $X_b$ of the $(g+1, g+1)$-form $\sigma \wedge \bar{\sigma} \wedge t^{g-1}$ on $X$, where $t$ is any $(1,1)$-form on $X$ whose restrictions to fibers give the polarization.

Conversely, given $B$ with coordinates $z_i$ and a holomorphic function $\mathcal{F}$, the above formulas produce an algebraically integrable system over the open subset of $B$ where $Im(\frac{\partial^2 \mathcal{F}}{\partial z_i \partial z_j}) > 0$ : the family of abelian varieties is determined by its periods $p_{ij}$, and the symplectic form is given by the cubic $c_{ijk}$. The slightly weaker notion of an analytically integrable system (just replace abelian varieties by polarized complex tori) is recovered over the open subset where $Im(\frac{\partial^2 \mathcal{F}}{\partial z_i \partial z_j})$ is invertible, cf. [**DM2**]. The Kahler "metric" in this case is still non-degenerate, but possibly indefinite.

We note that the $w_i$ play a role dual to that of the $z_i$: the $z_i$ give the flat structure on $B$ determined by integration over the $\alpha$ cycles, while the $w_i$ give the flat structure determined by integration over the $\beta$ cycles. We also note that, up to this point, the $z_i$ as well as the $w_i$ have been determined only up to arbitrary additive constants. In other words, what we have are the vector fields $\partial/\partial z_i, \partial/\partial w_i$, or the 1-forms $dz_i, dw_i$. We will discuss below the additional choice used in the SW setup to fix these additive constants.

Similar structures have arisen in various more specialized contexts, e.g. in the WDVV equations of conformal field theory and on moduli spaces of Calabi-Yau threefolds, as well as in the Seiberg-Witten setup described in §1. They often go under the title of "special geometry". The point is that these specialized contexts are not really needed: the structure of special geometry on $B$ (consisting of a covering by open subsets on which an appropriate prepotential, dual flat holomorphic coordinate systems, and a Kahler metric are defined), arises naturally on the base of any algebraically integrable system. The converse is also true, locally, as we have just seen. Globally, though, there is an integrability constraint: the monodromy action (say on the coordinates $z_i, w_i$) in an integrable system involves integral symplectic transformations plus (complex) translations, while special geometry allows the symplectic transformation to be real instead of integral. We refer the reader to [**F**] for a very clear differential-geometric discussion of special geometry and its relationship to integrable systems.

**Seiberg-Witten Differentials**

In the Seiberg-Witten picture of super Yang-Mills, there are the charges $a_i$ and $a_i^D$which are described locally as holomorphic functions on the quantum moduli space $B$. There is less ambiguity in choosing these $a_i, a_i^D$ than in the coordinates $z_i, w_i$ on the base of an algebraically integrable system: the group of their linear combinations $\Sigma n_i a_i + n_i^D a_i^D$ (with integer coefficients $n_i$) should be uniquely determined in the pure SYM case, and determined modulo the masses of the added particles in general. A natural way to get functions with precisely such ambiguity on the base of an integrable system $\pi : X \to B$ is to choose a meromorphic differential 1-form $\lambda$ on $X$ and let $a_i, a_i^D$ be its periods over a set of 1-cycles representing $\alpha_i, \beta_i$ respectively and avoiding the poles of $\lambda$. We want $a_i, a_i^D$ to be a possible choice of the coordinates $z_i, w_i$, which are defined up to translation (once we have fixed the $\alpha_i, \beta_i$). In other words, we want

$$da_i = dz_i, \quad da_i^D = dw_i.$$

Since $dz_i, dw_i$ were defined as the contraction of $\alpha_i, \beta_i$ respectively with the symplectic form $\sigma$, the condition becomes simply:

$$d\lambda = \sigma.$$

The behavior of the singularities of $\lambda$ is constrained by this condition. When restricted to a general fiber $X_b$, $\lambda$ will have poles along the union of some irreducible divisors $D_{b,j}$. One constraint is that the residue $\mathrm{Res}_j$ of $\lambda$ along $D_{b,j}$ is (locally in $B$) independent of $b$. A *Seiberg-Witten differential* is a meromorphic 1-form $\lambda$ satisfying $d\lambda = \sigma$ and $\mathrm{Res}_j = m_j$, the mass of the $j$-th particle. On an algebraically integrable system with a specified Seiberg-Witten differential, the local coordinates $z_i, w_i$ can be specified as the charges $a_i, a_i^D$, with just the right ambiguity.

One way to describe an integrable system (though by no means the only way!) is in terms of a family $C \to B$ of spectral curves $C_b$. In the simplest situation, the abelian varieties $X_b$ are just the Jacobians $J(C_b)$. More generally, $X_b$ may be realized as the Prym of $C_b$ with respect to an involution, or as the generalized Prym with respect to a correspondence, cf. [**D1**]. An example of such a system, due to Mukai [**Mu**], is obtained from the Jacobians of a complete linear system $C \to B$ of curves on a symplectic surface $(S, \bar{\sigma})$. The symplectic form $\sigma$ on $X$ then corresponds via Abel-Jacobi to the pullback to $C$ of $\bar{\sigma}$. Mukai studied the case where $S$ is a $K3$ surface. Taking $S = T^*E$ to be the cotangent bundle of some curve $E$ yields Hitchin's system on $E$. One obtains many more examples by allowing $S$ to have a symplectic form $\bar{\sigma}_0$ defined only on an open subset $S_0 \subset S$, and imposing appropriate restrictions on the intersection of the spectral curves $C_b$ with $S - S_0$. For example, we can take $S$ to be the total space of the line bundle $\omega_E(D)$ for some effective divisor $D$ on $E$, to recover Markman's system [**Mn**] parametrizing meromorphic Higgs bundles on $E$. Note that this $S$ has a natural meromorphic 1-form $\bar{\lambda}$, given away from $D$ (i.e. where $S$ can be identified with $T^*E$) as the action 1-form $\bar{\lambda} = pdq$, whose differential is $d\bar{\lambda} = \bar{\sigma}_0$. It follows that the "tautological" 1-form on $C$ (pullback of $\bar{\lambda}$) induces the Seiberg-Witten differential $\lambda$ on the integrable system $X = J(C/B) \to B$. The system discussed in §3 is of this form.

**Linearity: complexified Duistermaat-Heckman**

One feature of the Seiberg-Witten picture which we have not yet discussed is the linear dependence of the symplectic form on the particle masses. From one point of view, this linearity follows from existence of the Seiberg-Witten differential: the equation $\sigma = d\lambda$ implies that the cohomology class of $\sigma$ is a fixed linear combination of the residues of $\lambda$, which are just the masses. On the other hand, linearity can also be interpreted as arising from a complex analogue of the theorem of Duistermaat-Heckman [**DH**].

Let $(X, \sigma)$ be a complex symplectic manifold (i.e. the symplectic form $\sigma$ is holomorphic, of type (2,0)) equipped with a Hamiltonian action of a complex reductive group $G$. The Hamiltonian property of the action means that there is a moment map $\mu : X \to \mathfrak{g}^*$ to the dual of the Lie algebra. Ignoring bad loci, we will assume (or pretend) that there is a good non-singular quotient $X/G$. Let $G_a \subset G$ be the stabilizer of $a \in \mathfrak{g}^*$ under the coadjoint action. The symplectic quotient $X_a$ (or $X//_aG$) is defined as

$$X_a := \mu^{-1}(a)/G_a = \mu^{-1}(\mathcal{O}_a)/G,$$

where $\mathcal{O}_a$ is the (coadjoint) orbit of $a$. It is again a complex symplectic manifold, with form $\sigma_a$. As $a$ ranges over orbits of a fixed "type" in $\mathfrak{g}^*$, the topology of $X_a$ remains locally constant, so cohomologies of nearby $X_a$'s of fixed type can be identified. The complex Duistermaat-Heckman theorem says that, under this identification, the cohomology class $[\sigma_a] \in H^2(X_a)$ varies linearly with $a$.

Consider first the abelian case where $G = T$ is a torus $T = Hom(\Lambda, \mathbb{C}^*)$. The moment map then factors

$$X \to X/T \to \mathbf{t}^*,$$

and we assume both maps are bundles, at least over some open $\mathbf{t}_0^* \subset \mathbf{t}^*$. The $T$-bundle $X \to X/T$ has first chern class

$$c_1 \in H^2(X/T, \Lambda^*).$$

For each $a$ in $\mathbf{t}_0^*$, this restricts to a class (still denoted by $c_1$) in $H^2(X_a, \Lambda^*)$. The theorem in this case says that for $a, a_0 \in \mathbf{t}_0^* = \Lambda \otimes \mathbb{C}$,

$$[\sigma_a] - [\sigma_{a_0}] = \langle c_1, a - a_0 \rangle.$$

Because of possible monodromy of the fibers of $X/T$ over $\mathbf{t}_0^*$, both sides may be multivalued, so we should think of the linearity as being local in $a$.

Next, let $G$ be an arbitrary reductive group, and consider reduction at a regular semisimple element $a$. This means that the stabilizer $G_a$ is a maximal torus $T \subset G$. The coadjoint orbits of semisimple elements are parametrized by the quotient $\mathbf{t}^*/W$ of the dual Cartan by the Weyl group. Let $\mathfrak{g}_0^*, \mathbf{t}_0^*$ denote the open subsets of $\mathfrak{g}^*, \mathbf{t}^*$ respectively parametrizing regular semisimple elements. The parameter space for regular semisimple coadjoint orbits is then

$$\mathbf{t}_0^*/W \approx \mathfrak{g}_0^*/G.$$

The moment map in this nonabelian situation no longer factors. The closest we get to factorization is the commutative diagram:

$$\begin{array}{ccccc} & & X_0 & & \\ & \swarrow & & \searrow^{\mu} & \\ X_0/G & & & & \mathfrak{g}_0^* \\ & \searrow & & \swarrow & \\ & & \mathbf{t}_0^*/W = \mathfrak{g}_0^*/G & & \end{array}$$

with $X_0 := \mu^{-1}(\mathfrak{g}_0^*)$. Somewhat surprisingly, it is easy to reduce this situation to the abelian case, by considering $Y := \mu^{-1}(\mathbf{t}^*)$. The symplectic form of $X$ restricts to a symplectic form on $Y$, and the $T$-action on $Y$ is Hamiltonian with respect to this form, so there is a moment map $\mu_T : Y \to \mathbf{t}^*$ compatible with $\mu$ via the inclusions $Y \subset X$ and $\mathbf{t}^* \subset \mathfrak{g}^*$. Now for $a \in \mathbf{t}_0^*$ there is an obvious identification of symplectic manifolds $X//_aG \approx Y//_aT$, so the dependence of the symplectic class $[\sigma_a]$ of $X//_aG$ on $a \in t_0^*$ is again locally linear. (In the original case with $G$ compact, the multivaluedness is replaced by discontinuities across walls, resulting in $[\sigma_a]$ being linear on each Weyl chamber).

What happens when $a$ is allowed to degenerate? If $G$ is compact, $a$ remains semisimple (though not regular), and there is a fairly obvious linearity for all such $a$ in a given open face $\mathbf{f}_0$ of the Weyl chamber, i.e. a face $\mathbf{f}$ minus its proper subfaces. In the complex reductive case, there are more possibilities: the most natural orbit which corresponds to an $a$ in a complexified open face $\mathbf{f}_0$ consists of elements which are not semisimple but regular, in the sense that their centralizers have the minimal possible dimension. The closure of this orbit consists of a finite number of other orbits. The smallest of these orbits (the unique closed orbit, which is in the closure of each of the others) consists of the semisimple elements. Although we will not encounter non-regular elements in the integrable systems we need in the following section, it seems worthwhile to point out that linearity is indeed a general phenomenon applicable to all strata.

We identify $\mathfrak{g}, \mathbf{t}$ with $\mathfrak{g}^*, \mathbf{t}^*$ via an invariant form. Consider an arbitrary stratum $S \subset \mathfrak{g}$ corresponding to a face $\mathbf{f}_0 \subset \mathbf{f} \subset \mathbf{t}$. $S$ is a bundle over (a finite quotient of the complexified) $\mathbf{f}_0$, and the fibers are (co)adjoint orbits of a fixed type. This bundle has local sections, which can be taken of the form $n + \mathbf{f}_0$, where $n$ is some fixed nilpotent in the reductive subalgebra $\mathbf{l} = \mathbf{l}_{\mathbf{f}}$ (= common centralizer of all elements of $\mathbf{f}_0$). All elements $a = n + t$ of this section have the same centralizer $\mathbf{c}$ in $\mathfrak{g}$. ($\mathbf{c}$ is the centralizer in $\mathbf{l}$ of $n$). Since $\mathbf{f}$ is the center of $\mathbf{l}$, it is contained in $\mathbf{c}$, which actually decomposes as $\mathbf{c} = \mathbf{f} + \mathbf{d}$ with $\mathbf{d}$ semisimple. Let $D \subset G$ be the subgroup corresponding to $\mathbf{d}$. The reduction to the abelian case now proceeds as follows: We observe that $\mu^{-1}(n + \mathbf{f}_0)/D$ is symplectic, and has a residual Hamiltonian action of the subtorus $F \subset T$ whose algebra is $\mathbf{f}$. Let $\mu_F$ be the moment map to $\mathbf{f}$ (which we identify with $\mathbf{f}^*$). Then for $a = n + t$ as above, we have an isomorphism of symplectic manifolds:

$$X//_aG \approx \mu_F^{-1}(t)/F = (\mu^{-1}(n + \mathbf{f}_0)/D)//_tF.$$

Since the symplectic form on the right varies linearly with $t$ (which parametrizes $a = n + t$), we conclude that $[\sigma_a]$ varies linearly with $a$, as claimed.

In a sense, there is linearity even as we degenerate from one stratum to another. We illustrate this phenomenon with the following example.

The left action of $G$ on itself lifts to a Hamiltonian action of $G$ on $X := T^*G \approx G \times \mathfrak{g}^*$. The quotient map $X \to X/G$ is

$$\begin{aligned} G \times \mathfrak{g}^* &\to \mathfrak{g}^* \\ (g, a) &\mapsto a, \end{aligned}$$

while the moment map sends

$$(g, a) \mapsto ad_g^* a.$$

The symplectic quotients

$$X//_aG = \mathcal{O}_a$$

are the coadjoint orbits. These fit, as discussed above, into a finite number of continuous families, each indexed by the $W$-quotient of some open face. Inside each family we expect linearity. As we move between families, the cohomology (and even the dimension) jumps. Yet there is a way to express linearity uniformly for all $a \in \mathfrak{g}^*$. Let $G_a, \mathfrak{g}_a$ be the stabilizers of $a$ in $G, \mathfrak{g}$ respectively. Then $a$ restricts to $\overline{a} \in \mathfrak{g}_a^*$ which is $G_a$-invariant, so it gives an element of $(\mathfrak{g}_a^*)^{G_a}$, a bi-invariant 1-formon $G_a$. This gives a class in $H^1(G_a, \mathbb{C})$, which is then sent to $H^2(\mathcal{O}_a, \mathbb{C})$ by the transgression map (= the coboundary map $E_2^{01} \to E_2^{20}$ in the Leray spectral sequence of the fibration $G \to \mathcal{O}_a$, with fiber $G_a$, where $E_2^{pq} = H^p(\mathcal{O}_a, H^q(G_a, \mathbb{C}))$. The image is seen to be $[\sigma_a]$, and the construction is obviously linear on each $\mathbf{f}_0 \subset \mathbf{f} \subset \mathbf{t}$, since the stabilizers are actually constant there.

## §3. Which integrable system

We have seen that all the structures on the quantum moduli space of $N = 2$ SYM predicted by Seiberg and Witten can be realized in terms of algebraically integrable systems. The features of "special geometry" occur in any algebraically integrable system. The distinguished electric and magnetic charges appear as soon as we fix a Seiberg-Witten differential, and perhaps the most natural way to get this is as tautological 1-form on a family of curves contained in the total space of a line bundle $\omega_E(D)$ of meromorphic differentials on some fixed curve $E$. Since a quantum theory with $\beta = 0$ depends on a parameter $\tau = \tau_{cl}$, it is natural to take $E$ to be the elliptic curve corresponding to this $\tau$. Finally, the linearity properties suggest that our family of systems, parametrized by the masses, could arise as the family of symplectic reductions of a larger symplectic manifold.

There remains the technicality of identifying a specific integrable system for each type of theory, and testing that these theories have reasonable physical properties. In this section we first give several descriptions of the system [**DW**] which solves the $SU(n)$ theory with one adjoint, and then describe some of the tests which confirm the choice.

**Meromorphic Higgs bundles**

Fix a curve $E$, an effective divisor $D$ on it, and an integer $n \geq 2$. The total space of the system is the moduli space $X$ of equivalence classes of pairs $(V, \varphi)$, where $V$ is a rank-$n$ vector bundle on $E$ with trivial determinant, $\varphi : V \to V \otimes \omega_E(D)$ is a traceless endomorphism with values in $\omega_E(D)$, and the pairs are subject to an appropriate semistability condition. The base is

$$B := \oplus_{i=2}^{n} H^0(E, (\omega_E(D))^{\otimes i}),$$

and the map $\pi : X \to B$ sends a pair $(V, \varphi)$ to (the sequence of coefficients of) the characteristic polynomial

$$\det(t \cdot I - \varphi).$$

(It is best to think of $t$ as a fiber coordinate taking values in $\omega_E(D)$.)

The generic fibers of $\pi$ are abelian varieties. Consider the surface $S$ which is the total space of $\omega_E(D)$. The points $b = (b_2, \ldots, b_n) \in B$ parametrize a linear system of curves $C_b \subset S$, given by the equation $t^n + b_2 t^{n-2} + \cdots + b_n = 0$. Let $\lambda$ be the tautological meromorphic 1-form on $S$. Then the pair $(V, \varphi)$ determines a sheaf $L :=$ coker $(\varphi - \lambda \cdot Id)$ on $S$ whose support is the curve $C_b$. In fact, for $(V, \varphi)$ generic, $L$ is a line bundle on $C_b$. Thus we have a map $\pi^{-1}(b) \to Pic(C_b)$. If we were working with $GL(n)$-bundles rather than $SL(n)$-bundles (remove the restrictions $\det V \approx \mathcal{O}$, trace $(\varphi) = 0$), this would be an isomorphism of $\pi^{-1}(b)$ with a component of $Pic(C_b)$. The restrictions mean that we get instead a translate of $\text{Prym}(C_b/E)$.

This construction, modelled on Hitchin's system [**Hi**], works much more generally, cf. [**D2**]. In particular, the structure group $SL(n)$ can be replaced by any reductive $G$. The total space $X$ now parametrizes pairs $(V, \varphi)$ with $V$ a $G$-bundle over $E$, and $\varphi$ a section of $ad(V) \otimes \omega_E(D)$. The base is replaced by

$$B := \oplus_{i=1}^{r} H^0(E, (\omega_E(D))^{\otimes d_i}),$$

where $r$ is the rank of $G$ and $d_i$ are the degrees of the basic invariant polynomials on the Lie algebra $\mathfrak{g}$. Once we choose a representation $\rho$ of $G$, we get a family of spectral curves $\{C_{\rho,b} \mid b \in B\}$ in $S$ as before.

As shown in [**D2**], [**Fa**] and further references therein, the generic fiber of $\pi$ can be identified as (a translate of) a certain distinguished abelian variety $\text{Prym}(C_b)$ which is an isogeny summand in the Jacobian $J(C_{\rho,b})$ for each representation $\rho$. The most natural way to view this is to abandon the spectral curves $C_{\rho,b}$ altogether, in favor of the cameral cover $C_b \to E$, a Galois cover with group $W$, the Weyl group of $G$. This is the pullback to $E$, via the classifying map $E \dashrightarrow \mathbf{t}/W \approx \mathfrak{g}/G$ determined locally in $E$ by the point $b \in B$, of the $W$-cover $\mathbf{t} \to \mathbf{t}/W$. ($\mathbf{t}$ is the

Cartan subalgebra, and the classifying map is really a map from the total space of $(\omega_E(D))^*$ to $\mathbf{t}/W$. The local form above depends on the choice of a local section of the line bundle $(\omega_E(D))^*$, so is really defined only modulo homotheties). Now $J(C_b)$ decomposes according to the irreducible representations of the Galois group $W$, and the distinguished Prym, Prym $(C_b)$, is the component corresponding to the action of $W$ on the weights. (The question of the dependence of the spectral Jacobian on $\rho$ was raised in [**AM**] and studied, in the case of the Toda system, in [**Ml**]. A solution, under some conditions, was given in [**Ka**]. The case $E = \mathbb{P}^1$ was analyzed in [**AHH**] and [**B**], and the general case is in [**D2**], [**Me**]). In this section we will use only the $SL(n)$ system, postponing the appearance of other groups to §4.

The total space $X$ is not symplectic, but has a Poisson structure [**B**], [**Mn**]. In general, a Poisson structure on an algebraic manifold $X$ is a Lie bracket operation on $\mathcal{O}_X$ satisfying the properties of the usual Poisson bracket. Equivalently, it can be given by a section $\alpha$ of $\wedge^2 T_X$ satisfying an integrability condition. If $\alpha$ is non-degenerate on some open set $X_0$, then its inverse $\sigma = \alpha^{-1}$ is a two-form on $X$, and the integrability condition says that $\sigma$ is closed, i.e. $\sigma$ then gives a symplectic structure. In general, a Poisson manifold admits a natural foliation by submanifolds (whose conormal bundles consist of the null spaces of $\alpha$) which inherit a symplectic structure. A typical example is $\mathfrak{g}^*$, which has the natural (Kirillov-Kostant) Poisson bracket, the symplectic leaves being precisely the coadjoint orbits. More generally, a Hamiltonian $G$-action on a symplectic manifold $Y$ induces a Poisson structue on $X = Y/G$. This is the ordinary quotient; the symplectic quotients $X_a := Y//_a G$, for $a \in \mathfrak{g}^*$, are recovered as the symplectic leaves of $X$. (The Kirillov-Kostant Poisson structure on $\mathfrak{g}^*$ is recovered when we take $Y := T^*G$).

Markman's construction of the Poisson structure is based on the moduli space $\mathcal{M}_D$ of "bundles with a level-$D$ structure" on $E$ : $\mathcal{M}_D$ parametrizes equivalence classes of stable pairs $(V, \eta)$ where

$$\eta : V_D \xrightarrow{\sim} \mathcal{O}_D^n$$

is a trivialization of the fibers of $V$ at points of $D$ (sending the given volume element of $V_D$ to the standard one on $\mathcal{O}_D^n$, since our structure group is $SL(n)$). Consider the cotangent bundle $Y := T^*\mathcal{M}_D$. By elementary deformation theory, its points are (equivalence classes of) triples $(V, \eta, \varphi)$ with $(V, \eta) \in \mathcal{M}_D$ and $(V, \varphi) \in X$, a meromorphic Higgs bundle with poles on $D$. The symmetry group $G$ is the product of copies of $SL(n)$, one for each point of $D$. (More precisely, $G = \{g \in \mathrm{Aut}_D(\mathcal{O}_D^n) \mid \det g = 1\}$, allowing for possibly non-reduced $D$.) The $G$ action on $\mathcal{M}_D$ (with quotient $\mathcal{M} = \mathcal{M}_0$, the moduli of stable bundles on $E$) lifts to a Hamiltonian action on $Y$:

$$g : (V, \eta, \varphi) \mapsto (V, g \circ \eta, \varphi),$$

with quotient $X = Y/G$, so we get a Poisson structure on $X$. The moment map

$$\mu : Y \to \mathfrak{g}^* \approx \mathfrak{g}$$

sends $(V, \eta, \varphi)$ to $\varphi^\eta$, the element of $\mathrm{End}_D(\mathcal{O}_D^n)$ which is conjugate via $\eta$ to the residue $\mathrm{Res}_D\varphi \in \mathrm{End}_D(V_D)$. The symplectic leaf $X_a$, for $a \in \mathfrak{g}$, can therefore be identified as the moduli space of pairs $(V, \varphi) \in X$ such that $\mathrm{Res}_D\varphi$ is conjugate to the given $a$. In particular, each abelian variety $\pi^{-1}(b) \approx \mathrm{Prym}(C_b/E)$ is contained in one symplectic leaf $X_a$, and is a Lagrangian subvariety there.

A number of special cases of this construction are well known. When $D = 0$, we get Hitchin's system [**Hi**] on $T^*\mathcal{M}$. When $E$ has genus 0, we get the polynomial matrices system studied in [**AHH**] and [**B**]. For more details and examples, see [**Mn**] or [**DM1**].

For the SYM theory with adjoint matter, we take $E$ to be the elliptic curve whose $\tau$ parameter (in the halfplane $\mathbb{H}$) is the coupling constant of the theory. We are looking for a 1-parameter family of integrable systems (parametrized by mass $m$), each having an $(n-1)$-dimensional base and $(n-1)$-dimensional fibers. One checks easily that $\dim(X_a)$ equals the dimension of the adjoint orbit of $a$; so we are looking for orbits of the smallest possible dimension $2(n-1)$. There is essentially only one 1-parameter family of possibilities: $D$ consists of a single point $\infty \in E$, and $a$ is the conjugacy class of the diagonal matrix with eigenvalues $m(1, 1, \ldots, 1, 1-n)$. This conjugacy class consists of all diagonalizable $SL(n)$ matrices $A$ with

$$\operatorname{rank}(A - m \cdot Id) \leq 1.$$

(As $m \to 0$ there is one further, nilpotent, orbit of the right dimension: the conjugacy class of the elementary matrix $e_{12}$. But this turns out to give a trivial integrable system, i.e. the same as for $a = 0$: all the spectral curves are completely reducible).

Our system can be described rather explicitly. The moduli space of degree 0, rank $n$ vector bundles on $E$ can be identified with the symmetric product $S^nE$: each equivalence class contains a unique decomposable bundle $V = \oplus_{i=1}^n L_i$, with $L_i = \mathcal{O}_E(q_i - \infty) \in \operatorname{Pic}^0 E$. The moduli of $SL(n)$-bundles is the fiber over 0 of the Abel-Jacobi map $S^nE \to E$. It is a projective space

$$\{q_1 + \cdots + q_n | \sum q_i = n\infty\} = \mathbb{P}H^0(E, \mathcal{O}(n\infty)) \approx \mathbb{P}^{n-1}.$$

In terms of the decomposition $V = \oplus L_i$, any Higgs field $\varphi$ on $V$ can be written as a matrix with entries

$$\varphi_{ij} \in \Gamma(L_j \otimes L_i^{-1}(\infty)).$$

Let $A := \operatorname{Res}_\infty \varphi$. Its diagonal elements are all zeroes, and $A - m \cdot Id$ has rank $\leq 1$, so $A$ is conjugate, via a diagonal matrix, to $m$ times $e = \Sigma_{i \neq j} e_{ij}$. This means that each $(V, \varphi)$ in our system is equivalent to a pair $(V, \varphi)$ with $\operatorname{Res}_\infty(\varphi_{ij}) = m$ for $i \neq j$. For a given $V$, this condition uniquely determines the $\varphi_{ij}$ (as long as $L_i \neq L_i$). The remaining free parameters are the coordinates $q_i$ of $V = \oplus \mathcal{O}(q_i - \infty)$ and the diagonal entries $p_i = \varphi_{ii}$. Away from the diagonals $q_i = q_j$, these $(q_i, p_i)$ can be identified with the canonical coordinates on $T^*\mathcal{M}$.

**The spectral curves**

The family of spectral curves $C_b$, for the fundamental, $n$-dimensional representation of $SL(n)$, can be described as a linear system on the compactification $\overline{S} = \mathbb{P}(\omega_E(\infty) \oplus \mathcal{O}_E)$ of the surface $S$, or on various birationally equivalent surfaces. More precisely, it forms an affine open subspace of this linear system, the complement consisting of curves which contain the section at infinity, $Z_\infty := \overline{S} - S$, with some multiplicity $k > 0$. By removing this multiple component, we recover the spectral curves for the analogous problem with $n$ replaced by $n - k$. (If $A$ is an endomorphism of a vector space $V$, and $W$ is an invariant subspace, then

$$\operatorname{rank}_W(A - m \cdot Id) \leq \operatorname{rank}_V(A - m \cdot Id),$$

so the conjugacy class restricts correctly.)

We can write down an explicit equation

$$p_n(t) = t^n + \Sigma_{k=2}^n f_k t^{n-k} = 0$$

for the spectral curves. Here $t$ is a coordinate on the fibers of $\omega_E(\infty)$, and $f_k \in \Gamma(\mathcal{O}_E(k\infty))$. (We identify $\mathcal{O}_E \approx \omega_E$). Once we have such a $p_n$ for each $n$, the previous remark and the tracelessness condition imply that the general spectral curve has equation

$$C_u : p_n + \Sigma_{\ell=0}^{n-2} u_\ell p_\ell = 0,$$

with arbitrary constants $u_\ell$ which can therefore be interpreted as coordinates on the quantum moduli-space, the base of the integrable system.

The condition on $p_n$ which specifies the right conjugacy class at infinity is that when we substitute $t = t' + \xi^{-1}$, where $\xi$ is a local coordinate on $E$ at $\infty$, we get a function of $t'$ and $\xi$ with at most first order poles in $\xi$. Write the equation of $E$ in Weierstrass form,

$$E : y^2 = x^3 + bx - c.$$

The local coordinate $\xi$ determines a sequence $x_k, k = 2, 3, \ldots,$ of rational functions on $E$ which are regular on $E - \infty$ and such that near $\infty$,

$$x_k - \xi^{-k} = O(\xi^{-1}).$$

We may as well use the local coordinate $\xi = x^{-1/2}$, and then the $x_k$ can be read off the Taylor expansions:

$$y = \xi^{-3}(1 + b\xi^4 - c\xi^6)^{1/2} = \xi^{-3}(1 + \frac{b}{2}\xi^4 - \frac{c}{2}\xi^6 - \frac{b^2}{8}\xi^8 + \frac{bc}{4}\xi^{10} - \frac{2c^2 - b^3}{16}\xi^{12} + \ldots)$$

and

$$y^{-1} = \xi^3(1+b\xi^4-c\xi^6)^{-1/2} = \xi^3(1-\frac{b}{2}\xi^4+\frac{c}{2}\xi^6+\frac{3b^2}{8}\xi^8-\frac{3bc}{4}\xi^{10}+\frac{6c^2-5b^3}{16}\xi^{12}+\ldots).$$

Explicitly, we get:

$$\begin{aligned}
x_2 &= x \\
x_3 &= y \\
x_4 &= x^2 \\
x_5 &= xy \\
x_6 &= x^3 \\
x_7 &= x^2y - \frac{b}{2}y \\
x_8 &= x^4 \\
x_9 &= x^3y - \frac{b}{2}xy + \frac{c}{2}y \\
x_{10} &= x^5 \\
x_{11} &= x^4y - \frac{b}{2}x^2y + \frac{c}{2}xy + \frac{3b^2}{8}y \\
x_{12} &= x^6,
\end{aligned}$$

and so on. In terms of this basis, one checks by an easy induction argument that

$$p_n = t^n - \Sigma_{k=2}^n (-1)^k (k-1) \binom{n}{k} x_k t^{n-k},$$

so the first few polynomials are:

$$\begin{aligned}
&1\\
&t\\
&t^2 - x\\
&t^3 - 3xt + 2y\\
&t^4 - 6xt^2 + 8yt - 3x^2\\
&t^5 - 10xt^3 + 20yt^2 - 15x^2t + 4xy\\
&t^6 - 15xt^4 + 40yt^3 - 45x^2t^2 + 24xyt - 5x^3\\
&t^7 - 21xt^5 + 70yt^4 - 105x^2t^3 + 84xyt^2 - 35x^3t + 6(x^2y - \frac{b}{2}y).
\end{aligned}$$

This is a bit cleaner than the formulas in [**DW**], mostly because we set the sum of the roots in the Weierstrass equation of $E$ to 0. A more-or-less equivalent solution was given already in [**T**], where the generating function for the $p_n$ was written in terms of theta functions and exponentials on $E$. A slightly different (but equivalent) set of polynomials was given in [**IM1**].

**Elliptic solitons**

It is curious to note that the same system arises in several rather different contexts. Here is a brief outline of one of these, following the work of Treibich and Verdier [**T**], [**TV**].

Finite dimensional solutions of the KP hierarchy (= infinite dimensional integrable system) are well understood. They are parametrized by Krichever data, consisting of a curve $C$, a point $p \in C$, a line bundle $L$, and local trivializations. Such data determine a point in the total space of the hierarchy with the property that the infinite collection of commuting KP flows emanating from the point sweep out only a finite dimensional space, isomorphic to $J(C)$. The curve $C$, the point $p$ and the trivializations remain fixed, while $L$ moves freely in $J(C)$.

An elliptic soliton is a finite dimensional solution of KP in which the first flow, $KP_1$, already closes up, i.e. evolves on an elliptic curve inside the possibly larger Jacobian $J(C)$. The direction of $KP_1$, for points coming from Krichever data, is the tangent vector at $p$ to the Abel-Jacobi image of $C$ in $J(C)$. So the condition is that $J(C)$ should contain an elliptic curve $E$ which is tangent at $AJ(p)$ to the Abel-Jacobi image $AJ(C)$.

Start with an $n$-sheeted branched cover $\pi : C \to E$ and a point $p \in E$ with local coordinate $\xi$. Let $\pi^*(p) = q_0 + \cdots + q_{n-1}$, and let $\xi_i := \xi \circ \pi$ near $q_i$. The tangent vector $t_i := \partial/\partial\xi_i$ lives naturally in $\Gamma(\mathcal{O}_{q_i}(q_i))$. Its image

$$(AJ)_* t_i \in T_{AJ(q_i)} J(C)$$

can then be identified with the image of $t_i$ under the coboundary map $\delta$ for the sequence

$$0 \to \mathcal{O}_C \to \mathcal{O}_C(q_i) \to \mathcal{O}_{q_i}(q_i) \to 0.$$

Similarly, the tangent at $\pi^*p$ to the image $\pi^*E$ of $E$ in $J(C)$ is given by $\Sigma_{i=0}^{n-1}\delta(t_i) = \delta(\Sigma_{i=0}^{n-1}t_i)$. The tangency condition then says that $\delta(\Sigma_{i=0}^{n-1}t_i)$ is proportional to $\delta(t_0)$, and therefore equal (by the residue theorem) to $\delta(nt_0)$. This happens if and only if there is some $f \in \Gamma(\mathcal{O}(\pi^*p))$ with

$$\mathrm{Res}_{q_i}(f\pi^*dz) = \begin{cases} 1 & 1 \le i \le n-1 \\ 1-n & i = 0, \end{cases}$$

where $dz$ is a non-zero closed 1-form on $E$. This is equivalent to saying that $\pi : C \to E$ is one of our spectral curves, and $f\pi^*dz$ becomes the Seiberg-Witten differential.

**Tests**

What makes this description of SYM with adjoint matter convincing for a physicist, is that it can be "tested" by working out various limits and special cases, and that it yields results consistent with the "predictions" based on the expected behavior of the physical system. We outline some of these tests.

**Consistency with Seiberg-Witten**

For $n = 2$, with given $E$ and $m$, our system produces a 1-parameter family of double covers $C_u \to E$. Each $C_u$ has genus 2, and the integrable system is the family of 1-dimensional Pryms. Write the equation of $E$ as $y^2 = \Pi_{i=1}^3(x - e_i) = x^3 + bx - c$, with $\sum e_i = 0$ as before. The cover $C_u$ is given explicitly in terms of our polynomials $p_i$ as:

$$0 = p_2 + up_0 = t^2 - x + u.$$

Its Galois group over the $\mathbb{P}^1$ with coordinate $x$ is $\mathbb{Z}_2 \times \mathbb{Z}_2$, with involutions $\alpha : y \mapsto -y$, $,\beta : t \mapsto -t$, and $\alpha\beta$. The quotients are a rational curve (with coordinate $t$); the original $E$; and another elliptic curve $E_u$, respectively.

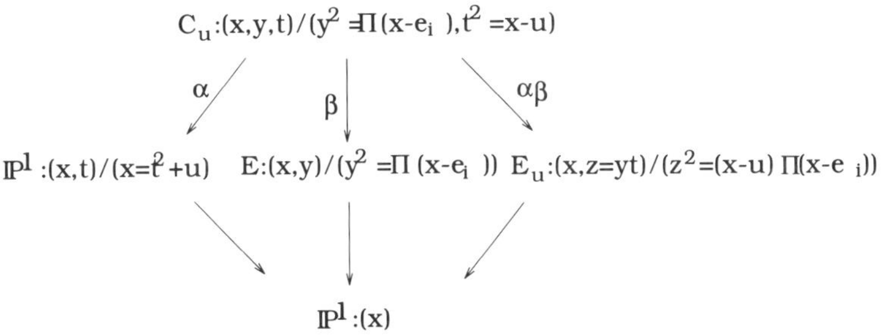


The Prym can then be identified with $E_u$.

The original Seiberg-Witten solution is given directly as the family of elliptic curves $E'_u$, double covers of $\mathbb{P}^1$ branched at $e_iu + e_i^2$ $(i = 1, 2, 3)$ and $\infty$ ([**SW2**], eq. (16.17)). These are indeed isomorphic to our $E_u$: the transformation

$$x \mapsto \frac{(u^2 + b)x - c}{u - x}$$

takes the $e_i$ and $u$ to $e_iu + e_i^2$ and $\infty$, respectively.

**Mass to zero: the N=4 limit**

The limit $m \to 0$ of $N = 2$ SYM with gauge group $G$ and adjoint matter is known to acquire a higher form of supersymmetry: it is the $N = 4$ SYM theory, with the same gauge group. (A physicist might rather say that two of the four supersymmetries of $N = 4$ SYM are broken when the theory is perturbed by the addition of a mass term to the Lagrangian). The extra symmetry of the $N = 4$ theory means that "everything" (i.e. the ingredients of the special geometry package encountered in §2: the quantum moduli space, the Kahler metric, the prepotential...) can be computed explicitly. The outcome matches perfectly to the corresponding quantities for Hitchin's system [**Hi**], i.e. the system of holomorphic (no poles) Higgs bundles on $E$. (This computation is in §2.3 of [**DW**]). So the physics predicts that our system should specialize, as $m \to 0$, to Hitchin's for $SU(n)$, which it of course does. (This is not really an honest "test", since we started out searching for deformations of Hitchin's system, so this limit was built in from the beginning).

**$\tau$ to $\infty$: the flow to pure N=2**

One of the most interesting limits occurs as both $m$ and $\tau$ go to infinity simultaneously, with $\tau$ proportional to $\log(m)$. Equivalently, after an obvious rescaling, we can keep $m = 1$ and let the elliptic curve degenerate ($\tau \to \infty$) while adjusting the coefficients $u_\ell$ in the spectral curves $C_u$.

The physics of this limit is easy to work out. Essentially, the mass going to infinity means that in this limit the particle is so heavy that it cannot be affected by the rest of the system; we get the pure N=2 theory, with no mass. This theory was solved in [AF] and [KLTY]. The solution does not involve $\tau$, as we saw in the section on adding matter: instead of $\tau$ having a limit (which was there called $\tau_{cl}$), it now goes to $\infty$. The spectral curves found in [AF] and [KLTY] have the explicit equation:

$$C'_b : w^2 = (z^n + b_2 z^{n-2} + \cdots + b_n)^2 + 1$$

in the $(z, w)$-plane, where the coefficients $b_k$ play a role analogous to our $u_\ell$.

In [DW], we show that these curves are indeed recovered as the appropriate limits of the spectral curves for the theory with adjoint mass. This is done by substituting the $\tau \to \infty$ limit into the explicit equations obtained above for the $C_u$, and performing the relevant rescalings. Instead of repeating the derivation here, we will describe the geometry of this degeneration.

The pure N=2 curve $C'_b$ is the fibre product over $\mathbb{P}^1$ (with coordinate $v$) of the rational two-sheeted cover $w^2 = v^2 + 1$ and the (again rational, with coordinate $z$) n-sheeted cover $v = z^n + b_2 z^{n-2} + \cdots + b_n$. In the adjoint theory, the analogue of the double cover is just the elliptic curve $E \to \mathbb{P}^1$ (which becomes rational in the limit $\tau \to \infty$), and $C_u$ is again an n-sheeted branched cover of $E$, but it is no longer the fiber product of $E$ with any $n$-sheeted cover of $\mathbb{P}^1$. The degree 2 map $\psi : C \to \mathbb{P}^1_z$ and the degree n map $\mathbb{P}^1_z \to \mathbb{P}^1_v$ arise only in the limit: $C$ becomes hyperelliptic.

The type of corank-1 degeneration involved in going from $C_u$ to $C'_b$ can be seen more generally. Consider a degree n branched cover $\pi : C \to E$ where $E$ is elliptic and $C$ has genus n. Up to isogeny, the Jacobian decomposes $J(C) \approx P \times E$, where $P$ is the Prym. There are the Abel-Jacobi and Abel-Prym maps:

$$AJ : C \to J(C)$$
$$AP : C \to P$$

and their derivatives

$$\phi : C \to \mathbb{P}^{n-1}$$

$$\psi : C \to \mathbb{P}^{n-2}.$$

Here $\phi$ is the canonical map and $\psi$ is the composition of $\phi$ with linear projection from the point $0 \in \mathbb{P}^{n-1}$ determined by the isogeny.

In the degeneration, we want each of $E, C$ to acquire a node: the singular curves are $E_0, C_0$, with normalizations $E', C'$, and singular points $p_0 \in E_0, q_0 \in C_0$ with branches $p_i, q_i, i = 1, 2$ in the normalizations. The Jacobians $J(C), J(E)$ become non-compact: $J(E_0)$ is a $C^*$, while $J(C_0)$ is a $C^*$-extension of $J(C')$; but the Prym $P$ remains compact. This type of degeneration occurs when $\pi' : C' \to E'$ is totally ramified above each of the $p_i$: ${\pi'}^{-1}(p_i) = nq_i$. In this case we still have the Abel-Jacobi and Abel-Prym maps, going to the compactified Jacobians. The derivative $\phi$ is still the canonical map of the singular $C_0$; its image has a node at $0 = \phi(q_0)$. On the other hand, $\psi$ is no longer a morphism on $C_0$; its restriction $\psi' : C' \to \mathbb{P}^{n-2}$ is now the canonical map of $C'$.

Within the $2n - 2$-dimensional family of all covers $\pi : C \to E$ we restrict attention to the $n$-dimensional family of our spectral curves. Here there exist points $p \in E, q \in C$ such that $AJ(C)$ at $q$ is tangent to $\pi^*(E)$ at $p$. This is equivalent to saying that $AP : C \to P(C/E)$ is ramified at $q$, or that the linear system giving $\psi$ has a base point at $q$, or that $\phi(q) = 0$.

Combining these two effects, we find that when the corank-1 degeneration described above occurs within the family of spectral curves, the two branches $\phi(q_1), \phi(q_2)$ are tangent at 0. Equivalently, the canonical map $\psi'$ satisfies $\psi(q_1) = \psi(q_2)$; so $C'$ is hyperelliptic, $\psi$ is two-to-one, and $\mathcal{O}(p_1 + p_2)$ is the hyperelliptic bundle. It follows immediately that the two maps, of degrees 2 and $n$ respectively, from $C'$ to $\mathbb{P}^1$ fit in a cartesian diagram. The corank-1 degeneration of the spectral curves (for the adjoint theory) therefore reproduces exactly the spectral curves of the pure $N = 2$ theory.

**Higgs to $\infty$: symmetry breaking**

This time we fix both the curve $E$ and the mass $m$, and instead we flow to the theory whose gauge group is some Levi subgroup $G \subset SU(n)$. Fix a diagonal matrix $\psi$, and consider the flow sending a Higgs bundle $(V, \varphi)$ in our system to $(V, \varphi_s)$ where $\varphi_s := \varphi + s\psi$. (As before we decompose $V = \oplus L_i$ and let $\psi$ act diagonally. The flow is well-defined only on the $n!$-sheeted cover of the system on which this decomposition exists). The physics predicts that the limit of the $SU(n)$ integrable system as $s \to \infty$ along this flow should be the $G$-system, where $G \subset SU(n)$ is the centralizer of $\psi$.

For generic $\psi$, with distinct eigenvalues, this is clear: After rescaling we have $\varphi_s \sim \varphi'_s := \psi + \frac{1}{s}\varphi$, so the limiting spectral curve is $\det(t - \psi) = 0$, which consists of $n$ copies of $E$ meeting at $\infty$. The Prym is then the product $J(E)^{(n-1)}$, independent of the initial $(V, \varphi)$. This trivial system corresponds to the Abelian structure group $U(1)^{n-1}$ which is the maximal torus of $SU(n)$ and the centralizer of $\psi$.

In general, $\psi$ will have, say, $k$ distinct eigenvalues with multiplicities $n_1, \ldots, n_k$, $\Sigma n_i = n$. The centralizer is

$$G = U(1)^{k-1} \times \Pi_{i=1}^{k} SU(n_i).$$

Now the limit of the spectral curves of $\varphi'_s$ is non-reduced, consisting of $k$ components, the $i$-th with multiplicity $n_i$. What we need is the limit of the Jacobians. This clearly has a factor (up to isogeny) from each component, and by rescaling around the $i$-th eigenvalue $\psi_i$ of $\psi$ (i.e. setting $t - \psi_i = s^{-1}\tilde{t}$) we see that this factor is the Jacobian of the spectral curve

$$\det(\tilde{t} - \varphi_i),$$

where $\varphi_i$ is the $n_i \times n_i$ dimensional $i$-th block of $\varphi$. The claim is then that this is one of the $SU(n_i)$ spectral curves, i.e. that the residue at $\infty$ of the $i$-th block is conjugate to $m \cdot$ diag $(1, 1, \ldots, 1, 1 - n_i)$. An equivalent condition is: rank $(\varphi_i - m) \leq 1$, and this condition is clearly inherited by any block in $\varphi$, as required.

**Singularities**

Much of the effort in [**DW**] was devoted to analyzing the discriminant, the hypersurface $D \subset B$ over which the spectral curves are singular. Physically, that is where all the interesting phenomena occur, and where many predictions can be made. For instance, as $\tau \to \infty$ ("at weak coupling"), the limit of $D$ should be reducible: this is a classical limit, where the mass spectrum can be analyzed directly from the classical Lagrangian. $D$ is the locus where some mass vanishes. This mass can belong to either of the two types (vector or hyper) of $N = 2$ SUSY multiplets mentioned in section 1 ("Adding matter"), and this gives rise to the two components. In fact, one knows the explicit equations in $B$ for these components. A Maple assisted computation, using the explicit equations obtained above for the spectral curves, confirmed the reducibility and produced the precise predicted factors for $SU(3)$.

For any $\tau$ (and any $n$), the most interesting locus in $B$ is the $(n-1)$-fold self-intersection of $D$, a finite set. Physically, this corresponds to the massive vacua, where the (low energy) gauge group is broken to its maximal torus by monopole condensation. A classification result due to 't Hooft predicts that the possible massive vacua should correspond to index $n$ subgroups of $\mathbb{Z}_n \times \mathbb{Z}_n$.

The corresponding result can be obtained directly from the integrable system. The general spectral curve has genus $n$, and we are imposing the maximal number, $n - 1$, of nodes. The normalized spectral curve is then of genus 1, and must be an unramified $n$-sheeted cover of the base $E$. These covers are of course classified by the subgroups as above. Conversely, given any $n$-sheeted cover $\pi : E' \to E$ and a point $\infty'$ above $\infty \in E$, there is a unique meromorphic 1-form on $E'$ with residue $m$ at $\pi^{-1}(\infty) \backslash \infty'$ and residue $(1 - n)m$ at $\infty'$ (modulo the constant form $\pi^* dz$), hence a unique map of $E'$ to the total space of $\omega_E(\infty)$ whose image is a spectral curve, as required. (Changing the choice of $\infty'$ amounts to a deck transformation, leaving the spectral curve unchanged).

We also analyzed, assisted with some massive and miraculous-looking Maple computations, the locus of cusps in the $SU(3)$ theory, again confirming some rather delicate predictions coming from the physics. It seems that there is some very beautiful geometry still waiting to be understood in the structure of the discriminant for the $SU(n)$ theories.

## §4. Other integrable systems

Having studied in some detail the system which solves one particular family of SYM theories, we proceed to consider what is known for some of the other families. Our discussion will include,

- The pure $N = 2$ SYM for arbitrary gauge group $G$, and its relation to the Toda systems.
- The situation when fundamental matter is added to the pure theory.

- The theory with adjoint matter for arbitrary $G$, and its relation to the elliptic Calogero-Moser system.
- A few words about the integrable systems arising in string theory.

**Pure N=2 SYM**

The solutions by Seiberg and Witten of the pure $N = 2$ SYM theory [**SW1**] and the theories with fundamental or adjoint matter [**SW2**] led rapidly to the creation of a whole literature of solutions to specific SYM theories. The general pattern was: postulate the general form of a family of curves which builds-in some of the expected symmetries of the theory, then adjust coefficients to enforce all other available information: further symmetries, limiting theories which were already known, internal consistencies. All cases which were worked out this way, e.g. [**KLTY**], [**AF**], [**DS**], [**HO**], [**APS**], [**BL**], [**AS**], [**H**], [**AAG**] and further references therein, were based on postulating curves which are hyperelliptic or closely related to hyperelliptics.

It was observed in [**GKMMM**] that the curves obtained in [**KLTY**], [**AF**] for the pure $G = SU(n)$ theory are precisely the spectral curves of the periodic Toda system. The analogue for all semisimple groups was found in [**MW**], which we follow here.

**The Toda system**

A simple way to describe the Toda system is via the construction of Adler-Kostant-Symes. Start with a decomposition $\mathfrak{g} = \mathbf{a} \oplus \mathbf{b}$ of a Lie algebra $\mathfrak{g}$ as vector space direct sum of two of its subalgebras $\mathbf{a}, \mathbf{b}$. This induces a dual decomposition $\mathfrak{g}^* = \mathbf{a}^* \oplus \mathbf{b}^*$, an injection $i := \mathbf{b}^* \hookrightarrow \mathfrak{g}^*$, and a projection $\pi : \mathfrak{g} \twoheadrightarrow \mathbf{b}$. Let $\mathcal{O}_x \subset \mathbf{b}^*$ denote the coadjoint orbit of some $x \in \mathbf{b}^*$. The image $i(\mathcal{O}_x) \subset \mathfrak{g}^*$ has, in general, nothing to do with the coadjoint orbit $\mathcal{O}_{i(x)}$ of $i(x)$ in $\mathfrak{g}^*$. So $i^*$ of a $G$-invariant function $f$ on $\mathfrak{g}^*$ can be a complicated, non-constant function on $\mathcal{O}_x$. The AKS claim is that any two such, $i^* f_1$ and $i^* f_2$, Poisson commute with respect to the natural (Kirillov-Kostant) symplectic structure $\sigma_{\mathcal{O}_x}$ on $\mathcal{O}_x$. Indeed, let $v_j \in \mathfrak{g}$ correspond to the differential $df_j|_{i(x)} \in \mathfrak{g}^{**} = \mathfrak{g}$, so $i^* df_j$ is given by $\pi(v_j)$. $G$-invariance of $f_j$ implies that $v_j$ is conormal to $\mathcal{O}_{i(x)}$, i.e. that

$$0 = \langle i(x), [\mathfrak{g}, v_j] \rangle.$$

We set $a_j := v_j - \pi(v_j) \in \mathbf{a}$, and compute:

$$\begin{aligned} \{i^* f_1, i^* f_2\}_{\mathcal{O}_x} &= \sigma_{\mathcal{O}_x}(\pi v_1, \pi v_2) \\ &= \langle x, [\pi(v_1), \pi(v_2)]_{\mathbf{b}} \rangle \\ &= \langle i(x), [\pi(v_1), \pi(v_2)]_{\mathfrak{g}} \rangle \\ &= \langle i(x), [v_1, \pi(v_2)]_{\mathfrak{g}} \rangle - \langle i(x), [a_1, \pi(v_2)]_{\mathfrak{g}} \rangle \\ &= \langle i(x), [v_1, \pi(v_2)] \rangle - \langle i(x), [a_1, v_2] \rangle + \langle i(x), [a_1, a_2] \rangle. \end{aligned}$$

We have just seen that the first two terms vanish when $f_1, f_2$ are $G$-invariant, and the third term vanishes since it pairs $i(x) \in \mathbf{b}^* = \mathbf{a}^\perp$ with $[a_1, a_2] \in \mathbf{a}$. This proves AKS.

A shifted version of AKS is very convenient in applications. Replace the inclusion $i : \mathbf{b}^* \hookrightarrow \mathfrak{g}^*$ by its shifted version $\epsilon + i$, where $\epsilon$ is an element of $\mathfrak{g}^*$ which annihilates $[\mathbf{a}, \mathbf{a}]$ and $[\mathbf{b}, \mathbf{b}]$. The above argument goes through verbatim. For example, take $\mathfrak{g} = \mathbf{sl}(n)$, $\mathbf{a} =$ strictly lower triangular matrices, $\mathbf{b} =$ upper triangulars, and (after identifying $\mathfrak{g}$ with $\mathfrak{g}^*$, so that $\mathbf{b}^*$ becomes the lower triangulars):

$$\epsilon = \begin{pmatrix} 0 & 1 & & & \\ & 0 & & & \\ & & \ddots & & \\ & & & 0 & 1 \\ & & & & 0 \end{pmatrix}, x = \begin{pmatrix} 0 & & & & \\ 1 & 0 & & & \\ & & \ddots & & \\ & & & 0 & \\ & & & 1 & 0 \end{pmatrix}.$$

The orbit $\mathcal{O}_x \subset \mathbf{b}^*$ consists of all matrices of the form:

$$\begin{pmatrix} b_1 & & 0 \\ a_1 & \ddots & \\ & \ddots & \\ 0 & & a_{n-1} b_n \end{pmatrix}$$

with $a_j \neq 0$ and $\Sigma b_j = 0$. The shifted orbit consists of tridiagonal matrices:

$$\begin{pmatrix} b_1 & 1 & & & 0 \\ a_1 & \cdot & \cdot & & \\ & \cdot & \cdot & \cdot & \\ & & \cdot & \cdot & 1 \\ 0 & & & a_{n-1} & b_n \end{pmatrix}.$$

This is a $2(n-1)$-dimensional symplectic manifold $(X, \sigma)$, isomorphic as symplectic manifold to the cotangent bundle of the maximal torus $T \approx (\mathbb{C}^*)^{n-1}$. The restriction to it of the symmetric polynomials of $SL(n)$ (= coefficients of the characteristic polynomial) gives $n-1$ independent, commuting Hamiltonians. This is the classical Toda system.

An equivalent system is obtained if $\epsilon$ is replaced by $m \cdot \epsilon$ with $m \in \mathbb{C}^*$. When $m = 0$, the Hamiltonians become the symmetric functions of the $b_i$. We get the cotangent bundle $X$ of $(\mathbb{C}^*)^{n-1}$, fibered by its moment map over $(\mathbb{C})^{n-1}$. Another variant is to take $\epsilon = 0$, but replace $\mathbf{a}$ by the algebra $\mathbf{so}(n)$ of skew symmetric matrices. Now $\mathbf{b}^*$ sits in $\mathbf{sl}(n)$ as the symmetric matrices, and the orbit consists of symmetric tridiagonal matrices

$$\begin{pmatrix} b_1 & c_1 & & & \\ c_1 & b_2 & & & \\ & & \ddots & & \\ & & & b_{n-1} & c_{n-1} \\ & & & c_{n-1} & b_n \end{pmatrix}.$$

This is conjugate to the previous matrix with $a_i = c_i^2$, so we obtain a new version $(\tilde{X}, \tilde{\sigma})$ of Toda which is just a $2^{n-1}$-sheeted cover of the previous, basic version, $(X, \sigma)$.

The Toda system for a semisimple Lie algebra $\mathfrak{g}$ is obtained with $\mathbf{b}$ a Borel subalgebra, $\mathbf{a}$ the nilpotent radical of the opposite Borel, $x = \Sigma e_\alpha$ and $\epsilon = \Sigma e_{-\alpha}$, where the sums are over the simple roots $\alpha$ with respect to $\mathbf{b}$, and the $e_{\pm\alpha}$ are normalized root vectors. The total space can again be identified with the cotangent bundle of the maximal torus, a typical element having the form $\sum_{\alpha \text{ simple}} (b_\alpha h_\alpha + a_\alpha e_\alpha + e_{-\alpha})$ with respect to a standard basis $h_\alpha, e_{\pm\alpha}$.

**Periodic Toda**

The periodic Toda systems are obtained when the semisimple algebra $\mathfrak{g}$ is replaced by a loop algebra $\mathfrak{g}^{(1)}$. This has an extended Dynkin diagram consisting of the simple roots $\alpha$ of $\mathfrak{g}$ plus the affine root $\alpha_0$, corresponding to the highest root of $\mathfrak{g}$. The general element of the periodic Toda system can thus be written as

$$\Sigma_\alpha(b_\alpha h_\alpha + a_\alpha e_\alpha + e_{-\alpha}) + ze_{\alpha_0} + z^{-1}a_0 e_{-\alpha_0},$$

where $z$ is the loop variable, and the product $a_0 \prod_\alpha a_\alpha = \mu$ is a fixed non-zero constant. (For $\mathfrak{g} = \mathbf{sl}(n)$, this is:

$$\begin{pmatrix} b_1 & 1 & & a_0 z^{-1} \\ a_1 & \ddots & \ddots & \\ & \ddots & \ddots & 1 \\ z & & a_{n-1} & b_n \end{pmatrix},$$

with $\prod a_i = \mu, \Sigma b_j = 0$).

As in the case of Higgs bundles, the spectral curves $C_{\rho,b}$ depend on the choice of a representation $\rho$ of $G$. These Toda curves and their relationships were first studied in [**Ml**] and [**MS**], with some clarifications added in [**MW1**]. Each $C_{\rho,b}$ is a branched cover of the $z$-line $\mathbb{P}^1_z$. The latter plays the role for periodic Toda which the base curve $E$ played for the meromorphic Higgs system in §2. The cameral cover in this situation has a strikingly simple description. The classifying map $\mathbb{P}^1_z \dashrightarrow \mathbf{t}/W$, which is usually only locally defined, is now a morphism for $z \neq 0, \infty$. It factors through the double cover

$$\mathbb{P}^1_z \to \mathbb{P}^1_w$$
$$z \mapsto w = z + \mu z^{-1}$$

which maps $\mathbb{P}^1_z \backslash \{0, \infty\}$ to $\mathbb{A}^1_w = \mathbb{P}^1_w \backslash \{\infty\}$, followed by an affine linear map

$$\mathbb{A}^1_w \mapsto \mathbf{t}/W$$

whose image in $\mathbf{t}/W$ is a straight line pointing in the direction of the invariant polynomial of highest degree, which is equal to the Coxeter number $h_\mathfrak{g}$ of $\mathfrak{g}$. More precisely, the algebra of polynomial functions on $\mathbf{t}/W$ is isomorphic to the algebra of invariant polynomials on $\mathbf{t}$, which is filtered by the degree. The equations of the image of $\mathbb{A}^1_w$ are obtained by fixing the values of all polynomials on $\mathbf{t}/W$ whose pullback to $\mathbf{t}$ has degree $< h_\mathfrak{g}$. The induced $W$-cover of $\mathbb{P}^1_w$ then has $r =$ rank($\mathfrak{g}$) branch points in $\mathbb{A}^1_w$, and one at $w = \infty$. The choice of a Borel labels the finite branch points $w_\alpha$ by the simple roots $\alpha$ in such a way that the monodromy at $w_\alpha$ is the reflection corresponding to $\alpha$. The monodromy at $\infty$ is then the product of these reflections, the Coxeter element. The Cameral cover is the pullback to $\mathbb{P}^1_z$. It can be described as the $W$-cover obtained from the trivial cover $W \times \mathbb{P}^1_z$ by making $r + 1$ cuts labelled by the extended set of roots $\{\alpha\} \cup \{\alpha_0\}$, each running from a point $z_\alpha$ to $\mu z_\alpha^{-1}$, and pasting across each cut according to the corresponding element of $W$.

In addition to the loop algebras $\mathfrak{g}^{(1)}$, there are twisted versions $\mathfrak{g}^{(k)}$ where $k$ is the order of an automorphism of the Dynkin diagram of $\mathfrak{g}$, and each of these also gives rise to a periodic Toda system. The basic observation of Martinec and Warner [**MW1**] is that each of the known families of pure $N = 2$ SYM curves arises as spectral curves for one of these twisted periodic Toda systems. The specific identification involves Langlands duality (Olive-Montonen duality, to a physicist) exchanging the direction of arrows in the extended diagram. The (untwisted) extended diagrams for types $A, D, E$ are self-dual, yielding ordinary Toda systems. The remaining cases yield twisted algebras and systems:

$$(B_r^{(1)})^\vee = A_{2r-1}^{(2)}, (C_r^{(1)})^\vee = D_{r+1}^{(2)}, (F_4^{(1)})^\vee = E_6^{(2)}, (G_2^{(1)})^\vee = D_4^{(3)}.$$

We will now describe explicitly the spectral curves for some of these systems.

**The spectral curves**

For the classical groups we can write the equation of the Toda spectral curves $C_{F,u}$ for the fundamental representation $F$ as

$$w = z + \mu z^{-1}, x^\epsilon w + p = 0$$

where

$$p = p(x) = x^{n+1} + u_2 x^{n-1} + \cdots + u_{n+1}$$

for type $A_n$ $(G = SL(n+1))$, and

$$p = p(x^2) = x^{2n} + u_2 x^{2n-2+} \ldots + u_{2n}$$

for types $B_n, C_n$, and $D_n$, and the $u_\ell$ are coordinates on the base. Here $\epsilon = 0$ for $A_n, B_n, C_n$ and $\epsilon = 2$ for $D_n$. In general, $\deg_x(p)$ is the dimension of the representation, while the $w$ term modifies the coefficient $u_{h_\mathfrak{g}}$ corresponding to the invariant of highest degree. For $A_n, B_n, C_n, D_n$, these Coxeter numbers are $n+ 1, 2n, 2n, 2n-2$. (The $B_n$ case is slightly different since 0 is a weight of $F$. So the actual "spectral" curve has equation $x(w + p(x)) = 0$. We may safely ignore the $x$ factor, since the distinguished Prym lives in the Jacobian of the other factor).

For $A_n$, the fundamental curve has genus $n$, so its Jacobian is the distinguished Prym. It can be described geometrically as the fiber product over $\mathbb{P}^1_w$ of the double cover $w = z+\mu z^{-1}$, branched at $w = \pm 2\sqrt{\mu}$, and the degree $n+1$ cover $w = -p(x)$, with total ramification over $w = \infty$ and $n$ simple branch points at the images of the roots of $p'(x) = 0$. The substitution $y = 2z + p(x)/2$, $4\mu = \Lambda^{2(n+1)}$ converts it to

$$y^2 = p^2(x) - \Lambda^{2(n+1)},$$

the form obtained in [**AF**] and [**KLTY**]. (Here we have reinstated the scale parameter $\Lambda$ of the asymptotically free theory. Previously this was set to 1).

The other spectral curves for $A_n$ are easily obtained from the fundamental one. The curve for $\wedge^2 F$ is of degree $\frac{n(n+1)}{2}$ over the $z$ line, and of genus $n(n-1)+1-[\frac{n+1}{2}]$. The obvious correspondence between it and $C_{F,u}$ produces a copy of $J(C_{F,u})$ in $J(C_{\wedge^2 F,u})$. The $A_3$ case is described in [**Ml**], [**MS**]: The tetragenal curve $C_{F,u}$ of genus 3 gives rise via Recillas' trigonal construction (cf. [**D3**]) to the degree 6, genus 5 curve $C_{\wedge^2 F,u}$, together with an involution whose quotient is a trigonal curve $C'$ of genus 2. The Jacobian of $C_{F,u}$ is isogenous to $\mathrm{Prym}(C_{\wedge^2 F,u}/C')$. (Actually, the degree 6 and 3 curves produced by the trigonal construction have genera 7 and 4 respectively, but each has 2 nodes, over $z = 0$ and $z = \infty$; the curves $C_{\wedge^2 F,u}$ and $C'$, of genera 5 and 3, are their respective desingularizations). A beautiful picture of the $g = 11$ curve $C_{\wedge^2 F,u}$ for $A_4$ is given in [**MW1**]. In contrast to these small curves, there is the gigantic cameral cover, of degree $n!$ over $\mathbb{P}^1_z$ and of genus $1 + \frac{n-1}{2(n+1)}(n+2)!$, whose Jacobian contains those of all (irreducible components of) spectral covers.

For $D_n, C_{F,u}$ has equation $x^2w + p(x^2) = 0$. This depends on $n$ parameters, but has genus $2n - 1$, too big. The quotient $C'$ by the involution $x \mapsto -x$ is hyperelliptic of genus $n-1$; the $n$-dimensional piece is Prym $(C_{F,u}/C')$, and it can be realized as the Jacobian of a curve $C''$, the quotient of $C_{F,u}$ by the involution $x \mapsto -x, z \mapsto \mu z^{-1}$. The substitution $y := x^2(z - \mu z^{-1})$, $4\mu = \Lambda^{4(n-1)}$ takes the equation of $C_{F,u}$ to

$$y^2 = p^2(x^2) - \Lambda^{4(n-1)}x^4,$$

the form of the $SO(2n)$ SYM curve obtained in [**BL**]. In terms of $s = xy, t = x^2$, the quotient $C''$ becomes:

$$s^2 = tp^2(t) - \Lambda^{4(n-1)}t^3.$$

For type $C_n$, the fundamental Toda curve $C_{F,u}$ has equation $w + p(x^2) = 0, w = z + \mu z^{-1}$. The substitution $y = 2z + p(x^2)$ changes this to

$$y^2 = p^2(x^2) - 4\mu.$$

Just as in the $D_n$ case, $C_{F,u}$ has automorphism group $\mathbb{Z}_2 \times \mathbb{Z}_2$ over the $t = x^2$ line:

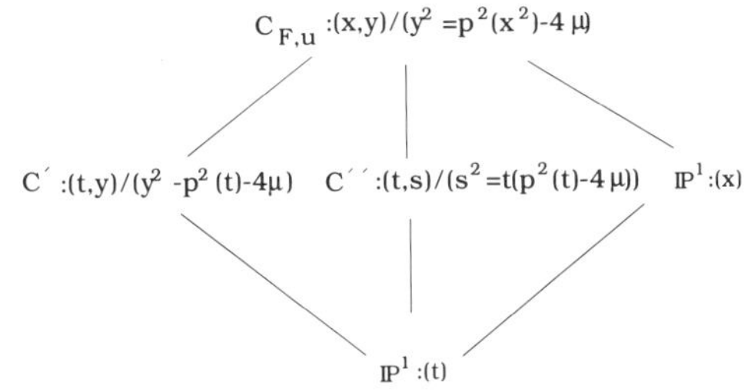


and the $n$-dimensional piece is $J(C'')$. But this is **not** the right abelian variety for the SYM theory. Rather, we should consider the Toda curves for $(C_n^{(1)})^\vee = D_{n+1}^{(2)}$. These have the form

$$w^2 + p(x^2) = 0, \; w = z + \mu z^{-1}.$$

The coordinates on the three $\mathbb{Z}_2$ quotients by the involutions $x \mapsto \pm x, z \mapsto \pm\mu z^{-1}$ are now:

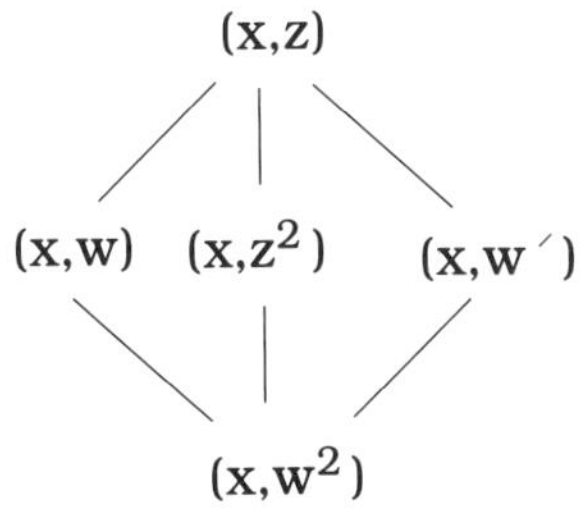

where $w' = w - 2\mu z^{-1} = z - \mu z^{-1}$. The corresponding genera are:

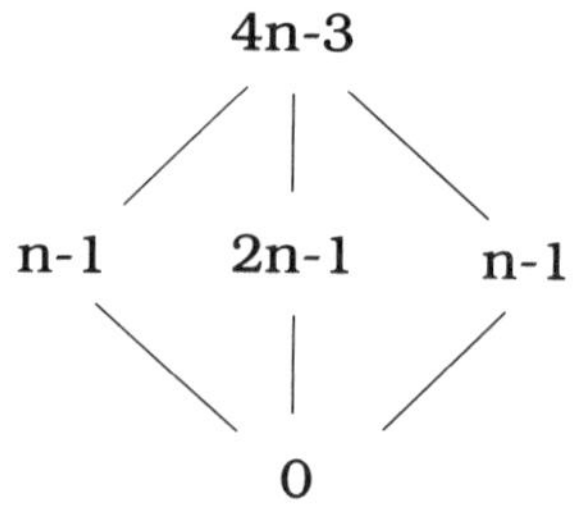


The middle curve has equation $v + \frac{\mu^2}{v} + p(x^2) - 2\mu = 0$, where $v = z^2$. It factors further: in fact, it looks just like the $C_n$-Toda curve, with $z, \mu, p$ replaced by $v, \mu^2, p - 2\mu$ respectively. So its Jacobian has an $n$-dimensional piece, which is the Jacobian of:

$$s^2 = t((p(t) - 2\mu)^2 - 4\mu^2) = tp(t)(p(t) - 4\mu).$$

This is essentially the $C_n$-curve obtained in [**AS**].

For $B_n$ the story is similar, but simpler. The dual Toda curve has equation $x(z+\mu z^{-1}) + p(x^2) = 0$. The substitution $y = 2xz + p(x^2)$, $4\mu = \Lambda^{2(2n-1)}$ converts this to the form obtained in [**DS**],

$$y^2 = p^2(x^2) - \Lambda^{2(2n-1)}x^2.$$

Again this is of genus $2n - 1$. The genus $n$ quotient is

$$s^2 = t(p^2(t) - \Lambda^{2(2n-1)}t).$$

For $G_2$, we need Toda curves for $(G_2^{(1)})^\vee = D_4^{(3)}$. These are given [**MW1**] as:

$$3(z - \mu z^{-1})^2 - x^8 + 2u_2x^6 - [u_2^2 + 6(z + \mu z^{-1})]x^4 + [u_4 + 2u_2(z + \mu z^{-1})]x^2 = 0.$$

The various quotients involve the functions:

$$\begin{aligned}
w &= z + \mu z^{-1} \\
s &= x^2 \\
u &= x(z - \mu z^{-1}) \\
v &= x(w - s^2 + \frac{u_2}{3}s).
\end{aligned}$$

Some quotient curves, indicated by the functions which generate their fields, are:

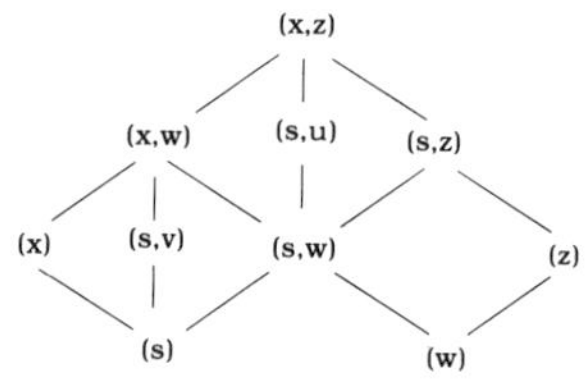

The corresponding genera are:

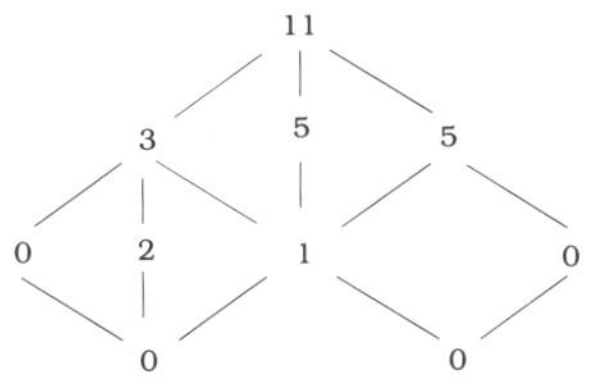


I have not checked whether the genus 2 curve with functions $(s, v)$ is the one corresponding to the $G_2$ theory. An explicit family of genus 2 curves was proposed in several works (see references in [**AAG**]) for the $G_2$ theory, but according to [**LPG**] this family does not have good physical properties, and does not match the Toda curves.

The $E_6$ curves were considered in [**LW**], and turn out to involve some beautiful geometry. The simplest curve is based on the 27-dimensional representation of $E_6$. The resulting 27-sheeted cover of the $z$ line behaves like the lines on a 1-parameter family of cubic surfaces. Each of the 6 simple roots corresponds to an ordinary double point acquired by the surface, so the local monodromy at each finite branch point is the product of 6 disjoint transpositions. Lerche and Warner study this genus-34 spectral curve, and go on to relate it to an integrable system coming from a string theory compactification on a Calabi-Yau threefold which degenerates to a fibration of the $E_6$ ALE singularity by a family of cubic surfaces.

**Fundamental matter**

The spectral curves for SYM with various gauge groups and numbers of particles ("quarks") in the fundamental representation have been determined by various means. A very incomplete list includes [**HO**], [**APS**], [**AS**], [**H**], [**MN**], [**KP**], [**DKP1**],[**DKP2**]. Each of these solutions is supported by substantial evidence, and yet the total picture seems far from complete. No really clear unifying principle is known, and there is no analogue of the Toda integrable system which is responsible for the various curves. There is also a certain level of disagreement among comparable solutions, see the discussion in [**MN**] and [**DKP1**]. (The curves involved in those solutions are the same, but the parametrizations differ. An integrable system would of course give a preferred parametrization).

**Adjoint matter**

As soon as [**MW**] and [**DW**] appeared, each with a class of integrable systems which solves its respective version of SYM, the question arose whether there was a common generalization: an integrable system for SYM with adjoint matter and arbitrary gauge group $G$, going to the periodic Toda system in the limit as $m$ and $\tau$ go together to $\infty$.

In [**Mc**], Martinec suggested that the elliptic Calogero-Moser system (cf. [**OP**]) may provide this common generalization. There is such a system for each semisimple $G$, and the case $G = SU(n)$ agrees exactly with the system in [**DW**]. There is also a direct computation [**I1**] showing that the $A_n$ and $D_n$ systems can degenerate to the respective Toda systems, and there is a deformed family of integrable systems depending on some additional parameters (the elliptic spin models, [**I2**]). A way of obtaining the $A_n$ system by symplectic reduction is given in [**GN**].

One way to try to understand the geometry of these systems is as follows. Fix an elliptic curve $E$, a semisimple group $G$, and its maximal torus $T$. Let

$\mathcal{M}, \mathcal{M}'$ denote the moduli spaces of semistable bundles on $E$ with structue groups $G, T$ respectively. A special feature of genus 1 is that the natural map $\mathcal{M}' \to \mathcal{M}$ is surjective, in fact it is a Galois cover with group $W$. (It is not true that the structure group of every semistable $G$-bundle on $E$ can be reduced to $T$; but every $S$-equivalence class of semistable bundles contains one whose structure group can be reduced). Now let $X$ be the moduli space of Higgs bundles on $E$ with first order pole at $\infty$, and let $X'$ be its finite cover induced by $\mathcal{M}' \to \mathcal{M}$. (A point of $X'$ is an equivalence class of pairs $(V, \varphi)$ where $V$ is a $T$-bundle on $E$ and $\varphi$ is a meromorphic Higgs field on $V \otimes_T G$). There is a "moment map"

$$\mu : X' \to \mathbf{t}^{\perp}/T,$$

where $\mathbf{t}^{\perp} \subset \mathfrak{g}^*$ is the annihilator of the Cartan $\mathbf{t}$, and the $T$-action is coadjoint. Now **all** the fibers of $\mu$ are symplectic, of the minimal dimension $2r$ (where $r = \text{rank}(\mathfrak{g})$), and at least for some of them (but certainly not for all!), the restrictions of all the $X$-Hamiltonians Poisson commute, giving a large family of small integrable systems. (Presumably, this involves an analogue of the AKS proof of integrability of the Toda system). As the elliptic curve degenerates to $\mathbb{P}^1$ with 0 glued to $\infty$, this construction specializes to the periodic Toda.

The question remains, though: of all the possible $T$-orbits, which one corresponds to SYM with adjoint matter? A necessary condition seems to be that the $T$ orbit $\mathcal{O}$ needs to be $W$-invariant; otherwise the limit as $m \to 0$ will not be the Hitchin system $T^*\mathcal{M}$, but its finite cover $T^*\mathcal{M}'$. (Or something in-between, if $\mathcal{O}$ is invariant under a subgroup of $W$.) For $SU(n)$, there is a unique such class (up to scalars), namely $\sum_{i \neq j} e_{ij}$. The natural analogue for arbitrary $\mathfrak{g}$ would be $\sum_\alpha e_\alpha$, the sum over all roots. Unfortunately, this is ill defined: the $e_\alpha$ can be normalized only up to sign, and different signs give distinct $T$ orbits. For the orthogonal groups, there does not seem to be any $W$-invariant $T$ orbit. David Lowe has pointed out [**L**] that for $Sp(n)$ there is an invariant orbit, namely the $T$ orbit of the defining matrix $J$. (It is a signed sum of the long root vectors). This does give an integrable system, but a rather boring one: it is the $n$-fold product of the $SU(2)$ system, which does not flow to the pure $N = 2$ $C_n$ spectral curves described earlier.

In sum, the elliptic Calogero-Moser systems (or their spin variants) seem to be a promising general class of systems for realizing the SYM theory with adjoint matter for arbitrary gauge group, but at the moment we lack a good geometric understanding of these systems, e.g. via spectral curves, except in the case of $SU(n)$.

**Stringy integrable systems?**

Quantum field theory is a limit of string theory compactified on a Calabi-Yau threefold, and the SW duality can be realized as the limiting form of various string dualities. A series of recent papers, such as [**MW2**], [**KLMVW**], [**GHL1**], [**GHL2**], [**LW**], suggest that much of the QFT structure studied in this paper - the spectral curves, integrable systems, SW differentials and so on - arises as limit of corresponding stringy structures. A possible framework for studying the transition from stringy integrable systems to the SW systems is proposed in [**F**], where SW-type integrable systems are shown to arise from rigid special geometry, while the stringy type arises from local special geometry. From a purely mathematical point of view, it should be interesting to understand the relations between the SW-type systems, whose basic input is an elliptic curve; the Mukai system on a $K3$ surface, which degenerates to Hitchin's on a curve ([**DEL**], [**DM1**]); and the Calabi-Yau systems constructed in [**DM2**].

## References


[AAG] M. R. Abolhasani, M. Alishahiha, A. M. Ghezelbash, hep-th 9606043.

[AHH] M. Adams, J. Harnad, J. Hurtubise, *Isospectral Hamiltonian flows in finite and infinite dimensions II: Integration of flows*, Commun. Math. Phys. (1990), 555-585.

[AM] M. Adler, P. van Moerbeke, *Completely integrable systems, Euclidean Lie algebras, and curves*, Advances in Math. **38** (1980), 267-379.

[AF] P. C. Argyres, A. Faraggi, *The vacuum Structure and Spectrum of $N = 2$ Supersymmetric $SU(N)$ Gauge Theory*, Phys. Rev. Lett. **74** (1995), hep-th 9411057.

[APS] P. C. Argyres, R. Plesser, A. D. Shapere,, *The Coulomb phase of $N = 2$ supersymmetric QCD*, Phys. Rev. Lett. **75:1699-1702** (1995), no. e-Print Archive: hep-th 9505100.

[AS] P. C. Argyres, A. D. Shapere,, *The vacuum structure of $N = 2$ super-QCD with classical gauge groups*, hep-th 9509175.

[B] A. Beauville,, *Jacobiennes des courbes spectrales et systemes hamiltoniens completement integrables*, Acta Math. **164** (1990), 211-235.

[Be] J. Bernstein, *Lectures on Supersymmetry, Lectures at IAS, Fall 1996*, Preprint available at http://www.math.ias.edu/drm/QFT.

[Bo] F. Bottacin, *Symplectic geometry on moduli spaces of stable pairs*, Orsay, 1992.

[BL] A. Brandhuber, K. Landsteiner, *On the Monodromies of $N = 2$ Supersymmetric Yang-Mills Theory with Gauge Group $SO(2n)$*, hep-th 9507008.

[DS] U. H. Danielsson, B. Sundborg, *The Moduli Space and Monodromies of $N = 2$ Supersymmetric $SO(2r + 1)$ Yang-Mills Theory*, hep-th 9504102.

[DKP1] E. D'Hoker, I. M. Krichever, D. H. Phong, hep-th 9609041.

[DKP2] E. D'Hoker, I. M. Krichever, D. H. Phong,, hep-th 9609145.

[D1] R. Donagi, *Decomposition of spectral covers*, in Journees de Geometrie Algebrique D'Orsay, Asterisque **218** (1993), 145-175.

[D2] R. Donagi, *Spectral Covers*, Current Topics in Complex Algebraic Geometry, (H. Clemens and J. Kollar, eds.), MSRI Pub., vol. 28, 1995, pp. 65-86, Alg Geom, 9505009.

[D3] R. Donagi, *The tetragonal construction*, Bull. Amer. Math. Soc. (N.S.) **4** (1981), 181-185.

[DEL] R. Donagi, L. Ein and R. Lazarsfeld, *A non-linear deformation of the Hitchin dynamical system*, AlgGeom 9504017.

[DM1] R. Donagi, E. Markman, *Spectral curves, algebraically completely integrable Hamiltonian systems, and moduli of bundles*, R. Donagi et al, Integrable systems and Quantum Groups, LNM 1620, (1996), 1-119 Alg Geom, 9507017.

[DM2] R. Donagi, E. Markman, *Cubics, Calabi-Yau threefolds, Integrable Systems and Mirrors*, Israel Math. Conf. Proc., vol. 9, AMS, 1995, AlgGeom 9408004.

[DW] R. Donagi, E. Witten, *Supersymmetric Yang-Mills theory and integrable systems*, Nucl. Phys B460 **299** (1996), hep-th, 9510101.

[DH] J. J. Duistermaat, G. J. Heckman, *On the Variation in the Cohomology of the Symplectic Form of the Reduced Phase Space*, Invent. Math. **69** (1982), 259.

[FS] L. D. Faddeev, A. A. Slavnov, *Gauge Fields: Introduction to Quantum Theory*, Benjamin/Cummings, 1980.

[Fa] G. Faltings, *Stable G-bundles and Projective Connections*, Jour. Alg. Geo **2** (1993), 507-568.

[F] D. S. Freed, *On special Kahler manifolds*, preprint.

[Fr] P. Freund, Introduction to supersymmetry, *Camb. Monographs on Math. Phys.*, Camb. U. Press, 1986.

[GHL1] C. Gomez, R. Hernandez, E. Lopez, hep-th 9604057.

[GHL2] C. Gomez, R. Hernandez, E. Lopez, hep-th 9608104.

[GKMMM] A. Gorsky, I. Krichever, A. Marshakov, A. Mironov, A. Morozov, hep-th 9505035.

[GN] A. Gorsky, N. Nekrasov, *Elliptic Calogero-Moser system from two-dimensional current algebra*, hep-th 9401021.

[H] A. Hanany, *On the quantum moduli space of vacua of $N = 2$ supersymmetric gauge theories*, hep-th 9509176.

[HO] A. Hanany, Y. Oz, *On the quantum moduli space of vacua of $N = 2$ supersymmetric $SU(N)$ gauge theories*, hep-th 9505075.

[Hi] N. Hitchin,, *Stable bundles and integrable systems*, Duke Math. J. **54** (1987), 91-114.

[I1] V. I. Inozemtsev, *The finite Toda lattices*, Comm. Math. Phys. **121** (1989), 629-38.

[I2] V. I. Inozemtsev, *Lax representation with spectral parameter on a torus for integrable particle systems*, J. Lett. Math. Phys. **17** (1989), 11-17.

[IM1] H. Itoyama, A. Morozov, hep-th 9511126.

[IM2] H. Itoyama, A. Morozov, hep-th 9512161.

[Ka] V. Kanev, *Spectral curves, simple Lie algebras and Prym-Tjurin varieties*, Proc. Symp. Pure Math. Part I **49** (1989), 627-645.

[KLMVW] A. Klemm, W. Lerche, P. Mayr, C. Vafa, N. Warner, hep-th 9604034.

[KLTY] A. Klemm, W. Lerche, S. Theisen, S. Yankielowicz, *Simple Singularities and $N = 2$ Supersymmetric Yang-Mills Theory*, Phys. Lett. **169** (1995), no. B344, hep-th 9411048.

[K] I. M. Krichever, hep-th 9205110.

[KP] I. M. Krichever, D. H. Phong, hep-th 9604199.

[LPG] K. Landsteiner, J. M. Pierre, S. B. Giddings,, hep-th 9609059.

[LW] W. Lerche, N. P. Warner, hep-th 9608183.

[Mn] E. Markman, *Spectral Curves and Integrable Systems*, Comp. Math. **93** (1994), 255-290.

[Mv] A. Marshakov, hep-th 9607159.

[Mc] E. Martinec, hep-th 9510204.

[MW1] E. Martinec, N. P. Warner, *Integrable Systems and Supersymmetric Gauge Theory*, hep-th 9509161.

[MW2] E. Martinec, N. P. Warner, hep-th 9511052.

[Ml] A. McDaniel, *Representations of $sl(n, \mathbb{C})$ and the Toda lattice*, Duke Math. J. **56** (1988), 47-99.

[MS] A. McDaniel, L. Smolinsky, *A Lie theoretic Galois Theory for the spectral curves of an integrable system* II, prep. (1994).

[Me] J. Y. Merindol, *Varietes de Prym d'un Revetement Galoisien*, prep. (1993).

[MN] J. A. Minahan, D. Nemeschansky, hep-th 9601059.

[MO] C. Montonen, D. Olive, *Magnetic Monopoles as Gauge Particles?*, Phys. Lett. **72B** (1977), 177.

[Mu] S. Mukai, *Symplectic structure of the moduli space of sheaves on an abelian or K3 surface*, Inv. Math. **77** (1984), 101-116.

[NT] T. Nakatsu, K. Takasaki, hep-th 9509162.

[N] N. Nekrasov, *Holomorphic bundles and many-body systems*, hep-th 9503157.

[OP] M. A. Olshanetski, A. M. Perelomov, *Classical integrable finite-dimensional systems related to Lie algebras*, Phys. Rep. **71C** (1981), 313-400.

[SW1] N. Seiberg, E. Witten, *Monopole Condensation, and Confinement in $N = 2$ Supersymmetric Yang-Mills Theory*, Nucl. Phys. **B426** (1994), 19, hep-th 9407087.

[SW2] N. Seiberg, E. Witten, *Monopoles, duality and chiral symmetry breaking in* $N = 2$ *supersymmetric QCD*, Nucl. Phys. **B431** (1994), 484-550, hep-th 9408099.

[T] A. Treibich, *Tangential polynomials and elliptic solitons*, Duke Math. J. **59** (1989), 611-627.

[TV] A. Treibich, J. L. Verdier, *Solitons Elliptiques*, The Grothendieck Festschrift III (1990), Birkhäuser, 437-480.

[W] E. Witten, *Monopoles and Four Manifolds*, hep-th/9411102.

[WB] J. Wess, J. Bagger, *Supersymmetry and supergravity*, Princeton U. Press, 1992.

DEPARTMENT OF MATHEMTICS,
UNIVERSITY OF PENNSYLVANIA,
PHILADELPHIA, PA 19104-6395

*Current address*: School of Mathematics,
Institute for Advanced Study,
Princeton, NJ 08540

Proceedings of Symposia in Pure Mathematics
Volume **62.2**, 1997

# Notes On Stable Maps And Quantum Cohomology

W. Fulton and R. Pandharipande

*Dedicated to the memory of Claude Itzykson*

CONTENTS



## 0. Introduction

**0.1. Overview.** The aim of these notes is to describe an exciting chapter in the recent development of quantum cohomology. Guided by ideas from physics (see [W]), a remarkable structure on the solutions of certain rational enumerative geometry problems has been found: the solutions are coefficients in the multiplication table of a quantum cohomology ring. Associativity of the ring yields non-trivial relations among the enumerative solutions. In many cases, these relations suffice to solve the enumerative problem. For example, let $N_d$ be the number of degree $d$, rational plane curves passing through $3d-1$ general points in $\mathbf{P}^2$. Since there is a unique line passing through 2 points, $N_1 = 1$. The quantum cohomology ring of $\mathbf{P}^2$ yields the following beautiful associativity relation determining all $N_d$ for $d \geq 2$:

$$N_d = \sum_{d_1+d_2=d,\ d_1,d_2>0} N_{d_1} N_{d_2} \left( d_1^2 d_2^2 \binom{3d-4}{3d_1-2} - d_1^3 d_2 \binom{3d-4}{3d_1-1} \right).$$

---

1991 *Mathematics Subject Classification.* Primary 14N10, 14H10; Secondary 14E99.

The first author was supported in part by NSF Grant DMS 9307922.
The second author was supported in part by an NSF Post-Doctoral Fellowship.

Similar enumerative formulas are valid on other homogeneous varieties. Viewed from classical enumerative geometry, the quantum ring structure is a complete surprise.

The path to quantum cohomology presented here follows the work of Kontsevich and Manin. The approach is algebro-geometric and involves the construction and geometry of a natural compactification of the moduli space of maps. The large and exciting conjectural parts of the subject of quantum cohomology are avoided here. We focus on a part of the story where the proofs are complete. We also make many assumptions that are not strictly necessary, but which simplify the presentation.

It should be emphasized that this is in no way a survey of quantum cohomology, or any attempt at evaluating various approaches. In particular, the symplectic point of view is not covered (see [R-T]). Another algebro-geometric approach, using a different compactification, can be found in [L-T 1].

These notes are based on a jointly taught course at the University of Chicago in which our main efforts were aimed at understanding the papers of Kontsevich and Manin. We thank R. Donagi for instigating this course. Thanks are due to D. Abramovich, P. Belorousski, I. Ciocan-Fontanine, C. Faber, T. Graber, S. Kleiman, A. Kresch, C. Procesi, K. Ranestad, H. Tamvakis, J. Thomsen, E. Tjøtta, and A. Vistoli for comments and suggestions. A seminar course at the Mittag-Leffler Institute has led to many improvements. Some related preprints that have appeared since the Santa Cruz conference are pointed out in footnotes.

**0.2. Notation.** In this exposition, for simplicity, we consider only homology classes of even dimension. To avoid doubling indices, we set, for a complete variety $X$,

$$A_d X = H_{2d}(X, \mathbb{Z}), \quad A^d X = H^{2d}(X, \mathbb{Z}).$$

When $X$ is nonsingular of dimension $n$, identify $A^d X$ with $A_{n-d} X$ by the Poincaré duality isomorphism

$$A^d X \xrightarrow{\sim} A_{n-d} X, \quad c \mapsto c \cap [X].$$

In particular, a closed subvariety $V$ of $X$ of pure codimension $d$ determines classes in $A_{n-d}X$ and $A^d X$ via the duality isomorphism. Both of these classes are denoted by $[V]$. For homogeneous varieties, which are our main concern, the Chow groups coincide with the topological groups. Hence $A_d X$ and $A^d X$ can be taken to be the Chow homology and cohomology groups for homogeneous varieties (see [F]).

If $X$ is complete, and $c$ is a class in the ring $A^* X = \bigoplus A^d X$, and $\beta$ is a class in $A_k X$, we denote by $\int_\beta c$ the degree of the class of the zero cycle obtained by evaluating $c_k$ on $\beta$, where $c_k$ is the component of $c$ in $A^k X$. When $V$ is a closed, pure dimensional subvariety of $X$, we write $\int_V c$ instead of $\int_{[V]} c$. We use cup product notation $\cup$ for the product in $A^* X$.

We will be concerned only with varieties over $\mathbb{C}$ since the relevant moduli spaces have not yet been constructed in positive characteristic. Let $[n]$ denote the finite set of integers $\{1, 2, \ldots, n\}$.

**0.3. Compactifications of moduli spaces.** An important feature of quantum cohomology is the use of intersection theory in a space of maps of curves into a variety, rather than in the variety itself. To carry this out, a good compactification of such a space is required. At least when $X$ is sufficiently positive, Kontsevich has constructed such a compactification. We start, in this section, by reviewing some related moduli spaces with similar properties. Kontsevich's space of stable maps will be introduced in section 0.4.

Algebraic geometers by now have become quite comfortable working with the moduli space $M_g$ of projective nonsingular curves of genus $g$, and its compactification $\overline{M}_g$, whose points correspond to projective, connected, nodal curves of arithmetic genus $g$, satisfying a stability condition (due to Deligne and Mumford) that guarantees the curve has only a finite automorphism group. These moduli spaces are irreducible varieties of dimension $3g-3$ if $g \geq 2$, smooth if regarded as (Deligne-Mumford) stacks, and with orbifold singularities if regarded as ordinary coarse moduli spaces.

Some related spaces have become increasingly important. The moduli space $M_{g,n}$ parametrizes projective nonsingular curves $C$ of genus $g$ together with $n$ distinct marked points $p_1, \ldots, p_n$ on $C$. $M_{g,n}$ has a compactification $\overline{M}_{g,n}$ whose points correspond to projective, connected, nodal curves $C$, together with $n$ distinct, nonsingular, marked points, again with a stability condition equivalent to the finiteness of automorphism groups. $\overline{M}_{g,1}$ is often called the universal curve over $\overline{M}_g$ (although, as coarse moduli spaces, this is a slight abuse of language).

The first remarkable feature of the space $\overline{M}_{g,n}$ is that it compactifies $M_{g,n}$ without ever allowing the points to come together. When points on a smooth curve approach each other, in fact, the curve sprouts off one or more components, each isomorphic to the projective line, and the points distribute themselves at smooth points on these new components, in a way that reflects the relative rates of approach.

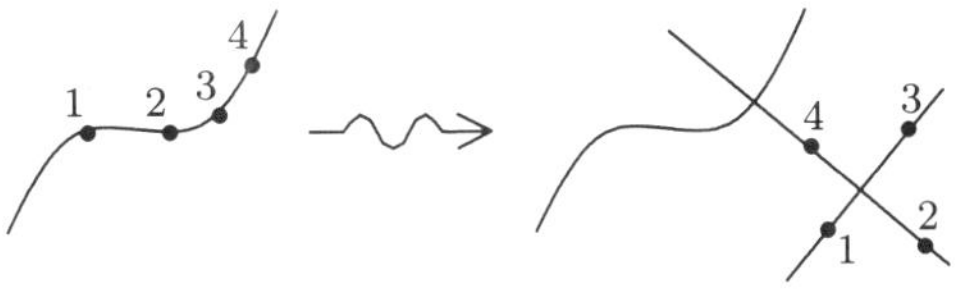


The spaces $\overline{M}_{g,n}$ again are smooth stacks, or orbifold coarse moduli spaces, of dimension $3g-3+n$, as long as this integer is nonnegative. The case of genus zero plays a prominent role in our story. In this case, $\overline{M}_{0,n}$ is a fine moduli space and a nonsingular variety. A point in $\overline{M}_{0,n}$ corresponds to a curve which is a tree of projective lines meeting transversally, with $n$ distinct, nonsingular, marked points; the stability condition is that each component must have at least three special points, which are either the marked points or the nodes where the component meets the other components.

For $n = 3$, of course, $M_{0,3} = \overline{M}_{0,3}$ is a point. For $n = 4$, $M_{0,4}$ parametrizes 4 distinct marked points on a projective line. Since, up to isomorphism, one can fix the first three of these points, say to be 0, 1, and $\infty$, $M_{0,4}$ is isomorphic to $\mathbf{P}^1 \setminus \{0, 1, \infty\}$. It is not hard to guess what $\overline{M}_{0,4}$ must be. In fact, the three added points are represented by the following three marked curves:

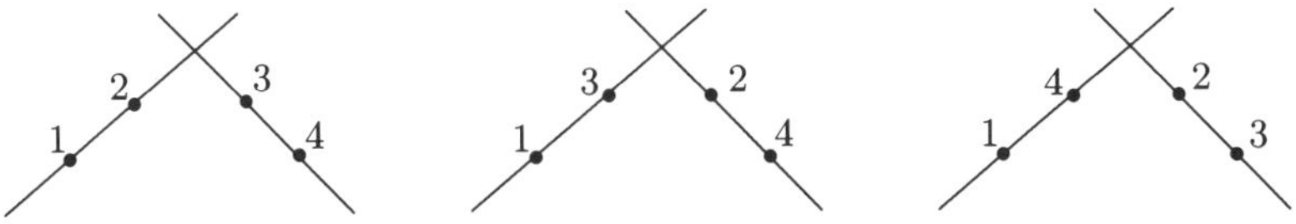


In general, the closures of the loci of trees of a given combinatorial type are smooth subvarieties of $\overline{M}_{0,n}$, and all such loci meet transversally. There is a divisor $D(A|B)$ in $\overline{M}_{0,n}$ for each partition of $[n]$ into two disjoint sets $A$ and $B$, each with at least two elements. A generic point of $D(A|B)$ is represented by two lines meeting transversally, with points labeled by $A$ and $B$ on each:

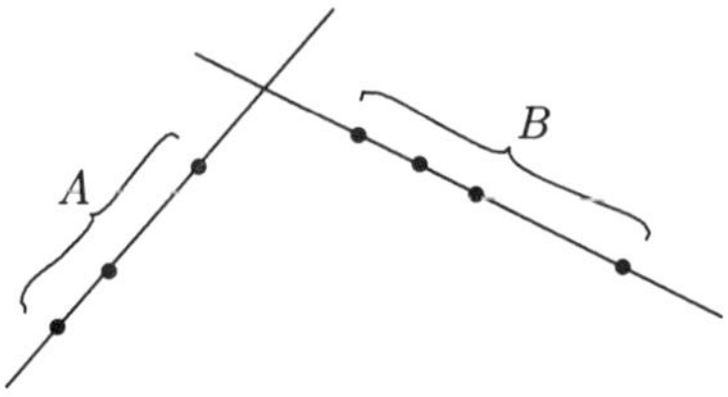


It is convenient to allow labeling by finite sets other than $[n]$; we write $\overline{M}_{g,A}$ for the corresponding moduli space where $A$ is the labeling set. Let $B \subset A$ (if $g = 0$, then let $|B| \geq 3$). It is a fundamental fact that there is a morphism $\overline{M}_{g,A} \to \overline{M}_{g,B}$ which "forgets" the points marked in $A \setminus B$. On the open locus $M_{g,n}$ this map is the obvious one, but it is more subtle on the boundary: removing some points may make a component unstable, and such a component must be collapsed. For example, the map from $\overline{M}_{0,5}$ to $\overline{M}_{0,4}$ forgetting the point labeled 5 sends

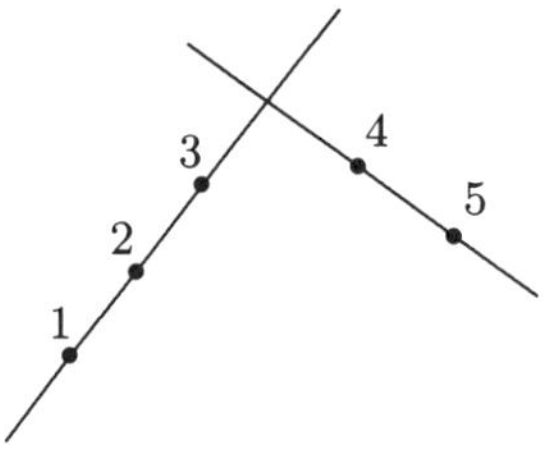


to

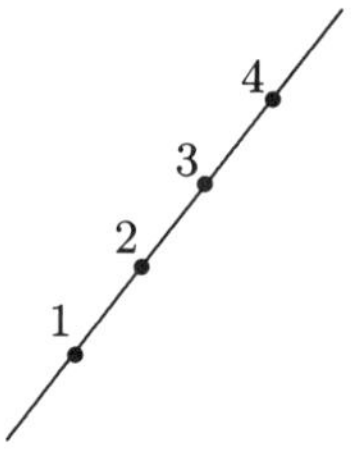


and

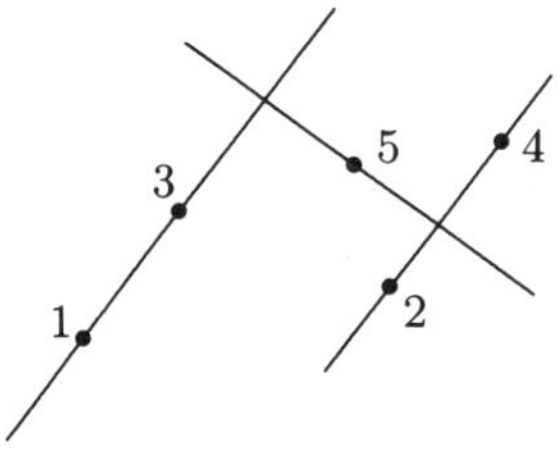


to

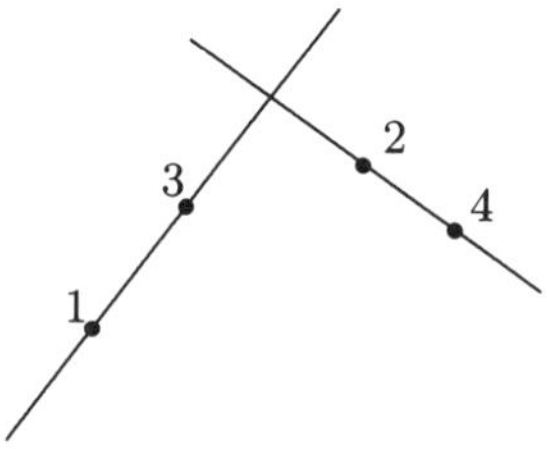


The algebra that shows this is a morphism is carried out in [Kn].

In particular, for any subset $\{i, j, k, l\}$ of four integers in $[n]$, we have a morphism from $\overline{M}_{0,n}$ to $\overline{M}_{0,\{i,j,k,l\}}$. The inverse image of the point $P(i, j \mid k, l)$

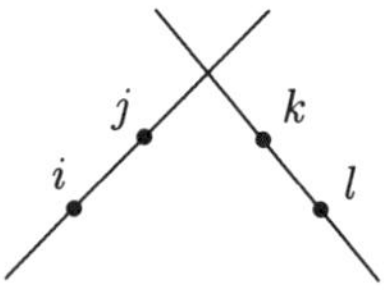


is a divisor on $\overline{M}_{0,n}$. This inverse image is a multiplicity-free sum of divisors $D(A|B)$: the sum is taken over all partitions $A \cup B = [n]$ satisfying $i, j \in A$ and $k, l \in B$. The fact that the three boundary points in $\overline{M}_{0,\{i,j,k,l\}} \cong \mathbf{P}^1$ are linearly equivalent implies their inverse images in $\overline{M}_{0,n}$ are linearly equivalent as well. Hence, the fundamental relation is obtained:

$$\sum_{i,j\in A \ k,l\in B} D(A|B) = \sum_{i,k\in A \ j,l\in B} D(A|B) = \sum_{i,l\in A \ j,k\in B} D(A|B) \tag{1}$$

in $A^1(\overline{M}_{0,n})$. Keel [Ke] has shown that the classes of these divisors $D(A|B)$ generate the Chow ring, and that the relations (1), together with the (geometrically obvious) relations $D(A|B)\cdot D(A'|B') = 0$ if there are no inclusions among the sets $A$, $B$, $A'$, $B'$, give a complete set of relations.

**0.4. The space of stable maps.** In the remainder of the introduction, the basic ideas and constructions in quantum cohomology are introduced. The goal here is to give a precise overview with no proofs. The ideas introduced here are covered carefully (with proofs) in the main sections of these notes.

Let $X$ be a smooth projective variety, and let $\beta$ be an element in $A_1X$. Let $M_{g,n}(X,\beta)$ be the set of isomorphism classes of pointed maps $(C, p_1, \ldots, p_n, \mu)$ where $C$ is a projective nonsingular curve of genus $g$, the markings $p_1, \ldots, p_n$ are distinct points of $C$, and $\mu$ is a morphism from $C$ to $X$ satisfying $\mu_*([C]) = \beta$. $(C, p_1, \ldots, p_n, \mu)$ is *isomorphic* to $(C', p'_1, \ldots, p'_n, \mu')$ if there is a scheme isomorphism $\tau : C \to C'$ taking $p_i$ to $p'_i$, with $\mu' \circ \tau = \mu$. Of course, if $\beta \neq 0$, $M_{g,n}(X,\beta)$ is empty unless $\beta$ is the class of a curve in $X$. There are also other problems. For example, if $g = 0$, which will be the case of interest to us, $M_{g,n}(X,\beta)$ is empty if $\beta \neq 0$ and $X$ contains no rational curves. To obtain a well-behaved space, one needs to make strong assumptions on $X$. In general, there is a compactification

$$M_{g,n}(X,\beta) \subset \overline{M}_{g,n}(X,\beta),$$

whose points correspond to stable maps $(C, p_1, \ldots, p_n, \mu)$ where $C$ a projective, connected, nodal curve of arithmetic genus $g$, the markings $p_1, \ldots, p_n$ are distinct nonsingular points of $C$, and $\mu$ is a morphism from $C$ such that $\mu_*([C]) = \beta$. Again, the stability condition (due to Kontsevich) is equivalent to finiteness of automorphisms of the map. Alternatively, $(C, p_1, \ldots, p_n, \mu)$ is a *stable* map if both of the following conditions hold for every irreducible component $E \subset C$:

(1) If $E \cong \mathbf{P}^1$ and $E$ is mapped to a point by $\mu$, then $E$ must contain at least three special points (either marked points or points where $E$ meets the other components of $C$).
(2) If $E$ has arithmetic genus 1 and $E$ is mapped to a point by $\mu$, then $E$ must contain at least one special point.

Condition (2) is relevant only in case $g = 1$, $n = 0$, and $\beta = 0$ (in other cases, (2) is automatically satisfied). From conditions (1) and (2), it follows that $\overline{M}_{1,0}(X,0) = \emptyset$. Thus, in practice, (1) is the important condition.

When $X$ is a point, so $\beta = 0$, one recovers the pointed moduli space of curves $\overline{M}_{g,n} \cong \overline{M}_{g,n}(\text{point}, 0)$. When $X \cong \mathbf{P}^r$ is a projective space, we write $\overline{M}_{g,n}(\mathbf{P}^r, d)$ in place of $\overline{M}_{g,n}(\mathbf{P}^r, d[\text{line}])$.

The simplest example is $\overline{M}_{0,0}(\mathbf{P}^r, 1)$, which is the Grassmannian $\mathbf{G}(\mathbf{P}^1, \mathbf{P}^r)$. If $n \geq 1$, $\overline{M}_{0,n}(\mathbf{P}^r, 1)$ is a locally trivial fibration over $\mathbf{G}(\mathbf{P}^1, \mathbf{P}^r)$ with the configuration space $\mathbf{P}^1[n]$ of [F-M] as the fiber. Let us look at the space $\overline{M}_{0,0}(\mathbf{P}^2, 2)$. An open set in this space is the space of nonsingular conics, since to each such conic $D$ there is an isomorphism $\mathbf{P}^1 \xrightarrow{\sim} D \subset \mathbf{P}^2$, unique up to equivalence. Singular conics $D$ that are the unions of two lines are similarly the isomorphic image $C \xrightarrow{\sim} D \subset \mathbf{P}^2$, where $C$ is the union of two projective lines meeting transversally at a point. This gives:

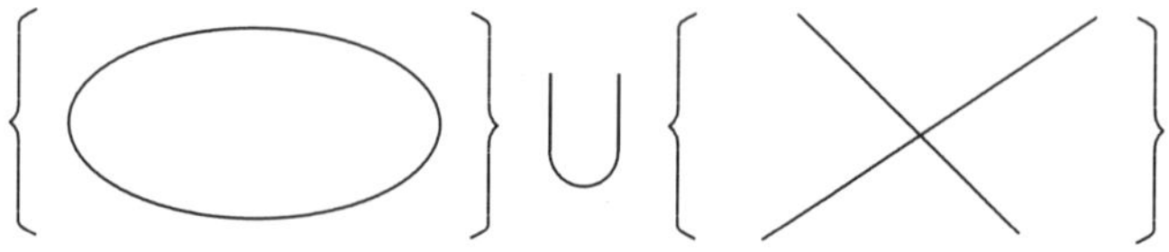

We also have maps from the same $C$ to $\mathbf{P}^2$ sending each line in the domain onto the same line in $\mathbf{P}^2$. To determine this map up to isomorphism, however, the point that is the image of the intersection of the two lines must be specified, so the data for this is a line in $\mathbf{P}^2$ together with a point on it. Finally, there are maps from $\mathbf{P}^1$ to a line in the plane that are branched coverings of degree two onto a line in the plane. These are determined by specifying the line together with the two branch points. The added points consist of:

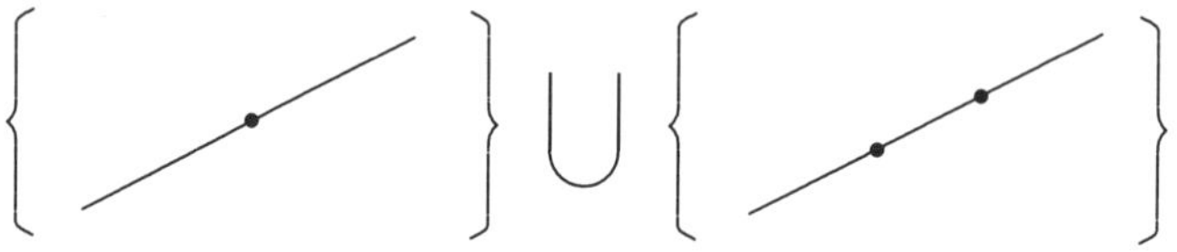

Thus, we recover the classical space of complete conics – but in quite a different realization from the usual one. The same discussion is valid when $\mathbf{P}^2$ is replaced by $\mathbf{P}^r$, but this time the space is not the classical space of complete conics in space. The classical space specifies a plane together with a complete conic contained in the plane; Kontsevich's space has blown down all the planes containing a given line.

Let $X$ be a complete nonsingular variety with tangent bundle $T_X$. $X$ is *convex* if, for every morphism $\mu : \mathbf{P}^1 \to X$,

$$H^1(\mathbf{P}^1, \mu^*(T_X)) = 0. \tag{2}$$

Convexity is a very restrictive condition on $X$. The main examples of convex varieties are homogeneous spaces $X = G/P$, where $G$ is a Lie group and $P$ is a parabolic subgroup. Since $G$ acts transitively on $X$, $T_X$ is generated by global sections. Hence, $\mu^*(T_X)$ is globally generated for every morphism of $\mathbf{P}^1$, and the vanishing (2) is obtained. Projective spaces, Grassmannians, smooth quadrics, flag varieties, and products of such varieties are all homogeneous. It is for homogeneous spaces that the theory of quantum cohomology takes its simplest form. The development of quantum cohomology in sections 7–10 is carried out only in the homogeneous case. Other examples of convex varieties include abelian varieties and projective bundles over curves of positive genus.

The genus 0 moduli space of stable maps is well-behaved in case $X$ is convex. In this case, $\overline{M}_{0,n}(X,\beta)$ exists as a projective nonsingular stack or orbifold coarse moduli space, containing $M_{0,n}(X,\beta)$ as a dense open subset. When $\overline{M}_{0,n}(X,\beta)$ is nonempty, its dimension is given by

$$\dim \overline{M}_{0,n}(X,\beta) = \dim X + \int_\beta c_1(T_X) + n - 3.$$

Here, $c_1(T_X)$ is the first Chern class of the tangent bundle to $X$. We assume always that the right side of this equation is nonnegative. In the stack or orbifold sense, this is a compactification with normal crossing divisors. When $X$ is projective space, $\overline{M}_{0,n}(X,d)$ is irreducible. These assertions are Theorems 1–3 in these notes and are established in sections 1–6.

We will also write $\overline{M}_{0,A}(X,\beta)$ when the index set is a set $A$ instead of $[n]$. These varieties also have forgetful morphisms $\overline{M}_{0,A}(X,\beta) \to \overline{M}_{0,B}(X,\beta)$ when $B$

is a subset of $A$. In addition, if $|A| \geq 3$, there are morphisms $\overline{M}_{0,A}(X,\beta) \to \overline{M}_{0,A}$ that forget the map $\mu$. In both these cases, as before, one must collapse components that become unstable.

When $X$ is convex, the spaces $\overline{M}_{0,n}(X,\beta)$ have fundamental boundary divisors analogous to the divisors $D(A|B)$ on $\overline{M}_{0,n}$. Let $n \geq 4$. Let $A \cup B$ be a partition of $[n]$. Let $\beta_1 + \beta_2 = \beta$ be a sum in $A_1 X$. There is a divisor in $\overline{M}_{0,n}(X,\beta)$ determined by:

$$(3) \qquad D(A, B; \beta_1, \beta_2) = \overline{M}_{0,A\cup\{\bullet\}}(X,\beta_1) \times_X \overline{M}_{0,B\cup\{\bullet\}}(X,\beta_2),$$

$$D(A, B; \beta_1, \beta_2) \subset \overline{M}_{0,n}(X,\beta).$$

A moduli point in $D(A, B, \beta_1, \beta_2)$ corresponds to a map with a reducible domain $C = C_1 \cup C_2$ where $\mu_*([C_1]) = \beta_1$ and $\mu_*([C_2]) = \beta_2$. The points labeled by $A$ lie on $C_1$ and points labeled by $B$ lie on $C_2$. The curves $C_1$ and $C_2$ are connected at the points labeled $\bullet$.

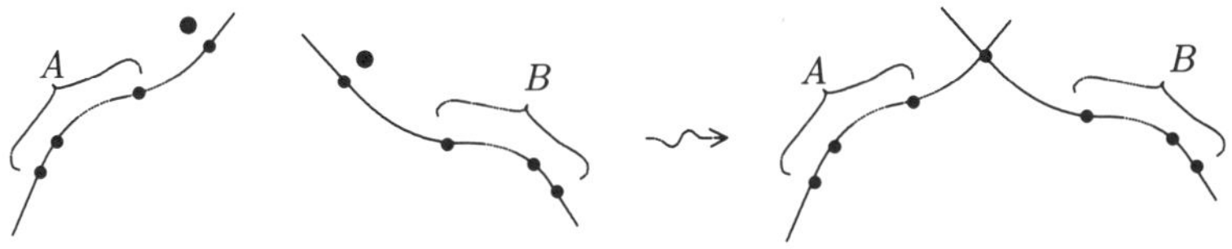


Finally, the fiber product in the definition (3) corresponds to the condition that the maps must take the same value in $X$ on the marked point $\bullet$ in order to be glued. This fiber product is defined via evaluation maps discussed in the next section.

For $i, j, k, l$ distinct in $[n]$, set

$$D(i, j \mid k, l) = \sum D(A, B; \beta_1, \beta_2).$$

The sum is over all partitions $A \cup B = [n]$ with $i, j \in A$ and $k, l \in B$ and over all classes $\beta_1, \beta_2 \in A_1 X$ satisfying $\beta_1 + \beta_2 = \beta$. Using the projection $\overline{M}_{0,n}(X,\beta) \to \overline{M}_{0,\{i,j,k,l\}} \cong \mathbf{P}^1$, the fundamental linear equivalences

$$(4) \qquad D(i, j \mid k, l) = D(i, k \mid j, l) = D(i, l \mid j, k)$$

on $\overline{M}_{0,n}(X,\beta)$ are obtained via pull-back of the the 4-point linear equivalences on $\overline{M}_{0,\{i,j,k,l\}}$ as in (1).

**0.5. Gromov-Witten invariants and quantum cohomology.** Let $X$ be a convex variety. For each marked point $1 \leq i \leq n$, there is a canonical *evaluation map*

$$\rho_i : \overline{M}_{0,n}(X,\beta) \to X$$

defined for $[C, p_1, \ldots, p_n, \mu]$ in $\overline{M}_{0,n}(X,\beta)$ by:

$$\rho_i([C, p_1, \ldots, p_n, \mu]) = \mu(p_i).$$

Given classes $\gamma_1, \ldots, \gamma_n$ in $A^*X$, a product is determined in the ring $A^*(\overline{M}_{0,n}(X,\beta))$ by:

$$(5) \qquad \rho_1^*(\gamma_1) \cup \cdots \cup \rho_n^*(\gamma_n) \in A^*(\overline{M}_{0,n}(X,\beta)).$$

If $\sum \operatorname{codim}(\gamma_i) = \dim(\overline{M}_{0,n}(X,\beta))$, the product (5) can be evaluated on the fundamental class of $\overline{M}_{0,n}(X,\beta)$. In this case, the *Gromov-Witten invariant* $I_\beta(\gamma_1 \cdots \gamma_n)$

is defined by:

$$I_\beta(\gamma_1 \cdots \gamma_n) = \int_{\overline{M}_{0,n}(X,\beta)} \rho_1^*(\gamma_1) \cup \cdots \cup \rho_n^*(\gamma_n). \tag{6}$$

The multiplicative notation in the argument of $I_\beta$ is used to indicate $I_\beta$ is a symmetric function of the classes $\gamma_1, \ldots, \gamma_n$.

Let $X$ be a homogeneous space. Poincaré duality and Bertini-type transversality arguments imply a relationship between the Gromov-Witten invariants and enumerative geometry. If $\gamma_i = [V_i]$ for a subvariety $V_i \subset X$, the Gromov-Witten invariant (6) equals the number of marked rational curves in $X$ with $i^{th}$ marked point in $V_i$, suitably counted. For example, when $X = \mathbf{P}^2$, $\beta = d[\text{line}]$, $n = 3d-1$, and each $V_i$ is a point,

$$N_d = I_d(\underbrace{[p] \cdots [p]}_{3d-1}).$$

The Gromov-Witten invariants are used to define the quantum cohomology ring. Associativity of this ring is established as a consequence of the 4-point linear equivalences (4). Associativity amounts to many equations among the Gromov-Witten invariants which often lead to a determination of all the invariants in terms of a few basic numbers.

Given $\gamma_1, \ldots, \gamma_n \in H^*X$ (not necessarily of even degrees), there are more general Gromov-Witten invariants in $H^*\overline{M}_{0,n}$ defined by

$$I^X_{0,n,\beta}(\gamma_1 \otimes \cdots \otimes \gamma_n) = \eta_*(\rho_1^*(\gamma_1) \cup \cdots \cup \rho_n^*(\gamma_n))$$

where $\eta : \overline{M}_{0,n}(X,\beta) \to \overline{M}_{0,n}$ is the projection. The set of multilinear maps

$$\{I^X_{0,n,\beta} : (H^*X)^{\otimes n} \to H^*\overline{M}_{0,n}\}$$

is called the *Tree-Level System of Gromov-Witten Invariants.* We will not need these generalities here.

The construction and proofs of the basic properties of $\overline{M}_{0,n}(X,\beta)$ are undertaken in sections 1–6. The theory of Gromov-Witten invariants and quantum cohomology for homogeneous varieties is presented in sections 7–10 with the examples of $\mathbf{P}^2$, $\mathbf{P}^3$, and a smooth quadric 3-fold $\mathbf{Q}^3$ worked out in detail. If Theorems 1–3 are taken for granted, sections 1–6 can be skipped. No originality is claimed for these notes except for some aspects of the proofs of Theorems 1–4. Constructions of Kontsevich's moduli space of stable maps can also be found in [A], [K], [B-M]. In [A], a generalization to the case in which the domain is a surface is analyzed.

**0.6. Calculation of $N_d$.** We end this introduction by sketching how these moduli spaces of maps can be used to calculate the number $N_d$ of degree $d$ rational plane curves passing through $3d-1$ general points in $\mathbf{P}^2$. The formula (7) will be recovered in section 9 from the general quantum cohomology results, but it may be useful now to see a direct proof.

For $d=1$, $N_1 = 1$ is the number of lines through 2 points. $N_d$ is determined for $d \geq 2$ by the *recursion formula*:

$$N_d = \sum_{d_1+d_2=d,\ d_1,d_2>0} N_{d_1}N_{d_2}\left(d_1^2d_2^2\binom{3d-4}{3d_1-2} - d_1^3d_2\binom{3d-4}{3d_1-1}\right). \tag{7}$$

For example, (7) yields[1] :

$$N_2 = 1,\ N_3 = 12,\ N_4 = 620,\ N_5 = 87304,\ N_6 = 26312976, \ldots$$

The strategy of proof is to utilize the fundamental linear relations (4) among boundary components of $\overline{M}_{0,n}(\mathbf{P}^2, d)$. Intersection of a curve $Y$ in this moduli space with the linear equivalence (4) will yield (7). We will take $n = 3d$ (not $3d-1$) with $d \geq 2$, so $n \geq 6$. Label the marked points by the set

$$\{1, 2, 3, \ldots, n-4, q, r, s, t\}.$$

The forgetful morphism $\overline{M}_{0,n}(\mathbf{P}^2, d) \to \overline{M}_{0,\{q,r,s,t\}}$ yields the relations (4) on $\overline{M}_{0,n}(\mathbf{P}^2, d)$:

$$D(q, r \mid s, t) = D(q, s \mid r, t). \tag{8}$$

Recall from section 0.4:

$$D(q, r \mid s, t) = \sum_{q,r\in A,\ s,t\in B,\ d_1+d_2=d} D(A, B; d_1, d_2).$$

The curve $Y \subset \overline{M}_{0,n}(\mathbf{P}^2, d)$ is determined by a selection of general points and lines in $\mathbf{P}^2$. More precisely, let $z_1, \ldots, z_{n-4}, z_s, z_t$ be $n-2$ general points in $\mathbf{P}^2$ and let $l_q, l_r$ be general lines. Let the curve $Y$ be defined by the intersection:

$$Y = \rho_1^{-1}(z_1) \cap \cdots \cap \rho_{n-4}^{-1}(z_{n-4}) \cap \rho_q^{-1}(l_q) \cap \rho_r^{-1}(l_r) \cap \rho_s^{-1}(z_s) \cap \rho_t^{-1}(z_t).$$

$\overline{M}_{0,n}(\mathbf{P}^2, d)$ is a nonsingular, fine moduli space on the open set of automorphism-free maps (see section 1.2). It is not difficult to show the locus of maps with non-trivial automorphisms in $\overline{M}_{0,n}(\mathbf{P}^2, d)$ is of codimension at least 2 if $(n, d) \neq (0, 2)$. Therefore, by Bertini's theorem applied to each evaluation map and the generality of the points and lines, we conclude $Y$ is a nonsingular curve contained in the automorphism-free locus which intersects all the boundary divisors transversally at general points of the boundary. It remains only to compute the intersection of $Y$ with each side of the linear equivalence (8).

The points of

$$Y \cap\ D(A, B; d_1, d_2)$$

correspond bijectively to maps $\mu : C = C_A \cup C_B \to \mathbf{P}^2$ satisfying:

(a) $C_A, C_B \cong \mathbf{P}^1$ and meet transversally at a point.
(b) The markings of $A$, $B$ lie on $C_A$, $C_B$ respectively.
(c) $\mu_*([C_A]) = d_1[\text{line}]$, $\mu_*([C_B]) = d_2[\text{line}]$.
(d) $\forall 1 \leq i \leq n-4$, $\mu(i) = z_i$.
(e) $\mu(q) \in l_q$, $\mu(r) \in l_r$, $\mu(s) = z_s$, $\mu(t) = z_t$.

Let $q, r \in A$ and $s, t \in B$. $Y \cap\ D(A, B; 0, d)$ is nonempty only when $A = \{q, r\}$. In this case, $C_A$ is required to map to the point $l_q \cap l_r$. The restriction $\mu : C_B \to \mathbf{P}^2$ must map the $3d-2$ markings on $C_B$ to the $3d-2$ given points, and in addition, $\mu$ maps the point $C_A \cap C_B$ to $l_q \cap l_r$. Therefore,

$$\#\, Y \cap\ D(\{q, r\}, \{1, \ldots, n-4, s, t\}; 0, d) = N_d.$$

[1]The number $N_3 = 12$ is the classically known number of singular members in a pencil of cubic curves through 8 given points. The number $N_4 = 620$ was computed by H. Zeuthen in [Z]. Z. Ran reports $N_4 = 620$ as well as the higher $N_d$ 's can be derived from his formulas in [R1]; see [R2] for a comparison of the two approaches. Some $N_d$'s are also computed in [C-H 2].

For $1 \leq d_1 \leq d-1$, $Y \cap D(A, B; d_1, d_2)$ is nonempty only when $|A| = 3d_1 + 1$. There are $\binom{3d-4}{3d_1-1}$ partitions satisfying $q, r \in A$, $s, t \in B$, and $|A| = 3d_1 + 1$. A simple count of maps satisfying (a)-(e) yields

$$\# Y \cap D(A, B; d_1, d_2) = N_{d_1} N_{d_2} d_1^3 d_2$$

for each partition. There are $N_{d_1}$ choices for the image of $C_A$ and $N_{d_2}$ choices for the image of $C_B$. The points labeled $q$ and $r$ map to any of the $d_1$ intersection points of $\mu(C_A)$ with $l_q$ and $l_r$ respectively. Finally, there are $d_1 d_2$ choices for the image of the intersection point $C_A \cap C_B$ corresponding to the intersection points of $\mu(C_A) \cap \mu(C_B) \subset \mathbf{P}^2$. The last case is simple: $Y \cap D(A, B; d, 0) = \emptyset$. Therefore,

$$\# Y \cap D(q, r \mid s, t) = N_d + \sum_{d_1 + d_2 = d,\ d_1 > 0,\ d_2 > 0} N_{d_1} N_{d_2} d_1^3 d_2 \binom{3d-4}{3d_1 - 1}.$$

Now consider the other side of the linear equivalence (8). Let the markings now satisfy $q, s \in A$ and $r, t \in B$. $Y \cap D(A, B; 0, d)$ and $Y \cap (A, B; d, 0)$ are both empty. For $1 \leq d_1 \leq d - 1$, $Y \cap (A \cup B, d_1, d_2)$ is nonempty only when $|A| = 3d_1$. There are $\binom{3d-4}{3d_1-2}$ such partitions and

$$\# Y \cap D(A, B; d_1, d_2) = N_{d_1} N_{d_2} d_1^2 d_2^2$$

for each. Therefore,

$$\# Y \cap D(q, s \mid r, t) = \sum_{d_1 + d_2 = d,\ d_1 > 0,\ d_2 > 0} N_{d_1} N_{d_2} d_1^2 d_2^2 \binom{3d-4}{3d_1 - 2}.$$

The linear equivalence (8) implies

$$\# Y \cap D(q, r \mid s, t) = \# Y \cap D(q, s \mid r, t).$$

The recursion (7) follows immediately.

In the general development of quantum cohomology described in sections 8 and 9, these numerical relations obtained by intersection with the basic linear equivalences arise as ring associativity relations.

## 1. Stable maps and their moduli spaces

**1.1. Definitions.** An $n$-pointed, genus $g$, complex, *quasi-stable* curve

$$(C,\ p_1, \ldots, p_n)$$

is a projective, connected, reduced, (at worst) nodal curve of arithmetic genus $g$ with $n$ distinct, nonsingular, marked points. Let $S$ be an algebraic scheme over $\mathbb{C}$. A *family* of $n$-pointed, genus $g$, quasi-stable curves over $S$ is a flat, projective map $\pi : \mathcal{C} \to S$ with $n$ sections $p_1, \ldots, p_n$ such that each geometric fiber $(\mathcal{C}_s,\ p_1(s), \ldots, p_n(s))$ is an $n$-pointed, genus $g$, quasi-stable curve. Let $X$ be an algebraic scheme over $\mathbb{C}$. A *family of maps* over $S$ from $n$-pointed, genus $g$ curves to $X$ consists of the data $(\pi : \mathcal{C} \to S, \{p_i\}_{1 \leq i \leq n}, \mu : \mathcal{C} \to X)$:

(i) A family of $n$-pointed, genus $g$, quasi-stable curves $\pi : \mathcal{C} \to S$ with $n$ sections $\{p_1, \ldots, p_n\}$.

(ii) A morphism $\mu : \mathcal{C} \to X$.

Two families of maps over $S$,

$$(\pi : \mathcal{C} \to S, \{p_i\}, \mu), \quad (\pi' : \mathcal{C}' \to S, \{p'_i\}, \mu'),$$

are *isomorphic* if there exists a scheme isomorphism $\tau : \mathcal{C} \to \mathcal{C}'$ satisfying: $\pi = \pi' \circ \tau$, $p'_i = \tau \circ p_i$, $\mu = \mu' \circ \tau$. When $\pi : C \to \mathrm{Spec}(\mathbb{C})$ is the structure map, $(\pi : C \to \mathrm{Spec}(\mathbb{C}), \{p_i\}, \mu)$ is written as $(C, \{p_i\}, \mu)$.

Let $(C, \{p_i\}, \mu)$ be a map from an $n$-pointed quasi-stable curve to $X$. The *special points* of an irreducible component $E \subset C$ are the marked points and the component intersections of $C$ that lie on $E$. The map $(C, \{p_i\}, \mu)$ is *stable* if the following conditions hold for every component $E \subset C$:

(1) If $E \cong \mathbf{P}^1$ and $E$ is mapped to a point by $\mu$, then $E$ must contain at least three special points.
(2) If $E$ has arithmetic genus 1 and $E$ is mapped to a point by $\mu$, then $E$ must contain at least one special point.

A family of pointed maps $(\pi : \mathcal{C} \to S, \{p_i\}, \mu)$ is *stable* if the pointed map on each geometric fiber of $\pi$ is stable.

If $X = \mathbf{P}^r$, stability can be expressed in the following manner. Let $\omega_{\mathcal{C}/S}$ denote the relative dualizing sheaf. A flat family of maps $(\pi : \mathcal{C} \to S, \ \{p_i\}, \mu)$ is *stable* if and only if $\omega_{\mathcal{C}/S}(p_1 + \ldots + p_n) \otimes \mu^*(\mathcal{O}_{\mathbf{P}^r}(3))$ is $\pi$-relatively ample.

Let $X$ be an algebraic scheme over $\mathbb{C}$. Let $\beta \in A_1 X$. A map $\mu : C \to X$ *represents* $\beta$ if the $\mu$-push-forward of the fundamental class $[C]$ equals $\beta$. Define a contravariant functor $\overline{\mathcal{M}}_{g,n}(X, \beta)$ from the category of complex algebraic schemes to sets as follows. Let $\overline{\mathcal{M}}_{g,n}(X, \beta)(S)$ be the set of isomorphism classes of stable families over $S$ of maps from $n$-pointed, genus $g$ curves to $X$ representing the class $\beta$.

**1.2. Existence.** Let $X$ be a projective, algebraic scheme over $\mathbb{C}$. Projective coarse moduli spaces of maps exist for general $g$. In the genus 0 case, if $X$ is a projective, nonsingular, convex variety, the coarse moduli spaces are normal varieties with finite quotient singularities.

THEOREM 1. *There exists a projective, coarse moduli space* $\overline{M}_{g,n}(X, \beta)$.

$\overline{M}_{g,n}(X, \beta)$ is a scheme together with a natural transformation of functors

$$\phi : \overline{\mathcal{M}}_{g,n}(X, \beta) \to \mathcal{H}om_{Sch}(*, \overline{M}_{g,n}(X, \beta))$$

satisfying properties:

(I) $\phi(\mathrm{Spec}(\mathbb{C})) : \overline{\mathcal{M}}_{g,n}(X, \beta)(\mathrm{Spec}(\mathbb{C})) \to \mathcal{H}om(\mathrm{Spec}(\mathbb{C}), \overline{M}_{g,n}(X, \beta))$ is a set bijection.
(II) If $Z$ is a scheme and $\psi : \overline{\mathcal{M}}_{g,n}(X, \beta) \to \mathcal{H}om(*, Z)$ is a natural transformation of functors, then there exists a unique morphism of schemes

$$\gamma : \overline{M}_{g,n}(X, \beta) \to Z$$

such that $\psi = \tilde{\gamma} \circ \phi$. ($\tilde{\gamma} : \mathcal{H}om(*, \overline{M}_{g,n}(X, \beta)) \to \mathcal{H}om(*, Z)$ is the natural transformation induced by $\gamma$.)

Let $(C, \{p_i\}, \mu)$ be a map of an $n$-pointed, quasi-stable curve to $X$. An automorphism of the map is an automorphism, $\tau$, of the curve $C$ satisfying

$$p_i = \tau(p_i), \quad \mu = \mu \circ \tau.$$

It is straightforward to check that $(C, \{p_i\}, \mu)$ is stable if and only if $(C, \{p_i\}, \mu)$ has a finite automorphism group. Let $\overline{M}^*_{g,n}(X,\beta) \subset \overline{M}_{g,n}(X,\beta)$ denote the open locus of stable maps with no non-trivial automorphisms.

A nonsingular variety $X$ is convex if for every map $\mu : \mathbf{P}^1 \to X, H^1(\mathbf{P}^1, \mu^*(T_X)) = 0$ (see section 0.4). The second and third theorems concern the convex, genus 0 case.

THEOREM 2. *Let $X$ be a projective, nonsingular, convex variety.*

(i) *$\overline{M}_{0,n}(X,\beta)$ is a normal projective variety of pure dimension*

$$\dim(X) + \int_\beta c_1(T_X) + n - 3.$$

(ii) *$\overline{M}_{0,n}(X,\beta)$ is locally a quotient of a nonsingular variety by a finite group.*

(iii) *$\overline{M}^*_{0,n}(X,\beta)$ is a nonsingular, fine moduli space (for automorphism-free stable maps) equipped with a universal family.*

In part (i), $\overline{M}_{0,n}(X,\beta)$ is not claimed in general to be irreducible (or even nonempty).

In fact, if the language of stacks is pursued, it can be seen that the moduli problem of stable maps from $n$-pointed, genus 0 curves to a nonsingular, convex space $X$ determines a complete, nonsingular, algebraic stack. For simplicity, the stack theoretic view is not taken in these notes; the experienced reader will see how to make the required modifications.

The *boundary* of $\overline{M}_{0,n}(X,\beta)$ is the locus corresponding to reducible domain curves. The boundary of the fine moduli space $\overline{M}_{0,n}$ is a divisor with normal crossings. In the coarse moduli spaces $\overline{M}_g$ and $\overline{M}_{g,n}$, the boundary is a divisor with normal crossings modulo a finite group. $\overline{M}_{0,n}(X,\beta)$ has the same boundary singularity type as these moduli spaces of pointed curves.

THEOREM 3. *Let $X$ be a nonsingular, projective, convex variety. The boundary of $\overline{M}_{0,n}(X,\beta)$ is a divisor with normal crossings (up to a finite group quotient).*

The organization of the construction is as follows. First $\overline{M}_{g,n}(\mathbf{P}^r, d)$ is explicitly constructed in sections 2–4. If $X \subset \mathbf{P}^r$ is a closed subscheme, it is not difficult to define a natural, closed subscheme

$$\overline{M}_{g,n}(X,d) \subset \overline{M}_{g,n}(\mathbf{P}^r, d)$$

of maps that factor through $X$. $\overline{M}_{g,n}(X,d)$ is a disjoint union of the spaces $\overline{M}_{g,n}(X,\beta)$ as $\beta$ varies in $A_1X$. By the universal property, it can be seen that the coarse moduli spaces $\overline{M}_{g,n}(X,\beta)$ do not depend on the projective embedding of $X$ (see section 5). The deformation arguments required to deduce Theorem 2 from the convexity assumption are covered in section 5. The boundary of the space of maps is discussed in section 6.

**1.3. Natural structures.** The universal property of the moduli space of maps immediately yields geometric structures on $\overline{M}_{g,n}(X,\beta)$. Consider first the marked points. The $n$ marked points induce $n$ canonical evaluation maps $\rho_1, \ldots, \rho_n$ on $\overline{M}_{g,n}(X,\beta)$. For $1 \le i \le n$, define a natural transformation

$$\theta_i : \overline{\mathcal{M}}_{g,n}(X,\beta) \to \mathcal{H}om(*, X)$$

as follows. Let $\zeta = (\pi : \mathcal{C} \to S,\ \{p_i\},\ \mu)$ be an element of $\overline{\mathcal{M}}_{g,n}(X,\beta)(S)$. Let

$$\theta_i(S)(\zeta) = \mu \circ p_i \in \mathcal{H}om(S, X).$$

$\theta_i$ is easily seen to be a natural transformation. By Theorem 1, $\theta_i$ induces a unique morphism of schemes $\rho_i : \overline{M}_{g,n}(X,\beta) \to X$.

By the universal properties of the moduli spaces $\overline{M}_{g,n}$ of $n$-pointed Deligne-Mumford stable genus $g$ curves (in case $2g-2+n>0$), each element $\zeta \in \overline{\mathcal{M}}_{g,n}(X,\beta)(S)$ naturally yields a morphism $S \to \overline{M}_{g,n}$ ([Kn]). Therefore, there exist natural forgetful maps $\eta : \overline{M}_{g,n}(X,\beta) \to \overline{M}_{g,n}$.

## 2. Boundedness and a quotient approach

**2.1. Summary.** In this section, the case $X = \mathbf{P}^r$ will be considered. The boundedness of the moduli problem of pointed stable maps is established. The arguments lead naturally to a quotient approach to the coarse moduli space. To set up the quotient approach, a result on equality loci of families of line bundles is required.

**2.2. Equality of line bundles in families.** Results on scheme theoretic equality loci are recalled. Let $\pi : \mathcal{C} \to S$ be a flat family of quasi-stable curves. By the theorems of cohomology and base change (cf. [H]), there is a canonical isomorphism $\mathcal{O}_S \cong \pi_*(\mathcal{O}_{\mathcal{C}})$. Hence, for any line bundle $\mathcal{N}$ on $S$, there is a canonical isomorphism $\mathcal{N} \cong \pi_*\pi^*(\mathcal{N})$. Suppose $\mathcal{L}$ and $\mathcal{M}$ are two line bundles on $\mathcal{C}$. The existence of a line bundle $\mathcal{N}$ on $S$ such that $\mathcal{L} \otimes \mathcal{M}^{-1} \cong \pi^*(\mathcal{N})$ is equivalent to the joint validity of (a) and (b):

(a) $\pi_*(\mathcal{L} \otimes \mathcal{M}^{-1})$ is locally free.

(b) The canonical map $\pi^*\pi_*(\mathcal{L} \otimes \mathcal{M}^{-1}) \to \mathcal{L} \otimes \mathcal{M}^{-1}$ is an isomorphism.

Let $\mathcal{L}_s$ be a line bundle on the geometric fiber $\mathcal{C}_s$ of $\pi$. The *multidegree* of $\mathcal{L}_s$ assigns to each irreducible component of $\mathcal{C}_s$ the degree of the restriction of $\mathcal{L}_s$ to that component.

PROPOSITION 1. *Let $\mathcal{L}$, $\mathcal{M}$ be line bundles on $\mathcal{C}$ such that the multidegrees of $\mathcal{L}_s$ and $\mathcal{M}_s$ coincide on each geometric fiber $\mathcal{C}_s$. Then, there is a unique closed subscheme $T \to S$ satisfying the following two properties:*

(I) *There is a line bundle $\mathcal{N}$ on $T$ such that $\mathcal{L}_T \otimes \mathcal{M}_T^{-1} \cong \pi^*(\mathcal{N})$.*

(II) *If $(R \to S,\ \mathcal{N})$ is a pair of a morphism from $R$ to $S$ and a line bundle on $R$ such that $\mathcal{L}_R \otimes \mathcal{M}_R^{-1} \cong \pi^*(\mathcal{N})$, then $R \to S$ factors through $T$.*

*Proof.* The proof of the Theorem of the Cube (II) in [M1] also establishes this proposition. The multidegree condition implies $\mathcal{L}_s \cong \mathcal{M}_s$ if and only if $h^0(\mathcal{C}_s, \mathcal{L}_s \otimes \mathcal{M}_s^{-1}) = 1$. The multidegree condition is required for $T$ to be a *closed* subscheme. □

**2.3. Boundedness.** Let $(C, \{p_i\}, \mu)$ be a stable map from an $n$-pointed, genus $g$ curve to $\mathbf{P}^r$. Let

$$\mathcal{L} = \omega_C(p_1 + \ldots + p_n) \otimes \mu^*(\mathcal{O}_{\mathbf{P}^r}(3)).$$

$\mathcal{L}$ is ample on $C$. A simple argument shows there exists an $f = f(g,n,r,d) > 0$ such that $\mathcal{L}^f$ is very ample on $C$ and $h^1(C, \mathcal{L}^f) = 0$, so

$$\text{degree}(\mathcal{L}^f) = f \cdot (2g-2+n+3d) = e,$$

$$h^0(C, \mathcal{L}^f) = e - g + 1.$$

Let $W \cong \mathbb{C}^{e-g+1}$ be a vector space. An isomorphism

$$W^* \xrightarrow{\sim} H^0(C, \mathcal{L}^f) \tag{9}$$

induces embeddings $\iota : C \hookrightarrow \mathbf{P}(W)$ and $\gamma : C \hookrightarrow \mathbf{P}(W) \times \mathbf{P}^r$ where $\gamma = (\iota, \mu)$. The $n$ sections $\{p_i\}$ yield $n$ points $(\iota \circ p_i, \mu \circ p_i)$ of $\mathbf{P}(W) \times \mathbf{P}^r$. Let $H$ be the Hilbert scheme of genus $g$ curves in $\mathbf{P}(W) \times \mathbf{P}^r$ of multidegree $(e, d)$. Let $P_i = \mathbf{P}(W) \times \mathbf{P}^r$ be the Hilbert scheme of a point in $\mathbf{P}(W) \times \mathbf{P}^r$. Via the isomorphism (9), a point in $H \times P_1 \times \ldots \times P_n$ is associated to the stable map $(C, \{p_i\}, \mu)$.

The locus of points in $H \times P_1 \times \ldots \times P_n$ corresponding to stable maps has a natural quasi-projective scheme structure. There is a natural closed incidence subscheme

$$I \subset H \times P_1 \times P_2 \times \ldots \times P_n$$

corresponding to the locus where the $n$ points lie on the curve. There is an open set $U \subset I$ satisfying the following:

(i) The curve $C$ is quasi-stable.
(ii) The natural projection $C \to \mathbf{P}(W)$ is a non-degenerate embedding.
(iii) The $n$ points lie in the nonsingular locus of $C$.
(iv) The multidegree of $\mathcal{O}_{\mathbf{P}(W)}(1) \otimes \mathcal{O}_{\mathbf{P}^r}(1)|_C$ equals the multidegree of

$$\omega_C^f(fp_1 + fp_2 + \ldots + fp_n) \otimes \mathcal{O}_{\mathbf{P}^r}(3f+1)|_C.$$

By Proposition 1, there exists a natural closed subscheme $J \subset U$ where the line bundles of condition (iv) above coincide. $J$ corresponds to the locus of stable maps. The natural $PGL(W)$-action on $\mathbf{P}(W) \times \mathbf{P}^r$ yields $PGL(W)$-actions on $H$, $P_i$, $I$, $U$, and $J$. To each stable map from an $n$-pointed, genus $g$ curve to $\mathbf{P}^r$, we have associated a $PGL(W)$-orbit in $J$. If two stable maps are associated to the same orbit, the two stable maps are isomorphic. The stability condition implies that a stable map has no infinitesimal automorphisms. It follows that the $PGL(W)$-action on $J$ has finite stabilizers.

**2.4. Quotients.** The moduli space of stable maps is $J/PGL(W)$. It may be possible to construct the quotient $J/PGL(W)$ via Geometric Invariant Theory. Another method will be pursued here. The quotient will be first constructed as a proper, algebraic variety by using auxiliary moduli spaces of pointed curves. Projectivity will then be established via J. Kollár's semipositivity approach.

## 3. A rigidification of $\overline{\mathcal{M}}_{g,n}(\mathbf{P}^r, d)$

**3.1. Review of Cartier divisors.** An effective Cartier divisor $D$ on a scheme $Y$ is a closed subscheme that is locally defined by a non-zero-divisor. An effective Cartier divisor determines a line bundle $\mathcal{L} = \mathcal{O}(D)$ together with a section $s \in H^0(Y, \mathcal{L})$ locally not a zero-divisor such that $D$ is the subscheme defined by $s = 0$. (As an invertible sheaf, $\mathcal{O}(D)$ can be constructed as the subsheaf of rational functions with at most simple poles along $D$ with $s$ equal to the function 1, see [M2].) Conversely, if the pair $(\mathcal{L}, s)$ satisfies:

(i) $\mathcal{L}$ is line bundle on $Y$.
(ii) $s \in H^0(Y, \mathcal{L})$ is a section locally not a zero divisor.

then the zero scheme of $s$ is an effective Cartier divisor on $Y$.

LEMMA 1. *Let the pairs* $(\mathcal{L}, s)$ *and* $(\mathcal{L}', s')$ *satisfy (i) and (ii) above. If the two pairs yield the same Cartier divisor, then there exists a unique isomorphism* $\mathcal{L} \to \mathcal{L}'$ *taking* $s$ *to* $s'$.

**3.2. Definitions.** We assume throughout the construction that $r > 0$, $d > 0$, and $(g, n, r, d) \neq (0, 0, 1, 1)$. If $r = 0$, the functor of stable maps to $\mathbf{P}^0$ is coarsely represented by $\overline{M}_{g,n}$. If $d = 0$, the functor $\overline{\mathcal{M}}_{g,n}(\mathbf{P}^r, 0)$ is coarsely represented by $\overline{M}_{g,n} \times \mathbf{P}^r$ and, $\overline{M}_{0,0}(1,1)$ is easily seen to be Spec($\mathbb{C}$). For all other values, the construction of $\overline{M}_{g,n}(X, \beta)$ will be undertaken.

Let $\mathbf{P}^r = \mathbf{P}(V)$. Then, $V^* = H^0(\mathbf{P}^r, \mathcal{O}_{\mathbf{P}^r}(1))$. Let $\bar{t} = (t_0, \ldots, t_r)$ span a basis of $V^*$. A *$\bar{t}$-rigid stable* family of degree $d$ maps from $n$-pointed, genus $g$ curves to $\mathbf{P}^r$ consists of the data

$$(\pi : \mathcal{C} \to S,\ \{p_i\}_{1 \leq i \leq n}\ ,\ \{q_{i,j}\}_{0 \leq i \leq r,\ 1 \leq j \leq d}\ ,\ \mu)$$

where:

(i) $(\pi : \mathcal{C} \to S,\ \{p_i\}, \mu)$ is a stable family of degree $d$ maps from $n$-pointed, genus $g$ curves to $\mathbf{P}^r$.

(ii) $(\pi : \mathcal{C} \to S,\ \{p_i\}, \{q_{i,j}\})$ is a flat, projective family of $n + d(r+1)$-pointed, genus $g$, Deligne-Mumford stable curves with sections $\{p_i\}$ and $\{q_{i,j}\}$.

(iii) For $0 \leq i \leq r$, there is an equality of Cartier divisors

$$\mu^*(t_i) = q_{i,1} + q_{i,2} + \ldots + q_{i,d}.$$

Condition (iii) implies each fibered map of the family intersects each hyperplane $(t_i) \subset \mathbf{P}^r$ transversally. Condition (ii) guarantees these hyperplane intersections are unmarked, nonsingular points.

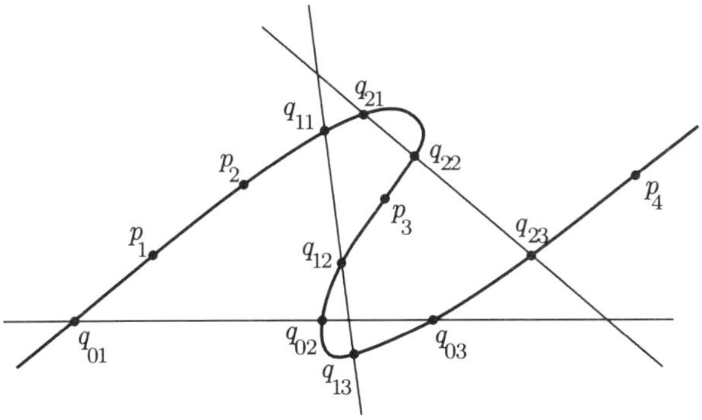


If $(g, n, r, d) = (0, 0, 1, 1)$, then $n + d(r+1) = 2$. There are no Deligne-Mumford stable 2-pointed genus 0 curves. This is why $(0, 0, 1, 1)$ is avoided.

Define a contravariant functor $\overline{\mathcal{M}}_{g,n}(\mathbf{P}^r, d, \bar{t})$ from the category of complex algebraic schemes to sets as follows. Let $\overline{\mathcal{M}}_{g,n}(\mathbf{P}^r, d, \bar{t})(S)$ be the set of isomorphism classes of $\bar{t}$-rigid stable families over $S$ of degree $d$ maps from $n$-pointed, genus $g$ curves to $\mathbf{P}^r$. Note that the functor $\overline{\mathcal{M}}_{g,n}(\mathbf{P}^r, d, \bar{t})$ depends only upon the spanning hyperplanes $(t_i) \subset \mathbf{P}^r$ and not upon the additional $\mathbb{C}^*$-choices in the defining equations $t_i$ of the hyperplanes. Nevertheless, it is natural for the following constructions to consider the equations of the hyperplanes $\bar{t} = (t_0, \ldots, t_r)$.

PROPOSITION 2. *There exists a quasi-projective coarse moduli space,*

$$\overline{M}_{g,n}(\mathbf{P}^r, d, \bar{t}),$$

*and a natural transformation of functors*

$$\psi : \overline{\mathcal{M}}_{g,n}(\mathbf{P}^r, d, \bar{t}) \to \mathcal{H}om(*, \overline{M}_{g,n}(\mathbf{P}^r, d, \bar{t}))$$

*satisfying the analogous conditions (I) and (II) of Theorem 1.*

The genus 0 case is simpler.

PROPOSITION 3. *$\overline{M}_{0,n}(\mathbf{P}^r, d, \bar{t})$ represents the functor $\overline{\mathcal{M}}_{0,n}(\mathbf{P}^r, d, \bar{t})$ and is a nonsingular algebraic variety.*

**3.3. Proofs.** A complete proof of Proposition 3 will be given. The proof of Proposition 2 is almost identical. Remarks indicating the differences will be made. The dependence of the coarse and fine moduli property on the genus in Propositions 2 and 3 is a direct consequence of the fact that $\overline{M}_{g,n}$ is a coarse moduli space for $g > 0$ and a fine moduli space for $g = 0$.

The idea behind the construction is the following. Let $m = n + d(r + 1)$. The data of the $\bar{t}$-rigid stable family immediately yields a morphism of the base $S$ to $\overline{M}_{g,m}$. In fact, the image of $S$ lies in a universal, locally closed subscheme of $\overline{M}_{g,m}$. This subscheme is denoted by $B$. The first step of the construction is to identify $B$. The morphism $S \to B$ does not contain all the data of the $\bar{t}$-rigid stable family. Consider the case in which the base $S$ is a point. The corresponding point in $B$ records the domain curve $C$, the marked points $\{p_i\}$, and the pull-back divisors under $\mu$ of the hyperplanes in $\mathbf{P}^r$ determined by $\bar{t}$. The map $\mu$ is determined by the pull-back divisors up to the diagonal torus action on $\mathbf{P}^r$. The torus information is recorded in the total space of $r$ tautological $\mathbb{C}^*$-bundles over $B$. The $\bar{t}$-rigid moduli space is expressed as the total space of these $r$ distinct $\mathbb{C}^*$-bundles over $B$. To canonically construct the universal family over the $\bar{t}$-rigid moduli space, the equations $t_i$ of the hyperplanes are needed. This is why the equations $t_i$ (rather than the spanning hyperplanes $(t_i)$) are explicitly chosen.

Proposition 3 is proved by an explicit construction of $\overline{M}_{0,n}(\mathbf{P}^r, d, \bar{t})$ together with a universal family of $\bar{t}$-rigid stable maps. Let $\overline{M}_{0,m}$ be the Mumford-Knudsen compactification of the moduli space of $m$-pointed, genus 0 curves. Let $\pi : \overline{U}_{0,m} \to \overline{M}_{0,m}$ be the universal curve with $m$ sections $\{p_i\}_{1 \le i \le n}$ and $\{q_{i,j}\}_{0 \le i \le r,\ 1 \le j \le d}$. Since $\overline{U}_{0,m}$ is nonsingular and the sections are of codimension 1, there are canonically defined line bundles:

$$\mathcal{H}_i = \mathcal{O}_{\overline{U}_{0,m}}(q_{i,1} + q_{i,2} + \ldots + q_{i,d}),$$

for $0 \le i \le r$. Let $s_i \in H^0(\overline{U}_{0,m}, \mathcal{H}_i)$ be the canonical section representing the Cartier divisor $(q_{i,1} + q_{i,2} + \ldots + q_{i,d})$.

For any morphism $\gamma : X \to \overline{M}_{0,m}$, consider the fiber product:

$$\begin{array}{ccc} X \times_{\overline{M}_{0,m}} \overline{U}_{0,m} & \xrightarrow{\overline{\gamma}} & \overline{U}_{0,m} \\ \downarrow \pi_X & & \downarrow \pi \\ X & \xrightarrow{\gamma} & \overline{M}_{0,m} \end{array}$$

We call the morphism $\gamma : X \to \overline{M}_{0,m}$ *$\mathcal{H}$-balanced* if

(a) For $1 \le i \le r$, $\pi_{X*}\overline{\gamma}^*(\mathcal{H}_i \otimes \mathcal{H}_0^{-1})$ is locally free.
(b) For $1 \le i \le r$, the canonical map

$$\pi_X^* \pi_{X*} \overline{\gamma}^*(\mathcal{H}_i \otimes \mathcal{H}_0^{-1}) \to \overline{\gamma}^*(\mathcal{H}_i \otimes \mathcal{H}_0^{-1})$$

is an isomorphism.

If $\gamma$ is $\mathcal{H}$-balanced, the line bundles $\overline{\gamma}^*(\mathcal{H}_i)$ are isomorphic on the fibers of $\pi_X$. Let $B \subset \overline{M}_{0,m}$ be the universal, locally closed subscheme satisfying the two following properties:

(i) The inclusion $\iota : B \hookrightarrow \overline{M}_{0,m}$ is $\mathcal{H}$-balanced.
(ii) Every $\mathcal{H}$-balanced morphism $\gamma : X \to \overline{M}_{0,m}$ factors (uniquely) through $B$.

By Proposition 1, $B$ exists. In fact, $B \subset \overline{M}_{0,m}$ is a Zariski open subscheme. In the $g > 0$ case, the above constructions exist over the stacks $\overline{M}_{g,m}$ and $\overline{U}_{g,m}$. $B_{g,m}$ is a locally closed substack of $\overline{M}_{g,m}$ of positive codimension.

Let $\mathcal{G}_i = \pi_{B*}\overline{\iota}^*(\mathcal{H}_i \otimes \mathcal{H}_0^{-1})$ for $1 \leq i \leq r$. Let $\tau_i : Y_i \to B$ be the total space of the canonical $\mathbb{C}^*$-bundle associated to $\mathcal{G}_i$. $Y_i$ is the affine bundle associated to $\mathcal{G}_i$ minus the zero section. The pull-back $\tau_i^*(\mathcal{G}_i)$ has a tautological non-vanishing section and hence is canonically trivial. Consider the product

$$Y = Y_1 \times_B \times Y_2 \times_B \ldots \times_B Y_r$$

equipped with projections $\rho_i : Y \to Y_i$ and a morphism $\tau : Y \to B$. Form the cartesian square:

$$\begin{array}{ccc} \mathcal{U} & \xrightarrow{\overline{\tau}} & \overline{U}_{0,m} \\ \downarrow{\scriptstyle \pi_Y} & & \downarrow{\scriptstyle \pi} \\ Y & \xrightarrow{\tau} & B \subset \overline{M}_{0,m}. \end{array}$$

The line bundles $\overline{\tau}^*(\mathcal{H}_i)$ for $1 \leq i \leq r$ are canonically isomorphic to $\mathcal{L} = \overline{\tau}^*(\mathcal{H}_0)$ on $\mathcal{U}$ since

$$\overline{\tau}^*(\mathcal{H}_i \otimes \mathcal{H}_0^{-1}) \stackrel{\sim}{=} \pi_Y^* \rho_i^* \tau_i^*(\mathcal{G}_i)$$

and $\tau_i^*(\mathcal{G}_i)$ is canonically trivial.

Via pull-back and the canonical isomorphisms, $\overline{\tau}^*(s_i)$ canonically corresponds to a section of $\mathcal{L}$. Since these $r+1$ sections do not vanish simultaneously, they define a morphism of $\mu : \mathcal{U} \to \mathbf{P}^r$. The canonical method of obtaining $\mu$ is as follows. Define a vector space map $V^* \to H^0(\mathcal{L})$ by sending $t_i$ to $\overline{\tau}^*(s_i)$. The induced surjection $V^* \otimes \mathcal{O} \to \mathcal{L}$ canonically yields a morphism

$$\mu : \mathcal{U} \to \mathbf{P}^r.$$

Note that the equations $t_i$ are used to define the morphism $\mu$. The sections $\{p_i\}$, $\{q_{i,j}\}$ pull back to sections of $\pi_Y$. We claim that the family

(10) $$(\pi_Y : \mathcal{U} \to Y, \{p_i\}, \{q_{i,j}\}, \mu)$$

is a universal family of $\overline{t}$-rigid stable maps, so $\overline{M}_{0,n}(\mathbf{P}^r, d, \overline{t}) = Y$.

The stability of the family of maps

(11) $$(\pi_Y : \mathcal{U} \to Y, \{p_i\}, \mu)$$

is straightforward. Each fiber $C$ of $\pi_Y$ is an $m$-pointed, genus 0 stable curve with markings $\{p_i\}$ and $\{q_{i,j}\}$. Let $E \subset C$ be an irreducible component. Suppose $\dim(\mu(E)) = 0$. By the transversality condition (iii), $E$ has no markings from the sections $\{q_{i,j}\}$. Since $C$ is a stable $m$-pointed curve and no $\{q_{i,j}\}$ markings lie on $E$, $\deg_E(\omega_C(p_1 + \ldots + p_n)) > 0$. Hence, condition (1) in the definition of map stability (section 1.1) holds for $E$. Therefore (11) is a stable family of maps. By construction, it is a $\overline{t}$-rigid stable family.

Finally, it must be shown (10) is universal. Let

$$(\pi : \mathcal{C} \to S, \{p_i\}, \{q_{i,j}\}, \nu) \tag{12}$$

be a family of $\bar{t}$-rigid stable maps. Since $(\pi : \mathcal{C} \to S, \{p_i\}, \{q_{i,j}\})$ is a flat family of $m$-pointed, genus 0 stable curves, there is an induced map $\lambda : S \to \overline{M}_{0,m}$ such that the pull-back family $S \times_{\overline{M}_{0,m}} \overline{U}_{0,m}$ is canonically isomorphic to $(\pi : \mathcal{C} \to S, \{p_i\}, \{q_{i,j}\})$.

First we show $\lambda$ is $\mathcal{H}_i$-balanced. The pair $(\overline{\lambda}^*(\mathcal{H}_i), \overline{\lambda}^*(s_i))$ yields the Cartier divisor $q_{i,1}+\ldots+q_{i,d}$ on $\mathcal{C}$. The map $\nu$ is induced by a vector space homomorphism $\psi : V^* \to H^0(\mathcal{C}, \nu^*(\mathcal{O}_{\mathbf{P}(V)}(1)))$. Let $z_i = \psi(t_i)$. By condition (iii) of $\bar{t}$-rigid stability, the pair $(\nu^*(\mathcal{O}_{\mathbf{P}(V)}(1)), z_i)$ yields the Cartier divisor $q_{i,1}+\ldots+q_{i,d}$ on $\mathcal{C}$. By Lemma 1, there are *canonical* isomorphisms

$$\overline{\lambda}^*(\mathcal{H}_i) \stackrel{\sim}{=} \nu^*(\mathcal{O}_{\mathbf{P}(V)}(1)) \tag{13}$$

for all $i$. Hence $\lambda$ is $\mathcal{H}_i$-balanced.

By the universal property of $B$, $\lambda$ factors through $B$: $\lambda : S \to B$. There are canonical isomorphisms

$$\pi_*(\overline{\lambda}^*(\mathcal{H}_i \otimes \mathcal{H}_0^{-1})) \stackrel{\sim}{=} \lambda^*(\mathcal{G}_i). \tag{14}$$

The canonical isomorphisms (13) yield canonical sections of $\overline{\lambda}^*(\mathcal{H}_i \otimes \mathcal{H}_0^{-1})$. The canonical isomorphisms (14) then yield nowhere vanishing sections of $\lambda^*(\mathcal{G}_i)$ over $S$. Hence there is a canonical a map $S \to Y$. It is easily checked the pull-back of the universal family over $Y$ yields a $\bar{t}$-rigid stable family of maps canonically isomorphic to (12).

## 4. The construction of $\overline{M}_{g,n}(\mathbf{P}^r, d)$

**4.1. Gluing.** While a given pointed stable map $\mu : C \to \mathbf{P}^r$ may not be rigid for a given basis $\bar{t}$ of $V^* = H^0(\mathbf{P}^r, \mathcal{O}_{\mathbf{P}^r}(1))$, the map will be rigid (by Bertini's theorem) for some choice of basis. The moduli space $\overline{M}_{g,n}(\mathbf{P}^r, d)$ is obtained by gluing together quotients of $\overline{M}_{g,n}(\mathbf{P}^r, d, \bar{t})$ for different choices of bases $\bar{t}$.

For notational convenience, set $\overline{M}(\bar{t}) = \overline{M}_{g,n}(\mathbf{P}^r, d, \bar{t})$. We write $(\pi : \mathcal{U} \to \overline{M}(\bar{t}), \{p_i\}, \{q_{i,j}\}, \mu)$ for the universal family of $\bar{t}$-rigid stable maps in the genus 0 case. If $g > 0$, more care is required.

Let $\mathfrak{S}_d$ denote the symmetric group on $d$ letters. The group

$$G = G_{d,r} = \mathfrak{S}_d \times \ldots \times \mathfrak{S}_d \quad (r+1 \text{ factors})$$

has a natural action on $\overline{M}(\bar{t})$ obtained by permuting the ordering in each of the $r+1$ sets of sections $\{q_{i,1}, \ldots, q_{i,d}\}$, $0 \leq i \leq r$. For any $\sigma \in G$, the family

$$(\pi : \mathcal{U} \to \overline{M}(\bar{t}), \{p_i\}, \{q_{i,\sigma(j)}\}, \mu) \tag{15}$$

is also a $\bar{t}$-rigid family over $\overline{M}(\bar{t})$. By the universal property, the permuted family (15) induces an automorphism of $\overline{M}(\bar{t})$. Since $\overline{M}(\bar{t})$ is quasi-projective and $G$ is finite, there is a quasi-projective quotient scheme $\overline{M}(\bar{t})/G$.

Let $\bar{t}$ and $\bar{t}'$ be distinct choices of bases of $V^*$. Let $\mu : \mathcal{U} \to \mathbf{P}^r$ be the universal family over $\overline{M}(\bar{t})$. Let

$$\overline{M}(\bar{t}, \bar{t}') \subset \overline{M}(\bar{t})$$

denote the open locus over which the divisors $\mu^*(t_0'), \ldots, \mu^*(t_r')$ are étale, disjoint, and disjoint from the sections $\{p_i\}$. The open set $\overline{M}(\bar{t}, \bar{t}')$ is certainly $G$-invariant. Let $\overline{M}(\bar{t}, \bar{t}')/G$ denote the quasi-projective quotient.

PROPOSITION 4. *There is a canonical isomorphism*

$$\overline{M}(\bar{t},\bar{t}')/G \cong \overline{M}(\bar{t}',\bar{t})/G.$$

*Proof.* The divisors $\mu^*(t_i')$ define an étale Galois cover $\mathcal{E}$ of $\overline{M}(\bar{t},\bar{t}')$ with Galois group $G$ over which a $\bar{t}'$-rigid stable family is defined. The fiber of $\mathcal{E}$ over $(C, \{p_i\}, \{q_{i,j}\}, \mu)$ is the set of orderings $\{q_{i,j}'\}$ of the points mapped by $\mu$ to the hyperplane $(t_i' = 0)$. Therefore there is a map

$$\mathcal{E} \to \overline{M}(\bar{t}') \tag{16}$$

which is easily seen be $G$-equivariant for the Galois $G$-action on $\mathcal{E}$ and the $\{q_{i,j}'\}$-permutation $G$-action on the $\overline{M}(\bar{t}')$. Moreover (16) factors through $\overline{M}(\bar{t}',\bar{t})$. Hence there exists a map of quotients

$$\overline{M}(\bar{t},\bar{t}') \cong \mathcal{E}/\text{Galois} \to \overline{M}(\bar{t}',\bar{t})/G. \tag{17}$$

The map (17) is $G$-invariant for the $\{q_{i,j}\}$-permutation action on $\overline{M}(\bar{t},\bar{t}')$. Therefore (17) descends to $\overline{M}(\bar{t},\bar{t}')/G \to \overline{M}(\bar{t}',\bar{t})/G$. The inverse is obtained by interchanging $\bar{t}$ and $\bar{t}'$ in the above construction. In fact, there is a natural action of $G \times G$ on $\mathcal{E}$ and canonical isomorphisms $\overline{M}(\bar{t},\bar{t}')/G \cong \mathcal{E}/(G \times G) \cong \overline{M}(\bar{t}',\bar{t})/G$. □

In case $g > 0$, the coarse moduli spaces $\overline{M}_{g,n}(\mathbf{P}^r, d, \bar{t})$ do not (in general) have universal families. The permutation action of $G$ can be defined on a Hilbert scheme or a stack and then descended to $\overline{M}_{g,n}(\mathbf{P}^r, d, \bar{t})$. The open sets $\overline{M}(\bar{t}',\bar{t})$ and $\overline{M}(\bar{t},\bar{t}')$ are well defined for $g > 0$ and still satisfy Proposition 4.

The cocycle conditions on triple intersections are easily established. Hence, the schemes $\overline{M}(\bar{t})/G$ canonically patch together along the open sets $\overline{M}(\bar{t},\bar{t}')/G$ to form the scheme $\overline{M}_{g,n}(\mathbf{P}^r, d)$. The results on boundedness show $\overline{M}_{g,n}(\mathbf{P}^r, d)$ is covered by a finite number of these open sets $\overline{M}(\bar{t})/G$. Hence, $\overline{M}_{g,n}(\mathbf{P}^r, d)$ is an algebraic scheme of finite type over $\mathbb{C}$. The universal properties of $\overline{M}_{g,n}(\mathbf{P}^r, d)$ are easily obtained from the universal properties of the moduli spaces of $\bar{t}$-rigid stable maps.

**4.2. Separation and completeness.** Let $(X, x)$ be a nonsingular, pointed curve. Let $\iota : X \setminus \{x\} = U \hookrightarrow X$. Let

$$(\pi : \mathcal{C} \to X,\ \{p_i\},\ \mu) \tag{18}$$

$$(\pi' : \mathcal{C}' \to X,\ \{p_i'\},\ \mu') \tag{19}$$

be two families over $X$ of stable maps to $\mathbf{P}^r = \mathbf{P}(V)$.

PROPOSITION 5. *An isomorphism between the families (18) and (19) over $U$ extends to an isomorphism over $X$.*

*Proof.* Choose a basis $\bar{t} = (t_0, \ldots, t_r)$ of $V^*$ that intersects the maps $\mu : \mathcal{C}_x \to \mathbf{P}^r$ and $\mu' : \mathcal{C}_x' \to \mathbf{P}^r$ transversally at unmarked, nonsingular points. Since it suffices to prove the isomorphism extends over a local étale cover of $(X, x)$, it can be assumed that the Cartier divisors $\mu^*(t_i)$ and $\mu'^*(t_i)$ split into sections $\{q_{i,j}\}$ and $\{q_{i,j}'\}$ of $\pi$ and $\pi'$. Then $\mathcal{C}$, $\mathcal{C}'$ are Deligne-Mumford stable $m = n + d(r+1)$ pointed curves. Therefore, by the separation property of the functor of Deligne-Mumford stable pointed curves, there exists an isomorphism (of pointed curves) $\tau : \mathcal{C} \to \mathcal{C}'$ over $X$. Since $\tau \circ \mu'$ and $\mu$ agree on an open set, $\tau \circ \mu' = \mu$. □

Proposition 5 and the valuative criterion show $\overline{M}_{g,n}(X,\beta)$ is a separated algebraic scheme.

Properness is also established by the valuative criterion. To complete 1 dimensional families of stable maps, semi-stable reduction techniques for curves are used (as in [K-K-M] and [Ha]).

PROPOSITION 6. *Let* $\mathcal{F} = (\pi : \mathcal{C} \to U,\ \{p_i\},\ \mu)$ *be a family of stable maps to* $\mathbf{P}^r$. *There exists a base change* $\gamma : (Y,y) \to (X,x)$ *satisfying:*

(i) $\gamma_W : Y \setminus \{y\} = W \to U$ *is étale.*

(ii) *The pull-back family* $\gamma_W^*(\mathcal{F})$ *extends to a stable family over* $(Y,y)$.

*Proof.* First, after restriction to a Zariski open subset of $U$, it can be assumed that the fibers $\mathcal{C}_\xi$ all have the same number of irreducible components. There may be non-trivial monodromy around the point $x \in X$ in the set of irreducible components of the fibers $\mathcal{C}_\xi$. After a base change (possibly ramified at $x$), this monodromy can be made trivial. It can therefore be assumed that $\mathcal{F}$ is a union of stable families $\mathcal{F}_j = (\pi_j : \mathcal{C}_j \to U,\ \{p_i^j\}, \{p_i^c\},\ \mu_j)$ where $\pi_j$ is family of *irreducible*, nodal, projective curves. The markings $\{p_i^j\}$ are the markings of $\mathcal{C}$ that lie on $\mathcal{C}_j$. The marking $\{p_j^c\}$ correspond to intersections of components in $\mathcal{F}$. It suffices to prove Proposition 6 separately for each stable family $\mathcal{F}_j$.

For technical reasons, it is convenient to consider families of nonsingular curves. After restriction, normalization, and base change of $\mathcal{F}_j$, a family

$$\tilde{\mathcal{F}}_j = (\tilde{\pi}_j : \tilde{\mathcal{C}}_j \to U, \{p_i^j\}, \{p_i^c\}, \{p_i^n\}, \tilde{\mu}_j) \tag{20}$$

can be obtained where $\tilde{\mathcal{F}}_j$ is a family of stable maps of irreducible, *nonsingular*, projective curves. The additional markings $\{p_i^n\}$ correspond to the nodes. Consider the nodal locus in $\mathcal{F}_j$. This locus consists of curves and isolated points. Via restriction of $U$ to a Zariski open set, it can be assumed the nodal locus (if non-empty) is of pure dimension 1. A normalization now separates the sheets along the nodal locus. A base change then may be required to make the separated points $\{p_i^n\}$ sections. If the normalized family $\tilde{\mathcal{F}}_j$ is completed, $\mathcal{F}_j$ can be completed by identifying the nodal markings on $\tilde{\mathcal{F}}_j$. This nodal identification commutes with the map to $\mathbf{P}^r$. It therefore suffices to prove Proposition 6 for these normalized families (20).

By the above reductions, it suffices to prove Proposition 6 for a family of stable maps of irreducible, nonsingular, projective curves. Let

$$(\pi : \mathcal{C} \to U, \{p_i\}, \mu) \tag{21}$$

be such a family. Let $\pi : \mathcal{E} \to X$ be a flat extension of $\pi : \mathcal{C} \to U$ over the point $x \in X$. After blow-ups in the special fiber of $\mathcal{E}$, it can be assumed the map $\mu : \mathcal{C} \to \mathbf{P}^r$ extends to $\mu : \mathcal{E} \to \mathbf{P}^r$. By Lemma 2 below applied to the flat extension $\pi : \mathcal{E} \to X$, there exists a base change $\gamma : (Y,y) \to (X,x)$ and a family of pointed curves

$$\pi_Y : \mathcal{C}_Y \to (Y,y)$$

satisfying conditions (i)–(iii) of Lemma 2. Via $\tau : \mathcal{C}_Y \to \mathcal{E}$, $\mu$ naturally induces a map

$$\mu_Y : \mathcal{C}_Y \to \mathbf{P}^r.$$

The family $(\pi_Y : \mathcal{C}_Y \to (Y,y), \{p_i\}, \mu_Y)$ is certainly an extension of the family over $Y \setminus \{y\}$ determined by the $\gamma$ pull-back of the stable family (21). The special fiber is a map of a pointed quasi-stable curve to $\mathbf{P}^r$. Unfortunately, the special fiber may not be stable. A stable family of maps is produced in two steps. First, unmarked, $\mu_Y$-collapsed, $-1$-curves in the special fiber are sequentially blow-down. A multiple of the line bundle

$$\omega_{\pi_Y}(\sum_i p_i) \otimes \mu_Y^*(\mathcal{O}_{\mathbf{P}^r}(3)) \tag{22}$$

is then $\pi_Y$- relatively basepoint free. Second, as in [Kn], the relative morphism determined by a power of the line bundle (22) blows-down the remaining destabilizing $\mathbf{P}^1$'s to yield a stable extension over $(Y, y)$. □

LEMMA 2. *Let $\pi_X : \mathcal{S}_X \to (X, x)$ be a flat, projective family of curves with $l$ sections $s_1, \ldots, s_l$ satisfying the following condition: $\forall \xi \neq x$, $\pi^{-1}(\xi) = \mathcal{C}_\xi$ is a projective nonsingular curve with $l$ distinct marked points $s_1(\xi), \ldots, s_l(\xi)$. There exists a base change $\gamma : (Y, y) \to (X, x)$ étale except possibly at $y$ with a family of $l$-pointed curves $\pi_Y : \mathcal{S}_Y \to (Y, y)$ and a diagram:*

$$\begin{array}{ccc} \mathcal{S}_Y & \xrightarrow{\tau} & \mathcal{S}_X \\ \downarrow{\scriptstyle \pi_Y} & & \downarrow{\scriptstyle \pi_X} \\ (Y, y) & \xrightarrow{\gamma} & (X, x) \end{array}$$

*satisfying the following properties:*

(i) *$\mathcal{S}_Y$ is a nonsingular surface. $\pi_Y : \mathcal{S}_Y \to (Y, y)$ is a flat, projective family of $l$-pointed quasi-stable curves.*
(ii) *For each marking $1 \leq i \leq l$, $\tau \circ s_i = s_i \circ \gamma$.*
(iii) *Over $W = Y \setminus \{y\}$, there is isomorphism $\mathcal{S}_W \xrightarrow{\sim} \gamma_W^*(\mathcal{S}_U)$, where $U = X \setminus \{x\}$. The morphism $\tau|_{\mathcal{S}_W}$ is the composition*

$$\mathcal{S}_W \xrightarrow{\sim} \gamma_W^*(\mathcal{S}_U) \to \mathcal{S}_U$$

*where the second map is the natural projection.*

*Proof.* The method is by standard semi-stable reduction (cf. [K-K-M], [Ha]). First, the singularities of $\mathcal{S}_X$ are resolved. Note that all singularities lie in the special fiber. Next, the surface $\mathcal{S}_X$ is blown-up sufficiently to ensure the reduced scheme supported on the special fiber has normal crossing singularities in $\mathcal{S}_X$. The required blow-ups have point centers in the special fiber. Finally, the resulting surface is blown-up further (at points in the special fiber) to ensure the marking sections $s_1, \ldots, s_l$ do not intersect each other and do not pass through nodes of the reduced scheme supported on the special fiber. Let $\hat{\pi} : \hat{\mathcal{S}}_X \to (X, x)$ be the resulting nonsingular surface. The singularities of the morphism $\hat{\pi}$ are locally of the form $z_1^\alpha z_2^\beta = t$ where $z_1$, $z_2$ are coordinates on $\hat{\mathcal{S}}_X$ and $t$ is a coordinate on $X$. Let $\{\alpha_j, \beta_j\}$ be the set of exponents that occur at the singularities of $\hat{\pi}$. Let $\gamma : (Y, y) \to (X, x)$ be a base change whose ramification index over $x$ is divisible by all $\alpha_j$ and $\beta_j$. Let $\mathcal{S}_Y$ be the normalization of $\gamma^*(\hat{\mathcal{S}}_X)$. A straightforward local analysis shows the family $\pi_Y : \mathcal{S}_Y \to (Y, y)$ has an $l$-pointed, reduced, nodal special fiber. The surface $\mathcal{S}_Y$ has singularities of the local form $z_1 z_2 - t^k$ in the special fiber. Blowing-up $\mathcal{S}_Y$ yields a nonsingular surface with the required properties (i)–(iii). □

By the valuative criterion, Propositions 5 and 6 prove $\overline{M}_{g,n}(\mathbf{P}^r, d)$ is a separated and proper complex algebraic scheme.

**4.3. Projectivity.** The projectivity of the proper schemes $\overline{M}_{g,n}(\mathbf{P}^r, d)$ is established here by a method due to J. Kollár ([Kol]). Proofs of the projectivity of $\overline{M}_{g,n}(\mathbf{P}^r, \beta)$ can also be found in [A] and [C]. Kollár constructs ample line bundles on proper spaces via sufficiently nontrivial quotients of semipositive vector bundles. A vector bundle $E$ on an algebraic scheme $S$ is *semipositive* if for every morphism of a projective curve $f : C \to S$, every quotient line bundle of $f^*(E)$ has nonnegative degree on $C$.

The first step is a semipositivity lemma. Let

$$\mathcal{F} = (\pi : \mathcal{C} \to S,\ \{p_i\},\ \mu) \tag{23}$$

be a stable family of maps over $S$ to $\mathbf{P}^r$. Let

$$E_k(\pi) = \pi_*\Big(\omega_\pi^k(\sum_{i=1}^{n} kp_i) \otimes \mu^*(\mathcal{O}(3k))\Big).$$

LEMMA 3. *$E_k(\pi)$ is a semipositive vector bundle on $S$ for $k \geq 2$.*

*Proof.* A slight perturbation of the arguments in [Kol] is required. It suffices to prove semipositivity in case the base is a nonsingular curve $X$. Let $\gamma : Y \to X$ be a flat base change. By map stability, Serre duality, and the base change theorems, it follows (for $k \geq 2$) $E_k$ commutes with pull-back:

$$E_k(\pi_Y) \stackrel{\sim}{=} \gamma^*(E_k(\pi_X))$$

where $\pi_Y$ is the pull-back family over $Y$. It therefore suffices to prove semipositivity after base change.

Using the methods of section 4.2, it can be assumed (after base change) that $\mathcal{F}$ is a union of component stable families $\mathcal{F}_j = (\pi_j : \mathcal{C}_j \to X,\ \{p_i^j\}, \{p_i^c\},\ \mu_j)$ where $\pi_j$ is family of stable maps and the generic element of $\mathcal{F}_j$ is a map of an *irreducible*, projective, nodal curve. The notation introduced in the proof of Proposition 6 is employed. After further base change and normalization of $\mathcal{F}_j$, it can be assumed that

$$\tilde{\mathcal{F}}_j = (\tilde{\pi}_j : \tilde{\mathcal{C}}_j \to X,\ \{p_i^j\}, \{p_i^c\}, \{p_i^n\},\ \tilde{\mu}_j) \tag{24}$$

is a family of stable maps where the generic element is a map of an irreducible, projective, *nonsingular* curve.

A semipositivity result for the family $\tilde{\mathcal{F}}_j$ is first established. Let $H_1$, $H_2$, $H_3 \subset \mathbf{P}^r$ be general hyperplanes. After base change, it can be assumed $\tilde{\mu}_j^*(H_l)$ is a union of $d$ reduced sections for each $l$. These $3d$ sections are distinct from the sections $\{p_i^j\}$, $\{p_i^c\}$, $\{p_i^n\}$. Therefore,

$$\omega_{\tilde{\pi}_j}^k\big(\sum kp_i^j + \sum(k-1)p_i^c + \sum(k-1)p_i^n\big) \otimes \tilde{\mu}_j^*(\mathcal{O}_{\mathbf{P}^r}(3k)) \stackrel{\sim}{\to} \omega_{\tilde{\pi}_j}^k\big(\sum \alpha_q X_q\big) \tag{25}$$

where $X_q$ are distinct sections of $\tilde{\pi}_j$ and $\alpha_q \leq k$. The surface $\tilde{\mathcal{C}}_j$ has finitely many singularities of the form $z_1 z_2 - t^\alpha$. These singularities are resolved by blow-up,

$$\tau : \mathcal{S}_j \to \tilde{\mathcal{C}}_j.$$

Since the relative dualizing sheaf of the family $\mathcal{S}_j$ is trivial on the exceptional $\mathbf{P}^1$'s of $\tau$, Lemma 4 below can be applied to deduce the semipositivity of $F_k(\tilde{\pi}_j)$ for $k \geq 2$ where

$$F_k(\tilde{\pi}_j) = \tilde{\pi}_{j*}\Big(\omega^k_{\tilde{\pi}_j}\big(\sum kp^j_i + \sum(k-1)p^c_i + \sum(k-1)p^n_i\big) \otimes \mu^*(\mathcal{O}(3k))\Big).$$

For $k \geq 2$, the restriction of the line bundle (25) to a fiber of $\tilde{\mathcal{C}}_j$ is equal to

$$\omega \otimes \omega^{k-1}\big(\sum(k-1)p^j_i + \sum(k-1)p^c_i + \sum(k-1)p^n_i\big) \otimes \tag{26}$$
$$\tilde{\mu}^*_j(\mathcal{O}_{\mathbf{P}^r}(3k-3)) \otimes \mu^*_j(\mathcal{O}_{\mathbf{P}^r}(3))\big(\sum_i p^j_i\big)$$

where $\omega$ is the dualizing sheaf of the fiber. By stability for the family $\tilde{\mathcal{F}}_j$, the product of the middle two factors in (26) is ample for $k \geq 2$. The last factor in (26) is certainly of non-negative degree. By Serre duality, for $k \geq 2$,

$$R^1\tilde{\pi}_{j*}\Big(\omega^k_{\tilde{\pi}_j}\big(\sum kp^j_i + \sum(k-1)p^c_i + \sum(k-1)p^n_i\big) \otimes \mu^*(\mathcal{O}(3k))\Big) = 0. \tag{27}$$

The semipositivity of $E_k(\pi)$ will be obtained from the semipositivity of $F_k(\tilde{\pi}_j)$. The $(k-1)$-multiplicities will naturally arise in considering dualizing sheaves on nodal and reducible curves.

Let $\tilde{\pi}_{\cup_j} : \bigcup_j \tilde{\mathcal{C}}_j \to X$ be the disjoint union of the families $\tilde{\mathcal{F}}_j$. There is natural morphism from the disjoint union to $\mathcal{C}$

$$\rho : \bigcup_j \tilde{\mathcal{C}}_j \to \mathcal{C}$$

obtained by identifying nodal marked points and gluing components along intersection marked points. Consider the natural sequence of sheaves on $\mathcal{C}$:

$$0 \to \rho_*(\omega_{\tilde{\pi}_{\cup_j}}) \to \omega_\pi \to K \to 0. \tag{28}$$

The quotient $K$ is easily identified as $\bigoplus_{p^c_i, p^n_i} \mathcal{O}_p$ where the sum is over all nodal and component intersection sections of the family $\mathcal{F}$. Tensoring (28) with the line bundle $\omega^{k-1}_\pi(\sum kp_i) \otimes \mu^*(\mathcal{O}_{\mathbf{P}(V)}(3k))$ yields the exact sequence:

$$0 \to \rho_*\Big(\omega^k_{\tilde{\pi}_{\cup_j}}\big(\sum kp_i + \sum(k-1)p^c_i + \sum(k-1)p^n_i\big) \otimes \mu^*(\mathcal{O}_{\mathbf{P}(V)}(3k))\Big) \to$$
$$\omega^k_\pi\big(\sum_i kp_i\big) \otimes \mu^*(\mathcal{O}_{\mathbf{P}(V)}(3k)) \to \bigoplus_{p^c_i, p^n_i} \mathcal{O}_p \otimes \mu^*(\mathcal{O}_{\mathbf{P}(V)}(3k)) \to 0.$$

Certainly $\tilde{\pi}_{\cup_j *} = \pi_*\rho_*$. Note the vanishing of $R^1$ determined in (27). These facts imply the $\pi$ direct image of the above sequence on $\mathcal{C}$ yields an exact sequence on $X$:

$$0 \to \bigoplus_j F_k(\tilde{\pi}_j) \to E_k(\pi) \to \bigoplus_{p^c_i, p^n_i} \mathcal{O}_X \otimes \mu^*(\mathcal{O}_{\mathbf{P}(V)}(3k)) \to 0.$$

Finally, since an extension of semipositive bundles is semipositive ([Ko1]), $E_k(\pi)$ is semipositive. $\square$

LEMMA 4. *Let $\pi : S \to X$ be a map from a nonsingular projective surface to a nonsingular curve. Assume the general fiber of $\pi$ is nonsingular. Let $X_q$ be a set of distinct sections of $\pi$. Then*

$$\pi_*(\omega^k_{S/X}(\sum \alpha_q X_q))$$

*is semipositive provided $k \geq 2$ and $\alpha_q \leq k$ for all $q$.*

*Proof.* This is precisely Proposition 4.7 of [Kol]. □

The second step is the construction of a non-trivial quotient. Let $\mathcal{F}$ be the family (23). Let $\mathbf{P}(E^*_k)$ be the projective bundle over $S$ obtained from the subspace projectivization of $E^*_k$. The condition of stability implies there is a canonical $S$-embedding $\iota : \mathcal{C} \to \mathbf{P}(E^*_f)$ for some $f = f(d, g, n, r)$ (see section 2.3). The morphism $\mu$ then yields a canonical $S$-embedding:

$$\gamma : \mathcal{C} \to \mathbf{P}(E^*_f) \times_{\mathbb{C}} \mathbf{P}^r.$$

The $n$ sections $\{p_i\}$ yield $n$ sections $\{(\iota \circ p_i, \mu \circ p_i)\}$ of $\mathbf{P}(E^*_f) \times \mathbf{P}^r$ over $S$. Let $S_i$ denote the subscheme of $\mathbf{P}(E^*_f) \times \mathbf{P}^r$ defined by the $i^{th}$ section. Denote the projection of $\mathbf{P}(E^*_f) \times \mathbf{P}^r$ to $S$ also by $\pi$. Let $\mathcal{M} = \mathcal{O}_{\mathbf{P}(E^*_f)}(1) \otimes \mathcal{O}_{\mathbf{P}^r}(1)$. $\mathcal{M}$ is an $\pi$-relatively ample line bundle. Note $\pi_*(\mathcal{M}^l) \cong \text{Sym}^l(E_f) \otimes \text{Sym}^l(\mathbb{C}^{r+1})$. By the stability of semipositivity under symmetric and tensor products ([Kol]) and Lemma 3, $\pi_*(\mathcal{M}^l)$ is semipositive. Fix a choice of $l$ (depending only on $d$, $g$, $n$, and $r$) large enough to ensure

$$\pi_*(\mathcal{M}^l) \oplus \bigoplus_{i=1}^{n} \pi_*(\mathcal{M}^l) \to \pi_*(\mathcal{M}^l \otimes \mathcal{O}_{\mathcal{C}}) \oplus \bigoplus_{i=1}^{n} \pi_*(\mathcal{M}^l \otimes \mathcal{O}_{S_i}) \to 0. \tag{29}$$

Such a choice of $l$ is possible by the boundedness established in section 2.3. Let $Q$ be the quotient in (29). By boundedness and the vanishing of higher direct images, the quotient $Q$ is a vector bundle for large $l$.

The quotient (29) is nontrivial in the following sense. Let $G = GL$ be the structure group of the bundle $E_f$. $G$ is naturally the structure group of $\pi_*(\mathcal{M}^l)$. Let $W$ be the $G$-representation inducing the bundle $\pi_*(\mathcal{M}^l) \oplus \bigoplus_1^n \pi_*(\mathcal{M}^l)$. Let $q$ be the rank of the quotient bundle of (29). The quotient sequence (29) yields a set theoretic classifying map to the Grassmannian:

$$\rho : S \to \mathbf{Gr}(q, W^*)/G.$$

LEMMA 5. *There exists a set theoretic injection*

$$\delta : \overline{M}_{g,n}(\mathbf{P}^r, d) \to \mathbf{Gr}(q, W^*)/G.$$

*Let $\lambda : S \to \overline{M}_{g,n}(\mathbf{P}^r, d)$ be the map induced by the stable family (23). There is a (set theoretic) factorization $\rho = \delta \circ \lambda$.*

*Proof.* For large $l$, the sequence (29) is equivalent to the data of a Hilbert point in $J$ (see section 2.3). Since the $G$ orbits of $J$ are exactly the stable maps, the lemma follows. □

LEMMA 6. *A stable map has a finite number of automorphisms.*

*Proof.* As simple consequence of the definition of stability, there are no infinitesimal automorphisms. The total number is therefore finite. □

Suppose the map to moduli $\lambda : S \to \overline{M}_{g,n}(\mathbf{P}^r, d)$ is a generically finite algebraic morphism. Then, in the terminology of [Kol], Lemmas 5 and 6 show the classifying map $\rho$ is *finite* on an open set of $S$.

PROPOSITION 7. *(Lemma 3.13, [Kol]) Let the base $S$ of (23) be a normal projective variety. Suppose the classifying map is finite on an open set of $S$. Then, the top self-intersection number of Det($Q$) on $S$ is positive.*

If $\overline{M}_{g,n}(\mathbf{P}^r, d)$ were a fine moduli space equipped with a universal family, Det$(Q)$ would be well defined and ample (by Proposition 7 and the Nakai-Moishezon criterion) on $\overline{M}_{g,n}(\mathbf{P}^r, d)$. Since $\overline{M}_{0,n}(\mathbf{P}^r, d)$ is expressed locally as a quotient of a fine moduli space by a finite group, it is easily seen Det$(Q)^k$ is a well defined line bundle on $\overline{M}_{0,n}(\mathbf{P}^r, d)$ for some sufficiently large $k$. The exponent $k$ is taken to trivialize the $\mathbb{C}^*$-representations at the fixed points. In the higher genus case, Det$(Q)$ is a well defined line bundle on the Hilbert scheme $J$ or the stack. Since the moduli problem has finite automorphisms, Det$(Q)^k$ is well defined on the coarse moduli space for some $k$.

Since the moduli spaces $\overline{M}_{g,n}(\mathbf{P}^r, d)$ are not fine, subvarieties are not equipped with stable families. Proposition (7) and the Nakai-Moishezon criterion do not directly establish the ampleness of Det$(Q)^k$. An alternative approach (due to J. Kollár) is followed. Recall the Hilbert scheme $J$ (of section 2.3) is equipped with a universal family and, therefore, a canonical map

$$J \to \overline{M}_{g,n}(\mathbf{P}^r, d).$$

Let $X \subset \overline{M}_{g,n}(\mathbf{P}^r, d)$ be a subvariety. Using $J$ and the finite automorphism property of a stable map, a morphism $Y \to X$ of algebraic schemes can be constructed satisfying

(i) $Y \to X$ is finite and surjective.
(ii) $Y$ is equipped with a stable family of maps such that $Y \to X$ is the corresponding morphism to moduli.

The existence of $Y \to X$ is exactly the conclusion of Proposition 2.7 in [Kol] under slightly different assumptions. Nevertheless, the argument is valid in the present setting. The construction of $Y$ is subtle. First $Y$ is constructed as an algebraic space. Then, a lemma of Artin is used to find an algebraic scheme $Y$. Since $Y$ has a universal family, Proposition 7 implies Det$(Q)^k$ has positive top intersection on $Y$ and therefore on $X$. The Nakai-Moishezon criterion can be applied to conclude the ampleness of Det$(Q)^k$ on $\overline{M}_{g,n}(\mathbf{P}^r, d)$.

**4.4. Automorphisms.** We use the notation of sections 3.2 and 4.1. In the genus 0 case, $\overline{M}(\bar{t})$ is nonsingular. Therefore, the space $\overline{M}_{0,n}(\mathbf{P}^r, d)$ is locally a quotient of a nonsingular variety by a finite group.

LEMMA 7. *Let $\xi \in \overline{M}(\bar{t})$ be a point at which the $G_{d,r}$ action is not free. Then $\xi$ corresponds to a stable map with nontrivial automorphisms.*

*Proof.* $G_{d,r}$ acts by isomorphism on the stable maps of the universal family over $\overline{M}(\bar{t})$. The $G_{d,r}$ action is not free at $\xi \in \overline{M}(\bar{t})$ if and only if there exists a $1 \neq \gamma \in$

$G_{d,r}$ fixing $\xi$. The element $\gamma$ induces an automorphism of the map corresponding to $\xi$. The automorphism is nontrivial on the marked points $\{q_{i,j}\}$. $\square$

Over the automorphism-free locus, the $G_{d,r}$-action on $\overline{M}(\overline{t})$ (and on the universal family over $\overline{M}(\overline{t})$) is free. It follows that the quotient over the automorphism-free locus is a nonsingular quasi-projective variety denoted by $\overline{M}^*_{0,n}(\mathbf{P}^r, d)$. A universal family over $\overline{M}^*_{0,n}(\mathbf{P}^r, d)$ is obtained by patching. Theorems 1 and 2 have been established in the case $X \cong \mathbf{P}^r$.

## 5. The construction of $\overline{M}_{g,n}(X, \beta)$

**5.1. Proof of Theorem 1.** Let $X$ be a projective algebraic variety. Existence of the coarse moduli space $\overline{M}_{g,n}(X, \beta)$ is established via a projective embedding $\iota : X \hookrightarrow \mathbf{P}^r$. Let $\iota_*(\beta)$ be $d$ times the class of a line in $\mathbf{P}^r$.

LEMMA 8. *There exists a natural closed subscheme*

$$\overline{M}_{g,n}(X, \beta, \overline{t}) \subset \overline{M}_{g,n}(\mathbf{P}^r, d, \overline{t})$$

*satisfying the following property. Let* $(\pi : \mathcal{C} \to S, \{p_i\}, \{q_{i,j}\}, \mu)$ *be a* $\overline{t}$*-rigid stable family of genus* $g$*,* $n$*-pointed, degree* $d$ *maps to* $\mathbf{P}^r$*. Then, the natural morphism* $S \to \overline{M}_{g,n}(\mathbf{P}^r, d, \overline{t})$ *factors through* $\overline{M}_{g,n}(X, \beta, \overline{t})$ *if and only if* $\mu$ *factors through* $\iota$ *and each geometric fiber of* $\pi$ *is a map to* $X$ *representing the homology class* $\beta \in A_1 X$*.*

*Proof.* The lemma is proved in case $g = 0$. If $g > 0$, then $\overline{M}_{g,n}(\mathbf{P}^r, d, \overline{t})$ is not a fine moduli space and the argument is more technical.

Let

$$(\pi_M : \mathcal{U} \to \overline{M}_{0,n}(\mathbf{P}^r, d, \overline{t}), \{p_i\}, \{q_{i,j}\}, \mu)$$

be the universal family over $\overline{M}_{0,n}(\mathbf{P}^r, d, \overline{t})$. On a genus 0 curve, any vector bundle generated by global sections has no higher cohomology. Therefore, by this cohomology vanishing and the base change theorems, $\pi_{M*}\mu^*(\mathcal{O}_{\mathbf{P}^r}(k))$ is a vector bundle for all $k > 0$. (This argument must be modified in the $g > 0$ case since $\pi_{M*}\mu^*(\mathcal{O}_{\mathbf{P}^r}(k))$ need not be a vector bundle even on the Hilbert scheme $J$ or the stack. Nevertheless, it is not hard to define the closed subscheme determined by $X$ on the Hilbert scheme $J$ or the stack and then descend it to the coarse moduli space.) Let $\mathcal{I}_X$ be the ideal sheaf of $X \subset \mathbf{P}^r$. Let $I_X(k) = H^0(\mathbf{P}^r, \mathcal{I}_X(k))$. Let $l >> 0$ be selected so that $\mathcal{I}_X(l)$ is generated by the global sections $I_X(l)$. These sections $I_X(l)$ yield sections of the vector bundle $\pi_{M*}\mu^*(\mathcal{O}_{\mathbf{P}^r}(l))$. Let $Z \subset \overline{M}_{0,n}(\mathbf{P}^r, d, \overline{t})$ be the scheme-theoretic zero locus of these sections. The restriction of $\mu$ to $\pi_M^{-1}(Z)$ factors though $\iota$. Since $Z$ is an algebraic scheme, $Z$ is a finite union of disjoint connected components. The homology class in $A_1(X) = H_2(X, \mathbb{Z})$ represented by a map with moduli point in $Z$ is a deformation invariant of the map. Therefore, the represented homology class is constant on each connected component of $Z$. Let $Z_\beta \subset Z$ be the union of components of $Z$ which consist of maps representing the class $\beta \in A_1 X$. Let $\overline{M}_{0,n}(X, \beta, \overline{t}) = Z_\beta$. The required properties are easily established. $\square$

By the functorial property, $\overline{M}_{g,n}(X, \beta, \overline{t})$ is invariant under the $G_{d,r} \cong \mathfrak{S}_d \times \cdots \times \mathfrak{S}_d$ action on $\overline{M}_{g,n}(\mathbf{P}^r, d, \overline{t})$. The quotient

$$\overline{M}_{g,n}(X, \beta, \overline{t})/G_{d,r}$$

is an open set of $\overline{M}_{g,n}(X,\beta)$. A patching argument identical to the patching argument of section 4.1 yields a construction of $\overline{M}_{g,n}(X,\beta)$ as a closed subscheme of $\overline{M}_{g,n}(\mathbf{P}^r,d)$. The functorial property of $\overline{M}_{g,n}(X,\beta)$ shows the construction is independent of the projective embedding of $X$. Projectivity of $\overline{M}_{g,n}(X,\beta)$ is obtained from the projectivity of $\overline{M}_{g,n}(\mathbf{P}^r,d)$. This completes the proof of Theorem 1.

**5.2. Proof of Theorem 2.** Let $g=0$. Let $X$ be a projective, nonsingular, convex variety. Theorem 2 is certainly true in case $\beta = 0$ since $\overline{M}_{0,n}(X,0) = \overline{M}_{0,n} \times X$. In general, a deformation study is needed to establish Theorem 2.

By the functorial property, the Zariski tangent space to the scheme $\overline{M}_{0,n}(X,\beta,\overline{t})$ at the point $(C,\{p_i\},\{q_{i,j}\},\mu : C \to X)$ is canonically isomorphic to the space of first order deformations of the pointed $\overline{t}$-stable map $(C,\{p_i\},\{q_{i,j}\},\mu : C \to X)$. The later deformation space corresponds bijectively to the space of first order deformations of the pointed stable map $(C,\{p_i\},\mu : C \to X)$.

Let $\text{Def}(\mu)$ denote the space of first order deformations of the pointed stable map $(C,\{p_i\},\mu : C \to X)$. Consider first the case in which $C \cong \mathbf{P}^1$. Let $\text{Def}_R(\mu)$ be the space of first order deformations of $(C,\{p_i\},\mu : C \to X)$ with $C$ held rigid. There is an natural exact sequence:

$$0 \to H^0(C,T_C) \to \text{Def}_R(\mu) \to \text{Def}(\mu) \to 0.$$

Stability of $\mu$ implies the left map is injective. Let $\text{Hom}(C,X)$ be the quasi-projective scheme of morphisms from $C$ to $X$ representing the class $\beta$. $\text{Hom}(C,X)$ is an open subscheme of the Hilbert scheme of graphs in $C \times X$. The Zariski tangent space to $\text{Hom}(C,X)$ is naturally identified:

$$T_{\text{Hom}(C,X)}([\mu]) \cong H^0(C,\mu^*T_X)$$

(see [Ko2]). There is an exact sequence:

$$0 \to \text{Ker} \to \text{Def}_R(\mu) \to H^0(C,\mu^*T_X) \to 0$$

where Ker corresponds to the deformations of the markings. Therefore, $\dim_{\mathbb{C}}\text{Ker} = n$. Since $X$ is convex, the above sequences suffice to compute the dimension of $\text{Def}(\mu)$:

$$\dim_{\mathbb{C}}\text{Def}(\mu) = \dim(X) + \int_\beta c_1(T_X) + n - 3.$$

The dimension of the tangent space to $\overline{M}_{0,n}(X,\beta,\overline{t})$ is established in case $C \cong \mathbf{P}^1$.

Before proceeding further, the following deformation result is needed. A proof can be found in [Ko2].

LEMMA 9. *Let $\mathcal{C}/S$ and $\mathcal{X}/S$ be flat, projective schemes over $S$. Let $s \in S$ be a geometric point. Let $C_s$, $X_s$ be the fibers over $s$ and let $f : C_s \to X_s$ be a morphism. Assume the following conditions are satisfied:*

(i) *$C_s$ has no embedded points.*
(ii) *$X_s$ is nonsingular.*
(iii) *$S$ is equidimensional at $s$.*

*Then, the dimension of every component of the quasi-projective variety $Hom_S(\mathcal{C},\mathcal{X})$ at the point $[f]$ is at least*

$$\dim_{\mathbb{C}} H^0(C_s, f^*T_{X_s}) - \dim_{\mathbb{C}} H^1(C_s, f^*T_{X_s}) + \dim_s S.$$

Again, let $(C \cong \mathbf{P}^1, \{p_i\}, \{q_{i,j}\}, \mu : C \to X)$ correspond to a point of the space $\overline{M}_{0,n}(X, \beta, \overline{t})$. By Lemma 9 and the convexity of $X$, every component of $\text{Hom}(C, X)$ at $[\mu]$ has dimension at least $\dim_{\mathbb{C}} H^0(C, \mu^* T_X)$. Therefore, every component of $\overline{M}_{0,n}(X, \beta, \overline{t})$ at $[\mu]$ has dimension at least $\dim(X) + \int_\beta c_1(T_X) + n - 3$. By the previous tangent space computation, it follows $[\mu]$ is a nonsingular point of $\overline{M}_{0,n}(X, \beta, \overline{t})$. Before attacking the reducible case, a lemma is required.

LEMMA 10. *Let $X$ be a nonsingular, projective, convex space. Let $\mu : C \to X$ be a morphism of a projective, connected, reduced, nodal curve of arithmetic genus 0 to $X$. Then,*

$$H^1(C, \mu^* T_X) = 0. \tag{30}$$

*and $\mu^* T_X$ is generated by global sections on $C$.*

*Proof.* Let $E \subset C$ be an irreducible component of $C$; $E \cong \mathbf{P}^1$. Let

$$\mu^* T_X|_E \cong \bigoplus \mathcal{O}_{\mathbf{P}^1}(\alpha_i).$$

Suppose there exists $\alpha_i < 0$. The composition of a rational double cover of $E$ with $\mu$ would then violate the convexity of $X$ . It follows that:

$$\forall i, \quad \alpha_i \geq 0. \tag{31}$$

We will prove the following statement by induction on the number of components of $C$:

$$H^1(C, \mu^* T_X \otimes \mathcal{O}_C(-p)) = 0 \tag{32}$$

for any nonsingular point $p \in C$. Equation (32) is true by condition (31) when $C \cong \mathbf{P}^1$ is irreducible. Assume now $C$ is reducible and $p \in E \cong \mathbf{P}^1$. Let $C = C' \cup E$; let $\{p'_1, \ldots, p'_q\} = C' \cap E$. Since $C$ is a tree, $C'$ has exactly $q$ connected components each intersecting $E$ in exactly 1 point. There is a component sequence:

$$0 \to \mu^* T_X|_{C'} \otimes \mathcal{O}_{C'}(-\sum_{j=1}^{q} p'_j) \to \mu^* T_X \otimes \mathcal{O}_C(-p) \to \mu^* T_X|_E \otimes \mathcal{O}_E(-p) \to 0.$$

Equation (32) now follows from the inductive assumptions on $C'$ and $E$. The inductive assumption (32) is applied to every connected component of $C'$.

We now prove $H^1(C, \mu^* T_X) = 0$. If $C \cong \mathbf{P}^1$, then the lemma is established by condition (31). Assume now $C = C' \cup E$ where $E \cong \mathbf{P}^1$. There is a component sequence

$$0 \to \mu^* T_X|_{C'} \otimes \mathcal{O}_{C'}(-\sum_{j=1}^{q} p'_j) \to \mu^* T_X \to \mu^* T_X|_E \to 0. \tag{33}$$

Equation (30) now follows from (32) applied to every connected component of $C'$.

Finally, an analysis of sequence (33) also yields the global generation result. $\mu^* T_X|_E$ is generated by global sections by (31). Sequence (33) is exact on global sections by (32). Hence $\mu^* T_X$ is generated by global sections on $E$. But, every point of $C$ lies on some component $E \cong \mathbf{P}^1$. □

In sections 7 and 8, the following related lemma will be required:

LEMMA 11. *Let $\mu : \mathbf{P}^1 \to X$ be a non-constant morphism to a nonsingular, projective, convex space $X$. Then $\int_{\mu_*[\mathbf{P}^1]} c_1(T_X) \geq 2$.*

*Proof.* Since $\mu$ is non-constant, the differential

$$d\mu : T_{\mathbf{P}^1} \to \mu^*(T_X)$$

is nonzero. Let $s \in H^0(\mathbf{P}^1, T_{\mathbf{P}^1})$ be a vector field with two distinct zeros $p_1, p_2 \in \mathbf{P}^1$. Then, $d\mu(s) \in H^0(\mathbf{P}^1, \mu^*(T_x)) \neq 0$ and $d\mu(s)$ vanishes (at least) at $p_1$ and $p_2$. By the proof of Lemma 10, $\mu^*(T_X) \stackrel{\sim}{=} \bigoplus \mathcal{O}_{\mathbf{P}^1}(\alpha_i)$ where $\alpha_i \geq 0$ for all $i$. The existence of $d\mu(s)$ implies that $\alpha_j \geq 2$ for some $j$. □

Let $C$ now be a reducible curve. $C$ must be a tree of $\mathbf{P}^1$'s. Let $q$ be the number of nodes of $C$. Again, let $\text{Def}(\mu)$ be the first order deformation space of the pointed stable map $\mu$. The *dual graph* of a pointed curve $C$ of arithmetic genus 0 consists of vertices and edges corresponding bijectively to the irreducible components and nodes of $C$ respectively. The *valence* of a vertex in the dual graph is the numbers of edges incident at that vertex. Let $\text{Def}_G(\mu) \subset \text{Def}(\mu)$ be the first order deformation space of the pointed stable map $\mu$ preserving the dual graph. $\text{Def}_G(\mu)$ is a linear subspace of codimension at most $q$. Let $\text{Def}_G(C)$ be the space of first order deformations of the curve $C$ which preserve the dual graph. A simple calculation yields

$$\dim_{\mathbb{C}} \text{Def}_G(C) = \sum_{|\nu| \geq 4} |\nu| - 3$$

where the sum is taken over vertices $\nu$ of the dual graph of valence at least 4.

The natural linear map $\text{Def}_G(\mu) \to \text{Def}_G(C)$ is now analyzed. Let $S$ be the nonsingular universal base space of deformations of $C$ preserving the dual graph. Let $\mathcal{C}$ be the universal deformation over $S$. Let $\mathcal{X} = X \times S$. Let $s_0 \in S$ correspond to $C$. By Lemmas 9 and 10, every component of $\text{Hom}_S(\mathcal{C}, \mathcal{X})$ at $[\mu]$ has dimension at least $\dim(X) + \int_\beta c_1(T_X) + \dim(S)$. The tangent space to the fiber of $\text{Hom}_S(\mathcal{C}, \mathcal{X})$ over $s_o$ at $[\mu]$ is canonically $H^0(C, \mu^* T_X)$. The latter space has dimension $\dim(X) + \int_\beta c_1(T_X)$. Hence, $\text{Hom}_S(\mathcal{C}, \mathcal{X})$ is nonsingular at $[\mu]$ of dimension $\dim(X) + \int_\beta c_1(T_X) + \dim(S)$ and the projection morphism to $S$ is smooth at $[\mu]$. Therefore, $\text{Def}_G(\mu) \to \text{Def}_G(C)$ is surjective.

The above definitions and results yield a natural exact sequence:

$$0 \to \text{Def}_C(\mu) \to \text{Def}_G(\mu) \to \text{Def}_G(C) \to 0$$

where $\text{Def}_C(\mu)$ is the space of first order deformations of the pointed stable map $\mu$ which restrict to the trivial deformation of $C$. As in the case where $C \stackrel{\sim}{=} \mathbf{P}^1$, $\text{Def}_C(\mu)$ differs from $\text{Def}_R(\mu)$ only by the tangent fields obtained from automorphisms:

$$0 \to H^0(C, T_C^{auto}) \to \text{Def}_R(\mu) \to \text{Def}_C(\mu) \to 0.$$

$H^0(C, T_C^{auto})$ is the space of tangent fields on the components of $C$ that vanish at all the nodes of $C$. Note $H^0(C, T_C^{auto}) = \sum_{|\nu| \leq 3} 3 - |\nu|$. Finally, there is an exact sequence containing $\text{Def}_R(\mu)$ and the tangent space to $\text{Hom}(C, X)$:

$$0 \to \text{Ker} \to \text{Def}_R(\mu) \to H^0(C, \mu^* T_X) \to 0.$$

From these exact sequences, Lemma 10, and some arithmetic, it follows

$$\dim_{\mathbb{C}} \text{Def}_G(\mu) = \dim(X) + \int_\beta c_1(T_X) + n - 3 - q. \tag{34}$$

Let $\mathcal{C}$ be a smoothing of the reducible curve $C$ over a base $S$ and let $\mathcal{X} = X \times S$. A simple application of Lemma 9 shows that $[\mu] \in \overline{M}_{0,n}(X, \beta, \overline{t})$ lies in the closure of the locus of maps with irreducible domains. Since the irreducible domain locus is pure dimensional of dimension $\dim(X) + \int_\beta c_1(T_X) + n - 3$,

$$\dim_{\mathbb{C}} \mathrm{Def}(\mu) \geq \dim(X) + \int_\beta c_1(T_X) + n - 3. \tag{35}$$

It follows from (34) and (35) that $\mathrm{Def}_G(\mu)$ is of maximal codimension $q$ in $\mathrm{Def}(\mu)$ and that the inequality in (35) is an equality. Since $\mathrm{Def}(\mu)$ is of dimension $\dim(X) + \int_\beta c_1(T_X) + n - 3$, $[\mu]$ is a nonsingular point of $\overline{M}_{0,n}(X, \beta, \overline{t})$. Since $\overline{M}_{0,n}(X, \beta, \overline{t})$ is nonsingular of pure dimension $\dim(X) + \int_\beta c_1(T_X) + n - 3$, parts (i) and (ii) of Theorem 2 are established. Part (iii) follows from the corresponding result in the case $X \cong \mathbf{P}^r$.

## 6. The boundary of $\overline{M}_{0,n}(X, \beta)$

**6.1. Definitions.** Let $X$ be nonsingular, projective, and convex. Let the genus $g = 0$. The *boundary* of $\overline{M}_{0,n}(X, \beta)$ is the locus corresponding to reducible domain curves. Boundary properties of the Mumford-Knudsen space $\overline{M}_{0,m}$ (where $m = n + d(r+1)$) are passed to $\overline{M}_{0,n}(\mathbf{P}^r, d)$ by the local quotient construction. The boundary locus of $\overline{M}_{0,m}$ is a divisor with normal crossings. Since $\overline{M}_{0,n}(\mathbf{P}^r, d, \overline{t})$ is a product of $\mathbb{C}^*$-bundles over an open set of $\overline{M}_{0,m}$, the boundary locus of $\overline{M}_{0,n}(\mathbf{P}^r, d, \overline{t})$ is certainly a divisor with normal crossing. $\overline{M}_{0,n}(\mathbf{P}^r, d)$ is locally the $G_{d,r}$-quotient of $\overline{M}_{0,n}(\mathbf{P}^r, d, \overline{t})$. The boundary of $\overline{M}_{0,n}(\mathbf{P}^r, d)$ is therefore a union of subvarieties of pure codimension 1. Over the automorphism-free locus, the boundary of $\overline{M}_{0,n}(\mathbf{P}^r, d)$ is a divisor with normal crossings.

Let $X$ be a nonsingular, projective, convex variety. The corresponding boundary results for $\overline{M}_{0,n}(X, \beta)$ are consequences of the deformation analysis of section 5.2. The boundary locus of $\overline{M}_{0,n}(X, \beta, \overline{t})$ is a divisor with normal crossing singularities. A pointed map $\mu : C \to X$ such that $C$ has $q$ nodes lies in the intersection of $q$ branches of the boundary. The dimension computation

$$\dim_{\mathbb{C}} \mathrm{Def}_G(\mu) = \dim \overline{M}_{0,n}(X, \beta, \overline{t}) - q$$

shows these branches intersect transversally at $[\mu]$. This completes the proof of Theorem 3. In particular $\overline{M}_{0,n}(X, \beta)$ has the same boundary singularity type as $\overline{M}_g$ and $\overline{M}_{g,n}$.

A class $\beta \in H_2(X, \mathbb{Z})$ is *effective* if $\beta$ is represented by some genus 0 stable map to $X$. If $n = 0$, the boundary of $\overline{M}_{0,0}(X, \beta)$ decomposes into a union of divisors which are in bijective correspondence with effective partitions $\beta_1 + \beta_2 = \beta$. For general $n$, the boundary decomposes into a union of divisors in bijective correspondence with data of weighted partitions $(A, B; \beta_1, \beta_2)$ where

(i) $A \cup B$ is a partition of $[n] = \{1, 2, \ldots, n\}$.
(ii) $\beta_1 + \beta_2 = \beta$, $\beta_1$ and $\beta_2$ are effective .
(iii) If $\beta_1 = 0$ (resp. $\beta_2 = 0$), then $|A| \geq 2$ (resp. $|B| \geq 2$).

$D(A, B; \beta_1, \beta_2)$, the divisor corresponding to the data $(A, B; \beta_1, \beta_2)$, is defined to be the locus of maps $\mu : C_A \cup C_B \to X$ satisfying the following conditions:

(a) $C$ is a union of two quasi-stable curves $C_A$, $C_B$ of genus 0 meeting in a point.
(b) The markings of $A$ (resp. $B$) lie on $C_A$ (resp. $C_B$).
(c) The map $\mu_A = \mu|_{C_A}$ (resp. $\mu_B$) represents $\beta_1$ (resp. $\beta_2$).

The deformation results of section 5 show the locus maps satisfying (a)–(c) and $C_A \cong C_B \cong \mathbf{P}^1$ is dense in $D(A, B; \beta_1, \beta_2)$. If $X = \mathbf{P}^r$, then it is easily seen that $D(A, B; \beta_1, \beta_2)$ is irreducible. In general, we do not claim the divisor $D(A, B; \beta_1, \beta_2)$ is irreducible, although that is the case in all the examples we have seen.

**6.2. Boundary divisors.** The boundary divisor of $\overline{M}_{0,n}$ corresponding to the marking partition $A \cup B = [n]$ is naturally isomorphic (by gluing) to the product

$$\overline{M}_{0,A\cup\{\bullet\}} \times \overline{M}_{0,B\cup\{\bullet\}}.$$

An analogous construction exists for the boundary divisor $D(A, B; \beta_1, \beta_2)$ of the space $\overline{M}_{0,n}(X, \beta)$.

Let $K = D(A, B; \beta_1, \beta_2)$ be a boundary divisor of $\overline{M}_{0,n}(X, \beta)$. Let $\overline{M}_A = \overline{M}_{0,A\cup\{\bullet\}}(X, \beta_1)$ and $\overline{M}_B = \overline{M}_{0,B\cup\{\bullet\}}(X, \beta_2)$. Let $e_A : \overline{M}_A \to X$ and $e_B : \overline{M}_B \to X$ be the evaluation maps obtained from the additional marking $\bullet$. Let $\tau_A$, $\tau_B$ be the projections of $\overline{M}_A \times \overline{M}_B$ to the first and second factors respectively. Let $\tilde{K} = \overline{M}_A \times_X \overline{M}_B$ be the fiber product with respect to the evaluation maps $e_A$, $e_B$. $\tilde{K} \subset \overline{M}_A \times_{\mathbb{C}} \overline{M}_B$ is the closed subvariety $(e_A \times_{\mathbb{C}} e_B)^{-1}(\Delta)$ where $\Delta \subset X \times X$ is the diagonal.

Properties of $\tilde{K}$ can be deduced from the local quotient constructions of $\overline{M}_A$ and $\overline{M}_B$. It will be shown that $\tilde{K}$ is a normal projective variety of pure dimension with finite quotient singularities. Let $\overline{M}_A(X, \bar{t}_A)$, $\overline{M}_B(X, \bar{t}_B)$ be the $\bar{t}_A$, $\bar{t}_B$-rigid moduli spaces. $\tilde{K}$ is the $G_A \times G_B$-quotient of the corresponding subvariety

$$\tilde{K}(X, \bar{t}_A, \bar{t}_B) \subset \overline{M}_A(X, \bar{t}_A) \times \overline{M}_B(X, \bar{t}_B),$$

$$\tilde{K}(X, \bar{t}_A, \bar{t}_B) = (e_A \times_{\mathbb{C}} e_B)^{-1}(\Delta).$$

The differential of $e_A$ at a point $[\mu]$ of $\overline{M}_A(X, \bar{t}_A)$ is determined in the following manner. The case in which the domain $C \cong \mathbf{P}^1$ is irreducible is most straightforward. Then, there are natural linear maps:

$$\text{Def}(\mu) \to H^0(\mu^* T_X / T_C(-p_\bullet)) \to T_X(\mu(p_\bullet)). \tag{36}$$

The first map in (36) is the natural surjection of $\text{Def}(\mu)$ onto the deformation space of the moduli problem obtained by forgetting all the markings except $\bullet$. The natural fiber evaluation $H^0(\mu^* T_X) \to T_X(\mu(p_\bullet))$ is well defined on the space $H^0(\mu^* T_X / T_C(-p_\bullet))$. This is the second map in (36). The composition of maps in (36) is simply the differential of $e_A$ at $[\mu]$. Since $\mu^* T_X$ is generated by global sections by Lemma 10, it follows that the differential of $e_A$ is surjective at $[\mu]$. A similar argument shows the differential of $e_A$ is surjective for each $[\mu] \in \overline{M}_A(X, \bar{t}_A)$. The differential of $e_B$ is therefore also surjective. The surjectivity of the differentials of $e_A$ and $e_B$ imply $\tilde{K}(X, \bar{t}_A, \bar{t}_B)$ is nonsingular. Thus $\tilde{K}$ is a normal projective variety of pure dimension with finite quotient singularities.

By gluing the universal families over $\overline{M}_A(\bar{t}_A)$ and $\overline{M}_B(\bar{t}_B)$ along the markings $\bullet$, a natural family of Kontsevich stable maps exists over $\tilde{K}(X, \bar{t}_A, \bar{t}_B)$. The induced map

$$\tilde{K}(X, \bar{t}_A, \bar{t}_B) \to K$$

is seen to be $G_A \times G_B$ invariant. Therefore, a natural map $\psi : \tilde{K} \to K$ is obtained.

LEMMA 12. *Results on the morphism* $\psi$ :

(i) *If* $A \neq \emptyset$ *and* $B \neq \emptyset$, *then* $\psi : \tilde{K} \to K$ *is an isomorphism.*
(ii) *If* $A \neq \emptyset$, *or* $B \neq \emptyset$, *or* $\beta_A \neq \beta_B$, *then* $\psi$ *is birational.*

(iii) *If* $A = B = \emptyset$ *(*$n = 0$*) and* $\beta_A = \beta_B = \beta/2$ *then* $\psi$ *is generically* 2 *to* 1.

*Proof.* First part (i) is proven. Let $q_A \in A$ and $q_B \in B$ be fixed markings (whose existence is guaranteed by the assumptions of (i)). Let $\mathcal{L}$ be a very ample line bundle on $X$ against which all degrees of maps are computed. Let $d_A$, $d_B$ be the degrees of $\beta_A$, $\beta_B$ respectively. Let $K = (A \cup B, \beta_A, \beta_B)$. Let $\mu : C \to \mathbf{P}^r$ correspond to a moduli point $[\mu] \in K$. Let $C = \bigcup C_i$ be the union of irreducible components. Let $q_A \in C_1$, $q_B \in C_l$ where $1 \neq l$ and let

$$C_1, C_2, \ldots, C_l$$

be the unique minimal path from $C_1$ to $C_l$ which exists since $C$ is a tree of components. For $1 \leq i \leq l-1$, let $x_i = C_i \cap C_{i+1}$. Each node $x_i$ divides $C$ into two connected curves

$$C = C_{A,i} \cup C_{B,i}$$

labeled by the points $q_A$, $q_B$. Let $d_i$ be the degree of $\mu$ restricted to $C_{A,i}$. The degrees $d_i$ increase monotonically. Since $[\mu] \in K$, $d_i = d_A$ for some $i$. Let $j$ be the minimal value satisfying $d_j = d_A$. If $d_{j+1} > d_A$, then $\psi^{-1}[\mu]$ is the unique point determined by cutting at the node $x_j$. If $d_{j+1} = d_A$, then the subcurve

$$C \setminus (C_{A,j} \cup C_{B,j+1})$$

must contain (by stability) a nonempty set of marked points $P_{j+1}$. Let $k$ be maximal index satisfying $d_{j+k} = d_A$. The analogously defined marked point sets

$$P_{j+1}, \ldots, P_{j+k}$$

are all nonempty. There must be a index $t$ satisfying $P_{j+t'} \subset A$ for $1 \leq t' \leq t$ and $P_{j+t'} \subset B$ for $t < t' \leq k$. $\psi^{-1}[\mu]$ is then the unique point determined by cutting at the node $x_{j+t}$. Therefore, $\psi$ is bijective in case $A$ and $B$ are nonempty.

Let $\overline{M}_{0,n}(X, \beta, \bar{t})$ be a locally rigidified moduli space containing the point $[\mu] \in K$. If $|A|, |B| \geq 1$, a similar argument shows the boundary components of $\overline{M}_{0,n}(X, \beta, \bar{t})$ lying over $K$ are *disjoint*. Therefore, $K$ is normal. In case $A$ and $B$ are nonempty, $\psi$ is a bijective morphism of normal varieties and hence an isomorphism.

Note, for example, that the component $K = D(\emptyset, \emptyset; 2, 3)$ of $\overline{M}_{0,0}(\mathbf{P}^r, 5)$ is not normal. $K$ intersects itself along the codimension 2 locus of moduli points $[\mu]$ of the form:

$$\mu : C_1 \cup C_2 \cup C_3 \to \mathbf{P}^r$$

with restricted degrees $d_1 = 2$, $d_2 = 1$, $d_3 = 2$. In this case, $\psi : \tilde{K} \to K$ is a normalization.

Parts (ii) and (iii) follow simply from the defining properties (a)–(c) of $K$. □

The fundamental relations among the Gromov-Witten invariants will come from the following linear equivalences among boundary components in $\overline{M}_{0,n}(X, \beta)$.

PROPOSITION 8. *For* $i, j, k, l$ *distinct in* $[n]$, *set*

$$D(i, j \mid k, l) = \sum D(A, B; \beta_1, \beta_2),$$

*the sum over all partitions such that* $i$ *and* $j$ *are in* $A$, $k$ *and* $l$ *are in* $B$, *and* $\beta_1$ *and* $\beta_2$ *are effective classes in* $A_1 X$ *such that* $\beta_1 + \beta_2 = \beta$. *Then, we have the linear equivalence of divisors*

$$D(i, j \mid k, l) \sim D(i, l \mid j, k)$$

*on* $\overline{M}_{0,n}(X,\beta)$.

*Proof.* The proof is obtained by examining the map

$$\overline{M}_{0,n}(X,\beta) \to \overline{M}_{0,n} \to \overline{M}_{0,\{i,j,k,l\}} \cong \mathbf{P}^1,$$

and noting that the divisor $D(i,j \mid k,l) \subset \overline{M}_{0,n}(X,\beta)$ is the multiplicity-free inverse image of the point $D(i,j \mid k,l) \in \overline{M}_{0,\{i,j,k,l\}}$. The deformation methods of section 5 can be used to prove that the inverse image of the point $D(i,j \mid k,l) \in \overline{M}_{0,\{i,j,k,l\}}$ is multiplicity-free. Since points are linearly equivalent on $\mathbf{P}^1$, the linear equivalence on $\overline{M}_{0,n}(X,\beta)$ is established. □

## 7. Gromov-Witten invariants

In sections 7–10, unless otherwise stated, $X$ will denote a homogeneous variety and the genus $g$ will be zero. Since the the tangent bundle of $X$ is generated by global sections, $X$ is convex. The moduli spaces $\overline{M}_{0,n}(X,\beta)$ are therefore available with the properties proved in sections 1–6. In addition, the cohomology of $X$ has a natural basis of algebraic cycles (classes of Schubert varieties), so $A^iX = H^{2i}X$ can be identified with the Chow group of cycle classes of codimension $i$. The effective classes $\beta$ in $A_1X$ (see section 6.1) are non-negative linear combinations of the Schubert classes of dimension 1. Each 1-dimensional Schubert class is represented by an embedding $\mathbf{P}^1 \subset X$.

The varieties $\overline{M}_{0,n}(X,\beta)$ come equipped with $n$ morphisms $\rho_1,\dots,\rho_n$ to $X$, where $\rho_i$ takes the point $[C,p_1,\dots,p_n,\mu] \in \overline{M}_{0,n}(X,\beta)$ to the point $\mu(p_i)$ in $X$. Given arbitrary classes $\gamma_1,\dots,\gamma_n$ in $A^*X$, we can construct the cohomology class

$$\rho_1{}^*(\gamma_1) \cup \cdots \cup \rho_n{}^*(\gamma_n)$$

on $\overline{M}_{0,n}(X,\beta)$, and we can evaluate its homogeneous component of the top codimension on the fundamental class, to produce a number, called a *Gromov-Witten invariant*, that we denote by $I_\beta(\gamma_1\cdots\gamma_n)$:

$$I_\beta(\gamma_1\cdots\gamma_n) = \int_{\overline{M}_{0,n}(X,\beta)} \rho_1{}^*(\gamma_1) \cup \cdots \cup \rho_n{}^*(\gamma_n). \tag{37}$$

If the classes $\gamma_i$ are homogeneous, this will be a nonzero number only if the sum of their codimensions is the dimension of $\overline{M}_{0,n}(X,\beta)$, that is,

$$\sum \operatorname{codim}(\gamma_i) = \dim X + \int_\beta c_1(T_X) + n - 3.$$

It follows from the definition that $I_\beta(\gamma_1\cdots\gamma_n)$ is invariant under permutations of the classes $\gamma_1,\dots,\gamma_n$.

The conventions of [K-M] require $n \geq 3$. However, it will be convenient for us to take $n \geq 0$. A 0-pointed invariant occurs when the moduli space $\overline{M}_{0,0}(X,\beta)$ is of dimension 0. In this case $I_\beta = \int_{\overline{M}_{0,0}(X,\beta)} 1$. By Lemma 11, $\overline{M}_{0,0}(X,\beta)$ is of dimension 0 if and only if $\dim(X) = 1$ and $\int_\beta c_1(X) = 2$. Hence, for homogeneous varieties, 0-pointed invariants only occur on $X \cong \mathbf{P}^1$. In this case, $I_1 = 1$ is the unique 0-pointed invariant.

Let $M^*_{0,n}(X,\beta) = M_{0,n}(X,\beta) \cap \overline{M}^*_{0,n}(X,\beta)$. We start with a simple lemma.

LEMMA 13. *If* $n \geq 1$, *then* $M^*_{0,n}(X,\beta) \subset \overline{M}_{0,n}(X,\beta)$ *is a dense open set.*

*Proof.* If $\beta = 0$, then $\overline{M}_{0,n}(X,0)$ is nonempty only if $n \geq 3$. The equality $\overline{M}^*_{0,n}(X,0) = \overline{M}_{0,n}(X,0)$ is deduced from the corresponding equality for $\overline{M}_{0,n}$. Assume $\beta \neq 0$. By Theorem 3, $M_{0,n}(X,\beta) \subset \overline{M}_{0,n}(X,\beta)$ is a dense open set. Let $(\mathbf{P}^1, \{p_i\}, \mu)$ be a point in $M_{0,n}(X,\beta)$. It suffices to show that $(\mathbf{P}^1, \{p'_i\}, \mu)$ is automorphism-free for general points $p'_1, \ldots, p'_n \in \mathbf{P}^1$. The automorphism group $A$ of the unpointed map $\mu : \mathbf{P}^1 \to X$ is finite since $\beta \neq 0$. There exists a (nonempty) open set of $\mathbf{P}^1$ consisting of points with trivial $A$-stabilizers. If $p'_1, \ldots, p'_n$ belong to this open subset, the pointed map $(\mathbf{P}^1, \{p'_i\}, \mu)$ is automorphism-free. □

Let $X = G/P$, so $G$ acts transitively on $X$. Let $\Gamma_1, \ldots, \Gamma_n$ be pure dimensional subvarieties of $X$. Let $[\gamma_i] \in A^*X$ be the corresponding classes (see our notational conventions in section 0.2). Assume

$$\sum_{i=1}^{n} \operatorname{codim}(\Gamma_i) = \dim(X) + \int_\beta c_1(T_X) + n - 3.$$

Let $g\Gamma_i$ denote the $g$-translate of $\Gamma_i$ for $g \in G$.

LEMMA 14. *Let $n \geq 0$. Let $g_1, \ldots, g_n \in G$ be general elements. Then, the scheme theoretic intersection*

$$\rho_1^{-1}(g_1\Gamma_1) \cap \cdots \cap \rho_n^{-1}(g_n\Gamma_n) \tag{38}$$

*is a finite number of reduced points supported in $M_{0,n}(X,\beta)$ and*

$$I_\beta(\gamma_1 \cdots \gamma_n) = \#\ \rho_1^{-1}(g_1\Gamma_1) \cap \cdots \cap \rho_n^{-1}(g_n\Gamma_n).$$

*Proof.* If $n = 0$, $I_1 = 1$ on $\mathbf{P}^1$ is the only case and the lemma holds since $\overline{M}_{0,0}(\mathbf{P}^1, 1)$ is a nonsingular point. Assume $n \geq 1$. $M^*_{0,n}(X,\beta) \subset \overline{M}_{0,n}(X,\beta)$ is a dense open set by Lemma 13. By simple transversality arguments (with respect to the $G$-action), it follows that the intersection (38) is supported in $M^*_{0,n}(X,\beta)$. By Theorem 2, $M^*_{0,n}(X,\beta)$ is nonsingular. An application of Kleiman's Bertini theorem ([Kl]) now shows that the intersection (38) is a finite set of reduced points. To see that the number of points in (38) agrees with the intersection number, consider the fiber diagram:

$$\begin{array}{ccc} \cap_{i=1}^n \rho_i^{-1}(g_i\Gamma_i) & \longrightarrow & \overline{M} \times \prod_{i=1}^n g_i\Gamma_i \\ \downarrow & & \downarrow \\ \overline{M} & \xrightarrow{\iota} & \overline{M} \times X^n \end{array} \tag{39}$$

where $\overline{M} = \overline{M}_{0,n}(X,\beta)$ and $\iota$ is the graph of the morphism $(\rho_1, \ldots, \rho_n)$. From (39), one sees that

$$\prod_{i=1}^{n} \rho_i^*[g_i\Gamma_i] \cap [\overline{M}] = \iota^*[\overline{M} \times \prod_{i=1}^{n} g_i\Gamma_i] = [\cap_{i=1}^n \rho_i^{-1}(g_i\Gamma_i)]$$

in $A_0(\overline{M})$, which is the required assertion. □

Lemma 14 relates the Gromov-Witten invariants to enumerative geometry. We see $I_\beta(\gamma_1 \cdots \gamma_n)$ equals the number of pointed maps $\mu$ from $\mathbf{P}^1$ to $X$ representing the class $\beta \in A_1X$ and satisfying $\mu(p_i) \in g_i\Gamma_i$. We will need three basic properties satisfied by the Gromov-Witten invariants:

(I) $\beta = 0$. In this case, $\overline{M}_{0,n}(X,\beta) = \overline{M}_{0,n} \times X$, and the mappings $\rho_i$ are all equal to the projection $p$ onto the second factor. Since

$$\rho_1{}^*(\gamma_1) \cup \cdots \cup \rho_n{}^*(\gamma_n) = p^*(\gamma_1 \cup \cdots \cup \gamma_n),$$

$$\begin{aligned} I_\beta(\gamma_1 \cdots \gamma_n) &= \int_{\overline{M}_{0,n} \times X} p^*(\gamma_1 \cup \cdots \cup \gamma_n) \\ &= \int_{p_*[\overline{M}_{0,n} \times X]} \gamma_1 \cup \cdots \cup \gamma_n. \end{aligned}$$

Note that $\overline{M}_{0,n}$ is empty if $0 \le n \le 2$. If $n > 3$, $p_*[\overline{M}_{0,n} \times X] = 0$, since the fibers of $p$ have positive dimension. The only way the number $I_\beta(\gamma_1 \cdots \gamma_n)$ can be nonzero is when $n = 3$, so that $\overline{M}_{0,n}$ is just a point. In this case, $I_\beta(\gamma_1 \cdot \gamma_2 \cdot \gamma_3)$ is the classical intersection number $\int_X \gamma_1 \cup \gamma_2 \cup \gamma_3$.

(II) $\gamma_1 = 1 \in A^0 X$. If $\beta \neq 0$, then the product $\rho_1{}^*(\gamma_1) \cup \cdots \cup \rho_n{}^*(\gamma_n)$ is the pullback of a class on $\overline{M}_{0,n-1}(X,\beta)$ by the map from $\overline{M}_{0,n}(X,\beta)$ to $\overline{M}_{0,n-1}(X,\beta)$ that forgets the first point. Since the fibers of this map have positive dimension, the evaluation $I_\beta(\gamma_1 \cdots \gamma_n)$ must vanish. Therefore, by (I), $I_\beta(\gamma_1 \cdots \gamma_n)$ vanishes unless $\beta = 0$ and $n = 3$. In this case, $I_0(1 \cdot \gamma_2 \cdot \gamma_3) = \int_X \gamma_2 \cup \gamma_3$.

(III) $\gamma_1 \in A^1 X$ and $\beta \neq 0$. In this case,

$$I_\beta(\gamma_1 \cdots \gamma_n) = \left( \int_\beta \gamma_1 \right) \cdot I_\beta(\gamma_2 \cdots \gamma_n). \tag{40}$$

For a map $\mu : C \to X$ with $\mu_*[C] = \beta$, there are $\left(\int_\beta \gamma_1\right)$ choices for the point $p_1$ in $C$ to map to a point in $\Gamma_1$, where $\Gamma_1$ is a variety representing $\gamma_1$. Equation (40) is therefore a consequence of Lemma 14.

For a formal intersection-theoretic proof of (40), consider the mapping

$$\psi : \overline{M}_{0,n}(X,\beta) \to X \times \overline{M}_{0,n-1}(X,\beta)$$

which is the product of $\rho_1$ and the map that forgets the first point. By the Künneth formula, we can write $\psi_*[\overline{M}_{0,n}(X,\beta)] = \beta' \times [\overline{M}_{0,n-1}(X,\beta)] + \alpha$, where $\beta'$ is a class in $A_1 X$, and $\alpha$ is some homology class that is supported over a proper closed subset of $\overline{M}_{0,n-1}(X,\beta)$. The class $\beta'$ can be calculated by restricting to what happens over a generic point of $\overline{M}_{0,n-1}(X,\beta)$. Representing such a point by $(C, p_2, \ldots, p_n, \mu)$ with $C \cong \mathbf{P}^1$, one sees that the fiber over this point is isomorphic to $C$ and $\beta' = \mu_*[C] = \beta$. Using the projection formula as in (I) and (II), it follows that

$$\begin{aligned} I_\beta(\gamma_1 \cdots \gamma_n) &= \int_{\beta \times [\overline{M}_{0,n-1}(X,\beta)]} \gamma_1 \times \rho_2{}^*(\gamma_2) \cup \cdots \cup \rho_n{}^*(\gamma_n) \\ &= \int_\beta \gamma_1 \cdot \int_{\overline{M}_{0,n-1}(X,\beta)} \rho_2{}^*(\gamma_2) \cup \cdots \cup \rho_n{}^*(\gamma_n), \end{aligned}$$

as asserted.

It should be noted that the generic element of $\overline{M}_{0,0}(X,\beta)$ may not be a *birational* map of $\mathbf{P}^1$ to $X$. This is seen immediately for $X \cong \mathbf{P}^1$ where the generic element of $\overline{M}_{0,0}(\mathbf{P}^1, d)$ is a $d$-fold branched covering of $\mathbf{P}^1$. This phenomenon occurs in higher dimensions. For example, let $X$ be the complete flag variety

$\mathbf{Fl}(\mathbb{C}^3)$ (the space of pairs $(p, l)$ satisfying $p \in l$ where $p$ and $l$ are a point and a line in $\mathbf{P}^2$). Let $\beta \in A_1\mathbf{Fl}(\mathbb{C}^3)$ be the class of the curve $\mathbf{P}^1 \subset \mathbf{Fl}(\mathbb{C}^3)$ determined by all pairs $(p, l)$ for a fixed line $l$. One computes $\int_\beta c_1(T_{\mathbf{Fl}(\mathbb{C}^3)}) = 2$, so the dimension of $\overline{M}_{0,0}(\mathbf{Fl}(\mathbb{C}^3), \beta)$ is $3 + 2 - 3 = 2$ by Theorem 2. Directly, one sees that $\overline{M}_{0,0}(\mathbf{Fl}(\mathbb{C}^3), \beta)$ is isomorphic to the space of lines in $\mathbf{P}^2$. In particular, $\overline{M}_{0,0}(\mathbf{Fl}(\mathbb{C}^3), \beta)$ has no boundary. As in the case of $\mathbf{P}^1$, it is seen that every element of $M_{0,0}(\mathbf{Fl}(\mathbb{C}^3), 2\beta)$ corresponds to a double cover of an element of $\overline{M}_{0,0}(\mathbf{Fl}(\mathbb{C}^3), \beta)$. The boundary of $\overline{M}_{0,0}(\mathbf{Fl}(\mathbb{C}^3), 2\beta)$ consists of degenerate double covers. Note also that every element of $\overline{M}_{0,0}(\mathbf{Fl}(\mathbb{C}^3), 2\beta)$ has a nontrivial automorphism. Since the space of image curves of maps in $\overline{M}_{0,0}(\mathbf{Fl}(\mathbb{C}^3), 2\beta)$ is only 2-dimensional, it follows that all Gromov-Witten invariants of $\mathbf{Fl}(\mathbb{C}^3)$ of the form $I_{2\beta}(\gamma_1 \cdots \gamma_n)$ vanish.

## 8. Quantum cohomology

We keep the notation of section 7. Let $T_0 = 1 \in A^0X$, let $T_1, \ldots, T_p$ be a basis of $A^1X$, and let $T_{p+1}, \ldots, T_m$ be a basis for the other cohomology groups. The classes of Schubert varieties form the natural basis for homogeneous varieties. The fundamental numbers counted by the Gromov-Witten invariants are the numbers

$$N(n_{p+1}, \ldots, n_m; \beta) = I_\beta({T_{p+1}}^{n_{p+1}} \cdots {T_m}^{n_m}) \tag{41}$$

for $n_i \geq 0$. The invariant (41) is nonzero only when $\sum n_i\,(\mathrm{codim}(T_i) - 1) = \dim X + \int_\beta c_1(T_X) - 3$. In this case, it is the number of pointed rational maps meeting $n_i$ general representatives of $T_i$ for each $i$, $p + 1 \leq i \leq m$.

Define the numbers $g_{ij}, 0 \leq i, j \leq m$, by the equations

$$g_{ij} = \int_X T_i \cup T_j. \tag{42}$$

(If the $T_i$ are the Schubert classes, then for each $i$ there is a unique $j$ such that $g_{ij} \neq 0$. For this $j$, $g_{ij} = 1$.)

Define $(g^{ij})$ to be the inverse matrix to the matrix $(g_{ij})$. Equivalently, the class of the diagonal $\Delta$ in $X \times X$ is given by the formula

$$[\Delta] = \sum_{e\,f} g^{ef}\, T_e \otimes T_f \tag{43}$$

in $A^*(X \times X) = A^*X \otimes A^*X$. The following equations hold:

$$T_i \cup T_j = \sum_{e,\,f} \left(\int_X T_i \cup T_j \cup T_e\right) g^{ef} T_f = \sum_{e,\,f} I_0(T_i{\cdot}T_j{\cdot}T_e) g^{ef} T_f. \tag{44}$$

The idea is to define a "quantum deformation" of the cup multiplication of (44) by allowing nonzero classes $\beta$. Here enters a key idea from physics – to write down a "potential function" that carries all the enumerative information.

Define, for a class $\gamma$ in $A^*X$,

$$\Phi(\gamma) = \sum_{n \geq 3} \sum_{\beta} \frac{1}{n!} I_\beta(\gamma^n), \tag{45}$$

where $\gamma^n$ denotes $\gamma \cdots \gamma$ ($n$ times).

LEMMA 15. *For a given integer $n$, there are only finitely many effective classes $\beta \in A_1X$ such that $I_\beta(\gamma^n)$ is not zero.*

*Proof.* Since $X$ is a homogeneous space, the effective classes in $A_1X$ are the non-negative linear combination of finitely many (nonzero) effective classes $\beta_1, \dots, \beta_p$. By Lemma 11, $\int_{\beta_i} c_1(T_X) \geq 2$. Hence, for a given integer $N$, there are only a finite number of effective $\beta$ for which $\int_\beta c_1(T_X) \leq N$. If $I_\beta(\gamma^n)$ is nonzero, then

$$\dim M_{0,n}(X, \beta) \leq n \cdot \dim X$$

which implies that $\int_\beta c_1(T_X) \leq (n-1) \cdot \dim X + 3 - n$. □

Let $\gamma = \sum y_i T_i$. By Lemma 15, $\Phi(\gamma) = \Phi(y_0, \dots, y_m)$ becomes a formal power series in $\mathbb{Q}[[y]] = \mathbb{Q}[[y_0, \dots, y_m]]$:

$$\Phi(y_0, \dots, y_m) = \sum_{n_0+\dots+n_m \geq 3} \sum_\beta I_\beta({T_0}^{n_0} \cdots {T_m}^{n_m}) \frac{{y_0}^{n_0}}{n_0!} \cdots \frac{{y_m}^{n_m}}{n_m!}. \tag{46}$$

Define $\Phi_{ijk}$ to be the partial derivative:

$$\Phi_{ijk} = \frac{\partial^3 \Phi}{\partial y_i \, \partial y_j \partial y_k}, \quad 0 \leq i, j, k \leq m. \tag{47}$$

A simple formal calculation, using (46), gives the following equivalent formula:

$$\Phi_{ijk} = \sum_{n \geq 0} \sum_\beta \frac{1}{n!} I_\beta(\gamma^n \cdot T_i \cdot T_j \cdot T_k). \tag{48}$$

Now we define a new "quantum" product $*$ by the rule:

$$T_i * T_j = \sum_{e,\, f} \Phi_{ije} \, g^{ef} \, T_f. \tag{49}$$

The product in (49) is extended $\mathbb{Q}[[y]]$-linearly to the $\mathbb{Q}[[y]]$-module $A^*X \otimes_{\mathbb{Z}} \mathbb{Q}[[y]]$, thus making it a $\mathbb{Q}[[y]]$-algebra. One thing is evident from this remarkable definition: this product is commutative, since the partial derivatives are symmetric in the subscripts.

It is less obvious, but not difficult, to see $T_0 = 1$ is a unit for the $*$-product. In fact, it follows from property (I) of section 7, together with (48), that

$$\Phi_{0jk} = I_0(T_0 \cdot T_j \cdot T_k) = \int_X T_j \cup T_k = g_{jk},$$

and from this we see that $T_0 * T_j = \sum g_{je} \, g^{ef} \, T_f = T_j$.

The essential point, however, is the associativity:

THEOREM 4. *This definition makes $A^*X \otimes \mathbb{Q}[[y]]$ into a commutative, associative $\mathbb{Q}[[y]]$-algebra, with unit $T_0$.*

We start the proof by writing down what associativity says:

$$(T_i * T_j) * T_k = \sum_{e,\, f} \Phi_{ije} \, g^{ef} \, T_f * T_k = \sum_{e,\, f} \sum_{c,\, d} \Phi_{ije} \, g^{ef} \, \Phi_{fkc} \, g^{cd} \, T_d,$$

$$T_i * (T_j * T_k) = \sum_{e,\, f} \Phi_{jke} \, g^{ef} \, T_i * T_f = \sum_{e,\, f} \sum_{c,\, d} \Phi_{jke} \, g^{ef} \, \Phi_{ifc} \, g^{cd} \, T_d.$$

Since the matrix $(g^{cd})$ is nonsingular, the equality of $(T_i * T_j) * T_k$ and $T_i * (T_j * T_k)$ is equivalent to the equation

$$\sum_{e,\, f} \Phi_{ije} \, g^{ef} \, \Phi_{fkl} = \sum_{e,\, f} \Phi_{jke} \, g^{ef} \, \Phi_{ifl}$$

for all $l$. If we set

$$F(i,j \mid k,l) = \sum_{e,f} \Phi_{ije}\, g^{ef}\, \Phi_{fkl}, \tag{50}$$

and use the symmetry $\Phi_{ifl} = \Phi_{fil}$, we see that the associativity is equivalent to the equation

$$F(i,j \mid k,l) = F(j,k \mid i,l). \tag{51}$$

It follows from (48) that

$$F(i,j \mid k,l) = \sum \frac{1}{n_1!\, n_2!} I_{\beta_1}(\gamma^{n_1}\cdot T_i\cdot T_j\cdot T_e)\, g^{ef}\, I_{\beta_2}(\gamma^{n_2}\cdot T_k\cdot T_l\cdot T_f), \tag{52}$$

where the sum is over all nonnegative $n_1$ and $n_2$, over all $\beta_1$ and $\beta_2$ in $A_1X$, and over all $e$ and $f$ from 0 to $m$. We need the following lemma. Recall from section 6, the divisor $D(A,B;\beta_1,\beta_2)$. In case $A$ and $B$ are nonempty,

$$D(A,B;\beta_1,\beta_2) = \overline{M}_{0,A\cup\{\bullet\}}(X,\beta_1) \times_X \overline{M}_{0,B\cup\{\bullet\}}(X,\beta_2).$$

LEMMA 16. *Let $\iota$ denote the natural inclusion of $D(A,B;\beta_1\beta_2)$ in the Cartesian product $\overline{M}_{0,A\cup\{\bullet\}}(X,\beta_1) \times \overline{M}_{0,B\cup\{\bullet\}}(X,\beta_2)$, and let $\alpha$ be the embedding of $D(A,B;\beta_1,\beta_2)$ as a divisor in $\overline{M}_{0,n}(X,\beta)$, with $\beta = \beta_1 + \beta_2$. Then for any classes $\gamma_1,\ldots,\gamma_n$ in $A^*X$,*

$$\iota_* \circ \alpha^*({\rho_1}^*(\gamma_1) \cup \cdots \cup {\rho_n}^*(\gamma_n)) =$$
$$\sum_{e,f} g^{ef} \left(\prod_{a\in A} {\rho_a}^*(\gamma_a)\cdot{\rho_\bullet}^*(T_e)\right) \times \left(\prod_{b\in B} {\rho_b}^*(\gamma_b)\cdot{\rho_\bullet}^*(T_f)\right).$$

*Proof.* Let $M_1 = \overline{M}_{0,A\cup\{\bullet\}}(X,\beta_1)$, $M_2 = M_{0,B\cup\{\bullet\}}(X,\beta_2)$, $M = \overline{M}_{0,n}(X,\beta)$, and $D = D(A,B;\beta_1\beta_2)$. From the identification of $D$ with $M_1 \times_X M_2$, we have a commutative diagram, with the right square a fiber square:

$$\begin{array}{ccccc} M & \xleftarrow{\ \alpha\ } & D & \xrightarrow{\ \iota\ } & M_1 \times M_2 \\ \downarrow{\scriptstyle \rho} & & \downarrow{\scriptstyle \eta} & & \downarrow{\scriptstyle \rho'} \\ X^n & \xleftarrow[\ p\ ]{} & X^{n+1} & \xrightarrow[\ \delta\ ]{} & X^{n+2} \end{array} \tag{53}$$

Here $\rho$ is the product of the evaluation maps denoted $\rho_i$, $\rho'$ is the product of maps $\rho_i$ and the two others denoted $\rho_\bullet$, $\delta$ is the diagonal embedding that repeats the last factor, and $p$ is the projection that forgets the last factor. Then we have

$$\begin{aligned} \iota_* \circ \alpha^*({\rho_1}^*(\gamma_1) \cup \cdots \cup {\rho_n}^*(\gamma_n)) &= \iota_* \circ \alpha^* \circ \rho^*(\gamma_1 \times \cdots \times \gamma_n) \\ &= \iota_* \circ \eta^* \circ p^*(\gamma_1 \times \ldots \times \gamma_n) \\ &= \iota_* \circ \eta^*(\gamma_1 \times \ldots \times \gamma_n \times [X]) \\ &= \rho'^* \circ \delta_*(\gamma_1 \times \ldots \times \gamma_n \times [X]) \\ &= \rho'^*(\gamma_1 \times \ldots \times \gamma_n \times [\Delta]) \\ &= \sum_{e,f} g^{ef} \rho'^*(\gamma_1 \times \ldots \times \gamma_n \times T_e \times T_f) \end{aligned}$$

$$= \sum_{e,f} g^{ef} \left(\prod_{a\in A} {\rho_a}^*(\gamma_a)\cdot{\rho_\bullet}^*(T_e)\right) \times \left(\prod_{b\in B} {\rho_b}^*(\gamma_b)\cdot{\rho_\bullet}^*(T_f)\right).$$

□

Fix $\beta \in A_1X$ and $\gamma_1, \dots, \gamma_n \in A^*X$, and fix four distinct integers $q, r, s$, and $t$ in $[n]$. Set

$$G(q,r \mid s,t) = \sum g^{ef} I_{\beta_1}\left(\prod_{a\in A} \gamma_a \cdot T_e\right) \cdot I_{\beta_2}\left(\prod_{b\in B} \gamma_b \cdot T_f\right), \tag{54}$$

where the sum is over all partitions of $[n]$ into two sets $A$ and $B$ such that $q$ and $r$ are in $A$ and $s$ and $t$ are in $B$, and over all $\beta_1$ and $\beta_2$ that sum to $\beta$, and over $e$ and $f$ between 0 and $m$. It follows from Lemma 16 that

$$G(q,r \mid s,t) = \sum \int_{D(A,B;\beta_1,\beta_2)} {\rho_1}^*\gamma_1 \cup \cdots \cup {\rho_n}^*\gamma_n,$$

the sum over $A$ and $B$ and $\beta_1$ and $\beta_2$ as above. Now Proposition 8 from section 7 implies

$$G(q,r \mid s,t) = G(r,s \mid q,t). \tag{55}$$

Apply (55) in the following case :

$$\gamma_i = \gamma, \quad \text{for } 1 \le i \le n-4,$$

$$\gamma_{n-3} = T_i, \quad \gamma_{n-2} = T_j, \quad \gamma_{n-1} = T_k, \quad \gamma_n = T_l,$$

$$q = n-3, \quad r = n-2, \quad s = n-1, \quad t = n.$$

Then (54) becomes

$$G(q,r \mid s,t) = \sum \binom{n-4}{n_1-2} g^{ef} I_{\beta_1}(\gamma^{n_1-2} \cdot T_i \cdot T_j \cdot T_e) \cdot I_{\beta_2}(\gamma^{n_2-2} \cdot T_k \cdot T_l \cdot T_f),$$

the sum over $n_1$ and $n_2$, each at least 2, adding to $n$, and $\beta_1$ and $\beta_2$ adding to $\beta$; the binomial coefficient is the number of partitions $A$ and $B$ for which $A$ has $n_1$ elements, and $B$ has $n_2$ elements. This can be rewritten

$$G(q,r \mid s,t) = n! \sum \frac{1}{n_1!\, n_2!} g^{ef} I_{\beta_1}(\gamma^{n_1} \cdot T_i \cdot T_j \cdot T_e) \cdot I_{\beta_2}(\gamma^{n_2} \cdot T_k \cdot T_l \cdot T_f), \tag{56}$$

the sum over nonnegative $n_1$ and $n_2$ adding to $n-4$, and $\beta_1$ and $\beta_2$ adding to $\beta$.

The required equality (51) then follows immediately from (55) and (56), together with (52). This completes the proof of Theorem 4.

While the definition of the quantum cohomology ring depends upon a choice of basis $T_0, \dots, T_m$ of $A^*X$, the rings obtained from different basis choices are canonically isomorphic. The variables $y_0, \dots, y_m$ should be identified with the dual basis to $T_0, \dots, T_m$. If $T'_0, \dots, T'_m$ is another basis of $A^*X$ and $T'_i = \sum a_{ij} T_j$ is the change of coordinates, let

$$y_i = \sum a_{ji} y'_j \tag{57}$$

be the dual coordinate change. Relation (57) yields an isomorphism of $\mathbb{Q}$-vector spaces

$$A^*X \otimes \mathbb{Q}[[y]] \cong A^*X \otimes \mathbb{Q}[[y']].$$

It is easy to check that the quantum products defined respectively on the left and right by the $T$ and $T'$ bases agree with this identification.

Let $V$ denote the underlying free abelian group of $A^*X$. Let $\mathbb{Q}[[V^*]]$ be the completion of the graded polynomial ring $\bigoplus_{i=0}^{\infty} Sym^i(V^*) \otimes \mathbb{Q}$ at the unique maximal graded ideal. The quantum product defines a canonical ring structure on the free $\mathbb{Q}[[V^*]]$-module $V \otimes_{\mathbb{Z}} \mathbb{Q}[[V^*]]$. Let $QH^*X = (V \otimes_{\mathbb{Z}} \mathbb{Q}[[V^*]], *)$ denote the quantum cohomology ring. There is a canonical injection of abelian groups

$$\iota : A^*X \hookrightarrow QH^*X$$

determined by $\iota(v) = v \otimes 1$ for $v \in V$. The injection $\iota$ is *not* compatible with the $\cup$ and $*$ products.

It is worth noting that the quantum cohomology ring $QH^*X$ is *not* in general a formal deformation of $A^*X$ over the local ring $\mathbb{Q}[[V^*]]$. It can be seen directly from the definitions that the $*$-product does not specialize to the $\cup$-product when the formal parameters are set to 0. At the end of section 9, a presentation of $QH^*\mathbf{P}^2$ shows explicitly the difference between $A^*\mathbf{P}^2$ and the specialization of $QH^*\mathbf{P}^2$. In section 10, a ring deformation of $A^*X$ will be constructed via a smaller quantum cohomology ring.

## 9. Applications to enumerative geometry

We write the potential function as a sum:

$$\Phi(y_0, \dots, y_m) = \Phi_{\text{classical}}(y) + \Phi_{\text{quantum}}(y).$$

The classical part has the terms for $\beta = 0$:

$$\Phi_{\text{classical}}(y) = \sum_{n_0+\ldots+n_m=3} \int_X (T_0^{\,n_0} \cup \cdots \cup T_m^{\,n_m}) \frac{y_0^{\,n_0}}{n_0!} \cdots \frac{y_m^{\,n_m}}{n_m!}.$$

Since the associativity equations involve only third derivatives, we can modify $\Phi$ by any terms of degree at most 2. Using properties (I)–(III) of section 7, we see that $\Phi_{\text{quantum}}(y)$ can be replaced by $\Gamma(y)$:

$$\Gamma(y) = \sum_{n_{p+1}+\ldots+n_m \geq 0} \sum_{\beta \neq 0} N(n_{p+1}, \dots, n_m; \beta) \prod_{i=1}^{p} e^{(\int_\beta T_i) y_i} \prod_{i=p+1}^{m} \frac{y_i^{\,n_i}}{n_i!},$$

where $N(n_{p+1}, \dots, n_m; \beta) = I_\beta(T_{p+1}^{\,n_{p+1}} \cdots T_m^{\,n_m})$. The partial derivatives of $\Phi_{\text{classical}}$ involve only the numbers $\int_X T_i \cup T_j \cup T_k$, while $\Gamma$ involves the interesting enumerative geometry numbers. From this form of $\Gamma$, it is easy to read off its partial derivatives.

Let us look again at the projective plane from this point of view. Take the obvious basis: $T_0 = 1$, $T_1$ the class of a line, and $T_2$ the class of a point. Note that $g_{ij}$ is 1 if $i + j = 2$, and 0 otherwise, so the same is true for $g^{ij}$. Therefore,

$$T_i * T_j = \Phi_{ij0} T_2 + \Phi_{ij1} T_1 + \Phi_{ij2} T_0.$$

For example,

$$\begin{aligned} T_1 * T_1 &= T_2 + \Gamma_{111} T_1 + \Gamma_{112} T_0, \\ T_1 * T_2 &= \Gamma_{121} T_1 + \Gamma_{122} T_0, \\ T_2 * T_2 &= \Gamma_{221} T_1 + \Gamma_{222} T_0. \end{aligned}$$

Therefore,

$$(T_1 * T_1) * T_2 = (\Gamma_{221} T_1 + \Gamma_{222} T_0) + \Gamma_{111}(\Gamma_{121} T_1 + \Gamma_{122} T_0) + \Gamma_{112} T_2,$$

$$T_1 * (T_1 * T_2) = \Gamma_{121}(T_2 + \Gamma_{111} T_1 + \Gamma_{112} T_0) + \Gamma_{122} T_1.$$

The fact that the coefficients of $T_0$ must be equal in these last two expressions gives the equation:

$$\Gamma_{222} = {\Gamma_{112}}^2 - \Gamma_{111}\Gamma_{122}. \tag{58}$$

If $\beta = d[\text{line}]$, the number $N(n, \beta)$ is nonzero only when $n = 3d - 1$, when it is the number $N_d$ of plane rational curves of degree $d$ passing through $3d - 1$ general points. So,

$$\Gamma(y) = \sum_{d\geq 1} N_d e^{dy_1} \frac{{y_2}^{3d-1}}{(3d-1)!}.$$

From this we read off the partial derivatives:

$$\begin{aligned}
\Gamma_{222} &= \sum_{d\geq 2} N_d e^{dy_1} \frac{{y_2}^{3d-4}}{(3d-4)!} \\
\Gamma_{112} &= \sum_{d\geq 1} d^2 N_d e^{dy_1} \frac{{y_2}^{3d-2}}{(3d-2)!} \\
\Gamma_{111} &= \sum_{d\geq 1} d^3 N_d e^{dy_1} \frac{{y_2}^{3d-1}}{(3d-1)!} \\
\Gamma_{122} &= \sum_{d\geq 1} d N_d e^{dy_1} \frac{{y_2}^{3d-3}}{(3d-3)!}.
\end{aligned}$$

Therefore,

$${\Gamma_{112}}^2 = \sum_{d\geq 2} \sum_{d_1+d_2=d} {d_1}^2 N_{d_1} {d_2}^2 N_{d_2} e^{dy_1} \frac{{y_2}^{3d-4}}{(3d_1-2)!\,(3d_2-2)!},$$

$$\Gamma_{111}\Gamma_{122} = \sum_{d\geq 2} \sum_{d_1+d_2=d} {d_1}^3 N_{d_1} d_2 N_{d_2} e^{dy_1} \frac{{y_2}^{3d-4}}{(3d_1-1)!\,(3d_2-3)!}.$$

In all these sums, $d_1$ and $d_2$ are positive. Equating the coefficients of

$$e^{dy_1} {y_2}^{3d-4}/(3d-4)!,$$

we get the identity ($d \geq 2$):

$$N_d = \sum_{d_1+d_2=d} N_{d_1} N_{d_2} \left[ {d_1}^2 {d_2}^2 \binom{3d-4}{3d_1-2} - {d_1}^3 d_2 \binom{3d-4}{3d_1-1} \right]. \tag{59}$$

Here a binomial coefficient $\binom{n}{m}$ is defined to be zero if any of $n$, $m$, or $n - m$ is negative. This is the recursion formula discussed in the introduction.

Note that the quantum formalism has removed any necessity to be clever. One simply writes down the associativity equations, and reads off enumerative information. One can organize the information in these associativity equations more systematically as follows (see [DF-I]). Let $F(i, j \mid k, l)$ be defined by (50). For $0 \leq i, j, k, l \leq m$, define:

$$\begin{aligned}
A(i,j,k,l) &= F(i,j \mid k,l) - F(j,k \mid i,l) \\
&= \sum_{e,f} \Phi_{ije} g^{ef} \Phi_{fkl} - \Phi_{jke} g^{ef} \Phi_{fil}.
\end{aligned}$$

Associativity (Theorem 4) amounts to the equations $A(i,j,k,l) = 0$ for all $i,j,k,l$. The symmetry of $\Phi_{ijk}$ in the subscripts and $g^{ef}$ in the superscripts and the basic facts about $\Phi_{0jk}$ imply:

(i) $A(k,j,i,l) = -A(i,j,k,l)$,
(ii) $A(l,k,j,i) = A(i,j,k,l)$,
(iii) $A(i,j,k,l) = 0$ if $i = k$ or $j = l$ or if any of the indices $i,j,k,l$ equals 0.

We consider equations equivalent if they differ by sign. For distinct $i,j,k,l$, the 24 possible equations divide into 3 groups of 8. The equation $A(i,j,k,l) = 0$ that says $F(i,j \mid k,l) = F(j,k \mid i,l)$ can be labelled by a duality diagram from topological field theory (see [DF-I]):

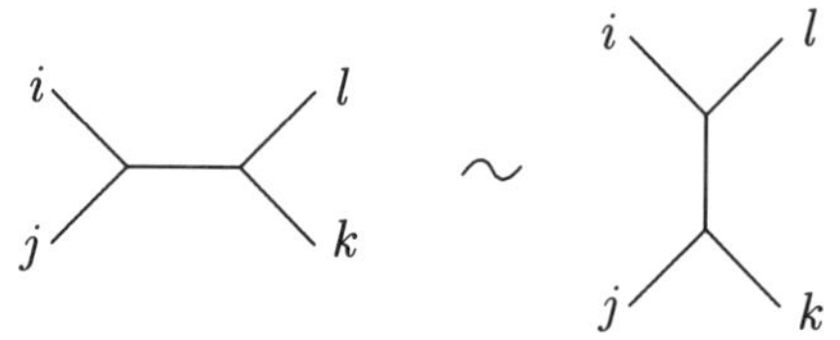


This diagram corresponds to the equations:

$$A(i,j,k,l) = A(j,i,l,k) = A(k,l,i,j) = A(l,k,j,i) = 0$$

$$-A(i,l,k,j) = -A(k,j,i,l) = -A(l,i,j,k) = -A(j,k,l,i) = 0.$$

To obtain the equations, read the labels around the left or right diagram (either clockwise or counterclockwise, but always reading two grouped together at an end first). The other sixteen equations correspond similarly to the diagrams:

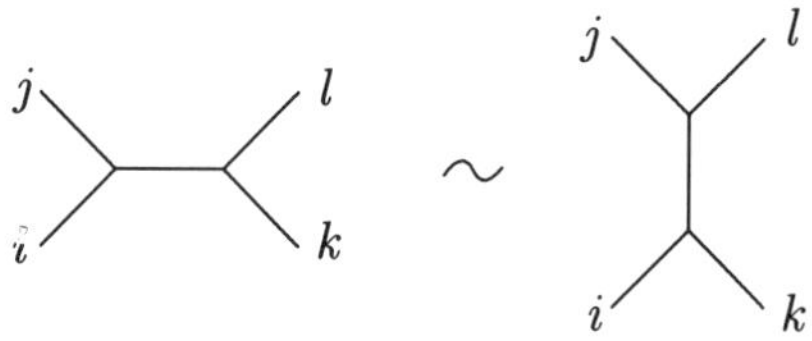


and

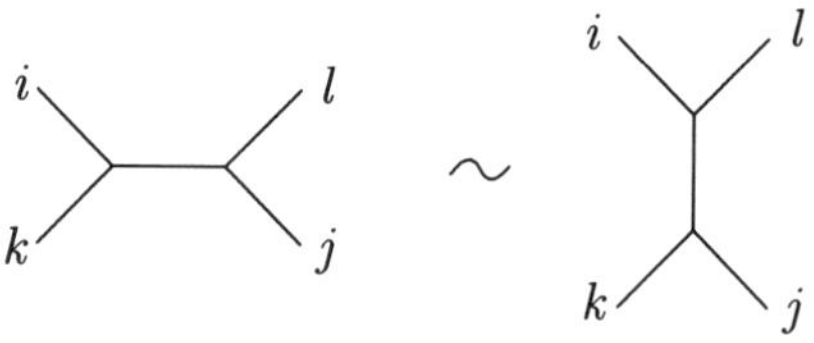


In practice, one only needs to write down one equation for each such diagram.

When 3 of the 4 labels are distinct, say $i,i,j,k$, there is only 1 equation up to sign (which occurs 8 times). It corresponds to:

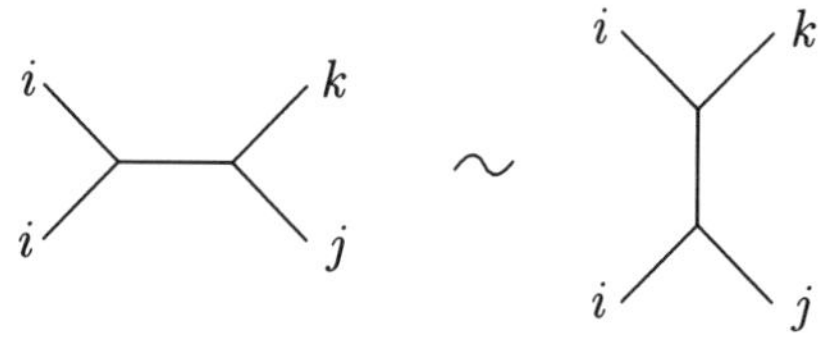


When two labels are distinct, there is again only 1 equation up to sign (occurring 4 times):

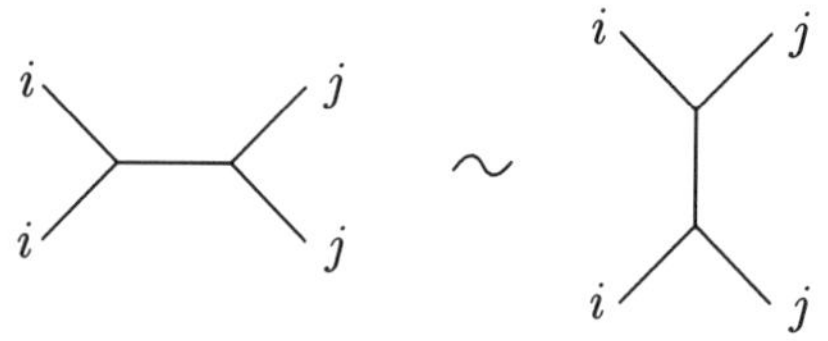


The symmetry in these diagrams reflects the symmetry in the equations. Taking just one equation for each diagram, one sees that the number $N(m)$ of equations for $\operatorname{rank}(A^*X) = m+1$ is

$$N(m) = 3\binom{m}{4} + m\binom{m-1}{2} + \binom{m}{2} = \frac{m(m-1)(m^2-m+2)}{8},$$

so $N(2) = 1$, $N(3) = 6$, $N(4) = 21$, $N(5) = 55$, $N(6) = 120$, and $N(7) = 231$. For the complete flag manifold $\mathbf{Fl}(\mathbb{C}^n)$, $m = n! - 1$. The number of equations for $\mathbf{Fl}(\mathbb{C}^4)$ is $N(23) = 30861$.

Let us work this out for the two varieties $X = \mathbf{P}^3$ and $X = \mathbf{Q}^3$ (a smooth quadric 3-fold), which have very similar classical cohomology rings. Each has a basis :

$$\begin{aligned} T_0 &= 1, \\ T_1 &= \text{hyperplane class}, \\ T_2 &= \text{line class}, \\ T_3 &= \text{point class}. \end{aligned}$$

The difference in the classical product is that $T_1 \cup T_1 = T_2$ for $\mathbf{P}^3$ but $T_1 \cup T_1 = 2T_2$ for $\mathbf{Q}^3$. Let $c = 1$ for $\mathbf{P}^3$ and $c = 2$ for $\mathbf{Q}^3$. The $N(3) = 6$ equations are:

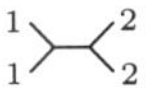


$$2\Gamma_{123} - c\Gamma_{222} = \Gamma_{111}\Gamma_{222} - \Gamma_{112}\Gamma_{122}$$

$$\Gamma_{133} - c\Gamma_{223} = \Gamma_{111}\Gamma_{223} - \Gamma_{113}\Gamma_{122}$$

$$c\Gamma_{233} = 2\Gamma_{113}\Gamma_{123} - \Gamma_{112}\Gamma_{133} - \Gamma_{111}\Gamma_{233}$$

$$\Gamma_{233} = \Gamma_{113}\Gamma_{222} - \Gamma_{112}\Gamma_{223}$$

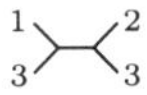


$$\Gamma_{333} = \Gamma_{123}^2 - \Gamma_{122}\Gamma_{133} + \Gamma_{113}\Gamma_{223} - \Gamma_{112}\Gamma_{233}$$

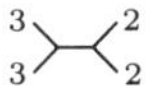


$$0 = \Gamma_{133}\Gamma_{222} - 2\Gamma_{123}\Gamma_{223} + \Gamma_{122}\Gamma_{233}$$

The function $\Gamma$ has the form:

$$\Gamma = \sum N_{a,b} e^{dy_1} \frac{y_2^a}{a!}\frac{y_3^b}{b!}. \tag{60}$$

For $\mathbf{P}^3$ the sum in (60) is over non-negative $a, b$ satisfying $a + 2b = 4d$, $d \geq 1$. A crucial difference is that for $\mathbf{Q}^3$, the sum in (60) is over $a + 2b = 3d$, $d \geq 1$ reflecting the fact that $c_1(T_{\mathbf{P}^3}) = 4T_1$ while $c_1(T_{\mathbf{Q}^3}) = 3T_1$. In each case, $N_{a,b}$ is the number of degree $d$ rational curves in $X$ meeting $a$ general lines and $b$ general points of $X$.

Each of the six differential equations above yields a recursion among the $N_{a,b}$:

(1) For $a \geq 3$, $b \geq 0$, $2dN_{a-2,b+1} - cN_{a,b} =$

$$\sum N_{a_1,b_1} N_{a_2,b_2} \binom{b}{b_1}\left(d_1^3\binom{a-3}{a_1} - d_1^2 d_2 \binom{a-3}{a_1-1}\right)$$

(2) For $a \geq 2$, $b \geq 1$, $dN_{a-2,b+1} - cN_{a,b} =$

$$\sum N_{a_1,b_1} N_{a_2,b_2} \binom{a-2}{a_1}\left(d_1^3\binom{b-1}{b_1} - d_1^2 d_2 \binom{b-1}{b_1-1}\right)$$

(3) For $a \geq 1$, $b \geq 2$, $cN_{a,b} =$

$$\sum N_{a_1,b_1} N_{a_2,b_2} \left(2d_1^2 d_2 \binom{a-1}{a_1}\binom{b-2}{b_1-1} - d_1^2 d_2 \binom{a-1}{a_1-1}\binom{b-2}{b_1}\right.$$

$$-d_1^3\binom{a-1}{a_1}\binom{b-2}{b_1}\Big)\Big)$$

(4) For $a \geq 3,\ b \geq 1,\quad N_{a-2,b+1} =$

$$\sum N_{a_1,b_1}N_{a_2,b_2}d_1^2\left(\binom{a-3}{a_1}\binom{b-1}{b_1-1} - \binom{a-3}{a_1-1}\binom{b-1}{b_1}\right)$$

(5) For $a \geq 2,\ b \geq 2,\quad N_{a-2,b+1} =$

$$\sum N_{a_1,b_1}N_{a_2,b_2}\Big(d_1d_2\binom{a-2}{a_1-1}\binom{b-2}{b_1-1} - d_1d_2\binom{a-2}{a_1-2}\binom{b-2}{b_1}$$
$$+d_1^2\binom{a-2}{a_1}\binom{b-2}{b_1-1} - d_1^2\binom{a-2}{a_1-1}\binom{b-2}{b_1}\Big)$$

(6) For $a \geq 3,\ b \geq 2,\quad 0 =$

$$\sum N_{a_1,b_1}N_{a_2,b_2}d_1\Big(\binom{a-3}{a_1}\binom{b-2}{b_1-2} - 2\binom{a-3}{a_1-1}\binom{b-2}{b_1-1}$$
$$+\binom{a-3}{a_1-2}\binom{b-2}{b_1}\Big)$$

In these formulas, the sum is over non-negative $a_1, a_2, b_1, b_2$ satisfying

(i) $a_1 + a_2 = a,\ b_1 + b_2 = b,$
(ii) $a + 2b = 4d,\ a_i + 2b_i = 4d_i,\ d_i > 0$ for $\mathbf{P}^3$,
$a + 2b = 3d,\ a_i + 2b_i = 3d_i,\ d_i > 0$ for $\mathbf{Q}^3$.

For $\mathbf{P}^3$, one starts with the $N_{0,2} = 1$ for the number of lines through two points. For $\mathbf{Q}^3$, $N_{1,1} = 1$ is not hard to compute directly. In each case, the six recursions are more than enough to solve for all the other $N_{a,b}$. These numbers for $\mathbf{P}^3$ include the classical results: there are $N_{4,0} = 2$ lines meeting 4 general lines, $N_{8,0} = 92$ conics meeting 8 general lines, and $N_{12,0} = 80160$ twisted cubics meeting 12 general lines. See [DF-I] for more of these numbers[2]. For $\mathbf{Q}^3$, computations yield:

$(d=1)$ $N_{1,1} = 1,\ N_{3,0} = 1$
$(d=2)$ $N_{0,3} = 1,\ N_{2,2} = 1,\ N_{4,1} = 2\ ,\ N_{6,0} = 5$
$(d=3)$ $N_{1,4} = 2,\ N_{3,3} = 5,\ N_{5,2} = 16,\ N_{7,1} = 59,\ N_{9,0} = 242$
$(d=4)$ $N_{0,6} = 6,\ N_{2,5} = 20,\ N_{4,4} = 74,\ N_{6,3} = 320,\ N_{8,2} = 1546,$
$N_{10,1} = 8148,\ N_{12,0} = 46230$
$(d=5)$ $N_{1,7} = 106,\ N_{3,6} = 448,\ N_{5,5} = 2180,\ N_{7,4} = 11910,$
$N_{9,3} = 71178,\ N_{11,2} = 457788,\ N_{13,1} = 3136284,$
$N_{15,0} = 22731810.$

The reader is invited to work out the equations for some other simple homogeneous spaces such as $\mathbf{P}^4$, $\mathbf{P}^1 \times \mathbf{P}^1$, $\mathbf{Gr}(2,4)$, or the incidence variety $\mathbf{Fl}(\mathbb{C}^3)$ of points on lines in the plane. For very pleasant excursions along these paths, see [DF-I].

There is a simple method of obtaining a presentation of $QH^*X$ from $\Phi$ and a presentation of $A^*X$. It will be convenient to consider $A^*X_{\mathbb{Q}} = H^*(X,\mathbb{Q})$, the cohomology ring of $X$ with rational coefficients. Following the notation of section 8, let $QH^*X = (V \otimes_Z \mathbb{Q}[[V^*]], *)$. There is a canonical embedding:

$$\iota_{\mathbb{Q}} : A^*X_{\mathbb{Q}} \hookrightarrow QH^*X$$

[2]The numbers $N_{a,b}$ given in [DF-I] are correct, although their version of equation (6) has a misprint.

of $\mathbb{Q}$-vector spaces. In the discussion below, $A^*X_{\mathbb{Q}}$ is viewed as a $\mathbb{Q}$-subspace of $QH^*X$ via $\iota_{\mathbb{Q}}$. The results relating presentations of $A^*X_{\mathbb{Q}}$ and $QH^*X$ are established in Propositions 9 and 10.

PROPOSITION 9. *Let $z_1, ..., z_r$ be homogeneous elements of positive codimension that generate $A^*X_{\mathbb{Q}}$ as a $\mathbb{Q}$-algebra. Then, $z_1, \dots, z_r$ generate $QH^*X$ as a $\mathbb{Q}[[V^*]]$-algebra.*

The proof requires a lemma. Note that for $\gamma \in \mathbb{Q}[[V^*]]$ there is a well-defined constant term $\gamma(0) \in \mathbb{Q}$.

LEMMA 17. *Let $T_0, \dots, T_m$ be any homogeneous $\mathbb{Q}$-basis of $A^*X_{\mathbb{Q}}$. Let $w_1, w_2 \in A^*X_{\mathbb{Q}}$ be homogeneous elements. Let*

$$w_1 \cup w_2 = \sum_{k=0}^{m} c_k T_k, \quad c_k \in \mathbb{Q},$$

$$w_1 * w_2 = \sum_{k=0}^{m} \gamma_k T_k, \quad \gamma_k \in \mathbb{Q}[[V^*]],$$

*be the unique expansions in $A^*X_{\mathbb{Q}}$ and $QH^*X$ respectively.*

(i) *If* $\text{codim}(T_k) > \text{codim}(w_1) + \text{codim}(w_2)$, *then* $\gamma_k(0) = 0$.
(ii) *If* $\text{codim}(T_k) = \text{codim}(w_1) + \text{codim}(w_2)$, *then* $\gamma_k(0) = c_k$.

*Proof.* By linearity of the $*$-product, it can be assumed that $w_1$ and $w_2$ are basis elements $T_i$ and $T_j$ respectively. In the basis $T_0, \dots, T_m$ of $A^*_{\mathbb{Q}}X$, the $*$-product is determined by:

$$T_i * T_j = T_i \cup T_j + \sum_{i=1}^{m} \Gamma_{ijl} g^{lk} T_k$$

where the dual coordinates $y_0, \dots, y_m$ are taken in $V^* \otimes \mathbb{Q}$. $\Gamma_{ijl}(0) = \sum_{\beta \neq 0} I_\beta(T_i \cdot T_j \cdot T_l)$. Therefore, if $\Gamma_{ijl}(0) \neq 0$, there must exist a nonzero effective class $\beta \in A_1X$ such that

$$\dim \overline{M}_{0,3}(X, \beta) = \text{codim}(T_i) + \text{codim}(T_j) + \text{codim}(T_l).$$

Since $X$ is homogeneous, $\int_\beta c_1(X) \geq 2$ by Lemma 11. By the dimension formula,

$$\text{codim}(T_i) + \text{codim}(T_j) + \text{codim}(T_l) \geq \dim(X) + 2. \tag{61}$$

Equation (61) yields $\text{codim}(T_l) \geq \dim(X) - \text{codim}(T_i) - \text{codim}(T_j) + 2$. For $g^{lk}$ to be nonzero, it follows that $\text{codim}(T_k) \leq \text{codim}(T_i) + \text{codim}(T_j) - 2$. The lemma is proven. □

We will apply Lemma 17 to products in a basis of $A^*X_{\mathbb{Q}}$ consisting of monomials $z^I = z_1^{i_1} \cup \dots \cup z_r^{i_r}$. Let

$$z^{*I} = \underbrace{z_1 * \dots * z_1}_{i_1} * \underbrace{z_2 * \dots * z_2}_{i_2} * \dots * \underbrace{z_r * \dots * z_r}_{i_r}$$

denote the corresponding monomial in $QH^*X$. Let

$$\{z^I \mid I \in \mathcal{S}\} \tag{62}$$

be a monomial $\mathbb{Q}$-basis of $A^*X_{\mathbb{Q}}$. Choose an ordering of the set $\mathcal{S}$ so that $\text{codim}(z^I) \leq \text{codim}(z^J)$ for $I < J$. Let

$$z^{*I} = \sum_{J \in \mathcal{S}} \gamma_{IJ} z^J, \quad \gamma_{IJ} \in \mathbb{Q}[[V^*]]$$

be the unique expansion in $QH^*X$. An inductive application of Lemma 17 yields:

(i) If $J > I$, then $\gamma_{IJ}(0) = 0$.

(ii) $\gamma_{II}(0) = 1$.

Therefore, the matrix $(\gamma_{IJ}(0))$ is invertible over $\mathbb{Q}$. It follows that the matrix $(\gamma_{IJ})$ is invertible over $\mathbb{Q}[[V^*]]$. In particular, $\{z^{*I} \mid I \in \mathcal{S}\}$ is a $\mathbb{Q}[[V^*]]$-basis of $QH^*X$. Proposition 9 is proved.

Let $K$ be the kernel of the surjection

$$\phi : \mathbb{Q}[Z] = \mathbb{Q}[Z_1, \ldots, Z_r] \to A^*X_{\mathbb{Q}}$$

determined by $\phi(Z_i) = z_i$. Let $K'$ be the kernel of the corresponding surjection

$$\phi' : \mathbb{Q}[[V^*]][Z] \to QH^*X$$

determined by $\phi'(Z_i) = z_i$ Using our choice (62) of monomial basis, there is a method of constructing elements of $K'$ from elements of $K$. Let $f \in K$. The polynomial $f$ is also an element of $\mathbb{Q}[[V^*]][Z]$. There is a unique expansion:

$$\phi'(f) = \sum_{I \in \mathcal{S}} \xi_I z^{*I}, \quad \xi_I \in \mathbb{Q}[[V^*]].$$

Then, $f' = f(Z_1, \ldots, , Z_r) - \sum_{I \in \mathcal{S}} \xi_I Z^I$ is in $K'$.

The ideal $K$ is homogeneous provided the degree of $Z_i$ is taken to be the codimension of $z_i$. We need the following fact.

LEMMA 18. *Let $f \in K$ be homogeneous of degree $d$ and let $I \in \mathcal{S}$. If* $\deg(Z^I) \geq d$, *then* $\xi_I(0) = 0$.

*Proof.* If $d > \dim(X)$, the statement is vacuous. Assume $d \leq \dim(X)$. Let $\phi'(f) = \sum_{I \in \mathcal{S}} \tilde{\xi}_I z^I$, $\tilde{\xi}_I \in \mathbb{Q}[[V^*]]$ be the unique expansion. Apply Lemma 17 repeatedly to the monomials of $f$ in the basis $\{z^I \mid I \in \mathcal{S}\}$ of $A^*X_{\mathbb{Q}}$. It follows that if $\deg(Z^I) \geq d$, then $\tilde{\xi}_I(0) = 0$. The change of basis relations (i) and (ii) for the $\mathbb{Q}[[V^*]]$-basis $\{z^{*I} \mid I \in \mathcal{S}\}$ now imply the lemma. □

Now suppose the elements $f_1, \ldots, f_s$ are homogeneous generators of $K$, so

$$A^*X_{\mathbb{Q}} = \mathbb{Q}[Z]/(f_1, \ldots, f_s)$$

is a presentation of the cohomology ring.

PROPOSITION 10. *The ideal $K'$ is generated by $f'_1, \ldots, f'_s$, so*

$$QH^*X = \mathbb{Q}[[V^*]][Z]/(f'_1, \ldots, f'_s)$$

*is a presentation of the quantum cohomology ring.*

*Proof.* Since we have a surjection

$$\mathbb{Q}[[V^*]][Z]/(f'_1 \ldots, f'_s) \to QH^*X$$

and $QH^*X$ is a free $\mathbb{Q}[[V^*]]$-module with basis $\{z^{*I} \mid I \in \mathcal{S}\}$, it suffices to show that the monomials $\{Z^I \mid I \in \mathcal{S}\}$ span the $\mathbb{Q}[[V^*]]$-module on the left. By Nakayama's lemma, it suffices to show that these monomials generate the $\mathbb{Q}$-vector space

$$\mathbb{Q}[[V^*]][Z]/(f'_1, \ldots, f'_s, \mathfrak{m}), \tag{63}$$

where $\mathfrak{m} \subset \mathbb{Q}[[V^*]]$ is the maximal ideal. Let $f'_i = f_i - \sum \xi_{iI} Z^I$. Define $\overline{f}'_i \in \mathbb{Q}[Z]$ by $\overline{f}'_i = f_i - \sum \xi_{iI}(0) Z^I$. The $\mathbb{Q}$-algebra (63) can be identified with

$$\mathbb{Q}[Z]/(\overline{f}'_1, \ldots, \overline{f}'_s).$$

By Lemma 18, all the terms $\xi_{iI}(0)Z^I$ have strictly lower degree than $f_i$. It is then a simple induction on the degree to see that the same monomials $\{Z^I\}$ that span modulo $(f_1, \ldots, f_s)$ will also span modulo $(\overline{f}'_1, \ldots, \overline{f}'_s)$. □

For example, let $X = \mathbf{P}^2$. Let $Z = Z_1$ and let $A^*_{\mathbb{Q}}\mathbf{P}^2 = \mathbb{Q}[Z]/Z^3$ be the standard presentation with the monomial basis $1, Z, Z^2$. A presentation of $QH^*\mathbf{P}^2$ is obtained:

$$QH^*\mathbf{P}^2 \xrightarrow{\sim} \mathbb{Q}[[y_0, y_1, y_2]][Z]/(Z^3 - \Gamma_{111}Z^2 - 2\Gamma_{112}Z - \Gamma_{122}) \tag{64}$$

where $\Gamma$ is the quantum potential of $\mathbf{P}^2$. By (64) and the determination of $\Gamma$,

$$QH^*\mathbf{P}^2 \otimes_{\mathbb{Q}[[V^*]]} \mathbb{Q}[[V^*]]/\mathfrak{m} = \mathbb{Q}[Z]/(Z^3 - 1).$$

Note that $QH^*\mathbf{P}^2$ does not specialize to $A^*\mathbf{P}^2$.

## 10. Variations

The algebra $QH^*X = A^*X \otimes \mathbb{Q}[[V^*]]$ may be regarded as the "big" quantum cohomology ring. There is also a "small" quantum cohomology ring, $QH^*_sX$, that incorporates only the 3-point Gromov-Witten invariants in its product. $QH^*_sX$ is obtained by restricting the $*$-product to the formal deformation parameters of the divisor classes. Most computations of quantum cohomology rings have been of this small ring, which is often easier to describe; the small ring is often denoted $QH^*X$.

It is simplest to define $QH^*_sX$ in the Schubert basis $T_0, \ldots, T_m$. Let

$$\overline{\Phi}_{ijk} = \Phi_{ijk}(y_0, y_1, \ldots, y_p, 0, \ldots, 0) = \int_X T_i \cup T_j \cup T_k \; + \; \overline{\Gamma}_{ijk}. \tag{65}$$

The modified quantum potential $\overline{\Gamma}_{ijk}$ is determined by

$$\overline{\Gamma}_{ijk} = \sum_{n\geq 0} \frac{1}{n!} \sum_{\beta\neq 0} I_\beta(\gamma^n \cdot T_i \cdot T_j \cdot T_k)$$

where $\gamma = y_1T_1 + \ldots + y_pT_p$. By the divisor property (III) of section 7,

$$\overline{\Gamma}_{ijk} = \sum_{\beta\neq 0} I_\beta(T_i \cdot T_j \cdot T_k) q_1{}^{\int_\beta T_1} \cdots q_p{}^{\int_\beta T_p}, \tag{66}$$

where $q_i = e^{y_i}$. Note that only 3-point invariants occur. Let $\mathbb{Z}[q] = [q_1, \ldots, q_p]$. By Theorem 4, the product

$$T_i * T_j = \sum_{e,f} \overline{\Phi}_{ije} g^{ef} T_f \; = T_i \cup T_j \; + \; \sum_{e,f} \overline{\Gamma}_{ije} g^{ef} T_f$$

then makes the $\mathbb{Z}[q]$-module $A^*X \otimes_{\mathbb{Z}} \mathbb{Z}[q]$ into a commutative, associative $\mathbb{Z}[q]$-algebra with unit $T_0$. From equation (66), it easily follows that the small quantum cohomology is a deformation of $A^*X$ is the usual sense: $A^*X$ is recovered by setting the variables $q_i = 0$.

For example, let $X = \mathbf{P}^r$. Then, $q = q_1$. If $T_i$ is the class of a linear subspace of codimension $i$ and $\beta$ is $d$ times the class of a line, then the number $I_\beta(T_i \cdot T_j \cdot T_k)$ can be nonzero only if $i + j + k = r + (r+1)d$; this can happen only for $d = 0$ or $d = 1$, and in each case the number is 1. It follows that,

(i) if $i + j \leq r$, then $T_i * T_j = T_{i+j}$;
(ii) if $r + 1 \leq i + j \leq 2r$, then $T_i * T_j = qT_{i+j-r-1}$.

Therefore the small quantum cohomology ring is:

$$QH_s^*\mathbf{P}^r = \mathbb{Z}[T,\, q]/(T^{r+1} - q),$$

where $T = T_1$ is the class of a hyperplane.

The following variation of Proposition 10 is valid for the small quantum cohomology ring (cf. [S-T]). As before let $z_1, \dots, z_r$ be homogenous elements of positive codimension that generate $A^*X$. (We use integer coefficients but rational coefficients could be used as well). Let $\mathbb{Z}[Z] = \mathbb{Z}[Z_1, \dots, Z_r]$, and let

$$A^*X = \mathbb{Z}[Z]/(f_1, \dots, f_s)$$

be a presentation with arbitrary homogeneous generators $f_1, \dots, f_s$ for the ideal of relations. Let $\mathbb{Z}[q, Z] = \mathbb{Z}[q_1, \dots, q_p, Z_1, \dots, Z_r]$. The variables $q_i, Z_j$ are graded by the following degrees: $\deg(q_i) = \int_{\beta_i} c_1(T_X)$ where $\beta_i$ is the class of the Schubert variety dual to $T_i$ and $\deg(Z_j) = \operatorname{codim}(z_j)$. Let $QH_s^*X = A^*X \otimes \mathbb{Z}[q]$ with the quantum product.

PROPOSITION 11. *Let $f_1', \dots, f_s'$ be any homogeneous elements in $\mathbb{Z}[q, Z]$ such that:*

(i) $f_i'(0, \dots, 0, Z_1, \dots, Z_r) = f_i(Z_1, \dots, Z_r)$ *in* $\mathbb{Z}[q, Z]$,
(ii) $f_i'(q_1, \dots, q_p, Z_1, \dots, Z_r) = 0$ *in* $QH_s^*X$.

*Then, the canonical map*

$$\mathbb{Z}[q, Z]/(f_1', \dots, f_s') \to QH_s^*X \tag{67}$$

*is an isomorphism.*

*Proof.* The proof is by a Nakayama-type induction. As the arguments are similiar to the proof of Proposition 10, we will be brief. The fact that each $q_i$ has positive degree implies the following statement. If $\psi : M \to N$ is a homogeneous map of finitely generated $\mathbb{Z}[q, Z]$-modules that is surjective modulo the ideal $(q) = (q_1, \dots, q_p)$, then $\psi$ is surjective. Hence, by (i), the map (67) is surjective. Similarly, if $\tilde{T}_0, \dots, \tilde{T}_m$ are homogeneous lifts to $\mathbb{Z}[Z]$ of a basis of $A^*X$, an easy induction shows that their images in $\mathbb{Z}[q, Z]/(f_1', \dots, f_s')$ generate this $\mathbb{Z}[q]$-module. Since $QH_s^*X$ is free over $\mathbb{Z}$ of rank $m+1$, the map (67) must be an isomorphism. □

A similar calculation, as in [S-T], yields the small quantum cohomology ring of the Grassmannian $X = \mathbf{Gr}(p, n)$ of $p$-dimensional subspaces of $\mathbb{C}^n$. Let $k = n - p$, let $0 \to S \to \mathbb{C}^n{}_X \to Q \to 0$ be the universal exact sequence of bundles on $X$, and let $\sigma_i = c_i(Q)$. Set $S_r(\sigma) = \det\left(\sigma_{1+j-i}\right)_{1 \le i,j \le r}$, and let $q = q_1$.

PROPOSITION 12. *The small quantum cohomology ring of* $\mathbf{Gr}(p, n)$ *is*

$$\mathbb{Z}[\sigma_1, \dots, \sigma_k, q]/\left(S_{p+1}(\sigma), S_{p+2}(\sigma), \dots, S_{n-1}(\sigma), S_n(\sigma) + (-1)^k q\right).$$

*Proof.* We use some standard facts about the Grassmannian. In particular, the cohomology has an additive basis of Schubert classes $\sigma_\lambda$, as $\lambda$ varies over partitions with $k \ge \lambda_1 \ge \dots \ge \lambda_p \ge 0$; $\sigma_\lambda = [\Omega_\lambda]$ is the class of a Schubert variety

$$\Omega_\lambda = \{L \in X : \dim L \cap V_{k+i-\lambda_i} \ge i \text{ for } 1 \le i \le p\},$$

where $V_1 \subset V_2 \subset \dots \subset V_n = \mathbb{C}^n$ is a given flag of subspaces. In $A^*(X)$, $S_r(\sigma)$ represents the $r^{th}$ Chern class of $S^\vee$, from which we have

$$A^*(X) = \mathbb{Z}[\sigma_1, \dots, \sigma_k]/\left(S_{p+1}(\sigma), \dots, S_n(\sigma)\right).$$

By Proposition 11, it suffices to show that the relations displayed in the proposition are valid in $QH_s^*X$.

Since $c_1(T_X) = n\sigma_1$, a number $I_\beta(\gamma_1 \cdot \gamma_2 \cdot \gamma_3)$ can be nonzero only if the sum of the codimensions of the $\gamma_i$ is equal to $\dim X + nd$, where $\beta$ is $d$ times the class of a line. If $d \geq 1$, all such numbers vanish when $\mathrm{codim}(\gamma_1) + \mathrm{codim}(\gamma_2) < n$. In particular, the relations $S_i(\sigma) = 0$ for $p < i < n$ remain valid in $QH_s^*X$. From the formal identity

$$S_n(\sigma) - \sigma_1 S_{n-1}(\sigma) + \sigma_2 S_{n-2}(\sigma) - \ldots + (-1)^k \sigma_k S_{n-k}(\sigma) = 0,$$

we therefore have $S_n(\sigma) = (-1)^{k-1}\sigma_k S_{n-k}(\sigma)$ in $QH_s^*X$. Since $S_{n-k}(\sigma) = \sigma_{(1^{n-k})}$, the proof will be completed by verifying that $\sigma_k * \sigma_{(1^{n-k})} = q$. Equivalently, when $\beta$ is the class of a line, we must show that

$$I_\beta(\sigma_k, \sigma_{(1^p)}, \sigma_{(k^p)}) = 1.$$

This is a straightforward calculation. First we have

$$\sigma_k = [\{L : L \supset A\}], \ \sigma_{(1^p)} = [\{L : L \subset B\}], \ \sigma_{(k^p)} = [\{L : L = C\}],$$

where $A$, $B$, and $C$ are linear subspaces of $\mathbb{C}^n$ of dimensions 1, $n-1$, and $p$ respectively. It is not hard to verify that any line in $X$ is a Schubert variety of the form $\{L : U \subset L \subset V\}$, where $U \subset V$ are subspaces of $\mathbb{C}^n$ of dimensions $p-1$ and $p+1$. Such a line will meet the three displayed Schubert varieties only if $V$ contains $A$ and $C$, and $U$ is contained in $B$ and $C$. For $A$, $B$, and $C$ general, there is only one such line, with $U = B \cap C$ and $V$ spanned by $A$ and $C$. □

This proposition was proved in another way by Bertram [Ber], where the beginnings of some "quantum Schubert calculus" can be found. For the small quantum cohomology ring of a flag manifold, following ideas of Bertram, Givental, and Kim, see [CF][3].

As with the big quantum cohomology ring, the small ring has a basis independent description. Let $\mathbb{Z}[A_1X]$ be the group algebra. The small $*$-product is naturally defined on the free $\mathbb{Z}[A_1X]$-module $A^*X \otimes_{\mathbb{Z}} \mathbb{Z}[A_1X]$. If $\beta_1, \ldots, \beta_p$ is a basis of $A_1X$ consisting of Schubert classes, then the dual Schubert classes $T_1, \ldots, T_p$ satisfy $\int_\beta T_i \geq 0$ for every effective class $\beta$. In this case, the small $*$-product on $A^*X \otimes_{\mathbb{Z}} \mathbb{Z}[A_1X]$ preserves the $\mathbb{Z}[q_1, \ldots, q_p]$-submodule:

$$A^*X \otimes_{\mathbb{Z}} \mathbb{Z}[q_1, \ldots, q_p] \subset A^*X \otimes_{\mathbb{Z}} \mathbb{Z}[A_1X].$$

Hence, in the Schubert basis, the small quatum cohomology ring can be taken to be $QH_s^*X = (A^*X \otimes_{\mathbb{Z}} \mathbb{Z}[q_1, \ldots, q_p], *)$.

The numbers $I_\beta(\gamma_1 \cdots \gamma_n)$ should not be confused with the numbers denoted by the expression $\langle \gamma_1, \ldots, \gamma_n \rangle_\beta$ which often occur in discussions of small quantum cohomology rings ([B-D-W], [Ber], [CF]). To define the latter, one *fixes* $n$ distinct points $p_1, \ldots, p_n$ in $\mathbf{P}^1$. Then, $\langle \gamma_1, \ldots, \gamma_n \rangle_\beta$ is the number of maps $\mu : \mathbf{P}^1 \to X$ satisfying: $\mu_*[\mathbf{P}^1] = \beta$ and $\mu(p_i) \in \Gamma_i$ for $1 \leq i \leq n$ (where $\Gamma_i$ is a subvariety in general position representing the class $\gamma_i$). For $n = 3$, the numbers agree: $I_\beta(\gamma_1 \cdot \gamma_2 \cdot \gamma_3) = \langle \gamma_1, \gamma_2, \gamma_3 \rangle_\beta$. For $n > 3$, the numbers $\langle \gamma_1, \ldots, \gamma_n \rangle_\beta$ and $I_\beta(\gamma_1 \cdots \gamma_n)$ are solutions to different enumerative problems. In fact, $\langle \gamma_1, \ldots, \gamma_n \rangle_\beta$ can be expressed

[3] This Schubert calculus is extended to flag manifolds in [F-G-P].

in terms of the 3-points numbers while $I_\beta(\gamma_1 \cdots \gamma_n)$ cannot. For $1 < k < n-1$, $\langle \gamma_1, \dots, \gamma_n \rangle_\beta =$

$$\sum_{\beta_1+\beta_2=\beta} \sum_{e,f} \langle \gamma_1, \dots, \gamma_k, T_e \rangle_{\beta_1} g^{ef} \langle T_f, \gamma_{k+1}, \dots, \gamma_n \rangle_{\beta_2} \tag{68}$$

Equation (68) can be seen geometrically by deforming $\mathbf{P}^1$ to a union of two $\mathbf{P}^1$'s meeting at a point with $p_1, \dots, p_k$ going to fixed points on the first line and $p_{k+1}, \dots, p_n$ going to fixed points on the second. Algebraically, in the small quantum cohomology ring,

$$\gamma_1 * \cdots * \gamma_n = \sum_\beta \sum_{e,f} q^\beta \langle \gamma_1, \dots, \gamma_n, T_e \rangle_\beta g^{ef} T_f.$$

Equation (68) amounts to the associativity of this product.

We conclude with a few general remarks to relate the discussion and notation here to that in [K-M 1].

The numbers that we have denoted $I_\beta(\gamma_1 \cdots \gamma_n)$ are part of a more general story. Let $\eta$ denote the forgetful map from $\overline{M}_{0,n}(X,\beta)$ to $\overline{M}_{0,n}$. For any cohomology classes $\gamma_1, \dots, \gamma_n$ on $X$, one can construct a class

$$I^X_{0,\,n,\,\beta}(\gamma_1 \otimes \cdots \otimes \gamma_n) = \eta_* \left( {\rho_1}^*(\gamma_1) \cup \cdots \cup {\rho_n}^*(\gamma_n) \right) \tag{69}$$

in the cohomology ring $H^*(\overline{M}_{0,n})$. These are called (tree-level, or genus zero) Gromov-Witten classes. The number we denoted $I_\beta(\gamma_1 \cdots \gamma_n)$ is the degree of the zero-dimensional component of this class, which they denote by $\langle I^X_{0,\,n,\,\beta} \rangle (\gamma_1 \otimes \cdots \otimes \gamma_n)$. The intersections with divisors that we have carried out on $\overline{M}_{0,n}(X,\beta)$ can be carried out with the corresponding divisors on $\overline{M}_{0,n}$; this has the advantage that the intersections take place on a nonsingular variety.

One of the main goals of [K-M 1] and especially [K-M 2] is to show how Gromov-Witten classes can be reconstructed from the numbers obtained by evaluating them on the fundamental classes. The idea is that a cohomology class in $H^*(\overline{M}_{0,n})$ is known by evaluating it on the classes of the closures of the strata determined by the combinatorial types of the labeled trees. As we saw and exploited for divisors, these numbers can be expressed in terms of the numbers $I_\beta$ for the pieces making up the tree.

Kontsevich and Manin also allow cohomology classes of odd degrees, in which case one has to be careful with signs and the ordering of the terms. For an interesting application to some Fano varieties, see [Bea].

Since the space $H = H^*(X, \mathbb{Q})$ can be identified with its dual by Poincaré duality, the maps $I^X_{0,\,n,\,\beta}$ can be regarded as maps

$$H_*(\overline{M}_{0,n+1}) \to \operatorname{Hom}(H^{\otimes n}, H). \tag{70}$$

Both of these, for varying $n$, have a natural operad structure, that on the first coming from all the ways to glue together labeled trees of $\mathbf{P}^1$'s to form new ones, and the second from all the ways to compose homomorphisms. Remarkably, the associativity (Theorem 4), is equivalent to the assertion that (70) is a morphism of operads.

The structure constants $g^{ij}$ put a metric on the cohomology space $H^*(X, \mathbb{C})$; with coordinates given by the basis for the cohomology, there is a (formal) connection given by the formula $A^k_{ij} = \sum \Phi_{ije} g^{ek}$. In this formalism of Dubrovin, the associativity translates to the assertion that this is a flat connection.

The numbers calculated here are part of a much more ambitious program described in [K-M 1] and [K]. The hope is to extend the story to varieties without the positivity assumptions made here, with some other construction of what should be the fundamental class of $\overline{M}_{g,n}(X, \beta)$. (For varieties whose tangent bundles are not as positive as those considered here, the definition of the potential function $\Phi$ is modified by multiplying the summands in (45) by $e^{-\int_\beta \omega}$, for a Kähler class $\omega$, in the hopes of making the power series converge on some open set of the cohomology space $H$.)

Even if this program is carried out, however – and associativity has been proved by symplectic methods [R-T] in some cases beyond those mentioned here[4] – the interpretation cannot always be in enumerative terms as simple as those we have discussed, cf. [C-M]. On the other hand, these ideas from quantum cohomology have inspired some recent work in enumerative geometry, even in cases where the associativity formalism does not apply directly, cf. [C-H 1] and [P][5].

## References


[A] V. Alexeev, *Moduli spaces $M_{g,n}$ for surfaces*, preprint 1994.

[Bea] A. Beauville, *Quantum cohomology of complete intersections*, preprint 1995.

[B] K. Behrend, *Gromov-Witten invariants in algebraic geometry*, preprint 1996.

[B-F] K. Behrend and B. Fantechi, *The intrinsic normal cone*, preprint 1996.

[B-M] K. Behrend and Yu. Manin, *Stacks of stable maps and Gromov-Witten invariants*, preprint 1995.

[Ber] A. Bertram, *Quantum Schubert calculus*, preprint 1994.

[B-D-W] A. Bertram, G. Daskalopoulos, and R. Wentworth, *Gromov-Witten invariants for holomorphic maps from Riemann surfaces to Grassmannians*, J. Amer. Math. Soc. **9** (1996), 529-571.

[C] M. Cornalba, *A simple proof of the projectivity of Kontsevich's space of maps*, preprint 1995.

[C-H 1] L. Caporaso and J. Harris, *Rational curves on rational ruled surfaces*, preprint 1996.

[C-H 2] L. Caporaso and J. Harris, *Degrees of Severi Varieties*, preprint 1996.

[CF] I. Ciocan-Fontanine, *Quantum cohomology of flag varieties*, Internat. Math. Res. Notices 1995, no. 6, 263-277.

[C-M] B. Crauder and R. Miranda, *Quantum cohomology of rational surfaces*, in *The moduli space of curves*, R. Dijkgraaf, C. Faber, and G. van der Geer, eds., Birkhauser, 1995, pp 33-80.

[D-M] P. Deligne and D. Mumford, *The irreducibility of the space of curves of given genus*, Inst. Hautes Études Sci. Publ. Math. **36** (1969), 75-110.

[DF-I] P. Di Francesco and C. Itzykson, *Quantum intersection rings*, in *The moduli space of curves*, R. Dijkgraaf, C. Faber, and G. van der Geer, eds., Birkhauser, 1995, pp 81-148.

[E-K] L. Ernström and G. Kennedy, *Recursive formulas for the characteristic numbers of rational plane curves*, preprint 1996.

[F-G-P] S. Fomin, S. Gelfand, and A. Postnikov, *Quantum Schubert polynomials*, preprint 1996.

[F] W. Fulton, *Intersection Theory*, Springer-Verlag, 1984.

[F-M] W. Fulton and R. MacPherson, *A Compactification of Configuration Spaces*, Ann. of Math. **130** (1994), 183-225.

[G] A. Gathmann, *Counting rational curves with multiple points and Gromov-Witten invariants of blow-ups*, preprint 1996.


---

[4]At this conference, J. Li lectured on his work with G. Tian (see [L-T 2]) using cones in vector bundles to construct a "virtual fundamental class" to use in place of $[\overline{M}_{g,n}(X, \beta)]$ in case $\overline{M}_{g,n}(X, \beta)$ has the wrong dimension. This approach has been clarified and extended by K. Behrend and B. Fantechi using the language of stacks ([B-F], [B]). Algebraic computations in the non-convex case can be found, for example, in [G], [G-P], [K].

[5]Also, [E-K], [K-Q-R].

[G-P] L. Göttsche and R. Pandharipande, *The quantum cohomology of blow-ups of* $\mathbf{P}^2$ *and enumerative geometry*, preprint 1996.

[Ha] J. Harris, *On the Severi problem*, Invent. Math. **84** (1986), 445-461.

[H] R. Hartshorne, *Algebraic Geometry*, Springer-Verlag, 1977.

[K-Q-R] S. Katz, Z. Qin, and Y. Ruan, *Composition law and nodal genus-2 curves in* $\mathbf{P}^2$, preprint 1996.

[Ke] S. Keel, *Intersection theory on moduli spaces of stable n-pointed curves of genus zero*, Trans. AMS **330** (1992), 545-574.

[K-K-M] G. Kempf, F. Knudsen, D. Mumford, and B. Saint-Donat, *Toroidal Embeddings I*, Springer Lecture Notes **339**, 1973.

[Kl] S. Kleiman, *The transversality of a general translate*, Compositio Math. **28** (1974), 287-297.

[Kn] F. Knudsen, *Projectivity of the moduli space of stable curves. II*, Math. Scand. **52** (1983), 1225-1265.

[K] M. Kontsevich, *Enumeration of rational curves via torus actions*, in *The moduli space of curves*, R. Dijkgraaf, C. Faber, and G. van der Geer, eds., Birkhauser, 1995, pp 335-368.

[K-M 1] M. Kontsevich and Yu. Manin, *Gromov-Witten classes, quantum cohomology, and enumerative geometry*, Commun. Math. Phys. **164** (1994), 525-562.

[K-M 2] M. Kontsevich and Yu. Manin, *Quantum cohomology of a product*, Invent. Math. **124** (1996), 313-339.

[Ko1] J. Kollár, *Projectivity of complete moduli*, J. Diff. Geom. **32** (1990), 235-268.

[Ko2] J. Kollár, *Rational curves on algebraic varieties*, Springer-Verlag, 1996.

[L-T 1] J. Li and G. Tian, *The quantum cohomology of homogeneous varieties*, preprint 1995.

[L-T 2] J. Li and G. Tian, *Virtual moduli cycles and Gromov-Witten invariants of algebraic varieties*, preprint 1996.

[M1] D. Mumford, *Abelian varieties*, Oxford University Press, 1970.

[M2] D. Mumford, *Lectures on curves on an algebraic surface*, Annals of Math. Studies **59**, Princeton Univ. Press, 1966.

[P] R. Pandharipande, *Intersections of* $\mathbb{Q}$*-divisors on Kontsevich's moduli space* $\overline{M}_{0,n}(\mathbf{P}^r, d)$ *and enumerative geometry*, *Counting elliptic plane curves with fixed j-invariant*, *The canonical class of* $\overline{M}_{0,n}(\mathbf{P}^r, d)$ *and enumerative geometry*, preprints 1995.

[R1] Z. Ran, *Enumerative geometry of singular plane curves*, Invent. Math. **97** (1989), 447-465.

[R2] Z. Ran, *On the quantum cohomology of the plane, old and new*, preprint 1995.

[R-T] Y. Ruan and G. Tian, *A mathematical theory of quantum cohomology*, J. Diff. Geom. **42** (1995), 259-367.

[S-T] B. Siebert and G. Tian, *On quantum cohomology of Fano manifolds and a formula of Vafa and Intrilligator*, preprint 1994.

[W] E. Witten, *Two-dimensional gravity and intersection theory on moduli space*, Surveys in Diff. Geom. **1** (1991), 243-310; *The Verlinde algebra and the cohomology of the Grassmannian*, in *Geometry, topology, and physics*, Intern. Press: Cambridge, MA, 1995, pp 357-422.

[Z] H. Zeuthen, *Almindelige Egenskaber ved Systemer af plane Kurver*, Danske Videnskabernes Selskabs Skrifter, Naturvidenskabelig og Mathematisk, Afd. **10** Bd. IV (1873), 286-393.

Department of Mathematics, University of Chicago, Chicago, Illinois, 60637
*E-mail address*: `fulton@math.uchicago.edu`

Department of Mathematics, University of Chicago, Chicago, Illinois, 60637
*E-mail address*: `rahul@math.uchicago.edu`

Proceedings of Symposia in Pure Mathematics
Volume **62.2**, 1997

# MAPPING CLASS GROUPS AND MODULI SPACES OF CURVES

RICHARD HAIN AND EDUARD LOOIJENGA

## CONTENTS



## 1. INTRODUCTION

It is classical that there is a very strong relation between the topology of $\mathcal{M}_g$, the moduli space of smooth projective curves of genus $g$, and the structure of the mapping class group $\Gamma_g$, the group of homotopy classes of orientation preserving diffeomorphisms of a compact orientable surface of genus $g$. The geometry of $\mathcal{M}_g$, the topology of $\mathcal{M}_g$, and the structure of $\Gamma_g$ are all intimately related. Until recently, the principal tools for studying these topics were Teichmüller theory (complex analysis and hyperbolic geometry), algebraic geometry, and geometric topology. Recently, a fourth cornerstone has been added, and that is physics which enters through the theories of quantum gravity and conformal field theory. Already these new ideas have had a remarkable impact on the subject through the ideas of Witten and the work of Kontsevich. In this article, we survey some recent developments in the understanding of moduli spaces. Some of these are classical (do not use physical ideas), while others are modern. One message we would like to convey is that algebraic geometers, topologists, and physicists who work on moduli spaces of curves may have a lot to learn from each other.

Having said this, we should immediately point out that, partly due to our own limitations, there are important developments that we have not included in this survey. Our most notable omission is the arithmetic aspect of the theory, much of which originates in Grothendieck's fundamental works [19], [20]. We direct readers

1991 *Mathematics Subject Classification.* Primary 14-02, 14H10, 32G15; Secondary 20F34.
First author supported in part by grants from the National Science Foundation.

to the volume [21] and to the recent papers of Ihara, Nakamura and Oda for other recent developments (see Nakamura's survey [71] for references). Other topics we have not covered include conformal field theory and recent work of Ivanov [38] and Ivanov and McCarthy [39] on homomorphisms from mapping class groups and arithmetic groups to mapping class groups. Of particular importance is Ivanov's version of Margulis rigidity for mapping class groups [38] which he obtains using some recent fundamental work of Kaimanovich and Masur [47] on the ergodic theory of Teichmüller space.

We shall denote the moduli space of $n$ pointed smooth projective curves of genus $g$ by $\mathcal{M}_g^n$. Knudsen, Mumford and Deligne constructed a canonical compactification $\overline{\mathcal{M}}_g^n$ of it. It is the moduli space of stable $n$ pointed projective curves of genus $g$. It is a projective variety with only finite quotient singularities. Perhaps the most important developments of the decade concern the Chow rings[1] of $\mathcal{M}_g^n$ and $\overline{\mathcal{M}}_g^n$.

The first Chern class of the relative cotangent bundle of the universal curve associated to the $i$th point is a class $\overline{\tau}_i$ in $\mathrm{CH}^1(\overline{\mathcal{M}}_g^n)$. One can consider monomials in the $\overline{\tau}_i$'s of polynomial degree equal to the dimension of some $\overline{\mathcal{M}}_g^n$. For such a monomial, one can take the degree of the monomial as a zero cycle on $\overline{\mathcal{M}}_g^n$ to obtain a rational number. These can be assembled into a generating function. Witten conjectured that this formal power series satisfies a system of partial differential operators. Kontsevich proved this using topological arguments, and thereby provided inductive formulas for these intersection numbers. These developments are surveyed in Section 6.

For each positive integer $i$, Mumford defined a *tautological class* $\overline{\kappa}_i$ in $\mathrm{CH}^i(\overline{\mathcal{M}}_g)$. The restrictions $\kappa_i$ of these classes to $\mathrm{CH}^\bullet(\mathcal{M}_g)$ generate a subalgebra of $\mathrm{CH}^\bullet(\mathcal{M}_g)$ which is called the *tautological algebra* of $\mathcal{M}_g$. Faber has conjectured that this ring has the structure of the $(p,p)$ part of the cohomology ring of a smooth complex projective variety of complex dimension $g-2$. That is, it satisfies Poincaré duality and has the "Hard Lefschetz Property" with respect to $\kappa_1$. Considerable evidence now exists for this conjecture, much of which is presented in Section 5. Other developments on the Chow ring, such as explicit computations in low genus, are also surveyed there.

In the early 80s, Harer proved that the cohomology in a given degree of $\mathcal{M}_g$ is independent of the genus once the genus is sufficiently large relative to the degree. These stable cohomology groups form a graded commutative algebra which is known to be free. The tautological classes $\kappa_i$ freely generate a polynomial algebra inside the stable cohomology ring. Mumford and others have conjectured that the stable cohomology of $\mathcal{M}_g$ is generated by the $\kappa_i$'s. Some progress has been made towards this conjecture which we survey throughout the paper. In Section 4 we consider the stabilization maps from an algebro-geometric point of view, and in Section 10 we survey Kontsevich's methods for constructing classes in the cohomology of the $\mathcal{M}_g^n$.

We have also tried to advertise the fecund work of Dennis Johnson on the Torelli groups. The Torelli group $T_g$ is the subgroup of the mapping class group $\Gamma_g$ consisting of those diffeomorphism classes that act trivially on the homology of the reference surface. This mysterious group, in some sense, measures the difference

[1]All Chow rings and cohomology groups in this paper are with $\mathbb{Q}$ coefficients except when explicit coefficients are used.

between curves and abelian varieties and appears to play a subtle role in the geometry of $\mathcal{M}_g$. Johnson proved that $T_g$ is finitely generated when $g \geq 3$ and computed its first integral homology group. These computations have direct geometric applications, especially when combined with M. Saito's work in Hodge theory — for example, they restrict the normal functions defined over $\mathcal{M}_g$ and its standard level covers. From this, one can give a computation of the Picard group of the generic curve with a level $l$ structure. Johnson's work and its applications is surveyed in Section 7.

Since $\Gamma_g$ is the orbifold fundamental group of $\mathcal{M}_g$, an algebraic variety, one should be able to apply Hodge theory and Galois theory to study its structure. In Section 9 we survey recent work on applications of Hodge theory to understanding the structure of the Torelli groups, mainly via Malcev completion. In Section 10 we combine this Hodge theory with recent results of Kawazumi and Morita to show that the cohomology of $\mathcal{M}_g$ constructed by Kontsevich using graph cohomology are, after stabilization, polynomials in the $\kappa_i$'s. Thus Hodge theory provides some evidence for Mumford's conjecture that the stable cohomology of the mapping class group is generated by the $\kappa_i$'s.

Some of the results we discuss have not yet appeared in the literature, at least not in the form in which we present them. Rather than mention all such results, we simply mention a few instances where we believe our presentation to be novel: the correspondences in Section 4.1, the rôle of the fundamental normal function for orbifold fundamental groups in Section 8.6, Theorem 9.11 and the contents of Section 10.3.

*Notation and Conventions.* All varieties will be defined over $\mathbb{C}$ unless explicitly stated to the contrary. Unless explicit coefficients are used, all (co)homology groups are with rational coefficients. We will often abbreviate *mixed Hodge structure* by MHS. The sub- or superscript *pr* on a (co)homology group will denote the primitive part in both the context of the Hard Lefschetz Theorem and in the context of Hopf algebras.

*Acknowledgements.* We would like to thank Carel Faber for his comments on part of an earlier version of this paper and Shigeyuki Morita for explaining to us some of his recent work. We also appreciate the useful comments by a referee. We gratefully acknowledge support by the AMS that enabled us to attend this conference.

## 2. Mapping Class Groups

Fix a compact connected oriented reference surface $S_g$ of genus $g$, and a sequence of distinct points $(x_0, x_1, x_2, \ldots)$ in $S_g$. Let us write $S_g^n$ for the open surface $S - \{x_1, \ldots, x_n\}$ and $\pi_g^n$ for its fundamental group $\pi_1(S_g^n, x_0)$. This group admits a presentation with generators $\alpha_{\pm 1}, \ldots, \alpha_{\pm g}, \beta_1, \ldots, \beta_n$ and relation

$$(\alpha_1, \alpha_{-1}) \cdots (\alpha_g, \alpha_{-g}) = \beta_1 \cdots \beta_n,$$

where $(x, y)$ denotes the commutator of $x$ and $y$.[2] The generators are represented by loops that do not meet outside the base point; $\beta_i$ is represented by a loop that follows an arc to a point close to $x_i$, makes a simple loop around $x_i$, and returns to the base point along the same arc.

Let $\mathrm{Diff}^+(S)_r^n$ denote the group of orientation preserving diffeomorphisms of $S$ that fix the $x_i$ for $i = 1, \ldots, n{+}r$, and are the identity on $T_{x_i}S$ for $i = n{+}1, \ldots, n{+}r$.

[2]For $n = 0$ the righthand side is to be interpreted as the unit element.

Although not really necessary at this stage, it is convenient to assume that $2g-2+n+2r>0$. In other words, we do not consider the cases where $(g,n,r)$ is $(0,0,0)$, $(0,1,0)$, $(0,0,1)$, $(0,2,0)$ or $(1,0,0)$. We will keep this assumption throughout the paper.

The *mapping class group* $\Gamma^n_{g,r}$ is defined to be the group of connected components of this group:

$$\Gamma^n_{g,r} = \pi_0 \operatorname{Diff}^+(S)^n_r.$$

We omit the decorations $n$ and $r$ when they are zero. The mapping class group $\Gamma^n_g$ acts on $\pi^n_g$ by outer automorphisms. A theorem that goes back to Baer (1928) and Nielsen (1927) [72] identifies $\Gamma_g$, via this representation, with the subgroup of $\operatorname{Out}(\pi_g)$ (of index two) that acts trivially on $H_2(\pi_g) \cong H_2(S_g)$.

When $n \geq 1$ we can consider the diagonal action of $\operatorname{Aut}(\pi^n_g)$ on $(\pi^n_g)^n$. Clearly, $\operatorname{Out}(\pi^n_g)$ acts on the set of orbits of $\pi^n_g$ (which acts by inner automorphisms on each component) in $(\pi^n_g)^n$. Now $\Gamma^n_g$ can be identified with the group of outer automorphisms of $\pi^n_g$ that preserve the image of $(\beta_1,\dots,\beta_n)$ in $\pi^n_g\backslash(\pi^n_g)^n$. If we choose $x_n$ as a base point, then a corresponding assertion holds: $\Gamma^n_g$ can be identified with a subgroup of $\operatorname{Aut}(\pi_1(S^{n-1}_g, x_n))$ that is characterized in a similar way. The evident homomorphism $\Gamma^n_g \to \Gamma^{n-1}_g$ is surjective and its kernel can be identified with $\pi_1(S^{n-1}_g, x_n)$ (acting by inner automorphisms). Ivanov and McCarthy [39] recently showed that the resulting exact sequence cannot be split.

2.1. **Generators and basic properties.** Although a lot is known about these groups they are still poorly understood. Let us quickly review some of their basic properties. Dehn proved in [10] that the mapping class groups are generated by the 'twists' that are now named after him: if $\alpha$ is a simple (unoriented) loop on $S^{n+r}_g$, then parameterize a regular neighborhood of $\alpha$ in $S^{n+r}_g$ by the cylinder $[0,1]\times S^1$ (preserving orientations) and define an automorphism of $S_g$ that on this neighborhood is given by $(t,z)\mapsto (t,e^{2\pi i t}z)$ and is the identity elsewhere. The isotopy class of this automorphism only depends on the isotopy class of $\alpha$ and is called the *Dehn twist* along $\alpha$. (Perhaps we should add that $\alpha$ is, in turn, already determined by its free homotopy class, in other words, by the associated conjugacy class in $\pi^n_g$.) The corresponding element of $\Gamma^n_{g,r}$ is the identity precisely when $\alpha$ bounds a disk in $S_g - \{x_{n+1},\dots,x_{n+r}\}$ which meets $\{x_1,\dots,x_n\}$ in at most one point. Several people have found a finite presentation for the mapping class groups. One with few generators was given by Waynryb [79]. From this presentation one sees that the mapping class groups considered here are perfect when $g\geq 3$ (a result due to Powell [75] in the undecorated case).

There is an obvious homomorphism $\Gamma_{g,n} \to \Gamma^n_g$. It is easy to see that it is surjective and that the kernel is generated by the Dehn twists around the points $x_1,\dots,x_n$. These Dehn twists generate a free abelian central subgroup of $\Gamma_{g,n}$ of rank $n$. Now recall that a central extension of a discrete group $G$ by $\mathbb{Z}$ determines an extension class in $H^2(G;\mathbb{Z})$; it has a geometric interpretation as a first Chern class. In the present case we have $n$ such classes $\tau_i \in H^2(\Gamma^n_g;\mathbb{Z})$, $i=1,\dots,n$.

Conversely, each subgroup of $H^2(G;\mathbb{Z})$ determines a central extension of $G$ by that subgroup. Harer proved that $H^2(\Gamma_{g,r};\mathbb{Z})$ is infinite cyclic if $g\geq 3$ [27], so that there is a corresponding central extension

$$0 \to \mathbb{Z} \to \widetilde{\Gamma}_{g,r} \to \Gamma_{g,r} \to 1.$$

Since $H_1(\Gamma_{g,r};\mathbb{Z})$ vanishes, this central extension is perfect (and universal). A nice presentation of it was recently given by Gervais [16]. The central extension by $\mathbb{Z}$ has a simple description which we will give in Section 3.

2.2. **Stable cohomology.** The mapping class groups $\Gamma^n_{g,r}$ turn up in a connected sum construction that we describe next. It is convenient to do this in a somewhat abstract setting. Suppose we are given a closed, oriented (but not necessarily connected) surface $S$, a finite subset $Y \subset S$, and a fixed point free involution $\iota$ of $Y$. Assume that $\iota$ has been lifted to an orientation reversing linear involution $\tilde\iota$ on the spaces of rays $\mathrm{Ray}(TS|Y)$. The *real oriented blow up* $S_Y \to S$ is a surface with boundary canonically isomorphic to $\mathrm{Ray}(TS|Y)$. So $\tilde\iota$ defines an orientation reversing involution of this boundary. Welding the boundary components of $S_Y$ by means of this involution produces a closed surface $S(\tilde\iota)$. Some care is needed to give it a differentiable structure inducing the given one on $S_Y$. Although there is no unique way to do this, all natural choices lie in the same isotopy class. If $S$ happens to have a complex structure, then each choice of a real ray $L$ in $T_pS \otimes_{\mathbb{C}} T_{\iota(p)}S$ determines a lift of $\iota$ over the pair $\{p, \iota(p)\}$: if $l$ is a ray in $T_pS$, then $\tilde\iota(l)$ is determined uniquely by the condition $l \otimes_{\mathbb{C}} \tilde\iota(l) = L$.

If $S(\tilde\iota)$ is connected, then each finite subset $X$ of $S - Y$ determines a natural homomorphism from the mapping class group which is perhaps best denoted by $\Gamma(S)^X_Y$ (a product of groups of the type $\Gamma^n_{g,r}$) to the mapping class group $\Gamma(S(\tilde\iota))^X$. The image of this homomorphism is simply the stabilizer of the simple loops indexed by $Y/\iota$ that are images of boundary components of $S_Y$. Its kernel is a free abelian group whose generators can be labeled by a system of representatives $R$ of $\iota$ orbits in $Y$. Indeed, for each element $y$ of $R$, take the composite of the Dehn twist around $y$ and the inverse of the Dehn twist around $\iota(y)$. These maps appear in the stability theorems and are at the root of the recent operad theoretic approaches to the study of the cohomology of mapping class groups.

**Theorem 2.1** (Stability theorem, Harer [29]). *There exists a positive constant $c$ with the following property. If $S(\tilde\iota)$ is connected and $S'$ is a connected component of $S$ and $X$ a finite subset of $S' \setminus Y$, then the homomorphism*

$$\Gamma(S')^X_{Y\cap S'} \to \Gamma(S(\tilde\iota))^X$$

*induces an isomorphism on integral cohomology in degree $\leq c.\mathrm{genus}(S')$.*

The constant $c$ appearing in this theorem was $1/3$ in Harer's original paper. It was later improved to $1/2$ by Ivanov in [35]. Most recently, Harer [31] has showed that we can take $c$ to be about $2/3$ and that this is the minimal possible value. There is also a version for twisted coefficients, due to Ivanov [36].

Harer's theorem says essentially that the $k$th cohomology group of $\Gamma^n_{g,r}$ depends only on $n$, provided that $g$ is large enough. These stable cohomology groups are the cohomology of a single group, namely the group $\Gamma^n_\infty$ of compactly supported mapping classes of a surface $S_\infty$ of infinite genus (with one end, say) that fix a given set of $n$ distinct points.

Among the homomorphisms defined above are maps $\Gamma_{g,1} \times \Gamma_{g',1} \to \Gamma_{g+g'}$. These stabilize and define homomorphisms of $\mathbb{Q}$ algebras

$$\mu : H^\bullet(\Gamma_\infty) \to H^\bullet(\Gamma_\infty) \otimes H^\bullet(\Gamma_\infty).$$

This defines a coproduct on $H^\bullet(\Gamma_\infty)$. Together with the cup product, this gives $H^\bullet(\Gamma_\infty)$ the structure of a connected graded-bicommutative Hopf algebra. The classification of such Hopf algebras implies that $H^\bullet(\Gamma_\infty)$ is free as a graded algebra and is generated by its set of primitive elements

$$H^\bullet_{\mathrm{pr}}(\Gamma_\infty) := \{x \in H^+(\Gamma_\infty) : \mu(x) = x \otimes 1 + 1 \otimes x\}.$$

For each $i > 0$, Mumford [69] and Morita [63] independently found a class $\kappa_i$ in $H^{2i}_{\mathrm{pr}}(\Gamma_\infty)$ (we shall recall the definition in Section 4) and Miller [61] and Morita [63] independently showed that each $\kappa_i$ is nonzero. So the $\kappa_i$'s generate a polynomial subalgebra of the stable cohomology. Mumford conjectured that they span all of $H^\bullet_{\mathrm{pr}}(\Gamma_\infty)$. This has been verified by Harer in a series of papers [27], [30], [32] in degrees $\leq 4$.[3]

The first Chern class $\tau_i \in H^2(\Gamma^n_g; \mathbb{Z})$ stabilizes also and we may think of it as an element of $H^\bullet(\Gamma^n_\infty; \mathbb{Z})$ $(i = 1, \dots, n)$. The forgetful map $\Gamma^n_\infty \to \Gamma_\infty$ gives $H^\bullet(\Gamma^n_\infty)$ the structure of a module over this Hopf algebra. From the stability theorem one can deduce:

**Theorem 2.2** (Looijenga [58]). *The algebra $H^\bullet(\Gamma^n_\infty; \mathbb{Z})$ is freely generated by the classes $\tau_1, \dots, \tau_n$ as a graded-commutative $H^\bullet(\Gamma_\infty; \mathbb{Z})$ algebra.*

## 3. Moduli Spaces

A conformal structure and an orientation on $S_g$ determine a complex structure on $S_g$. The *Teichmüller space* $\mathcal{X}^n_{g,r}$ is the space of conformal structures on $S_g$ (with some reasonable topology) up to isotopies that fix $\{x_1, \dots, x_{n+r}\}$ pointwise and act trivially on the tangent spaces $T_{x_i}S$ for $i = n+1, \dots, n+r$. It is, in a natural way, a complex manifold of dimension $3g - 3 + n + 2r$. As a real manifold it is diffeomorphic to a cell. The group $\Gamma^n_{g,r}$ acts naturally on it. This action is properly discontinuous and a subgroup of finite index acts freely.

If $\Gamma$ is any subgroup of $\Gamma^n_{g,r}$ that acts freely, then the orbit space $\Gamma\backslash\mathcal{X}^n_{g,r}$ is a classifying space for $\Gamma$ and so its singular integral cohomology coincides with $H^\bullet(\Gamma; \mathbb{Z})$. This is even true with twisted coefficients: if $V$ is a $\Gamma$ module, then the trivial sheaf over $\mathcal{X}^n_{g,r}$ with fiber $V$ comes with an obvious (diagonal) action of $\Gamma$. Passing to $\Gamma$ orbits yields a locally constant sheaf $\mathbb{V}$ on $\Gamma\backslash\mathcal{X}^n_{g,r}$. The cohomology of this sheaf equals $H^\bullet(\Gamma; V)$. For an arbitrary subgroup $\Gamma$ of $\Gamma^n_{g,r}$ these statements still hold as long as we take our coefficients to be $\mathbb{Q}$ vector spaces (but $\mathbb{V}$ need no longer be locally constant). For $\Gamma = \Gamma^n_{g,r}$, we denote the orbit space by $\mathcal{M}^n_{g,r}$.

The space $\mathcal{M}^n_{g,r}$ is, in a natural way, a normal analytic space and the obvious forgetful maps such as $\mathcal{M}^n_{g,r} \to \mathcal{M}^n_g$ are analytic. An interpretation as a coarse moduli space makes it possible to lift this analytic structure to the algebraic category. To see this, we first choose a nonzero vector in each tangent space $T_{x_i}S_g$. Each triple $(C; x, v)$, where $C$ is a connected nonsingular complex projective curve $C$ of genus $g$, $x$ an injective map $x : \{1, \dots, n+r\} \to C$, and $v$ a nowhere zero section of $TC$ over $\{n+1, \dots, n+r\}$, determines an element of $\mathcal{M}^n_{g,r}$. This point depends only on the isomorphism class of $(C, x, v)$ with respect to the obvious notion of isomorphism. Since each conformal structure on $S$ gives $S$ the structure of a nonsingular complex projective curve, $\mathcal{M}^n_{g,r}$ can be identified can be identified with the space of isomorphism classes of such triples. From the work of Knudsen, Mumford and Deligne, we know that $\mathcal{M}^n_g$ is, in a natural way, a quasi-projective orbifold. Recall that they also constructed a projective completion $\overline{\mathcal{M}}^n_g$ of $\mathcal{M}^n_g$, the

[3]He also tells us that he has checked that there are no stable primitive classes in degree 5.

*Deligne-Mumford completion* [12], that also admits the interpretation of a coarse moduli space. Its points parameterize the connected stable $n$ pointed curves $(C, x)$ of arithmetic genus $g$, where we now allow $C$ to have ordinary double points, but still require $x$ to map to the smooth part of $C$ and the automorphism group of $(C, x)$ to be finite. The *Deligne-Mumford boundary* $\overline{\mathcal{M}}_g^n - \mathcal{M}_g^n$ is a normal crossing divisor in the orbifold sense. There is a projective morphism $\overline{\mathcal{M}}_g^{n+1} \to \overline{\mathcal{M}}_g^n$, defined by forgetting the last point. It comes with $n$ sections $x_1, \dots, x_n$. The fibers of this morphism are stable $n$ pointed curves (modulo finite automorphism groups) and the morphism can be regarded as the universal stable $n$ pointed curve (in an orbifold sense). Let $\omega$ denote the relative dualizing sheaf of this morphism, considered as a line bundle in the orbifold sense. We can then think of $\mathcal{M}_{g,r}^n$ as the set of $(v_{n+1}, \dots, v_{n+r})$ in the total space of $x_{n+1}^*\omega \oplus \dots \oplus x_{n+r}^*\omega$ restricted to $\mathcal{M}_g^{n+r}$ that have each component nonzero. So $\mathcal{M}_{g,r}^n$ is also quasi-projective.

Each finite quotient group $G$ of $\Gamma_g^n$ determines, in an obvious way, a Galois cover $\mathcal{M}_g^n[G] \to \mathcal{M}_g^n$. The Deligne-Mumford completion $\overline{\mathcal{M}}_g^n[G]$ of this cover is, by definition, the normalization of $\overline{\mathcal{M}}_g^n$ in $\mathcal{M}_g^n[G]$.

**Theorem 3.1** (Looijenga [55]). *There exists a finite group $G$ such that $\overline{\mathcal{M}}_g[G]$ is smooth with a normal crossing divisor as Deligne-Mumford boundary.*

This has been extended by De Jong and Pikaart [74] to arbitrary characteristic, and by Boggi and Pikaart (independently) to the $n$-pointed case. (They show that it also can be arranged that each irreducible component of the Deligne-Mumford boundary of $\overline{\mathcal{M}}_g^n[G]$ is smooth.) This makes it relatively easy to define the Chow algebra of $\overline{\mathcal{M}}_g^n$: if $\overline{\mathcal{M}}_g^n[G]$ is smooth, then define $\mathrm{CH}^\bullet(\overline{\mathcal{M}}_g^n)$ to be the $G$ invariant part of $\mathrm{CH}^\bullet(\overline{\mathcal{M}}_g^n[G])$ (we take algebraic cycles modulo rational equivalence with coefficients in $\mathbb{Q}$). It is easy to see that this is independent of the choice of $G$.

The central extension of $\Gamma_g$ by $\mathbb{Z}$ $(g \geq 3)$ discussed in Section 2.1 takes the geometric form of a complex line bundle over Teichmüller space with $\Gamma_g$ action and hence yields an orbifold line bundle over $\mathcal{M}_g$. According to Arbarello-Cornalba [3] it is the determinant bundle of the direct image of the relative dualizing sheaf of $\mathcal{M}_g^1 \to \mathcal{M}_g$ (this is a rank $g$ vector bundle). The orbifold fundamental group of the associated $\mathbb{C}^\times$ bundle is just the central extension of $\Gamma_g$ by $\mathbb{Z}$ mentioned in Section 2.1.

Since $\Gamma_g$ is perfect when $g \geq 3$, we have $H^1(\Gamma_g) = 0$. Ivanov has asked the following question:

*Question 3.2* (Ivanov). Is it true that $H^1(\Gamma)$ vanishes for all finite index subgroups $\Gamma$ of $\Gamma_g$, at least when $g$ is sufficiently large?

This would imply that the Picard group of each finite unramified cover of $\mathcal{M}_g$ (in the orbifold sense) is finitely generated. The answer to Ivanov's questions is affirmative, for example, for subgroups of finite index of $\Gamma_g$, $g \geq 3$, that contain the Torelli group — see (7.4).

## 4. Algebro-Geometric Stability

The Deligne-Mumford completion $\overline{\mathcal{M}}_g^n$ comes with a natural stratification into orbifolds, with each stratum parameterizing stable $n$ pointed curves of a fixed topological type $T$. Denote this stratum by $\mathcal{M}(T)$. It has codimension equal to the number of singular points of $T$. The normalization of the topological type $T$ is an oriented closed surface $S$ that comes with $n$ distinct numbered points $X = \{x_1, \dots, x_n\}$ and a finite subset $Y$ of $S - X$ with a fixed point free involution $\iota$, so that $T$ is recovered by identifying the points of $Y$ according to $\iota$. These topological data define a moduli space $\mathcal{M}(S)^{X \cup Y}$ of the same type (we hope that the notation is self-explanatory) and there is a natural morphism $\mathcal{M}(S)^{X\cup Y} \to \mathcal{M}(S/\iota)^X$ that is a Galois cover of orbifolds. This morphism extends to a finite surjective morphism from the Deligne-Mumford completion $\overline{\mathcal{M}}(S)^{X\cup Y}$ to the closure of $\mathcal{M}(T)$ in $\overline{\mathcal{M}}_g^n$. The resulting morphism $\overline{\mathcal{M}}(S)^{X\cup Y} \to \overline{\mathcal{M}}_g^n$ has only self-intersections of normal crossing type and so carries a normal bundle in the orbifold sense. This normal bundle is a direct sum of line bundles with one summand for each $\iota$ orbit $\{p, p'\}$, namely $p^*\omega^{-1} \otimes p'^*\omega^{-1}$. (To see this, notice that the restriction of the universal curve to $\overline{\mathcal{M}}(T)$ has a quadratic singularity along the locus defined by the pair $\{p, p'\}$. Associating to a local defining equation its hessian determines a natural isomorphism between $p^*\omega^{-1} \otimes p'^*\omega^{-1}$ and the normal bundle of a divisor in the Deligne-Mumford boundary passing through $\overline{\mathcal{M}}(T)$.) We now see before us an algebro-geometric incarnation of the map that appears in the stability theorem: the set of normal vectors that point towards the interior $\mathcal{M}_g^n$ is the restriction to $\mathcal{M}(S)$ of the total space of the direct sum of $\mathbb{C}^\times$ bundles in this normal bundle. So $\mathcal{M}(S)_Y^X$ maps to the latter space, and although we do not have a morphism $\mathcal{M}(S)_Y^X \to \mathcal{M}_g^n$, the map on cohomology behaves as if there were. In particular, the map $H^\bullet(\mathcal{M}_g^n) \to H^\bullet(\mathcal{M}(S)_Y^X)$ is a MHS morphism. So the stability theorem implies:

**Theorem 4.1** (Algebro-geometric stability). *Suppose that the finite set $X$ is contained in a connected component $S'$ of $S$ of genus $g'$, so that $\mathcal{M}(S')_{Y\cap S'}^X$ appears as a factor of $\mathcal{M}(S)_Y^X$. Choose points in the remaining factors so that we have an inclusion of $\mathcal{M}(S')_{Y\cap S'}^X$ in $\mathcal{M}(S)_Y^X$. Then for $k \le cg'$ the composite map*

$$H^k(\mathcal{M}_g^n) \to H^k(\mathcal{M}(S)_Y^X) \to H^k(\mathcal{M}(S')_{Y\cap S'}^X)$$

*is an isomorphism and so is the map*

$$H^k(\mathcal{M}_{g'}^n) \to H^k(\mathcal{M}(S')_{Y\cap S'}^X)$$

*induced by the forgetful morphism $\mathcal{M}(S')_{Y\cap S'}^X \to \mathcal{M}(S')^X \cong \mathcal{M}_{g'}^n$. These maps are also MHS morphisms.*

So the stable rational cohomology $H^\bullet(\Gamma_\infty^n)$ comes with a natural MHS. A geometric consequence of this result is that each stable rational cohomology class of $\mathcal{M}_g^n$ (that is, a class whose degree is in the stability range) extends across the open part of the blow up of $\overline{\mathcal{M}}(T)$ parameterizing the normal directions pointing towards the interior. Pikaart showed that these partial extensions can be made to come from a single extension to $\overline{\mathcal{M}}_g^n$, at least if $g$ is large compared with $k$. But then it is not hard to show that if this is possible for large $g$, then it is possible in the stable range and so the conclusion is:

**Theorem 4.2** (Pikaart [73]). *The restriction map $H^k(\overline{\mathcal{M}}_g^n) \to H^k(\mathcal{M}_g^n)$ is surjective in the stable range. Consequently, the MHS on $H^k(\Gamma_\infty^n)$ is pure of weight $k$.*

Mumford's Conjecture, if known, would imply this result, and so Pikaart's Theorem is evidence for the truth of this conjecture.

We illustrate this theorem with the known stable classes. We have seen in the previous section that $\overline{\mathcal{M}}_g^n$ comes with $n$ orbifold line bundles $x_i^*\omega$, $i = 1, \ldots, n$. Let $\overline{\tau}_{n,i}$ denote the first Chern class of this line bundle, regarded as an element of $\mathrm{CH}^1(\overline{\mathcal{M}}_g^n)$. The restriction of this class to $\mathrm{CH}^1(\mathcal{M}_g^n)$ is a pull-back of the restriction of $\overline{\tau}_{n-1,i}$ to $\mathcal{M}_g^{n-1}$ (when $n \geq 1$) and so we denote that restriction simply by $\tau_i$. The underlying cohomology class of $\tau_i$ in $H^{2i}(\mathcal{M}_g^n) \cong H^{2i}(\Gamma_g^n)$ is what we denoted earlier by that symbol, in particular, it is stable.

For the definition of the tautological classes of $\mathcal{M}_g^n$, we shall not use Mumford's original definition, but a modification proposed by Arbarello-Cornalba. This might begin with the observation that the "functor" which associates to an $(n+1)$-pointed stable genus $g$ curve $(C; x_1, \ldots, x_n, x)$ the cotangent space $T_x^*C$ defines an orbifold line bundle over $\overline{\mathcal{M}}_g^{n+1}$. It is not quite the same as the relative dualizing sheaf $\omega$ of the forgetful map $\overline{\mathcal{M}}_g^{n+1} \to \overline{\mathcal{M}}_g^n$: a little computation shows that it is in fact $\omega(\sum_{i=1}^n (x_i))$. This is perhaps a more natural bundle to consider than $\omega$. In any case, we denote the direct image of $c_1(\omega(\sum_{i=1}^n (x_i))^{i+1} \in \mathrm{CH}^{i+1}(\overline{\mathcal{M}}_g^{n+1})$ under the projection $\overline{\mathcal{M}}_g^{n+1} \to \overline{\mathcal{M}}_g^n$ by $\overline{\kappa}_{n,i} \in \mathrm{CH}^i(\mathcal{M}_g^n)$ and its restriction to $\mathcal{M}_g^n$ by $\kappa_{n,i}$. The cohomology class underlying $\kappa_{n,i}$ can be regarded as an element of $H^{2i}(\Gamma_g^n)$ (of Hodge bidegree $(i,i)$). These cohomology classes stabilize and, for $n = 0$, they define the nonzero primitive elements of degree $2i$ alluded to in 2.2.

We regard (for $k = 0, 1, \ldots, n$) $\mathrm{CH}^\bullet(\overline{\mathcal{M}}_g^n)$ as a $\mathrm{CH}^\bullet(\overline{\mathcal{M}}_g^k)$-algebra via the obvious forgetful morphism, and view the classes $\overline{\kappa}_{k,i}$ as elements of $\mathrm{CH}^\bullet(\overline{\mathcal{M}}_g^n)$ when appropriate. The class $\overline{\kappa}_{n,i}$ is then not equal to $\overline{\kappa}_{n-1,i}$, but according to formula (1.10) of [4] we have:

$$\overline{\kappa}_{n,i} = \overline{\kappa}_{n-1,i} + (\overline{\tau}_{n,n})^i.$$

As Arbarello-Cornalba explain, the classes $\overline{\kappa}_{n,i}$ possess a nice property not enjoyed by Mumford's classes. First recall that every stratum of $\overline{\mathcal{M}}_g^n$ is the image of a finite map $\overline{\mathcal{M}}(S)^{X \cup Y} \to \overline{\mathcal{M}}_g^n$ and that $\overline{\mathcal{M}}(S)^{X \cup Y}$ is a product of varieties of the type $\overline{\mathcal{M}}_{g_\alpha}^{n_\alpha}$. The pull-back of $\overline{\kappa}_{n,i}$ along this map is the sum of the classes $\overline{\kappa}_{n_\alpha,i}$ (pulled back along the projection $\overline{\mathcal{M}}(S)^{X \cup Y} \to \overline{\mathcal{M}}_{g_\alpha}^{n_\alpha}$). Carel Faber pointed out to us that a similar property is enjoyed by the divisor class of the Deligne-Mumford boundary, but we know of no other examples. Since this behaviour is reminiscent of that of a primitive element in a Hopf algebra under the coproduct, we ask:

*Question 4.3.* What other collections $\left\{\mu_{g,n} \in \mathrm{CH}^k(\mathcal{M}_g^n)\right\}_{g,n}$ have this property?

4.1. **Correspondences between moduli spaces.** There is an altogether different way to relate the cohomology of the moduli spaces $\mathcal{M}_g^n$ for different values of $g$. This involves certain Hecke type correspondences. For simplicity we shall restrict ourselves to the undecorated case $n = 0$. We return to the reference surface $S_g$ and suppose that we are given a subgroup $\pi$ of $\pi_g$ of finite index $d$, say. (For what follows only its conjugacy class will matter.) This subgroup determines an unramified finite covering $\tilde{S} \to S$ of closed oriented surfaces. The genus $\tilde{g}$ of $\tilde{S}$ is then equal to $d(g-1)+1$. Consider the group of pairs $(\tilde{h}, h) \in \mathrm{Diff}^+(\tilde{S}) \times \mathrm{Diff}^+(S)$ such that $\tilde{h}$ is a lift of $h$. Let $\Gamma_g(\pi)$ be its group of connected components. The projection $\Gamma_g(\pi) \to \Gamma_g$ has as kernel the group of covering transformations of $\tilde{S} \to S$ (so is finite) and its image consists of the outer automorphisms of $\pi_g$ that come from an automorphism which preserves the subgroup $\pi$ (so is of finite index, $e$, say). There is a corresponding finite covering of moduli spaces $p_1 : \mathcal{M}_g(\pi) \to \mathcal{M}_g$, where $\mathcal{M}_g(\pi)$ is simply the coarse moduli space of finite unramified coverings of nonsingular complex projective curves $\tilde{C} \to C$ topologically equivalent to $\tilde{S} \to S$. There is also a finite map $p_2 : \mathcal{M}_g(\pi) \to \mathcal{M}_{\tilde{g}}$. Together they define a one-to-finite correspondence $p_2 p_1^{-1}$ from $\mathcal{M}_g$ to $\mathcal{M}_{\tilde{g}}$. This extends over the Deligne-Mumford compactifications: if $p_1 : \overline{\mathcal{M}}_g(\pi) \to \overline{\mathcal{M}}_g$ denotes the normalization of $\overline{\mathcal{M}}_g$ in $\mathcal{M}_g(\pi)$, then $p_2$ extends to a finite morphism $p_2 : \overline{\mathcal{M}}_g(\pi) \to \overline{\mathcal{M}}_{\tilde{g}}$. We have an induced map

$$T_\pi := e^{-1} p_{1*} p_2^* : \mathrm{CH}^\bullet(\overline{\mathcal{M}}_{\tilde{g}}) \to \mathrm{CH}^\bullet(\overline{\mathcal{M}}_g)$$

and likewise on cohomology. A computation shows that any monomial in the tautological classes is an "eigen class" for such correspondences:

**Proposition 4.4.** *The map $T_\pi$ sends $\overline{\kappa}_{i_1}\overline{\kappa}_{i_2}\cdots\overline{\kappa}_{i_r}$ to $d^r\overline{\kappa}_{i_1}\overline{\kappa}_{i_2}\cdots\overline{\kappa}_{i_r}$.*

This proposition suggests the consideration, for given positive integers $r$ and $s$, of sequences of classes $(x_g \in \mathrm{CH}^s(\overline{\mathcal{M}}_g))_{g\geq 2}$ of fixed degree that have the property that $T_\pi(x_g) = d^r x_{(g-1)d+1}$ for each index $d$ subgroup $\pi$ of $\pi_g$.

*Question 4.5.* Is for such a system the image of $x_g$ in $H^\bullet(\mathcal{M}_g)$ stable? Is it in fact a polynomial of degree $r$ in primitive stable classes?

An affirmative answer would give us a notion of stability for the Chow groups of the moduli spaces $\mathcal{M}_g$.

## 5. Chow Algebras and the Tautological Classes

We have already encountered some of the basic classes on $\overline{\mathcal{M}}_g^n$: the first Chern classes $\overline{\tau}_i \in \mathrm{CH}^1(\overline{\mathcal{M}}_g^n)$ $(i = 1, \dots, n)$ and the tautological classes $\overline{\kappa}_i$ $(i = 1, 2, \dots)$. More such classes come from the boundary: if $\prod_i \overline{\mathcal{M}}_{g_i}^{n_i} \to \overline{\mathcal{M}}_g$ is a Galois covering of a stratum of the boundary as in Section 4, then we can add to these the push-forwards along this map of the exterior products of the corresponding classes on the factors. Let us call the subalgebra of $\mathrm{CH}^\bullet(\overline{\mathcal{M}}_g^n)$ generated by all these classes the *tautological subalgebra* and denote it by $\mathcal{R}^\bullet(\overline{\mathcal{M}}_g^n)$. The image of this algebra in $\mathrm{CH}^\bullet(\mathcal{M}_g^n)$ is denoted by $\mathcal{R}^\bullet(\mathcal{M}_g^n)$; it is generated by the classes $\kappa_{n,i}$ $(i = 1, 2, \dots)$ and $\tau_i$ $(i = 1, \dots, n)$. It is possible that these classes generate the rational Chow ring of $\overline{\mathcal{M}}_g^n$ modulo homological equivalence, but this is of course unknown. In any case, these subalgebras are preserved under pull-back and push-forward along the natural maps that we have met so far.

The first computations were done by Mumford [69] who found a presentation of $\mathrm{CH}^\bullet(\overline{\mathcal{M}}_2)$. Subsequently Faber [13] calculated $\mathrm{CH}^\bullet(\overline{\mathcal{M}}_2^1)$, $\mathrm{CH}^\bullet(\overline{\mathcal{M}}_3)$ and obtained partial results on $\mathrm{CH}^\bullet(\overline{\mathcal{M}}_4)$. In all these cases the tautological algebra is the whole Chow algebra.

This is also the case for $\overline{\mathcal{M}}_0^n$, whose Chow algebra was computed by Keel. This is a very remarkable algebra which appears in other contexts. Because of this, we describe it explicitly. We first introduce notation for the divisor classes on

$\overline{\mathcal{M}}_0^n$. The boundary divisor $\overline{\mathcal{M}}_0^n - \mathcal{M}_0^n$ parameterizes all singular stable $n$ pointed rational curves. Its components correspond to the topological types of $n$ pointed stable rational curves with exactly one singular point. Such curves have exactly two irreducible components. By collecting the points $x_i$ lying on the same component, we obtain a partition $P$ of $\{1,\dots,n\}$ into two subsets. The stability property implies that both members of $P$ have at least two elements. We denote the corresponding class in $\mathrm{CH}^1(\overline{\mathcal{M}}_0^n)$ by $D(P)$.

**Theorem 5.1** (Keel [49]). *The Chow algebra* $\mathrm{CH}^\bullet(\overline{\mathcal{M}}_0^n)$ *coincides with* $H^\bullet(\overline{\mathcal{M}}_0^n)$ *and, as a* $\mathbb{Q}$ *algebra, is generated by the* $D(P)$*'s subject to the following relations:*

(i) *If* $\{i,j,k\}$ *are distinct integers in* $\{1,\dots,n\}$*, then the sum of the* $D(P)$*'s for which* $P$ *separates* $i$ *from* $\{j,k\}$ *is independent of* $j$ *and* $k$ *(and equals* $\overline{\tau}_i$*).*
(ii) $D(P)\cdot D(P')=0$ *if* $P$ *and* $P'$ *are independent in the sense that the partition they generate has four nonempty members.*

The relations (ii) are geometrically obvious since the divisors $D(P)$ and $D(P')$ do not meet if $P$ and $P'$ are independent. The additive relations (i) are not difficult to see either: if $C$ is a stable $n$ pointed rational curve, then a moment of thought shows that there is a unique morphism $z: C\to\mathbb{P}^1$ that is an isomorphism on one irreducible component, constant on the other irreducible components, and is such that $z(x_i)=1$, $z(x_j)=0$ and $z(x_k)=\infty$. The differential $z^{-1}dz$ restricted to $x_i$ defines a section of $x_i^*\omega$. The image of $z^{-1}dz$ in $T^*_{x_i}C$ vanishes precisely when $z$ collapses the irreducible component containing $x_i$.

In [59], Manin derives a formula for the Poincaré polynomial of $\overline{\mathcal{M}}_0^n$. Such a formula was independently found by Getzler [17] who also obtained the $\mathcal{S}_n$ equivariant Poincaré polynomial of $H^\bullet(\overline{\mathcal{M}}_0^n)$. That is, he determined the character of the $\mathcal{S}_n$ representations $H^k(\overline{\mathcal{M}}_0^n)$, $k\geq 0$. Kaufmann [53] recently gave a formula for the intersection number of classes of strata of complementary dimension.

We now turn to the Chow and cohomology algebras of the moduli spaces $\mathcal{M}_g$. First we list some results about the Chow algebras.

$\mathrm{CH}^\bullet(\mathcal{M}_1^n)=\mathbb{Q}$ for $n=1,2$ (folklore)
$\mathrm{CH}^\bullet(\mathcal{M}_2)=\mathbb{Q}$ (folklore),
$\mathrm{CH}^\bullet(\mathcal{M}_2^1)=\mathbb{Q}[\tau]/(\tau^2)$ (Mumford [69]),
$\mathrm{CH}^\bullet(\mathcal{M}_3)=\mathbb{Q}[\kappa_1]/(\kappa_1^2)$ (Faber[13]),
$\mathrm{CH}^\bullet(\mathcal{M}_3^1)=\mathbb{Q}[\kappa_1,\tau]/(\kappa_1^2,4\tau^2-\tau\kappa_1)$ (Faber[13]),
$\mathrm{CH}^\bullet(\mathcal{M}_4)=\mathbb{Q}[\kappa_1]/(\kappa_1^3)$ (Faber [13]),
$\mathrm{CH}^\bullet(\mathcal{M}_5)=\mathbb{Q}[\kappa_1]/(\kappa_1^4)$ (Izadi [40] combined with Faber [14]).

The reason that such computations can be made is that, when $g$ and $n$ are both small, the moduli space $\mathcal{M}_g^n$ has a concrete description. For example, when $g=2$, each curve is hyperelliptic and therefore given by configuration of 6 points on the projective line. In the case $g=3$ a nonhyperelliptic curve is realized by its canonical system as a quartic curve in $\mathbb{P}^2$. The double cover of the projective plane along this curve is a Del Pezzo surface of degree 2, i.e., is obtained by blowing up 7 points in the plane in general position. General curves of genus 4 and 5 can be described as complete intersections of multidegrees $(2,3)$ (in $\mathbb{P}^3$) and $(2,2,2)$ (in $\mathbb{P}^4$), respectively.

5.1. **The tautological algebra of $\mathcal{M}_g$ and Faber's Conjecture.** On the basis of numerous calculations, Faber, around 1993, made the following conjecture.

**Conjecture 5.2** (Faber[1]). *The tautological algebra $\mathcal{R}^\bullet(\mathcal{M}_g)$ is a graded Frobenius algebra with socle in degree $g-2$. That is, $\dim \mathcal{R}^{g-2}(\mathcal{M}_g) = 1$, and the intersection product defines a nondegenerate bilinear form $\mathcal{R}^i(\mathcal{M}_g) \times \mathcal{R}^{g-2-i}(\mathcal{M}_g) \to \mathcal{R}^{g-2}(\mathcal{M}_g)$ $(i = 0, \dots, g-2)$. Moreover, $\kappa_1$ has the Lefschetz property in $\mathcal{R}^\bullet(\mathcal{M}_g)$ in the sense that multiplication by $(\kappa_1)^{g-2-2i}$ maps $\mathcal{R}^i(\mathcal{M}_g)$ isomorphically onto $\mathcal{R}^{g-2-i}(\mathcal{M}_g)$ for $0 \le i \le (g-2)/2$.*

Since the conjecture was made, evidence for it has been growing. For example:

**Theorem 5.3** (Looijenga [57]). *The algebra $\mathcal{R}^\bullet(\mathcal{M}_g)$ is trivial in degree $> g-2$ and $\mathcal{R}^{g-2}(\mathcal{M}_g)$ is generated by the class of the hyperelliptic locus (a closed irreducible variety of codimension $g-2$).*

In particular $\kappa_1^{g-1} = 0$. Since $\kappa_1$ is ample on $\mathcal{M}_g$, we recover a theorem of Diaz [11] which asserts that every complete subvariety of $\mathcal{M}_g$ must be of dim $\le g-2$.

Actually, in [57] a stronger result is proven, which, among other things, implies that $\mathcal{R}^k(\mathcal{M}_g^n) = 0$ for $k > g-2+n$. An induction argument then shows that $\mathcal{R}^{3g-3+n}(\overline{\mathcal{M}}_g^n)$ is spanned by the classes of the zero dimensional strata. But zero dimensional strata can be connected by one dimensional strata and the one dimensional strata are all rational. This shows that $\mathcal{R}^{3g-3+n}(\overline{\mathcal{M}}_g^n) \cong \mathbb{Q}$.

Faber recently proved that the tautological class $\kappa_{g-2}$ is nonzero. To describe his result, we find it convenient to introduce a compactly supported version of the tautological algebra: let $\mathcal{R}_c^\bullet(\mathcal{M}_g^n)$ be defined as the set of elements in $\mathcal{R}^\bullet(\overline{\mathcal{M}}_g^n)$ that restrict trivially to the Deligne-Mumford boundary. This is a graded ideal in $\mathcal{R}^\bullet(\overline{\mathcal{M}}_g^n)$ and the intersection product defines a map

$$\mathcal{R}^\bullet(\mathcal{M}_g^n) \times \mathcal{R}_c^\bullet(\mathcal{M}_g^n) \to \mathcal{R}_c^\bullet(\mathcal{M}_g^n)$$

that makes $\mathcal{R}_c^\bullet(\mathcal{M}_g^n)$ a $\mathcal{R}^\bullet(\mathcal{M}_g^n)$-module. Notice that every complete subvariety of $\mathcal{M}_g$ of codimension $d$ whose class is in $\mathcal{R}^\bullet(\overline{\mathcal{M}}_g^n)$ defines a nonzero element of $\mathcal{R}_c^d(\mathcal{M}_g^n)$ (but it is by no means clear that such elements span $\mathcal{R}_c^\bullet(\mathcal{M}_g^n)$). A somewhat stronger form of the first part of Faber's Conjecture is:

**Conjecture 5.4.** *The intersection pairings*

$$\mathcal{R}^k(\mathcal{M}_g) \times \mathcal{R}_c^{3g-3-k}(\mathcal{M}_g) \to \mathcal{R}_c^{3g-3}(\mathcal{M}_g) \cong \mathbb{Q}, \quad k = 0, 1, 2, \dots$$

*are perfect (Poincaré duality) and $\mathcal{R}_c^\bullet(\mathcal{M}_g)$ is a free $\mathcal{R}^\bullet(\mathcal{M}_g)$ module of rank one.*

Faber [14] finds a compactly supported class $I_g$ in $\mathcal{R}_c^{2g-1}(\mathcal{M}_g)$ with $\kappa_{g-2} \cdot I_g \neq 0$. So $\mathcal{R}_c^\bullet(\mathcal{M}_g)$ should be the ideal generated by this element. Faber verified his conjecture for genera $\le 15$ by writing down many relations in $\mathcal{R}^\bullet(\mathcal{M}_g)$ (this evidently gives an upper bound) and using the nonvanishing of $\kappa_{g-2}$ (this gives a surprisingly strong lower bound).

A refined form of Conjecture 5.4 (which we shall not state here) also takes care of the Lefschetz property.

*Question 5.5.* Does the tautological ring of $\overline{\mathcal{M}}_g^n$ satisfy Poincaré duality? Does it have the Lefschetz property with respect to $\overline{\kappa}_{n,1}$? (It is known that $\overline{\kappa}_{n,1}$ is ample [8].)

5.2. **Cohomology of some moduli spaces.** As may be expected, even less is known about the cohomology algebras. Here is an incomplete list of special results. In genus 0 we have that the Chow algebra of $\overline{\mathcal{M}}_0^n$ maps isomorphically onto its rational cohomology algebra. The cohomology of $\mathcal{M}_0^n$ is easily computed if we start out from the observation that this space is the projective arrangement of type $A_{n-2}$. It then follows for instance, that its cohomology in degree $p$ is of type $(p,p)$. There are similar descriptions of the moduli spaces of $n$-pointed hyperelliptic curves of genus $g$ when $n = 0,1,2$ that involve arrangements of type $A$ or $D$. These again should enable us to determine their rational cohomology ring, but it seems that this hasn't been done yet. In the same spirit arrangements of various types (among them $E_6$ and $E_7$) were used in [54] to prove that

$$H^\bullet(\mathcal{M}_3) = \mathrm{CH}^\bullet(\mathcal{M}_3) + \mathbb{Q}u,$$

where $u$ is a class of degree 6 of Hodge bidegree $(6,6)$ and

$$H^\bullet(\mathcal{M}_3^1) = \mathrm{CH}^\bullet(\mathcal{M}_3^1) + \mathbb{Q}u + \mathbb{Q}u\tau + \mathbb{Q}u\kappa_1 + \mathbb{Q}v,$$

where $v$ is a class of degree 7 and of Hodge bidegree $(6,6)$.

*Question 5.6.* The image of the tautological algebra in $H^{2p}(\mathcal{M}_g^n)$ consists of classes of type $(p,p)$. Are all such classes of this form?

A version of the Hodge conjecture asserts that the rational classes in degree $2p$ of type $(p,p)$ are in the image of the Chow algebra, so modulo this conjecture we are asking whether every Chow class on $\mathcal{M}_g^n$ is homologically equivalent to a tautological class.

## 6. The Ribbon Graph Picture

Around 1981 Thurston, Mumford and Harer observed that partial completions of the Teichmüller spaces $\mathcal{X}_g^n$ with $n > 0$ possess two natural $\Gamma_g^n$ equivariant triangulations. One is based on the hyperbolic geometry of $S_g^n$ (Thurston) and the other based on the singular euclidean geometry of $S_g^n$ (Mumford, Harer). The last approach was actually a direct, but very powerful application of work that Jenkins and Strebel had done 10–20 years earlier. It is this approach that we shall explain.

The basic notion is that of a *ribbon graph*. This is a finite graph[4] $G$ together with a cyclic order on the set of oriented edges[5] emanating from each vertex. As we shall see, there is a canonical construction of a surface that contains $G$ and of which $G$ is a deformation retract. This construction should explain the name. We first give a somewhat more abstract characterization of ribbon graphs which is very useful in some applications.

Let $X(G)$ be the set of oriented edges of $G$. Let $\sigma_1$ be the involution of $X(G)$ that reverses the orientation of each edge. The set $X_1(G)$ of $\sigma_1$ orbits can be identified with the set of edges of $G$. The cyclic orderings define another permutation $\sigma_0$ of $X(G)$ as follows. Each oriented edge $e$ has an initial vertex $\mathrm{in}(e)$ and a terminal vertex $\mathrm{term}(e)$. Define $\sigma_0(e)$ to be the successor of $e$ with respect to the given cyclic order on the set of oriented edges that have $\mathrm{in}(e)$ as their initial vertex. The set of orbits $X_0(G)$ of $\sigma_0$ can be identified with the set of vertices of $G$. Put $\sigma_\infty := (\sigma_1\sigma_0)^{-1} = \sigma_0^{-1}\sigma_1$. Call an orbit of this permutation a *boundary cycle*.

[4]For us a graph is a cell complex of pure dimension one; its zero cells are called *vertices* and its one cells *edges*. So it has no isolated vertices.

[5]An oriented edge of a graph is an edge together with an orientation of it.

(Draw a picture to see why.) The set of boundary cycles wil be denoted $X_\infty(G)$. These data form a complete invariant of $G$, for we can reverse the construction and associate to a finite nonempty set $X$ endowed with a fixed point free involution $\sigma_1$ and a permutation $\sigma_0$ of $X$, a ribbon graph $G(X, \sigma_0, \sigma_1)$ whose oriented edges are indexed by $X$ and such that $\sigma_0$ and $\sigma_1$ are the permutations defined above.

For every oriented edge $e$ of $G$ we form the one point compactification $\Delta_e$ of the half strip $e \times [0, \infty)$; this is just a 2-simplex, parameterized in an unusual way. We make identifications along the boundaries of these simplices with the help of $\sigma_0$ and $\sigma_\infty$: $e \times \{0\}$ is identified with $\sigma_1 e \times \{0\}$ and and $\{\mathrm{term}(e)\} \times [0, \infty)$ with $\{\mathrm{in}(\sigma_\infty e)\} \times [0, \infty)$ (in either case, the identification map is essentially the identity). This is easily seen to be a compact, triangulated surface $S(G)$ that contains $G$ as a subcomplex. Its vertex set can be identified with the disjoint union of the vertex set of $G$ (so $X_0(G)$) and $X_\infty(G)$. We call vertices of the latter type *cusps*. Notice that $G$ is a deformation retract of $S(G) - X_\infty(G)$ and that the surface is canonically oriented if we insist that the cyclic orderings of the edges emanating from each vertex are induced by the orientation. Let us say that the ribbon graph $G$ is $n$-*pointed* if we are given a injection $y : \{1, \dots, n\} \hookrightarrow X_\infty(G) \cup X_0(G)$ whose image contains $X_\infty(G)$ and the vertices of valency $\leq 2$.

Suppose that we are given a *metric* $l$ on $G$. That is, a function that assigns to every (unoriented) edge of $X$ a positive real number. Give $[0, \infty)$ the standard metric and every half strip $e \times [0, \infty)$ the product metric. This defines (at least locally) a metric on $S(G)$. This metric is euclidean except possibly at the vertices. However, it is not difficult to show that the underlying conformal structure extends across all the vertices of $S(G)$ so that we end up with a compact Riemann surface $C(G, l)$. Notice that each cusp has a "circumference" — this is the length of the associated boundary cycle. It is clear that we get the same complex structure if $l$ is replaced by a positive multiple of it and so we may just as well assume that the total length of $G$ is 1. With this convention, the sum of the circumferences of the cusps is 2. The work of Jenkins and Strebel shows that all compact Riemann surfaces arise in this way:

**Theorem 6.1** (Strebel [78]). *Let $(C; x : \{1, \dots, n\} \hookrightarrow C)$ be an $n$-pointed connected Riemann surface (so that the complement of the image of $x$ has negative Euler characteristic as usual) and let $c_1, \dots, c_n$ be nonnegative real numbers, not all zero. Then there exists an $n$-pointed metrized ribbon graph $(G, y, l)$, with $y(i)$ a cusp of $G$ of circumference $c_i$ when $c_i > 0$ and a vertex of $G$ otherwise, such that $(C(G, l), y)$ and $(C, x)$ are isomorphic as $n$-pointed Riemann surfaces. Moreover, $(G, y, l)$ is unique up to the obvious notion of isomorphism.*

The results of Strebel also include a continuity property: a continuous variation of the complex structure on $C$ corresponds to a continuous variation of $(G, y, l)$ in a sense that we make precise. Denote by $\mathcal{RG}_g^n$ the set of isomorphism classes of $n$-pointed ribbon graphs $(G, y)$ that are marked in the sense that we are given an isotopy class of homeomorphisms $h : S_g \to S(G)$ with $h(x_i) = y(i)$, $i = 1, \dots, n$. On this set $\Gamma_g^n$ acts, and it is easy to see that the number of orbits of markings is finite.

Suppose that $(G, y, [h])$ represents an element of $\mathcal{RG}_g^n$. Denote the geometric realization of the abstract simplex on the set $X_1(G)$ by $\Delta(G)$. Notice that the metrics $l$ on $G$ that give $G$ unit length are parameterized by the interior of $\Delta(G)$. The circumferences of the cusps add up to two, so half the circumferences are the barycentric coordinates of a simplicial projection $\lambda : \Delta(G) \to \Delta^{n-1}$. Let $s$ be an edge of $G$ that is not a loop and does not connect two vertices in the image of $y$. Then collapsing that edge yields a member $(G/s, y/s, [h]/s)$ of $\mathcal{RG}_g^n$. We can regard $\Delta(G/s)$ as a face of $\Delta(G)$. Making these identifications produces a simplicial complex which we will denote by $\widehat{\mathbb{X}}_g^n$. It comes with a simplicial map $\lambda : \widehat{\mathbb{X}}_g^n \to \Delta^{n-1}$. We have a simplicial action of $\Gamma_g^n$ on $\widehat{\mathbb{X}}_g^n$ which preserves the fibers of $\lambda$. The union of relative interiors of simplices of $\widehat{\mathbb{X}}_g^n$ indexed by the elements of $\mathcal{RG}_g^n$ is an open subset $\mathbb{X}_g^n$ of $\widehat{\mathbb{X}}_g^n$. The results of Strebel can be strengthened to:

**Proposition 6.2** (cf. [56]). *The above construction defines a $\Gamma_g^n$ equivariant homeomorphism of $\mathbb{X}_g^n$ onto $\mathcal{X}_g^n \times \Delta^{n-1}$.*

Now consider the quotient space

$$\widehat{\mathbb{M}}_g^n := \Gamma_g^n \backslash \widehat{\mathbb{X}}_g^n.$$

This is a finite simplicial orbicomplex that is equipped with a simplicial map $\lambda : \widehat{\mathbb{M}}_g^n \to \Delta^{n-1}$. We regard this complex as a compactification of its open subset $\mathbb{M}_g^n := \Gamma_g^n \backslash \mathbb{X}_g^n$. According to the above theorem, the latter is canonically homeomorphic with $\mathcal{M}_g^n \times \Delta^{n-1}$. This raises the question of how this compactification compares to that of Deligne-Mumford. The answer is essentially due to Kontsevich:

**Theorem 6.3** (Kontsevich [50], see also [56]). *The simplicial orbicomplex $\widehat{\mathbb{M}}_g^n$ is a quotient space of $\overline{\mathcal{M}}_g^n \times \Delta^{n-1}$. Moreover, the part of $\widehat{\mathbb{M}}_g^n$ where $\lambda_i > 0$ carries an oriented piecewise linear circle bundle (in the orbifold sense) whose pull-back to $\overline{\mathcal{M}}_g^n \times \{\lambda \in \Delta^{n-1} | \lambda_i > 0\}$ is the oriented circle bundle coming from the standard line bundle $\overline{\tau}_i$. In particular, the part of $\widehat{\mathbb{M}}_g^n$ lying over the interior of $\Delta^{n-1}$ carries the tautological cohomology classes underlying $\overline{\tau}_{n,i}$, $i = 1, \dots, n$.*

The defining equivalence relation on $\overline{\mathcal{M}}_g^n \times \Delta^{n-1}$ is a little subtle and we refer to [56] for details regarding both statement and proof.[6]

This compactification of $\mathcal{M}_g^n \times \Delta^{n-1}$ plays a crucial rôle in Kontsevich's proof of the Witten conjectures. There are however earlier applications. These include Harer's stability theorem we met before, the computation of the Euler characteristic of $\mathcal{M}_g^n$, and the proof that $\Gamma_g^n$ is a virtual duality group of dimension $4g - 4 + n$. We shall not explain the relation with stability here, but we will briefly touch on the other applications.

There is also a remarkable arithmetic aspect of ribbon graphs that is presently under intense investigation, but which we merely mention in passing. This is the observation, made by Grothendieck in a research proposal [20], that for a metrized ribbon graph $(G, l)$ all of whose edges have equal length, the corresponding Riemann surface $C(G, l)$ is, in a canonical way, a ramified covering of the Riemann sphere $\mathbb{P}^1$ with ramification locus contained in $\{0, 1, \infty\}$. The graph $G$ appears here as the preimage of the interval $[0, 1]$, its vertex set as the preimage of 0 and the set of cusps as the preimage of $\infty$. The preimage of 1 consists of the midpoints of the edges. At these points we have simple ramification. A covering of this type is naturally an algebraic curve defined over some number field. Conversely, every connected covering of the Riemann sphere of this type arises in this manner. The

[6] The space used by Kontsevich is not quite $\widehat{\mathbb{M}}_g^n$, but basically the part lying over the interior of $\Delta^{n-1}$ times a half line. In this case the circumference map $\lambda$ has image $(0, \infty)^n$.

absolute Galois group of $\mathbb{Q}$ acts on the collection of isomorphism types of such coverings, and thus also on each finite set $\mathcal{RG}_g^n$. It is very difficult to come to grips with this action. For more information we refer to the collection [21] and to Grothendieck's manuscripts [19] and [20]. Since these metrized ribbon graphs represent the barycenters of the simplices of $\mathbb{M}_g^n$, one can also think of this as an action of the absolute Galois group on the simplices of $\mathbb{M}_g^n$, but the significance of this is not clear to us.

6.1. **Virtual duality and virtual Euler characteristic.** We first make some observations about simplicial complexes. Let $K$ be a simplicial complex, $L$ a subcomplex. Set $U := K - L$. Then $U$ admits a canonical deformation retraction onto the union of the closed simplices of the barycentric subdivision of $K$ that lie in $U$. This is a subcomplex, called the *spine* of $U$, whose $k$-simplices correspond to strictly increasing chains $\sigma_0 \subset \sigma_1 \subset \cdots \subset \sigma_k$ of simplices of $K$ not in $L$. Further, if $\Gamma$ is a group of automorphisms of $K$ that preserves $L$, and if

(i) $U$ is contractible,
(ii) a subgroup of $\Gamma$ of finite index acts freely on $U$,

then $\Gamma\backslash U$ is a simplicial 'orbicomplex' that is also a virtual classifying space for $\Gamma$.[7] It has $\Gamma\backslash\operatorname{spine}(U)$ as deformation retract, and so the dimension of this spine is an upper bound for the virtual homological dimension of $\Gamma$.

We apply this in the situation where $K$ is the preimage of the first vertex of $\Delta^{n-1}$ in $\widehat{\mathbb{X}}_g^n$ under $\lambda$ and $U = K \cap \mathbb{X}_g^n$. Notice that $U \cong \mathcal{X}_g^n$. The simplices meeting $U$ are indexed by the elements of $\mathcal{RG}_g^n$ with a single boundary cycle. A simple calculation shows that when $g \geq 1$, the number of edges of such a graph is at most $6g-5+2n$ and at least $2g-1+n$. For $g=0$ these numbers are $2n-5$, resp. $n-2$. So the spine of $U$ has dimension $\leq 4g-4+n$, resp. $n-3$. One can verify that this is, in fact, an equality. It follows that $U$ admits a subcomplex of this dimension as an equivariant deformation retract. Hence:

**Theorem 6.4** (Harer [28]). *If $n \geq 1$, then for every level structure, the moduli space $\mathcal{M}_g^n[G]$ contains a subcomplex of dimension $4g-4+n$ (when $g>0$) or $n-3$ (when $g=0$) as a deformation retract.*

From this he deduces a similar result for the case when $n=0$: $\mathcal{M}_g[G]$ has the homotopy type of complex of dimension $4g-5$.

*Problem 6.5.* Is there a Lefschetz type of proof of this fact? For instance, the Lefschetz property would follow if one can find an orbifold stratification of $\mathcal{M}_g^n$ with all strata affine subvarieties of codimension $\leq g$ ($n \geq 1$) or $\leq g-1$ ($n=0$). That would also show that the cohomological dimension of $\mathcal{M}_g^n$ for quasicoherent sheaves is $\leq g-1$ ($n \geq 1$) or $\leq g-2$ ($n=0$).

Let us return to the general situation considered earlier and suppose, in addition, that

(iii) $\Gamma\backslash K$ is a finite complex, and
(iv) $U$ is a simplicial manifold of dimension $d$, say.

These conditions are satisfied in the case at hand. It is then natural (and standard) to assign to each simplex of $K$ the weight that is the reciprocal of the order of its

[7] This means there is a normal subgroup $\Gamma_1 \subset \Gamma$ of finite index such that a $\Gamma/\Gamma_1$-cover of this space classifies $\Gamma_1$.

$\Gamma$ stabilizer. This weighting is constant on orbits. Wall's Euler characteristic of $\Gamma$ is simply the usual alternating sum of the number of $\Gamma$ orbits of simplices not in $L$, except that each is counted with its weight. Equivalently, it is the orbifold Euler characteristic of the quotient $\Gamma\backslash K$.

In the present case, a ribbon graph $G$ defining a member of $\mathcal{RG}_g^n$ gives a contribution $|\operatorname{Aut}(G)|^{-1}(-1)^{|X_1(G)|}$ to the virtual Euler characteristic. The computation of the resulting sum is a combinatorial problem that was first solved by Harer and Zagier. Kontsevich [50] later gave a shorter proof. The answer is:

**Theorem 6.6** (Harer-Zagier [34]). *The orbifold Euler characteristic of $\mathcal{M}_g^n$ equals*

$$(-1)^{n-1}\frac{(2g+n-3)!}{(2g-2)!}\zeta(1-2g).$$

*Here $\zeta$ denotes the Riemann zeta function.*

Harer and Zagier also find formulae for the actual Euler characteristics of $\mathcal{M}_g^1$ and $\mathcal{M}_g$. These are often negative so that there must be lot of cohomology in odd degrees.

For the discussion of virtual duality we go back to the general situation and assume that beyond the four conditions already imposed we have:

(v) $L$ has the homotopy type of a bouquet of $r$-spheres.

Then the theory of Bieri-Eckmann can be invoked in a virtual setting: if we set $D := \tilde{H}_r(L;\mathbb{Z})$ and regard $D$ as a $\Gamma$ module in an obvious way, then $H_{d-r-1}(\Gamma;D)$ is of rank one and for any $\Gamma$-module $V$ with rational coefficients the cap products

$$\cap : H^k(\Gamma;V)\otimes H_{d-r-1}(\Gamma;D)\to H_{d-r-1-k}(\Gamma;V\otimes D),\quad k=0,1,2,\dots$$

are isomorphisms. One calls $D$ the *Steinberg module* of $\Gamma$.

Harer [28] proves that in the present case hypothesis (v) is satisfied: $L$ is a subcomplex of dimension $2g-3-n$, resp. $n-4$ which is $(2g-4-n)$-connected, resp. $(n-5)$-connected when $g>0$, resp. $g=0$.

We shall call the corresponding orbifold local system $\mathbb{D}$ over $\mathcal{M}_g^n$ the *Steinberg sheaf*. The homology group $H_{4g-4+n}(\mathcal{M}_g^n;\mathbb{D})$ is of rank one. For every orbifold local system $\mathbb{V}$ of rational vector spaces on $\mathcal{M}_g^n$, cap product with a generator of this homology group defines isomorphisms

$$H^k(\mathcal{M}_g^n;\mathbb{V})\xrightarrow{\sim} H_{4g-4+n-k}(\mathcal{M}_g^n;\mathbb{V}\otimes\mathbb{D}),\quad k=0,1,2,\dots$$

when $g>0$ and $n>0$ (and similar isomorphisms in the remaining cases). In particular, taking $\mathbb{V}$ to be $\mathbb{Q}$, we see that $H_\bullet(\mathcal{M}_g^n;\mathbb{D})$ has a canonical MHS. This suggests that $\mathbb{D}$ has some Hodge theoretic significance. Unfortunately it is not of finite rank, yet we wonder:

*Question 6.7.* Is the Steinberg sheaf motivic? In particular, does it have natural completions that carry (compatible) Hodge and étale structures?

6.2. **Intersection numbers on the Deligne-Mumford completion.** The intersection numbers in question are those defined by monomials in the $\overline{\tau}_i$'s. To be precise, define for every such monomial $\overline{\tau}_1^{d_1}\dots\overline{\tau}_n^{d_n}$ (with all $d_i\geq 0$) the intersection number $\int_{\overline{\mathcal{M}}_g^n}\overline{\tau}_1^{d_1}\dots\overline{\tau}_n^{d_n}$ where $g$ is chosen in such a way that this has a possibility of being nonzero: $3g-3+n=d_1+\cdots+d_n$. A physics interpretation suggests that we should combine these numbers into the generating function

$$\sum_{n=1}^{\infty} \frac{1}{n!} \sum_{g>1-\frac{1}{2}n} \sum_{d_1+\cdots+d_n=3g-3+n} t_{d_1} \ldots t_{d_n} \int_{\overline{\mathcal{M}}_g^n} \overline{\tau}_1^{d_1} \cdots \overline{\tau}_n^{d_n}.$$

Now pass to a new set of variables $T_1, T_3, T_5, \ldots$ by setting

$$t_i = 1.3.5. \cdots (2i+1) T_{2i+1}.$$

The resulting expansion $F(T_1, T_3, T_5, \ldots)$ encodes all these intersection numbers.

Witten [80] conjectured two other characterizations of this function, both of which allow computation of its coefficients. These were proved by Kontsevich in his celebrated paper [50]. Perhaps the most useful characterization is the one which says that $F$ is killed by a Lie algebra of differential operators isomorphic to the Lie algebra of polynomial vector fields in one variable. This Lie algebra comes with a basis $(L_k)_{k\geq -1}$ corresponding to the vector fields $(z^k\partial/\partial z)_{k\geq -1}$ and Witten verified the identities $L_k(F) = 0$ for $k = -1, 0$ within the realm of algebraic geometry. However no such proof is known for $k \geq 1$. Kontsevich's strategy is to represent the classes $\overline{\tau}_1^{d_1} \cdots \overline{\tau}_n^{d_n}$ by piecewise differential forms on the ribbon graph model that can actually be integrated. This allows him to convert the intersection numbers into weighted sums over ribbon graphs. This leads to a new characterization of the generating function that is more manageable. Still a great deal of ingenuity is needed to complete the proof of Witten's Conjecture.

## 7. Torelli Groups and Moduli

In the early 80s, Dennis Johnson published a series of pioneering papers [42, 43, 44] on the Torelli groups. Although this work is in geometric topology, it has several interesting applications to algebraic geometry. Here we review some of his work.

First a remark on notation. In the remainder of the paper we will write $V_g$ for the symplectic vector space $H_1(S_g)$ and $Sp_g(\mathbb{Z})$ for the group $\mathrm{Aut}(H_1(S_g,\mathbb{Z}), \langle\ ,\ \rangle)$; this does not really clash with standard notation, since a choice of a symplectic basis of $H_1(S_g;\mathbb{Z})$ identifies this with the standard integral symplectic group of genus $g$. Likewise, $Sp_g$ will stand for the algebraic $\mathbb{Q}$-group defined by the symplectic transformations of $V_g$; so its group of $\mathbb{Q}$-points, $Sp_g(\mathbb{Q})$, is just the group of symplectic automorphisms of $V_g$.

The mapping class group $\Gamma_{g,r}^n$ acts on the homology of the reference surface $S_g$. Since each of its elements preserves the orientation of $S_g$, we have a homomorphism

(1) $$\Gamma_{g,r}^n \to Sp_g(\mathbb{Z}).$$

which is surjective. The *Torelli group* $T_{g,r}^n$ is defined to be its kernel[8] so that we have an extension

$$1 \to T_{g,r}^n \to \Gamma_{g,r}^n \to Sp_g(\mathbb{Z}) \to 1.$$

The homology groups of $T_{g,r}^n$ are therefore $Sp_g(\mathbb{Z})$ modules.

The simplest kind of element of $T_{g,r}^n$ is a Dehn twist along a simple loop in $S_g^{n+r}$ that separates $S$ into two connected components. We call such a loop a *separating simple loop*. Another type of element of $T_{g,r}^n$ is determined by a *separating pair of simple loops*. This is a pair of two disjoint nonisotopic loops $\alpha_1, \alpha_2$ on $S_g^{n+r}$ that together separate $S$ into two connected components. The Dehn twist along

[8]Note that there is no general agreement on the definition of $T_{g,r}^n$ when $r + n > 1$.

$\alpha_1$ composed with the inverse of the Dehn twist along $\alpha_2$ is in $T_{g,r}^n$. The first of Johnson's results is:

**Theorem 7.1** (Johnson [42, 43, 44]). *When $g \geq 3$, $T_{g,r}^n$ is generated by elements associated to a finite number of separating simple loops and a finite number of separating pairs of simple loops. If $[S_g] \in \wedge^2 H_1(S_g;\mathbb{Z})$ corresponds to the fundamental class of $S_g$, then there are natural $Sp_g(\mathbb{Z})$ equivariant surjective homomorphisms*

$$\tau_g^1 : T_g^1 \to \wedge^3 H_1(S_g;\mathbb{Z}) \text{ and } \tau_g : T_g \to \wedge^3 H_1(S_g;\mathbb{Z})/([S_g] \wedge H_1(S_g;\mathbb{Z})).$$

*In both cases, the kernel of $\tau$ is the subgroup generated by the elements associated to simple separating loops. Finally, the kernels of the induced homomorphisms*

$$H_1(T_g^1;\mathbb{Z}) \to \wedge^3 H_1(S_g;\mathbb{Z}) \text{ and } H_1(T_g^1;\mathbb{Z}) \to \wedge^3 H_1(S_g;\mathbb{Z})/([S_g] \wedge H_1(S_g;\mathbb{Z}))$$

*are both 2-torsion.*

Johnson also finds an explicit description of this 2-torsion. We will give it in a moment, but first we want to point out an algebro-geometric consequence of this theorem. Let $\widetilde{\mathcal{M}}_g \subset \overline{\mathcal{M}}_g$ be the complement of the irreducible divisor whose generic point parametrizes irreducible singular stable curves, and let $\widetilde{\mathcal{M}}_g^1$ be its preimage in $\overline{\mathcal{M}}_g^1$.

**Corollary 7.2.** *When $g \geq 3$, the orbifold fundamental group of $\widetilde{\mathcal{M}}_g$ (resp. $\widetilde{\mathcal{M}}_g^1$) is isomorphic to an extension of $Sp_g(\mathbb{Z})$ by $\wedge^3 H_1(S_g;\mathbb{Z})/([S_g] \wedge H_1(S_g;\mathbb{Z}))$ (resp. $\wedge^3 H_1(S_g;\mathbb{Z})$).*

Johnson's theorem shows that the $Sp_g(\mathbb{Z})$ action on $H_1(T_{g,r}^n)$ is the restriction of a representation of the algebraic group $Sp_g$. We shall see shortly the importance of this property. Let $\lambda_1, \lambda_2, \ldots, \lambda_g$ be a fundamental set of weights of $Sp_g$ so that $\lambda_j$ corresponds to the $j$th fundamental representation of $Sp_g$. This last representation can be realized as the natural $Sp_g$ action on the primitive part of $\wedge^j V_g$.

The next result follows from Johnson's Theorem by standard arguments.

**Corollary 7.3.** *For each $g \geq 3$, there is a natural $Sp_g(\mathbb{Z})$ equivariant isomorphism*

$$\tau_{g,r}^n : H^1(T_{g,r}^n) \xrightarrow{\sim} V(\lambda_3) \oplus V(\lambda_1)^{\oplus(r+n)}.$$

A theorem of Ragunathan [76] implies that when $g \geq 2$, the first cohomology of each finite index subgroup of $Sp_g(\mathbb{Z})$ with coefficients in a rational representation of $Sp_g(\mathbb{Q})$ vanishes. So Johnson's computation also gives:

**Corollary 7.4.** *If $g \geq 3$, then every finite index subgroup of $\Gamma_{g,r}^n$ that contains $T_{g,r}^n$ has zero first Betti number.*

The situation is very different when $g < 3$. The Torelli groups $T_1$ and $T_1^1$ are trivial, while Geoff Mess [60] proved that when $g = 2$, $T_2$ is a countably generated free group. He also computed $H_1(T_2;\mathbb{Z})$. It is the $Sp_2(\mathbb{Z})$ module obtained by inducing the trivial representation up to $Sp_2(\mathbb{Z})$ from the stabilizer $(\mathbb{Z}/2) \ltimes (SL_2(\mathbb{Z}) \times SL_2(\mathbb{Z}))$ of a decomposition of $H_1(S_2;\mathbb{Z})$ into two symplectic modules each of rank 2. (We shall sketch a proof in the next subsection.) It is still unknown whether, for any $g \geq 3$, $T_g$ is finitely presented.

*Problem 7.5.* Determine whether $T_g$ is finitely presented when $g$ is sufficiently large.

Next, we describe Johnson's computation of the torsion in $H_1(T_g;\mathbb{Z})$. Denote the field of two elements by $\mathbb{F}_2$. Recall that an $\mathbb{F}_2$ quadratic form on $H_1(S_g;\mathbb{F}_2)$ associated to the mod two symplectic form $\langle\ ,\ \rangle$ on $H_1(S_g;\mathbb{F}_2)$ is a function $\omega : H_1(S_g;\mathbb{F}_2)\to\mathbb{F}_2$ satisfying

$$\omega(a+b)=\omega(a)+\omega(b)+\langle a,b\rangle.$$

The difference between any two such is an element of $H^1(S_g;\mathbb{F}_2)$. This makes the set $\Omega_g$ of such quadratic forms an affine space over the $\mathbb{F}_2$ vector space $H^1(S_g;\mathbb{F}_2)$. Denote the algebra of $\mathbb{F}_2$ valued functions on $\Omega_g$ by $S\,\Omega_g$. All such functions are given by polynomials and so we have a filtration

$$\mathbb{F}_2=S_0\Omega_g\subset S_1\Omega_g\subset S_2\Omega_g\subset\cdots\subset S\,\Omega_g,$$

where $S_d\Omega_g$ denotes the space of polynomial functions of degree $\le d$. Since $f=f^2$ for each $f\in S\,\Omega_g$, the associated graded algebra is naturally isomorphic to the exterior algebra $\wedge^\bullet H_1(S_g;\mathbb{F}_2)$. The algebra $S\,\Omega_g$ has as a distinguished element which is called the *Arf invariant*, denoted here by arf. If $a_1,\dots,a_g,b_1,\dots,b_g$ is a symplectic basis of $H^1(S;\mathbb{F}_2)$, then arf is defined by

$$\mathrm{arf}:\omega\mapsto\sum_i\omega(a_i)\omega(b_i).$$

It is an element of $S_2\Omega_g$, and its zero set $\Psi_g$ is an affine quadric in $\Omega_g$. Let $S_d\Psi_g$ denote the image of $S_d\Omega_g$ in the set of $\mathbb{F}_2$ valued functions on $\Psi_g$.

**Theorem 7.6** (Johnson [44]). *There are natural isomorphisms*

$$\sigma_{g,1}:H_1(T_{g,1};\mathbb{F}_2)\xrightarrow{\sim}S_3\Omega_g,\quad \sigma_g^1:H_1(T_g^1;\mathbb{F}_2)\xrightarrow{\sim}S_3\Omega_g/\mathbb{F}_2\,\mathrm{arf},$$
$$\sigma_g:H_1(T_g;\mathbb{F}_2)\xrightarrow{\sim}S_3\Psi_g$$

*which are equivariant with respect to the* $Sp_g(\mathbb{F}_2)$*-action. These induce natural isomorphisms*

$$H_1(T_{g,1};\mathbb{Z})_{\mathrm{tor}}\cong S_2\Omega_g,\quad H_1(T_g^1;\mathbb{Z})_{\mathrm{tor}}\cong H_1(T_g;\mathbb{Z})_{\mathrm{tor}}\cong S_2\Psi_g.$$

*Moreover, the natural isomorphisms*

$$\phi_{g,r}^n:H_1(T_{g,r}^n;\mathbb{F}_2)/H_1(T_{g,r}^n;\mathbb{Z})_{\mathrm{tor}}\xrightarrow{\sim}\left[H_1(T_{g,r}^n;\mathbb{Z})/\mathrm{torsion}\right]\otimes\mathbb{F}_2$$

*correspond, under the isomorphisms* $\sigma_{g,r}^n$ *and* $\tau_{g,r}^n$*, to the obvious isomorphisms*

$$\phi_{g,1}:S_3\Omega_g/S_2\Omega_g\xrightarrow{\sim}\wedge^3H_1(S_g;\mathbb{F}_2),\quad \phi_g^1:S_3\Omega_g/(\mathbb{F}_2\,\mathrm{arf}+S_2\Omega_g)\xrightarrow{\sim}\wedge^3H_1(S_g;\mathbb{F}_2),$$
$$\phi_g:S_3\Psi_g/S_2\Psi_g\xrightarrow{\sim}\wedge^3H_1(S_g;\mathbb{F}_2)/([S_g]\wedge H_1(S_g;\mathbb{F}_2)).$$

The homomorphisms $\tau_g^1$ and $\tau_g$ admit direct conceptual definitions that we will give later. Here we give a formula for the image of the standard generators of $T_{g,1}$ in $\wedge^3H_1(S_g;\mathbb{Z})$ and in $S_3\Omega_g$ under $\tau_{g,1}$ and $\sigma_{g,1}$, respectively.

Let $(\alpha_1,\alpha_2)$ be a separating pair of simple loops. Let $t$ be the corresponding element of $T_{g,1}$ — recall that this is the product of the Dehn twist about $\alpha_1$ and the *inverse* of the Dehn twist about $\alpha_2$. The two loops decompose $S_g$ into two pieces $S'$ and $S''$, say, where we suppose that $S'$ contains the point $x_1$. We orient $\alpha_1$ and $\alpha_2$ as boundary components of $S''$. The resulting cycles are opposite in $H_1(S'';\mathbb{Z})$: $[\alpha_2]=-[\alpha_1]$, and each spans the radical of the intersection pairing on this group. So there is a well-defined element in $\wedge^2H_1(S'';\mathbb{Z})/[\alpha_1]\wedge H_1(S'';\mathbb{Z})$ representing the intersection pairing on $H_1(S'';\mathbb{Z})$. Its wedge with $[\alpha_1]$ can be regarded as an element of $\wedge^3H_1(S'';\mathbb{Z})$. Since the inclusion $S''\subset S_g$ induces an injection on first homology, we can also view the latter as an element of $\wedge^3 H_1(S_g;\mathbb{Z})$. This is the element $\tau_{g,1}(t)$; it is clear that it only depends on the image of $t$ in $T_g^1$.

Next we associate to $t$ a function $\sigma_t : \Omega_g \to \mathbb{F}_2$ as follows. If $\omega \in \Omega_g$ takes the value 1 on $[\alpha]$, then we put $\sigma_t(\omega) = 0$; if it takes the value 0 on $[\alpha]$, then the restriction of $\omega$ to $H_1(S'';\mathbb{F}_2)$ factors through a nondegenerate quadratic function on $H_1(S'';\mathbb{F}_2)/\mathbb{F}_2[\alpha]$. Then $\sigma_{g,1}(t)(\omega)$ is its Arf invariant. It can be shown that $\sigma_{g,1}(t)$ lies in $S_3\Omega$.

Now suppose that $t$ is the element of $T_{g,1}$ associated to a separating simple loop $\alpha$. Denote the pieces $S'$ and $S''$ as before. In this case, $\tau_{g,1}(t)$ is trivial and $\sigma_{g,1}(t)$ is the element of $S\Omega_g$ that assigns to $\omega$ the Arf invariant of its restriction to $H_1(S'';\mathbb{F}_2)$. Notice that if $\alpha$ is a simple loop around $x_1$, then $\sigma_{g,1}(t)$ is just the function arf. (This explains why we mod out by this function when passing from $T_{g,1}$ to $T_g^1$.)

Without a base point there is no way of telling $S'$ and $S''$ apart. It is because of this ambiguity that we have to restrict functions to $\Psi_g$ in order to obtain a well defined function.

A diffeomorphism of $S_g$ onto a smooth projective curve $C$ determines a natural isomorphism between $\Omega_g$ and the space of theta characteristics of $C$ (i.e., square roots of the canonical bundle $K_C$; see for instance Appendix B of [2]). This suggests that Johnson's computation should have an algebro-geometric interpretation, if not interesting applications to the geometry of curves.

*Problem 7.7.* Give an algebro-geometric construction of the epimorphism $T_g \to S_3\Omega$.

Van Geemen has suggested such a construction (unpublished).

7.1. **Torelli space and period space.** The group $T_g$ acts freely on $\mathcal{X}_g$. The quotient $\mathcal{T}_g$ is therefore a complex manifold. It is called *Torelli space.* According to the discussion at the beginning of Section 3, $\mathcal{T}_g$ is then a classifying space for $T_g$ so that there is a canonical isomorphism $H_\bullet(T_g;\mathbb{Z}) \cong H_\bullet(\mathcal{T}_g;\mathbb{Z})$. Torelli space has a moduli interpretation; it is the moduli space of smooth projective curves $C$ of genus $g$ together with a symplectic isomorphism

$$\gamma : H_1(S_g;\mathbb{Z}) \to H_1(C;\mathbb{Z}).$$

There are also decorated versions $\mathcal{T}_{g,r}^n$ of Torelli space. Their points are points of $\mathcal{M}_{g,r}^n$ together with a symplectic isomorphism $\gamma$ of $H_1(S;\mathbb{Z})$ with the first homology of the curve corresponding to the point of $\mathcal{M}_g$. It is clear that the map $\mathcal{T}_{g,r}^n \to \mathcal{M}_{g,r}^n$ is Galois with Galois group $Sp_g(\mathbb{Z})$.

Denote the Siegel space associated to $V_g$ by $\mathfrak{h}_g$. To be precise, $\mathfrak{h}_g$ is the set of pure Hodge structures on $V_g$ with Hodge numbers $(-1,0)$ and $(0,-1)$, polarized by the intersection form. This is a contractible complex manifold of dimension $g(g+1)/2$ on which the group $Sp_g(\mathbb{R})$ acts properly and transitively. We can also regard $\mathfrak{h}_g$ as the moduli space of pairs consisting of a $g$ dimensional principally polarized abelian variety $A$ plus a symplectic isomorphism

$$\gamma : H_1(S;\mathbb{Z}) \to H_1(A;\mathbb{Z}).$$

This interprets the $Sp_g(\mathbb{Z})$ orbit space of $\mathfrak{h}_g$ as the moduli space of principally polarized abelian varieties of dimension $g$, $\mathcal{A}_g$. We regard $\mathcal{A}_g$ as an orbifold with orbifold fundamental group $Sp_g(\mathbb{Z})$, although $Sp_g(\mathbb{Z})$ does not act faithfully on $\mathfrak{h}_g$. The kernel of this action is $\{\pm 1\}$.

Assigning to a smooth projective curve the Hodge structure on its first homology group defines a map $\mathcal{T}_g \to \mathfrak{h}_g$, the *period map* for $T_g$. It is an isomorphism in genus 1, an open imbedding when $g = 2$, and 2:1 with ramification along the hyperelliptic locus when $g \geq 3$.[9] The reason for this is that for all abelian varieties we have the equality

$$[A; \gamma] = [A; -\gamma]$$

of points of $\mathfrak{h}_g$ as $-\operatorname{id}$ is an automorphism of each abelian variety. On the other hand, we have the equality

$$[C; \gamma] = [C; -\gamma]$$

of points of $\mathcal{T}_g$ if and only if $C$ is hyperelliptic.

Mess's result (mentioned at the beginning of the section) can now be deduced from this: $\mathcal{T}_2$ is the complement in $\mathfrak{h}_2$ of the locus of principally polarized abelian varieties that are products of two elliptic curves. The locus of such reducible abelian varieties is a countable disjoint union of copies of $\mathfrak{h}_1 \times \mathfrak{h}_1$. The group $Sp_2(\mathbb{Z})$ permutes them transitively, and each is stabilized by a product of two copies of $SL_2(\mathbb{Z})$ and an involution that switches the two copies of the upper half plane. Mess's result follows easily using the stratified Morse theory of Goresky and MacPherson — use distance from a generic point of $\mathfrak{h}_2$ as the Morse function. Since each component of $\mathfrak{h}_2 - \mathcal{T}_2$ is a totally geodesic divisor, the distance function has a unique critical point (necessarily a minimum) on each stratum. It follows that $\mathcal{T}_2$ has the homotopy type of a wedge of circles, one for each component of $\mathfrak{h}_2 - \mathcal{T}_2$.

The period map gives, after passage to $Sp_g(\mathbb{Z})$ orbit spaces, a morphism $\mathcal{M}_g \to \mathcal{A}_g$, the period mapping for $\mathcal{M}_g$. This period mapping extends to the partial completion $\widetilde{\mathcal{M}}_g$ of $\mathcal{M}_g$ and the resulting map $\widetilde{\mathcal{M}}_g \to \mathcal{A}_g$ is proper.

Now assume $g \geq 3$ and denote the image of the period map $\mathcal{T}_g \to \mathfrak{h}_g$ by $\mathcal{S}_g$. This space is the quotient of $\mathcal{T}_g$ by the subgroup $\{\pm 1\}$ of $Sp_g(\mathbb{Z})$. Consequently

$$H^\bullet(\mathcal{S}_g) \cong H^\bullet(\mathcal{T}_g)^{\{\pm 1\}}.$$

Observe that $\mathcal{S}_g$ is a locally closed analytic subvariety of $\mathfrak{h}_g$, but not closed. The $\{\pm 1\}$ cover $\mathcal{T}_g \to \mathcal{S}_g$ extends as a $\{\pm 1\}$ cover $\overline{\mathcal{T}}_g \to \overline{\mathcal{S}}_g$ over the closure of $\mathcal{S}_g$ in $\mathfrak{h}_g$, and the $\{\pm 1\}$ action on the total space is the restriction of an $Sp_g(\mathbb{Z})$ action. Both $\overline{\mathcal{T}}_g$ and $\overline{\mathcal{S}}_g$ are rather singular along the added locus (which has one component of codimension 2 and others of codimension 3). If we pass to $Sp_g(\mathbb{Z})$ orbit spaces, then the natural map

$$\widetilde{\mathcal{M}}_g \to Sp_g(\mathbb{Z}) \backslash \overline{\mathcal{T}}_g \cong Sp_g(\mathbb{Z}) \backslash \overline{\mathcal{S}}_g$$

resolves these singularities in an orbifold sense. A resolution of a normal analytic variety always induces a surjection on fundamental groups and so it follows from (7.1) that the fundamental group of $\overline{\mathcal{T}}_g$ is abelian and is $Sp_g(\mathbb{Z})$ equivariantly a quotient of $\Lambda^3 H_1(S_g; \mathbb{Z})/([S_g] \wedge H_1(S_g; \mathbb{Z}))$.

*Problem 7.8.* Understand the topology of $\mathcal{S}_g$ and its closure $\overline{\mathcal{S}}_g$ in $\mathfrak{h}_g$. In particular, how close is $\mathcal{S}_g$ to being a finite complex? (Observe that if it has a finite 2-skeleton, then $T_g$ is finitely presented.)

Related, but formally independent of this problem, is the question of whether the cohomology of $T_g$ stabilizes in a suitable sense:

[9]It is stated incorrectly in [24] that $\mathcal{T}_2 \to \mathfrak{h}_2$ is an unramified 2:1 map onto its image.

*Question 7.9.* Is $H^k(T_g)$ expressible as an $Sp_g(\mathbb{Z})$ module in a manner that is independent of $g$ if $g$ is large enough? For example, from Johnson's Theorem, we know that $H^1(T_g)$ is the third fundamental representation of $Sp_g$ for all $g \geq 3$.

7.2. **The Johnson homomorphism.** The proof of Johnson's Theorem is nontrivial and uses geometric topology, but the homomorphism $\tau_g^1$ is easily described.

Since $T_g$ is torsion free, the projection $\mathcal{T}_g^1 \to \mathcal{T}_g$ defines the universal curve over $\mathcal{T}_g$. Denote the corresponding bundle of jacobians by $\mathcal{J}_g \to \mathcal{T}_g$. Since the local system of first homology groups associated to the universal curve is canonically framed, this jacobian bundle $\mathcal{J}_g \to \mathcal{T}_g$ is analytically trivial as a bundle of Lie groups: we have a natural trivializing projection $p : \mathcal{J}_g \to \operatorname{Jac} S_g$, where $\operatorname{Jac} S_g := H_1(S_g; \mathbb{R}/\mathbb{Z})$ is the "jacobian" of the reference surface.

The usual Abel-Jacobi map, which assigns to an ordered pair of points $(x, y)$ on a smooth curve $C$ the divisor class of $(x) - (y)$, induces a morphism

$$\mathcal{T}_g^1 \times_{\mathcal{T}_g} \mathcal{T}_g^1 \to \mathcal{J}_g.$$

over $\mathcal{T}_g$. This provides a correspondence

$$\begin{array}{ccccc} \mathcal{T}_g^1 \times_{\mathcal{T}_g} \mathcal{T}_g^1 & \longrightarrow & \mathcal{J}_g & \xrightarrow{p} & \operatorname{Jac} S_g \\ \big\downarrow{\scriptstyle pr_2} & & & & \\ \mathcal{T}_g^1 & & & & \end{array}$$

from $\mathcal{T}_g^1$ to $\operatorname{Jac} S_g$. It induces homomorphisms

$$H_k(T_g^1) \cong H_k(\mathcal{T}_g^1) \to H_{k+2}(\operatorname{Jac} S_g).$$

The first of these is the Johnson homomorphism

$$\tau_g^1 : H_1(T_g^1) \to H_3(\operatorname{Jac} S_g)$$

for $T_g^1$. Since $\mathcal{T}_g^1 \to \mathcal{T}_g$ is a fibration of Eilenberg-MacLane spaces, we have an exact sequence of fundamental groups:

$$1 \to \pi_g \to T_g^1 \to T_g \to 1.$$

This induces an exact sequence

$$H_1(S_g; \mathbb{Z}) \to H_1(T_g^1; \mathbb{Z}) \to H_1(T_g; \mathbb{Z}) \to 0$$

on homology. Since $\operatorname{Jac} S_g$ is a topological group with torsion free homology, its integral homology has a product — the *Pontrjagin product.* It is not difficult to check that the composite

$$H_1(S_g; \mathbb{Z}) \to H_1(T_g^1; \mathbb{Z}) \to H_3(\operatorname{Jac} S_g; \mathbb{Z})$$

is the map given by Pontrjagin product with the class $[S_g]$. It follows that there is a natural homomorphism

$$H_1(T_g; \mathbb{Z}) \to H_3(\operatorname{Jac} S_g; \mathbb{Z}) / \left([S_g] \times H_1(S_g; \mathbb{Z})\right).$$

This is the Johnson homomorphism $\tau_g$ for $T_g$.

Johnson's Theorem, alone and in concert with Saito's theory of Hodge modules, has several interesting applications to the geometry of moduli spaces of curves as we shall see in subsequent sections.

7.3. **Monodromy of roots of the canonical bundle.** In this subsection we assume that $g \geq 2$. Suppose that $C$ is a smooth projective curve of genus $g$. Since its canonical bundle $K_C$ is of degree $2g-2$ and since $\operatorname{Pic}^0 C$ is a divisible group, $K_C$ has $n$th roots whenever $n$ divides $2g-2$. Any two such $n$th roots will differ by an $n$ torsion point of $\operatorname{Pic}^0 C$. Because of this, $n$th roots of $K_C$ are rigid under deformation. It follows that they form a locally constant sheaf (in the orbifold sense) $\mathrm{Root}^n$ over $\mathcal{M}_g$. The fiber over $C$, denoted $\mathrm{Root}^n C$, is a principal homogenous space over $H_1(C;\mathbb{Z}/n)$, the group of $n$ torsion points of $\operatorname{Pic}^0 C$.

Choose a conformal structure on $S_g$. Denote the corresponding algebraic curve by $C$. Sipe [77] determined the monodromy representation

$$\rho^n : \Gamma_g \to \operatorname{Aut}\mathrm{Root}^n(C).$$

of this sheaf. Before giving it, we make some remarks. Since the Torelli group acts trivially on the $n$ torsion of $\operatorname{Pic}^0 C$, it follows that the restriction of $\rho^n$ to $T_g$ factors through a representation $T_g \to H_1(S;\mathbb{Z}/(2g-2)) \to H_1(S;\mathbb{Z}/n)$. However, the action of $\Gamma_g$ on the set $\mathrm{Root}^2 C$ of square roots of $K_C$ (the set of theta characteristics of $C$) factorizes through $Sp_g(\mathbb{Z})$ also (even through $Sp_g(\mathbb{F}_2)$) — this is because there is a canonical correspondence between square roots of $K_C$ and $\mathbb{F}_2$ quadratic forms on $H_1(S;\mathbb{F}_2)$ associated to the intersection form. It follows that the image of the monodromy representation $\rho^n$ will be contained in an extension of $Sp_g(\mathbb{Z}/n)$ by a subgroup of $2\cdot H^1(S_g;\mathbb{Z}/n)$. In fact, it is all of this group.

**Theorem 7.10** (Sipe [77]). *The monodromy group of* $\mathrm{Root}^n$ *is an extension of* $Sp_g(\mathbb{Z}/n)$ *by the subgroup* $2\cdot H^1(S_g;\mathbb{Z}/n)$ *of* $H_1(S_g;\mathbb{Z}/n)$.

The subgroup $2\cdot H^1(S_g;\mathbb{Z}/n)$ appears as a quotient of the Torelli group $T_g$. In [24] it is shown that the restriction of the monodromy representation to $T_g$ is the composite of the Johnson homomorphism with a natural surjection

$$\wedge^3 H_1(S_g;\mathbb{Z})/([S_g]\wedge H_1(S_g;\mathbb{Z}) \to H_1(S_g;\mathbb{Z}/(g-1)) \to 2\cdot H_1(S_g;\mathbb{Z}/n).$$

7.4. **Picard groups of level covers.** Denote the moduli space of smooth projective genus $g$ curves with a level $l$ structure by $\mathcal{M}_g[l]$. This is convenient shorthand for the notation $\mathcal{M}_g[Sp_g(\mathbb{Z}/l)]$ introduced in Section 3. Denote the kernel of the reduction mod $l$ map

$$Sp_g(\mathbb{Z}) \to Sp_g(\mathbb{Z}/l)$$

by $Sp_g(\mathbb{Z})[l]$, and its full inverse image in $\Gamma_g$ by $\Gamma_g[l]$. Then $\mathcal{M}_g[l]$ is the quotient of Teichmüller space $\mathcal{X}_g$ by $\Gamma_g[l]$. As in the case of $\mathcal{M}_g$, there is a canonical isomorphism

$$H^\bullet(\mathcal{M}_g[l]) \cong H^\bullet(\Gamma_g[l]).$$

This holds with rational coefficients for all $l$, and arbitrary coefficients whenever $\Gamma_g[l]$ is torsion free, which holds whenever $Sp_g(\mathbb{Z})[l]$ is torsion free — $l \geq 3$.

We know from (7.4) that

$$H^1(\Gamma_g[l]) \cong H^1(\mathcal{M}_g[l]) = 0$$

when $g \geq 3$. By standard arguments (cf. [24, §5]), this implies that

$$c_1 : \operatorname{Pic}\mathcal{M}_g[l]\otimes\mathbb{Q} \to H^2(\mathcal{M}_g[l])$$

is injective, and therefore that $\operatorname{Pic}\mathcal{M}_g[l]$ is finitely generated when $g \geq 3$.

The stable cohomology of an arithmetic group depends only on the ambient real algebraic group [5]. Based on this, one might expect that the natural map

$$H^k(\Gamma_g) \to H^k(\Gamma_g[l])$$

is an isomorphism for all $l \geq 0$, once the genus $g$ is sufficiently large compared to the degree $k$. It follows from Johnson's work that this is true when $k = 1$ (cf. [24]), but the only evidence for it when $k > 1$ is Harer's computation of the second homology of the spin mapping class groups [33], and Foisy's theorem from which Harer's computation now follows:

**Theorem 7.11** (Foisy [15]). *For all $g \geq 3$, the natural map $H^2(\Gamma_g) \to H^2(\Gamma_g[2])$ is an isomorphism. Consequently, $\operatorname{Pic} \mathcal{M}_g[2]$ is finitely generated of rank 1.*

*Question 7.12.* Is $\operatorname{Pic} \mathcal{M}_g[l]$ rank 1 for all $g \geq 3$ and all $l \geq 1$?

This would be the case if we knew that the $Sp_g(\mathbb{Z})$ action on $H^2(T_g)$ extended to an algebraic action of $Sp_g$, for we could then invoke Borel's computation of the stable cohomology of arithmetic groups [5].

7.5. **Normal Functions.** Each rational representation $V$ of $Sp_g$ gives rise to an orbifold local system $\mathbb{V}$ over $\mathcal{M}_g[l]$. Such a local system underlies an admissible variation of Hodge structure. First, if $V$ is irreducible, then $V$ underlies a variation of Hodge structure unique up to Tate twist ([24, (9.1)]). Every polarized $\mathbb{Q}$ variation of Hodge structure whose monodromy representation comes from a rational representation of $Sp_g$ has the property that each of its isotypical components is an admissible variation of Hodge structure of the form $A_\lambda \otimes \mathbb{V}(\lambda)$, where $A_\lambda$ is a Hodge structure and $\mathbb{V}(\lambda)$ is a variation of Hodge structure corresponding to the $Sp_g$ module with highest weight $\lambda$ — cf. [24, (9.2)].

For a Hodge structure $V$ of weight $-1$ one defines the corresponding *intermediate jacobian* $JV$ by

$$JV = V_{\mathbb{C}}/(F^0V + V_{\mathbb{Z}}).$$

Its interest comes from the fact that it parametrizes the extensions of $\mathbb{Z}$ by $V$ in the MHS category: if $E$ is an extension of the $\mathbb{Z}$ (with its trivial Hodge structure of weight zero) by $V$, then choose an integral lift $e \in E$ of 1 and consider the image of $e$ in

$$E_{\mathbb{C}}/(F^0E + V_{\mathbb{Z}}) \cong V_{\mathbb{C}}/(F^0V + V_{\mathbb{Z}}).$$

This is independent of the lift and yields a complete invariant of the extension. There is an inverse construction that makes $JV$ support a variation of mixed Hodge structure $\mathbb{E}$ that is universal as an extension of the trivial Hodge structure $\mathbb{Z}$ by the constant Hodge structure $V$:

$$0 \to \mathbb{V}_{JV} \to \mathbb{E} \to \mathbb{Z}_{JV} \to 0$$

(see [7]). This immediately generalizes to a relative setting: if $\mathbb{V}$ is an admissible variation of $\mathbb{Z}$ Hodge structure of weight $-1$ over a smooth variety $X$, then we have a corresponding bundle $\pi : \mathcal{J}\mathbb{V} \to X$ of intermediate jacobians over $X$ supporting a universal extension

$$0 \to \pi^*\mathbb{V} \to \mathbb{E} \to \pi^*\mathbb{Z}_X \to 0.$$

A section $\sigma$ of $\mathcal{J}\mathbb{V}$ over $X$ determines an extension of Hodge structures:

$$0 \to \mathbb{V} \to \sigma^*\mathbb{E} \to \mathbb{Z}_X \to 0.$$

A *normal function* is a section of $\mathcal{J}\mathbb{V}$ such that the corresponding extension $\mathbb{E}$ is an admissible variation of mixed Hodge structure. The normal functions arising from algebraic cycles are normal functions in this sense — cf. [24, §6].

We briefly recall Griffiths' construction of a normal function associated to a family of homologically trivial algebraic cycles. First we consider the case where the base is a point. Suppose that $X$ is a smooth projective variety. A homologically trivial algebraic $d$-cycle $Z$ in $X$ canonically determines an extension of $\mathbb{Z}$ by $H_{2d+1}(X;\mathbb{Z}(-d))$ by pulling back the exact sequence

$$0 \to H_{2d+1}(X;\mathbb{Z}(-d)) \to H_{2d+1}(X,|Z|;\mathbb{Z}(-d)) \to H_{2d}(|Z|;\mathbb{Z}(-d)) \to \cdots$$

of MHSs along the inclusion

$$\mathbb{Z} \to H_{2d}(|Z|;\mathbb{Z}(-d))$$

that takes 1 to the class of $Z$. So an integral lift of 1 is given by an integral singular $2d+1$ chain $W$ in $X$ whose boundary is $Z$. Integration identifies $JH_{2d+1}(X;\mathbb{Z}(-d))$ with the *Griffiths intermediate Jacobian*

$$J_d(X) := \operatorname{Hom}_{\mathbb{C}}(F^d H^{2d+1}(X);\mathbb{C}(-d))/H^{2d+1}(X;\mathbb{Z}(-d)),$$

and under this isomorphism the extension class in question is just given by integration over $W$.

Families of homologically trivial cycles give rise to normal functions: Suppose that $\mathcal{X} \to T$ is a family of smooth projective varieties over a smooth base $T$ and that $\mathcal{Z}$ is an algebraic cycle in $\mathcal{X}$ which is proper over $T$ of relative dimension $d$. Then the local system whose fiber over $t \in T$ is $H_{2d+1}(X_t;\mathbb{Z}(-d))$ naturally underlies a variation of Hodge structure $\mathbb{V}$ over $T$ of weight $-1$ so that we can form the $d$th *relative intermediate jacobian* $\mathcal{J}_d(\mathcal{X}/T) \to T$, whose fiber over $t \in T$ is $J_d(X_t)$. The family of cycles $\mathcal{Z}$ defines a section of this bundle which is a normal function.

**Theorem 7.13** (Hain [24]). *Suppose that $\mathbb{V}$ is an admissible variation of Hodge structure of weight $-1$ over $\mathcal{M}_g[l]$ whose monodromy representation factors through a rational representation of $Sp_g$. If $g \geq 3$, then the space of normal functions associated to $\mathbb{V}$ is finitely generated of rank equal to the number of copies of the variation $\mathbb{V}(\lambda_3)$ of weight $-1$ that occur in $\mathbb{V}$.*

The theorem implies that, up to torsion and multiples, there is only one normal function over $\mathcal{M}_g$ associated to a variation of Hodge structure whose monodromy factors through a rational representation of $Sp_g$. So what is the generator of these normal functions?

To answer this question, recall that if $C$ is a smooth projective curve of genus $g$ and $x \in C$, we have the Abel-Jacobi morphism

$$C \to \operatorname{Jac} C, \quad y \mapsto (y) - (x).$$

Denote the image 1-cycle in $\operatorname{Jac} C$ by $C_x$ and the cycle $i_* C_x$ by $C_x^-$, where $i : \operatorname{Jac} C \to \operatorname{Jac} C$ takes $u$ to $-u$. The cycle $C_x - C_x^-$ is homologous to zero, and therefore defines a point $\nu^1(C,x)$ in $J_1(\operatorname{Jac} C)$. Pontrjagin product with the class of $C$ induces a homomorphism

$$A : \operatorname{Jac} C \to J_1(\operatorname{Jac} C).$$

We call the cokernel of $A$ the *primitive first intermediate Jacobian* $J_1^{\mathrm{pr}}(\operatorname{Jac} C)$ of $\operatorname{Jac} C$. The family of such primitive intermediate jacobians over $\mathcal{M}_g$ is the unique one (up to isogeny) associated to the variation of Hodge structure of weight $-1$ over $\mathcal{M}_g$ whose associated $\Gamma_g$ module is $V(\lambda_3)$. It is not difficult to show that

$$\nu^1(C,x) - \nu^1(C,y) = 2A(x-y).$$

It follows that the image of $\nu^1(C,x)$ in $J_1^{\mathrm{pr}}(\operatorname{Jac} C)$ is independent of $x$. This is the value of the normal function associated with $C - C^-$ over $[C]$.

We can do better and realize half of this generator by a generalized normal function as follows. Let $A$ be a principally polarized abelian variety of dimension $g \geq 3$. The polarization determines a distinguished element $\omega$ of $H_2(A;\mathbb{Z})$. If $Z$ and $Z'$ are two piecewise smooth cycles representing $\omega$, then their difference is the boundary of a piecewise smooth 3-chain $W$ on $A$. Represent the dual of $H_3(A;\mathbb{R})$ by translation invariant 3-forms on $A$. Then integrating these forms over $W$ determines an element of $H_3(A;\mathbb{R})$. Another choice of $W$ gives a class that differs from this one by an element of $H_3(A;\mathbb{Z})$, and so we have a well-defined element $[Z-Z']$ of $H_3(A;\mathbb{R}/\mathbb{Z})$. Notice that the latter torus is naturally identified with the first intermediate jacobian $J_1(A)$ of $A$. We declare $Z$ and $Z'$ to be equivalent if $[Z-Z'] = 0$ and denote the space of piecewise smooth cycles representing $\omega$ modulo this equivalence relation by $D(A)$. This is clearly a torsor of $J_1(A)$ and so it has a natural complex structure. In view of its connection with Deligne cohomology, we call it the *Deligne torsor* of $A$. This torsor contains naturally a subtorsor $D(A)[2]$ of the 2-torsion in $J_1(A)$, $J_1(A)[2] \cong H_3(A;\frac{1}{2}\mathbb{Z}/\mathbb{Z})$: Let $a = (a_1, a_{-1}, \dots, a_g, a_{-g})$ be a symplectic basis of $H_1(A;\mathbb{Z})$. Each basis element $a_i$ is uniquely represented by a homomorphism $\alpha_i : S^1 \to A$ and so $\omega$ is represented by the 2-cycle $\sum_{i=1}^g \alpha_i \times \alpha_{-i}$. This cycle defines an element $z(a) \in D(A)$. It is easily verified that $z(a)$ only depends on the mod two reduction of $a$ and that if $a$ runs over all symplectic bases, $z(a)$ runs over an entire orbit $D(A)[2]$ of $J_1(A)[2]$. (So $J_1(A)[2]\backslash D(A)$ has a canonical point which identifies it with $J_1(A)$.) The group $Sp(H_1(A;\mathbb{Z}))$ acts on $D(A)$ as an affine transformation group in a way that is easily made explicit. The lifts of these transformations to a universal covering of $D(A)$ form a group of affine symplectic transformations. It is an extension of $Sp(H_1(A;\mathbb{Z}))$ by $H_3(A;\mathbb{Z})$ which splits if we enlarge the extension to $H_3(A;\frac{1}{2}\mathbb{Z})$.

The Pontrjagin product with $\omega$ defines a homomorphism $A \to J_1(A)$ which gives rise to corresponding primitive notions: the *primitive Deligne torsor* $D^{\mathrm{pr}}(A) := A\backslash D(A)$ is a torsor of the primitive intermediate Jacobian $J_1^{\mathrm{pr}}(A) := A\backslash J_1(A)$. We have corresponding universal Deligne torsors over $\mathcal{A}_g$ which we denote $\mathcal{D}_g \to \mathcal{A}_g$ and $\mathcal{D}_g^{\mathrm{pr}} \to \mathcal{A}_g$. By the above argument, these torsors become trivial on the Galois cover of $\mathcal{A}_g$ representing principally polarized abelian varieties with a level 2 structure. The torsors themselves are nontrivial, for it can be shown that the orbifold fundamental groups of these torsors are nonsplit extensions of the integral symplectic group of genus $g$

For $C$ a nonsingular projective curve of genus $g \geq 3$ and $x \in C$, the Abel-Jacobi morphism $C \to \operatorname{Jac} C$ defined by $y \mapsto (y) - (x)$ defines a cycle in the homology class of the natural polarization of $\operatorname{Jac} C$ and so we get an element $[(C,x)]$ of $D(\operatorname{Jac} C)$. Its image in $D^{\mathrm{pr}}(\operatorname{Jac} C)$ is independent of $x$ and so can be denoted by $[C]$. Universally this produces holomorphic lifts of the period map:

$$\nu_g^1 : \mathcal{M}_g^1 \to \mathcal{D}_g \text{ and } \nu_g : \mathcal{M}_g \to \mathcal{D}_g^{\mathrm{pr}}.$$

We call $\nu_g$ the *fundamental normal function* on $\mathcal{M}_g$.

7.6. **Picard group of the generic curve with a level $l$ structure.** The classification of normal functions (7.13) implies that there are no sections of $\mathrm{Pic}^0$ of infinite order defined over $\mathcal{M}_g[l]$ when $g \geq 3$. This, combined with Sipe's computation (7.10) of the monodromy of roots of the canonical bundle allows one to determine the Picard group of the generic point of $\mathcal{M}_g[l]$. The case $l = 1$ was the subject of the Franchetta Conjecture which was deduced from Harer's computation of $\Gamma_g$ by Beauville (unpublished) and by Arbarello and Cornalba [3].

**Theorem 7.14** (Hain [25]). *The Picard group of the generic curve of genus $g \geq 3$ with a level $l$ structure is of rank 1, has torsion subgroup isomorphic to the $l$ torsion points $H_1(\mathrm{Jac}\, S_g; \mathbb{Z}/l)$, and, modulo torsion, is generated by the canonical bundle if $l$ is odd, and a theta characteristic if $l$ is even.*

## 8. Relative Malcev Completion

Fundamental groups of smooth algebraic varieties are quite special as we know from the work of Morgan [62] and others. The least trivial restrictions on these groups come from Hodge theory and Galois theory. Since $\Gamma_g$ is the (orbifold) fundamental group of $\mathcal{M}_g$, a smooth orbifold, Hodge theory and Galois theory should have something interesting to say about its structure. To put a MHS on a group one needs to linearize it. One way to do this is to replace the group by some kind of algebraic envelope and put a MHS on the coordinate ring of this (pro)algebraic group.

In this section we introduce these linearizations and use them to establish a relation between the fundamental normal function and a remarkable central extension that is hidden in a quotient of the mapping class group. Here the impact of mixed Hodge theory is not yet felt, but we are setting the stage for Section 9 where it is omnipresent.

8.1. **Classical Malcev completion.** Suppose that $\pi$ is a finitely generated group. The classical Malcev (or unipotent) completion of $\pi$ consists of a prounipotent group $\mathcal{U}(\pi)$ (over $\mathbb{Q}$) and a homomorphism $\pi \to \mathcal{U}(\pi)$. It is characterized by the following universal mapping property: if $U$ is a unipotent group, and $\phi : \pi \to U$ is a homomorphism, there is a unique homomorphism of prounipotent groups $\mathcal{U}(\pi) \to U$ through which $\phi$ factors. There are several well known constructions of the unipotent completion, which can be found in [23], for example. Each (pro)unipotent group $U$ is isomorphic to its Lie algebra $\mathfrak{u}$, a (pro)nilpotent Lie algebra via the exponential map. Thus, to give the Malcev group $\mathcal{U}(\pi)$ associated to $\pi$ it suffices to give its associated pronilpotent Lie algebra $\mathfrak{u}(\pi)$. This Lie algebra is called the *Malcev Lie algebra associated to* $\pi$. It comes with a natural descending filtration whose $k$th term $\mathfrak{u}^{(k)}(\pi)$ is the closed ideal of $\mathfrak{u}(\pi)$ generated by its $k$-fold commutators ($k = 1, 2, \dots$) and it is complete with respect to this filtration. We will refer to this filtration as the *Malcev filtration.*

When $\pi$ is the fundamental group $\pi_1(X, x)$ of a smooth complex algebraic variety, $\mathfrak{u}(\pi)$ has a canonical MHS which was first constructed by Morgan [62]. If $X$ is also complete, or more generally, when $H_1(X)$ has a pure Hodge structure of weight $-1$, then the weight filtration is the Malcev filtration:

$$W_{-k}\mathfrak{u}(\pi_1(X, x)) = \mathfrak{u}^{(k)}(\pi_1(X, x)).$$

Alternatively, this MHS determines and is determined by a MHS on the coordinate ring $\mathcal{O}(\mathcal{U}(\pi))$ of the associated Malcev group.

We shall denote the Malcev completion of $\pi_g^n = \pi_1(S_g^n, x_0)$ by $\mathfrak{p}_g^n$.

8.2. **Relative Malcev completion.** The Malcev completion of a group $\pi$ is trivial when $H_1(\pi)$ vanishes, for then $\pi$ has no non-trivial unipotent quotients. Since the first homology of $\Gamma_g$ vanishes for all $g$, its Malcev completion will be trivial. Deligne has defined the notion of Malcev completion of a group $\pi$ relative to a Zariski dense homomorphism $\rho : \pi \to S$, where $S$ is a reductive algebraic group defined over a base field $F$ (that we assume to be of characteristic zero).

The *Malcev completion of* $\pi$ *relative to* $\rho : \pi \to S$ is a a proalgebraic $F$-group $\mathcal{G}(\pi, \rho)$, which is an extension

$$1 \to \mathcal{U} \to \mathcal{G}(\pi, \rho) \to S \to 1$$

of $S$ by a prounipotent group, together with a lift $\tilde{\rho} : \pi \to \mathcal{G}(\pi, \rho)$ of $\rho$.[10] It is characterized by the following universal mapping property: if $G$ is an $F$-group which is an extension of $S$ by a unipotent group $U$, and if $\phi : \pi \to G$ is a homomorphism, then there is a unique homomorphism $\mathcal{G}(\pi, \rho) \to G$ through which $\phi$ factors:

$$\phi : \pi \xrightarrow{\tilde{\rho}} \mathcal{G}(\pi, \rho) \to G.$$

Since $S$ is reductive, we should think of $\mathcal{U}$ as the prounipotent radical of $\mathcal{G}(\pi, \rho)$. One can show, for instance, that $\mathcal{U}$ has a Levi supplement so that $\mathcal{G}(\pi, \rho)$ is a semidirect product of $S$ and $\mathcal{U}$. The Lie algebra $\mathfrak{g}(\pi, \rho)$ of $\mathcal{G}(\pi, \rho)$ also comes with a Malcev filtration with respect to which it is complete: $\mathfrak{g}(\pi, \rho)^{(0)} = \mathfrak{g}(\pi, \rho)$, and for $k \geq 1$, $\mathfrak{g}(\pi, \rho)^{(k)}$ is the closed ideal generated by $k$-fold commutators in the Lie algebra of $\mathcal{U}$.

We will often write $\mathcal{G}(\pi)$ instead of $\mathcal{G}(\pi, \rho)$ when the representation $\rho$ is clear from the context. We shall denote the completion of the (orbifold) fundamental group of a pointed orbifold $(X, x)$ with respect to a Zariski dense reductive representation $\rho : \pi_1(X, x) \to S$ by $\mathcal{G}(X, x; \rho)$, or simply $\mathcal{G}(X, x)$ when $\rho$ is clear from the context.

When $S$ is trivial, we recover the classical Malcev completion. The universal property of the Malcev completion of $\ker \rho$ yields a natural homomorphism of proalgebraic $F$-groups $\mathcal{U}(\ker \rho)(F) \to \mathcal{G}(\pi)$. In general, it is neither surjective nor injective as the following two examples show.

**Example 8.1.** The fundamental group of the symplectic Lie group $Sp_g(\mathbb{R})$ is infinite cyclic and hence so is its universal cover $\widehat{Sp}_g(\mathbb{R}) \to Sp_g(\mathbb{R})$. This universal cover is not an algebraic group (which follows for instance from the fact that the complexification of $Sp_g(\mathbb{R})$, $Sp_g(\mathbb{C})$, is simply connected). The preimage $\widehat{Sp}_g(\mathbb{Z})$ of $Sp_g(\mathbb{Z})$ in this covering contains the universal central extension of $Sp_g(\mathbb{Z})$ by $\mathbb{Z}$. Now take for $\pi$ this central extension and for $\rho$ its natural homomorphism to $Sp_g(\mathbb{C})$. The corresponding relative Malcev completion is then reduced to $Sp_g(\mathbb{C})$ itself, so that the homomorphism from $\mathcal{U}(\mathbb{Z})(\mathbb{C})$ (which is just the abelian group $\mathbb{C}$) to $\mathcal{G}(\widehat{Sp}_g(\mathbb{Z}))$ is trivial. We will see that this example is realized inside a quotient of the mapping class group.

[10]In many cases the completion of $\pi$ over an algebraic closure $\bar{F}$ of $F$ is the set of $\bar{F}$ points of the completion of $\pi$ over $F$. This is the case for the mapping class groups when $g \geq 3$, but we do not know whether this is true in general, except when $S$ is trivial.

**Example 8.2.** In this example, $\ker \rho$ is trivial, but $\mathcal{U}(\pi)$ is not. The basic fact we need (see [25, (10.3)]) is that there is always a natural $S$ equivariant isomorphism

$$H_1(\mathfrak{u}(\pi)) \cong \prod_{\alpha \in \check{S}} H_1(\pi; V_\alpha) \otimes V_\alpha^*,$$

where $V_\alpha$ denotes a representation with highest weight $\alpha$. For $\pi$ we take $\Gamma$, a finite index subgroup of $SL_2(\mathbb{Z})$, for $S$ we take $SL_2(\mathbb{Q})$, and for $\rho$ we take the natural inclusion. Denote the $n$th power of the fundamental representation of $SL_2$ by $S^nV$. For all such $\Gamma$, there is an infinite number of integers $n \geq 0$ such that $H^1(\Gamma; S^nV)$ is non-trivial.[11] It follows that $\mathcal{U}(\Gamma)$ has an infinite dimensional $H_1$, even though $\ker \rho$ is trivial.

This example suggests the following problem:

*Problem 8.3.* Investigate the relationship between the theory of modular forms associated to a finite index subgroup $\Gamma$ of $SL_2(\mathbb{Z})$ and the completion of $\Gamma$ relative to the inclusion $\Gamma \hookrightarrow SL_2(\mathbb{Q})$.

8.3. **The relative Malcev completion of $\Gamma_g$.** The natural homomorphism $\rho : \Gamma_{g,r}^n \to Sp_g$ has Zariski dense image. Denote the completion of $\Gamma_{g,r}^n$ relative to $\rho$ by $\mathcal{G}_{g,r}^n$, its prounipotent radical by $\mathcal{U}_{g,r}^n$ and their Lie algebras by $\mathfrak{g}_{g,r}^n$ and $\mathfrak{u}_{g,r}^n$.

The following theorem indicates the presence of essentially one copy of the universal central extension of $Sp_g(\mathbb{Z})$ in quotients of each mapping class group of genus $g$ when $g \geq 3$.

**Theorem 8.4** (Hain [23]). *When $g \geq 2$, the homomorphism*

$$\mathcal{U}(T_{g,r}^n) \to \mathcal{U}_{g,r}^n \tag{2}$$

*is surjective. When $g \geq 3$, its kernel is a central subgroup isomorphic to the additive group.*

This phenomenon is intimately related to the cycle $C - C^-$ and its normal function as we shall now explain.

8.4. **The central extension.** The existence of the central extension has both a group theoretic and a geometric explanation. It is also related to the Casson invariant through the work of Morita [65, 66]. We begin with the group theoretic one.

The group analogue of the Malcev filtration for the Torelli group $T_g$ is the most rapidly descending central series of $T_g$ with torsion free quotients:

$$T_g = T_g^{(1)} \supset T_g^{(2)} \supset T_g^{(3)} \supset \cdots$$

Note that $T_g^{(1)}/T_g^{(2)}$ is the maximal torsion free abelian quotient of $T_g$, which is

$$V(\lambda_3)_g\mathbb{Z} := \Lambda^3 H_1(S_g;\mathbb{Z})/\left([S_g] \times H_1(S_g;\mathbb{Z})\right)$$

by Johnson's Theorem (7.1). The group $\Gamma_g/T_g^{(3)}$ can be written as an extension

$$1 \to T_g^{(2)}/T_g^{(3)} \to \Gamma_g/T_g^{(3)} \to \Gamma_g/T_g^{(2)} \to 1. \tag{3}$$

It turns out that this sequence contains a multiple of the universal central extension of $Sp_g(\mathbb{Z})$ by $\mathbb{Z}$.

[11] This is easily seen when $\Gamma$ is free, for example. In general it is related to the theory of modular forms.

Since $V_g(\lambda_3)$ is a rational representation of $Sp_g$, and since the surjection

$$\wedge^2 V_g(\lambda_3)_{\mathbb{Z}} \to T_g^{(2)}/T_g^{(3)}$$

induced by the commutator is $Sp_g(\mathbb{Z})$ equivariant, it follows that $T_g^{(2)}/T_g^{(3)} \otimes \mathbb{Q}$ is also a rational representation of $Sp_g$. Because $V_g(\lambda_3)$ is an irreducible symplectic representation, there is exactly one copy of the trivial representation in $\wedge^2 V_g(\lambda_3)$. This copy of the trivial representation survives in $T_g^{(2)}/T_g^{(3)} \otimes \mathbb{Q}$ [23] so that there is an $Sp_g(\mathbb{Z})$ equivariant projection $T_g^{(2)}/T_g^{(3)} \to \mathbb{Z}$. Pushing the extension (3) out along this map gives an extension

$$0 \to \mathbb{Z} \to E \to \Gamma_g/T_g^{(2)} \to 1 \tag{4}$$

Note that $E$ is a quotient of $\Gamma_g$. We will manufacture a multiple of the universal central extension of $Sp_g(\mathbb{Z})$ from this group that turns out to be the obstruction to the map $\mathcal{U}(T_g) \to \mathcal{U}_g$ being injective. (Full details can be found in [23].)

The group $\Gamma_g/T_g^{(2)}$ can be written as an extension

$$0 \to V_g(\lambda_3)_{\mathbb{Z}} \to \Gamma_g/T_g^{(2)} \to Sp_g(\mathbb{Z}) \to 1. \tag{5}$$

Morita [67] showed that this extension is *semisplit*, that is, if we replace $V_g(\lambda_3)_{\mathbb{Z}}$ by $\frac{1}{2}V_g(\lambda_3)_{\mathbb{Z}}$, it splits. (This can also be seen using the normal function of $C - C^-$.)

**Theorem 8.5** (Morita [66], Hain [23]). *The extension of $Sp_g(\mathbb{Z})$ by $\mathbb{Z}$ obtained by pulling back the extension (4) along a semisplitting of (5) contains the universal central extension of $Sp_g(\mathbb{Z})$.*

The geometric picture uses the fundamental normal function $\nu_g$. The lifted period maps $\nu_g$ and $\nu_g^1$ to the Deligne torsors are easily seen to extend over the partial completions $\widetilde{\mathcal{M}}_g$ resp. $\widetilde{\mathcal{M}}_g^1$:

$$\tilde{\nu}_g : \widetilde{\mathcal{M}}_g \to \mathcal{D}_g^{\mathrm{pr}}, \quad \tilde{\nu}_g^1 : \widetilde{\mathcal{M}}_g^1 \to \mathcal{D}_g.$$

The orbifold fundamental group of $\mathcal{D}_g^{\mathrm{pr}}$, resp. $\mathcal{D}_{g,1}$, is an extension of $Sp_g(\mathbb{Z})$ by $V_g(\lambda_3)_{\mathbb{Z}}$, resp. $\wedge^3 V_{g,\mathbb{Z}}$, as both the base and fiber are Eilenberg MacLane spaces with these groups as orbifold fundamental groups. But by 7.2 the orbifold fundamental group of $\widetilde{\mathcal{M}}_g$, resp. $\widetilde{\mathcal{M}}_g^1$, also has such a structure. Indeed:

**Theorem 8.6.** *For $g \geq 3$, the normal functions $\tilde{\nu}_g : \widetilde{\mathcal{M}}_g \to \mathcal{D}_g^{\mathrm{pr}}$ and $\tilde{\nu}_g^1 : \widetilde{\mathcal{M}}_g^1 \to \mathcal{D}_g$ induce an isomorphism on orbifold fundamental groups. (The former can be identified with $\Gamma_g/T_g^{(2)}$ and the latter with $\Gamma_g^1/(T_g^1)^{(2)}$.)*

From this theorem we recover the fact that (5) is semisplit, not split. But we get more, since it should also lead to a description of that extension.

The extension (4) can also be realized geometrically.

**Proposition 8.7** (Hain [23]). *There is a canonical (locally homogeneous) line bundle $\mathcal{B}_g$ over the bundle $\mathcal{D}_{g,1}^{\mathrm{pr}} \to \mathcal{A}_g$ that realizes the central extension (4) via the isomorphism of the previous proposition as an extension of orbifold fundamental groups. In particular, both $\tilde{\nu}^*\mathcal{B}_g$ and $\nu^*\mathcal{B}_g$ have nonzero rational first Chern class. The bundle $\nu^*\mathcal{B}_g$ is canonically metrized and its square is isomorphic (as a metrized line bundle) to the metrized line bundle associated to the archimedean height of the cycle $C - C^-$.*

## 9. Hodge Theory of the Mapping Class Group

One reason that mixed Hodge theory is so powerful is that the MHS category is abelian. In many situations this turns out to have topological implications for algebraic varieties that are difficult, if not impossible, to obtain directly. A somewhat related (but less exploited) property is that a MHS is canonically split over $\mathbb{C}$. This implies that the weight filtration (which often has a topological interpretation) splits in a way that is compatible with all the algebraic structure naturally present. So, for many purposes, there is no loss of information regarding this algebraic structure if we pass to the corresponding weight graded object. For example, the Malcev filtration on the Malcev Lie algebra of a smooth projective variety is minus the weight filtration, and it therefore splits over $\mathbb{C}$ in a natural way. This splitting is natural in the sense that it respects the Lie algebra structure and is preserved under all base point preserving morphisms. But if we vary the complex structure on $X$ or the base point $x$, then the splitting will, in general, vary with it.

A basic example is the Malcev Lie algebra $\mathfrak{p}_g^1$ of $\pi_g^1 = \pi_1(S_g^1, x_0)$. The group $\pi_g^1$ is free on $2g$ generators and it is a classical fact that the graded of $\mathfrak{p}_g^1$ with respect to the Malcev filtration is just the free Lie algebra generated by $V_g$. If $S_g$ is given a conformal structure, then $V_g$ has a pure Hodge structure of weight $-1$ and the weight filtration of $\mathfrak{p}_g^1$ is minus the Malcev filtration. The splitting allows us to identify $\mathfrak{p}_g^1 \otimes \mathbb{C}$ with the completion of $\mathrm{Lie}(V_g) \otimes \mathbb{C}$.

We shall come back to this example in Section 10.3. But for now we will focus on the relative Malcev completions introduced in the previous section.

9.1. **Hodge theory of $\mathcal{G}_{g,r}^n$.** A choice of a conformal structure on $S_g$ and nonzero tangent vectors at $x_{n+1}, \dots, x_{n+r}$ determines a point $x_o$ of the moduli space $\mathcal{M}_{g,r}^n$. We can thus identify $\Gamma_{g,r}^n$ with the orbifold fundamental group of $(\mathcal{M}_{g,r}^n, x_o)$. This induces an isomorphism of $\mathcal{G}_{g,r}^n$ with $\mathcal{G}(\mathcal{M}_{g,r}^n, x_o)$, the completion of $\pi_1(\mathcal{M}_{g,r}^n, x_o)$ with respect to the standard symplectic representation. We shall write $\mathcal{G}_{g,r}^n(x_o)$ for $\mathcal{G}(\mathcal{M}_{g,r}^n, x_o)$ and denote its prounipotent radical by $\mathcal{U}_{g,r}^n(x_o)$. There is a general Hodge de Rham theory of relative Malcev completion [25]. Applying it to $(\mathcal{M}_{g,r}^n, x_o)$, one obtains the following result:

**Theorem 9.1** (Hain [26]). *For each choice of a base point $x_o$ of $\mathcal{M}_{g,r}^n$, there is a canonical MHS on the coordinate ring $\mathcal{O}(\mathcal{G}_{g,r}^n(x_o))$ which is compatible with its Hopf algebra structure. Consequently, the Lie algebra $\mathfrak{g}_{g,r}^n(x_o)$ of $\mathcal{G}_{g,r}^n(x_o)$ and the Lie algebra $\mathfrak{u}_{g,r}^n(x_o)$ of its prounipotent radical both have a natural MHS.*

Denote the Malcev Lie algebra of the subgroup of $\pi_1(\mathcal{M}_{g,r}^n, x_o)$ corresponding to the Torelli group $T_{g,r}^n$ by $\mathfrak{t}_{g,r}^n(x_o)$. The normal function of $C - C^-$ can be used to lift the MHS from $\mathfrak{u}_{g,r}^n(x_o)$ to $\mathfrak{t}_{g,r}^n(x_o)$.

**Theorem 9.2** (Hain [26]). *For each $g \geq 3$ and for each choice of a base point $x_o$ of $\mathcal{M}_{g,r}^n$, there is a canonical MHS on $\mathfrak{t}_{g,r}^n(x_o)$ which is compatible with its bracket. Moreover, the canonical central extension*

$$0 \to \mathbb{Q}(1) \to \mathfrak{t}_{g,r}^n(x_o) \to \mathfrak{u}_{g,r}^n(x_o) \to 0$$

*is an extension of MHSs, and the weight filtration equals the Malcev filtration.*

9.2. **A presentation of $\mathfrak{t}_g$.** We denote the Malcev Lie algebra of $T_{g,r}^n$ by $\mathfrak{t}_{g,r}^n$. The existence of a MHS on $\mathfrak{t}_{g,r}^n(x_o)$ implies that, after tensoring with $\mathbb{C}$, there is a canonical isomorphism

$$\mathfrak{t}^n_{g,r}(x_o) \otimes \mathbb{C} \cong \prod_m \mathrm{Gr}^W_{-m} \mathfrak{t}^n_{g,r}(x_o) \otimes \mathbb{C}.$$

Since the left hand side is (noncanonically) isomorphic to $\mathfrak{t}^n_{g,r} \otimes \mathbb{C}$, to give a presentation of $\mathfrak{t}^n_{g,r} \otimes \mathbb{C}$, it suffices to give a presentation of its associated graded. It follows from Johnson's Theorem (7.1) that each graded quotient of the lower central series of $\mathfrak{t}_g$ is a representation of the algebraic group $Sp_g$. We will give a presentation of $\mathrm{Gr}^W_\bullet \mathfrak{t}_g$ in the category of representations of $Sp_g$. Recall that $\lambda_1, \dots, \lambda_g$ is a set of fundamental weights of $Sp_g$. For a nonnegative integral linear combination of the fundamental weights $\lambda = \sum_{i=1}^g n_i \lambda_i$ we denote by $V_g(\lambda)$ the representation of $Sp_g$ with highest weight $\lambda$.

For all $g \geq 3$, the representation $\Lambda^2 V_g(\lambda_3)$ contains a unique copy of $V_g(2\lambda_2) + V_g(0)$. Denote the $Sp_g$ invariant complement of this by $R_g$. Since the quadratic part of the free Lie algebra $\mathrm{Lie}(V_g)$ is $\Lambda^2 V_g$, we can view $R_g$ as being a subspace of the quadratic elements of $\mathrm{Lie}(V_g(\lambda_3))$.

As mentioned earlier, it is unknown whether any $T_g$ is finitely presented when $g \geq 3$. But the following theorem says that its de Rham incarnation is:

**Theorem 9.3** (Hain [26]). *For all $g \geq 3$, $\mathfrak{t}_g$ is isomorphic to the completion of its associated graded $\mathrm{Gr}^W_\bullet \mathfrak{t}_g$. When $g \geq 6$, this has presentation*

$$\mathrm{Gr}^W_\bullet \mathfrak{t}_g = \mathrm{Lie}(V_g(\lambda_3))/(R_g),$$

*where $R_g$ is the set of quadratic relations defined above. When $3 \leq g < 6$, the relations in $\mathrm{Gr}^W_\bullet \mathfrak{t}_g$ are generated by the quadratic relations $R_g$, and possibly some cubic relations. In particular, $\mathfrak{t}^n_{g,r}$ is finitely presented whenever $g \geq 3$.*

Note that this, combined with (9.2) gives a presentation of $\mathrm{Gr}^W_\bullet \mathfrak{u}_g$ when $g \geq 6$:

**Corollary 9.4.** *For all $g \geq 3$, $\mathfrak{u}_g$ is isomorphic to the completion of its associated graded $\mathrm{Gr}^W_\bullet \mathfrak{u}_g$. When $g \geq 6$, this has quadratic presentation*

$$\mathrm{Gr}^W_\bullet \mathfrak{u}_g = \mathrm{Lie}(V_g(\lambda_3))/(R_g + V_g(0)),$$

*where $R_g$ is the set of quadratic relations defined above and where $V_g(0)$ is the unique copy of the trivial representation in $\Lambda^2 V_g(\lambda_3)$. When $3 \leq g < 6$, the relations in $\mathrm{Gr}^W_\bullet \mathfrak{u}_g$ are generated by the quadratic relations $R_g + V_g(0)$, and possibly some cubic relations. In particular, $\mathfrak{u}^n_{g,r}$ is finitely presented whenever $g \geq 3$.*

The proof that the relations in the presentation of $\mathfrak{t}_g$ are generated by quadratic relations when $g \geq 6$ and quadratic and cubic ones when $g \geq 3$ is not topological, but uses deep Hodge theory and, surprisingly, intersection homology. The key ingredients are a result of Kabanov, which we state below, and M. Saito's theory of Hodge modules.

We define the *Satake compactification* $\overline{\mathcal{M}}_g^{\mathrm{sat}}$ of $\mathcal{M}_g$ as the closure of $\mathcal{M}_g$ inside the (Baily-Borel-)Satake compactification of $\mathcal{A}_g$.

**Theorem 9.5** (Kabanov [46]). *For each irreducible representation $V$ of $Sp_g$, the natural map*

$$IH^2(\overline{\mathcal{M}}_g^{\mathrm{sat}}; \mathbb{V}) \to H^2(\mathcal{M}_g; \mathbb{V})$$

*is an isomorphism when $g \geq 6$. Here $\mathbb{V}$ denotes the generically defined local system corresponding to $V$.*

Such a local system $\mathbb{V}$ is, up to a Tate twist, canonically a variation of Hodge structure. Saito's purity theorem then implies that $H^2(\mathcal{M}_g;\mathbb{V})$ is pure of weight 2+ the weight of $\mathbb{V}$ when $g \geq 6$. It is this purity result that forces $H^2(\mathfrak{t}_g)$ to be of weight 2, and implies that no higher order relations are needed.

9.3. **Understanding $\mathfrak{t}_g$.** Even though we have a presentation of $\mathfrak{t}_g$, we still do not have a good understanding of its graded quotients, either as vector spaces or as $Sp_g$ modules. There is an exact sequence

$$0 \to \mathfrak{p}_g \to \mathfrak{t}_g^1 \to \mathfrak{t}_g \to 0$$

of Lie algebras (recall that $\mathfrak{p}_g$ stands for the Malcev Lie algebra of $\pi_g = \pi_1(S, x_0)$).

It is the de Rham incarnation of the exact sequence of fundamental groups associated to the universal curve. Fix a conformal structure on $(S, x_0)$. Then this sequence is an exact sequence of MHSs. Since $\mathrm{Gr}^W$ is an exact functor, and since $\mathrm{Gr}^W_\bullet\, \mathfrak{p}_g$ is well understood, it suffices to understand $\mathrm{Gr}^W_\bullet\, \mathfrak{t}_g^1$.

There is a natural representation

$$\mathfrak{t}_g^1 \to \operatorname{Der} \mathfrak{p}_g \tag{6}$$

It is a morphism of MHS, and therefore determined by the graded Lie algebra homomorphism

$$\mathrm{Gr}^W_\bullet\, \mathfrak{t}_g^1 \to \operatorname{Der} \mathrm{Gr}^W_\bullet\, \mathfrak{p}_g.$$

One can ask how close it is to being an isomorphism. Since this map is induced by the natural homomorphism

$$\Gamma_g^1 \to \varprojlim \operatorname{Aut} \mathbb{C}\pi_g/I^m, \tag{7}$$

the homomorphism (6) factors through the projection $\mathfrak{t}_g^1 \to \mathfrak{u}_g^1$, and therefore cannot be injective. On the other hand, we have the following (reformulated) result of Morita:

**Theorem 9.6** (Morita [68]). *There is a natural Lie algebra surjection*

$$Tr_M : W_{-1} \operatorname{Der} \mathrm{Gr}^W_\bullet\, \mathfrak{p}_g \to \oplus_{k\geq 1} S^{2k+1} H_1(S)$$

*onto an abelian Lie algebra whose composition with (6) is trivial. Here $S^m$ denotes the $m^{th}$ symmetric power.*

One may then hope that the sequence

$$0 \to \mathbb{C} \to \mathrm{Gr}^W_\bullet\, \mathfrak{t}_g^1 \to W_{-1} \operatorname{Der} \mathrm{Gr}^W_\bullet\, \mathfrak{p}_g \to \oplus_{k\geq 1} S^{2k+1} H_1(S) \to 0$$

is exact. However, there are further obstructions to exactness at $W_{-1} \operatorname{Der} \mathrm{Gr}^W_\bullet\, \mathfrak{p}_g$ which were discovered by Nakamura [70]. They come from Galois theory and use the fact that $\mathcal{M}_g^1$ is defined over $\mathbb{Q}$.[12] On the other hand, one can ask:

*Question 9.7.* Is the map $\mathfrak{u}_g^1 \to \operatorname{Der} \mathfrak{p}_g$ injective? Equivalently, is $\mathcal{G}_g^1$ the Zariski closure of the image of the representation (7)?

A good understanding of $\mathfrak{t}_g$ may help in understanding the stable cohomology of $\Gamma_g$ as we shall explain in the next subsection.

[12] Actually, Nakamura proves his result for a corresponding sequence for $\mathfrak{t}_{g,1}$, but his obstructions most likely appear in this case too.

9.4. **Torelli Lie algebras and the cohomology of $\Gamma_g$.** Each Malcev Lie algebra $\mathfrak{g}$ can be viewed as a complete topological Lie algebra. A basis for the neighbourhoods of 0 being the terms $\mathfrak{g}^{(k)}$ of the Malcev filtration. One can define the continuous cohomology of such a $\mathfrak{g}$ to be

$$H^\bullet(\mathfrak{g}) := \varinjlim H^\bullet(\mathfrak{g}/\mathfrak{g}^{(k)}).$$

If $\mathfrak{g}$ has a MHS, then so will $H^\bullet(\mathfrak{g})$. The continuous cohomology of $\mathfrak{t}_{g,r}^n$, $\mathfrak{u}_{g,r}^n$, etc. each has an action of $Sp_g$. The general theory of relative Malcev completion [25] gives a canonical homomorphism

$$H^\bullet(\mathfrak{u}_{g,r}^n)^{Sp_g} \to H^\bullet(\mathcal{M}_{g,r}^n) \tag{8}$$

One can ask how much of the cohomology of $\mathcal{M}_{g,r}^n$ is captured by this map.

Fix a base point $x_o$ of $\mathcal{M}_{g,r}^n$. Then $\mathfrak{t}_{g,r}^n$, etc. all have compatible MHSs, and these induce MHSs on their continuous cohomology groups. These groups have the property that the weights on $H^k$ are $\geq k$.

**Theorem 9.8** (Hain [26]). *The map (8) is a morphism of MHS.*

Since $H_1(T_g)$ is a quotient of $\mathfrak{u}_g$, there is an induced map

$$H^\bullet(H_1(T_g)) \to H^\bullet(\mathfrak{u}_g). \tag{9}$$

This is also a morphism of MHS. The following result follows directly from [22, (9.2)], the presentation (9.3) of $\mathfrak{t}_g$, and the existence of the MHS on $\mathfrak{u}_g$.

**Proposition 9.9.** *If $g \geq 3$, the map (9) surjects onto the lowest weight subring*

$$\oplus_{k\geq 0} W_k H^k(\mathfrak{u}_g)$$

*of $H^\bullet(\mathfrak{u}_g)$, and the kernel is generated by the ideal generated by the unique copy of $V_g(2\lambda_2)$ in $H^2(H_1(T_g))$.*

Similar results hold when $r + n > 0$ — cf. [26, §14.6].

The following result of Kawazumi and Morita tells us that the image of the lowest weight subring of $H^\bullet(\mathfrak{u}_g)^{Sp_g}$ contains no new cohomology classes.

**Theorem 9.10** (Kawazumi-Morita [48]). *The image of the natural map*

$$H^\bullet(H_1(T_g))^{Sp_g} \to H^\bullet(\mathcal{M}_g)$$

*is precisely the subring generated by the $\kappa_i$'s.*

If we combine this with the previous two results and Pikaart's Purity Theorem (4.2), we obtain the following strengthening of the theorem of Kawazumi and Morita (and obtained independently by Morita, building on our work):

**Theorem 9.11.** *When $k \leq g/2$, the image of $H^k(\mathfrak{u}_g)^{Sp_g} \to H^k(\mathcal{M}_g)$ is the degree $k$ part of the subring generated by the $\kappa_i$'s.*

To continue the discussion further, it seems useful to consider cohomology with symplectic coefficients.

9.5. **Cohomology with symplectic coefficients.** The irreducible representations of $Sp_g$ are parametrized by Young diagrams with $\leq g$ rows (and no indexing of the boxes), in other words, by nonincreasing sequences of nonnegative integers whose terms with index $> g$ are zero. So any such sequence $\alpha = (\alpha_1, \alpha_2, \dots)$ defines an irreducible representation of $Sp_h$ for all $h \geq g$. We will denote the representation of $Sp_g$ corresponding to $\alpha$ by $V_{g,\alpha}$, and the corresponding (orbifold) local system over $\mathcal{M}_g$ by $\mathbb{V}_{g,\alpha}$.

A theorem of Ivanov [36] (that in fact pertains to more general local systems) implies that, when $r \geq 1$, the group $H^k(\Gamma^n_{g,r}; V_{g,\alpha})$ is independent of $g$ once $g$ is large enough. In the case at hand we have a more explicit result that we state here for the undecorated case (a case that Ivanov actually excludes).

**Theorem 9.12** (Looijenga [58]). *Let $\alpha = (\alpha_1, \alpha_2, \dots)$ be a nonincreasing sequence of nonnegative integers that is eventually zero, and let $c_1, c_2, \dots$ be weighted variables with $\deg(c_i) = 2i$. Put $|\alpha| := \sum_{i\geq 1} \alpha_i$. Then there exists a finitely generated, evenly graded $\mathbb{Q}[c_1, \dots, c_{|\alpha|}]$-module $A^\bullet_\alpha$ (that can be described explicitly) and a graded homomorphism of $H^\bullet(\Gamma_\infty)$ modules*

$$A^\bullet_\alpha[-|\alpha|] \otimes H^\bullet(\Gamma_\infty) \to H^\bullet(\Gamma_g; V_{g,\alpha})$$

*that is an isomorphism in degree $\leq cg - |\alpha|$. It is also a MHS morphism if we take $A^{2k}_\alpha$ to be pure of type $(k,k)$. In particular, we have*

$$A^\bullet_\alpha[-|\alpha|] \otimes H^\bullet(\Gamma_\infty) \cong H^\bullet(\Gamma_\infty; V_\alpha)$$

*both as MHSs and as graded $H^\bullet(\Gamma_\infty)$ modules. So, by (4.2), $H^k(\Gamma_\infty; V_\alpha)$ is pure of weight $k + |\alpha|$.*

It is useful to try to understand all cohomology groups with symplectic coefficients at the same time. To do this we take a leaf out of the physicist's book and consider the "generating function"

$$\oplus_\alpha H^\bullet(\Gamma_g; V^*_\alpha) \otimes V_\alpha \tag{10}$$

where $\alpha$ ranges over all partitions with $\leq g$ rows, and $*$ denotes dual. This is actually a graded commutative ring as the Peter-Weyl Theorem implies that the coordinate ring $\mathcal{O}_g$ of $Sp_g$ is

$$\mathcal{O}_g = \oplus_\alpha (\operatorname{End} V_\alpha)^* \cong \oplus_\alpha V^*_\alpha \otimes V_\alpha.$$

The mapping class group acts on $\mathcal{O}_g$ by composing the right translation action of $Sp_g$ on $\mathcal{O}_g$ with the canonical representation $\Gamma_g \to Sp_g$. The corresponding cohomology group $H^\bullet(\Gamma_g; \mathcal{O})$ is then the "generating function" (10). Note that $\mathcal{O}_g$ is a variation of Hodge structure of weight 0, so the group $H^k(\Gamma_g; \mathcal{O}_g)$ is stably of weight $k$ by the above theorem.

There is a canonical algebra homomorphism

$$H^\bullet(\mathfrak{u}_g) \to H^\bullet(\Gamma_g; \mathcal{O}_g)$$

whose existence follows from the de Rham theory of relative completion suggested by Deligne — cf. [25]. The map (8) of the previous subsection is just its invariant part. This map is a MHS morphism for each choice of complex structure on $S$. The $\alpha$ isotypical part of both sides stabilizes as $g$ increases. It is natural to ask:

*Question 9.13.* Is this map stably an isomorphism?

This has been verified by Hain and Kabanov (unpublished) in degrees $\leq 2$ for all weights, and in degree 3 and weight 3. If the answer is yes, or even if one has surjectivity, then it will follow from the theorem of Kawazumi and Morita (9.10) that the stable cohomology of $\mathcal{M}_g$ is generated by the $\kappa_i$'s. A consequence of injectivity and Pikaart's Purity Theorem would be that for each $k$, $H^k(\mathfrak{u}_g)$ is pure of weight $k$ once the genus is sufficiently large. This is equivalent to the answer to the following question being affirmative.

*Question 9.14.* Are $H^\bullet(\mathfrak{u}_g)$ and $U\,\mathrm{Gr}^W_\bullet\,\mathfrak{u}_g$ stably Koszul dual?

Note that $U\,\mathrm{Gr}^W_\bullet\,\mathfrak{u}_g$ and the lowest weight subalgebra of $H^\bullet(\mathfrak{u}_g)$ have dual quadratic presentations.

## 10. Algebras Related to the Cohomology of Moduli Spaces of Curves

The ribbon graph description is the root of a number of ways of constructing (co)homology classes on moduli spaces of curves from certain algebraic structures. These constructions have in common that they actually produce cellular (co)chains on $\mathbb{M}_g^n$, and so they are recipes that assign numbers to 'oriented' ribbon graphs. The typical construction, due to Kontsevich [52], goes like this: assume that we are given a complex vector space $V$, a symmetric tensor $p \in V \otimes V$, and linear forms $T_k : V^{\otimes k} \to \mathbb{C}$ that are cyclically invariant. If $\Gamma$ is a ribbon graph, then the decomposition of $X(G)$ into $\sigma_1$ and $\sigma_0$-orbits gives isomorphisms

$$\otimes_{s\in X_1(\Gamma)} V^{\otimes\,\mathrm{or}(s)} \cong V^{\otimes X(G)} \cong \otimes_{v\in X_0(\Gamma)} V^{\otimes\,\mathrm{out}(v)},$$

where $\mathrm{or}(s)$ stands for the two-element set of orientations of the edge $s$ and $\mathrm{out}(v)$ for the set of oriented edges that have $v$ as initial vertex. Now $p^{\otimes X_1(\Gamma)}$ defines a vector of the lefthand side and a tensor product of certain $T_k$'s defines a linear form on the righthand side. Evaluation of the linear form on the vector gives a number, which is clearly an invariant of the ribbon graph. Since this invariant does not depend on an orientation on the set of edges of $G$, it cannot be used directly to define a cochain on the combinatorial moduli spaces. To this end we need some sign rules so that, for instance, the displayed isomorphisms acquire a sign. The tensors $p$ and $T_k$ are sometimes referred to as the *propagator* and the *interactions*, respectively, to remind us of their physical origin.

If $p$ is nondegenerate, then we may use it to identify $V$ with its dual. In this case $T_k$ defines a linear map $V^{\otimes(k-1)} \to V$. The properties one needs to impose on propagator and interactions in order that the above recipe produce a cocycle on $\mathbb{M}_g^n$ is that they define a $\mathbb{Z}/2$ graded $A_\infty$ algebra with inner product. A similar recipe assigns cycles on $\mathbb{M}_g^n$ to certain $\mathbb{Z}/2$ graded differential algebras. The cocycles can be evaluated on the cycles and this, in principle, gives a method of showing that some of the classes thus obtained are nonzero. We shall not be more precise, but instead refer to [52] or [1] for an overview. A simple example is to take $V = \mathbb{C}$, $p := 1 \otimes 1$ and $T_k(z^k)$ arbitrary for $k \geq 3$ odd, and zero otherwise. Kontsevich asserts that the classes thus obtained are all tautological.

10.1. **Outer space.** In Section 6 we encountered a beautiful combinatorial model for a virtual classifying space of the mapping class group $\Gamma_g^n$. There is a similar, but simpler, combinatorial model that does the same job for the outer automorphism group of a free group.

We fix an integer $r \geq 2$ and consider connected graphs $G$ with first Betti number equal to $r$ and where each vertex has degree $\geq 3$. Let us call these graphs *r-circular graphs.* The maximal number of edges (resp. vertices) such a graph can have is $3r-3$ (resp. $2r-2$). These bounds are realized by all trivalent graphs of this type. Notice that an $r$-circular graph $G$ has fundamental group isomorphic to the free group on $r$ generators, $F_r$. We say that $G$ is *marked* if we are given an isomorphism $\phi : F_r \to \pi_1(G, \text{base point})$ up to inner automorphism. The group $\operatorname{Out}(F_r)$ permutes these markings simply transitively. There is an obvious notion of isomorphism for marked $r$-circular graphs. We shall denote the collection of isomorphism classes by $\mathcal{G}_r$.

Let $(G,[\phi])$ represent an element of $\mathcal{G}_r$. The metrics on $G$ that give $G$ total length 1 are parameterized by the interior of a simplex $\Delta(G)$. We fit these simplices together in a way analogous to the ribbon graph case: if $s$ is an edge of $G$ that is not a loop, then collapsing it defines another element $(G/s,[\phi]/s)$ of $\mathcal{G}_r$. We may then identify $\Delta(G/s)$ with a face of $\Delta(G)$. After we have made these identifications we end up with a simplicial complex $\widehat{\mathbb{O}}_r$. The union of the interiors of the simplices $\Delta(G)$ (indexed by $\mathcal{G}_r$) will be denoted by $\mathbb{O}_r$; it is the complement of a closed subcomplex of $\widehat{\mathbb{O}}_r$. This construction is due to Culler-Vogtmann [9]. We call $\mathbb{O}_r$ the *outer space of order r* for reasons that will become apparent in a moment. Observe that $\widehat{\mathbb{O}}_r$ comes with a simplicial action of $\operatorname{Out}(F_r)$. We denote the quotient of $\widehat{\mathbb{O}}_r$ (resp. $\mathbb{O}_r$) by $\operatorname{Out}(F_r)$ by $\widehat{\mathbb{G}}_r$ (resp. $\mathbb{G}_r$). It is easy to see that $\widehat{\mathbb{G}}_r$ is a finite orbicomplex. The open subset $\mathbb{G}_r$ is the moduli space of metrized $r$-circular graphs. It has a spine of dimension $2r-3$.

**Theorem 10.1** (Culler-Vogtmann [9], Gersten). *The outer space of order r is contractible and a subgroup of finite index of* $\operatorname{Out}(F_r)$ *acts freely on it. Hence* $\mathbb{G}_r$ *is a virtual classifying space for* $\operatorname{Out}(F_r)$ *and* $\operatorname{Out}(F_r)$ *has virtual homological dimension* $\leq 2r-3$.

In contrast to the ribbon graph case, $\mathbb{O}_r$ is not piecewise smooth. If we choose $2g-1+n$ free generators for the fundamental group of our reference surface $S_g^n$, then each ribbon graph without vertices of degree $\leq 2$ determines an element of $\mathbb{G}_{2g-1+n}$: simply forget the ribbon structure. The ribbon data is finite and it is therefore not surprising that forgetting the ribbons defines a finite map

$$\hat{f} : \mathcal{S}_n \backslash \widehat{\mathbb{M}}_g^n \to \widehat{\mathbb{G}}_{2g-1+n}$$

of orbicomplexes. Here $\mathcal{S}_n$ stands for the symmetric group, which acts in the obvious way on $\widehat{\mathbb{M}}_g^n$. Following Strebel's theorem, the preimage of $\mathbb{G}_{2g-1+n}$ can be identified with $\mathcal{S}_n \backslash (\mathcal{M}_g^n \times \operatorname{int} \Delta^{n-1})$. We denote the resulting map by

$$f : \mathcal{S}_n \backslash (\mathcal{M}_g^n \times \operatorname{int} \Delta^{n-1}) \to \mathbb{G}_{2g-1+n}.$$

It induces the evident map

$$f_* : H_k(\Gamma_g^n)_{\mathcal{S}_n} \to H_k(\operatorname{Out} F_{2g-1+n})$$

on rational homology. It is unclear whether there is such an interpretation for the induced map on cohomology with compact supports. We remark that $\mathcal{M}_g^n \times \operatorname{int} \Delta^{n-1}$ is canonically oriented, but that its $\mathcal{S}_n$-orbit space is not (since transpositions reverse this orientation). Poincaré duality therefore takes the form

$$H_k(\mathcal{M}_g^n)_{\mathcal{S}_n} \cong H_k(\mathcal{S}_n \backslash (\mathcal{M}_g^n \times \operatorname{int} \Delta^{n-1}); \epsilon) \cong H_c^{6g-7+3n-k}(\mathcal{S}_n \backslash (\mathcal{M}_g^n \times \operatorname{int} \Delta^{n-1}); \epsilon),$$

where $\epsilon$ is the signum representation of $\mathcal{S}_n$. If $\delta$ denotes the (signum) character of $\mathrm{Out}(F_r)$ on $\wedge^r H_1(F_r)$, then the adjoint of $f_*$ is a map

$$H_c^{6g-7+3n-k}(\mathbb{G}_{2g-1+n};\delta) \to H_c^{6g-7+3n-k}(\mathcal{S}_n\backslash(\mathcal{M}_g^n \times \operatorname{int}\Delta^{n-1});\epsilon) \cong H_k(\mathcal{M}_g^n)_{\mathcal{S}_n}.$$

So, when $m \geq 1$, we have maps

$$H_c^{i+m-1}(\mathbb{G}_{m+1};\delta) \to \oplus_{2g-2+n=m} H_{2m-i}(\mathcal{M}_g^n)_{\mathcal{S}_n} \to H_{2m-i}(\mathbb{G}_{m+1}), \quad i = 0,1,\ldots.$$

There is a remarkable interpretation of this sequence that we will discuss next.

10.2. **Three Lie algebras.** We describe Kontsevich's three functors from the category of symplectic vector spaces to the category of Lie algebras and their relation with the cohomology of the moduli spaces $\mathcal{M}_g^n$. The basic references are [51] and [52].

We start out with a finite dimensional $\mathbb{Q}$ vector space $V$ endowed with a nondegenerate antisymmetric tensor $\omega_V \in V \otimes V$. Let $\mathrm{Ass}(V)$ be the tensor algebra (i.e., the free associative algebra) generated by $V$. We grade it by giving $V$ degree $-1$. The Lie subalgebra generated by $V$ is free and so we denote it by $\mathrm{Lie}(V)$. It is well-known that $\mathrm{Ass}(V)$ may be identified with the universal enveloping algebra of $\mathrm{Lie}(V)$. If we mod out $\mathrm{Ass}(V)$ by the two-sided ideal generated by the degree $\leq -2$ part of $\mathrm{Lie}(V)$, we obtain the symmetric algebra $\mathrm{Com}(V)$ of $V$. Define $\mathfrak{g}_{\mathrm{ass}}(V)$ (resp. $\mathfrak{g}_{\mathrm{lie}}(V)$) to be the Lie algebra of derivations of $\mathrm{Ass}(V)$ (resp. $\mathrm{Lie}(V)$) of degree $\leq 0$ that kill $\omega_V$. Since each derivation of $\mathrm{Lie}(V)$ extends canonically to its universal enveloping algebra, we have an inclusion $\mathfrak{g}_{\mathrm{lie}}(V) \subset \mathfrak{g}_{\mathrm{ass}}(V)$. There is also a corresponding Lie algebra $\mathfrak{g}_{\mathrm{com}}(V)$ of derivations of degree $\leq 0$ of $\mathrm{Com}(V)$ that kill $\omega_V$. Here we regard the latter as a two-form on the affine space $\operatorname{Spec}\mathrm{Com}(V)$. This Lie algebra is a quotient of $\mathfrak{g}_{\mathrm{ass}}(V)$. All three Lie algebras are graded and have as degree zero summand the Lie algebra $\mathfrak{sp}(V)$ of the group $\mathrm{Sp}(V)$ of symplectic transformations of $V$. A simple verification shows that the degree $-1$ summands have as $\mathfrak{sp}(V)$ representations the following natural descriptions:

$$\mathfrak{g}_{\mathrm{com}}(V)_{-1} \cong S^3(V), \quad \mathfrak{g}_{\mathrm{ass}}(V)_{-1} \cong S^3(V) \oplus \wedge^3 V, \quad \mathfrak{g}_{\mathrm{lie}}(V)_{-1} \cong \wedge^3 V.$$

These Lie algebras are functorial with respect to symplectic injections $(V,\omega_V) \hookrightarrow (W,\omega_W)$. Note that $\mathrm{Sp}(V)$ acts trivially on this cohomology of the Lie algebra in question because $\mathfrak{sp}(V) \subset \mathfrak{g}_*(V)$. This implies that $H^k(\mathfrak{g}_*(V))$, $* \in \{\mathrm{lie}, \mathrm{ass}, \mathrm{com}\}$, depends only on $\dim V$. We form the inverse limit:

$$H^k(\mathfrak{g}_*) := \varprojlim{}_V H^k(\mathfrak{g}_*(V)).$$

The sum over $k$, $H^\bullet(\mathfrak{g}_*)$, has the structure of a connected graded bicommutative Hopf algebra; the coproduct comes from the direct sum operation on symplectic vector spaces. It is actually bigraded: apart from the cohomological grading there is another coming from the grading of the Lie algebras. Notice that the latter grading has all its degrees $\geq 0$. The primitive part $H^\bullet_{\mathrm{pr}}(\mathfrak{g}_*)$ inherits this bigrading. Furthermore, the natural maps

$$H^\bullet(\mathfrak{g}_{\mathrm{com}}) \to H^\bullet(\mathfrak{g}_{\mathrm{ass}}) \to H^\bullet(\mathfrak{g}_{\mathrm{lie}})$$

are homomorphisms of bigraded Hopf algebras. Consequently, we have induced maps between the bigraded pieces of their primitive parts.

**Theorem 10.2** (Kontsevich [51], [52]). *For* $* \in \{\text{lie}, \text{ass}, \text{com}\}$ *we have*

$$H^k_{\text{pr}}(\mathfrak{g}_*)_0 = H^k_{\text{pr}}(\mathfrak{sp}_\infty) \cong \begin{cases} \mathbb{Q} & \text{for } k = 3, 7, 11, \dots ; \\ 0 & \text{otherwise.} \end{cases}$$

*Furthermore,* $H^k_{\text{pr}}(\mathfrak{g}_*)_l = 0$ *when* $l$ *is odd and, when* $m > 0$, *we have a natural diagram*

$$\begin{array}{ccccc} H^k_{\text{pr}}(\mathfrak{g}_{\text{com}})_{2m} & \longrightarrow & H^k_{\text{pr}}(\mathfrak{g}_{\text{ass}})_{2m} & \longrightarrow & H^k_{\text{pr}}(\mathfrak{g}_{\text{lie}})_{2m} \\ \cong\downarrow & & \cong\downarrow & & \cong\downarrow \\ H_c^{k+m-1}(\mathbb{G}_{m+1}, \delta) & \xrightarrow{f_c^*} & \oplus_{2g-2+n=m} H_{2m-k}(\mathcal{M}_g^n)_{\mathcal{S}_n} & \xrightarrow{f_*} & H_{2m-k}(\mathbb{G}_{m+1}) \end{array}$$

*which commutes up to sign and whose rows are complexes. The maps in the top row are the natural maps and the bottom row is the sequence defined in Section 10.1.*

The proof is an intelligent application of classical invariant theory. For each of the three Lie algebras one writes down the standard complex. The subcomplex of invariants with respect to the symplectic group is quasi-isomorphic to the full complex. Weyl's invariant theory furnishes a natural basis for this subcomplex. Kontsevich then observes that this makes the subcomplex naturally isomorphic to a cellular chain (or cochain) complex of one of the cell complexes $\mathbb{G}_*$ and $\mathbb{M}_*^*$ whose (co)homology appears in the bottom row.

The diagram in this theorem suggests that the sequence of natural transformations Lie $\to$ Ass $\to$ Com is self dual in some sense. This can actually be pinned down by looking at the corresponding operads: Ginzburg and Kapranov [18] observed that these operads have "quadratic relations" and they proved the self duality of the operad sequence in a Koszul sense.

However, our main reason for displaying this diagram is that it pertains to the cohomology of the moduli spaces of curves in two apparently unrelated ways. The first one is evident. The $\mathcal{S}_n$ coinvariants of the homology of $\mathcal{M}_g^n$ features in the middle column, but the righthand column has something to do with the cohomology of a 'linearization' of $\Gamma_\infty$: we will see that $\mathfrak{g}_{\text{lie}}$ is intimately related to the Lie algebra of the relative Malcev completions discussed in Section 8. We explain this in the next subsection after a giving a restatement of Kontsevich's Theorem.

In this restatement the Lie algebra cohomology of $\mathfrak{g}_*(V)$ is replaced by the relative Lie algebra cohomology of the pair $(\mathfrak{g}_*(V), \mathfrak{k}(V))$, where $\mathfrak{k}(V)$ is a maximal compact Lie subalgebra of $\mathfrak{g}_*(V)_0$, and therefore of $\mathfrak{g}_*(V)$ ($\mathfrak{k}(V)$ is a unitary Lie algebra of rank $\dim V/2$). As above, the Lie algebra cohomology $H^k(\mathfrak{g}_*(V), \mathfrak{k}(V))$ depends only on the dimension of $V$ and stabilizes once $\dim V$ is sufficiently large. We denote the inverse limit of these groups by $H^\bullet(\mathfrak{g}_*, \mathfrak{k}_\infty)$. Likewise, we denote the stable cohomology of the pair $(\mathfrak{sp}_g, \mathfrak{k}_g)$ by $H^\bullet(\mathfrak{sp}_\infty, \mathfrak{k}_\infty)$. By a theorem of Borel [5], this is naturally isomorphic to the stable cohomology of $\mathcal{A}_g$ and is a polynomial algebra generated by classes $c_1, c_3, c_5, \dots$, where $c_k$ has degree $2k$.

Combining Kontsevich's Theorem 10.2 with Borel's computation and an elementary spectral sequence argument, we obtain the following result. (Use the fact that $(\mathfrak{sp}_g, \mathfrak{k}_g)$ is both a sub and a quotient of $(\mathfrak{g}_*(V), \mathfrak{k}(V))$.)

**Corollary 10.3.** *We have*

$$H^k_{\text{pr}}(\mathfrak{g}_*, \mathfrak{k}_\infty)_0 = H^k_{\text{pr}}(\mathfrak{sp}_\infty, \mathfrak{k}_\infty) \cong \begin{cases} \mathbb{Q} & \text{for } k = 2, 6, 10, \dots ; \\ 0 & \text{otherwise.} \end{cases}$$

*Furthermore, when $m > 0$, the natural maps $H^\bullet(\mathfrak{g}_*, \mathfrak{k}_\infty)_m \to H^\bullet(\mathfrak{g}_*)_m$ are isomorphisms.*

10.3. **Relation with the relative Malcev completion.** We begin with an observation. For a symplectic vector space $V$ and $* \in \{\text{lie}, \text{ass}, \text{com}\}$, denote by $\mathfrak{g}_*^\flat(V)$ the subalgebra of $\mathfrak{g}_*(V)$ generated by its summands of weight 0 and $-1$. Kontsevich's computation shows:

**Proposition 10.4.** *The graded cohomology groups $H^k(\mathfrak{g}_*^\flat(V), \mathfrak{k}(V))_l$ stabilize and the sum of the stable terms is a bigraded bicommutative Hopf algebra $H^\bullet(\mathfrak{g}_*^\flat, \mathfrak{k}_\infty)_\bullet$. In addition, the restriction map $H^k_{\text{pr}}(\mathfrak{g}_*, \mathfrak{k}_\infty)_l \to H^k_{\text{pr}}(\mathfrak{g}_*^\flat(V), \mathfrak{k}_{\text{lie}}(V))_l$ is an isomorphism when $l \leq k$.*

The case of interest here is that of lie where $\mathfrak{g}_{\text{lie}}^\flat(V)_0 = \mathfrak{sp}(V)$ and $\mathfrak{g}_{\text{lie}}^\flat(V)_{-1} \cong \Lambda^3 V$. Denote by $z_m$ the element of $H^{2m}_{\text{pr}}(\mathfrak{g}_{\text{lie}})_{2m}$ that corresponds, via Theorem 10.2, to $1 \in H_0(\mathbb{G}_{m+1})$. The preceding proposition yields:

**Corollary 10.5.** *We have a natural isomorphism of bigraded Hopf algebras*

$$\sum_{l \leq k} H^k(\mathfrak{g}_{\text{lie}}^\flat, \mathfrak{k}_\infty)_l \cong H^\bullet(\mathfrak{sp}_\infty, \mathfrak{k}_\infty)[z_1, z_2, \dots] \cong \mathbb{C}[c_1, c_3, c_5, \dots, z_1, z_2, z_3, \dots]$$

*where each $z_i$ and $c_j$ is primitive.*

The graded Lie algebra $\mathfrak{g}_{\text{lie}}^\flat(V)$ is the semi-direct product of $\mathfrak{sp}(V)$ and its elements of positive weight, which we shall denote by $\mathfrak{u}_{\text{lie}}^\flat$. Consequently, there are natural inclusions

$$H^\bullet(\mathfrak{u}_{\text{lie}}^\flat(V))^{Sp} \hookrightarrow H^\bullet(\mathfrak{g}_{\text{lie}}^\flat(V), \mathfrak{k}(V)) \text{ and } H^\bullet(\mathfrak{sp}(V), \mathfrak{k}(V)) \hookrightarrow H^\bullet(\mathfrak{g}_{\text{lie}}^\flat(V), \mathfrak{k}(V)).$$

Together these induce an algebra homomorphism

$$H^\bullet(\mathfrak{sp}(V), \mathfrak{k}(V)) \otimes H^\bullet(\mathfrak{u}_{\text{lie}}^\flat(V))^{Sp} \to H^\bullet(\mathfrak{g}_{\text{lie}}^\flat(V), \mathfrak{k}(V))$$

which is compatible with stabilization.

**Proposition 10.6.** *Upon stabilization, these maps induce an isomorphism*

$$H^\bullet(\mathfrak{sp}_\infty, \mathfrak{k}_\infty) \otimes H^\bullet(\mathfrak{u}_{\text{lie}}^\flat)^{Sp} \to H^\bullet(\mathfrak{g}_{\text{lie}}^\flat, \mathfrak{k}_\infty).$$

Next we relate the graded Lie algebra $\mathfrak{g}_{\text{lie}}^\flat$ to the filtered Lie algebra $\mathfrak{g}_{g,1}$ of the relative Malcev completion $\mathcal{G}_{g,1}$ of $\Gamma_{g,1} \to Sp_g$. Recall from Section 2 that $\pi_g^1$ is freely generated by $2g$ generators named $\alpha_\pm, \dots, \alpha_{\pm g}$ so that the commutator $\beta := (\alpha_1, \alpha_{-1}) \cdots (\alpha_g, \alpha_{-g})$ represents a simple loop around $x_1$. Using Latin letters for the logarithms of the images of elements of $\pi_g^1$ in its Malcev completion, we find that

$$b \equiv [a_1, a_{-1}] + \cdots + [a_g, a_{-g}] \mod (\mathfrak{p}_g^1)^{(3)}.$$

So the image of $b$ in $\operatorname{Gr}^2 \mathfrak{p}_g^1 \cong \Lambda^2 V_g$ is the symplectic form $\omega_S$.

The obvious homomorphism $\Gamma_{g,1} \to \operatorname{Aut}(\pi_g^1)$ induces a Lie algebra homomorphism

$$\mathfrak{g}_{g,1} \to \operatorname{Der} \mathfrak{p}_g^1. \tag{11}$$

whose image we denote by $\bar{\mathfrak{g}}_{g,1}$.[13]

[13] It is possible that this map is injective so that $\mathfrak{g}_{g,1} \cong \bar{\mathfrak{g}}_{g,1}$. Note that it is not surjective — see [68] and [70], and Section 9.3.

Notice that $\overline{\mathfrak{g}}_{g,1}$ is contained in the subalgebra $\mathrm{Der}(\mathfrak{p}_{g,1}, b)$ consisting of those derivations that kill $b$. Since (11) is (Malcev) filtration preserving, it induces Lie algebra homomorphisms

$$\mathrm{Gr}^\bullet \mathfrak{g}_{g,1} \to \mathrm{Gr}^\bullet \overline{\mathfrak{g}}_{g,1} \to \mathrm{Der}^\bullet(\mathrm{Gr}\,\mathfrak{p}_g^1, \omega_S).$$

Notice that the last term is just $\mathfrak{g}_{\mathrm{lie}}(V_g)$.

In view of (10.6), to construct an algebra homomorphism

$$H^\bullet(\mathfrak{g}_{\mathrm{lie}}^\flat, \mathfrak{k}_\infty) \to H^\bullet(\Gamma_\infty)$$

it suffices to construct an algebra homomorphism $H^\bullet(\mathfrak{u}_{\mathrm{lie}}^\flat)^{Sp_g} \to H^\bullet(\Gamma_\infty)$.

At this stage we need Hodge theory. Choose a conformal structure on $S_g$. Then, by (9.1), there are natural MHSs on $\mathfrak{g}_{g,1}$ and $\mathfrak{p}_{g,1}$ whose weight filtrations are the Malcev filtrations and such that (11) is a MHS morphism. Hence the image $\overline{\mathfrak{g}}_{g,1}$ has a natural MHS. Since $\mathrm{Gr}^\bullet \mathfrak{g}_{g,1}$ is generated by its summands in degree 0 and 1, the same is true for $\mathrm{Gr}^\bullet \overline{\mathfrak{g}}_{g,1}$. On the other hand, the summands in degree 0 and 1 of $\mathrm{Gr}^\bullet \overline{\mathfrak{g}}_{g,1}$ are equal to the summands of weight 0 and $-1$ of $\mathfrak{g}_{\mathrm{lie}}(V_g)$, and so the graded Lie algebra $\mathrm{Gr}^\bullet \overline{\mathfrak{g}}_{g,1}$ may be identified with $\mathfrak{g}_{\mathrm{lie}}^\flat(V_g)_\bullet$ (except that the indexing of the summands differs by sign).

Denote the pronilpotent radical $W_{-1}\overline{\mathfrak{g}}_{g,1}$ of $\overline{\mathfrak{g}}_{g,1}$ by $\overline{\mathfrak{u}}_{g,1}$. We know from Section 9 that the homomorphisms

$$H^\bullet(\overline{\mathfrak{u}}_{g,1})^{Sp_g} \to H^\bullet(\mathfrak{u}_{g,1})^{Sp_g} \to H^\bullet(\Gamma_{g,1}) \tag{12}$$

are morphisms of MHS. After weight grading these become bigraded algebra homomorphisms

$$H^\bullet(\mathfrak{u}_{\mathrm{lie}}^\flat(V_g)_\bullet)^{Sp_g} \to H^\bullet(\mathrm{Gr}_\bullet^W \mathfrak{u}_{g,1})^{Sp_g} \to \mathrm{Gr}_\bullet^W H^\bullet(\Gamma_{g,1}).$$

The sequence (12) stabilizes with $g$ to a sequence of Hopf algebras in the MHS category. The corresponding weight graded sequence is

$$H^\bullet((\mathfrak{u}_{\mathrm{lie}}^\flat)_\bullet)^{Sp_g} \to H^\bullet(\mathrm{Gr}_\bullet^W \mathfrak{u}_{\infty,1})^{Sp_g} \to \mathrm{Gr}_\bullet^W H^\bullet(\Gamma_\infty).$$

Each term in this sequence is a Hopf algebra and each map a Hopf algebra homomorphism. But by Pikaart's Purity Theorem we know that the last term is pure of weight $k$ in degree $k$, so that we can replace it by $H^\bullet(\Gamma_\infty)$ and obtain a map $H^\bullet(\mathfrak{u}_{\mathrm{lie}}^\flat)^{Sp_g} \to H^\bullet(\Gamma_\infty)$. We therefore have a Hopf algebra homomorphism

$$H^\bullet(\mathfrak{g}_{\mathrm{lie}}^\flat, \mathfrak{k}_\infty) \cong H^\bullet(\mathfrak{sp}_\infty, \mathfrak{k}_\infty) \otimes H^\bullet(\mathfrak{u}_{\mathrm{lie}}^\flat)^{Sp_g} \to H^\bullet(\Gamma_\infty).$$

If we compose the natural restriction map $H^\bullet(\mathfrak{g}_{\mathrm{lie}}, \mathfrak{k}_\infty) \to H^\bullet(\mathfrak{g}_{\mathrm{lie}}^\flat, \mathfrak{k}_\infty)$ with the above maps we get a homomorphism

$$H^\bullet(\mathfrak{g}_{\mathrm{lie}}, \mathfrak{k}_\infty) \to H^\bullet(\Gamma_\infty).$$

Kontsevich asked (at the end of [52]) about the meaning of this map. This can now be answered by invoking the theorem of Kawazumi and Morita (9.10), or rather a weaker form, which says that $z_i$ is mapped to a nonzero multiple of $\kappa_i$. This result was obtained with Kawazumi and Morita.

**Theorem 10.7.** *There is a natural Hopf algebra homomorphism*

$$H^\bullet(\mathfrak{g}_{\mathrm{lie}}, \mathfrak{k}_\infty) \to H^\bullet(\Gamma_\infty).$$

*The left hand side is a polynomial algebra generated by primitive elements $z_1, z_2, \dots$ and $c_1, c_3, \dots$ where $z_i$ has degree $2i$ and $c_j$ has degree $2j$. The image of this homomorphism is precisely the subalgebra generated by the $\kappa_i$s. The kernel is generated by elements of the form*

$$c_{2k+1} - a_k z_{2k+1} - P_{2k+1}, \quad k \in \{0, 1, 2, \dots\}$$

*where $P_{2k+1}$ is a polynomial in the $z_i$ and $c_j$ with no linear terms , and $a_k$ is a non zero rational number.*

Here we have used the fact, due to Mumford [69], that the image of $c_{2k+1}$ in $H^\bullet(\mathcal{M}_{g,1})$ is a polynomial in the odd $\kappa_i$'s. The theorem indicates that no new stable classes are to be expected from Hodge theory — that is, a de Rham version of the Mumford conjecture holds. Recent work of Kawazumi and Morita attempts to explain the kernel of the homomorphism $H^\bullet(\overline{\mathfrak{g}}_\infty) \to H^\bullet(\Gamma_\infty)$ in terms of secondary characteristic classes of surface bundles. The first element of the kernel is the difference $c_1 - 12\, z_1$. Its restriction to the Torelli group can be interpreted as the Casson Invariant (cf. [65].)

## References


[1] *Amsterdam-Utrecht seminar on moduli spaces of curves*, C. Faber and E. Looijenga, editors, A. and K. Peters Inc., to appear.

[2] E. Arbarello, M. Cornalba, P. Griffiths and J. Harris: *Geometry of Algebraic Curves I* Grundlehren Math. Wiss. 267, Springer Verlag (1985).

[3] E. Arbarello and M. Cornalba: *The Picard group of the moduli space of curves*, Topology 26 (1987), 153–171.

[4] E. Arbarello and M. Cornalba: *Combinatorial and algebro-geometric cohomology classes on the moduli spaces of curves*, to appear in the J. of Alg. Geom.

[5] A. Borel: *Stable real cohomology of arithmetic groups*, Ann. Sci. Ecole Norm. Sup. 7 (1974), 235–272.

[6] A. Borel: *Stable real cohomology of arithmetic groups II*, Manifolds and Lie Groups, Papers in Honor of Yozo Matsushima, Progress in Mathematics 14 , Birkhauser, Boston, 1981, 21–55.

[7] J. Carlson: *The geometry of the extension class of a mixed Hodge structure*, in *Algebraic Geometry Bowdoin, 1985*, Proc. Symp. Pure Math. 46 (1987), 199–222.

[8] M. Cornalba: *On the projectivity of the moduli spaces of curves*, J. reine angew. Math. 443 (1993), 11–20.

[9] M. Culler and K. Vogtmann: *Moduli of graphs and automorphisms of free groups*, Invent. Math. 84 (1986), 91–119.

[10] M. Dehn: *Die Gruppe der Abbildungsklassen*, Acta Math. 69 (1938), 135–206.

[11] S. Diaz: *A bound on the dimensions of complete subvarieties of $\mathcal{M}_g$*, Duke Math. J. 51 (1984), 405–408.

[12] P. Deligne and D. Mumford: *The irreducibility of the space of curves of given genus*, Inst. Hautes Études Sci. Publ. Math. 36 (1969), 75–109.

[13] C. Faber: *Chow rings of moduli spaces of curves I, II*, Ann. of Math. 132 (1990), 331–449.

[14] C. Faber: *The tautological ring of $\mathcal{M}_g$ is one-dimensional in degree $g-2$*, preprint Univ. van Amsterdam (1995).

[15] J. Foisy: *The second homology group of the level 2 mapping class group of an orientable surface*, Ph.D. Thesis, Duke University, Spring, 1996. (Available from `http://www.math.duke.edu/~foisy/`)

[16] S. Gervais: *Presentation and central extensions of mapping class groups*, Trans. Am. Math. Soc. 348 (1996), 3097–3132.

[17] E. Getzler: *Operads and moduli spaces of genus 0 Riemann surfaces*, in *The moduli space of curves*, R. Dijkgraaf, C. Faber, G. van der Geer, editors, Birkhäuser, 1995, 199–230.

[18] V. Ginzburg and M. Kapranov: *Koszul duality for operads*, Duke Math. J. 76 (1994), 203–272.

[19] A. Grothendieck: *La longue marche à travers de la théorie de Galois*, Manuscript written in 1981, to appear. (First few chapters obtainable from `malgoire@math.univ-montp2.fr` — see also `http://math.univ-montp2.fr/`)

[20] A. Grothendieck: *Esquisse d'un programme*, 1984 manuscript, will appear in a book edited by L. Schneps and P. Lochak and to be published by Cambridge UP.

[21] A. Grothendieck: *The Grothendieck theory of dessins d'enfants*, L. Schneps, editor, London Math. Soc. Lecture Note Series 200 (1994), Cambridge University Press.

[22] R. Hain: *Algebraic cycles and extensions of variations of mixed Hodge structure*, Proc. Symp. Pure Math. 53 (1991), 175–221.

[23] R. Hain: *Completions of mapping class groups and the cycle $C - C^-$*, in *Mapping Class Groups and Moduli Spaces of Riemann Surfaces*, C.-F. Bödigheimer and R. Hain, editors, Contemp. Math. 150 (1993), 75–105.

[24] R. Hain: *Torelli groups and Geometry of Moduli Spaces of Curves*, in *Current Topics in Complex Algebraic Geometry* (C. H. Clemens and J. Kollar, eds.) MSRI publications no. 28, Cambridge University Press, 1995, 97–143.

[25] R. Hain: *The Hodge de Rham Theory of Relative Malcev Completion*, Ann. Sci. École Norm. Sup. (4), to appear.

[26] R. Hain: *Infinitesimal presentations of the Torelli groups*, J. Amer. Math. Soc. to appear.

[27] J. Harer: *The second homology group of the mapping class group of an orientable surface*, Invent. Math. 72 (1983), 221–239.

[28] J. Harer: *The virtual cohomological dimension of the mapping class group of an orientable surface*, Invent. Math. 84 (1986), 157–176.

[29] J. Harer: *Stability of the homology of the mapping class groups of orientable surfaces*, Ann. of Math. 121 (1985), 215–129.

[30] J. Harer: *The third homology group of the moduli space of curves*, Duke J. Math. 63 (1992), 25–55.

[31] J. Harer: *Improved stability for the homology of the mapping class groups of surfaces*, Duke University preprint, 1993. (Available from `http://www.math.duke.edu/preprints/`)

[32] J. Harer: *The fourth homology group of the moduli space of curves*, Math. Ann., to appear. (Available from `http://www.math.duke.edu/preprints/1993.html`)

[33] J. Harer: *The rational Picard group of the moduli spaces of Riemann surfaces with spin structure*, in *Mapping Class Groups and Moduli Spaces of Riemann Surfaces*, C.-F. Bödigheimer and R. Hain, editors, Contemp. Math. 150 (1993), 107–136.

[34] J. Harer and D. Zagier: *The Euler characteristic of the moduli space of curves*, Invent. Math. 85 (1986), 457–485.

[35] N. Ivanov: *Stabilization of the homology of Teichmüller groups*, Leningrad Math. J. 1 (1989), Engl. transl., 675–691.

[36] N. Ivanov: *On the homology stability for Teichmüller modular groups: closed surfaces and twisted coefficients*, in *Mapping Class Groups and Moduli Spaces of Riemann Surfaces*, C.-F. Bödigheimer and R. Hain, editors, Contemp. Math. 150 (1993), 149–194.

[37] N. Ivanov: Mapping class groups, to appear in *Handbook in Geometric Topology*, (R. J. Daverman and R. B. Sher, eds.) North-Holland, 1997.

[38] N. Ivanov: *Homomorphisms from arithmetic to modular groups*, in preparation.

[39] N. Ivanov, R. McCarthy: *On injective homomorphisms between Teichmüller modular groups*, Preprint, 1994, 85 pp.

[40] E. Izadi: *The Chow ring of the moduli space of curves of genus 5*, in *The moduli space of curves*, R. Dijkgraaf, C. Faber, G. van der Geer, editors, Birkhauser, 1995, 267–304.

[41] D. Johnson: *An abelian quotient of the mapping class group $\mathcal{I}_g$*, Math. Ann. 249 (1980), 225–242.

[42] D. Johnson: *The structure of the Torelli group I: A finite set of generators for $\mathcal{I}$*, Ann. of Math. 118 (1983), 423–442.

[43] D. Johnson: *The structure of the Torelli group—II: A characterization of the group generated by twists on bounding curves*, Topology 24 (1985), 113–126.

[44] D. Johnson: *The structure of the Torelli group—III: The abelianization of $\mathcal{I}$*, Topology 24 (1985), 127–144.

[45] D. Johnson: *A survey of the Torelli group*, Contemp. Math. 20 (1983), 165–179.

[46] A. Kabanov: *The second cohomology of moduli spaces of Riemann surfaces with twisted coefficients*, Thesis, Duke University, 1995. (Available from `http://www.mth.msu.edu/~kabanov/`)

[47] V. Kaimanovich, H. Masur: *The Poisson boundary of the mapping class group*, Invent. Math. 125 (1996), 221–264.

[48] N. Kawazumi, S. Morita: *The primary approximation to the cohomology of the moduli space of curves and cocycles for the stable characteristic classes*, Hokkaido University preprint, # 343, August, 1996.
[49] S. Keel: *Intersection theory of moduli spaces of stable n-pointed curves of genus zero*, Trans. Amer. math. Soc. 330 (1992), 545–574.
[50] M. Kontsevich: *Intersection theory on the moduli space of curves and the matrix Airy function*, Comm. Math. Phys. 147 (1992), 1–23.
[51] M. Kontsevich: *Formal (non-)commutative symplectic geometry*, in The Gelfand mathematical seminars 1990–1992, 173-188, Birkhäuser Verlag, Basel (1993).
[52] M. Kontsevich: *Feynman diagrams and low-dimensional topology*, in: Proc. First Eur. Math. Congr. at Paris (1992), Vol. II, 97–121, Birkhäuser Verlag, Basel (1994).
[53] M. Kontsevich and Yu. Manin (with an appendix by R. Kaufmann): *Quantum cohomology of a product*, Invent. Math. 124 (1996), 313–340.
[54] E. Looijenga: *Cohomology of* $\mathcal{M}_3$ *and* $\mathcal{M}_3^1$, in *Mapping Class Groups and Moduli Spaces of Riemann Surfaces*, C.-F. Bödigheimer and R. Hain, editors, Contemp. Math. 150 (1993), 205–228.
[55] E. Looijenga: *Smooth Deligne-Mumford compactifications by means of Prym-level structures*, J. Alg. Geom., 3 (1994), 283–293.
[56] E. Looijenga: *Cellular decompositions of compactified moduli spaces of pointed curves*, in *The moduli space of curves*, R. Dijkgraaf, C. Faber, G. van der Geer, editors, Birkhauser, 1995, 369–400.
[57] E. Looijenga: *On the tautological ring of* $\mathcal{M}_g$, Invent. Math. 121 (1995), 411–419.
[58] E. Looijenga: *Stable cohomology of the mapping class group with symplectic coefficients and of the universal Abel-Jacobi map*, J. Alg. Geom. 5 (1996), 135–150.
[59] Yu. Manin: *Generating functions in algebraic geometry and sums over trees* in *The moduli space of curves*, R. Dijkgraaf, C. Faber, G. van der Geer, editors, Birkhäuser, 1995, 401–417.
[60] G. Mess: *The Torelli groups for genus 2 and 3 surfaces*, Topology 31 (1992), 775–790.
[61] E. Miller: *The homology of the mapping class group*, J. Diff. Geom. 24 (1986), 1–14.
[62] J. Morgan: *The algebraic topology of smooth algebraic varieties*, Publ. Math. IHES 48 (1978), 137–204; correction, Publ. Math. IHES 64 (1986), 185.
[63] S. Morita: *Characteristic classes of surface bundles*, Invent. Math. 90 (1987), 551–577.
[64] S. Morita: *On the structure and the homology of the Torelli group*, Proc. Japan Acad. 65 (1989), 147–150.
[65] S. Morita: *Casson's invariant for homology 3-spheres and characteristic classes of surface bundles I*, Topology 28 (1989), 305–323.
[66] S. Morita: *On the structure of the Torelli group and the Casson invariant*, Topology 30 (1991), 603–621.
[67] S. Morita: *The extension of Johnson's homomorphism from the Torelli group to the mapping class group*, Invent. Math. 111 (1993), 197–224.
[68] S. Morita: *Abelian quotients of subgroups of the mapping class group of surfaces*, Duke Math. J. 70 (1993), 699–726.
[69] D. Mumford *Towards an enumerative geometry of the moduli space of curves*, in Arithmetic and Geometry, M. Artin and J. Tate, editors (1983), Birkhaüser Verlag (Basel) 271–328
[70] H. Nakamura: *Coupling of universal monodromy representations of Galois-Teichmüller modular groups*, Math. Ann. 304 (1996), 99–119.
[71] H. Nakamura: *Galois rigidity of profinite fundamental groups*, Sugaku 47 (1995), 1–17 (in Japanese). English translation to appear in *Sugaku Exposition* (AMS).
[72] J. Nielsen: *Untersuchungen zur Topologie der geschlossenen zweiseitigen Flächen I*, Acta Math. 50 (1927), 189–358.
[73] M. Pikaart: *An orbifold partition of* $\overline{\mathcal{M}}_g^n$, in *The moduli space of curves*, R. Dijkgraaf, C. Faber, G. van der Geer, editors, Birkhäuser, 1995, 467–482.
[74] M. Pikaart and A. de Jong: *Moduli of curves with non-abelian level structure*, in *The moduli space of curves*, R. Dijkgraaf, C. Faber, G. van der Geer, editors, Birkhäuser, 1995, 483–509.
[75] J. Powell: *Two theorems on the mapping class group of a surface*, Proc. Amer. Math. Soc. 68 (1978), 347–350.
[76] M. Ragunathan: *Cohomology of arithmetic subgroups of algebraic groups: I*, Ann. of Math. 86 (1967), 409–424.

[77] P. Sipe: *Roots of the canonical bundle of the universal Teichmüller curve and certain subgroups of mapping class groups*, Math. Ann. 260 (1982), 67–92.
[78] K. Strebel: *Quadratic differentials*, Springer, 1984.
[79] B. Waynryb: *A simple presentation for the mapping class group of an oriented surface*, Isr. J. Math. 45 (1983), 157–174.
[80] E. Witten: *Two dimensional gravity and intersection theory on moduli space*, Surveys in Diff. Geom. 1 (1991), 243–310.

DEPARTMENT OF MATHEMATICS, DUKE UNIVERSITY, DURHAM, NC 27708-0320, USA
*E-mail address*: hain@math.duke.edu

FACULTEIT WISKUNDE EN INFORMATICA, UNIVERSITY OF UTRECHT, POSTAL BOX 80.010, NL-3508 TA UTRECHT, THE NETHERLANDS
*E-mail address*: looijeng@math.ruu.nl

Proceedings of Symposia in Pure Mathematics
Volume **62.2**, 1997

# Algebraic and Symplectic geometry of Gromov-Witten invariants

Jun Li[1] and Gang Tian[2]

This is an expanded version of the talk given by the first author in 1995 Santa Cruz conference on algebraic geometry. In the talk, he reported on the joint work of the authors in which they constructed virtual moduli cycles for a large class of moduli spaces by applying Fulton-MacPherson's normal cone construction to the tangent-obstruction complexes of moduli functors, and constructed Gromov-Witten invariants for projective manifolds using those cycles. Soon after that, the authors proved that this construction commutes with the refined Gysin maps. This is crucial in establishing relations among invariants defined via virtual cycles. For instance, the composition law (associativity law) of Gromov-Witten invariants follows from this property [LT1]. After hearing the talk, Behrend and Fantechi started and later succeeded in giving an alternative construction of the virtual moduli cycles [BF]. As an application, Behrend defined the Gromov-Witten invariants using such cycles [Be]. On another front, the authors and, independently, Fukaya-Ono developed an intersection theory on moduli spaces in the $C^\infty$-category. In particular, they constructed Gromov-Witten invariants for all symplectic manifolds [FO, LT2]. Different constructions were given later by Siebert [Si] and Ruan [Ru2]. Previously, in late 1993, Ruan and the second author constructed all Gromov-Witten invariants and proved their composition law for semi-positive symplectic manifolds ([RT1,RT2]), which include all symplectic manifolds of complex dimension less than four.

In this note, we will first briefly sketch the construction of these virtual cycles, following the construction carried out in [LT1]. We will also describe the relation between this construction and refined Gysin maps. As an application, we will show how this can be applied to prove the composition law for Gromov-Witten invariants. This constitutes the first two sections of this note. In section 3, we will compute the degree of the virtual moduli cycle on a moduli space of rational curves in a special class of Calabi-Yau varieties. The interest in this computation is that the degree of this cycle is negative, and it verifies mathematically the prediction [HK] based on the Mirror symmetry conjecture. In the last part, we will describe the construction of Gromov-Witten invariants for all symplectic manifolds following [LT2] and will

1991 *Mathematics Subject Classification.* 14C17, 14D20.
[1]Supported in part by NSF grant, A. Sloan fellowship and Terman fellowship.
[2]Supported in part by NSF grant.

sketch the idea of the proof that the algebraic and analytic constructions of GW-invariants lead to identical invariants [LT3].

Let us first recall the need to find virtual moduli cycles. Given a variety $X$, one can form moduli spaces of certain objects associated to $X$. Let us denote by $\mathcal{M}$ one such moduli space. $\mathcal{M}$ carries with it a collection of tautological topological classes, say $\alpha_1, \cdots, \alpha_k \in H^*(\mathcal{M})$. Let us assume $\mathcal{M}$ is complete. Then we obtain a functional[1]

$$\mathcal{F}(t_1, \cdots, t_k) = \int_{\mathcal{M}} \exp(t_1\alpha_1 + \cdots + t_k\alpha_k). \tag{1}$$

This functional is meaningful and well-behaved when deformation of objects in $\mathcal{M}$ is unobstructed, i.e. when $\mathcal{M}$ is smooth and has the expected dimension. When $\mathcal{M}$ has the expected dimension but is not necessarily reduced, we can replace the integration by an operational cohomological theory of schemes:

$$\mathcal{F}(t_1, \cdots t_k) = \deg\bigl(\exp(t_1\alpha_1 + \cdots t_k\alpha_k)[\mathcal{M}]\bigr)_0. \tag{2}$$

(Here we implicitly assume that the $\alpha_i \in A^*X$ are algebraic cocycles.) This still defines the desired functional. For instance, when $X$ is an algebraic surface and $\mathcal{M}$ the moduli spaces of rank two semistable sheaves, this is the Donaldson polynomial invariant[2]. When $X$ is a smooth projective variety and $\mathcal{M}$ the moduli spaces of stable morphisms from marked curves to $X$, this defines the Gromov-Witten invariants of $X$. However, when $\mathcal{M}$ does not have the expected dimension, we run into serious trouble if we keep the formula (2). For one, when $X$ is a member of a smooth family of varieties, the associated moduli spaces may not necessarily form a flat family, and hence the resulting functional will be likely to change as $X$ varies. Also, many of these functionals can be defined using (1) with $\mathcal{M}$ replaced by moduli of corresponding analytic objects, yielding interesting invariants of the topological space $X$ coupled with some auxiliary structure. These invariants mostly will differ from (2) when $\mathcal{M}$ does not have the expected dimension. The solution to this difficulty is to find a cycle in $\mathcal{M}$ whose dimension is equal to the expected dimension of $\mathcal{M}$, and to define the functional using (2) by replacing $[\mathcal{M}]$ with this cycle. Morally speaking, one should view the moduli space $\mathcal{M}$ as defined by certain relations $\Psi$. When $\mathcal{M}$ has bigger than expected dimension, it means that these relations are not in general position. The cycle we seek should be the cycle defined in the spirit of excess intersection theory. In the following, we will call such cycles virtual moduli cycles, and will denote them by $[\mathcal{M}]^{\mathrm{vir}}$, after they have been constructed. The correct formula for the mentioned functionals then takes the form

$$F(t_1, \cdots, t_k) = \deg\bigl(\exp(t_1\alpha_1 + \cdots + t_k\alpha_k)[\mathcal{M}]^{\mathrm{vir}}\bigr)_0. \tag{3}$$

The search for virtual moduli cycles was first called upon in the study of Donaldson invariants of algebraic surfaces. Donaldson invariants of algebraic surfaces are such functionals with $\mathcal{M}$ the moduli of semistable sheaves on that surface [Mo, Li1]. In case one encounters the situation where moduli spaces have bigger than expected dimensions, one needs to use virtual moduli cycles to give an algebraic definition of these invariants. This can be done either using rational embeddings

[1] $t_1, \cdots, t_k$ are formal variables and $\exp(\cdot)$ is the formal power series of the exponential function.

[2] We need to assume $p_g > 0$ here, otherwise it depends on the choice of polarization used to define the stability of sheaves.

of moduli spaces [FM1], or using the obstruction bundle in some special cases [Ba, MO, FM2]. The virtual moduli cycle then is constructed by either using excess intersection theory, in the first approach, or taking the Euler class of the obstruction bundle, in the second approach. The notion of virtual moduli cycles received much attention in the study of mirror symmetry of Calabi-Yau manifolds. In [AM, Vo], Aspinwall-Morrison and Voisin have determined the virtual moduli cycle of the moduli of degree $d$ maps $\mathbf{P}^1 \to \mathbf{P}^1 \subset X$, where $X$ is a Calabi-Yau threefold and $\mathbf{P}^1 \subset X$ is an isolated smooth rational curve with normal bundle $N_{\mathbf{P}^1/X} \cong \mathcal{O}(-1)^{\oplus 2}$. Katz has worked on constructing virtual moduli cycles and obtained some partial results in special cases [Kz]. Kontsevich used the moduli of stable maps and embedding of the moduli spaces into smooth moduli spaces functorially to compute virtual numbers of rational curves in quintic threefolds [Ko]. We should mention that the analytic approach to such problems is to perturb the parameters of the elliptic operators so that the solution spaces are smooth and have expected dimensions [Do, RT1].

From this incomplete account of previous research on virtual moduli cycles, it is evident that there are generally two approaches in constructing virtual moduli cycles: one by finding a "functorial" embedding of the moduli space and then applying excess intersection theory, and the other by taking the Euler class of the obstruction sheaf, when it is locally free. The first approach is satisfactory in some cases, although finding such an embedding is more the exception than the rule, thus restricting its usage to a very limited class of moduli problems. On the other hand, by working out the deformation-obstruction theory of the moduli problem, one usually can explicitly write down its obstruction sheaf. Unless the moduli space is smooth, in which case the obstruction sheaf will be locally free and the virtual cycle will be its Euler class, there is no apparent notion of the Euler class of a sheaf suitable for our purposes.

The virtual cycle, which will be constructed shortly, depends upon the combination of the tangent complex and obstruction sheaf (obstruction theory) of the moduli space, which we call the tangent-obstruction complex. To put it simply, the tangent-obstruction complex is a global deformation-obstruction theory of a moduli functor. The following observation explains why the tangent-obstruction complex as a whole is needed to determine the virtual moduli cycle. Let $p \in \mathcal{M}$ be a point and $O_p$ be the obstruction space to the deformations of $p \in \mathcal{M}$. Let $m = \dim T_p\mathcal{M}$ and $n = \dim O_p$. Then there is a Kuranishi map

$$\mathcal{K}_p : \hat{\mathbf{A}}_m \longrightarrow \hat{\mathbf{A}}_n,$$

where $\hat{\mathbf{A}}_m$ is the germ of $0 \in \mathbf{A}_m$, such that the germ of $\mathcal{M}$ at $p$ is isomorphic to $\mathcal{K}_p^{-1}(0)$. The two extreme cases are when $\dim \mathcal{K}^{-1}(0) = m - n$, in which case the effective obstruction directions span $O_p$, and when $\dim \mathcal{K}^{-1}(0) = m$, in which the obstruction is zero. They correspond to the case where $\mathcal{M}$ has the expected dimension at $p$ and where the obstruction sheaf is locally free at $p$, respectively. The general case is that the obstruction sheaf is partially effective, in which case the virtual moduli cycle, roughly speaking, should be the "Euler" class of the part of the obstruction sheaf that is ineffective. As we will see, such information is contained in the global obstruction theory of the moduli space.

## 1. Virtual cycles and Gromov-Witten invariants

To make the exposition short and easy to comprehend, we shall only consider the Gromov-Witten invariants, and shall indicate the main ingredients that make this construction work for more general moduli spaces.

We begin with Gromov-Witten invariants. Let $X$ be a fixed smooth complex projective variety, $\alpha \in H_2(X,\mathbb{Z})$ a homological class and $g, n$ integers. We first recall the notion of stable morphisms, which is due to Kontsevich [Ko]. We say a map

$$f : D \subset C \longrightarrow X$$

is stable if $C$ is a nodal curve, $D \subset C$ consists of $n$ ordered points away from singularities of $C$ and there are no non-trivial vector fields on $C$ fixing $D$ and $\mathrm{Sing}(C)$ that lie in the kernel of $f_*$. We form the moduli space of stable morphisms from $n$-pointed (arithmetic) genus $g$ curves to $X$ such that they send fundamental classes of curves to $\alpha$. We denote this moduli space by $\mathcal{M}^X_{\alpha,g,n}$. By work of Alexeev [Al], $\mathcal{M}^X_{\alpha,g,n}$ is projective. $\mathcal{M}^X_{\alpha,g,n}$ carries a collection of tautological cohomology classes: let $\mathrm{ev}_i : \mathcal{M}^X_{\alpha,g,n} \to X$ be the evaluation morphism sending $\{f : D \subset C \to X\} \in \mathcal{M}^X_{\alpha,g,n}$ to $f(x_i)$, where $D = \{x_1, \dots, x_n\}$, and let $\pi_{g,n} : \mathcal{M}^X_{\alpha,g,n} \to \mathcal{M}_{g,n}$ be the obvious morphism that send the above data to the stable contraction of the marked curve $D \subset C$. Then we have the tautological class

$$ev_1^* \cup \cdots \cup \mathrm{ev}_n^* \cup \pi_{g,n}^* : H^*(X)^{\times n} \times H^*(\mathcal{M}_{g,n}) \longrightarrow H^*(\mathcal{M}^X_{\alpha,g,n}).$$

(Unless otherwise is mentioned, all (co)homology take rational coefficients.) Next, by working out the deformation-obstruction theory of this moduli problem [LT1], one sees that at $\xi \in \mathcal{M}^X_{\alpha,g,n}$ represented by $f : D \subset C \longrightarrow X$, the Zariski tangent space $T_\xi \mathcal{M}^X_{\alpha,g,n}$ is

$$\mathrm{Ext}^1_C([f^*\Omega_X \to \Omega_C(D)], \mathcal{O}_C)$$

and the obstruction to deformations of $\xi$ in $\mathcal{M}^X_{\alpha,g,n}$ lies in

$$\mathrm{Ext}^2_C([f^*\Omega_X \to \Omega_C(D)], \mathcal{O}_C),$$

where the complex $[f^*\Omega_X \to \Omega_C(D)]$ is induced by the obvious $f^*\Omega_X \to \Omega_C$ and is indexed at $-1$ and 0. Using the Riemann-Roch theorem, the virtual dimension of $\mathcal{M}^X_{\alpha,g,n}$ is

$$\begin{aligned}\rho(\alpha, g, n) &\triangleq \sum_{i=1}^{2} (-1)^{i-1} \dim \mathrm{Ext}^i_C([f^*\Omega_X \to \Omega_C(D)], \mathcal{O}_C) \\ &= (3 - \dim X)(1 - g) + n + \alpha \cdot c_1(X).\end{aligned}$$

Witten proposed the notion of quantum cohomology in his study of two dimensional non-linear sigma model, which is fundamental in understanding the phenomenon of mirror symmetry for Calabi-Yau manifolds. The more general version of quantum cohomology is the GW-invariants, whose algebro-geometric version is a collection of multilinear maps

$$\phi^X_{\alpha,g,n} : H^*(X)^{\times n} \times H^*(\mathcal{M}_{g,n}) \longrightarrow \mathbb{Q}$$

defined by

$$\phi^X_{\alpha,g,n}(\alpha_1, \cdots, \alpha_n, \beta) = \deg\bigl([\mathrm{ev}_1^*(\alpha_1) \cup \cdots \cup \mathrm{ev}_n^*(\alpha_n) \cup \pi_{g,n}^*(\beta)][\mathcal{M}^X_{\alpha,g,n}]^{\mathrm{vir}}\bigr)_0,$$

where $[\mathcal{M}^X_{\alpha,g,n}]^{\mathrm{vir}}$ is the virtual moduli cycle that will be constructed shortly.

We first recall the notion of tangent-obstruction complexes. We begin with the the moduli space of stable morphisms $\mathcal{M}^X_{\alpha,g,n}$. Given any closed point $w \in \mathcal{M}^X_{\alpha,g,n}$ representing the map $f: D \subset C \to X$, it is known that the first order deformations of $f$ is parameterized by $\mathrm{Ext}^1_C([f^*\Omega_X \to \Omega_C(D)], \mathcal{O}_C)$ and the obstruction to deformations of $f$ lies in $\mathrm{Ext}^2_C([f^*\Omega_X \to \Omega_C(D)], \mathcal{O}_C)$. Such notion can be generalized to family version. Let $S$ be an affine scheme and

$$F : D \subset \mathcal{X} \longrightarrow X$$

be a family of stable morphisms from $n$-pointed curves $D \subset \mathcal{X}$ over $S$ to $X$. Let $\pi: \mathcal{X} \to S$ be the projection and let $\mathcal{C}^\bullet$ be the complex $[F^*\Omega_X \to \Omega_{\mathcal{X}/S}(D)]$, indexed at $-1$ and $0$.

LEMMA. *1. Given any sheaf $\mathcal{F}$ of $\mathcal{O}_S$-modules, we let $\tilde{S}$ be the split infinitesimal extension of the scheme $S$ by the sheaf $\mathcal{F}$. Namely, $\mathcal{O}_{\tilde{S}}$ is the split extension of $\mathcal{O}_S$ by $\mathcal{F}$. Then the set of all families of stable morphisms*

$$\tilde{F} : \tilde{D} \subset \tilde{\mathcal{X}} \longrightarrow X$$

*over $\tilde{S}$ extending the family $F$ is isomorphic to the group $\Gamma(S, \mathcal{E}xt^1_\pi(\mathcal{C}^\bullet, \pi^*\mathcal{F}))$.*
*2. Let $S \to S_0 \to \tilde{S}$ be inclusion (closed immersions) of affine $S$-schemes, let $\mathfrak{m}$ (resp. $\mathcal{I}$) be the ideal sheaf of $S \subset \tilde{S}$ (resp. $S_0 \subset \tilde{S}$). Suppose $\mathcal{I} \cdot \mathfrak{m} = 0$ in $\mathcal{O}_{\tilde{S}}$ and $F_0$ is a family over $S_0$ extending the family $F$, then there is a class*

$$ob(S, S_0, F_0) \in \Gamma(S, \mathcal{E}xt^2_\pi(\mathcal{C}^\bullet, \pi^*\mathcal{I})),$$

*canonical under base change, such that its vanishing is the necessary and sufficient condition for $F_0$ to be extendable to a family over $\tilde{S}$. We call such class the obstruction class to extending $F_0$ to $\tilde{S}$.*

We call such assignment $\mathcal{F} \Rightarrow [\mathcal{E}xt^\bullet_\pi(\mathcal{C}^\bullet, \pi^*\mathcal{F})]$ the tangent-obstruction complex to the moduli functor associated to the moduli space $\mathcal{M}^X_{\alpha,g,n}$. Note that for this complex there is a complex of locally free sheaves of $\mathcal{O}_S$-modules, say $\mathcal{E}^\bullet = [\mathcal{E}_1 \to \mathcal{E}_2]$, such that the sheaf $\mathcal{E}xt^i_\pi(\mathcal{C}^\bullet, \pi^*\mathcal{F})$ is the $i$-th sheaf cohomology of $\mathcal{E}^\bullet \otimes \mathcal{F}$. As usual, we call such complex a perfect tangent-obstruction complex, and call $\operatorname{rank} \mathcal{E}_2 - \operatorname{rank} \mathcal{E}_1$ the virtual dimension of $\mathcal{M}^X_{\alpha,g,n}$.

For the construction of virtual cycle to work for an arbitrary moduli space $\mathcal{M}$, what is needed is a perfect tangent-obstruction complex $[\mathcal{T}^1 \to \mathcal{T}^2]$ of $\mathcal{M}$.

DEFINITION. *Let $\mathfrak{F}$ be a moduli functor. A perfect tangent-obstruction complex of $\mathfrak{F}$ consists of two functors $\mathcal{T}^1$ and $\mathcal{T}^2$ as follows: let $S$ be any affine scheme and $\eta \in \mathfrak{F}(S)$.*

1. *For any sheaf $\mathcal{F}$ of $\mathcal{O}_S$-modules, $\mathcal{T}^1(\eta)(\mathcal{F})$ is the sheaf of $\mathcal{O}_S$-modules so that $\Gamma(\mathcal{T}^1(\eta)(\mathcal{F}))$ is the group that is the preimage of $\eta$ of the restriction map $\mathfrak{F}(\tilde{S}) \to \mathfrak{F}(S)$, where $\tilde{S}$ is the splitting extension of $S$ by $\mathcal{F}$;*
2. *$\mathcal{T}^2(\eta)(\mathcal{I}_{S_0 \subset \tilde{S}})$ is the sheaf of obstruction. Namely, for triple $S \to S_0 \to \tilde{S}$ of $S$-schemes as in 2) of the Lemma and $\eta_0 \in \mathfrak{F}(S_0)$ restricting to $\eta$, there is a natural obstruction class $ob(S, S_0, \eta_0) \in \Gamma_S(\mathcal{T}^2(\eta)(\mathcal{I}_{S_0 \subset \tilde{S}}))$ to extending $\eta_0$ to $\tilde{\eta} \in \mathfrak{F}(\tilde{S})$;*
3. *There is a two-term complex of locally free sheaves of $\mathcal{O}_S$-modules $\mathcal{E}^\bullet = [\mathcal{E}_1 \to \mathcal{E}_2]$ such that $\mathcal{T}^\bullet$ is the sheaf cohomology of $\mathcal{E}^\bullet$. Namely, suppose $i$:*

*$S' \to S$ is any morphism and $\eta' \in \mathfrak{F}(S')$ is the pull back of $\eta$, then $\mathcal{T}^1(\eta')(\mathcal{F})$ and $\mathcal{T}^2(\eta')(\mathcal{F})$ are the kernel and the cokernel of $i^*\mathcal{E}_1 \otimes \mathcal{F} \to i^*\mathcal{E}_2 \otimes \mathcal{F}$.*

We remark that the choice of tangent-obstruction complex is not unique, since one can choose the complex $\mathcal{E}^\bullet \oplus [0 \to \mathcal{L}]$, where $\mathcal{L}$ is any locally free sheaf, to be a different tangent-obstruction complex. In the end, the virtual cycle constructed will depend on the choice of tangent-obstruction complex.

Let us first look at one simple example that is key to our construction.

Let $W$ be any smooth scheme of dimension $m$ and let $V \to W$ be a rank $k$ vector bundle with a section $s : W \to V$. Let $Z = s^{-1}(0)$. Then the tangent complex $\mathcal{T}^1$ of $W$ is

$$\mathcal{T}^1(\mathcal{F}) = \ker\{\mathcal{O}_Z(TW) \otimes \mathcal{F} \xrightarrow{ds\otimes 1_{\mathcal{F}}} \mathcal{O}_Z(V) \otimes \mathcal{F}\}.$$

Let $\mathcal{T}^2$ be the cokernel of $ds$. Then $\mathcal{T}^\bullet = [\mathcal{T}^1 \to \mathcal{T}^2]$ is the tangent-obstruction complex of $Z$ associated to the defining equation $s$. In this setting, the virtual cycle $[Z]^{\mathrm{vir}}$ obviously should be

$$[Z]^{\mathrm{vir}} = e(V) \in A_{m-k}W.$$

Using the normal cone construction of Fulton-MacPherson, $e(V)$ is the image of the cycle of the normal cone to $Z$ in the graph $\Gamma_s$, denoted by $C_{Z/\Gamma_s}$, under the Gysin map

$$i_V^! : A_*V \longrightarrow A_{*-k}Z,$$

where $i_V$ is the zero section of $V$. Note that in this example, the obstruction sheaf is the cokernel of $\mathcal{O}_Z(TW) \to \mathcal{O}_Z(V)$. The virtual cycle $[Z]^{\mathrm{vir}} \in A_{m-k}Z$ is defined by first *constructing a cone $C_{Z/\Gamma_s} \subset V$, hence a cycle $[C_{Z/\Gamma_s}] \in A_mV$*, and then *intersecting $[C_{Z/\Gamma_s}]$ with the zero section of $V$, i.e. $[Z]^{\mathrm{vir}} = i_V^![C_{Z/\Gamma_s}]$.*

The key to the construction of virtual cycles is that one can reconstruct the normal cone cycle $[C_{Z/\Gamma_s}]$ provided that one knows the tangent-obstruction complex of $Z$. To illustrate this scheme, let us give an explicit description of the obstruction theory to deformations of $z$ in $Z$ as above, and investigate how such theory is related to the cone $C_{Z/\Gamma_s}$. Let $p \in Z$ be any point and let $m_1 = \dim T_pZ$. We choose a local coordinate $(z_1, \cdots, z_m)$ of $p \in W$ and a frame $\{e_1, \cdots, e_k\}$ of $V$ near $p$ such that if we let $f_i$ be the $i$-th component of $s : W \to V$, then $f_i \in \mathbb{C}[z_1, \cdots, z_{m_1}]$ for $i \le i \le k_1$, where $k_1 = k - (m - m_1)$, and $f_{k_1+i} = z_{m_1+i}$ for $1 \le i \le k - k_1$. Polynomials $\{f_1, \cdots, f_{m_1}\}$ provide the obstruction theory to deformations of $p$ with values in $T_2 \triangleq V_p/T_pZ$ as follows: let $B \to B_0$ be Artin rings and quotient homomorphism such that its kernel $I \subset B$ is annihilated by the maximal ideal $\mathfrak{m}$ of $B$. Let $\hat{A} = \mathbb{C}[[z_1, \cdots, z_{m_1}]]$. For any homomorphism $\varphi_0 : \hat{A} \to B_0$, we choose a lifting $\varphi : \hat{A} \to B$. Then the obstruction to $\varphi$ factor through $\hat{A}/(f_1, \cdots, f_{m_1})$ is

$$ob(B, B_0, \varphi_0) = (\varphi(f_1), \cdots, \varphi(f_{m_1})) \in I \otimes_{\mathbb{C}} T_2.$$

(Since $f_i$ has no linear terms, this is independent of the choices of $\varphi$.) Obviously, this is the obstruction theory to deformations of $z \in Z$ with values in $T_2$. Let $T_1 = T_pZ$. Then the induced homomorphism

$$f = (f_1, \cdots, f_{m_1}) : \hat{\mathrm{Sym}}(T_2^\vee) \longrightarrow \hat{\mathrm{Sym}}(T_1^\vee)$$

is a Kuranishi map of this obstruction theory. Here $\hat{\mathrm{Sym}}(T_2^\vee)$ is the formal completion of the symmetric algebra of $T_2^\vee$ with respect to the ideal $(T_2^\vee)$. Now let us investigate how to construct the cone $C_{Z/\Gamma_s}$ from $f$. We let $\hat{X} = \operatorname{Spec}\hat{\operatorname{Sym}}(T_1^\vee)$ and $\hat{Y} = \operatorname{Spec}\hat{\operatorname{Sym}}(T_2^\vee)$. Let $\Gamma_f \subset \hat{X} \times \hat{Y}$ be the graph of $f$ and let $C_{f^{-1}(0)/\Gamma_f}$ be the normal cone to $f^{-1}(0) \triangleq \Gamma_f \cap \hat{X} \times \{0\}$ in $\Gamma_f$. Let $\hat{p}$ be the formal completion of $Z$ along $p$. Then $\hat{p} \cong f^{-1}(0)$ and $C_{f^{-1}(0)/\Gamma_f}$ is a cone in $f^{-1}(0) \times T_2$. Obviously, the restriction of $C_{Z/\Gamma_f}$ to $V|_{\hat{p}}$ is the pull-back of $C_{f^{-1}(0)/\Gamma_f}$ under an appropriate cone isomorphism $V|_{\hat{p}} \to \hat{f}^{-1}(0) \times T_2$ that extends $V_p \to T_2$.

The construction of virtual moduli cycles will follow this scheme. We assume $S \subset \mathcal{M}^X_{\alpha,g,n}$ is an affine open set so that there is a tautological family of stable morphisms over $S$, say $F : D \subset \mathcal{X} \to X$. Let $\pi : \mathcal{X} \to S$ be the projection and let $\mathcal{K}$ be the sheaf $\mathcal{E}xt^2_\pi([F^*\Omega_X \to \Omega_{\mathcal{X}/S}(D)], \mathcal{O}_{\mathcal{X}})$. Let $\mathcal{E}$ be a locally free sheaves of $\mathcal{O}_S$-modules so that $\mathcal{K}$ is a quotient sheaf of $\mathcal{E}$, and let $E$ be the vector bundle on $S$ such that $\mathcal{O}_S(E) = \mathcal{E}$. We will also view $E$ as a cone over $S$. Let $x \in S$ be any closed point and let $\mathcal{E} \otimes k(x) \to \mathcal{K} \otimes k(x)$ be the restriction of $\mathcal{E} \to \mathcal{K}$. We let $T_2 = \mathcal{K} \otimes k(x)$, considered as a $\mathbb{C}$-vector space, and let $T_1 = T_x S$. By assumption, there is an obstruction theory to deformations of $x$ in $\mathcal{M}^X_{\alpha,g,n}$ with values in $T_2$. Let

$$f : \hat{\operatorname{Sym}}(T_2^\vee) \longrightarrow \hat{\operatorname{Sym}}(T_1^\vee)$$

be a Kuranishi map, following [La]. Let

$$\hat{R} = \operatorname{Spec}\hat{\operatorname{Sym}}(T_1^\vee) \quad \text{and} \quad \hat{x} = \operatorname{Spec}(\hat{\operatorname{Sym}}(T_1^\vee) \otimes_{\hat{\operatorname{Sym}}(T_2^\vee)} k).$$

$\hat{x} \subset \hat{R}$ and is isomorphic to the formal completion of $\mathcal{M}^X_{\alpha,g,n}$ along $x \in \mathcal{M}^X_{\alpha,g,n}$. (Here we assume the morphism associated to $x$ has no non-trivial automorphisms.) Let $\Gamma_f$ be the graph of $f$ which lies in $\hat{R} \times \operatorname{Spec}\hat{\operatorname{Sym}}(T_2^\vee)$. Then the normal cone to $\hat{x} \times \{0\}$ in $\Gamma_f$ is a purely $\dim T_1$-dimensional subcone in $\hat{x} \times T_2$. We denote this cone by $C_{\hat{x}/\Gamma_f}$ and denote by $N^f_x \triangleq [C_{\hat{x}/\Gamma_f}]$ the corresponding cycle in $Z_*(\hat{x} \times T_2)$. $N^f_x$ is independent of the choices of Kuranishi maps $f$ because of the following invariance Lemma.

Invariance Lemma. *1. Let $f$, $g : \hat{\operatorname{Sym}}(T_2^\vee) \to \hat{\operatorname{Sym}}(T_1^\vee)$ be two Kuranishi maps associated to the same obstruction theory. We fix a basis $\{z_1, \cdots, z_n\}$ of $T_1^\vee$ and $\{w_1, \cdots, w_m\}$ of $T_2^\vee$. Then there is an automorphism $A$ of $\hat{\operatorname{Sym}}(T_1^\vee \oplus T_2^\vee)$ of the form $z_i \mapsto \phi_i(z)$ and $w_i \mapsto \sum \phi_{ij}(z) w_j$, where*

$$\phi_i(z) \equiv z_i \mod \mathfrak{m}^2 \qquad \phi_{ij}(z) \equiv \delta_{ij} \mod \mathfrak{m} \quad \textit{and} \quad \mathfrak{m} = (z_1, \cdots, z_n),$$

*such that the induced isomorphism $\bar{A}$ of $\operatorname{Spec}(\hat{\operatorname{Sym}}(T_1^\vee \oplus T_2^\vee))$ induces an isomorphism of the graph of $f$ and the graph of $g$;*
*2. Let $f$ and $g$ be as in 1) and let $N^f_x$ and $N^g_x \in Z_*(\hat{x} \times T_2)$ be cycles defined using Kuranishi maps $f$ and $g$ respectively. Then there is an automorphism $\eta$ of vector bundle $\hat{x} \times T_2 \to \hat{x}$ such that $\eta|_x = id$ and $\eta^{-1}(N^f_x) = N^g_x$.*

The technical part of the construction is to show that there is a global cone in $E$ whose restriction to $\hat{x}$ are exactly the pull backs of cones $N^f_x$.

Basic Lemma. *There is a cone cycle $[C^{\mathcal{E}}] \in Z_*E$ of pure dimension* $\operatorname{rank} E + v.\dim \mathcal{M}^X_{\alpha,g,n}$ *of which the following holds: for each $x \in S$, there is a quotient homomorphism $\eta : E|_{\hat{x}} \to \hat{x} \times T_2$, where $T_2 = \mathcal{K} \otimes k(x)$, that extends the given quotient homomorphism $E|_x \to T_2$ such that $\eta^{-1}(N^f_x) = [C^{\mathcal{E}}]|_{\hat{x}} \in Z_*(E|_{\hat{x}})$.*

Once the basic lemma is established, we obtain a global cone cycle in $Z_*E$ of pure dimension $\operatorname{rank} E + v.\dim \mathcal{M}^X_{\alpha,g,n}$ whose restriction to germs $\hat{x}$ are the cones associated to Kuranishi maps at $x$. By invariance lemma, such cone cycles are unique. $[C^{\mathcal{E}}]$ are virtual normal cones to $\mathcal{M}^X_{\alpha,g,n}$ with respect to the tangent-obstruction complex chosen.

For the purpose of defining the virtual moduli cycle, we need to investigate the dependence of $[C^{\mathcal{E}}]$ on the choices of quotient homomorphisms $\mathcal{E} \to \mathcal{K}$. Let $\mathcal{F}$ be another locally free sheaf of $\mathcal{O}_S$-modules such that $\mathcal{K}$ is a quotient sheaf of $\mathcal{F}$. Then it is easy to see that there is a locally free sheaf of $\mathcal{O}_S$-modules $\mathcal{G}$ so that both $\mathcal{E}$ and $\mathcal{F}$ are quotient sheaves of $\mathcal{G}$ and that the diagram

$$\begin{array}{ccc} \mathcal{G} & \xrightarrow{\eta} & \mathcal{E} \\ {\scriptstyle\delta}\downarrow & & \downarrow \\ \mathcal{F} & \longrightarrow & \mathcal{K} \end{array}$$

is commutative. Then because of the invariance lemma and the basic lemma, the cone $[C^{\mathcal{G}}]$ is isomorphic to the pull-back of $[C^{\mathcal{E}}]$ as well as the pull-back of $[C^{\mathcal{F}}]$.

Now we come to our main task: defining Gromov-Witten invariants of smooth projective varieties. To accomplish this we need to overcome one technical difficulty: namely there are possibly maps in $\mathcal{M}^X_{\alpha,g,n}$ that have non-trivial automorphism groups. To this end, a few words on moduli of objects with finite automorphisms is in order. To avoid introducing more notation, we will speak of moduli of stable morphisms only. Given a point $z \in \mathcal{M}^X_{\alpha,g,n}$, associated to the morphism $z = \{f : D \subset \mathcal{X} \to X\}$, there are two germs of analytic varieties crucial to understanding deformations of $z$: one is the germ parameterizing the semi-universal family of $z$, and the other is the germ of $\mathcal{M}^X_{\alpha,g,n}$ at $z$. We denote the former by $U$ and the later by $V$. $U$ and $V$ are isomorphic if $G_z = \operatorname{Aut}(z) = \{1\}$. Otherwise, $G_z$ acts naturally on $U$ and $V \cong U/G_z$. (The automorphism groups we are dealing with are all finite and reduced.) The property of $V$, thus the local property of $\mathcal{M}^X_{\alpha,g,n}$ at $z$, is best understood by going through $U$, because there is a semi-universal family, say $\xi = \{F_U : D_U \subset \mathcal{X}_U \to X\}$, over $U$. For instance, the tangent-obstruction complex of $U$ is naturally

$$\mathcal{T}^\bullet(\xi)(\mathcal{F}) = \left[\mathcal{E}xt^\bullet_{\mathcal{X}_U/U}\left([F_U^*\Omega_X \to \Omega_{\mathcal{X}_U}(D_U)], \pi_U^*\mathcal{F}\right)\right].$$

Hence a common scheme in studying the geometry of $\mathcal{M}^X_{\alpha,g,n}$, following the work of Mumford [Mu], will be to first study the corresponding structure using local semi-universal families and then to analyze their descents under the automorphism groups. Of course for the informed readers, this is an example of a stack (or $\mathbb{Q}$-schemes). Since the automorphism groups involved are all finite, we prefer to treat them as schemes with certain structures locally expressible over Galois covers.

We now elaborate on this by working out the $\mathbb{Q}$-tangent-obstruction complex of $\mathcal{M}^X_{\alpha,g,n}$. Here we view $\mathcal{M}^X_{\alpha,g,n}$ as a $\mathbb{Q}$-schemes. Namely, there is a collection of finite groups $G_i$, affine $G_i$-schemes $U_i$, étale morphisms $U_i/G_i \xrightarrow{\pi_i} \mathcal{M}^X_{\alpha,g,n}$ and $G_i$-families $\xi_i = \{F_i : D_i \subset \mathcal{X}_i \to X\}$ over $U_i$ such that $\xi_i$ is a local semi-universal family for each of its members, and such that over some étale covering $U_{ij} \to U_i \times_{\mathcal{M}^X_{\alpha,g,n}} U_j$ there are isomorphisms

$$\varphi_{ij} : p_i^*\xi_i \cong p_j^*\xi_j,$$

where $p_i$ is the projection of $U_{ij}$ to $U_i$, satisfying $p_{ij}^*\varphi_{ij} \circ p_{jk}^*\varphi_{jk} = p_{ik}^*\varphi_{ik}$ over $U_{ijk}$, where $U_{ijk}$ is an étale cover of the product $U_i \times_{\mathcal{M}^X_{\alpha,g,n}} U_j \times_{\mathcal{M}^X_{\alpha,g,n}} U_k$ and $p_{ij}$

is the projection to $U_{ij}$. (See for instance [FP] in this volume.) Accordingly, the collection of tangent-obstruction complexes

$$\mathcal{T}^\bullet(\xi_i) = \left[\mathcal{E}xt^\bullet_{\mathcal{X}_i/U_i}([F_i^*\Omega_X \to \Omega_{\mathcal{X}_i/U_i}(D_i)], \mathcal{O}_{\mathcal{X}_i})\right]$$

satisfies

$$p_i^*\mathcal{T}^\bullet(\xi_i) \cong p_j^*\mathcal{T}^\bullet(\xi_j).$$

We call such a collection the $\mathbb{Q}$-tangent-obstruction complex of $\mathcal{M}^X_{\alpha,g,n}$, following [Mu].

Next we construct a $\mathbb{Q}$-complex of sheaves $\mathcal{E}^\bullet$ such that its sheaf cohomology $\mathfrak{H}^\bullet(\mathcal{E}^\bullet)$ is $\mathcal{T}^\bullet$. Since this will be needed later, we will give the detailed account here. We pick a sufficiently ample invertible sheaf $\mathcal{L}$ on $X$ and form a locally free sheaf of $\mathcal{O}_{\mathcal{X}}$-modules

(4) $$\mathcal{W}_1(\xi_i) = \pi_{U_i}^*\pi_{U_i*}\big(\omega_{\mathcal{X}_i/U_i}(D_i)^{\otimes 5} \otimes F_i^*(\mathcal{L}\otimes\Omega_X)\big) \otimes \big(\omega_{\mathcal{X}_i/U_i}(D_i)^{\otimes 5} \otimes F_i^*\mathcal{L}\big)^{-1}.$$

Let $\mathcal{W}_1(\xi_i) \to F_i^*\Omega_X$ be the obvious homomorphism and let $\mathcal{W}_2(\xi_i)$ be its kernel, which will be locally free. We then form complexes

$$\mathcal{B}^\bullet(\xi_i) = [\mathcal{W}_1(\xi_i) \to \Omega_{\mathcal{X}_i/U_i}(D_i)] \quad \text{and} \quad \mathcal{C}^\bullet(\xi_i) = [F_i^*\Omega_X \to \Omega_{\mathcal{X}_i/U_i}(D_i)],$$

indexed $-1$ and 0. Then $\mathcal{A}^\bullet(\xi_i) = [\mathcal{W}_2(\xi_i) \to 0]$ fits into the short exact sequence of complexes

(5) $$0 \longrightarrow \mathcal{A}^\bullet(\xi_i) \longrightarrow \mathcal{B}^\bullet(\xi_i) \longrightarrow \mathcal{C}^\bullet(\xi_i) \longrightarrow 0.$$

Let

$$\mathcal{E}_1(\xi_i) = \mathcal{E}xt^1_{\mathcal{X}_i/U_i}(\mathcal{B}^\bullet(\xi_i), \mathcal{O}_{\mathcal{X}_i}) \quad \text{and} \quad \mathcal{E}_2(\xi_i) = \mathcal{E}xt^1_{\mathcal{X}_i/U_i}(\mathcal{A}^\bullet(\xi_i), \mathcal{O}_{\mathcal{X}_i}),$$

both of which are locally free, and let

(6) $$\mathcal{E}^\bullet(\xi_i) = \left[\mathcal{E}_1(\xi_i) \to \mathcal{E}_2(\xi_i)\right]$$

be part of the long exact sequence induced from the short exact sequence (5). Because $\mathcal{L}$ is sufficiently ample, the sheaf cohomology $\mathfrak{H}^\bullet(\mathcal{E}^\bullet(\xi_i)) = \mathcal{T}^\bullet(\xi_i)$. Following the construction, it is apparent that the complex $\mathcal{E}^\bullet(\xi_i)$ admits a $G_i$-linearization that is compatible with $\mathfrak{H}^\bullet(\mathcal{E}^\bullet(\xi_i)) = \mathcal{T}^\bullet(\xi_i)$, and over $U_{ij}$ we have $p_i^*\mathcal{E}^\bullet(\xi_i) \cong p_j^*\mathcal{E}^\bullet(\xi_j)$ whose induced sheaf cohomology isomorphism

$$p_i^*\mathfrak{H}^\bullet(\mathcal{E}^\bullet(\xi)) \cong p_j^*\mathfrak{H}^\bullet(\mathcal{E}^\bullet(\xi_j))$$

coincides with $p_i^*\mathcal{T}^\bullet(\xi_i) \cong p_j^*\mathcal{T}^\bullet(\xi_j)$ above. We call such complexes $\mathbb{Q}$-complexes.

Now we construct the virtual moduli cycle $[\mathcal{M}^X_{\alpha,g,n}]^{\mathrm{vir}}$. Over each $U_i$, since $\mathcal{T}^2(\xi_i)$ is a quotient sheaf of $\mathcal{E}_2(\xi_i)$, by the basic lemma there is a unique cone cycle $[C^{\mathcal{E}_2(\xi_i)}]$ in $Z_*(E_2(\xi_i))$, where $E_2(\xi_i)$ is the vector bundle on $U_i$ such that $\mathcal{O}_{U_i}(E_2(\xi_i)) = \mathcal{E}_2(\xi_i)$. Clearly, both $E_2(\xi_i)$ and $[C^{\mathcal{E}_2(\xi_i)}]$ are $G_i$-equivariant. A standard descent argument shows that the collection of cones $E_2(\xi_i)/G_i$ patch together to form a cone $E_2$ over $\mathcal{M}^X_{\alpha,g,n}$, and the quotient cycles $[C^{\mathcal{E}_2(\xi_i)}]/G_i \in Z_*\big(E_2(\xi_i)/G_i\big)$ patch together to form a cycle $[C^{\mathcal{E}_2}]$ in $Z_*(E_2)$. Let $i\colon \mathcal{M}^X_{\alpha,g,n} \to E_2$ be the zero section. Since $E_2$ is a $\mathbb{Q}$-vector bundle, the Gysin map

$$i^!_{E_2} : A_*(E_2) \longrightarrow A_{*-\mathrm{rk}(E_2)}\mathcal{M}^X_{\alpha,g,n}$$

is well-defined over $\mathbb{Q}$. Hence, we define the virtual cycle $[\mathcal{M}^X_{\alpha,g,n}]^{\mathrm{vir}}$ to be

$$[\mathcal{M}^X_{\alpha,g,n}]^{\mathrm{vir}} = i^!_{E_2}[C^{\mathcal{E}_2}] \in A_*\mathcal{M}^X_{\alpha,g,n} \otimes_{\mathbb{Z}} \mathbb{Q}.$$

Of course we need to check that such cycles do not depend on the choices of the $\mathbb{Q}$-sheaves $\mathcal{E}_2(\xi_\bullet)$ chosen. Assume $\mathcal{F}_2$ is another $\mathbb{Q}$-sheaf so that the obstruction sheaf $\mathcal{T}^2(\xi_\bullet)$ is a quotient sheaf of $\mathcal{F}_2$. Then we can find a $\mathbb{Q}$-sheaf $\mathcal{G}$ and quotient homomorphisms $f:\mathcal{G} \to \mathcal{E}_2$ and $g:\mathcal{G} \to \mathcal{F}_2$ so that the diagram

$$\begin{array}{ccc} \mathcal{G} & \longrightarrow & \mathcal{E}_2 \\ \downarrow & & \downarrow \\ \mathcal{F}_2 & \longrightarrow & \mathcal{T}^2(\xi_\bullet) \end{array}$$

is commutative. Let $E_2$, $F_2$ and $G$ be cones associated to $\mathcal{E}_2$, $\mathcal{F}_2$ and $\mathcal{G}$ respectively, and let $\tilde{f} : G \to E_2$ and $\tilde{g} : G \to F_2$ be the induced maps. Then following the construction,

$$\tilde{f}^*[C^{\mathcal{E}_2}] = [C^{\mathcal{G}}] = \tilde{g}^*[C^{\mathcal{F}_2}] \in Z_*G.$$

Hence

$$i^!_{E_2}[C^{\mathcal{E}_2}] = i^!_G\tilde{f}^*[C^{\mathcal{E}_2}] = i^!_G\tilde{g}^*[C^{\mathcal{F}_2}] = i^!_{F_2}[C^{\mathcal{F}_2}] \in A_*\mathcal{M}^X_{\alpha,g,n}.$$

This completes the task of constructing the virtual moduli cycles $[\mathcal{M}^X_{\alpha,g,n}]^{\mathrm{vir}}$.

For completeness, we now sketch the construction of the cone $[C^{\mathcal{E}}]$ in the basic lemma. Since the lemma is local, we only need to consider the case where $S$ is one of the $U_i$ mentioned in defining the $\mathbb{Q}$-structure of $\mathcal{M}^X_{\alpha,g,n}$. For simplicity, we consider the case where $n = 0$. Let $F_0 : C_0 \to X$ be the family over $S$ and $[\mathcal{E}_1 \xrightarrow{\sigma} \mathcal{E}_2]$ be the complex of locally free sheaves of $\mathcal{O}_S$-modules constructed in (6) based on the family $F_0$. Our strategy to prove the lemma is as follows. Let $E_1$ and $E_2$ be the vector bundles over $S$ so that $\mathcal{O}_S(E_i) \cong \mathcal{E}_i$ and let $\hat{E}_1$ and $\hat{E}_2$ be the formal completions of the total spaces of $E_1$ and $E_2$ along their zero sections. Let $\pi_S : \hat{E}_1 \times_S \hat{E}_2 \to S$ be the projection. We will construct a graph $\Gamma \subset \hat{E}_1 \times_S \hat{E}_2$ so that over each closed $x \in S$, $\Gamma_x \triangleq \pi^{-1}(x) \cap \Gamma$ is the graph of a Kuranishi map of the obstruction theory to deformations of $x$ in $\mathcal{M}^X_{\alpha,g,n}$. In other words, if we view $\hat{E}_1$ as the image scheme of the obvious immersion $\hat{E}_1 \to \hat{E}_1 \times_S \hat{E}_2$, then $\Gamma \cap \hat{E}_1$ will be the formal completion of $S \times S$ along its diagonal and $\Gamma$ will be the graph of a relative Kuranishi map. Then we take $\mathcal{N}$ to be the normal cone to $\Gamma \cap \hat{E}_1$ in $\Gamma$. $\mathcal{N}$ will be a cone in $p^*E_2$, where $p:\Gamma \cap \hat{E}_1 \to S$ is the obvious projection. The invariance property of the cones associated to Kuranishi maps ensures that $\mathcal{N}$ will provide an algebraic cone cycle $[C^{\mathcal{E}}]$ as desired. Indeed, the support of $[C^{\mathcal{E}}]$ will be the support of $\mathcal{N}$. This will prove the basic lemma.

Before we continue, let us first introduce some notations. Let $B$ be the sheaf of algebra $\mathrm{Sym}_S(\mathcal{E}_1^\vee)$, $\mathfrak{m} \subset B$ be the ideal generated by $\mathcal{E}_1^\vee$, $B_k = B/\mathfrak{m}^{k+1}$ and $\hat{B} = \lim B_k$. Hence $\hat{E}_1 = \mathrm{Spec}_S(\hat{B})$. In the following, whenever there is a family of stable morphisms $F:C \to X$, we will use $\mathcal{W}_1(F)$ to denote the sheaf constructed in (4) so that $F^*\Omega_X$ is the quotient sheaf of $\mathcal{W}_1(F)$, and denote by $\mathcal{W}_2(F)$ the kernel of $\mathcal{W}_1(F) \to F^*\Omega_X$. Let $\sigma$ be the homomorphism in the complex $[\mathcal{E}_1 \to \mathcal{E}_2]$.

From the construction of the sheaf $\mathcal{E}_1$, there is a tautological family of curves $C_{B_1}$ over $\mathrm{Spec}_S B_1$, extending $C_0$ over $S$, and a tautological homomorphism $\varphi_1$

fitting the diagram of sheaves of $\mathcal{O}_{C_0}$-modules

$$\begin{array}{ccc} \mathcal{W}_1(F_0) & = & \mathcal{W}_1(F_0) \\ {\scriptstyle\varphi_1}\downarrow & & \downarrow \\ \Omega_{C_{B_1}/S} \otimes_{\mathcal{O}_{C_{B_1}}} \mathcal{O}_{C_0} & \xrightarrow{\beta_1} & \Omega_{C_0/S}. \end{array}$$

$J_1$ is exactly the ideal of the largest subscheme in $\mathrm{Spec}_S\, B_1$ over which $\varphi_1$ factor through $F_0^*\Omega_X$. It can be seen as follows. Because $\mathfrak{m}/\mathfrak{m}^2$ is the kernel of $\beta_1$, the composition of $\mathcal{W}_2(F_0) \to \mathcal{W}_1(F_0)$ with $\varphi_1$ lifts to

$$\xi_1 \in \mathrm{Hom}_{C_0}(\mathcal{W}_2(F_0), \pi^*(\mathfrak{m}/\mathfrak{m}^2)) = \Gamma_S(\mathcal{E}_1^\vee \otimes_{\mathcal{O}_S} \mathcal{E}_2),$$

where $\pi : C_0 \to S$ is the projection. Clearly, $\varphi_1$ factor through $F_0^*\Omega_X$ if and only if $\xi_1$ vanishes. We call $\xi_1$ the first obstruction class. It is easy to see that $\xi_1 = \sigma$. For simplicity, we denote by $I(\xi_1)$, called the vanishing ideal of $\xi_1$, the ideal in $B_1$ generated by the components of $\xi_1 : \mathcal{E}_2^\vee \to \mathcal{E}_1^\vee$. Now let $A_1 = B_1/J_1$ and let $C_{A_1}$ be the restriction of $C_{B_1}$ to $\mathrm{Spec}_S\, A_1 \subset \mathrm{Spec}_S\, B_1$. Then the fact that we have a diagram

$$\begin{array}{ccc} F_0^*\Omega_X & = & F_0^*\Omega_X \\ \downarrow & & \downarrow \\ \Omega_{C_{A_1}/S} \otimes_{\mathcal{O}_{C_{A_1}}} \mathcal{O}_{C_0} & \longrightarrow & \Omega_{C_0/S} \end{array}$$

means that the left arrow is induced by a morphism $F_1 : C_{A_1} \to X$ extending $F_0$.

Our next step is to find a second obstruction class $\xi_2$ and its vanishing ideal $J_2 = I(\xi_2) \subset R_2 \triangleq B_2/J_1 \cdot \mathfrak{m}$ so that $J_2/\mathfrak{m}^2 = J_1$ and $J_2$ defines the largest subscheme in $\mathrm{Spec}_S(B_2/J_1 \cdot \mathfrak{m})$ over which $F_1$ can be extended. We fix a family $C_{R_2}$ over $\mathrm{Spec}_S\, R_2$ extending $C_{B_1}$, and let $\phi_1 : \mathcal{W}_1(F_1) \to \Omega_{C_{A_1}/S}$ be induced by $F_1^*\Omega_X \to \Omega_{C_{A_1}/S}$. Because of the ampleness of the sheaf $\mathcal{W}_1(F_1)^{-1}$, $\phi_1$ lifts to $\varphi_2$ as shown in the diagram

$$\begin{array}{ccc} \mathcal{W}_1(F_1) & = & \mathcal{W}_1(F_1) \\ {\scriptstyle\varphi_2}\downarrow & & \downarrow \\ \Omega_{C_{R_2}/S} \otimes_{\mathcal{O}_{C_{R_2}}} \mathcal{O}_{C_{A_1}} & \xrightarrow{\beta_1} & \Omega_{C_{A_1}/S}. \end{array} \tag{7}$$

Let $\xi_2' \in \mathrm{Hom}(\mathcal{W}_2(F_1), \mathcal{I})$ be the element induced by the above diagram, similar to how $\xi_1$ is derived, where $\mathcal{I}$ is the ideal sheaf of $C_{A_1} \subset C_{R_2}$. We can choose $\varphi_2$ so that $\xi_2'$ restricted to $\xi_1$ under $\mathcal{O}_{C_{R_2}}(-C_{A_1}) \to \mathcal{O}_{C_{B_1}}(-C_{A_1})$. Using the inclusion $\mathcal{O}_{C_{R_2}}(-C_{A_1}) \subset \mathcal{O}_{C_{R_2}}$, $\xi_2'$ induces our second obstruction class

$$\xi_2 \in \mathrm{Hom}_{C_0}(\mathcal{W}_2(F_0), \pi^*(\mathcal{I}_{A_1 \subset R_2})) \subset \Gamma_S(\mathcal{E}_2 \otimes_{\mathcal{O}_S} R_2).$$

Let $I(\xi_2) \subset R_2$ be the ideal generated by components of $\xi_2$. Then $I(\xi_2) \subset R_2$ defines the subscheme of $\mathrm{Spec}_S\, R_2$ over which the above diagram induces an extension $F_2 : C_{A_2} \to X$ of $F_1$, where $A_2 = R_2/I(\xi_2)$.

Inductively, after constructing the $k$-th extension $F_k$ over $A_k$, we can find a diagram of sheaves of $\mathcal{O}_{C_{A_k}}$-modules, similar to (7), which in turn induces the $(k+1)$-th obstruction class $\xi_{k+1} \in \Gamma(\mathcal{E}_2 \otimes_{\mathcal{O}_S} \mathcal{I}_{A_R \subset R_{k+1}})$ whose vanishing locus is exactly where $F_k$ can be extended to $F_{k+1}$, according to the diagram given. Finally, we can choose $f: \mathcal{E}_2^\vee \to \hat{B}$ so that its restriction to $\mathcal{E}_2^\vee \to R_k$ are $\xi_k$. Then $f$ is the relative Kuranishi map we want to construct. Let $\Gamma \subset \hat{E}_1 \times_S \hat{E}_2$ be the graph of $f$.

From the construction, there is a tautological family of morphisms over $\Gamma \cap \hat{E}_1$ that extends the original family $F_0$. Let $\mathcal{N}$ be the normal cone to $\Gamma \cap \hat{E}_1$ in $\Gamma$. Of course, the choices of $\{\xi_k\}$ and $f$ are not unique. However, one shows that the cycles $[\mathcal{N}]$ do not depend on their choices. $[\mathcal{N}]$ is a cycle in $Z_*(\pi^* E_2)$, where $\pi : \Gamma \cap \hat{E}_1 \to S$ is induced by projection $\hat{E}_1 \to S$. Finally, one shows that there is a (cone) cycle $[C^{\mathcal{E}}] \in Z_* E_2$ of which the following holds: let $x \in S$ be any closed point, let $\widehat{x}$ be the formal completion of $\Gamma \cap \hat{E}_1$ along $x$ and let $p: \widehat{x} \to S$ be induced by the tautological family on $\Gamma \cap \hat{E}_1$. Then there is a homomorphism $(\pi^* E)|\widehat{x} \to p^*(E_2|_{\hat{x}})$, where $\hat{x}$ is the formal completion of $S$ along $x$, such that $p^{-1}([C^{\mathcal{E}}]|_{\hat{x}}) = \mathcal{N}|\widehat{x}$. Therefore, $\mathcal{N}$ provides a cone $[C^{\mathcal{E}}]$ as desired.

We end this section with a few words on Donaldson polynomial invariants.

EXAMPLE. (Donaldson invariants) Let $S$ be a smooth algebraic surface. We fix data $\xi = (r, I, d) \in \mathbf{N} \times \mathbf{Pic}(S) \times H^4(X, \mathbb{Z})$, and form the moduli space of semistable sheaves $\mathcal{F}$ (with respect to a fixed polarization $H$) satisfying $\operatorname{rk} \mathcal{F} = r$, $\det \mathcal{F} = I$ and $c_2(\mathcal{F}) = d$. We denote this moduli space by $\mathcal{M}_\xi$. When $r, I \cdot H$ and $d$ are coprime, $\mathcal{M}_\xi$ admits a universal family $\mathcal{E}$. The natural choice of tangent-obstruction complex of $\mathcal{M}_\xi$ is

$$\mathcal{D}^\bullet = \big[\mathcal{E}xt^1_{S\times\mathcal{M}_\xi/\mathcal{M}_\xi}(\mathcal{E},\mathcal{E})^0 \longrightarrow \mathcal{E}xt^2_{S\times\mathcal{M}_\xi/\mathcal{M}_\xi}(\mathcal{E},\mathcal{E})^0\big], \tag{8}$$

where the superscript 0 denotes the traceless part of the relative extension sheaves [At]. It can be shown that there are complexes of locally free sheaves $\mathcal{E}^\bullet = [\mathcal{E}_1 \to \mathcal{E}_2]$ so that $\mathfrak{H}^\bullet(\mathcal{E}^\bullet) = \mathcal{D}^\bullet$. Therefore, we obtain the virtual moduli cycle

$$[\mathcal{M}_\xi]^{\mathrm{vir}} \in A_{\mathrm{vir.\,dim}} \mathcal{M}_\xi,$$

following the outlined construction. Let $\mu_i : H_{2i}(S, \mathbb{R}) \to H^{4-2i}(\mathcal{M}_\xi, \mathbb{R})$ be induced by slant product with $c_2(\mathcal{E})$. We define

$$\phi_{r,I} : H_2(S, \mathbb{R}) \oplus H_4(S, \mathbb{R}) \longrightarrow \mathbb{R}$$

via $\phi_{r,I}(\alpha) = \deg\big(e^\alpha [\mathcal{M}_\xi]^{\mathrm{vir}}\big)_0$. When $r = 2$, this is the usual Donaldson polynomial. (It is independent of the choice of polarization when $p_g > 0$.[3] The story is more complicated, but understood, when $p_g = 0$.) For moduli of high rank stable sheaves, they are also smooth invariants (for $p_g > 0$), following the general construction of Euler classes of Fredholm bundles [LT2].

## 2. Virtual cycles and refined Gysin maps

Refined Gysin maps is an important tool in intersection theory. Given a square

$$\begin{array}{ccc} X \times_Z W & \longrightarrow & W \\ \big\downarrow & & \big\downarrow f \\ X & \xrightarrow{\ i\ } & Z, \end{array}$$

[3]Other than the case of K3-surfaces [HG], the only proof of this statement is using the moduli of ASD connections.

where $i: X \to Z$ is a local complete intersection immersion of codimension $k$, the refined Gysin map

$$i^! : A_* W \longrightarrow A_{*-k} X \times_Z W$$

is a homomorphism sending any class $B \in A_* W$ to $f_*(B) \cdot [X]$, viewed as a cycle in $X \times_Z W$. In case $B$ intersects $X$ properly, $i^! B$ is $[B \cap X]$, counting multiplicity.

The virtual cycle constructed commutes with the refined Gysin maps in the following sense. Let $Z$ be a smooth scheme, and let $\mathcal{M}_Z$ be a moduli space projective over $Z$, with a perfect ($\mathbb{Q}$-)tangent-obstruction complex $\mathcal{T}_Z^\bullet$. Hence we have the virtual cycle $[\mathcal{M}_Z]^{\mathrm{vir}}$. Now assume $i: X \to Z$ is an immersion of smooth subscheme. Let $\mathcal{M}_X = X \times_Z \mathcal{M}_Z$ and let $\mathcal{T}_X^\bullet$ be the ($\mathbb{Q}$)-tangent-obstruction complex of $\mathcal{M}_X$. Assume that there is an exact sequence

$$0 \longrightarrow \mathcal{T}_X^1(\mathcal{F}) \longrightarrow \mathcal{T}_Z^1(\mathcal{F}) \longrightarrow \pi^* \mathcal{N}_{X/Z} \otimes \mathcal{F} \longrightarrow \mathcal{T}_X^2(\mathcal{F}) \longrightarrow \mathcal{T}_Z^2(\mathcal{F}) \longrightarrow 0,$$

where $\mathcal{F}$ is any sheaf of $\mathcal{O}_{\mathcal{M}_X}$-modules, $\pi: \mathcal{M}_X \to X$ is the projection and $\mathcal{N}_{X/Z}$ is the sheaf of normal bundle to $X$ in $Z$, and assume further that the obstructions to deformations in $\mathcal{M}_Z$ are the images of the obstructions to deformations in $\mathcal{M}_X$ under $\mathcal{T}_X^2 \to \mathcal{T}_Z^2$, then the two virtual cycles $[\mathcal{M}_Z]^{\mathrm{vir}}$ and $[\mathcal{M}_X]^{\mathrm{vir}}$ are related by the refined Gysin maps.

THEOREM. *Let the notation be as above. Then*

$$i^! [\mathcal{M}_Z]^{\mathrm{vir}} \sim_{\mathrm{rat}} [\mathcal{M}_X]^{\mathrm{vir}}.$$

The immediate corollary of this theorem is that the GW-invariants and the Donaldson invariants are deformation invariants.

COROLLARY. *Let $X_t$ be a smooth family of complex projective varieties over a connected base $T$. Then for any $t, t' \in T$, and $\alpha, \alpha' \in H_i(X_t, \mathbb{Z})$ $(= H_i(X_{t'}, \mathbb{Z}))$,*

$$\phi_{\alpha,g,n}^{X_t} = \phi_{\alpha',g,n}^{X_{t'}}.$$

Another application of this principle is the proof of a set of relations satisfied by the GW-invariants. Here we will discuss the composition law for GW-invariants. The composition law is a set of relations among GW-invariants that are derived by applying Gysin maps to divisors of $\mathcal{M}_{g,n}$. Recall the two clutching morphisms of moduli spaces of stable curves constructed by Knudsen. Let $g = g_1 + g_2$ and $n = n_1 + n_2$ be any partitions of $g$ and $n$. Then

$$\tau_{g_1 g_2, n_1 n_2} : \mathcal{M}_{g_1, n_1+1} \times \mathcal{M}_{g_2, n_2+1} \longrightarrow \mathcal{M}_{g,n}$$

is the morphism mapping any pair of curves to the nodal curve obtained by adjoining the last marked point of the first curve to the first marked point of the second curve while keep the order of the remaining marked points. Another clutching morphism is

$$\tau_{g-1,n+2} : \mathcal{M}_{g-1,n+2} \longrightarrow \mathcal{M}_{g,n}$$

induced from identifying the last two marked points of curves in $\mathcal{M}_{g-1,n+2}$.

COMPOSITION LAW. *Let $X$ be any smooth projective variety. Assume $\tau_i$, $\tilde{\tau}_i \in H^*(X)$ are elements so that $[\Delta]^\vee = \sum_{i=1}^k \tau_i \otimes \tilde{\tau}_i$ is the Kunneth decomposition of the Poincare dual of the class $[\Delta]$, where $\Delta \subset X \times X$ is the diagonal. Then*

1. *for any $h_1 \in H_*(\mathcal{M}_{g_1,n_1+1})$ and $h_2 \in H_*(\mathcal{M}_{g_2,n_2+1})$, we have*

$$\begin{aligned}&\psi^X_{\alpha,g,n}(\xi_1,\cdots,\xi_n,\tau_{g_1g_2,n_2n_2\,*}(h_1\times h_2)^\vee)\\ &=\sum_{\alpha_1+\alpha_2=\alpha}\sum_{i=1}^k\psi^X_{\alpha_1,g_1,n_1+1}(\xi_1,\cdots,\xi_{n_1},\tau_i,h_1^\vee)\cdot\psi^X_{\alpha_2,g_2,n_2+1}(\xi_{n_1+1},\cdots,\xi_n,\tilde{\tau}_i,h_2^\vee).\end{aligned}$$

2. *for any $h \in H_*(\mathcal{M}_{g-1,n+2})$, we have*

$$\psi^X_{\alpha,g,n}(\xi_1,\cdots,\xi_n,\tau_{g-1,n+2*}(h)^\vee)=\sum_{i=1}^k\psi^X_{\alpha,g-1,n+2}(\xi_1,\cdots,\xi_n,\tau_i,\tilde{\tau}_i,h^\vee).$$

We now explain the geometry behind these two laws. Let $\alpha_1+\alpha_2=\alpha$ be a partition of $\alpha$ and let $W_{\alpha_1\alpha_2}$ be defined by the Cartesian product

$$\begin{array}{ccc} W_{\alpha_1\alpha_2} & \longrightarrow & \mathcal{M}^X_{\alpha_1,g_1,n_1+1}\times\mathcal{M}^X_{\alpha_2,g_2,n_2+1}\\ \downarrow & & \downarrow{\scriptstyle (e_{n_1+1},e_1)}\\ X & \xrightarrow{\Delta} & X\times X.\end{array}$$

Note that when $n_1=n_2=0$ and $g_1=g_2$, $\mathbb{Z}_2$ acts on $\cup_{\alpha_1+\alpha_2=\alpha}W_{\alpha_1\alpha_2}$ induced by $W_{\alpha_1\alpha_2}\to W_{\alpha_2\alpha_1}$, which is obtained by interchanging factors of pairs of maps. For convenience, we let $G$ be $\mathbb{Z}_2$ when $n_1=n_2=0$ and $g_1=g_2$ and $\{1\}$ otherwise. The construction of the clutching morphism $\tau_{g_1g_2,n_1n_2}$ can be adopted to define a similar morphism

$$\tilde{\tau}_{g_1g_2,n_1n_2}:(\bigcup_{\alpha_1+\alpha_2=\alpha}W_{\alpha_1\alpha_2})/G\longrightarrow\mathcal{M}^X_{\alpha,g,n},$$

which we still call clutching morphism. $\tilde{\tau}_{g_1g_2,n_1n_2}$ is finite but is not necessary an immersion. We also define the subscheme $Z_\alpha\subset\mathcal{M}^X_{\alpha,g,n}$ via the Cartesian product

$$\begin{array}{ccc} Z_\alpha & \longrightarrow & \mathcal{M}^X_{\alpha,g,n}\\ \downarrow & & \downarrow{\scriptstyle \pi}\\ \mathcal{M}_{g_1,n_1+1}\times\mathcal{M}_{g_2,n_2+1}/G & \xrightarrow{\tau_{g_1g_2,n_1n_2}} & \mathcal{M}_{g,n}.\end{array}$$

The morphism $\tilde{\tau}_{g_1g_2,n_1n_2}$ factor through $Z_\alpha$, denoted

$$\eta:(\bigcup_{\alpha_1+\alpha_2=\alpha}W_{\alpha_1\alpha_2})/G\longrightarrow Z_\alpha.$$

The composition law I follows from

COMPOSITION LAW I. *Let the notation be as above. Then the composition law I follows from*

$$\eta_*((\sum_{\alpha_1+\alpha_2=\alpha}\Delta^![\mathcal{M}^X_{\alpha_1,g_1,n_1+1}\times\mathcal{M}^X_{\alpha_2,g_2,n_2+1}]^{\mathrm{vir}})/G)=\tau^!_{g_1g_2,n_1n_2}[\mathcal{M}^X_{\alpha,g,n}]^{\mathrm{vir}}.$$

The composition law II is equivalent to a similar statement about cycles. Let $W_1$ and $W_2$ be defined by fiber products

$$\begin{array}{ccc} W_2 & \longrightarrow & \mathcal{M}^X_{\alpha,g,n} \\ \downarrow & & \downarrow \\ \mathcal{M}_{g-1,n+2} & \stackrel{\theta}{\longrightarrow} & \mathcal{M}_{g,n} \end{array} \quad \text{and} \quad \begin{array}{ccc} W_1 & \longrightarrow & \mathcal{M}^X_{\alpha,g-1,n+2} \\ \downarrow & & \downarrow{\scriptstyle (e_{n+1},e_{n+2})} \\ X & \stackrel{\Delta}{\longrightarrow} & X\times X. \end{array}$$

There is an obvious finite morphism $\zeta : W_1 \to W_2$.

COMPOSITION LAW II. *Let the notation be as above. Then the second composition law follows from*

$$\zeta_* \Delta^! [\mathcal{M}^X_{\alpha,g-1,n+2}]^{\mathrm{vir}} = \theta^! [\mathcal{M}^X_{\alpha,g,n}]^{\mathrm{vir}}.$$

Now we sketch the strategy to prove the composition law I. I should mention that in case all "relevant" schemes have expected dimensions, then the above two identities follow from properties of ordinary refined Gysin maps. One class of varieties of which $\mathcal{M}^X_{\alpha,0,n}$ are unobstructed are homogeneous varieties. This was discussed in details in [LT4] and in this volume by Fulton-Pandharipande [FP].

Let us look at the first composition law. Following the principle that virtual moduli cycles construction commutes with refined Gysin maps, one immediately sees that if one let $W_{\alpha_1\alpha_2}$ and $Z_\alpha$ be endowed with the obvious $\mathbb{Q}$-tangent-obstruction complex, then should $\eta : \cup W_{\alpha_1\alpha_2}/G \to Z_\alpha$ be an isomorphism, then the composition law would follow if one can show that the $\mathbb{Q}$-tangent-obstruction complex of $\cup W_{\alpha_1\alpha_2}/G$ is isomorphic to the $\mathbb{Q}$-tangent-obstruction complex of $Z_\alpha$. However, in general $\eta$ is only finite. There are possibly morphisms $f$ so that $\eta^{-1}(f)$ are not unique. Such morphisms are of the following forms: let $\{f_1 : C_1 \to X\} \in \mathcal{M}^X_{\beta_1,g_1,n_1+1}$ be a morphism with marked points $\{x_i\}$, let $\{f_2 : C_2 \to X\} \in \mathcal{M}^X_{\beta_2,g_2,n_2+1}$ be a morphism with marked points $\{y_i\}$ and let $\{f_0 : C_0 \to X\} \in \mathcal{M}^X_{\beta_0,0,2}$ with marked points $\{z_1, z_2\}$ so that $f_1(x_{n_1+1}) = f_0(z_1)$ and $f_0(z_2) = f_2(y_1)$. Assume $\alpha = \beta_0 + \beta_1 + \beta_2$. Then we obtain a new morphism $f \in \mathcal{M}^X_{\alpha,g,n}$ whose domain is the result of identifying $x_{n_1+1}$ with $z_1$ and identifying $z_2$ with $y_1$. Since $C_0$ has genus 0, $f$ lies in the image of $Z_{\beta_1+\beta_0,\beta_2}$ and in the image of $Z_{\beta_1,\beta_0+\beta_2}$. The key observation is that though $\cup W_{\alpha_1\alpha_2}/G \to Z_\alpha$ is not an isomorphism, the individual $W_{\alpha_1\alpha_2} \to Z_\alpha$ (or $W_{\alpha_1\alpha_2}/\mathbb{Z}_2 \to Z_\alpha$ when $\alpha_1 = \alpha_2$) are closed immersions. Further, for any partitions $\alpha_1 + \alpha_2 = \alpha$ there are pairs $(L_{\alpha_1\alpha_2}, f_{\alpha_1\alpha_2})$, where $L_{\alpha_1\alpha_2}$ are $\mathbb{Q}$-line bundles on $\mathcal{M}^X_{\alpha,g,n}$ and $f_{\alpha_1\alpha_2}$ are global sections of $L_{\alpha_1\alpha_2}$, that have the following significance: 1) $f^{-1}_{\alpha_1\alpha_2}(0)$ is the image scheme of the locally closed immersion $W_{\alpha_1\alpha_2} \to \mathcal{M}^X_{\alpha,g,n}$; 2) the $\mathbb{Q}$-tangent-obstruction complex of $f^{-1}_{\alpha_1\alpha_2}(0)$ induced by the defining equation $f_{\alpha_1\alpha_2}$ is isomorphic to the $\mathbb{Q}$-tangent-obstruction complex of $W_{\alpha_1\alpha_2}$; 3) let $L = \otimes_{\alpha_1+\alpha_2=\alpha} L_{\alpha_1\alpha_2}$ and $f = \Pi_{\alpha_1+\alpha_2=\alpha} f_{\alpha_1\alpha_2}$, then the pair $(L, f)$ is the pull-back of the pair $(L_0, f_0)$ defining the image $\mathbb{Q}$-divisor $\mathcal{M}_{g_1,n_1+1} \times \mathcal{M}_{g_2,n_2+1} \to \mathcal{M}_{g,n}$. Namely, $L_0$ is a $\mathbb{Q}$-line bundle on $\mathcal{M}_{g,n}$ and $f_0$ is a section of $L_0$ so that $f_0^{-1}(0)$ is the image scheme. Let $[W_{\alpha_1\alpha_2}]^{\mathrm{vir}}$, $[f^{-1}_{\alpha_1\alpha_2}(0)]^{\mathrm{vir}}$ and $[f^{-1}(0)]^{\mathrm{vir}}$ be the respective virtual moduli cycles constructed based on the respective $\mathbb{Q}$-tangent-obstruction complexes. Then applying the theorem in this section,

$$\begin{aligned}\tau^{!}_{g_1g_2,n_1n_2}[\mathcal{M}^X_{\alpha,g,n}]^{\mathrm{vir}} &= c_1(L)\cdot[\mathcal{M}^X_{\alpha,g,n}]^{\mathrm{vir}} = \sum_{\alpha_1+\alpha_2=\alpha} c_1(L_{\alpha_1\alpha_2})\cdot[\mathcal{M}^X_{\alpha,g,n}]^{\mathrm{vir}}\\ &= \sum_{\alpha_1+\alpha_2=\alpha}[f^{-1}_{\alpha_1\alpha_2}(0)]^{\mathrm{vir}} = \eta_*\Big(\sum_{\alpha_1+\alpha_2=\alpha}[W_{\alpha_1\alpha_2}]^{\mathrm{vir}}/G\Big)\\ &= \eta_*\Delta^!\Big(\sum_{\alpha_1+\alpha_2=\alpha}[\mathcal{M}^X_{\alpha_1,g_1,n_1+1}\times\mathcal{M}^X_{\alpha_2,g_2,n_2+1}]^{\mathrm{vir}}/G\Big).\end{aligned}$$

Here we view $c_1(L)\cdot[\mathcal{M}^X_{\alpha,g,n}]^{\mathrm{vir}}$ as a cycle class in $A_*(f^{-1}(0))\cong A_*Z_\alpha$. Thus to prove the composition law I, it suffices to show that the $\mathbb{Q}$-tangent-obstruction complex of $W_{\alpha_1\alpha_2}$ and that of $Z_{\alpha_1\alpha_2}\triangleq f^{-1}_{\alpha_1\alpha_2}(0)$ are isomorphic.

We now sketch its proof. For simplicity, we will give the description of the tangent and the obstruction spaces of any $w\in W_{\alpha_1\alpha_2}$ and the tangent and the obstruction spaces of $z=\tilde\tau_{g_1g_2,n_1n_2}(w)\in Z_{\alpha_1\alpha_2}=f^{-1}_{\alpha_1\alpha_2}(0)$. We will restrict ourselves to the case $n=0$ and $g_1\neq g_2$. (Hence $G=\{1\}$.) Let $w\in W_{\alpha_1\alpha_2}$ be any point associated to a pair of morphisms $f_1:C_1\to X$ and $f_2:C_2\to X$ with marked points $x_1\in C_1$ and $x_2\in C_2$ respectively. Then its image $z\in Z_{\alpha_1\alpha_2}$ associates to the morphism $f:C\to X$, where $C$ is the result of identifying $x_1$ with $x_2$. We denote by $x\in C$ the nodal point where $C_1$ and $C_2$ are adjoined. For simplicity, we assume $f$ has no non-trivial automorphisms. We claim that the tangent space of $Z_{\alpha_1\alpha_2}$ at $z$ is $\mathrm{Ext}^1_C([f^*\Omega_X\to\Omega^0_C],\mathcal{O}_C)$, where $\Omega^0_C$ is the quotient of $\Omega_C$ by its torsion at $x$. This is true because $df_{\alpha_1\alpha_2}:T_z\mathcal{M}^X_{\alpha,g,n}\to\mathbb{C}$ is the obvious homomorphism

$$\mathrm{Ext}^1_C([f^*\Omega_X\to\Omega_C],\mathcal{O}_C)\longrightarrow\mathrm{Ext}^1_{\hat x}(\Omega_{\hat x},\mathcal{O}_{\hat x}),$$

where $\hat x$ is the formal completion of $C$ along $x$ [DM]. A standard obstruction argument shows that the obstruction space to deformations of $z$ in $Z_{\alpha_1\alpha_2}$ is the cokernel of

$$\mathrm{Ext}^1_C(\Omega^0_C,\mathcal{O}_C)\longrightarrow\mathrm{Ext}^1_C(f^*\Omega_X,\mathcal{O}_C),$$

which is $\mathrm{Ext}^2_C([f^*\Omega_X\to\Omega^0_C],\mathcal{O}_C)$. Note that the obstruction space to deformations of $z$ in $\mathcal{M}^X_{\alpha,g,n}$ is the cohomology group $\mathrm{Ext}^2_C([f^*\Omega_X\to\Omega_C],\mathcal{O}_C)$. One further checks that the obstruction to deformations of $w\in Z_{\alpha_1\alpha_2}$ in $\mathcal{M}^X_{\alpha,g,n}$ is the image of the obstruction to deformations in $Z_{\alpha_1\alpha_2}$ under the homomorphism

$$\mathrm{Ext}^2_C([f^*\Omega_X\to\Omega^0_C],\mathcal{O}_C)\longrightarrow\mathrm{Ext}^2_C([f^*\Omega_X\to\Omega_C],\mathcal{O}_C).$$

Hence by the theorem at the beginning of this section,

$$c_1(L_{\alpha_1\alpha_2})\cdot[\mathcal{M}^X_{\alpha,g,n}]^{\mathrm{vir}}=[Z_{\alpha_1\alpha_2}]^{\mathrm{vir}}.$$

As to the tangent and the obstruction spaces of $W_{\alpha_1\alpha_2}$ at $w$, we first observe that the tangent space $T_wW_{\alpha_1\alpha_2}$ is the kernel of

$$\mathrm{Ext}^1_{C_1}([f_1^*\Omega_X\to\Omega_{C_1}(x_1)],\mathcal{O}_{C_1})\oplus\mathrm{Ext}^1_{C_2}([f_2^*\Omega_X\to\Omega_{C_2}(x_2)],\mathcal{O}_{C_2})\xrightarrow{(p_1,p_2)}T_{f(x)}X.$$

$p_i$ is defined as follows: given $v\in\mathrm{Ext}^1_{C_1}([f_1^*\Omega_X\to\Omega_{C_1}(x_1)],\mathcal{O}_{C_1})$, $v$ associates to a flat extension $\tilde f_1:\tilde x_1\subset\tilde C_1\to X$ of $f_1$ over $\mathrm{Spec}(\mathbb{C}[t]/(t^2))$. Then $p_1(v)$ is the tangent direction of $\tilde{f}(\tilde{x}_1)$.We now describe $p_1(v)$ explicitly. By definition, $\tilde{f}_1$ is constructed based on the diagram

$$\begin{array}{ccccccccc} & & & & f_1^*\Omega_X & = & f_1^*\Omega_X & & \\ & & & & \downarrow & & \downarrow & & \\ 0 & \longrightarrow & \mathcal{O}_{C_1} & \longrightarrow & \mathcal{B} & \longrightarrow & \Omega_{C_1}(x_1) & \longrightarrow & 0. \end{array}$$

Let $\mathcal{B}'$ be defined by the push-forward diagram

$$\begin{array}{ccc} \mathcal{O}_{C_1}(-x_1) & \longrightarrow & \mathcal{B}(-x_1) \\ \downarrow & & \downarrow \\ \mathcal{O}_{C_1} & \longrightarrow & \mathcal{B}', \end{array}$$

Then $f_1^*\Omega_X(-x_1) \to \mathcal{B}(-x_1) \to \mathcal{B}'$ lifts to $f_1^*\Omega_X \to \mathcal{B}'$ and $p_1$ is

$$T_{w_1}\mathcal{M}^X_{\alpha_1,g_1,n_1+1} = \operatorname{Ext}^1_{C_1}\left([f_1^*\Omega_{C_1} \to \Omega_{C_1}(x_1)], \mathcal{O}_{C_1}\right) \xrightarrow{p_1} e_1^*\Omega_X^\vee \equiv T_{f(x)}X$$

that sends $v$ to the composite

$$e_1^*\Omega_X \to \mathcal{B}'|_{x_1} \longrightarrow \operatorname{coker}\{\mathcal{B}(-x_1) \to \mathcal{B}'\} \equiv \mathbb{C}.$$

$p_2$ is defined similarly. Therefore, the induced homomorphism $(e_1, e_1)_*$ on tangent spaces is

$$(e_1, e_1)_* = (p_1, p_2) : T_w(\mathcal{M}^X_{\alpha_1,g_1,n_1+1} \times \mathcal{M}^X_{\alpha_2,g_2,n_2+1}) \longrightarrow (e_1, e_1)^*\Omega^\vee_{X\times X}.$$

Consequently, its composition with

$$(e_1, e_1)^*\Omega^\vee_{X\times X} \longrightarrow (e_1, e_1)^*\mathcal{N}_{\Delta(X)/X\times X}$$

is $p_1 - p_2$, after identifying the normal bundle $N_{\Delta(X)/X\times X}$ with $T_{\Delta(X)}$.

Finally, we combine these with the long exact sequence of cohomologies. Let $\mathcal{R}^\bullet$ be the complex $[f^*\Omega_X \to \Omega^0_C]$ and $\mathcal{R}^\bullet_j = [f_j^*\Omega_X \to \Omega_{C_j}(x_j)]$. Then the short exact sequence

$$0 \longrightarrow \mathcal{O}_C \longrightarrow \mathcal{O}_{C_1} \oplus \mathcal{O}_{C_2} \longrightarrow \mathcal{O}_x \longrightarrow 0$$

induces the long exact sequence

$$\longrightarrow \operatorname{Ext}^i_C(\mathcal{R}^\bullet, \mathcal{O}_C) \longrightarrow \oplus_{j=1}^2 \operatorname{Ext}^i_C(\mathcal{R}^\bullet, \mathcal{O}_{C_j}) \longrightarrow \operatorname{Ext}^i_C(\mathcal{R}^\bullet, \mathcal{O}_x) \longrightarrow$$

which simplifies to

$$\begin{aligned} 0 \longrightarrow \operatorname{Ext}^1_C(\mathcal{R}^\bullet, \mathcal{O}_C) \longrightarrow \oplus_{j=1}^2 \operatorname{Ext}^1_{C_j}(\mathcal{R}^\bullet_j, \mathcal{O}_{C_j}) \xrightarrow{\delta} f^*\Omega^\vee_X|_x \longrightarrow \\ \longrightarrow \operatorname{Ext}^2_C(\mathcal{R}^\bullet, \mathcal{O}_C) \longrightarrow \oplus_{j=1}^2 \operatorname{Ext}^2_{C_j}(\mathcal{R}^\bullet_j, \mathcal{O}_{C_j}) \longrightarrow 0. \end{aligned}$$

The above is the long exact sequence of the tangent spaces and obstruction spaces required by the theorem. Also, it is direct to check that the obstruction classes are compatible under these maps. Hence by the theorem

$$\sum [Z_{\alpha_1\alpha_2}]^{\mathrm{vir}} = \eta_*(\sum [W_{\alpha_1\alpha_2}]^{\mathrm{vir}}/G) = \eta_*(\sum \Delta^! [\prod_{i=1}^2 \mathcal{M}^X_{\alpha_i,g_i,n_i+1}]^{\mathrm{vir}}/G).$$

We end this section with a simple test of the second composition law. The number we are interested is the virtual number of degree three nodal rational curves in $\mathbf{P}^2$ passing through eight general points. It is known that there are 12 such curves. However, the virtual number of such curves is different due to the contribution of reducible curves. Let $H \in H_2(\mathbf{P}^2)$ be the hyperplane class. The invariant which we will compute is $\phi^{\mathbf{P}^2}_{3H,1,8}([p]^8, A)$, where $A \in H_*(\mathcal{M}_{1,8})$ is the cycle consisting of nodal rational curves (their geometric genus are 1) that is the image cycle $\frac{1}{2}\tau_{0,10\,*}([\mathcal{M}_{0,10}])$. Using composition law II and knowledge of $\phi^{\mathbf{P}^2}_{\bullet,0,10}$, we have $\phi^{\mathbf{P}^2}_{3H,1,8}([p]^8, A) = \frac{1}{2} \times 108 = 54$. Now we calculate $\phi^{\mathbf{P}^2}_{3H,1,8}([p]^8, A)$ directly using intersection theory on $\mathcal{M}^{\mathbf{P}^2}_{3H,1,8}$. We fix 8 general points on $\mathbf{P}^2$, say $p_1, \cdots, p_8$. There are 12 degree 3 nodal rational curves passing through these points. Let them be $C_1, \cdots, C_{12}$. Let $\tilde{C}_i$ be the normalizations of $C_i$ and let $f_i : \tilde{C}_i \to \mathbf{P}^2$ be the induced morphisms. Consider the subscheme

$$W = \left(\mathcal{M}^{\mathbf{P}^2}_{3H,1,8} \times_{\mathcal{M}_{1,8}} \mathcal{M}_{0,10}\right) \bigcap \left(\bigcap_{i=1}^{8} e_i^{-1}(p_i)\right),$$

where $e_i : \mathcal{M}^{\mathbf{P}^2}_{3H,1,8} \to \mathbf{P}^2$ is the i-th evaluation morphism. $W = W_1 \cup W_2$, where $W_1$ consists of 12 smooth points each corresponding to one of the curves $C_i$ and $W_2$ consists of 12 components each isomorphic to one of $\tilde{C}_i$ as follows. Let $C$ be the nodal rational curve with one nodal point and one marked point $x$. For each $z \in \tilde{C}_i$, we form the curve $D_{i,z}$ that is obtained by identifying $z \in \tilde{C}_i$ with $x \in C$. We define $g_{i,z} : D_{i,z} \to \mathbf{P}^2$ by $g_{i,z}|_{\tilde{C}_i} = f_i$ and $g_{i,z}|_C = \text{const.}$. Note that the virtual dimension of $W$ is zero. The virtual degree of component of $W_2$ is 7. Thus

$$\phi^{\mathbf{P}^2}_{3H,1,8}([p]^8, A) = \deg([W]^{\text{vir}}) = 12 + \frac{1}{2} \times 12 \times 7 = 54.$$

Here the factor $\frac{1}{2}$ accounts for the fact that morphisms $g_{i,z}$ have non-trivial automorphism groups $\mathbb{Z}_2$.

## 3. Example

One important application of Gromov-Witten invariants and their composition laws is that they provide us a new ring structure, called the quantum ring, on the group $H^*(X, \mathbb{R})$. Since this has been discussed in this volume [FP] (see also [KM, RT1]) we will not reproduce it here, only to point out that the binary relation $\times_q$ is defined using Gromov-Witten invariants $\phi^X_{\bullet,0,3}$, and the associativity of $\times_q$ follows from the first composition law of Gromov-Witten invariants.

The example we consider now is the degree of the virtual cycles of moduli of maps from rational curves of certain degree to a specific class of Calabi-Yau threefolds. The interest in this example is that the virtual degree is negative, and coincide with the number obtained using the mirror symmetry conjecture [HK].

The threefolds we consider are smooth Calabi-Yau's that contain smooth projective planes $\mathbf{P}^2 \subset Y$. Such $Y$ can be constructed as minimal resolutions of degree 18 hypersurfaces in the weighted projective space $\mathbf{P}^4(w)$, where $w = (6, 3, 1, 1, 1)$. A general degree 18 hypersurface $\tilde{Y} \subset \mathbf{P}^4(w)$ is smooth away from two isolated quotient singular points. After resolving these singularities, we obtain a smooth Calabi-Yau threefold $Y$ that contains two $\mathbf{P}^2$'s, which are exceptional divisors of the resolution $Y \to \tilde{Y}$. We denote one of $\mathbf{P}^2$'s by $W \subset Y$. Note that the normal bundle $N_{W/Y}$ has degree $-3$. Now let $h$ be the hyperplane class in $W$ and let $\mathcal{M}_{kh,0,0}$, in short $\mathcal{M}_k$, be the moduli of stable morphisms from rational curves to $Y$ with homology class $kh$. Evidently, all such morphisms factor through $W \subset Y$.

Hence $\mathcal{M}_1$ is isomorphic to $\mathbf{P}^2$, the space of lines in $W$, and $\mathcal{M}_2$ is the blowing-up of $\mathbf{P}^5$, the space of all conics in $W$, along the $\mathbf{P}^2$ that is the locus of singular conics. Note that

$$\dim \mathcal{M}_1 = 2 \quad \text{and} \quad \dim \mathcal{M}_2 = 5.$$

However, since $Y$ is Calabi-Yau, the virtual dimension of $\mathcal{M}_k$ is 0. Now the question arises: what is the degree of the virtual moduli cycle $[\mathcal{M}_k]^{\mathrm{vir}}$? We shall compute $\deg[\mathcal{M}_1]^{\mathrm{vir}}$ and $\deg[\mathcal{M}_2]^{\mathrm{vir}}$ in this note using the virtual moduli cycle constructed. We mention that the degrees of these two virtual moduli cycles were computed using the mirror symmetry conjecture, and are 3 and $-\frac{45}{8}$ respectively [HK].

We now carry out the computation. First we note that since the normal bundle to $W$ in $Y$ has negative degree, $\mathcal{M}_k$ is smooth, except for possibly quotient singularities due to non-trivial automorphism groups of maps in $\mathcal{M}_k$. Therefore, the $\mathbb{Q}$-obstruction sheaf is locally free and its fiber at any $\xi \in \mathcal{M}_k$, represented by $f: C \to W \subset X$, is

$$\mathrm{Ext}^2_C([f^*\Omega_X \to \Omega_C], \mathcal{O}_C) \cong H^1_C(f^*\mathcal{O}_W(-3)),$$

since the normal bundle to $W$ in $X$ is a line bundle of degree $-3$. Hence the virtual normal cone will be the total space of the obstruction bundle.

The computation of

$$\deg[\mathcal{M}_1]^{\mathrm{vir}} = 3$$

is standard, using the tautological family of lines in $\mathbf{P}^2$. Note that there are no maps in $\mathcal{M}_1$ that have non-trivial automorphism groups.

The evaluation of $\deg[\mathcal{M}_2]^{\mathrm{vir}}$ is more tricky since there are maps in $\mathcal{M}_2$ that have non-trivial automorphism groups $\mathbb{Z}_2$. The method we will use is the Bott fixed point formula. We begin with an $S^1$-action on $\mathbf{P}^2$, $\mu_t : \mathbf{P}^2 \to \mathbf{P}^2$ given by $\mu_t[x_0, x_1, x_2] = [t^a x_0, t^b x_1, t^c x_2]$, where $a, b, c$ are integers. For generic chosen $(a, b, c)$, $\mu_t$ has three fixed points: $\xi_0 = [1, 0, 0]$, $\xi_1 = [0, 1, 0]$ and $\xi_2 = [0, 0, 1]$, and three invariant lines $L_{ij}$, containing $\xi_i$ and $\xi_j$. Let $\tilde{\mu}_t : \mathcal{M}_2 \to \mathcal{M}_2$ be the induced action. Given any stable morphism $f: C \to \mathbf{P}^2$ of degree 2, it is fixed by $\mu_t$ if there are automorphisms $\eta_t \in \mathrm{Aut}(C)$ such that $\mu_t \cdot f = f \circ \eta_t$. Since $C$ is a rational nodal curve and $f$ has degree 2, this is possible only if $f$ falls into one of the following three groups:

*Group I*: $C \cong \mathbf{P}^1$ and $f: C \to \mathbf{P}^2$ is a double cover over a line $L_{ij}$ branched over $\xi_i$ and $\xi_j$;

*Group II*: $C \cong \mathbf{P}^1 \cup \mathbf{P}^1$, a nodal rational curve, and $f: C \to \mathbf{P}^2$ has image the union of two distinct lines $L_{ij}$ and $L_{ik}$;

*Group III*: $C \cong \mathbf{P}^1 \cup \mathbf{P}^1$, $f: C \to \mathbf{P}^2$ has image a single line $L_{ij}$ and $f$ sends the nodal point of $C$ to $\xi_i$.

Now let $p \in \mathcal{M}_2$ be any point fixed by the induced action $\tilde{\mu}_t : \mathcal{M}_2 \to \mathcal{M}_2$. Let $f_U : C_U \to X$ over $U$ be a semi-universal family of $p$, invariant under $\tilde{\mu}_t$. $\tilde{\mu}_t$ induces a linear transformation $L_{\mathcal{M}_2,p}: T_pU \to T_pU$, the Lie derivative. After fixing an $S^1$-linearization of the sheaf $\mathcal{O}_{\mathbf{P}^2}(-3)$ on $\mathbf{P}^2$, we obtain a canonical $S^1$-linearization of the obstruction sheaf,

$$R^1\pi_{U*}(f_U^*\mathcal{O}_{\mathbf{P}^2}(-3)) \xrightarrow{\cong} \tilde{\mu}_t^* R^1\pi_{U*}(f_U^*\mathcal{O}_{\mathbf{P}^2}(-3)).$$

Restricting to $p \in U$, and taking derivative at $t = 1$, we obtain a linear transformation

$$L_{E,p} : H^1_C(f^*\mathcal{O}_{\mathbf{P}^2}(-3)) \longrightarrow H^1_C(f^*\mathcal{O}_{\mathbf{P}^2}(-3)).$$

We denote by

$$m_p \triangleq \frac{1}{\#\operatorname{Aut}(f)} \frac{\det(L_{E,p})}{\det(L_{\mathcal{M}_2,p})}$$

the Bott residue at the fixed point $p \in \mathcal{M}_2$. The Bott fixed point formula states that

$$\deg[\mathcal{M}_2]^{\mathrm{vir}} = \sum_{p \text{ fixed points of } \mathcal{M}_2} m_p.$$

We now evaluate the Bott residues at each fixed points. We will give the detailed computation for group III maps and leave the computation for the remaining groups to the reader. We first fix an isomorphism

$$\mathcal{O}_{\mathbf{P}^2}(1) \xrightarrow{\cong} \mu_t^* \mathcal{O}_{\mathbf{P}^2}(1).$$

Since $\mu_t(\xi_0) = \xi_0$, it suffices to fix an isomorphism

$$\mathcal{O}_{\mathbf{P}^2}(1) \otimes k(\xi_0) \xrightarrow{\cong} \mu_t^* \mathcal{O}_{\mathbf{P}^2}(1) \otimes k(\xi_0) = \mathcal{O}_{\mathbf{P}^2}(1) \otimes k(\xi_0),$$

which we choose to be the multiplication by $t^{-a}$. That is, after fixing an element $1 \in \mathcal{O}_{\mathbf{P}^2}(1) \otimes k(\xi_0)$, then the above isomorphism sends 1 it to $t^{-a} \cdot 1$. Also, if $\eta_t$ acts on $\mathbf{P}^1$ by sending $[z_0, z_1]$ to $[w_0, w_1] = [t^\alpha z_0, t^\beta z_1]$, then the canonical isomorphism $\eta_t^* \omega_{\mathbf{P}^1} \cong \omega_{\mathbf{P}^1}$ is given by $dw_0 = t^{\alpha-\beta} dz_0$ and $dw_1 = t^{\beta-\alpha} dz_1$. Finally, if $C$ is a nodal curve and $c \in C$ is a nodal point, and if $\eta_t$ acts on $C$ fixing $c$, then the restriction of the canonical isomorphism $\eta_t^* \omega_C \xrightarrow{\cong} \omega_C$ to $c$ is the identity map.

Now consider $\xi \in \operatorname{fix}(\mathcal{M}_2)$ given by the map $f: C \to \mathbf{P}^2$, where $C$ is a nodal curve consisting two irreducible components $C_1$ and $C_2$ each isomorphic to $\mathbf{P}^1$, that is given by

$$f(x_1) = [1, x_1, 0] \quad \text{and} \quad f(y_1) = [1, y_1, 0],$$

where $x_1$ and $y_1$ are affine coordinates of $C_1$ and $C_2$ respectively centered at the nodal point of $C$. The automorphism $\eta_t$ of $C$ satisfying $\mu_t \cdot f = f \circ \eta_t$ is given by multiplying $\eta_t(x_1) = t^{b-a} x_1$ and $\eta_t(y_1) = t^{b-a} y_1$. Hence the isomorphism

$$L_{E_2,\xi} : f^* \mathcal{O}_{\mathbf{P}^2}(3) \otimes \omega_C \xrightarrow{\cong} f^* \mu_t^* \mathcal{O}_{\mathbf{P}^2}(3) \otimes \eta_t^* \omega_C$$

is given by multiplying $t^{-3a}$ at the nodal point. Hence the isomorphism

$$H^0(f^* \mathcal{O}_{\mathbf{P}^2}(3) \otimes \omega_C) \xrightarrow{\cong} H^0(f^* \mu_t^* \mathcal{O}_{\mathbf{P}^2}(3) \otimes \eta_t^* \omega_C)$$

is given by the diagonal matrix

$$\operatorname{diag}\{t^{-3a}, t^{-3a-(b-a)}, t^{-3a-2(b-a)}, t^{-3a-(b-a)}, t^{-3a-2(b-a)}\}$$

under the basis $\{1, x_1, x_1^2, y_1, y_1^2\}$ of $H^0(f^* \mathcal{O}_{\mathbf{P}^2}(3) \otimes \omega_C)$. Hence,

$$\det L_{E_2,\xi} = 3a(a+2b)^2(2a+b)^2.$$

Here the extra $-$ sign is necessary since we are computing the dual of the obstruction sheaf and the rank of the obstruction sheaf is odd.

Next, we compute $L_{\mathcal{M}_2,\xi}$. Since $\xi$ has automorphism group $\mathbb{Z}_2$, there is a $\mathbb{Z}_2$-invariant smooth analytic variety $0 \in S$ such that $S/\mathbb{Z}_2$ is an analytic neighborhood of $\xi \in \mathcal{M}_2$. Let $S_0 \subset S$ be the locus corresponding to singular maps, i.e. their images are singular conic curves. A local coordinate of $s \in S_0$ is given by families

$$f_{t_\bullet}(x_1) = [1, x_1 + t_3, t_1 x_1 + t_4], \quad f_{t_\bullet}(y_1) = [1, y_1 + t_3, t_2 y_1 + t_4],$$

where $t_\bullet = (t_1, \cdots, t_4)$. Then the induced action $\tilde{\mu}_t : S_0 \to S_0$ is given by

$$\tilde{\mu}_t(t_1, \cdots, t_4) = (t^{c-b}t_1, t^{c-b}t_2, t^{b-a}t_3, t^{c-a}t_4).$$

Hence the contribution to $L_{\mathcal{M}_2,\xi}$ from $TS_0$ is $(c-b)^2(b-a)(c-a)$. Now we choose a family normal to $S_0 \subset S$ at $0 \in S_0$. Let $h_s$ be a family of rational maps

$$h_s : ([z,z_1], s) - - \longrightarrow [z_0z_1, sz_0^2 + z_1^2, 0].$$

One checks that $h_0$ maps $\mathbf{P}^1$ to the line $z_2 = 0$, and is undefined at [1,0]. After blowing up $\mathbf{P}^1 \times \mathbf{A}^1$ at $\theta = ([1,0],0)$, $h$ lifts to a morphism $F : \mathrm{bl}_\theta(\mathbf{P}^1 \times \mathbf{A}^1) \to \mathbf{P}^2$ that has the following property: $F_0$ sends the central fiber of $\mathrm{bl}_\theta(\mathbf{P}^1 \times \mathbf{A}^1)$ to the line $z_2 = 0$, and sends the node of the central fiber to $[1,0,0]$. Thus it is a family normal to $S_0 \subset S$ at $0 \in S_0$. Obviously, $\mu_t F_s = F_{st^{2b-2a}}$. Hence the contribution to $L_{\mathcal{M}_2,\xi}$ from the normal direction is $2b-2a$. In conclusion, $L_{\mathcal{M}_2,\xi} = 2(b-a)^2(c-b)^2(c-a)$ and the Bott residue is

$$m_\xi = \frac{1}{2} \cdot \frac{3a(2a+b)^2(a+2b)^2}{2(b-a)^2(c-b)^2(c-a)}. \tag{9}$$

For the other two groups of fixed points, we will give the formulas for their Bott residues and leave the details to the reader. For the type II fixed point given by a map $f : \mathbf{P}^1 \cup \mathbf{P}^1 \to \mathbf{P}^2$ that sends $[x_0, x_1]$ to $[x_0, 0, x_1]$ and sends $[y_0, y_1]$ to $[0, y_0, y_1]$, its Bott residue is

$$m_\xi = -\frac{3c(a+2c)(2a+c)(b+2c)(2b+c)}{(a+b-2c)(a-b)^2(a-c)(b-c)}. \tag{10}$$

Note that it is symmetric with respect to $a$ and $b$. For the type I fixed point given by the map $f : \mathbf{P}^1 \to \mathbf{P}^2$ that sends $[z_0, z_1]$ to $[z_0^2, z_1^2, 0]$, the Bott residue is

$$m_\xi = -\frac{1}{2}\frac{(a+5b)(2a+4b)(3a+3b)(4a+2b)(5a+b)}{2^5(a-b)^2(c-a)(c-b)(c-\frac{a+b}{2})}. \tag{11}$$

Finally, we sum the Bott residues at all the fixed points. We fix an $S^1$-action that is given by $a = 0$, $b = 1$ and $c = 3$. For group I fixed points, there are three of them. Their Bott residues are given by (11) with $(a, b, c)$ taking values $(0, 1, 3)$, $(1, 3, 0)$ and $(3, 0, 1)$. Their values are $-2^{-3}$, $2^2 \cdot 5 \cdot 7$ and $-2^{-3} \cdot 3^4 \cdot 5$ respectively. For group II fixed points, we shall evaluate (10) with $(a, b, c)$ taking values $(0, 1, 3)$, $(1, 3, 0)$ and $(3, 0, 1)$ respectively. Their values are $3^3 \cdot 7$, 0 and $3^{-1} \cdot 5 \cdot 7$ respectively. Finally, for type III fixed points, we shall evaluate formula (9) with $(a, b, c)$ taking values $(0, 1, 3)$, $(0, 3, 1)$, $(1, 3, 0)$, $(3, 1, 0)$, $(1, 0, 3)$ and $(3, 0, 1)$, and the results are 0, 0, $-2^{-4} \cdot 3^{-1} \cdot 5^2 \cdot 7^2$, $-2^{-4} \cdot 3 \cdot 5^2 \cdot 7^2$, $2^{-1}3^{-1}$ and $-2^{-1}3^4$. respectively. Adding them together, we obtain the number:

$$-\frac{45}{8} = -6 + \frac{3}{8}.$$

As explained in [KM], this says that the virtual number of non-degenerate degree 1 and degree 2 maps are 3 and -6 respectively. Note that this coincides with the prediction from the mirror symmetry conjecture.

## 4. Comparison of Gromov-Witten invariants

In the last part of this paper, we will discuss issues related to the symplectic geometry of GW-invariants of projective manifolds. The constructions of GW-invariants for general symplectic manifolds were completed by the authors [LT2]. Different constructions were given by Fukaya-Ono at the same time [FO], and later by Siebert and Ruan [Si, Ru]. The idea of our approach is parallel to the algebraic construction of the invariants, namely, to find virtual moduli cycles of the moduli spaces of pseudo-holomorphic maps. The fundamental question we shall address in this section is: *given a projective manifold $X$ with Hodge class $\omega$, do the algebraic GW-invariants $\phi^{X,\mathrm{alg}}_{\bullet,g,n}$ coincide with the symplectic GW-invariants $\phi^{X,\mathrm{symp}}_{\bullet,g,n}$?* We shall first sketch the symplectic construction of GW-invariants for general symplectic manifolds, following [LT2]. After that, we will sketch the proof to the affirmative answer to the above question.

COMPARISON THEOREM. *Let $X$ be any smooth complex projective variety and let $\omega$ be a Kähler form on $X$. Then the algebraic GW-invariants $\phi^{X,\mathrm{alg}}_{\bullet,g,n}$ of $X$ coincide with the symplectic GW-invariants $\phi^{X,\mathrm{symp}}_{\bullet,g,n}$ of the symplectic manifold $(X,\omega)$.*

One immediate corollary is that the notion of uniruled is symplectic invariant.

COROLLARY. *(Kollár) Let $X_1$ and $X_2$ be smooth complex projective varieties with Kähler forms $\omega_1$ and $\omega_2$ respectively. Assume that $X_1$ is uniruled and that $(X_1,\omega_1)$ is sympletomorphic to $(X_2,\omega_2)$. Then $X_2$ is also uniruled.*

First, let us say a few words on the definition of GW-invariants of symplectic manifolds. Let $X$ be a smooth complex projective manifold and $J$ its holomorphic structure. As usual, we fix an $\alpha \in H_2(X,\mathbb{Z})$ and integers $g$ and $n$. We let $\mathcal{W}_{\alpha,g,n}$ be the topological space of all stable smooth maps from $n$-pointed genus $g$ curves $D \subset C$ to $X$ with fundamental class $\alpha$. Here we say $f: D \subset C \to X$ are smooth if $D \subset C$ are connected pointed nodal curves as before, $f: C \to X$ are continuous, are smooth when restricted to each irreducible components and $f_*([C]) = \alpha$. We say $f$ are stable if for any unstable components $\Sigma \subset (D,C)$, $f_*([\Sigma]) \neq 0$. $\mathcal{W}_{\alpha,g,n}$ is an infinite dimensional topological space, and is a Banach orbifold over the dense open subset consisting of all maps whose domains are regular. Let $\mathcal{M}^J_{\alpha,g,n}$ be the locus of holomorphic maps. $\mathcal{M}^J_{\alpha,g,n}$ is compact and is homeomorphic to $\mathcal{M}^X_{\alpha,g,n}$. Thus we can define GW-invariants if $\mathcal{M}^J_{\alpha,g,n}$ is smooth and is of the right dimension. Otherwise, we need to find a homological class $[\mathcal{M}^J_{\alpha,g,n}]^{\mathrm{symp}}$, of the expected dimension, in $H_*(\mathcal{W}_{\alpha,g,n})$. Note that $\mathcal{W}_{\alpha,g,n}$ admits evaluation maps and stable contraction, extending corresponding maps on $\mathcal{M}^X_{\alpha,g,n}$. Unlike in algebraic cases where $\mathcal{M}^X_{\alpha,g,n}$ is defined as the representative of the moduli functor, $\mathcal{M}^J_{\alpha,g,n}$ is defined as the zero locus of a section of an infinite rank vector bundle over the infinite dimensional space $\mathcal{W}_{\alpha,g,n}$. Let $\xi \in \mathcal{W}_{\alpha,g,n}$ be any point associated to a smooth map $f: D \subset C \to X$ with automorphism group $G_\xi$. We associate to it an infinite dimensional vector space (or its quotient)

$$E_\xi = C^\infty(\Omega^{0,1}_C(f^*TX))/G_\xi.$$

Here when $C_1,\cdots,C_l$ are irreducible components of $C$,

$$\Omega^{0,1}_C(f^*TX) = \oplus_{j=1}^l \Omega^{0,1}_{C_j}(f^*TX),$$

where $\Omega^{0,1}_{C_j}$ is the sheaf of smooth sections of (0,1)-forms on $C_j$ with values in appropriate vector bundles. Let $E$ be the union of $E_\xi$ for all $\xi \in \mathcal{W}_{\alpha,g,n}$ and $\pi: E \to \mathcal{W}_{\alpha,g,n}$ be the projection. $E$ is naturally a topological space and the fibers of $\pi$ are vector spaces or their finite quotients whose linear structures vary continuously with $\xi \in \mathcal{W}_{\alpha,g,n}$. We call such data infinite dimensional $\mathbb{Q}$-vector bundles. There is a continuous section $\Phi: \mathcal{W}_{\alpha,g,n} \to E$ that assigns to each $\xi \in \mathcal{W}_{\alpha,g,n}$ the value $\bar\partial f$. The simple observation is that

LEMMA. *The locus $\Phi^{-1}(0)$ is homeomorphic to $\mathcal{M}^J_{\alpha,g,n}$. Further, for any $\xi \in \mathcal{M}^J_{\alpha,g,n}$ the linearization*

$$d\Phi(\xi) : T_\xi\mathcal{W}_{\alpha,g,n} \longrightarrow E_\xi$$

*is Fredholm whose index is the expected dimension of $\mathcal{M}^J_{\alpha,g,n}$.*[4]

The symplectic class $[\mathcal{M}^J_{\alpha,g,n}]^{\mathrm{symp}} \in H_*(\mathcal{W}_{\alpha,g,n})$ should be the [0 – section] of $E$ intersecting with the [graph of $\Phi$]. Since we are working with $C^\infty$-category, such an intersection can be defined by choosing a perturbation $\tilde\Phi$ of $\Phi$, obeying certain finiteness condition governed by the Fredholm property of $d\Phi$, and then define the intersection class to be $[\tilde\Phi^{-1}(0)]$. For the moduli space $\mathcal{W}_{\alpha,g,n}$, the main difficulty is that $E$ is not a Banach vector bundle over a smooth Banach manifold in the usual sense. What is true is that $\mathcal{W}_{\alpha,g,n}$ is a weakly smooth orbifold near $\mathcal{M}^J_{\alpha,g,n}$ in the following sense.

DEFINITION-PROPERTY. *A weakly smooth orbifold structure of $\mathcal{W}_{\alpha,g,n}$ along $\mathcal{M}^J_{\alpha,g,n}$ is a collection $\Lambda$ of subsets $F \subset E$ that satisfies the following properties.*

1. *Each $F \in \Lambda$ is a finite rank $\mathbb{Q}$-subbundle of $E$ over an open subset $U_F \subset \mathcal{W}_{\alpha,g,n}$.*
2. *Let $Y_F = \Phi^{-1}(F)$, then $Y_F$ is an orbifold, $F|_{Y_F}$ is a smooth $\mathbb{Q}$-vector bundle, the lifting $\phi_F: Y_F \to F|_{Y_F}$ is smooth and $\dim Y_F - \mathrm{rk}(F|_{Y_F}) = r$, which is the index of $d\Phi$.*
3. *For any $\eta \in \mathcal{M}^J_{\alpha,g,n}$, we define a partial ordering $\succ_\eta$ of $\Lambda$ by saying $F_1 \succ_\eta F_2$ if there is a neighborhood $O$ of $\eta \in \mathcal{W}_{\alpha,g,n}$ such that $O \subset U_{F_1} \cap U_{F_2}$, $F_2|_O$ is a subbundle of $F_1|_O$, $Y_{F_2} \cap O \subset Y_{F_1} \cap O$ is a suborbifold, $F_2|_{Y_2\cap O} \subset F_1|_{Y_2 \cap O}$ is a smooth $\mathbb{Q}$-subbundle and at each $\xi \in \mathcal{M}^J_{\alpha,g,n} \cap O$ the induced homomorphism $T_\xi Y_{F_1}/T_\xi Y_{F_2} \to (F_1/F_2)|_\xi$ is an isomorphism. Then for any $F_1$, $F_2 \in \Lambda$ and $\eta \in Y_{F_1} \cap Y_{F_2} \cap \mathcal{M}^J_{\alpha,g,n}$, there is an $F_3$ such that $F_3 \succ_\eta F_1$ and $F_3 \succ_\eta F_2$.*
4. *$\cup_{F\in\Lambda} Y_F \supset \mathcal{M}^J_{\alpha,g,n}$ and $\lim_{F\in\Lambda} F_\xi = E_\xi$ for each $\xi \in \mathcal{M}^J_{\alpha,g,n}$.*

The main body of [LT2] is devoted to show that such collections do exist. Once we have such collections, then at each $\xi \in \mathcal{M}^J_{\alpha,g,n}$ we can pick an $F \in \Lambda$ that contains $\xi$ and then perturb the graph of the lifting $\phi_F: Y_F \to F|_{Y_F}$. One technicality is that $F|_{Y_F}$ is a $\mathbb{Q}$-bundle. Hence we will use piecewise smooth (simplicial) cycles with rational coefficients that are homotopic to the original graphs. In order for this scheme to work, we need to choose the perturbations so that they are compatible over their overlaps. This require some arguments since usually $Y_{F_1} \cap Y_{F_2}$ are not open in $Y_{F_1}$ or $Y_{F_2}$. This is where 3) of the definition-proposition is needed. 4) will be used to show that the homology classes constructed are independent of the constructions.

[4]To be rigorous, one needs to use completion of these vector spaces under certain norms. We refer the details to [LT2].

We now sketch the construction of $[\mathcal{M}^J_{\alpha,g,n}]^{\text{symp}}$. Since $\mathcal{M}^J_{\alpha,g,n} \approx \mathcal{M}^X_{\alpha,g,n}$, there is a simplicial decomposition of $\mathcal{M}^J_{\alpha,g,n}$. Hence we can speak of piecewise smooth subsets of $\mathcal{M}^J_{\alpha,g,n}$ and the real dimensions of these sets. Let $N = \dim \mathcal{M}^J_{\alpha,g,n}$. Let $\Lambda$ be the collection given in the definition-proposition. Without loss of generality, we can assume that for any $F \in \Lambda$, which is a $\mathbb{Q}$-bundle over $U_F \subset \mathcal{W}_{\alpha,g,n}$, and any open $O \subset U_F$, $F|_O$ is also in $\Lambda$. We first pick a finite subcollection $\Lambda_1 \subset \Lambda$ so that the union of $\{Y_F\}_{F\in\Lambda_1}$ contains $\mathcal{M}^J_{\alpha,g,n}$. To simplify the notation, for $\alpha \in \Lambda$ (by which we mean $F_\alpha \in \Lambda$) we will denote by $Y_\alpha$ the set $Y_{F_\alpha}$ and denote by $\phi_\alpha$ the lifting $\phi_{F_\alpha}$. We choose an open $O_\alpha \subset U_\alpha$ for each $\alpha \in \Lambda_1$ of which the following holds: $\overline{O_\alpha} \subset U_\alpha$, $\{O_\alpha \cap \mathcal{M}^J_{\alpha,g,n}\}_{\alpha\in\Lambda_1}$ are mutually disjoint and $T_1 \triangleq \mathcal{M}^J_{\alpha,g,n} - \cup_{\alpha\in\Lambda_1} O_\alpha$ is a piecewise smooth set and has dimension at most $N-1$. We next choose a finite collection $\Lambda_2 \subset \Lambda$ so that in addition to $\cup_{\alpha\in\Lambda_2} Y_\alpha$ covers $T_1$, we have $F_\beta \succcurlyeq_y F_\alpha$ for any $F_\beta \in \Lambda_2$, $F_\alpha \in \Lambda_1$ and $y \in Y_\alpha \cap Y_\beta \cap O_\alpha$. Since $T_1$ is compact, by 3) of the definition-proposition, $\Lambda_2$ does exist. Similar to $\Lambda_1$, we choose an open $O_\alpha \subset U_\alpha$ for each $\alpha \in \Lambda_2$ so that $\overline{O_\alpha} \subset U_\alpha$, $\{O_\alpha \cap \mathcal{M}^J_{\alpha,g,n}\}$ are mutually disjoint and $T_2 \triangleq T_1 - \cup_{\alpha\in\Lambda_2} O_\alpha$ is a piecewise smooth set and has dimension at most $N-2$. We will pick collections $\Lambda_k$ inductively according to this scheme. Note that $T_{N-1}$ will be a discrete set and hence we can assume $T_N = \emptyset$.

Now we perturb the graphs of the zero-sections. Without loss of generality, we can assume that for each $k$ the dimensions of $Y_\alpha \in \Lambda_k$ are identical. (They are different for different $k$.) Let $Z_\alpha = Y_\alpha \cap O_\alpha$ and let

$$S_k = \cup_{\alpha\in\Lambda_k} \overline{Z_\alpha}, \quad W_k = \cup_{\alpha\neq\beta\in\Lambda_k} \overline{Z_\alpha} \cap \overline{Z_\beta}, \quad \text{and} \quad T_k = T_{k-1} - \cup_{\alpha\in\Lambda_k} Z_\alpha.$$

We agree $T_0 = \mathcal{M}^J_{\alpha,g,n}$. We denote by $F_{S_k}$ the union of $F_\alpha|_{\overline{Z_\alpha}}$ for $\alpha \in \Lambda_k$. We fix a sufficiently small neighborhood $V_1$ of $T_1 \subset \mathcal{W}_{\alpha,g,n}$. Let $\delta_0 \subset F_{S_1}$ be the restriction to $S_1$ of the graph of the zero section of $E \to \mathcal{W}_{\alpha,g,n}$. Since away from $T_1$ the space $F_{S_1}$ is a $\mathbb{Q}$-vector bundle over orbifold near $\mathcal{M}^J_{\alpha,g,n}$, it is not difficult to show that there is a simplicial cycle $\delta_1$ with rational coefficients in $F_{S_1}$ such that 1) $\delta_1 = \delta_0$ when restricted to $T_1$ and $\partial S_1$; 2) the cycle $\delta_1$ intersects the graph $\Gamma_\Phi$ transversally[5] in $F_{S_1}$ along fibers over $\mathcal{M}^J_{\alpha,g,n} \cap \text{Int}(S_1 - V_1)$; 3) $\delta_1$ is homotopic to $\delta_0$ relative to the boundary $\partial S_1$[6]. By choosing $V_1$ small enough, we can assume $\cup_{\alpha\in\Lambda_2}\overline{Z_\alpha}$ covers $V_1 \cap \mathcal{M}^J_{\alpha,g,n}$. Using inclusion $F_{S_1}|_{S_1\cap S_2} \subset F_{S_2}|_{S_1\cap S_2}$, $\delta_1|_{S_1\cap S_2}$ is a closed subset of $F_{S_2}$. We next choose sufficiently small neighborhood $V_2$ of $T_2$. We choose $\delta_2$ to be a simplicial cycle in $F_{S_2}$ so that $\delta_2|_{S_2}$ is homotopic to the 0-section relative to the boundary $\partial S_2$, $\delta_2 = \delta_1$ restricted to $(S_2 - V_1) \cap S_1$ and $T_2$, and away from $V_2$ the cycle $\delta_2$ intersects transversally with the graphs $\Gamma_\Phi$ in $F_{S_2}$ along fibers over $\mathcal{M}^J_{\alpha,g,n} \cap \text{Int}(S_2)$. We let $\Gamma_2$ be $\delta_2$ when restricted to $S_2$ and be $\delta_1$ when restricted to $S_1 - S_2$. $\Gamma_2$ has the property that away from $V_2 \supset T_2$ it intersects transversally with the graph $\Gamma_\Phi$ in $F_{S_2}$ and in $F_{S_1}$ along fibers over $\mathcal{M}^J_{\alpha,g,n} \cap (\text{Int}(S_2) \cup \text{Int}(S_1))$. Inductively, by slightly perturbing $\Gamma_k$ near $T_k$ we can extend it to $\Gamma_{k+1}$ over $S_{k+1}$, etc.. Eventually, we obtain $\Gamma_N$ in the union of $F_{S_k}$ for all $k$. Restricted to each $S_k - S_{k+1}$, $\Gamma_N$ is transversal to $\Gamma_\Phi$ in $F_{S_k}$ along fibers over $\mathcal{M}^J_{\alpha,g,n} \cap \text{Int}(S_k - S_{k+1})$. Then the intersection $\Gamma_N \cap \Gamma_\Phi$ is a piecewise smooth, compact subset of $E$ that has pure dimension $r$ = index of $d\Phi$. We denote this intersection by $C^\Lambda_{\text{symp}}$. Let

[5]By which we mean that they intersect transversally as stratified sets.

[6]By this we mean that there is a cycle $\Sigma$ so that $\partial\Sigma - (\delta_0 - \delta_1)$ is contained in the fibers over $\partial S_1$

$\pi_*[C^{\Lambda}_{\text{symp}}] \in H_*(\mathcal{W}_{\alpha,g,n})$ be the image class, where $\pi: E \to \mathcal{W}_{\alpha,g,n}$ is the projection. It is direct to check that this class does not depend on the choices of perturbations. We define this class to be the symplectic class, denoted by $[\mathcal{M}^J_{\alpha,g,n}]^{\text{symp}}$.

Since $\mathcal{M}^X_{\alpha,g,n}$ is homeomorphic to $\mathcal{M}^J_{\alpha,g,n}$, the virtual moduli cycle $[\mathcal{M}^X_{\alpha,g,n}]^{\text{vir}} \in A_{r/2}\mathcal{M}^X_{\alpha,g,n}$ assigns a homological class $\tau_*[\mathcal{M}^X_{\alpha,g,n}]^{\text{vir}} \in H_r(\mathcal{W}_{\alpha,g,n})$ via the obvious homomorphism $\tau_*: A_{r/2}\mathcal{M}^X_{\alpha,g,n} \to H_r(\mathcal{W}_{\alpha,g,n})$.

COMPARISON THEOREM. *Let the notation be as above, then*

$$\tau_*[\mathcal{M}^X_{\alpha,g,n}]^{\text{vir}} = [\mathcal{M}^J_{\alpha,g,n}]^{\text{symp}} \in H_*(\mathcal{W}_{\alpha,g,n}).$$

We now sketch the main ingredient of the proof. To simplify the notation, we will deal with the case where $n = 0$. Namely, there are no marked points. Also, to keep the technical details at minimum, we will only deal with morphisms with trivial automorphism groups. Let $\xi \in \mathcal{M}^J_{\alpha,g,n}$ be any point corresponding to the holomorphic map $f: C \to X$. Our first observation is that

$$\operatorname{Coker} d\Phi(\xi) \cong \operatorname{Ext}^1_C([f^*\Omega_X \to \Omega_C], \mathcal{O}_C).$$

This is obvious when $C$ is smooth. Since then $T_\xi\mathcal{W}_{\alpha,g,n}$ is $C^\infty(f^*T_X) \oplus \mathcal{H}^1_{\bar\partial}(T_C)$ and

$$d\Phi(\xi): T_\xi\mathcal{W}_{\alpha,g,n} \cong C^\infty(f^*T_X) \oplus \mathcal{H}^1_{\bar\partial}(T_C) \longrightarrow E_\xi \cong C^\infty(\Omega^{0,1}_C(f^*T_X))$$

is the map sending $(\varphi, \eta)$ to $\bar\partial\varphi + f_*\eta$. Here $\mathcal{H}^1_{\bar\partial}(T_C)$ is the space of harmonic forms and $f_*$ is the obvious homomorphism $\Omega^{0,1}(T_C) \to \Omega^{0,1}(f^*T_X)$ induced by $T_C \to f^*T_X$. The required identity follows from comparing $d\Phi(\xi)$ with the exact sequence

$$\operatorname{Ext}^1_C(\Omega_C, \mathcal{O}_C) \longrightarrow \operatorname{Ext}^1_C(f^*\Omega_X, \mathcal{O}_C) \longrightarrow \operatorname{Ext}^1_C([f^*\Omega_X \to \Omega_C], \mathcal{O}_C) \longrightarrow 0.$$

Our next observation is that $\mathcal{M}^J_{\alpha,g,n}$ is an analytic scheme isomorphic to the underlying analytic scheme of $\mathcal{M}^X_{\alpha,g,n}$. This can be seen by the following lemma.

LEMMA. *Let $\Lambda$ be a collection given in the Definition-Proposition. We can enlarge $\Lambda$ if necessary so that for any $\xi \in \mathcal{M}^J_{\alpha,g,n}$ there is a member $F \in \Lambda$ of which the following holds:*

1. *$Y_F$ is a complex orbifold, $F|_{Y_F}$ is a holomorphic $\mathbb{Q}$-vector bundle and $\phi_F: Y_F \to F|_{Y_F}$ is holomorphic;*
2. *$\phi_F$ is (a finite quotient of) a Kuranishi map of $\xi$.*

Because of 2) of the lemma, the germ of $\xi \in \mathcal{M}^X_{\alpha,g,n}$ is isomorphic to the germ of $\xi \in \mathcal{M}^J_{\alpha,g,n}$ defined by the equation $\{\phi_F = 0\}$ (or its quotient by the automorphism group).

The proof of 1) amounts to choose a special subbundle $F$ over a neighborhood of $\xi \in \mathcal{W}_{\alpha,g,n}$. We will not discuss the construction here since it is quite technical. The proof of 2) is a straightforward application of the isomorphism of Čech cohomology and Dolbeault cohomology. Recall that given an Artin ring and its quotient ring $\bar B \to B$ with kernel $I$ annihilated by the maximal ideal $\mathfrak{m} \subset \bar B$, and given a family $f: C_B \to X$ of stable morphisms over $B$ extending $f_0: C \to X$, there is a natural obstruction class to the extension of $f$ to $\bar f$ over $\bar B$. One construction of this class is using Čech cohomology. Let $C_{\bar B}$ be an extension of $C_B$ and $\{U_i\}$ be an affine covering of $C_{\bar B}$. We fix extensions $f_i: U_i \to X$ of $f|_{C_B \cap U_i}$. Then the collection $(f_i - f_j)|_{U_i \cap U_j}$ defines an element in $\mathrm{Ext}^1_C(f_0^*\Omega_X, \mathcal{O}_C) \otimes_{\mathbb{C}} I$ whose image in $\mathrm{Ext}^1_C([f_0^*\Omega_X \to \Omega_C], \mathcal{O}_C) \otimes_{\mathbb{C}} I$ is the obstruction class to extending $f$ over $\bar{B}$. We call this class algebraic obstruction class. We can also construct an analytic obstruction class using Dolbeault cohomology. Instead of picking local extensions $f_i$, we let $g : C_{\bar{B}} \to X$ be a smooth map extending $f$. Then $\bar{\partial} g$ is an element in $H^1_{\bar{\partial}}(f_0^* T_X) \otimes_{\mathbb{C}} I$ whose image in Coker $(d\Phi(\xi))$ is the (analytic) obstruction to extending $f$ over $\bar{B}$. Using isomorphism of Čech and Dolbeault cohomologies, one checks directly that the two obstruction classes coincide under the given isomorphism Coker $(d\Phi(\xi)) \cong \mathrm{Ext}^1_C([f_0^*\Omega_X \to \Omega_C], \mathcal{O}_C)$. This proves 2).

Let $F_\beta \in \Lambda$ be any members in the lemma, which are finite rank $\mathbb{Q}$-bundles over $U_\beta \subset \mathcal{W}_{\alpha,g,n}$. Let $Y_\beta = \Phi^{-1}(F_\beta)$ and let $R_\beta = Y_\beta \cap \mathcal{M}^J_{\alpha,g,n}$. Using the liftings $\phi_\beta : Y_\beta \to F_\beta|_{Y_\beta}$, we can construct cones $C^{F_\beta} \subset F_\beta|_{R_\beta}$ that are limits $\lim_{t\to\infty} \Gamma_{t\phi_\beta}$, where $\Gamma_{t\phi_\beta}$ are the graphs of $t\phi_\beta$. Here we view $C^{F_\beta}$ as the limits of oriented currents. Since $\phi_\beta$ are holomorphic, such limits do exist. Following [Fu,p88], $C^{F_\beta}$ are the analytic analogue of the normal cones to $R_\beta$ in $\Gamma_{\phi_\beta}$. Because of the previous lemma, these cones are exactly the virtual normal cones to $\mathcal{M}^J_{\alpha,g,n}$ with respect to the tangent-obstruction complexes

$$d\Phi|_{R_\beta} : TY_\beta|_{R_\beta} \longrightarrow F_\beta|_{R_\beta}.$$

Now we construct another cycle representative of $[\mathcal{M}^J_{\alpha,g,n}]^{\mathrm{symp}}$ using these cones. We fix the collection $\Lambda_1, \cdots, \Lambda_N$ that was used in constructing $\Gamma_N$ and the symplectic class $[\mathcal{M}^J_{\alpha,g,n}]^{\mathrm{symp}}$. Because of the above lemma, we can assume without loss of generality that each member of $\Lambda_h \triangleq \cup\Lambda_k$ are holomorphic. We can also assume that to each $F_\beta \in \Lambda_h$ the cycle $\Gamma_N$ intersects the cone $C^{F_\beta}$ transversally as subsets of $F_\beta|_{Y_\beta}$. The collection $\{\Gamma_N \cap C^{F_\beta}\}$ will form a purely $r$-dimensional closed current, denoted by $C^{\Lambda_h}_{\mathrm{alg}}$. Let $\pi_*[C^{\Lambda_h}_{\mathrm{alg}}] \in H_*(\mathcal{W}_{\alpha,g,n})$. Because $C^{F_\beta}$ are the limits of $\Gamma_{t\phi_\beta}$,

$$\pi_*[C^{\Lambda_h}_{\mathrm{alg}}] = \pi_*[C^{\Lambda_h}_{\mathrm{symp}}] \in H_*(\mathcal{W}_{\alpha,g,n}).$$

It remains to show that $\pi_*[C^{\Lambda_h}_{\mathrm{alg}}]$ is $\tau_*[\mathcal{M}^X_{\alpha,g,n}]^{\mathrm{vir}}$. To accomplish this, we need to relate local complexes $[TY_\beta|_{R_\beta} \to F_\beta|_{R_\beta}]$ to the global complex $[E_1 \to E_2]$ used in constructing $[\mathcal{M}^X_{\alpha,g,n}]^{\mathrm{vir}}$ in section 1. We achieve this by constructing a collection of vector bundles $G_\beta$ over $R_\beta$, $\beta \in \Lambda_h$, of which the following holds: 1) both $E_2|_{R_\beta}$ and $F_\beta|_{R_\beta}$ are subbundles of $G_\beta$; 2) over $R_{\beta\beta'} = R_\beta \cap R_{\beta'}$, if $F_{\beta'}|_{R_{\beta\beta'}}$ is a subbundle of $F_\beta|_{R_{\beta\beta'}}$ then $G_{\beta'}|_{R_{\beta\beta'}}$ is a subbundle of $G_\beta|_{R_{\beta\beta'}}$ and further the diagram

$$\begin{array}{ccccc} E_2|_{R_{\beta\beta'}} & \longrightarrow & G_\beta|_{R_{\beta\beta'}} & \longleftarrow & F_\beta|_{R_{\beta\beta'}} \\ \| & & \uparrow & & \uparrow \\ E_2|_{R_{\beta\beta'}} & \longrightarrow & G_{\beta'}|_{R_{\beta\beta'}} & \longleftarrow & F_{\beta'}|_{R_{\beta\beta'}} \end{array}$$

is commutative. In $G_\beta$, we construct oriented currents $D_\beta \subset G_\beta$ so that $D_\beta \cap (E_2|_{R_\beta}) = C^{\mathcal{E}_2}|_{R_\beta}$ and $D_\beta \cap (F_\beta|_{R_\beta}) = C^{\Lambda_h}_{\mathrm{alg}}|_{R_\beta}$, and these intersections are transversal. Therefore, we must have

$$\pi_*[C^{\Lambda_h}_{\mathrm{alg}}] = \tau_*\eta^![C^{\mathcal{E}_2}] = \tau_*[\mathcal{M}^X_{\alpha,g,n}]^{\mathrm{vir}} \in H_*(\mathcal{W}_{\alpha,g,n}),$$

where $\eta$ is the zero section of $E_2$. This proves the comparison theorem.

## References


[Al] V. Alexeev, *Moduli spaces $M_{g,n}(W)$ for surfaces*, preprint.

[AM] P. Aspinwall, D. Morrison, *Topological field theory and rational curves*, Comm. Math. Phy. **151** (1993), 245-262.

[At] I. V. Artamkin, *On deformation of sheaves*, Math. USSR, Izv. **32 (no. 3)** (1989), 663-668.

[Ba] S. Bauer, *Diffeomorphism types of elliptic surfaces with $p_g = 1$*, Warwick preprint #34 (1992).

[BDW] A. Bertram, G. Daskalopoulos and R. Wentworth, *Gromov-Witten invariants for holomorphic maps from Riemann surfaces to Grassmannians*, J. Amer. Math. Soc. **9** (1996), 529-571.

[Be] K. Behrend, *Gromov-Witten invariants*, preprint (1996).

[BF] K. Behrend and B. Fantechi, *The intrinsic normal cone*, preprint (1996).

[Br] A. Bertram, *Quantum Schubert Calculus*, to appear in Journal of Amer. Math. Soc..

[CF] I. Ciocan-Fontanine, *Quantum cohomology of flag manifolds*, Intern. Math. Res. Notices **6** (1995), 263-277.

[CM] B. Crauder and R. Miranda, *Quantum Cohomology of Rational Surfaces*, The moduli space of curves (R. Dijkgraaf, C. Faber, G. Van der Geer, eds.), Progress in Mathematics **129**.

[Do] S. Donaldson, *Polynomial invariants for smooth four manifolds*, Topology **29** (1990), 257-315.

[DM] P. Deligne and D. Mumford, *The irreducibility of the space of curves of given genus*, Publ. I.H.E.S. **45**, 101-145.

[FGP] S. Fomin, S. Gelfand and A. Postnikov, *Quantum Schubert polynomials*, preprint (1996).

[FM1] R. Friedman and J. W. Morgan, *Smooth foru-manifolds and complex surfaces*, Ergebinisse der Mathematik und ihreh Grenzgebirte 3. Folge 27, Springer, Berlin Heidelberg New York, (1994).

[FM2] R. Friedman and J. W. Morgan, *Obstruction bundles, semiregularity, and Seiberg-Witten invariants*, preprint (1996).

[FO] K. Fukaya and K. Ono, *Arnold conjecture and Gromov-Witten invariants*, preprint (1996).

[Fu] W. Fulton, *Intersection theory*, Ergebnisse der Math. und ihrer Grenzgebiete 3. Folge Band 2, 1984.

[FP] W. Fulton and R. Pandharipande, *Notes on stable maps and quantum cohomology*, This volume.

[GH] L. Göttsche and D. Huybrechts, *preprint* (1996).

[GP] L. Göttsche and R. Pandharipande, *The quantum cohomology of blow-ups of $\mathbf{P}^2$ and enumerative geometry*, preprint (1996).

[HK] S. Hosono, A. Klemm, S. Theisen and S-T. Yau, *Mirror symmetry, mirror map and application to Calabi-Yau hypersurfaces*, Comm. Math. Phys. **167** (1995), 301-350.

[KM1] M. Kontsevich and Y. Manin, *GW classes, Quantum cohomology and enumerative geometry*, Comm. Math. Phys. **164** (1994), 525-562.

[KM2] Kontsevich and Y. Manin, *Quantum cohomology of product*, Invent. Math. **124** (1996), 313-339.

[Ko] M. Kontsevich, *Enumeration of rational curves via torus actions*, The Moduli Space of Curves, Editors R.Dijkgraaf, C.Faber, G. van der Geer, Progress in Mathematics vol.129, Birkhauser 1995, 335-368..

[Kz] S. Katz, *Lectures at 1993 enumerative geometry conference in Dyrkolbotn, Norway.*

[La] O. A. Laudal, *Foraml Moduli of Algebraic Structures.*, Lecture Notes in Mathematics, Vol. 754. Berlin Heidelberg New York, Springer 1979.

[Li] J. Li, *Algebraic geometric interpretation of Donaldson's polynomial invariants*, J. Diff. Geom. **37** (1993), 417-466.

[LT1] J. Li and G. Tian, *Virtual moduli cycle and Gromov-Witten invariants of algebraic varieties*, preprint (1996).

[LT2] J. Li and G. Tian, *Virtual moduli cycles and Gromov-Witten invariants of general symplectic manifolds*, preprint (1996).

[LT3] J. Li and G. Tian, *In preparation.*

[LT4] J. Li and G. Tian, *The quantum cohomology of homogeneous varieties*, to appear in J. of Alg. Geom..

[Mo] J. Morgan, *Comparison of the Donaldson polynomial invariants with their algebro-geome -tric analogues*, Topology **32 (no 3)** (1993), 449-488.

[MO] J. W. Morgan and K. O'Grady, *Differential Topology of complex surfaces*, Lecture Notes in Mathematics **1545** (1993).

[Mu] D. Mumford, *Towards an enumerative geometry of the moduli space of curves*, Arithmetic and Geometry II, Progress in Mathematics 36 (1983), 271-326.

[Ra] Z. Ran, *On the quantum cohomology of plane, old and new*, preprint.

[Ru1] Y. Ruan, *Topological Sigma model and Donaldson type invariants in Gromov theory*, to appear in Duke J. Math..

[Ru2] Y. Ruan, *Virtual neighborhoods and pseudo-holomorphic curves*, preprint (1996).

[RT1] Y. Ruan and G. Tian, *A mathematical theory of quantum cohomology*, J. Diff. Geom. **42 no. 2** (1995).

[RT2] Y. Ruan and G. Tian, *Higher genus symplectic invariants and sigma model coupled with gravity*, preprint (1995).

[Si] B . Siebert, *Gromov-Witten invariants for general symplectic manifolds*, preprint (1996).

[ST] B. Siebert and G. Tian, *On quantum cohomology of Fano manifolds and a formula of Vafa and Intrilligator*, preprint (1994).

[Vi] A. Vistoli, *Intersection theory on algebraic stacks and on their moduli spaces*, Invent. Math. **97** (1989), 613-670.

[Vo] C. Voisin, *A mathematical proof of Aspinwall-Morrison formula*, preprint.

[W1] E. Witten, *Topological sigma models*, Comm. Math. Phys. **118** (1988).

[W2] E. Witten, *Two dimensional gravity and intersection theory on moduli space*, Surveys in Diff. Geom. **1** (1991), 243-310.

[1]Stanford University
*E-mail address*: `jli@gauss.stanford.edu`

[2]M.I.T.
*E-mail address*: `tian@math.mit.edu`

# VI. Fundamental Groups and Non-Abelian Hodge Theory

Proceedings of Symposia in Pure Mathematics
Volume **62.2**, 1997

# On the Shafarevich maps

L. Katzarkov

Contents



## 1. Introduction

The Shafarevich conjecture says that the universal cover $\widetilde{X}$ of a smooth projective variety is holomorphically convex, meaning that for every infinite sequence without limit points in $\widetilde{X}$ there exists a holomorphic function unbounded on this sequence. By the Poincaré uniformization theorem, the conjecture is true for algebraic curves.

In the last five years there was major progress towards understanding the Shafarevich conjecture. The aim of this article is to provide an overview of these developments.

In section 2 we start with one of the key works of the period - the results of Kollár [**40**], [**41**], which show that the Shafarevich maps exist birationally. Similar results were proved by Campana [**6**].

Section 3 is devoted to certain approximations of the Shafarevich conjecture for special groups. The simplest example is a theorem of Jost-Yau [**30**] and Siu [**60**] which says that if the fundamental group of a curve of genus $g$ is a quotient of $\pi_1(X)$, then $X$ has a morphism onto a curve of genus at

1991 *Mathematics Subject Classification.* Primary 14C30, 14D07, Secondary 14D20, 53C55.

least $g$. There are many similar results for other groups and in some cases these can be used to prove the Shafarevich conjecture.

Nonabelian Hodge theory is considered in section 4. The original techniques are due to Corlette, Hitchin and Simpson [**8, 27, 54**]. A $p$-adic version of the ideas of Simpson and Corlette became available through the works of Gromov and Schoen [**23**]. All these combined with the precise analysis of the obstacles for holomorphic convexity done in the pioneering works of Lasell and Ramachandran [**42**] and Napier [**48**] allow us to prove the Shafarevich conjecture for surfaces with linear fundamental group.

In section 5 we propose an example which reduces the Shafarevich conjecture to some hard and unsolved group theoretic questions.

The last section contains various open problems and speculations related to the Shafarevich conjecture.

## 2. Results of János Kollár

One can easily notice that the Shafarevich conjecture is equivalent to the following two conditions:

1. There exists a normal analytic space $\mathbf{Sh}(\widetilde{X})$ with no compact complex subspaces and a proper morphism with connected fibers $\mathbf{Sh} : \widetilde{X} \longrightarrow \mathbf{Sh}(\widetilde{X})$. This morphism is known as the Cartan-Remmert reduction of $\widetilde{X}$.
2. $\mathbf{Sh}(\widetilde{X})$ is a Stein space.

Kollár observed in [**40**] and [**41**] that 1) is equivalent to:

**(∗):** There exists a normal variety $\mathbf{Sh}(X)$ and a proper morphism with connected fibers $\mathbf{Sh} : X \longrightarrow \mathbf{Sh}(X)$, which contracts exactly the subvarieties $Z$ in $X$ having the property that $\mathrm{im}[\pi_1(\mathrm{Res}(Z)) \longrightarrow \pi_1(X)]$ is finite. Here $\mathrm{Res}(Z)$ denotes any desingularization of $Z$.

One can consider also a relative version of condition $(*)$. Let $H \triangleleft \pi_1(X)$ be a normal subgroup. We will say that a subgroup $\Gamma \subset \pi_1(X)$ is almost contained in $H$ if the intersection $\Gamma \cap H$ has finite index in $\Gamma$ and we will write $\Gamma \lesssim H$. We have the following condition.

**$(*)^H$:** There exists a normal variety $\mathbf{Sh}^H(X)$ and a proper map with connected fibers $\mathbf{Sh}^H : X \longrightarrow \mathbf{Sh}^H(X)$, which contracts exactly the subvarieties $Z$ in $X$ having the property that $\mathrm{im}[\pi_1(\mathrm{Res}(Z)) \longrightarrow \pi_1(X)] \lesssim H$. Again $\mathrm{Res}(Z)$ denotes a desingularization of $Z$.

Notice that condition $(*)$ is a particular case of condition $(*)^H$ with $H = \{e\}$. The main difficulty in handling either condition lies in the existence of $\mathbf{Sh}(X)$. In [**40**], [**41**] J. Kollár suggested a relaxation of the conditions $(*)$ and $(*)^H$ that makes possible an explicit geometric construction approximating the variety $\mathbf{Sh}(X)$.

THEOREM 2.1 ( Kollár,[**40**],[**41**]). *For any smooth projective variety $X$ and any normal subgroup $H \triangleleft \pi_1(X)$ there exists a normal variety $Sh^H(X)$, a Zariski open set $X_0$ in $X$ and a proper morphism with connected fibers*

$$Sh^H : X_0 \longrightarrow Sh^H(X),$$

*such that if $Z \subset X$ passes through a very general point in $X$, $Sh^H$ contracts $Z$ iff* $\mathrm{im}[\pi_1(\mathrm{Res}(Z)) \longrightarrow \pi_1(X)] \lesssim H$. *(For the precise notion of a very general point see Definition 2.4 below.)*

We explain the main ideas of the proof of this theorem. What makes the proof work is putting the problem in the proper context - the formalism of normal cycles. This is also the the reason why the above theorem has far-going applications even though it does not answer the Shafarevich question completely. We will briefly explain the relevant notions and constructions.

DEFINITION 2.1. Let $X$ be a normal variety. By a normal cycle we mean a normal variety $W$ together with a finite morphism $w : W \longrightarrow X$ which is birational to its image.

DEFINITION 2.2. Let $X$ be a normal variety. By a family of normal cycles in $X$ we mean a diagram

$$\begin{array}{ccc} U & \xrightarrow{u} & X \\ {\scriptstyle p}\downarrow & & \\ S & & \end{array}$$

where

1. every connected component of $U$ is of finite type and reduced;
2. $p$ is flat with irreducible, geometrically reduced fibers;
3. for every $s \in S$ we have that

$$u_s : U_s \longrightarrow X$$

   is a normal cycle.

We will say that the family $u : U \to X$ is dominant if $u$ is dominant.

DEFINITION 2.3. Let $X$ be a normal variety. By a locally topologically trivial family of normal cycles we will mean a family of normal cycles such that

$$U \longrightarrow S.$$

is a topological fiber bundle.

We will say that $U \longrightarrow S$ is weakly complete if for every other locally topologically trivial family of normal cycles $W \longrightarrow T$ we have that there exists an open set $T_0$ in $T$ and a unique morphism $t_0 : T_0 \longrightarrow S$ such that pull back of $U \longrightarrow S$ by $t_0$ is isomorphic to the restriction of $W \longrightarrow T$ to $T_0$.

DEFINITION 2.4. Let $X$ be a normal variety. We will say that the point $x \in X$ is a very general point if for every normal cycle

$$w_0 : W_0 \longrightarrow X$$

with $x \in \operatorname{im}(w)$ there exists a dominant and locally topologically trivial family of normal cycles

$$S \longleftarrow W \longrightarrow X$$

and a closed point $s$ in $S$ such that the restriction of $S \leftarrow W \rightarrow X$ to $s$ is isomorphic to $w_0 : W_0 \longrightarrow X$.

It is not hard to see using Lemma 2.1 that the set of very general points contains the intersection of countably many Zariski dense subsets. In particular, there are very general points.

PROPOSITION 2.1. *( Kollár) Let $X$ be a normal variety and let $H$ be a normal subgroup of $\pi_1(X)$. Consider the maps*

$$T \xrightarrow{i} W \xrightarrow{w} Z \xrightarrow{n} X,$$

*and*

$$p : W \longrightarrow V$$

*Where all schemes $T, W, Z$ and $X$ are irreducible and normal and $p \circ i$ and $w$ are dominant. Choose a general point $o \in T$ and let $W_o$ be the fiber of $p$ containing $i(o)$. Assume also that $\operatorname{im}[\pi_1(T) \longrightarrow \pi_1(X)] \lesssim H$ and $\operatorname{im}[\pi_1(W_o) \longrightarrow \pi_1(X)] \lesssim H$, then*

$$\operatorname{im}[\pi_1(Z) \longrightarrow \pi_1(X)] \lesssim H.$$

Similar result was obtained by Campana ( see [**6**]).

SKETCH OF PROOF. Since $W$ and $Z$ are normal and $w$ is a dominant morphism we have that $\operatorname{im}[\pi_1(W) \longrightarrow \pi_1(Z)]$ has finite index in $\pi_1(Z)$. So what we need to show is that

$$\operatorname{im}(\pi_1(W) \longrightarrow \pi_1(X)) \lesssim H.$$

Replacing $W$ by $W \times_V T$ and $V$ by $T$ we can think of $i$ as a section $i : V \longrightarrow W$. Let us choose an open set $V^0$ in $V$ such that the following exact sequence splits

$$\pi_1(W_o) \longrightarrow \pi_1(W^0) \rightleftharpoons \pi_1(V^0) \longrightarrow 1,$$

where $W^0 = p^{-1}(V^0)$.

By carefully tracing all images from the splitting of the above sequence we get that

$$\operatorname{im}[\pi_1(W^0) \longrightarrow \pi_1(X)] \lesssim H.$$

We finish the proof by observing that $\pi_1(W^0)$ surjects onto $\pi_1(W)$. Q.E.D.

We proceed by observing the following corollary from the above proposition.

COROLLARY 2.1. *Let $X$ be a normal variety, $H$ is a normal subgroup in $\pi_1(X)$ and $x$ is a very general point in $X$. Let*

$$w_0 : W_0 \longrightarrow X$$

*be a normal cycle. Assume that*

*1) $x \in im(w_0)$;*

*2) $\mathrm{im}[\pi_1(W_0 \longrightarrow \pi_1(X)] \lesssim H$.*

*3) $\dim(W_0)$ is the greatest possible with properties 1) and 2).*

*Then $w_0$ is unique. That is, for every other cycle $w_0'$ satisfying 2), either $\mathrm{im}(w_0)$ is disjoint with $\mathrm{im}(w_0')$ or $\mathrm{im}(w_0') \subset \mathrm{im}(w_0)$.*

Now we are ready to prove Theorem 2.1. We start first with the following lemma:

LEMMA 2.1. *In the assumptions of the above corollary there exists a weakly complete locally topologically trivial family of normal cycles*

$$S(X) \longleftarrow W_X \longrightarrow X$$

*where $S(X)$ is a smooth scheme.*

The proof of the above lemma uses a standard argument with the universal family over the Hilbert scheme of $X$ and the universal property of the latter.

Now choose a normal cycle

$$w_0 : W_0 \longrightarrow X$$

that has the properties from the Corollary 2.1. The above lemma provides us with a weakly complete locally topologically trivial family of normal cycles

$$S(X) \longleftarrow W_X \longrightarrow X$$

which contains $W_0$. We need to find an open set $S_0$ in $S$ such that

$$w : s^{-1}(S_0) \longrightarrow X$$

is an open immersion. Furthermore, Corollary 2.1 gives that for $s' \neq s$, $s \in S(X)$, $x$ is not contained in $\mathrm{im}(w_{s'})$. Thus we conclude that a very general point $x$ of $X$ has only one preimage in $W$. Therefore $w$ is birational. Of course we need to take out $\cup D_i$ the union of all closed subvarieties over which $w$ is not an immersion.

Now let us define $sh_X^H = p \circ (w^0)^{-1}$ and $X^0 = w^0(W^0)$. Here $W^0$ and $w^0$ are defined as in the above corollary. This is the map we need. In other words Proposition 2.1 gives us the surjectivity of $w_0$ and corollary 2.1 provides the birationality.

Even though Theorem 2.1 does not answer completely the questions connected with the Shafarevich conjecture the technique developed for its proof is very powerful and rich in ideas and applications. Sometimes it even gives us the opportunity to avoid the Shafarevich conjecture.

The main idea is that as long as the information contained in the fundametal group is the essential one we can restrict ourselves to two big classes of varieties:

1. Varieties with finite fundamental group;
2. Varieties with large fundamental group. Recall that $X$ has large fundamental group if $\mathrm{im}[\pi_1(\mathrm{Res}(Z)) \to \pi_1(X)]$ is an infinite group for every irreducible and positive dimensional subvariety $Z$ of $X$, where $\mathrm{Res}(Z)$ denotes any resolution of $X$.

The idea to use only the information coming from the fundamental group resembles some rigidity results in differential geometry. This idea though seems bizarre in algebraic geometry. One of the things we are trying to do in this survey is to show that this might not be so strange.

The Shafarevich varieties are varieties with generically large fundamental groups.

The following conjecture was made by János Kollár

CONJECTURE 2.1. The varieties with large fundamental groups are fibrations of abelian varieties over varieties of general type.

If the Shafarevich conjecture was true then the universal coverings of the Shafarevich varieties will have lots of non-trivial holomorphic functions. Later on, after we discuss when the Shafarevich maps can be shown to be morphisms, we will go back to these questions. Now we would like to demonstrate how the techniques developed above helped J. Kollár to prove some very strong nonvanishing theorems for varieties with large fundamental groups. Unfortunately one needs an additional condition - residually finite fundamental group - which seems quite artificial. At the end of the paper we will try to explain why this actually might be the borderline requirement.

So we would like to discuss the following:

THEOREM 2.2. *Let $X$ be a smooth projective variety and let $\pi_1(X)$ be a residually finite infinite group. Let $D$ be a big Cartier divisor on $X$. If for every subvariety $Z$ of $X$, $\mathrm{im}[\pi_1(\mathrm{Res}(Z)) \to \pi_1(X)]$ is infinite, then:*

$$h^0(X, K_X + D) > 0,$$

*where $K_X$ is the canonical line bundle on $X$.*

Let us recall a definition:

DEFINITION 2.5. A line bundle $L$ on $X$ is called big if there is a dense open set $U \subset X$ such that if $x_1, x_2 \in U$ and $m \gg 1$ then there is a section $f \in H^0(X, L^m)$ such that $f(x_1) = 0$ and $f(x_2) \neq 0$. A line bundle $L$ is called nef if $deg_C L \geq 0$ for every compact irreducible curve $C \subset X$. A Cartier divisor is called big (respectively nef) if the line bundle $L = \mathcal{O}_X(D)$ is big (respectively nef).

We start the proof of the theorem by indicating the proof of the following:

LEMMA 2.2. *Let $X$ be a normal proper variety with generically large residually finite fundamental group. Let $D$ be a nef and big divisor on $Y$*

*and let $M > 0$. Then there exists an étale cover $m : X' \longrightarrow X$ such that if*

$$Z \longleftarrow U \xrightarrow{u} X'$$

*is a dominant family of normal cycles on $X'$ then*

$$\deg_D U_{gen} := (m^* D)^{\dim(U/Z)} . U_{gen} \geq M.$$

PROOF. Let $m : X' \longrightarrow X$ be an étale covering that corresponding to a normal subgroup $G$ in $\pi_1(X)$. It is clear that we can obtain a new dominant family of normal cycles on $X'$ by taking $U(X') = U(X) \times_X X'$.

We have that $\deg(U'_{i,s}/U_{i,s}) = [\pi_1(U'_{i,s}) : u^*(G)]$ and these implies that

$$\deg(U'_i)_{gen} = [\pi_1(U'_{i,s}) : u^*(G)]^{\dim(U_{i,s})} \cdot \deg_D(U_i)_{gen}.$$

Since $D$ is ample, there are only finitely many $i$ for which $\deg_D(U_i/S_i) < M$. Let $I$ be the corresponding index set. Therefore there exists an integer $r$ such that

$$\min_i \deg_D(U_i/S_i) \geq M/r.$$

Since $\pi_1(X)$ is residually finite and infinite we can chose a normal subgroup $G$ in $\pi_1(X)$ such that for every $i \in I$

$$\deg(U'_{i,s}/U_{i,s}) = [\pi_1(U'_{i,s}) : u^*(G)] \geq r.$$

This proves the lemma. □

Now put $N = K_X + D$, $D$ as before. It is well known that for the étale covering $m : X' \longrightarrow X$ we have

$$\begin{aligned} h^0(X', m^*(N)) &= \chi(X', m^*(N)) \\ &= \deg(m)\chi(X, N) = \deg(m) h^0(X, N). \end{aligned}$$

Now using a result of Kollár's the only thing we need to show is that for some etale covering $m : X' \longrightarrow X$ we have that

$$(m^* D)^{\dim(U/Z)} . U_{gen} \geq (1/2(\dim(X')^2 + \dim(X') + 1))^{\dim U/Z}.$$

According to the lemma above this can be achieved.

The existence of the Shafarevich map has many other applications. But from now on we will concentrate on the ones that are related with the source of all of these activities - the Shafarevich conjecture.

The Riemann uniformization theorem gives us the Shafarevich conjecture for a Riemann surface. Assume now that $\pi_1(X)$ surjects to the fundamental group of a Riemann surface of genus greater than two. Then by a theorem proved by Jost and Yau [**30**], Siu [**60**] we have a holomorphic map to a Riemann surface. Therefore we can define nonconstant holomorphic function on the universal covering of $X$. This fact is very suggestive. Observe that the fundamental group of a Riemann surface is a reductive group and that it also embeds in its unipotent completion. It is interesting to try to understand what is happening for the simplest possible case - an algebraic surface that maps to a Riemann surface. In this paper we try to show that even this case is quite hard and very rich in terms of the varieties of phenomena that it can offer. This paper deals exclusively with surfaces. We show in section 4 that for surfaces with nice fundamental groups we get the Shafarevich conjecture. In section 5 we show that for special surfaces one can expect some pathologies. The point we would like to make now is that the technique developed by Kollár suggests that at least for general enough families of curves the answer of the Shafarevich conjecture should be positive.

CONJECTURE 2.2. For every curve that passes through a very general point (excluding countable many closed subsets) in the moduli space of curves of genus $g$ with level structure the total space of the corresponding family $F$ of curves satisfies the Shafarevich conjecture.

Of course this is true for general enough hyperplane section of the compactified moduli space of curves of genus $g$ with level structure $L$. It follws from the fact that for general enough curve $C$ in the moduli space of curves of genus $g$ with level structure $L$ we have that $\pi_1(C\backslash\{P_1, \dots, P_k\})$ surjects onto $M_g^L$, the fundamental group of the moduli space of curves of genus $g$ with level structure $L$. Here $\{P_1, \dots, P_k\}$ are the points on $C$ that correspond to the singular fibers of our fibration $F$. The fact that $\pi_1(C \setminus \{P_1, \dots, P_k\})$ surjects onto $M_g^L$ means that the monodromy kills the image of the fundamental group of the generic fiber in $F$ in $\pi_1(F)$. Therefore the universal covering of $F$ maps properly to the universal covering of $C$ - the disc, so it is holomorphically convex.

Some evidence supporting Conjecture 2.2 is provided by the following:

PROPOSITION 2.2 (J. Kollár) ([**40**], Proposition 2.11). *Let* $f : X \to Y$ *be a morphism between irreducible normal varieties. Assume that* $\mathrm{im}(f)$ *contains a very general point in* $Y$. *Then the normalizer of* $\mathrm{im}[\pi_1(X) \to \pi_1(Y)]$ *is a finite index subgroup in* $\pi_1(Y)$.

With this proposition in mind let us consider the map from $C \setminus (P_1, \dots, P_k)$ to the moduli space of curves of genus $g$ with level structure $L$. We have the following sequence:

$$0 \longrightarrow T_g \longrightarrow M_g^L \longrightarrow Sp(2g, \mathbb{Z}) \longrightarrow 0,$$

where $T_g$ is the corresponding Torelli group.

Now let $N$ be a finitely generated normal subgroup of $M_g^L$ that maps to a subgroup of finite index in $Sp(2g, \mathbb{Z})$. We think of $N$ as the image of $\pi_1(C \setminus \{P_1, \dots, P_k\})$ in $M_g^L$. Following the above proposition we know that the normalizer of the image of $\pi_1(C \setminus \{P_1, \dots, P_k\}$ in $M_g^L$ is a subgroup of finite index in $M_g^L$. If it was the case that a group with finite index normalizer in $M_g^L$ is finite or with finite index itself we would had the conjecture. Unfortunately the Torelli group $T_g$ does not satisfy this property. Therefore we need to refine the question. Namely we first introduce $N$ a finitely generated normal subgroup of $M_g^L$ that maps to a subgroup of finite index in $Sp(2g, \mathbb{Z})$. After that to prove the above conjecture we need a positive answer of the following question.

QUESTION 2.1 (J. Kollár). Is such $N$ a subgroup of finite index in $M_g^L$?

Before we move to the case when the Shafarevich morphism actually exists we would like to discuss a geometrical result that as the above techniques shows follows from the Shafarevich conjecture.

THEOREM 2.3. *Let $X$ be a smooth projective variety for which the Shafarevich conjecture is correct and such that* $\operatorname{rank}\operatorname{Pic}(X) = 1$. *Then for every subvariety $Z$ in $X$ we have that* $\operatorname{im}[\pi_1(Z) \longrightarrow \pi_1(X)]$ *is infinite.*

This result follows from the following reasoning. Assume that the family of normal cycles $u : U \longrightarrow V$ from the Shafarevich conjecture exists. Assume also that $\dim(V) > 0$. Take now an effective divisor $H_V$ in $V$ and denote by $H_X = u^{-1}(H_V)$ the closure of $H_V$. On one hand $H_X = u^{-1}(H_V)$ is disjoint from every fiber of $g$. On other hand since $\operatorname{rank}\operatorname{Pic}(X) = 1$ every effective divisor on $X$ is ample so $H_X$ intersects every positive dimensional variety. Therefore we conclude that $\dim(V) = 0$.

For the detailed proofs of the statements of this section see [**40**], [**41**].

## 3. The Shafarevich morphism

**3.1. Factorization theorems.** In some cases one can show that the Shafarevich map is actually a morphism, which provides the first half of the Shafarevich conjecture. One obvious case is when the variety $X$ has abelian fundamental group. Then the Shafarevich morphism is just the Albanese map of $X$. This can be taken one step further. In this section we will survey some results which show that if the fundamental group of the variety $X$ admits a faithful linear representation to a complex Lie group $L$ then the Shafarevich morphism exists. We will also give some support of Conjecture 2.1. Many of the ideas in discussed bellow were influenced by J. Kollár, T. Pantev. M. Ramachandran. The main theorem of this section is the following:

THEOREM 3.1. *Let $X$ be a smooth projective variety which has a representation of its fundamental group to $GL(n, \mathbb{C})$ with a kernel $H$. Then the Shafarevich morphism $\boldsymbol{Sh}^H : X \to \boldsymbol{Sh}^H(X)$ exists.*

The proof will be explained in two steps. The complete proof can be found in [**39**].

**Step 1.**

First we consider the case when $\rho$ has reductive image. Put $S$ for the Zariski closure of $\rho(\pi_1(X))$ in $GL(n, \mathbb{C})$. For the moment we will make two simplifying assumptions:

- $S$ is semisimple;
- $\rho(\pi_1(X)) \subset S \cap GL(n, \overline{\mathbb{Q}})$

These assumptions together with the fact that $\pi_1(X)$ is a finitely generated group imply that $\rho : \pi_1(X) \to S$ is defined over a finite extension $E$ of $\mathbb{Q}$. Let $O$ be the ring of integers in $E$. Put $O_\mathfrak{p}$ for the localization of $O$ at some prime ideal $\mathfrak{p}$ in $O$ and $\widehat{O}_\mathfrak{p}$ for the completion of $O_\mathfrak{p}$ with respect to the natural discrete valuation. Denote by $E_\mathfrak{p}$ the field of fractions of $\widehat{O}_\mathfrak{p}$ and let $S(E_\mathfrak{p})$ be the group of $E_\mathfrak{p}$-points of $S$. Since $E_\mathfrak{p}$ is a local field we can use the Bruhat-Tits theory to attach to $S(E_\mathfrak{p})$ an affine building $B_\mathfrak{p}$. Similarly we consider the symmetric space $\mathrm{Symm}_\nu = S(E_\nu)/(\text{maximal compact subgroup})$ at an archimedean place $\nu$. Due to our assumptions $\rho$ induces a representation $\rho' : \pi_1(X) \to S(E_\mathfrak{p})$ with $\ker(\rho) = \ker(\rho')$. Here we introduce a definition:

DEFINITION 3.1. We call the representation $\rho' : \pi_1(X) \to S(E_\mathfrak{p})$ unbounded if its image is not contained in a maximal compact subgroup of $S(E_\mathfrak{p})$.

We are in a position to apply the formalism of the non-abelian Hodge theory. We introduce the technique of spectral coverings to show the existence of non-abelian analogues of the Albanese map. This technique is based on the existence of equivariant pluriharmonic maps to symmetric space and buildings proved by Corlette [**8**] in the case of symmetric spaces and by Gromov and Schoen [**23**] in the case of buildings.

The following theorem was proved in [**39**].

THEOREM 3.2. *Let $X$ be a smooth projective variety which has a Zariski dense unbouned representation $\rho' : \pi_1(X) \to S(E_\mathfrak{p})$. Then the Shafarevich morphism $\boldsymbol{Sh}^H : X \to \boldsymbol{Sh}^H(X) = Y$ exists and* $\dim Y \leq \operatorname{rank} S(E_\mathfrak{p})$.

For the proof of this theorem one needs to utilize the algebraic geometric information contained in the harmonic map data. This is achieved by constructing the so called spectral cover, which we proceed to describe.

Using the existence results in [**23**] we assign to every unbounded representation $\rho' : \pi_1(X) \to S(E_\mathfrak{p})$ a nontrivial $\rho'$-equivariant harmonic map $U : \widetilde{X} \to B_\mathfrak{p}$ to the affine Bruhat-Tits building of $S(E_\mathfrak{p})$. In every apartment $Ap$ of $B_\mathfrak{p}$ we have a canonical choice of coordinates modulo the action of the affine Weyl group $\widetilde{W}$. This means that locally we can choose one coordinate $z_1$ and consider also $e_1(z_1), \ldots, e_w(z_1)$, where $e_1, \ldots, e_w$ are all the elements of the usual Weyl group. Let $U^*(T^*Ap)$ be the pullback of the cotangent bundle of $Ap$ and let $d$ denote the exterior differentiation on $Ap$. Then we can define, modulo the action of $W$, the differential forms $de_1(z_1), \ldots, de_w(z_1)$ globally on the whole building. (Observe that after taking differentials we can make everything invariant under the translation part of the affine Weyl group.) Consider the complexified differentials $U^*de_1(z_1), \ldots, U^*de_w(z_1)$. Due to the harmonicity of $U$ the (1,0) part of each of the forms $U^*de_1(z_1), \dots, U^*de_w(z_1)$ is holomorphic. This follows from the fact that by theorem 7.3 [**23**] $U$ is actually pluriharmonic outside the singular set. We will denote these new holomorphic one forms again by $U^*de_1(z_1), \dots, U^*de_w(z_1)$. Since $U$ is an equivariant map these holomorphic one forms descend to multivalued forms on $X$. Choose a regular weight $\alpha$ for $S(E_{\mathfrak{p}})$. It determines a $W$-invariant $\mathbb{C}[x]$-valued function $P_\alpha : \mathfrak{h}\otimes\mathbb{C} \to \mathbb{C}[x]$, $P_\alpha(x) := \prod_{\mu\in W\alpha}(x-\mu)$ on the Cartan algebra $\mathfrak{h}$ of $S(E_{\mathfrak{p}})$. Let $\kappa_1, \dots, \kappa_w$ be the coefficients of $P_\alpha$. The $\kappa_i$'s are $W$-invariant polynomials on $\mathfrak{h}\otimes\mathbb{C}$. By using the standard identification of the complexifications of apartment $Ap$ and the Cartan algebra $\mathfrak{h}$ we can evaluate $\kappa_1, \dots, \kappa_r$ on the multi-valued forms $U^*de_1(z_1), \dots, U^*de_w(z_1)$ to obtain sections $h_i$ of $\operatorname{Symm}^i \Omega^1_X$, $i = 1, \dots, w$ well defined and holomorphic outside of the singularities of the harmonic map $U$. Furthermore, according to [**23**] the map $U$ is Lipschitz and therefore we get that the multi-holomorphic differentials $h_1, \dots, h_w$ are bounded. Using Theorem 3.2 [**39**] we conclude that $h_1, \dots, h_w$ extend to holomorphic differentials on the whole $X$.

Let $p : \mathcal{X} \to X$ be the total space of the cotangent bundle $T^*X$ and let $\lambda \in H^0(\mathcal{X}, p^*\Omega^1_X)$ be the tautological section. We define a spectral covering corresponding to the elements $\{h_i \in H^0(X, \operatorname{Symm}^i \Omega^1_X)\}_{i=1}^w$ to be the zero scheme $\mathfrak{S}$ of the section

$$\lambda^w + p^*(h_1)\lambda^{w-1} + \dots + p^*(h_w) \in H^0(\mathcal{X}, p^* \operatorname{Symm}^w \Omega^1_X).$$

Notice that different choices of a regular weight $\alpha$ result in birationally isomorphic $\mathfrak{S}$. $\mathfrak{S}$ is a projective subscheme of $\mathcal{X}$ and by construction the natural morphism $\mathfrak{S} \to X$ is a Galois cover with Galois group $W$. For more detailed account on spectral coverings see [**15**] and [**14**].

The statement of theorem 3.2 now follows from the abundance of holomorphic one forms on the spectral covering $S$. We consider the map $s : S \to ALB(S)/B$, where $ALB(S)$ is the Albanese variety of $S$ and $B$ is the subvariety of $ALB(S)$ generated by the image in $ALB(S)$ of the degeneracy set of the linearly independent holomorphic one forms on $S$. Denote by $k$ the dimension of the image $R$ of $S$ in $ALB(S)/B$. If the degeneracy set of the linearly independent holomorphic one forms on $R$ is a divisor in $R$. Then we are in a position to apply a version of Castelnuovo de Francis theorem see [**46**]. If the degeneracy set is of codimension at least two then by the theorem of Napier and Ramachandran ( theorem 6.3 section 6) we can find a curve $M$ in $R$ that avoids this set and such that the image of $\pi_1(M)$ in $\pi_1(R)$ is a subgroup of finite index. In this situation one easily gets a contradiction with the Zariski denseness of the representation $\rho' : \pi_1(X) \to S(E_{\mathfrak{p}})$. Morally the above argument says that the information about the representation $\rho' : \pi_1(X) \to S(E_{\mathfrak{p}})$ is contained in the degeneracy set of the linearly independent holomorphic one forms on $S$ or in other words in the ramification divisor of the spectral covering.

Finally let us mention that one gets a factorization not of $X$ but of a blow up of some finite nonramified covering $X'$ of $X$. To get the existence of the Shafarevich morphism $\mathbf{Sh}^H : X \to \mathbf{Sh}^H(X) = Y$ one argues as follows. By the above Castelnuovo de Franchis argument we get that there exists a normal projective variety $Y$ and a map $f : X \to Y$ with connected fibers so that for some finite etale cover $X' \to X$ the variety $\mathrm{Sh}^H(X')$ exists and the Shafarevich map $\mathrm{sh}^H : X' \to \mathrm{Sh}^H(X')$ is just the Stein factorization of the composition $X' \to X \to Y$. The blowups of $X'$ mentioned above are not essential here. They will be needed later to make the variety $\mathrm{Sh}^H(X')$ smooth, which is used in the Hodge theoretic arguments later when we use the theory of MHS in the proof. The finite cover $X'$ though is essential since we need to deal with the orbi-phenomenon. Now it only remains to observe that $Y$ must be $\mathrm{Sh}^H(X)$ because any subvariety in $X$ has a finite étale cover that is a subvariety in $X'$ and because the map $\mathrm{Sh}^H(X') \to Y$ is finite. Observe that there is no difference between $\mathrm{Sh}^H(X')$ and $Y$ from the view point of the Shafarevich map. The difference becomes visible only from the view point of the representation $\rho$. Namely $\pi_1(X') \to S(E_{\mathfrak{p}})$ factors completely through $\pi_1(\mathrm{Sh}^H(X'))$ but $\rho' : \pi_1(X) \to S(E_{\mathfrak{p}})$ factors only through an orbi-extension of $\pi_1(Y)$. In other words the subvarieties of $X$ whose fundamental groups are commensurable with the kernel of $\rho$ do not see this orbistructure.

LEMMA 3.1. *There exists at least one place $\nu$ for which the representation $\rho' : \pi_1(X) \to S(E_\nu)$ is unbounded.*

PROOF. If we assume that for all places $\nu$ the corresponding representations are bounded then we get that our initial representation $\rho : \pi_1(X) \longrightarrow S$ is finite. Indeed, only finitely many among the maps $U_{\mathfrak{p}}$ and $U_\nu$ are nontrivial since $\rho'(\pi_1(X))$ is finitely generated and hence only finitely many elements of $O$ occur in the denominators of the traces of its generators. Combined with the fact that $E_{\mathfrak{p}}$ is discrete in $\prod_{\mathfrak{p}} E_{\mathfrak{p}} \times \prod_\nu E_\nu$ this gives that the image of $\rho'$ acts discretely on a finite product $\prod_{\mathfrak{p}} B_{\mathfrak{p}} \times \prod_\nu \mathrm{Symm}_\nu$ via the diagonal action (for the proof of discreteness see [**5**]). Thus $\rho'(\pi_1(X))$ is simultaneously discrete and compact, hence finite, which contradicts the Zariski denseness of $\rho$. □

If we have an unbounded representation at the infinite place then we can use the results of Zuo [**62**] and Mok [**45**] (in case of discrete representation ) to get a factorization. Next we remove two simplifying assumtions we made at the beginning of Step 1. The above arguments show that we get the existence of the Shafarevich morphism for every semisimple representation defined over $\overline{\mathbb{Q}}$. To treat a general semisimple $\rho$ we would like to deform $\rho : \pi_1(X) \longrightarrow S$ to a representation that is defined over $\overline{\mathbb{Q}}$. Observe that existence of a nontrivial deformation of $\rho : \pi_1(X) \longrightarrow S$ is not an issue since the rigid representations are always defined over $\overline{\mathbb{Q}}$.

LEMMA 3.2. *In a small neighborhood of $\rho : \pi_1(X) \longrightarrow S$ we can find a Zariski dense representation defined over $\overline{\mathbb{Q}}$.*

The proof of this lemma can be found in [**42**]. Observe that deforming we might enlarge the kernel of the original representation.

Now we need the following:

LEMMA 3.3. *If the Shafarevich morphism exists for every representation defined over $\overline{\mathbb{Q}}$ in a neighborhood of $\rho : \pi_1(X) \longrightarrow S$ in the moduli space of representations then it exists for $\rho : \pi_1(X) \longrightarrow S$.*

PROOF. See [**39**].

□

The passage from semisimple to a reductive representation is almost straightforward. Indeed up to a finite index every reductive group is a product of a semisimple group and a torus. If this torus is trivial we go to Step 1. If it is not this means that we have a nontrivial Albanese variety $Alb(X)$. So we consider the map $X \to Y \times Alb(X)$ , where $Y$ is the factorization variety for the semisimple representation. The image of $X$ under this map will be the Shafarevich variety for our reductive representation.

**Step 2.**

Now we move to the case of representations to a linear group $L$. This case requires some new ideas. Every linear group $L$ is an extension of its nilpotent radical $N$ and its reductive quotient $G$

$$1 \to N \to L \to G \to 1.$$

If $G$ is a trivial group then we just use the Albanese map (see section 4). If $G$ is a nontrivial group using theorem 3.2 and the existence of a MHS on the relative completion of $\pi_1(X)$ described in section 4 one can prove the following (see [**39**]) :

THEOREM 3.3. *Let $X$ be a smooth projective variety such that it admits an infinite representation $\rho : \pi_1(X) \to L$ to a complex linear group with kernel $H$. Assume that the representation $\rho : \pi_1(X) \to G$ factors through a representation of the fundamental group of a variety $Y$. Then $\rho : \pi_1(X) \to L$ factors through a representation of the fundamental group of a variety $Z$ which admits a finite nonramified covering $Z' \to Z$ birationally isomorphic to a fibration of abelian varieties over $Y$.*

We will return to the proof of this theorem in section 4 when we introduce a MHS on the relative completion of $\pi_1(X)$.

This finishes the proof of theorem 3.1. We would like to comment on some properties of the Shafarevich varieties and show some support for Conjecture 2.1 we also comment on some further applications of the above techniques.

**3.2. Some applications of the technique of spectral coverings.** We first show that in the case of a representation to semisimple group $S$ $\mathbf{Sh}(X')$ is of general type. This is the answer of Conjecture 2.1 in a very special case when a Zariski dense almost faithful representation $\rho$ exists. Since the fundamental group of $X'$ is a group of finite index in the fundamental group of $X$ we get a Zariski dense almost faithful representation of $\pi_1(X')$, which we use to construct spectral covering $S'$ for $X'$ in the same way as we have constructed $S$.

THEOREM 3.4. *Let $X$ be a smooth variety which has an unbounded Zariski dense almost faithful representation $\rho : \pi_1(X) \to S(E_{\mathfrak{p}})$ of its fundamental group. Then* $\mathbf{Sh}(X')$ *is of general type.*

PROOF. We can restrict ourselves to the case when $\dim X = \dim \mathbf{Sh}(X')$. Observe that by [**47**] the Kodaira dimension of the spectral covering $S'$, $\kappa(S')$, is equal to $\kappa(X')$. This can be seen as follows.

Since $\kappa(X') \leq \kappa(S')$ is obvious, it is enough to show that

$$\kappa(S') \leq \kappa(X').$$

By construction there is a holomorphic one form $\alpha$ on $S'$ s.t. the collection of one forms $\{w^*\alpha\}_{\{w\in W\}}$ spans a rank$(U)$ dimensional subspace $V_x$ in $\Omega^1_{S',x}$ for a generic point $x \in S$. In particular, since the rank of $U$ is maximal we have $\dim S' = \operatorname{rank}(U) = \dim V_x$ and hence $\sigma = \otimes_{w\in W} w^* \det V$ is a well defined non-identically zero section in some power of the pull back of the canonical bundle $K'_X$ of $X'$.

The next step is to notice that the ramification divisor $R$ of $S'$ is defined locally by linear dependence conditions imposed on the one forms spanned by $\{w^*\alpha\}_{\{w\in W\}}$. This means that some power of $\sigma$ will vanish on $R$, i.e. $p^*K^i_{X'} \supseteq \mathcal{O}'_S(R)$ and hence

$$K'_S = p^*K_{X'} \otimes \mathcal{O}'_S(R) \subseteq p^*K^{N+1}_{X'},$$

which gives us the required inequality.

We need to show that $S'$ is of general type. Suppose it is not. But then by Ueno's theorem (see [**61**]) there exists a map $S' \longrightarrow N$ with generic fiber an abelian variety $A$ of positive dimension. Now using the fact that $\rho$ is a Zariski dense representation in a simple Lie group and the standard homotopy sequence

$$0 \longrightarrow \pi_1(A) \longrightarrow \pi_1(U) \longrightarrow \pi_1(N) \longrightarrow 0$$

we conclude that we get a factorization through $N$. (Here $U$ is an open set in $S'$.) But $\dim N < \dim \mathbf{Sh}(X')$ we get a contradiction with our main factorization theorem. □

Stronger result was proved by Zuo see [**63**].

We can actually say more about the characterization of $Y$. Namely some kind of hyperbolicity properties can be seen from the following:

PROPOSITION 3.1. *Let $X$ be a smooth projective variety with Zariski dense faithful representation $\rho$ of its fundamental group to a Lie group $G$ defined over a local field $K$ and such that* $\dim_{\mathbb{C}} \mathbf{Sh}(X) = \operatorname{rank}_K G$. *Then* $\mathbf{Sh}(X')$ *does not contain rational and elliptic curves.*

PROOF. The case of a rational curve follows from the definition of Shafarevich variety. We can reduce our considerations to the case when $\mathbf{Sh} : X' \longrightarrow \mathbf{Sh}(\mathbf{X'})$ is a finite covering. In this case we will think that they coincide. The first observation is that since $\mathbf{Sh}(X')$ has a large fundamental group we have an infinite abelian subgroup $\operatorname{im}[\pi_1(E) \to \pi_1(\mathbf{Sh}(X')) = \pi_1(X')]$ in the fundamental group of $\mathbf{Sh}(X')$. Furthermore, since the spectral covers are functorial under pullback of representations it follows that the restriction of the spectral cover $S'$ to $E$ is a spectral cover for $E$ and hence is either etale or non-reduced. Since $\mathbf{Sh}(X')$ is projective it follows that $E$ intersects the branch locus of $S'$ and thus the only possibility is that $E$ is contained in the branch locus and $S'_{|E}$ is non-reduced. Now, this means that there is a copy of $E$ inside $S'$. But we know that $S'$ embeds in its Albanese ( by construction the rank of the harmonic map is maximal) and hence $\mathrm{Alb}(S')$ contains an elliptic curve, i.e. it is reducible. Finally, if we denote by $Y$ the image of the composition map $S' \to \mathrm{Alb}(S') \to \mathrm{Alb}(S')/E$ it follows that the kernel of the homomorphism $\pi_1(S') \to \pi_1(Y)$ is abelian and in particular $\rho$ factors through $Y$ - a contradiction since the kernel should be finite. □

The same statement can be proved in case $\rho : \pi_1(X) \to S$ is a real unbounded Zariski dense representation ( see [**45**]). Combining this results with theorem 3.3 one gets that the variety $Z'$ from theorem 3.3 is a fibration of Abelian varieties over a variety of general type which is again in support of Conjecture 2.1. In other words we have proved:

THEOREM 3.5. *Let $X$ be a smooth projective variety with large linear fundamental group. Then there exists a finite nonramified covering $X'$ of $X$ which is a fibration of Abelian varieties over a variety of general type.*

The technique of spectral covering allows to analyze even further varieties with Zariski dense linear representations. We believe it has a forgoing applications. The first one is a generalization of the following theorem of Bloch.

THEOREM 3.6. *Let $X$ be a smooth projective variety such that irregularity of $X$ , $q(X) > \dim X$. Then every entire holomorphic map $f : \mathbb{C} \to X$ is algebraically degenerate meaning that the image of $f$ is contained in a proper subvariety in $X$.*

The following conjecture can be seen as a Nonabelian variant or Bloch's theorem.

CONJECTURE 3.1. Let $X$ be a smooth projective variety such that it admits representation $\rho : \pi_1(X) \to GL(n,\mathbb{C})$ with a reductive nontrivial

Zariski closure. Then any entire holomorphic map $f : \mathbb{C} \to X$ is algebraically degenerate.

So far we are able to give a proof of this conjecture under additional assumption that $\operatorname{rank} G > \dim X - 1$ and $\rho : \pi_1(X) \to GL(n, \mathbb{C})$ is nonrigid representation. (see [**35**]). The proof uses the above construction of a spectral covering. The idea is to work with holomorphic one multiforms instead with holomorphic one forms and use some results of value distribution theory. Observe that already theorem 3.1 suggests that the image of $f : \mathbb{C} \to X$ should be contained in the leaves of the foliation given by the Shafarevich morphism.

Similar result should be true in case of quasiprojective varieties $X \setminus D$ if we require that the representation $\rho : \pi_1(X \setminus D) \to G$ is quasi-unipotent at infinity.

**3.3. First applications to the Shafarevich Conjecture.** The factorization theorems can be applied to questions connected with the Shafarevich conjecture. The first one is the following theorem proved by Katzarkov and Ramachandran [**42**].

THEOREM 3.7 ([**37**]). *Let $X$ be a smooth projective surface which has a Zariski dense almost faithful representation $\rho$ of its fundamental group to a simple complex Lie group $G$. Then the Shafarevich conjecture is true for $X$.*

REMARK 3.1. Before this theorem some results were known in the case of compact Kähler surfaces. Gurjar and Shastri [**24**] have shown that Shafarevich conjecture is true for elliptic surfaces. T. Napier has studied the question when (ramified, infinite) coverings of holomorphically convex surface are holomorphically convex. He has obtained several partial results [**48**]. It follows from his work that the basic obstruction to the holomorphic convexity is the existence of connected noncompact analytic sets all of whose irreducible components are compact. Throughout the paper we will call such an analytic set an infinite chain.

SKETCH OF PROOF. Before explaining the idea of the proof we would like to comment on some basic properties of the holomorphically convex manifolds:

1) Let $f : X \to Y$ be a proper holomorphic map between complex manifolds. Then $Y$ is holomorphically convex iff $X$ is.

2) This is a theorem of Cartan-Remmert. If $\widetilde{X}$ is holomorphically convex and noncompact then there exists a proper map with connected fibers $\mathbf{Sh} : \widetilde{X} \to \widetilde{\mathbf{Sh}(X)}$ onto a Stein space $\widetilde{\mathbf{Sh}(X)}$. This map is called the Cartan-Remmert reduction of $\widetilde{X}$ and it is exactly the map we have introduced in the previous section.

The work of Napier we have mentioned above indicates that in case of complex surfaces the only obstruction to the holomorphic convexity of the universal cover might be the existence of infinite chains. Also Napier and Ramachandran had proved the following theorem in [**49**].

THEOREM 3.8. *Let $X$ be a compact Kähler surface. Then any intermediate regular covering $\widetilde{X} \longrightarrow X' \longrightarrow X$, which does not have two ends and admits a real analytic plurisubharmonic exhaustion function is holomorphically convex.*

The proof of the theorem 3.7 is a combination of this criterion with some nonabelian Hodge theoretic considerations that are used to rule out the possibility of existence of infinite chains. The argument is based on a theorem proved in [**42**] by Lasell and Ramachandran which provides the necessary link making the nonabelian Hodge theory suitable for questions connected with the Shafarevich conjecture. Before we formulate this theorem let us give some explanations.

Start with an algebraic surface $X$ and let $Y = \cup Y_i \hookrightarrow X$ be a divisor in $X$ that is a connected reducible curve such that for every component $Y_i$ of $Y$ the image $\mathrm{im}[\pi_1(Y_i) \to \pi_1(X)]$ is finite. We can produce a counterexample to the Shafarevich conjecture if we can arrange that $\mathrm{im}[\pi_1(Y) \longrightarrow \pi_1(X)]$ is infinite. The easiest thing to try is to show that

$$\mathrm{im}[H_1(Y,\mathbb{Q}) \longrightarrow H_1(X,\mathbb{Q})] \neq 0.$$

Indeed the above conditions will allow $Y$ to open up to an infinite chain in $\widetilde{X}$ and then by picking a point on every component of this infinite chain we can construct a discrete sequence that will by maximum principle violate the definition of holomorphic convexity. This idea however is too optimistic. The strictness of the geometric morphism of mixed Hodge structures [**9**] implies that for any subvariety $Y \subset X$ one has

$$\mathrm{im}[H_1(Y,\mathbb{Q}) \longrightarrow H_1(X,\mathbb{Q})] =$$

$$\mathrm{im}[H_1(\mathrm{Res}(Y),\mathbb{Q}) \longrightarrow H_1(X,\mathbb{Q})] =$$

$$\sum \mathrm{im}[H_1(\mathrm{Res}(Y_i),\mathbb{Q}) \longrightarrow H_1(X,\mathbb{Q}).$$

(Here $\mathrm{Res}(Y)$ and $\mathrm{Res}(Y_i)$ denote desingularizations of $Y$ and $Y_i$ respectively.)

The theorem of Lasell and Ramachandran [**42**] can be seen as a nonabelian analog of the above result of Deligne's. It says that if we have an algebraic surface $X$ a Zariski dense representation of $\pi_1(X)$ to a complex reductive Lie group $\rho : \pi_1(X) \longrightarrow G$ and $Y = \cup D_i \hookrightarrow X$ a divisor in $X$ such that restricted representations $\rho : \pi_1(D_i) \longrightarrow G$ are trivial then $\rho : \pi_1(U) \longrightarrow G$ factors through a finite group $\delta_n$ the order of which depends only on the complex rank of $G$. In particular for surfaces with reductive fundamental group this theorem ensures the absence of inifinite chains in the universal cover.

The second place in which we need the nonabelian Hodge theory is to construct a real analytic plurisubharmonic function which enables us to use the theorem of Napier and Ramachandran. Let us start with the case when $\rho : \pi_1(X) \longrightarrow G$ is a representation defined over $\overline{\mathbb{Q}}$. We do the localization procedure described in subsection 3.1 and get $G^{\mathfrak{p}}$ to be the new group defined over $E_{\mathfrak{p}}$. Since $E_{\mathfrak{p}}$ is a local field we can use the Bruhat-Tits theory to attach to $G^{\mathfrak{p}}$ a building $B_{\mathfrak{p}}$ at a nonarchimedian place $\mathfrak{p}$ or a symmetric space $Symm_\nu$ at an archimedian place $\nu$. It follows from [**5**] the new representation $\rho'(\pi_1(X))$ acts discretely on $T = \prod B_p \times Symm$ via the diagonal action. Now define $\varphi : \widetilde{X} \longrightarrow \mathbb{R}$ to be $\varphi = U' \circ \mathrm{dist}(., x)^2$, where $\mathrm{dist}(., x)$ is the distance to a fixed point $x$ on $T$ and $U : \widetilde{X} \to T$ is the pluriharmonic map given to us by the Nonabelian Hodge Theory. The facts that $\mathrm{dist}(., x)^2$ is convex and $U'$ is pluriharmonic give us that $\varphi = U' \circ \mathrm{dist}(., x)^2$ is a plurisubharmonic function on $\widetilde{Y}$. The fact that $\rho'(\pi_1(X))$ acts discretely on $T = \prod B_p \times Symm$ makes this function an exhaustion function. The function $\varphi = U' \circ \mathrm{dist}(., x)^2$ is well defined since only finitelly many of the harmonic maps to $B_p$ are going to be nontrivial.

This construction of adelic type uses a representation defined over $\overline{\mathbb{Q}}$. Of course $\rho : \pi_1(X) \longrightarrow G$ does not need to be defined over $\overline{\mathbb{Q}}$. To put ourselves in such favorable situation we need to deform this representation as in the proof of theorem 3.1. Deforming we might create a kernel. But according to the work of Napier [**48**] we have mentioned above the only thing we need to worry about is the existence of infinite chains of compact curves. The theorem of Lasell and Ramachandran [**42**] says that in the case of Zariski dense faithful $\rho : \pi_1(X) \to G$ such an infinite chain of compact curves does not exists.

Observe that the above harmonic maps might drop rank. In this case we use theorem 3.1. Namely in some cases the Shafarevich variety might be a curve. Therefore we get a proper map from $\widetilde{X}$ to a disc so $\widetilde{X}$ is holomophically convex. Q.E.D.

As an immediate consequence of the proof of theorem 3.7 one obtains a version for intermediate covers.

THEOREM 3.9 ([**37**]). *Let $X$ be a compact Kähler surface. Let $\rho : \pi_1(X) \to G$ be a reductive Zariski dense representation for which the intermediate regular covering*

$$\widetilde{X} \longrightarrow \widetilde{X}/\ker \rho \longrightarrow X,$$

*does not have two ends. Then $\widetilde{X}/\ker \rho$ is holomorphically convex.*

## 4. Hodge theory

In this section we develop the Hodge theory necessary for the proof of the Shafarevich conjecture for linear groups. First we start with:

**4.1. Abelian mixed Hodge considerations.** To demonstrate how the mixed Hodge theory works for the Shafarevich conjecture we prove:

THEOREM 4.1 ([**38**]). *Let $X$ be a smooth projective variety with a virtually nilpotent fundamental group. Then the Shafarevich conjecture is true for $X$.*

Recall that $\pi_1(X)$ is nilpotent if its lower central series has finitely many terms. We will assume that $\pi_1(X)$ is infinite since if $\pi_1(X)$ is finite the universal covering of $X$ is compact and the statement is obvious. Since finite nonramified coverings do not affect holomorphic convexity we will assume that $\pi_1(X)$ is nilpotent.

The proof of the above theorem is based on the following lemma:

LEMMA 4.1. *If $Z$ is a compact nodal curve and $f : Z \longrightarrow X$ is a map to a variety such that the MHS on $H^1(X)$ is of weights $> 0$ then the map*

$$f_* : L(Z, x) \longrightarrow L(X, f(x))$$

*is trivial if and only if the map*

$$f^* : H^1(X, \mathbb{Q}) \longrightarrow H^1(Z, \mathbb{Q})$$

*is trivial. Here $L(Z, x)$ and $L(X, f(x))$ are the corresponding Lie algebras of the unipotent completions $\pi_{un}(Z, x)$ and $\pi_{un}(X, f(x))$ of the fundamental groups $\pi_1(Z, x)$ and $\pi_1(X, f(x))$ respectively and $x$ is a point in $Z$.*

Having this we prove:

LEMMA 4.2. *Let $X$ be a smooth projective variety with a nilpotent fundamental group $\pi_1(X)$. Then for any algebraic curve $Z \subset X$ the fact* $\operatorname{im}[H_1(Z, \mathbb{Q}) \longrightarrow H_1(X, \mathbb{Q})]$ *is trivial is equivalent to the fact that* $\operatorname{im}[\pi_1(Z) \longrightarrow \pi_1(X)]$ *is finite.*

PROOF. Since we can always find a nodal curve $\widetilde{Z}$ with a proper map $\widetilde{Z} \to Z$ inducing a surjection $\pi_1(\widetilde{Z}, \tilde{x}) \to \pi_1(Z, x)$ it follows from the previous lemma that the map

$$f_* : L(Z, x) \longrightarrow L(X, f(x))$$

is the zero map. Furthermore, if $\pi_1(X)$ is a torsion free nilpotent group then by definition it embeds in $\pi_{un}(X, f(x))$. It is easy to see that torsion elements of a nilpotent group generate a finite group and hence

$$\pi_1(X, f(x)) \longrightarrow \pi_{un}(X, f(x))$$

is an embedding up to torsion which proves the lemma. □

Now we reduce the case of arbitrary dimensional fiber of the Albanese map to the curve case by invoking the following:

LEMMA 4.3. *Let $F$ be a connected subvariety in $X$ then we can find a curve $Z \subset F$ such that $\pi_1(Z)$ surjects on $\pi_1(F)$.*

Now we finish the proof of Theorem 4.1. We start with the Stein factorization of the Albanese map for $X$

$$Alb : X \longrightarrow S \longrightarrow \mathrm{im}(Alb) \subset Alb(X).$$

The universal covering $\widetilde{S}$ of $S$ is a fiber product of the universal covering $\widetilde{Alb(X)}$ of $Alb(X)$ and $S$. Therefore the map

$$\widetilde{S} \longrightarrow \widetilde{Alb(X)}$$

is a finite map and since $\widetilde{Alb(X)}$ is a Stein manifold we conclude that $\widetilde{S}$ is a Stein space. It follows from the definition of the Albanese morphism that the fibers of the map

$$Alb : X \longrightarrow S$$

are all subvarieties $F$ in $X$ for which the map $H_1(F, \mathbb{Q}) \longrightarrow H_1(X, \mathbb{Q})$ is trivial. We will show that if $\pi_1(X)$ is nilpotent the fact that $H_1(F, \mathbb{Q}) \longrightarrow H_1(X, \mathbb{Q})$ is trivial implies that the map $\pi_1(F) \longrightarrow \pi_1(X)$ is finite.

We have shown this in Lemma 4.2 when $F$ is an algebraic curve. Now if $\dim_{\mathbb{C}}(F) > 1$ we apply Lemma 4.3 to find a curve $Z$ in $F$ such that $\pi_1(Z)$ surjects on $\pi_1(F)$. The argument of Lemma 2.2 implies that $\pi_1(F)$ goes to zero in $\pi_1(X)$ since $\pi_1(Z)$ goes to zero in $\pi_1(X)$. Observe that the curve $Z$ is also contained in the fiber $F$ of the map

$$Alb : X \longrightarrow S.$$

Therefore the map $H_1(Z, \mathbb{Q}) \longrightarrow H_1(X, \mathbb{Q})$ is also trivial.

To finish the proof of theorem 4.1 we need to observe that $S$ satisfies the conditions for being the Shafarevich variety of $X$, $S = \mathbf{Sh}(X)$. Namely

1) There exists a holomorphic map with connected fibers $X \longrightarrow S$, which contracts only the subvarieties $Z$ in $X$ with the property that $\mathrm{im}[\pi_1(Z) \longrightarrow \pi_1(X)]$ is finite.

2) $\widetilde{S}$ is a Stein space.

REMARK 4.1. The above theorem proves theorem 3.1 in case the group $L$ is nilpotent.

There is a differrent way of showing the above result which came from a discusion with Ramachandran. One can use the higher Albanese varieties. These are defined in the following way. Denote by $G_s$ the complex simply connected group $\pi_{un}(X)/\Gamma^{s+1}$. The corresponding Lie algebra $g_s$ has MHS. Denote by $F^0G_s$ the closed subgroup in $G_s$ group that corresponds to $F^0 g_s$. Here $\pi_{un}(X)/\Gamma^{s+1}$ is the prounipotent completion of $\pi_1(X)/\Gamma^{s+1}$ and $F^0G_s$ is closed subgroup in $G_s$ that preserves the filtration on $G_s$.

Since the group $\pi_{un}(X)/\Gamma^{s+1}$ is unipotent then as it is easy to see we have a free action of $G_s(\mathbb{Z})$ on $G'_s/F^0G'_s$. Here we take the subgroups of $G'_s$ and $F^0G'_s$ defined over $\mathbb{Q}$.

Therefore we can obtain an $\pi_1(X)$ equivariant proper holomorphic map $X \longrightarrow G'_s/F^0G'_s$.

One can try to prove a solvable variant of the theorem 1.1. The maximum we were able to say is how much the solvable coverings differ from the nilpotent ones. We show in [**38**] the following theorem that is a slight generalization of a theorem of Arapura and Nori.

THEOREM 4.2. *If $\Gamma$ is a quotient of a Kähler group $\pi_1(X)$ so that $\Gamma$ is a complex linear solvable group, then there are two possibilities - either $\Gamma$ is virtually nilpotent or $\pi_1(X)$ surjects onto the fundamental group of a curve of genus bigger than zero.*

The above theorem gives some restrictions on the Kähler fundamental groups. In particular it means that any group $\Gamma$ with no surjective homomorphisms to a nonabelian free group possessing a solvable linear quotient defined over $\mathbb{Q}$ that is not virtually nilpotent cannot be Kähler. In case the fundamental group is solvable itself then the theorem of Arapura and Nori tells that it is actually a virtually nilpotent group so the Shafarevich conjecture is true in this case as well.

In the above argument we use all unipotent representations. The next step is following the ideas of Grothendieck, Deligne [**12**] and Simpson [**54**], [**55**], [**56**], [**57**], [**58**], [**59**] to use linear representations - in all ranks over different fields. Namely we would like to work with the proalgebraic completion of $\pi_1(X)$. Here we start our

**4.2. Nonabelian mixed Hodge considerations.** In what follows we skip the explanations of some of the technical Hodge theoretic statements and concentrate on the applications to the Shafarevich conjecture. The results in this and the next section are a joint work with T.Pantev ( for details see [**33**], [**39**]).

Let $X$ be a quasi-projective variety, $x \in X$ a base point and let $\rho : \pi_1(X, x) \to G$ be a Zariski dense representation to a complex reductive group $G$.

DEFINITION 4.1. Let $X$ be smooth. We say that $\rho$ arises from a *complex variation of Hodge structures* (CVHS) of weight $k$ if the following conditions are satisfied:

**(i)** There exists a compactification $\overline{X}$ of $X$ with $X_\infty := \overline{X} - X$ - a divisor with simple normal crossings, such that $\rho$ is quasi-unipotent around $X_\infty$.

**(ii)** There exists a faithful representation $G \subset GL(V)$ such that for the corresponding $\mathcal{C}^\infty$ local system $(\mathcal{V}, D)$ one has a $\mathcal{C}^\infty$ decomposition $\mathcal{V} = \oplus_{r+s=k}\mathcal{V}^{r,s}$ with respect to which $D$ satisfies Griffiths' transversality condition

$$D : \mathcal{V}^{r,s} \to A^{0,1}(\mathcal{V}^{r+1,s+1}) \oplus A^{1,0}(\mathcal{V}^{r,s}) \oplus A^{0,1}(\mathcal{V}^{r,s}) \oplus A^{1,0}(\mathcal{V}^{r-1,s+1}).$$

**(iii)** There exists a $D$-parallel hermitian form on $\mathcal{V}$ which makes the above decomposition orthogonal and which on $\mathcal{V}^{r,s}$ is positive definite if $r$ is even and negative definite if $r$ is odd.

The condition on the monodromy at infinity in the definition is an a-priori condition on the representation $\rho$ and is somewhat independent from the conditions **(ii)** and **(iii)**. In the case of real VHS it is automatically satisfied in view of Borel's theorem. Furthermore, it is straightforward to check that the quasi-unipotency does not depend on the particular choice of a compactification $\overline{X}$.

The above definition can be extended in the usual way to the singular case by means of hypercoverings.

DEFINITION 4.2. For any variety $X$ we say that $\rho$ arises from a *complex variation of Hodge structures* (CVHS) of weight $k$ if there exists a smooth hypercovering $X_\bullet$ of $X$ such that for any element $p : Z \to X$ in $X_\bullet$ we have that $p^*\rho$ arises from a CVHS of weight $k$.

REMARK 4.2. For any morphism $f : X \to Y$ with $X$ - an arbitrary variety and $Y$ - smooth and projective and for any representation $\rho: \pi_1(Y, f(x)) \to G$ arising from a CVHS of weight $k$ the natural pullback representation $f^*\rho : \pi_1(X, x) \to G$ arises from a CVHS of weight $k$ by default.

Let now $X$ be any variety and let $\rho : \pi_1(X) \to G$ be a Zariski dense representation in a complex reductive group. The Malcev completion of $\pi_1(X, x)$ relative to $\rho$ is the projective limit

$$\hat{\pi}_\rho(X,x) := \varprojlim \left\{ \begin{array}{c} 1 \longrightarrow R^u M \longrightarrow M \longrightarrow G \longrightarrow 1 \\ \uparrow \nearrow_{\rho} \\ \pi_1(X,x) \end{array} \left| \begin{array}{l} M \text{ - complex} \\ \text{linear algebraic} \\ \text{group, } R^u M \text{ -} \\ \text{the unipotent} \\ \text{radical of } M. \end{array} \right. \right\}.$$

One has the following fundamental fact.

PROPOSITION 4.1. *Let $\rho : \pi_1(X, x) \to G$ be a Zariski dense representation arising from a CVHS $\mathbb{V} := (\mathcal{V}, D)$. Then the Lie algebra* $\operatorname{Lie} \hat{\pi}_\rho(X, x)$ *admits a canonical functorial complex MHS in the sense that it is equipped with three opposed filtrations $(W_\bullet, F^\bullet, \overline{F}^\bullet)$ such that for every $k$ the graded quotient $Gr_k(W_\bullet)$ corresponds to a CVHS of weight $k$ on $X$ which belongs to the full tensor subgategory $\mathcal{C}_{\mathbb{V}}$ in the category of CVHS on $X$ that is generated by $\mathbb{V}$ and the Tate twists.*

REMARK 4.3. The functoriality of the MHS in the Proposition 4.1 means that for any morphism of varieties $f : (Z, z) \to (X, f(z))$ the map induced on completions

$$\hat{\pi}_{f^*\rho}(Z, z) \longrightarrow \hat{\pi}_\rho(X, f(z))$$

is a morphism of MHS of type $(0, 0)$. Also one expects that for a smooth projective $X$ the weight filtration on $\operatorname{Lie} \hat{\pi}_\rho(X, x)$ is non-zero only in weights $\leq \operatorname{weight}(\mathbb{V})$.

The main geometric application of the MHS on $\operatorname{Lie} \hat{\pi}_\rho(X, x)$ for us will be the following lemma.

LEMMA 4.4. *Let $X$ be a smooth projective variety and let $f : (Z, z) \to (X, x)$ be a (based) morphism from a reduced curve $Z$. Let $\rho : \pi_1(X, x) \to G$ be a Zariski dense representation arising from a CVHS $\mathbb{V}$ of weight $k$. Then the map*

$$f_* : \hat{\pi}_{f^*\rho}(Z, z) \longrightarrow \hat{\pi}_\rho(X, x)$$

*is trivial if and only if the maps*

$$f^* : H^1(X, \mathbb{U}) \longrightarrow H^1(Z, f^*\mathbb{U})$$

*are trivial for every irreducible $\mathbb{U} \in \mathrm{Ob}(\mathcal{C}_\mathbb{V})$.*

The above lemma is an analogue of lemma 4.2 and is just a different guise of the strictness property for morphisms of relative completions.

Next we will need the following general statement

THEOREM 4.3. *Let $r$ be a local system underlying a complex variation of MHS. There is a MHS on the completion of the local ring $\mathcal{O}_r$ of the moduli space of all representation of fixed rank equal to those of $r$ as well as on every quotient ring $\widehat{\mathcal{O}}_r/m^k$. Moreover $\widehat{\mathcal{O}}_r/m^k$ is a mixed Hodge representation of the relative completion of $\pi_1(X)$ w.r.t. $r$.*

The second statement is a consequence of the fact that the weight filtration goes by powers of $m^k$. This was first conjectured by Deligne. The proof follows from the work of Hain [**25**]. In [**25**] the above statement is proved for a real VHS. Adding to $r$ its complex conjugate we put ourselves in a position to apply Hain's result. A different proof can be found in [**34**]. The idea belongs to Simpson, who has developed in [**59**] more general technique.

PROPOSITION 4.2. *Let $\pi_1(X)$ be a linear group. Then there exists a CVHS $\mathbb{V} \to X$ such that $\pi_1(X)$ embeds in the relative completion with respect to $\mathbb{V}$.*

PROOF. Let $\rho : \pi_1(X) \to GL(n, \mathbb{C})$ be a faithful representation. Denote by $R$ an irreducible component of the the space of all rank $n$ representations of $\pi_1(X)$ containing $\rho$. Every element $a \in \pi_1(X)$ can be interpreted as a $GL(n, \mathbb{C})$-valued function on $R$: $a : R \to GL(n, \mathbb{C})$, $\mu \mapsto \mu(a)$. Since $\rho$ is faithful such an $a$ is uniquely determined by its value at $\rho$. On the other hand according to Corollary 6.12 of [**56**] the Higgs bundle corresponding to the semi-simplification of $\rho$ can be deformed (in its irreducible component) to a Higgs bundle which is fixed under the action of $\mathbb{C}^\times$. We need to worry a bit about the change of the complex structure when we move from the moduli space of representations to the moduli space of Higgs bundles. It could be that we have a fixed under the action of $\mathbb{C}^\times$ CVHS on every component of the moduli space of Higgs bundles but not on every component of the moduli space of representations ( or moduli space of of local systems).

One can show that this does not happen, namely in every component the moduli space of semisimple local systems of rank $n$ on $X$ there exists a local system underlying a complex variation of Hodge structures. Indeed the moduli space $Z$ of $\lambda$-connections realizes the moduli of Higgs bundles as a deformation of the moduli space of semisimple local systems on $X$. The irreducible components of $Z$ are the closures of the products of the irreducible components of the moduli of local systems with $\mathbb{C}^\times$ and every such component intersects the fiber of $Z$ over 0 (the moduli of Higgs bundles) in a union of components. But the natural $\mathbb{C}^\times$-action on $Z$ restricts to the dilation action on the Higgs bundles and this yields the statement. We can thus conclude that $R$ contains a point $\mathbb{V}$ which underlies a complex variation of Hodge structures.

Assume now $a \in \pi_1(X)$ be an element that goes to the identity in the relative completion of $\pi_1(X)$ w.r.t. $\mathbb{V}$. Then due to Theorem 4.3 $a$ goes to the identity in $GL(n, \widehat{\mathcal{O}}_{\mathbb{V}}/m^k)$ for every $k$ and hence the function $a$ equals the identity in $GL(n, \widehat{\mathcal{O}}_{\mathbb{V}})$. In particular $a = e$ on $R$ and thus $\rho(a) = \rho(e)$ which is a contradiction. □

We are ready now to prove the Shafarevich conjecture for surfaces with linear fundamental groups.

**4.3. Linear fundamental groups.** The following results are part of a joint work with T. Pantev and M. Ramachandran [**33**], [**39**].

LEMMA 4.5. *Let $X$ be an algebraic surface. If $\pi_1(X)$ is linear there are no infinite chains in $\widetilde{X}$.*

PROOF. The definition of an infinite chain was given in remark 3.1. To check that $X$ does not contain infinite chains we need to show that for every reducible curve $Z = \cup Z_i \subset X$ for which $\mathrm{im}[\pi_1(Z_i \to \pi_1(X)]$ is finite it follows that $\mathrm{im}[\pi_1(Z) \to \pi_1(X)]$ is also finite. Let $Z = \cup Z_i$ be such a curve. Since $\pi_1(X)$ is linear it follows by Malcev's theorem that it is residually finite and hence we can find a finite index subgroup $\Gamma \subset \pi_1(X)$ which intersects trivially each of the finite groups $\mathrm{im}[\pi_1(Z_i) \to \pi_1(X)]$. Therefore replacing $X$ with the finite cover of $X$ defined by $\Gamma$ we may assume that all $\mathrm{im}[\pi_1(Z_i) \to \pi_1(X)]$ are trivial. We have shown in the previous section that if $\pi_1(X)$ is linear then there exists a CVHS $\mathbb{V}$ such that $\pi_1(X)$ embeds in the relative completion $\hat{\pi}_{\mathbb{V}}(X)$. In particular $\mathrm{im}[\pi_1(Z) \to \hat{\pi}_{\mathbb{V}}(X)]$ is the same as $\mathrm{im}[\pi_1(Z) \to \pi_1(X)]$. On the other hand according to lemma 4.4 the map

$$f_* : \hat{\pi}_{f^*\mathbb{V}}(Z, z) \longrightarrow \hat{\pi}_{\mathbb{V}}(X, x)$$

is trivial if and only if the maps

$$f^* : H^1(X, \mathbb{U}) \longrightarrow H^1(Z, f^*\mathbb{U})$$

are trivial for every irreducible $\mathbb{U} \in \mathrm{Ob}(\mathcal{C}_{\mathbb{V}})$. But every $\mathbb{U}$ is a complex variation of Hodge structures and hence $H^1(X, \mathbb{U})$ and $H^1(Z, \mathbb{U})$ have canonical mixed Hodge structures. The usual strictness statement for morphisms of MHS induced by morphisms of varieties implies then that

$$f^* : H^1(X, \mathbb{U}) \longrightarrow H^1(Z, f^*\mathbb{U})$$

will be trivial if

$$f^{\nu *} : H^1(X, \mathbb{U}) \longrightarrow H^1(Z^\nu, f^{\nu *}\mathbb{U})$$

is trivial where $f^\nu : Z^\nu \to X$ is the map from the normalization of $Z$. But $Z^\nu = \coprod Z_i^\nu$ and thus

$$H^1(Z^\nu, f^{\nu *}\mathbb{U}) = \oplus_i H^1(Z_i^\nu, f^{\nu *}\mathbb{U})).$$

Finally the assumption that $\mathrm{im}[\pi_1(Z_i^\nu) \to \pi_1(X)]$ is trivial yields that $H^1(X, \mathbb{U}) \to H^1(Z_i^\nu, f^{\nu *}\mathbb{U})$ is the zero map and we get the statement of the lemma. □

Now we move to the proof of the following theorem:

THEOREM 4.4. *Let $X$ be an algebraic surface. If $\pi_1(X)$ is linear then $\widetilde{X}$ is holomorphically convex.*

PROOF. The proof is combination of the theorems in the nilpotent case and in the reductive case through the above lemma.

Notice first that the validity of the Shafarevich conjecture for $X$ will be unaffected if we replace $X$ with a blow-up of an étale cover of $X$. We will often use this fact in the subsequent arguments.

According to proposition 4.2 there exists a complex variation of Hodge structures $\mathbb{V}$ such that $\pi_1(X)$ embedds in $\hat{\pi}_{\mathbb{V}}(X)$. Let $G$ be the monodromy group of $\mathbb{V}$, i.e. the Zariski closure of the monodromy representation of $\pi_1(X)$ corresponding to $\mathbb{V}$. The group $G$ is a complex reductive group and without a loss of generality we may assume that $\mathrm{im}[\pi_1(X) \to G]$ does not contain a finite index subgroup isomorphic to $\mathbb{Z}$:

CLAIM 4.1. *The CVHS $\mathbb{V}$ from proposition 4.2 can be chosen so that its monodromy group is not commensurable to $\mathbb{Z}$.*

PROOF. Indeed, if there exists a $\mathbb{Z} \subset G$ with $\mathbb{Z} \sim \mathrm{im}[\pi_1(X) \to G]$, then $G \cong \mathbb{C}^\times$ because the image of $\pi_1(X)$ is Zariski dense. Since $X$ is Kähler it follows that the rank of the abelianization of $\pi_1(X)$ is at least two. Hence there exists a unitary character $\chi : \pi_1(X) \to U(1) \subset \mathbb{C}^\times$ that is linearly independent with the character $\pi_1(X) \to G$. The local system $\mathbb{V} \oplus \mathbb{C}_\chi$ is again a complex variation of Hodge structures and we can consider the relative completion $\hat{\pi}_{\mathbb{V}\oplus\mathbb{C}_\chi}(X)$ instead of $\hat{\pi}_{\mathbb{V}}(X)$. By construction the image of $\pi_1(X)$ in the monodromy group $G \times \mathbb{C}^\times$ of $\mathbb{V} \oplus \mathbb{C}_\chi$ contains a subgroup of finite index isomorphic to $\mathbb{Z} \times \mathbb{Z}$. Furthermore $\hat{\pi}_{\mathbb{V}\oplus\mathbb{C}_\chi}(X) \to \hat{\pi}_{\mathbb{V}}(X) \times \hat{\pi}_{\mathbb{C}_\chi}(X)$ has a finite kernel and thus $\pi_1(X)$ almost embedds in $\hat{\pi}_{\mathbb{V}\oplus\mathbb{C}_\chi}(X)$. To finish the argument we just need to replace $X$ with a finite étale cover. □

Put $\rho : \pi_1(X) \to G$ for the monodromy representation of $\mathbb{V}$ and let

$$\mathbf{Sh}^{\ker \rho} : X \longrightarrow \mathbf{Sh}^{\ker \rho}(X)$$

be the corresponding Shafarevich morphism. Denote by $a : X \to a(X) \to \mathrm{im}(\mathit{Alb}) \subset \mathit{Alb}(X)$ the Stein factorization of the Albanese map for $X$. We have the following possibilities:

(i.) $\dim a(X) = 2$;
(ii.) $\dim a(X) < 2$ but $\dim \mathbf{Sh}^{\ker\rho}(X) = 2$;
(iii.) $\dim \mathbf{Sh}^{\ker\rho}(X) = 1$ and $\dim a(X) = 1$;
(iv.) $\dim \mathbf{Sh}^{\ker\rho}(X) = 0$.

The case $\dim \mathbf{Sh}^{\ker\rho}(X) = 1$ and $\dim a(X) = 0$ is excluded for the following reason. If $\mathbf{Sh}^{\ker\rho}(X)$ is a curve, then the representation $\rho$ factors through an orbigroup corresponding to a suitable orbistructure on $\mathbf{Sh}^{\ker\rho}(X)$. Therefore we can find a smooth connected branched cover $C$ of $\mathbf{Sh}^{\ker\rho}(X)$ and a blow up $X'$ of an étale cover of $X$ so that the representation $\rho : \pi_1(X') \to G$ factors through $\pi_1(C)$. On the other hand since $\dim \mathbf{Sh}^{\ker\rho}(X) > 0$ it follows that $\mathrm{im}[\pi_1(X) \to G]$ is infinite and since $\pi_1(X')$ has finite index in $\pi_1(X)$ we conclude that the genus of $C$ is bigger than zero. Thus, replacing $X$ by $X'$ we may assume that $\dim \mathbf{Sh}^{\ker\rho}(X) = 1$ implies $\dim a(X) = 1$.

Next we show that in each of the cases listed above $\widetilde{X}$ is holomorphically convex.

Indeed, in case (i.) this follows from lemma 4.5 and a theorem of T.Napier [**48**], Theorem 6.1, asserting that for a surface $X$ the universal cover $\widetilde{X}$ is holomorphically convex provided that the Albanese image is two dimensional and there are no infinite chains in $\widetilde{X}$. Case (ii.) follows essentially from theorem 3.9. Indeed, since due to claim 4.1 $\mathrm{im}[\pi_1(X) \to G]$ is not commensurable with $\mathbb{Z}$ the covering $X' = X/\ker\rho$ does not have two ends and hence is holomorphically convex by theorem 3.9. Furthermore the Cartan-Remmert reduction of $X'$ surjects onto a cover of $\mathbf{Sh}^{\ker\rho}(X)$ and is thus of dimension two. Finally, according to a theorem of T.Napier [**48**], Theorem 6.2, any covering of a non-compact holomorphically convex surface with two dimensional Remmert reduction is also holomorphically convex provided that the covering does not have any infinite chains. Therefore lemma 4.5 implies the holomorphic convexity of $\widetilde{X}$. For case (iv.) notice that $\dim \mathbf{Sh}^{\ker\rho}(X) = 0$ implies that $\mathrm{im}\,\rho$ is is a finite group and thus $\pi_1(X)$ is virtually nilpotent. Therefore $\widetilde{X}$ is holomorphically convex by theorem 4.1.

Consider now case (iii.). Denote for simplicity $\mathbf{Sh}^{\ker\rho}(X)$ by $C$ and let $\pi : X \to C$ be the factorizing map. By construction the map $\pi$ contracts precisely all subvarieties $Z \subset X$ with the property $\mathrm{im}[\pi_1(Z) \to \pi_1(X)] \lesssim \ker\rho$. Moreover the representation $\rho$ factors through $\pi_1(C)$. Let $F$ be a smooth fiber of $\pi$ and let $i : F \hookrightarrow X$ be the corresponding inclusion. Since both $X$ and $C$ are smooth it is not hard to see that $\ker[\pi_1(X) \to \pi_1(C)] = \mathrm{im}[\pi_1(F) \to \pi_1(X)]$. There are two possibilities for $F$:

(a) For all $\mathbb{U} \in \mathrm{Ob}(\mathcal{C}_{\mathbb{V}})$ the map on the level of cohomology

$$i^* : H^1(X, \mathbb{U}) \longrightarrow H^1(F, \mathbb{U}_{|F})$$

is trivial.

(b) There exists an irreducible $\mathbb{U} \in \mathrm{Ob}(\mathcal{C}_{\mathbb{V}})$ for which the map

$$i^* : H^1(X, \mathbb{U}) \longrightarrow H^1(F, \mathbb{U}_{|F})$$

is not identically zero.

In alternative (a) due to lemma 4.4 the image $\mathrm{im}[\pi_1(F) \to \pi_1(X)] = \mathrm{im}[\pi_1(F) \to \hat{\pi}_{\mathbb{V}}(X)]$ must be trivial and therefore $\pi : X \to C$ must be the Shafarevich morphism for $X$. Since the universal cover of $C$ is Stein we conclude that $\widetilde{X}$ is holomorphically convex. Let $\mathbb{U}$ be the CVHS satisfying alternative (b). Since $\rho$ factors through $\pi_1(C)$ there is a CVHS $\mathbb{W} \to C$ so that $\mathbb{U} = \pi^*\mathbb{W}$. Put $\Sigma \subset C$ for the discriminant locus of $\pi$ and let $C^0 = C \setminus \Sigma$, $X^0 = X \setminus \pi^{-1}(\Sigma)$ and $\pi^0 : X^0 \to C^0$.

Claim 4.2. *The complex variation of Hodge structures $\mathbb{W}^\vee \to C^0$ is a direct summand in the $\mathbb{Z}$-variation of Hodge structures $R^1\pi^0_*\mathbb{C}$.*

Proof. Since $\pi^0 : X^0 \to C^0$ is a locally trivial topological fibrartion with fiber $F$ the restriction-inflation exact sequence for cohomology with coefficients in a local system gives

$$0 \longrightarrow H^1(C^0, \mathbb{W}) \longrightarrow H^1(X^0, \pi^{0*}\mathbb{W}) \xrightarrow{i^*} H^1(F, \pi^{0*}\mathbb{W}).$$

By assumption $i^*$ is not identically zero and therefore the inclusion $H^1(C^0, \mathbb{W}) \hookrightarrow H^1(X^0, \pi^{0*}\mathbb{W})$ is not an isomorphism. On the other hand by the Lerray spectral sequence

$$0 \longrightarrow H^1(C^0, \mathbb{W}) \longrightarrow H^1(X^0, \pi^{0*}\mathbb{W}) \longrightarrow H^0(C^0, R^1\pi^0_*\pi^{0*}\mathbb{W}),$$

and hence by the projection formula there is a non-zero homomorphism $\mathbb{W}^\vee \to R^1\pi^0_*\mathbb{C}$. Since $\mathbb{W}$ is irreducible and the monodromy of $R^1\pi^0_*\mathbb{C}$ is reductive [**9**] we conclude that $\mathbb{W}^\vee$ splits off as a direct summand which proves the claim. □

This claim together with the fact that $\pi_1(C)$ is finitely generated imply that there is a number field $K/\mathbb{Q}$ so that in the equivalence class of the monodromy representation of $\mathbb{W}$ there exists a representative defined over $K$ and having integral traces. In other words there is a representation $\tau : \pi_1(C) \to \mathrm{GL}(r, K)$ that is conjugate to the monodromy representation of $\mathbb{W}$ and such that for every $\gamma \in \pi_1(C)$ the number $\mathrm{Tr}(\tau(\gamma))$ belongs to the ring of integers of $K$. Furthermore by an argument of M.Larsen [**54**], Lemma 4.8 one can choose the realization $\tau$ so that $K$ is a purely imaginary quadratic extension of a totally real field. Also for each embedding $\sigma : K \to \mathbb{C}$ we get a local system $\mathbb{W}_\sigma$ which is also a direct summand in $R^1\pi^0_*\mathbb{C}$ since the monodromy group of the latter is defined over $\mathbb{Z}$. In particular $\mathbb{W}_\sigma$ underlies a CVHS of weight one for each $\sigma$. Adding all of the $\mathbb{W}_\sigma$'s in the same manner as in the proof of Theorem 5 of [**54**] we obtain a polarized $\mathbb{Q}$-variation of Hodge structures $\mathbb{S}$ whose monodromy group preserves an integral lattice $\mathbb{S}_{\mathbb{Z}}$. Since $\mathbb{W}$ is a restriction from a local system on $C$ it follows that $(\mathbb{S}, \mathbb{S}_{\mathbb{Z}})$ is a restriction from a VHS on $C$. By construction the weights of $\mathbb{S}$ can be chosen to be $(1,0)$ and $(0,1)$ and hence $\mathbb{S}/\mathbb{S}_{\mathbb{Z}}$ is a smooth family of abelian varieties over $C$. Denote the total space of this family by $A$. We have constructed the abelian scheme $A/C^0$ as a quotient of an abelian scheme of the form $\mathrm{Alb}(X^0/C^0) \times_{C^0} \mathrm{Alb}(X^0/C^0) \times_{C^0} \ldots \times_{C^0} \mathrm{Alb}(X^0/C^0)$ where $\mathrm{Alb}(X^0/C^0)$ is the relative albanesian for the smooth family of curves $\pi^0 : X^0 \to C^0$. Therefore the relative Albanese map provides a rational map

$$\begin{array}{ccc} X & \overset{f}{\dashrightarrow} & A \\ & \searrow \quad \swarrow & \\ & C & \end{array}$$

which we may assume a morphism since replacing $X$ by a blow-up of $X$ does not affect the holomorphic convexity of the universal cover. The fact that the Abel-Jacobi image of a curve generates the whole Jacobian implies that the smooth fibers of $\pi$ are not contracted to points in $A$ and thus the Stein factorization $Q$ of the image of $f$ is two dimensional. The universal cover of $Q$ is a finite cover of a closed subset in a product of a disk and a vector space and is therefore Stein. Finally, an irreducible curve $Z \subset X$ will be contracted by $f$ only if it is contained in a fiber of $\pi$ and if $H^1(X, \pi^*\mathbb{W}) \to H^1(Z, \pi^*\mathbb{W})$ is zero. Together with lemma 4.4 this implies that $\mathrm{im}[\pi_1(Z) \to \pi_1(X)]$ must be trivial. Consequently $Q$ is the Shafarevich variety for $X$ and $\widetilde{X}$ is holomorphically convex. This finishes the proof of the theorem. □

REMARK 4.4. The proof of the previous theorem can be adapted to yield a proof of theorem 3.1. Let $X$ be any smooth projective variety and let $\rho : \pi_1(X) \to GL(n, \mathbb{C})$ be an infinite linear representation. Due to Proposition 4.2 there exists a CVHS $\mathbb{V}$ with the property $\ker[\pi_1(X) \to \hat{\pi}_{\mathbb{V}}(X)] = \ker(\rho)$. Consider the Shafarevich variety $Y$ for the reductive local system $\mathbb{V}$. In the same way as above we can construct a $\mathbb{Z}$-VHS on $Y$ and a morphism $f$ from $X$ to the corresponding family of abelian varieties. Again Lemma 4.4 implies that $f$ contracts precisely the subvarieties in $X$ whose fundamental group is commensurable with $\ker(\rho)$. This fibration is exactly the Shafarevich variety from theorem 3.1.

As an immediate consequence of the proof of the above theorem we get:

THEOREM 4.5. *Let $X$ be an algebraic surface such that $\rho : \pi_1(X) \to G$ is a Zariski dense representation to a complex linear group $G$. Then $\widetilde{X}$ admits nonconstant holomorphic functions.*

There are known examples of surfaces with linear fundamental groups. By a theorem of Kollár (theorem 6.6.3 [**40**]) every hyperplane section of family of abelain varieties over Riemann surface has this property. M.Ramachandran informs me that there are examples of linear fundamental groups that are nontrivial extensions of two step nilpotent group by the fundamental group of a Riemann surface.

We will show in the last section how one can try to push the above methods to get the Shafarevich conjecture for surfaces with residually finite fundamental groups.

We would like to mention the following interesting direction which seems worth investigating. In the way we have assigned a classifying space of all unipotent variations of MHS to the prounipotent completion we can assign a classifying space to the relative completion with respect to a fixed CVHS. This space is nothing else but a fibration $F$ of nillmanifolds over the symmetric space $S$ that corresponds to the CVHS. In case we have just an usual representation that is a CVHS we have a harmonic map $h : \widetilde{X} \to S$. The question is what is the analog of $h : \widetilde{X} \to S$ for $\widetilde{X}$ and $F$. One can imagine that this is some kind of extensions of harmonic maps. In any case it looks like we have much more chances to embed the fundamental group of $X$ in its relative completions so the above question seems worth pursuing.

REMARK 4.5. Some recent unpublished results of Eliashberg and Gompf allow us to give topological characterization of the universal coverings of surfaces with a large linear fundamental group - they are bounded by a handelbody and have a Morse functions with indexes at their critical values less or equal to two.

## 5. Some advanced group theory

In the previous section we have tried to explain why the Shafarevich conjecture looks plausible for algebraic surfaces with residually finite fundamental groups. The main reason is the Hodge structure we have constructed on the proalgebraic completions of these fundamental groups. At the moment to us the case of surfaces with nonresidually finite fundamental groups seems out of reach. Recently some examples of interesting fundamental groups were suggested by F.Bogomolov and L. Katzarkov [**3**]. These examples are potential examples of nonresidually finite fundamental groups. It was shown in [**3**] that these examples are related to some group theoretic conjectures whose positive answer will give a couterexample to the Shafarevich conjecture. For completeness we give a brief description of the construction here.

We first describe the construction locally. We begin with some well known results on degeneration of curves. Let $X_D$ be a smooth complex surface fibered over a disc $D$. We assume that fibers over a punctured disc $D^* = D - 0$ are smooth curves of genus $g$ and the projection $t : X_D \to D$ is a complex Morse function. In particular the fiber $X_0$ over $0 \in D$ has only quadratic singular points and it has no multiple components. Denote by $P$ the set of singular points of $X_0$ and by $T : \pi_1(X_t) \to \pi_1(X_t)$ the monodromy transformation acting on fundamental group $\pi_1(X_t)$ of a generic fiber $X_t$. This monodromy action can be described in terms of Dehn twists. Obviously it induces an action on $H^1(X_t, \mathbb{Z})$. The following proposition describes completely the topology of $X_D$ and the projection $t : X_D \to D$.

We define a natural topological contraction $cr : X_D \to X_0$. Observe that $cr^{-1}(P)$ is an union $S$ of circles $S_i$. The restriction of $cr$ on $X_t - S$ is an isomorphism.

DEFINITION 5.1. We will call the loop $S_i$ in a first homotopy group of $X_t$ a geometric vanishing cycle.

We need the following easy lemma.

LEMMA 5.1. *1)The monodromy transformation $T$ acts via a unipotent transformation $T_H$ on the homology group $H_1(X_t, \mathbb{Z})$.*
*2) $(1 - T_H)^2 = 0$.*
*3) $(1 - T_H^N) = 0(modN)$ for any $N$.*

The geometric vanishing cycles consist of two different types:
1) First type includes homologically nontrivial vanishing classes.
They are all equivalent under the mapping class group $Map(g)$. The latter can be described either as group of connected components of the orientation preserving homeomorphisms of the Riemannian surface of genus $g$ or as a similar group of exterior automorphisms of the group $\pi_g$. The vanishing cycle from this class is a primitive element of $\pi_g$ which means it can be included into a set of generators of $\pi_g$ satisfying standard relation defining the fundamental group of the curve.
2) Second class consists of elements in $\pi_g$ which are homologous to zero. Any vanishing cycle of this type cuts the Riemann surface $X_t$ into two pieces and the number of handles in these pieces is the only invariant which distinguish the type of a cycle under the action of $Map(g)$.

DEFINITION 5.2. We shall denote the vanishing cycles of the first type as $NZ$-cycles and the second type by $Z$-cycles.

Vanishing $Z$-cycles correspond to the singular points of the singular fiber which divide this fiber into two components and $NZ$-cycles to the ones which don't. Each $Z$-cycle defines a primitive element in the center of the quotient $\pi_g/[[\pi_g, \pi_g], \pi_g]$, while each $NZ$-cycle defines a primitive element in the abelian quotient $\pi_g/[\pi_g, \pi_g] = \mathbb{Z}^{2g}$. Assume now that we made a change of the variable $t = u^N$ and consider the induced family of curves over $D$ with a coordinate $u$. Denote the resulting family as $X_U$. It is a singular surface and the singular set can be identified with the set $P$ of the singular points of the fiber $X_0$. The new monodromy transformation $T^U = T^N$ and therefore acts trivially on $H_1(X_t, \mathbb{Z}_N), t \neq 0$ by the lemma 5.1.

LEMMA 5.2. *The surface $X_U$ contracts to the central fiber $X_0$.*

Indeed the fiberwise contraction of $X_D$ to $X_0$ can be lifted into a contraction of $X_U$.

REMARK 5.1. The fundamental group $\pi_1(X_D - P) = \pi_1(X_D) = \pi_1(X_0)$ since the singular points of the fiber $X_0$ are nonsingular points of $X_D$. The analogous statement is not true however for $X_U$.

The following theorem follows from lemma 5.1 and plays an essential role in our construction. The proof can be found in [**3**].

THEOREM 5.1. *The fundamental group $\pi_1(X_U - P)$ is equal to the quotient of $\pi_1(X_t) = \pi_g$ by a normal subgroup generated by the elements $s_i^N$.*

Let us denote by $G_X$ the fundamental group of $X_U - P$ and by $\widetilde{X}$ its universal covering. The description of the group $G_X$ can be obtained in purely geometric terms. Namely the vanishing cycles $S_i$ don't intersect and therefore we can first contract $Z$-cycles to obtain the union of smooth Riemannian surfaces $X_i$ with normal intersections. The graph corresponding to this system of surfaces is tree since every point corresponding to $Z$-cycle splits it into two components. Remaining $NZ$-cycles lie on different surfaces $X_j$ and constitute a finite isotropic subset of primitive elements in $\pi_1(X_j)$. The vanishing cycle $S_i$ which contracts to the point of $X_j$ defines a map of a free group onto $\pi_1(X_j)$ with $S_i$ corresponding to standard relations in $\pi_1(X_j)$.

We have the following lemma.

LEMMA 5.3. *If $N > 2$ then the group $G_X$ has a natural surjective projection on the quotient group $\pi_g/[[\pi_g, \pi_g], \pi_g]$ with additional relation $x^N = 1$. If $N = 2$ then $G_X$ maps surjectively onto the group $UC_g$, which is a central extension of $\mathbb{Z}_2^{2g}$ by $\mathbb{Z}_2$ and the images of all $s_i$ have order 2.*

PROOF. See [**3**]. □

Here we start to develop the idea of the construction. Namely we show that we can work with open surfaces and get results concerning the universal coverings of the closed surfaces. The advantage is that we get bigger variety of fundamental groups.

THEOREM 5.2. *There is a natural $G_X$-invariant imbedding of $\widetilde{X}$ into a smooth surface $\widetilde{X}_U$ with $\widetilde{X}_U/G_X = X_U$. The complement of $\widetilde{X}$ in $\widetilde{X}_U$ consists of a discrete subset of points.*

PROOF. See [**3**]. □

Now we globalize the construction. We will show that we can do the construction described above globally in a compact surface.

Let $X$ be smooth surface with a proper map to a smooth projective curve $C$. We assume that the map $f : X \to C$ is described locally by a set of holomorphic Morse functions. Thus the generic fiber is a smooth curve $X_t, t \in C$ of genus $g$. As in the previous section we denote the fundamental group of $X_t$ by $\pi_g$. Abusing the notations now we denote by $P$ the set of all singular points of the fibers and by $P_C$ the set of points in $C$ corresponding to the singular fibers. Thus $f(P) = P_C$ and all the points in $P$ are singular double points. The principal difference of the global situation lies in the presence of the global monodromy group which is the image of $\pi_1(C - P_C)$ in the mapping class group $Map(g)$. We denote this group by $M_X$.

Let us select now an integer $N$ and consider a base change $h : R \to C$ where $R$ such that the map $h$ is $N$-ramified at all the preimages of the points from $P_C$ in $R$. Now consider a surface $S$ obtained via a base change $h : R \to C$. We have the finite map $h' : S \to X$ defined via $h$ and the projection $z : S \to R$ with a generic fiber $S_t = X_{h(t)}$.

The surface $S$ is singular with the set of singular points equal to $h^{-1}(P) = Q$ and the set of singular fibers over the points of $h^{-1}P_C = P_R$. The monodromy group $M_S$ of the family $S$ is subgroup of finite index of the group $M_X$.

THEOREM 5.3. *The fundamental group $\pi_1(S - Q)$ is an extension of $\pi_1(R)$ by the quotient of $\pi_g$ by a normal subgroup generated by the orbits of $M_S(s_i^N)$.*

PROOF. Indeed we have a natural surjection of $\pi_g$ on the kernel of the projection onto $\pi_1(R)$ since there are no multiple fibers in the projection $z$. A standard Van Kampen type of an argument reduces all relations to the local ones and finishes the proof of the theorem.

□

Now we move to the second step of our construction, modifying $S - Q$ so that we get a smooth compact surface $S^N$ with almost the same fundamental group. Let us assume that $N \geq 2$.

LEMMA 5.4. *There exists a smooth projective surface $S^N$ with a finite map $f : S^N \to S$ such that the image of the homomorphism $f_* : \pi_1(S^N) \to \pi_1(S - Q)$ is a subgroup of finite index in $\pi_1(S - Q)$.*

PROOF. We have allready shown that there exists a projection of $\pi_1(X_g)$ to a finite group $UC_g$ which is invariant under $Map(g)$. This projection factors through the fundamental group of a small neighborhood of any singular fiber of $S - Q$. Therefore, if the map $h : S \to R$ has a topological section we obtain a finite fiberwise covering of $S$ which is ramified only over the singular set $Q$. Hence the resulting surface coincides locally with the smooth surface described in theorem 5.2. We denote this covering by $S^N$. As it follows from theorem 5.2 this covering is smooth and the map $f$ is finite.

Since the map $f$ is finite the preimage $f^{-1}(Q)$ consists of a finite number of smooth points and therefore $\pi_1(S^N) = \pi_1(S^N - f^{-1}(Q))$. This also proves the fact that the image of the homomorphism $f_* : \pi_1(S^N) \to \pi_1(S - Q)$ is a subgroup of finite index in $\pi_1(S - Q)$. Observe that we can always get a topological section $h : S \to R$ via additional base change.

□

This is the end of the construction and we suggest some possible applications. To find a couterexample to the Shafarevich conjecture we need to control the existence of an infinite connected chain of compact curves in the universal covering of $S^N$. In other words we need to control the image of the fundamental groups of the irreducible componnents of the reducible singular fibers in the fundamental group of the whole surface $S^N$. The above construction shows that we need to look at the image of the fundamental groups of the open irreducible componnents of the reducible singular fibers in the fundamental group of the open surface $S - Q$.

Following the discussion in section 2 ( page 10 ) we can construct a smooth family of curves $S \to R$ with the following properties:

1) The fundamental group of $R_0$ surjects onto the fundamantal group of the moduli space $M_g^L$ - the mapping class group $Map(g)$. Here $R_0$ is the open curve that parametrizes only the smooth fibers.

2) $S \to R$ has a section $s : R \to S$.

3) All singular fibers have singularities that are ordinary double points.

It easy to arrange so that in the above family of curves of genus $g = 2k$ there is a fiber which consists of two components each of genus $k$ that intersect in one point.

DEFINITION 5.3. Denote by $\pi_g/(x^N = 1)$ the quotient of the fundamental group of a Riemann surface of genus $g$ by the group generated by the $N-th$ powers of the images of the primitive elements in $\pi_g$ under the action of a subgroup of a finite index in $Map(g)$.

The free group on $2k$ generators $\mathbb{F}^{2k}$ embeds in a stadard way in the fundamental group of a Riemann surface of genus $2k$ punctured at one point.

DEFINITION 5.4. We will denote by $P(2k)^N$ the quotient of $\mathbb{F}^{2k}$ by the subgroup generated in $\mathbb{F}^{2k}$ by the orbits of the $N$ powers of embedded loops in the fundamental group of a Riemann surface of genus $2k$ punctured in one point.

Applying the above construction we get that the image of the fundamental group of every component in the fundamental group of $S - Q$ is equal to $P^N(k)$. The image of the fundamental group of a generic fiber in the fundamental group of $S - Q$ is equal to $\pi_g/(x^N = 1)$. We formulate the following question.

QUESTION 5.1. Are there such a $2k$ and $N$ such that $P^N(2k)$ is a finite group and $\pi_{2k}/(x^N = 1)$ is an infinite group?

If the answer of the above question is affirmative then we get a contraexample to the Shafarevich conjecture.

Observe that irreducible components in the construction above could be made of different genus. This gives us much bigger variety of examples. Each of these examples produces hard ( according to Olshanskii and Zelmanov) group theoretic question simillar to the one we have formulated above. The Shafarevich conjecture in particular implies that all these questions have negative answer. It is also worth noticing that the surfaces $S^N$ described above most probably have nonresisually finite fundamental groups as it was conjectured by Zelmanov.

We believe that the constructions we have discuss might have other applications. One can try to explore a combination of this construction with the famous examples of Serre's of conjugate varieties with different fundamental groups ( see e.g. [**52**], [**13**] as well as the last section of this paper). For better understanding of the monodromy action one should probably try to define and study the Picard-Lefschetz transformation on the fundamental group and a nonabelian variant of the Clemens-Schmidt sequence.

One can also try to use simillar ideas to construct examples of fundamental groups of projective or quasiprojective varieties without any infinite linear representations (see e.g. [**1, 2**]).

## 6. Further questions

This section contains some speculations and some open questions for future research.

**6.1. Nonabelian Hodge theory.** We start with the sketch of a program that could lead to a proof of the Shafarevich conjecture for surfaces with residually finite fundamental groups. This program is closely related to some very basic results in the nonabelian mixed Hodge theory. The main problem at the moment is that residually finite groups without infinite linear complex representations are invisible to Hodge theoretic considerations.

We would like to formulate some more nonabelian Hodge theoretic statements which might help to overcome this problem.

First we define (see [**34**]) a non-abelian mixed Hodge structure on a finitely generated group $\Gamma$. Denote by $\mathfrak{L}$ the free Lie algebra over $\mathbb{C}$ on generators $D^{i,j}$ for $i, j < 0$. There is a unique bigrading $\mathfrak{L} = \oplus \mathfrak{L}^{i,j}$ having the property $D^{i,j} \in \mathfrak{L}^{i,j}$. The $m$-weight subspace $W_m(\mathfrak{L}) := \oplus_{i+j \leq m} \mathfrak{L}^{i,j}$ for this bigrading is obviously an ideal in $\mathfrak{L}$ and the quotient $\mathfrak{L}/W_m$ is a nilpotent Lie algebra. In particular we get a projective system of unipotent algebraic groups $\exp(\mathfrak{L}/W_m(\mathfrak{L}))$. Denote by $\mathfrak{U}$ its projective limit. The group $\mathfrak{U}$ is the unipotent tail of the Galois group of the Tannakian category of all complex mixed Hodge structures [**13**]. The natural action of $\mathbb{C}^\times$ on $\mathfrak{L}$ for which the element $\lambda$ acts as a multiplication by $\lambda^{-i-j}$ on $\mathfrak{L}^{i,j}$ clearly preserves the filtration $W_\bullet$ and thus induces an action of $\mathbb{C}^\times$ on $\mathfrak{U}$. Denote by $U(1)^\delta$ the group $U(1)$ considered as a discrete group.

DEFINITION 6.1 (NMHS). Non-abelian mixed Hodge structure will be a pair $\Gamma \subset \mathcal{G}$ consisting of a pro-algebraic group $\mathcal{G}$ defined over $\mathbb{R}$ together with a finitely generated subgroup $\Gamma \subset \mathcal{G}_{\mathbb{R}}$ and an action of the group $U(1)^\delta \ltimes \mathfrak{U}$ on $\mathcal{G}_{\mathbb{R}}$ so that:

1. $\Gamma$ is Zariski dense in $\mathcal{G}$;
2. $(U(1)^\delta \ltimes \mathfrak{U}) \times \Gamma \longrightarrow \mathcal{G}^{\text{an}}$ is continuous, where $\mathcal{G}^{\text{an}}$ is $\mathcal{G}$ considered as pro-analytic group.
3. the element $-1 \in U(1)$ acts as a Cartan involution on the pro-reductive part $\mathcal{G}/R^u\mathcal{G}$ of $\mathcal{G}$.

REMARK 6.1. This definition is a simplification of the one given in [**34**]. It is modified to serve the purposes of this paper. It has bad properties if one needs to work with deformations. Similarly the pro-algebraic completion is not the best object to work if one needs to consider deformations. A better way is suggested in [**34**] where instead of working with the proalgebraic completion we work with the completion with respect to all linear groups with coefficients in finitely generated algebras over $\mathbb{C}$.

Denote by $\varpi^{alg}$ the algebraic completion of $\pi_1(X)$ and by $\varpi^{hod}$ the completion of $\pi_1(X)$ with respect to all complex variations of mixed Hodge structures. The following proposition is proven in [**34**]. The proof is based on ideals of Deligne and Simpson [**54**], [**55**], [**56**], [**57**].

PROPOSITION 6.1. *$\varpi^{alg}$ and $\varpi^{hod}$ admit NMHS and the images* $\mathrm{im}[\pi_1(X) \to \varpi^{alg}]$ *and* $\mathrm{im}[\pi_1(X) \to \varpi^{hod}]$ *are the same.*

This proposition should be considered as a nonabelian strictness statement. We hope that a modification of this statement will allow us to use the full strength of the result of Lasell and Ramachandran we have mentioned above. In this way one could hope to prove the Shafarevich conjecture for surfaces with fundamental groups that do not have infinitely many elements of finite order modulo conjugation. As a preparatory step towards such a proof one needs to study deformations of non-abelian Hodge structures.

Simpson [**55**], [**56**], [**57**] defined a nonabelian variant of the Gauss-Manin connection. He showed that for any smooth projective morphism $f : X \longrightarrow S$ the corresponding family of relative moduli space of local systems has a natural structure of crystal over $S$, i.e. is a $\mathcal{D}_S$-scheme in a natural way. This structure is the non-abelian Gauss-Manin connection (NGMC).

We formulate the Nonabelian version of the $(p, p)$ - class theorem for a very simple case a smooth fibration of curves over a curve.

THEOREM 6.1. *Let $X_t$ be a smooth family of curves and let $M_t$ be the corresponding family of moduli spaces of representations. Let $s$ be a section in $M_t$ which is CVHS in one point and $s$ is a flat section with respect to NGMC . Then $s$ gives a CVHS for any $t$.*

We also formulate the following version:

THEOREM 6.2. *Let $X_t$ be a smooth family of curves and let $M_t$ be the corresponding family of moduli spaces of representations. Let $s$ be a section in $M_t$ which is an unitary connection in one point and $s$ is a flat section with respect to NGMC. Then $s$ gives an unitary connection for any $t$.*

The proof of these theorems can be found in [**34**].

Next, we formulate a conjectural nonabelian version of the invariant cycle theorem for a very simple case a smooth fibration of curves over a curve.

CONJECTURE 6.1. Let $X_t$ be a smooth family of curves and let $M'_t$ be the corresponding family of moduli spaces of Higgs bundles. Then any invariant under the monodromy family is a subfamily of Hyperkähler varieties in $M'_t$.

An interesting question will be if this family is a subfamily of integrable systems. Some version of the nonabelian Hodge conjecture will be to ask whether this integrable subsystem comes from a curve to which the general fiber of our family $X_t$ maps finite to one.

Even if the answer of the Shafarevich conjecture turns out to be negative we believe that the Hodge theoretic methods we have discussed above could help us to prove the following:

CONJECTURE 6.2. *Let $X$ be a compact Kähler manifold such that $\rho : \pi_1(X) \to G$ is a Zariski dense representation to a complex linear group $G$. Then the universal covering of $X$ admits nonconstant holomorphic functions.*

At the moment it seems that using Hodge theoretic methods one can show that if $X$ is a smooth algebraic surface with large algebraic fundamental group with finite torsion then its universal covering is holomorphically convex.

The above construction of relative completion of $\pi_1(X)$ has other applications. The results of Simpson's (see [**54**] ) imply that the group $L$

$$1 \to K \to L \to G \to 1.$$

where $L$ is a split extension of an arbitrary group $K$ and $G$ is a lattice in a group of non Hodge type e.g. $SL(n, \mathbb{Z}), n > 2$ cannot be a Kähler group.

The prounipotent radical of any relative completion of $\pi_1(X)$ with respect to any variation of Hodge structures is quadratically presented (meaning that the corresponding Lie algebra is quadratically presented, see e.g. [**7**] or [**25**]). Examples of groups with these properties were constructed by Kapovich and Millson [**32**]. Their construction is based on the study of the moduli space of linkages. Here are some other examples we would like to suggest:

Take $G$ to be a lattice in a high rank group of Hodge type. Then the direct product of $G$ with any nonquadratically presented nilpotent Lie group is not Kähler. The reason is that the prounipotent radical of the relative with respect to $G$ completion of the direct product is $G$ itself since it has no infinite unipotent representations.

One can try to modify this example. Let $G$ act on some vector space $V$ ( one needs that $H^1(G, \rho) > 0$, where $\rho$ a representation of $G$ in $End(V)$). Consider the free Lie algebra $L = Lie(V)$ on $V$. Consider also the nilpotent Lie algebra $n = L/Lk, k > 3$ and $Lk$ is the homogeneous ideal generated by $k$ powers. We can find a nilpotent lattice $N$ in $n$ using the Baker-Campbell-Hausdorff formula. Then the semidirect product

$$1 \to N \to S \to G \to 1.$$

with the action described above will not be a Kähler group since the corresponding prounipotent radical of $S$ with respect to $G$, $N$ is not quadratically presented.

The above example can be used to construct non Kähler groups in a different way too. One can easily observed that the prounipotent radical of

$S$ has a MHS and all odd weight pieces in the prounipotent radical must have even dimension over $\mathbb{Q}$. This is probably well known for the usual prounipotent completion.

**6.2. $L_2$ methods.** Recently T. Napier and M. Ramachandran [**50**] have developed a technique that can be used to prove the following very strong generalization of the Lefschetz hyperplane section theorem. Let $Y \subset X$ be two varieties and $\mathcal{F}$a coherent sheaf on $X$. We denote by $\widehat{Y}$ the formal neighborhood of $Y$ in $X$ and by $\widehat{\mathcal{F}}$ the completion of $\mathcal{F}$ along $Y$. The following theorem follows from [**50**].

THEOREM 6.3. *(Napier, Ramachandran) Let $X$ be a smooth quasiprojective surface and $Y$ a curve in $X$. Assume that there exists an ample line bundle $L \to X$ such that $H^0(\widehat{Y}, \widehat{L}) < \infty$. Then* $\mathrm{im}[\pi_1(Y) \to \pi_1(X)]$ *has finite index in $\pi_1(X)$).*

One can try to use the above theorem in some of the examples described in section 5. Unfortunately for interesting curves $Y$ in these examples, the union of the base curve in $S - Q$ with $K_1$ and $K_2$, one gets that $H^0(\widehat{Y}, \widehat{L}) = \infty$.

It is interesting to relate results describedin this paper to the results of Gromov, Henkin and Shubin ([**22**]) where they study the existence of nonzero $L_2$ holomorphic functions and forms on strictly pseudoconvex domains. A combination of their result with ours implies the following:

PROPOSITION 6.2. *Let $X$ be a smooth projective surface with linear fundamental group and $L$ be the total space of a $\mathbb{C}^\times$ fibration over its universal covering. Then $L$ admits nonzero $L_2$ holomorphic forms.*

It is interesting if one can get existence of nonzero $L_2$ holomorphic functions on $L$ and to try combine the methods. Namely to try to define what is a $L_2$ variation of Hodge structures. It is tempting to say that the $L_2$ theory can be used as an analog of the Hodge theory for residually finite groups. The condition that corresponds to the residual finiteness is the big growth condition. Some preliminary steps in this direction were taken in [**39**] where the Shafarevich conjecture was proved for Kähler hyperbolic threefolds under additional conditions.

A conjectural universal approach to all of these statements is discussed in [**36**]. The idea is to look for a differential graded algebra $A_X$ computing the cohomology with coefficients in all local systems over $X$ together with their Hodge structure. For example, when the homotopy type of $X$ is nilpotent, then such a dga exists and is the one coming from the prounipotent completion of $\pi_1(X)$. The dga $A_X$ will have many other applications. A formality statement for $A_X$, for instance, will imply that the prounipotent radical of the relative completion of $\pi_1(X)$ with respect to some variation of Hodge structure is quadratically presented. This also follows form the work of Hain [**25**]. The latter can be used to obtain restrictions on Kähler fundamental groups.

**6.3. Conjugate varieties.** The dga $A_X$ seems to be the maximal object that stays unchanged under conjugations. This opens another sequence of questions that relate the Shafarevich conjecture to the action of $\mathrm{Aut}(\mathbb{C}/\mathbb{Q})$ on the set of all projective varieties. The importance of $A_X$ for the geometry can be seen on some examples. It gives a kind of nonabelian Torelli theorem. For example if $X$ is a smooth projective variety with the same fundamental group as a ball quotient and with the same cohomology for all local systems then $X$ is actually byholomorphic to the ball quotient.

Let $X$ be a variety defined over finitely generated extension $k$ of $\mathbb{Q}$. For each extension of scalars $s : k \to \mathbb{C}$ we get a different variety $X_s$. We call these varieties conjugate varieties. It is natural to expect that $A_{X_s}$ is independent of $s$.

The following question came out from a discussion with D. Reed. Let $X_{s'}$ and $X_{s"}$ are two conjugate varieties. What can one one say about the topology of $\widetilde{X_{s'}}$ and $\widetilde{X_{s"}}$? Are $\widetilde{X_{s'}}$ and $\widetilde{X_{s"}}$ topologically different even in the case when $\pi_1(X_{s'})$ and $\pi_1(X_{s'})$ are isomorphic?

One can ask a different question. Let $X_{s'}$ and $X_{s"}$ are two conjugate varieties that are holomorphic families of isomorphic fibers over the same base. Are the monodromy actions the same? It seems that combining the ideas from section 5 with the works of Deligne [**9**], Reed [**51**] and Serre [**52**] one can construct interesting examples in this direction.

**6.4. Symplectic versus Kähler geometry.** In their seminal paper [**16**] Eliashberg and Gromov define the symplectic analogue of a Stein manifold, a Weinstein manifold. They show that a Weinstein manifold admits a compatible Stein structure. In this direction one can try to complete the following program:

QUESTION 6.1. Let $X_0$ be a variety with a large fundamental group. Consider a small deformation of $X$. Can one always find a small neighborhood $U$ around $X$ such that all varieties in this neighborhood $U$ are with large fundamental groups?

Assume further that the universal covering of $X_0$ is Stein.

QUESTION 6.2. Are the universal coverings of all varieties in $U$ Stein?

In another direction one can try to look for varieties $X_1$ and $X_2$ in $U$ such that they have Stein universal coverings and the Eliashberg invariants ( see [**17**]) for these two varieties are not the same. The Eliashberg invariant measures when two Stein manifolds can be deformed to each other via Stein manifolds. If the Eliashberg invariants for $X_1$ and $X_2$ are different then this will be a counterexample to the Shafarevich conjecture since this will mean that we will have a non Stein space in $U$. Observe that if the universal covering of a variety with a large fundamental group is holomorphically convex then it is Stein. Of course one has to be able to show that all this deformations can be done in the class of universal coverings of varieties. If some of the above questions has a negative answer this will mean that even in case of surfaces the holomorphic convexity of the universal covering does not depend only on the properties of the fundamental group.

At the moment it looks more plausible that one should be able to carry this program for symplectic then for Kähler manifolds. Of course one can speculate and ask for a variant of the Shafarevich conjecture for symplectic manifolds. The first question then to be answered is:

QUESTION 6.3. What is the correct notion of holomorphic convexity in the symplectic category?

If one can introduce such a notion then a natural question is what is the analogue of the Shafarevich conjecture for symplectic four manifolds. The previous section suggests that this is hard to expect in general. We will restrict ourselves to a special case. Let us introduce a definition.

DEFINITION 6.2. Let $X$ is a symplectic fourfold. We will say that $X$ has a large fundamental group if for every symplectic twofold $Y \subset X$ we have that $Im(\pi_1(Y) \to \pi_1(X))$ is infinite.

A reasonable question as some examples ( see e.g. [**19**]) suggest is the following:

QUESTION 6.4. Is the universal covering of a symplectic fourfold with a large fundamental group Weinstein?

As Gompf has shown [**19**] every finitely presented group can be a fundamental group of a symplectic manifold. So the symplectic form does not seem to put any restrictions on the fundamental groups. It is reasonable to ask if it puts any restrictions on the universal coverings. The question above is a very partial case of such a possible restriction. The techniques available at the moment - Seiberg-Witten invariants and some partial cases of the Bochner formula for almost complex manifolds seem not to be strong enough to draw any conclusions even in some partial cases.

Observe also that the construction defined in section 5 extends to the symplectic category where it could lead to interesting examples of symplectic four dimensional manifolds. Using this construction Bogomolov and Katzarkov have defined in [**4**] an obstruction to a symplectic Lefschetz pencil being a Kähler Lefschetz pencil. Let us begin with a symplectic fourfold $X$. Consider the corresponding Lefschetz pencil with reducible fiber and apply to it the construction from section 5. It is shown in [**4**] that the construction goes through and for a fix integer $N$ we get a symplectic fourfold $S^N$.

Let $\rho$ be a representation $\rho : \pi_1(S^N) \to GL(n, \mathbb{C})$ such that $Im(\rho)$ is not virtually equal to $\mathbb{Z}$. Denote by $Y_i$ the componnents of the preimage of the reducible fiber of $S$ in $S^N$ and denote by $F$ the general fiber of $S^N$. Denote by $\Gamma$ the image of $\pi_1(F)$ in $\pi_1(S^N)$ and by $\Gamma_i$ the images of the fundamental groups of $Y_i$ in $\pi_1(S^N)$.

If the restrictions of $\rho$ on $\Gamma$ and $\Gamma_i$ for all $i$ are both finite or infinite we will say that the obstruction $O(X)^{N,n}$ is equal to zero and to one otherwise.

PROPOSITION 6.3. *If $X$ is a Kähler surface then $O(X)^{N,n}$ is trivial for every pair $N, n$.*

Indeed otherwise we will have that covering of $S^N$ with linear Galois group contains infinite chain of compact curves so it cannot be holomorphically convex. This contradicts theorem 4.4.

REMARK 6.2. If the Shafarevich conjecture happens to be correct for Kähler surfaces we can drop in the above definition the representation $\rho$ and compare the cardinalities only of $\Gamma$ and $\Gamma_i$.

REMARK 6.3. The above obstruction does not distinguish in general Kähler surfaces from symplectic fourfolds. It distinguishes only symplectic from Kähler Lefschetz pencils.

In [**4**] we have constructed examples of symplectic fourfolds with nontrivial $O(X)^{N,n}$. These examples suggest that a positive answer of 6.2 might not exist.

We would like to indicate how our obstruction can be used to distinguish two sympletic Lefschetz pencils $X_1$ and $X_2$ with similar local monodromies. According to [**44**] two $C^\infty$ Lefschetz pencils $X_1$ and $X_2$ are diffeomorphic if their global monodromies are the same. Now following [**20**] we see that iff the general fiber is a Riemann surface of genus greater than one then $X_1$ and $X_2$ are symplectic Lefschetz pencils and they are symplectomorphic as symplectic Lefschetz pencils iff their global monodromies are the same. So every invariant that distinguishes $X_1$ and $X_2$ as symplectic Lefschetz pencils need to come from their global monodromies. The obstruction $O(X)^{N,n}$ can be used for defining s uch an invariant in the following way.

Denote by $W(X_i), i = 1, 2$ the smallest $N$ for which $O(X_i)^{N,n}$ is not trivial. In case all $O(X_i)^{N,n}$ are trivial we will say that $W(X_i) = 0$.

PROPOSITION 6.4. *If $W(X_1) \neq W(X_2)$ then $X_1$ and $X_2$ are not symplectomorphic with the considered symplectic Lefschetz pencil structure.*

Indeed as it follows from the construction above this means that the global monodromies of $X_1$ and $X_2$ are different.

All these suggest that holomorphic convexity of the universal coverings of projective surfaces could have a motivic character so in the next subsection we formulate an arithmetic variant of the Shafarevich conjecture.

**6.5. Arithmetic version of the holomorphic convexity conjecture.** The considerations in [**3**] lead us to consider the following number theory problem.

Let $K$ be a number field with a ring of integers $O_K$. Start with a projective semi-stable curve $C_K$ over $K$. Since $K$ is not algebraically closed we can find a nontrivial map $C \to \mathrm{Spec}(O_K)$ whose generic fiber is isomorphic to $C_K$. Due to the semi-stable reduction theorem we may assume (after passing to a finite extension of $K$ if necessary) that $C$ is also semi-stable. This means that $C$ is a normal variety with semi-stable fibers consisting of normally intersecting divisors. The function field $K(C)$ is regular and has dimension one over $K$. Any maximal ideal $\nu$ in $O_K$ defines a subring $O_\nu$ consisting of the elements in $K(C)$ regular along any of the components of the preimage of $\nu$ in $C$. The ring $O_\nu$ contains the ideal of elements which are trivial on the preimage of $\nu$. Denote this ideal by $M_\nu$. The quotient ring $O_\nu/M_\nu$ is a finite sum of fields of rational functions on the components of the fiber of $C$ over $\nu$. Consider the maximal nonramified extension $K(C)^{nr}$ of $K(C)$. It is a Galois extension with a profinite Galois group. The field $K(C)^{nr}$ contains the maximal nonramified extension $K^{nr}$ of the field $K$ as a subfield. Any maximal ideal $\mu$ in the ring of integers $O_{K^{nr}}$ contains a unique maximal ideal $\nu$ of $O_K$. We can now define the subring $A_\mu$ as the integral closure of $O_\nu$ in $K^{nr}(C)$. For any finite nonramified extension $L$ of $K(C)$ we have a model $C_L$ with a finite map onto $C$. As it follows from the above discussion the fibers of this new model $C_L$ are uniquely determined by $C$.

We formulate an arithmetic version of the Shafarevich conjecture.

CONJECTURE 6.3. There exists a constant $J(K(C))$ such that the number of components of any fiber of $C_L$ is bounded by $J(K(C))$ for any finite extension $L$ of $K(C)$ contained in $K(C)^{nr}$.

There exists some evidence for the above arithmetic conjecture - it is true if we consider only abelian coverings. Bogomolov proved that the torsion group of an abelian variety $A$ is finite if the latter is defined over an infinite nonramified extension of $K$ (see e.g. R.Coleman's paper in Duke Math. Journal, 54 (1987), p. 615).

If the above conjecture is correct for every curve over every finite extension of $\mathbb{Q}$ then we get that there aren't any infinite chains of compact curves on the universal covering of a projective surface with a residually finite fundamental group. According to a conjecture of M. Ramachandran this is the only obstruction to holomorphic convexity for the universal coverings of a projective surfaces.

**Acknowledgements:** We would like to thank A. Beilinson, F. Bogomolov, J. Carlson, H. Clemens, K. Corlette, P. Deligne, R. Donagi, Y. Eliashberg, M. Fried, M. Gromov, S. Gersten, R. Hain, S. Ivanov, M. Kapovich, J. Kollár, M. Nori, A. Olshanskii, T. Pantev, M. Ramachandran, C. Simpson, Y.T. Siu, R. Stern, G. Tian, D. Toledo, S. Weinberger, E. Zelmanov and S.T. Yau for the useful conversations and constant attention to the work. Special thanks to J. Kollár, T. Pantev and D. Toledo for carefully reading the text, fixing parts of it and helping me with organising the exposition. I am indebted to M. Ramachandran for explaining to me a lots of facts about the Shafarevich conjecture and to C. Simpson for teaching me nonabelian Hodge theory.

## References

[1] **H.Amoros, K.Corlette, D.Kotschick and D.Toledo,** *Fundamental groups of Käler manifolds,* Mathematical Surveys and Monographs **44**, American Mathematical Society, 1996.

[2] **I.Bauer,** *Irrational pencils on non-compact algebraic manifolds.* to appear in Int. Journal of Mathematics.

[3] **F.Bogomolov and L.Katzarkov** *Complex projective surfaces and infinite groups* , preprint, 1995.

[4] **F. Bogomolov and L.Katzarkov** *Symplectic four manifolds and projective surfaces* , preprint, 1996.

[5] **A.Borel and J.P.Serre** *Cohomologie d'iemmeubles et de S-groupe arithmétiques,* Topology **15** 1976, 211-232.

[6] **F.Campana,***Coreductions algebrique d'un espace analytic faiblements Kähleriennes compact,* Inv. Math. **63** 1981, 187-223.

[7] **J.Carlson and D.Toledo,** *Quadratic Presentations and Nilpotent Kähler Groups,* J. Geom. Anal. **5** 1995, 359-377.

[8] **K.Corlette,** *Nonabelian Hodge theory,* Proc. Symp. Pure. Math. **54**, part 2, 1993, 125-140.

[9] **P. Deligne,** *Theorie de Hodge II.* Publ. Math. I.H.E.S. **40** 1971, 5-58.

[10] **P.Deligne and D. Mumford,** *Irreducibility of the moduli space of curves,* Publ. Math. I.H.E.S. **6** 1969 75-109.

[11] **P.Deligne, P.Griffiths, J.Morgan and D.Sullivan** *Real Homotopy theory of Kähler manifolds,* Inv. Math. **29** 1975, 245-274.

[12] **P.Deligne, J.Milne, A.Ogus and K.Shih,** *Hodge cycles, motives and Shimura varieties,* Springer LNM, v. 900, 1982.

[13] **P.Deligne,** *Structures de Hodge mixtes réelles,* Proceedings of Symposia in Pure Mathematics, vol. **55** (1994), Part I, 509-514.

[14] **R.Donagi** *Spectral covers,* in Current topics in complex algebraic geometry (Berkeley, CA, 1992/93), MSRI Publ. **28** 1996, Cambridge Univ. Press, 65-86.

[15] **R.Donagi** *Decomposition of Spectral Covers,* Journ. de Algebr. Geometrie, Orsay, Asterisque **218** 1993, 145-175.

[16] **Y. Eliashberg, M. Gromov** *Convex Symplectic Manifolds,* Proc. Sym. in Pure Mathematics , vol. 52, part. 1, (1991), 135-162.

[17] **Y.Eliashberg,** in preparation.

[18] **H.Grauert, Th.Peternell and R.Remmert (Eds.)** *Several Complex Variebles 7,* Encyclopedia of Mathematical Sciences, vol. 74, Sringer-Verlag, 1994.

[19] **R. Gompf,** *A new construction of symplectic manifolds,* Annals of Math. **142** 1995, 527-595.

[20] **R. Gompf,** in preparation.

[21] **M. Gromov ,** *Hyperbolic groups,* Essays on Group theory, MSRI Publ. **8** 1987, 75-265.

[22] **M.Gromov, G.Henkin and M.Shubin** $L_2$ *holomorphic functions on pseudo convex coverings,* Preprint IHES, 1995.

[23] **M.Gromov, R.Schoen** *Harmonic maps into singular spaces and p-adic superrigidity for lattices in groups of rank one,* Publ. Math. I.H.E.S. **76** 1992, 165-246.

[24] **R.V.Gurjar,** *Coverings of algebraic surfaces,* in Algebraic geometry, Sendai, Adv. Stud. in Pure Math. vol. 10, 1987, 179-184.

[25] **R.Hain,** *The Hodge de Rham theory of the relative completion,* preprint, 1995.

[26] **R.Hain and S.Zucker,** *Unipotent variations of mixed Hodge structures,* Invent. Math. **88** (1987) 83-124.

[27] **N.J.Hitchin,** *The self-duality equations on a Riemann surface.* Proc. Lond. Math. Soc., (3) vol.55, 1987, 59-126.

[28] **H.Hironaka,** *Resolution of singularities of algebraic varieties over fields of characteristic zero I, II.* Ann. of Math., (2), v. 79, (1964), 109-326.

[29] **J.Jost and K.Zuo,** *Harmonic maps into Tits buildings, Factorization of non p-adic integral and nonrigid representations of* $\pi_1$ *of algebraic varieties,* preprint, 1994.

[30] **J.Jost and S.T.Yau,** *Applications of quasilinear PDE to algebraic geometry and arithmetic lattices,* Algebraic geometry and related topics: Proceedings of of Internatinal Symposium, Inchoen, Korea, Vol. 1 (1993), pp. 169-1990.

[31] **J.Jost and S.T.Yau,** *Harmonic maps and Kähler manifolds.* Math. Ann. **262** 1983, 145-166.

[32] **M. Kapovich and J. Millson** *On the deformation theory of representations of fundamental groups of compact hyperbolic 3-manifolds,* Topology **35** 1996, 1085-1106.

[33] **L.Katzarkov and T.Pantev,** *Representations of fundamental groups whose Higgs bundles are pullbacks,* J. Differential Geom. **39** 1994, 103-123.

[34] **L.Katzarkov and T.Pantev,** *Non abelian mixed Hodge fiction,* in preparation, 1995.

[35] **L.Katzarkov, T. Pantev,** ... *Non abelian Bloch theorem,* in preparation, 1995.

[36] **L.Katzarkov and T.Pantev** *On the Torelli dga,* in preparation, 1995.

[37] **L.Katzarkov and M.Ramachadran** *On the universal coverings of algebraic surfaces,* preprint 1994.

[38] **L.Katzarkov** *Nilpotent groups and universal coverings of smooth projective varieties.* to appear in J. Differential Geom..

[39] **L.Katzarkov, T.Pantev and M.Ramachandran,** *Geometric factorization of linear fundamental groups,* in preparation.

[40] **J.Kollár,** *Shafarevich maps and plurigenera of algebraic varieties,* Invent. Math. **113** 1993, 165-215.

[41] **J.Kollár,** *Shafarevich maps and automorphic forms,* Princeton University Press, (1995).

[42] **B.Lasell and M.Ramachandran,** *Observations on harmonic maps and singular varieties,* Ann. Sci. École Norm. Sup. **29**, 1996, 135-148.

[43] **A.I.Malcev,** *On a class of homogenious spaces,* Izv. Acad. Nauk. USSR , Ser. MATH. **13**, 1949, 9-32.

[44] **Y. Matsumoto** *Lefschetz fibrations of genus two-a topological aproach,* preprint 1995.

[45] **N.Mok,** *Factorization of semisimple discrete representation of Kähler groups,* Invent. Math. **110** 1992, 557-614.

[46] **N.Mok,** *Nonabelian Castelnuovo-deFrancis theorem,* preprint 1995.

[47] **S.Mori,** *Classification of higher dimensional varieties,* Proc. Symp. of Pure Math., vol.46, 1987.

[48] **T.Napier,** *Convexity properties of coverings of smooth projective varieties,* Math. Ann. **286** 1990, 433-480.

[49] **T.Napier and M.Ramachandran** *Structure theorems for complete Kähler manifolds and applications to a Lefshetz type of theorems,* GAFA **5** 1995, 809-851.

[50] **T.Napier and M.Ramachandran,** $L^2(\bar{\partial})$ *methods, weak Lefschetz theorems and topology of Káhller manifolds,* preprint 1996.

[51] **D.Reed,** *The topology of conjugate varieties,* Math. Ann. **305** 1996), 287-309.

[52] **J.P.Serre,** *Examples de Varietie Conjugees nonhomeomorphes,* Ouevres. Vol. 2 pp. 246-249. ( Springer 1986).

[53] **I.R. Shafarevich,** *Algebraic surfaces,* Trudi MIAN AN USSR (in Russian), 1965.

[54] **C.T.Simpson** *Higgs bundles and local systems,* IHES - Publications Mathematiques, 75, (1992), 5-95.

[55] **C.T.Simpson,** *Moduli spaces of representations 1,* IHES - Publications Mathematiques, 79, (1995).

[56] **C.T.Simpson,** *Moduli spaces of representations 2.* IHES - Publications Mathematiques, 80, (1995).

[57] **C.T.Simpson,** *Nonabelian Hodge theory* ., International congress of Mathematics Kyoto, 1990, Proceedings, Springer-Tokyo, (1991), pp. 747-756.

[58] **C.T.Simpson,** *The Hodge filtration on the non-abelian cohomology.* see these proceedings.

[59] **C.T.Simpson,** *Mixed twistor structures.* alg-geom/9705006.

[60] **Y.T. Siu,** *Complex analyticity of harmonic maps and strong rigidity of complex Kähler manifolds.* Ann. of Math, 112, 1980, pp. 73-110.

[61] **K. Ueno,** *Classification theory of algebraic varieties and compact complex spaces* . Springer LNM, 439, 1975.

[62] **K. Zuo,** *Factorization of nonrigid Zariski dense representations of* $\pi_1$ *of projective manifolds.* Invent. Math. **118** 1994, 37-46.

[63] **K. Zuo,** *Kodaira dimension and Chern hyperbolicity of the Shafarevich maps for representation of* $\pi_1$ *of compact Kähler manifolds,* J. Reine Angew. Math. **472** 1996, 139-156.

Department of Mathematics, UC Irvine, Irvine, CA, 92697
*E-mail address*: lkatzark@math.uci.edu

Proceedings of Symposia in Pure Mathematics
Volume **62.2**, 1997

# The Hodge filtration on nonabelian cohomology

Carlos Simpson

## 1. Introduction

Whereas usual Hodge theory concerns mainly the usual or abelian cohomology of an algebraic variety—or eventually the rational homotopy theory or nilpotent completion of $\pi_1$ which are in some sense obtained by extensions—nonabelian Hodge theory concerns the cohomology of a variety with nonabelian coefficients. Because of the basic fact that homotopy groups in higher dimensions are abelian, and since cohomology theories can generally be interpreted as spaces of maps into classifying (or Eilenberg-MacLane) spaces, nonabelian cohomology occurs essentially only in degree 1. There are certainly some degree 2 aspects which are as of yet totally untouched; and the same goes for the degree 1 case with twisted coefficient systems. (And see [**76**] for a direction of development combining the nonabelian coefficients in degree 1 with abelian coefficients in higher degrees—we discuss this briefly in a subsection of §5). If we leave these aside, we are left with the case of $H^1(X,G)$ for $G$ a nonabelian group. It is most natural to interpret this cohomology as a groupoid, or, when $G$ is a group-scheme, to interpret $H^1(X,G)$ as a stack. It is the stack of flat principal $G$-bundles on $X$. Recall from the usual abelian case that in order to obtain a Hodge structure, we must consider cohomology with complex coefficients. The analogue in the nonabelian case is that we must take as coefficient group a group-scheme $G$ over the complex numbers (and in fact it should be affine too). This then is the domain of application of the work that has been done in nonabelian Hodge theory: the study of properties and additional structure on the moduli stack $\mathcal{M}(X,G) := H^1(X,G)$ which are the analogues in an appropriate sense of the main structures or properties of abelian cohomology.

By its nature, the first nonabelian cohomology is an invariant of the fundamental group $\pi_1(X)$. The study of nonabelian Hodge theory may thus be thought of as the study of fundamental groups of algebraic varieties or compact Kähler manifolds. It is important to note, specially in light of Toledo's examples of $\pi_1(X)$ not residually finite, that the study of $\pi_1(X)$ via its nonabelian cohomology, i.e. via the spaces of homomorphisms $\pi_1(X)\to G$, will only "see" a certain part of $\pi_1(X)$ and in particular will not at all see the intersection of subgroups of finite index. It is an interesting question to try to understand what Hodge-theoretic methods could say about this more mysterious part of the fundamental group.

---

1991 *Mathematics Subject Classification.* Primary: 14C30, 14D20, 53C55.

We start in §2 by reviewing Corlette's nonabelian Hodge theorem [**11**] (cf also [**19**] and [**18**]) which is actually a generalization of the theorem of Eells and Sampson [**20**]. This theorem allows us to choose a prefered metric on any flat bundle. By a Bochner technique on Kähler manifolds ([**79**] [**41**] [**11**]) a harmonic metric is in fact pluriharmonic, and we recover in this way the holomorphic data of a *Higgs bundle*. The correspondence in the other direction characterizes exactly which Higgs bundles arise in this way [**36**] [**69**].

We then mention (refering to [**62**], [**75**] for proofs) the existence of moduli spaces for all of the objects in question. The space $M_{DR}(X,G)$ is the moduli scheme for principal holomorphic $G$-bundles with integrable connection. It is a coarse moduli space for the moduli stack $\mathcal{M}_{DR}(X,G)$; it is this moduli stack which should be thought of as the nonabelian de Rham cohomology, and the moduli space is a convenient scheme-theoretic version. The space $M_{Dol}(X,G)$ is the moduli scheme for semistable principal Higgs bundles with vanishing rational Chern classes; again it is a coarse moduli space for $\mathcal{M}_{Dol}(X,G)$, the moduli stack which is what should be thought of as the nonabelian Dolbeault cohomology. The harmonic metric construction and the Bochner technique (together with the inverse construction) give a homeomorphism $M_{DR}(X,G) \cong M_{Dol}(X,G)$ which is $\mathcal{C}^\infty$ on the smooth points.

The above work is the result of a long series of generalizations of the original work of Narasimhan and Seshadri. Without being exhaustive, I should at least mention the names of Mumford, Gieseker, Maruyama, Mehta and Ramanathan for the constructions of moduli spaces; and Donaldson, Uhlenbeck, Yau, Deligne and Beilinson for the inverse construction to the harmonic map construction. See the introductions and references of [**71**], [**75**] for more detailed historical references.

* * *

These constructions (or their predecessors in work of Eells and Sampson [**20**] [**67**] and Siu [**79**]) are the starting point for much of the work which has been done in nonabelian Hodge theory in the past several years. As many aspects are covered by other lectures (and their corresponding articles), and in any case many of the papers on the subject contain survey-like introductions, I will not try to survey all of these topics in the main part of the paper but will just mention some of them here in the introduction.

Hitchin was interested from the beginning in the completely integrable holomorphic hamiltonian system given by the moduli space of Higgs bundles [**35**]. This direction of research has branched off toward the Verlinde formula, quantization and so forth. I won't try to give references as this gets away from our principal concern of Hodge theory.

One of the principal applications of Hodge theory has always been to give restrictions on the topological type of varieties and their subvarieties. The nonabelian version presents this same feature. The existence of all of the various structures described above on the nonabelian cohomology $H^1(X,G)$ and various related considerations give restrictions on which groups can be fundamental groups of compact Kähler manifolds, or more generally on which homotopy types can arise. Some of these results such as [**8**] [**41**] [**67**] [**79**] [**81**] pre-date the general Hodge-theoretic point of view given above, being based on harmonic map considerations à la Eells-Sampson and Siu. Others come directly from the full correspondence between Higgs bundles and local systems and the subsidiary fact that Higgs bundles invariant under the natural action of $\mathbf{G}_m = \mathbf{C}^*$ correspond to variations of Hodge structure [**71**]. The restrictions on higher homotopy types are generally of two sorts: either one uses information about a homotopy type to obtain additional information on the harmonic map (such as its rank) and then concludes that such a harmonic map cannot exist ([**81**] [**8**] [**12**]); or one uses various notions about the cohomology of local systems to rule out higher homotopy types (these are the restrictions coming from work of Green-Lazarsfeld [**24**], Beauville [**3**], Arapura [**1**], recently Hironaka [**34**], also [**72**]—the idea of using these results to get restrictions will be explored in §3 below since many of these papers don't explicitly mention the aspect "restrictions on homotopy types" which comes out of their results).

Gromov has an $L^2$-Hodge theoretic argument to rule out free (and certain amalgamated) products of groups [**25**]. This is particularly interesting in relation to the theory we sketch here, because it allows one to "see" the whole fundamental group (for example, the free product of two groups with no subgroups of finite index is ruled out, which would evidently be impossible to do by looking at representations into linear groups). Gromov and Schoen [**26**] have also developed a generalization of the harmonic map theory to cover harmonic maps into negatively curved Euclidean buildings. Coupled with a Lefschetz technique [**73**] this gives results about fundamental groups and in particular seems to give an alternative proof of the result about amalgamated products. This technique is generalized in [**42**] [**44**].

With all of these restrictions on fundamental groups coming from Hodge theory, one might well wonder if there are any interesting fundamental groups at all. Toledo's example of a non residually finite fundamental group ([**83**]—there have since been several other generalizations) shows that the family of groups which can occur is, on the contrary, quite complicated. We can still ask whether the part "seen" by nonabelian Hodge theory with linear groups as coefficients can be nontrivial, for example, are there nontrivial ways of obtaining positive dimensional moduli spaces (other than easily known ways using curves and abelian varieties)? The answer here is affirmative too, and in fact the Higgs bundle picture is essential for calculating what happens to obtain examples [**74**].

One of the main types of results has been the *factorization theorem.* This type of result relates the fundamental group and the geometry of $X$. The typical type of statement is that if $\rho : \pi_1(X) \to \Gamma$ is a certain type of representation then it must *factor:* there is a morphism of varieties $X \to Y$ such that $\rho$ factors through $\pi_1(Y)$. Perhaps the original result of this type is that of Siu [**79**] (cf also Beauville's appendix to [**9**]) stating that when $\Gamma$ is the fundamental group of a Riemann surface, then any $\rho$ must factor through a Riemann surface $Y$. Gromov's result of [**25**] also passes through a similar type of statement: certain $L^2$ cohomology classes on the universal cover of $X$ must factor through maps to a Riemann surface. One of the first statements involving factorization through a higher dimensonal variety is that of Zuo [**86**]; and we now have a fairly complete picture of this type of result (cf [**42**] [**43**] [**44**] [**47**] [**86**] [**87**]): any nonrigid representation to a linear group $G$ must factor through a variety $Y$ of dimension less than or equal to the rank of $G$. Note that the example of [**74**] shows that we do not always get factorization through a curve...but apart from this we do not know for sure if the bound rank of $G$ is sharp. See also [**57**] and various generalizations for additional geometric information on the factorization variety $Y$.

Using the theory of harmonic maps to buildings mentioned above, one can extend these factorization theorems to the case of representations into linear groups over $p$-adic fields not going into a maximal compact subgroup [**42**] [**44**] [**87**].

These factorization techniques obviously have a certain application to the Shafarevich conjecture [**50**]. This has been pursued by Katzarkov, Lasell, Napier, Ramachandran ([**46**] [**49**] [**53**] [**60**] [**61**]), see also Zuo [**87**]. The main problem is that only the part of the fundamental group seen by a given linear representation can be treated. They obtain a full proof of the Shafarevich conjecture for surfaces whose fundamental group injects into a linear group.

Another recent development worth mentioning is Reznikov's proof of the Bloch conjecture that the Chern-Simons classes of flat bundles on Kähler manifolds are torsion [**64**].

The last principal area of work I would like to mention, one where there is still a fair amount to be done, is the noncompact (quasiprojective) case. The problem is to do the analogues of everything which we discuss in the compact Kähler (or projective algebraic) case, in the case of a quasiprojective variety. This problem becomes much more difficult in dimension $\geq 2$. Aside from the dimension distinction, the problem can be divided into several parts.

The first part is to obtain the analogue of Corlette's theorem. The main difficulty here is to get a starting point for a heat equation minimization process, that is to say an equivariant map of finite energy. If the eigenvalues of the monodromy at infinity are not of norm one, this becomes impossible and the problem is more difficult. Modulo this difficulty, the problem has been solved by Corlette [**13**] and Jost and Zuo [**42**].

The next problem is to obtain the analogues of the Bochner results, yielding a Higgs bundle. This is discussed to some extent in [**42**]. There may be a problem with Chern classes in general. The other aspect of this problem is that the appropriate Higgs bundle notion must include some data at infinity, namely a *parabolic structure*. The problem of associating a parabolic Higgs bundle with nice properties, to a harmonic bundle, is treated in Biquard [**5**] in the case when the divisor at infinity is smooth.

Biquard also provides the converse construction: given a parabolic Higgs bundle satisfying appropriate conditions, he gets back a Yang-Mills connection and hence a representation. This provides a relatively complete generalization in the case of smooth divisor at infinity. What is left open for the moment is to treat the case where the divisor at infinity has normal crossings.

A fourth aspect of the problem is to construct moduli spaces. This is now well understood for Higgs bundles and the like, due to work of Yokogawa and Maruyama [**56**] [**84**]. I think there is still a little work (probably not too hard) left to be done on the side of *filtered local systems*, which are the general representation-like objects which correspond to parabolic Higgs bundles.

It remains to be seen how all of these aspects fit together, and then to proceed with the generalizations of all of the further structures inherited by the moduli spaces in the compact case (i.e. the structures we will discuss in the present paper).

* * *

Rather than going into further detail on all of these applications and developments, I would like in the body of the paper to concentrate on a fundamental aspect—the nonabelian Hodge filtration [**70**]. We will discuss many of the basic subjects surrounding the Hodge filtration, such as the quaternionic structure and twistor spaces. And we give proofs of the results announced in [**70**], in particular the compactification of $M_{DR}$ which is a consequence of (and practically equivalent to) the Hodge filtration. The goal will be in the last section to introduce an open problem, that of studying degenerations of nonabelian Hodge structure coming from a degenerating family of varieties.

After our discussion of Corlette's theorem, its converse and the moduli spaces in §§1-2, we turn in §3 to a discussion of Hitchin's quaternionic structure on the moduli space for representations [**36**] [**22**]. We give an application to cohomology jump-loci retrieving the results of Green and Lazarsfeld via an argument of Deligne and along the way see how these results give restrictions on the higher homotopy type of non-simply connected Kähler manifolds.

In §4 we look at Deligne's complex analytic construction of the twistor space corresponding to the quaternionic structure [**14**]. He obtains the twistor space by glueing two copies of a family $M_{Hod}$ deforming between the moduli space $M_{DR}$ of vector bundles with integrable connection and the moduli space $M_{Dol}$ of Higgs bundles. This deformation (parametrized by $\mathbf{A}^1$) is the moduli space of *vector bundles with $\lambda$-connections*; over $\lambda = 0$ a $\lambda$-connection is just a Higgs field, whereas over $\lambda \neq 0$ a $\lambda$-connection is $\lambda$ times an integrable connection.

In §5 we explain the analogy with Rees modules (similar to the interpretation of [**17**]) which allows us to interpret the space $M_{Hod}$ as the Hodge filtration on $M_{DR}$.

As justification we establish in §6 the relationship with the Morgan-Hain Hodge filtration on the nilpotent completion of the fundamental group [**58**] [**30**].

We then proceed with certain results about the Hodge filtration in the nonabelian case, notably Griffiths transversality for its variation in a family, and regularity of the Gauss-Manin connection at singular points of a family (§8). In order to do this, we first introduce in §7 the notion of *formal groupoid* [**4**] [**39**]. This provides a general framework for looking at connections, Higgs fields and so forth, and in particular allows us to define these types of things for families of schemes rather than just for vector bundles.

In a detour §9 we investigate Goldman-Millson theory [**23**] for the local structure of $M_{Hod}$. The *isosingularity principle* which says that the singularities of $M_{Dol}$ are the same as those of $M_{DR}$, generalizes to give a trivalization of $M_{Hod}$ formally along prefered sections. This allows us to conclude, for example, that $M_{Hod}$ is flat over $\mathbf{A}^1$.

Then we come to a properness property of $M_{Hod}$; the limits of $\mathbf{G}_m$-orbits always exist (§10). This is the analogue of the classical property of the Hodge filtration, that $F^0$ is the whole space.

This weight property allows us to obtain a compactification of $M_{DR}$ in §11, by taking the quotient of an open set in $M_{Hod}$ by the action of $\mathbf{G}_m$. This compactification was announced without proofs in [**70**] so we take this opportunity to provide a complete version of the argument. (Drinfeld recently informed me that some of his students have obtained a compactification of the moduli space of logarithmic connections on $\mathbf{P}^1$ with singularities at a finite set of points, also using the method of $\lambda$-connections but independently of [**70**].)

At the end of the section we revisit Griffiths transversality in terms of this compactification: it says that the Gauss-Manin connection on $M_{DR}$ has poles of order 1 at infinity in the compactification. This gives a picture of a compact space

$\overline{M_{DR}}(X/S, G)$ with a lift of a frame vector field on $S$ to a vector field having a simple pole at infinity. A similar interpretation holds for the regularity of the Gauss-Manin connection.

In the penultimate section 12 we define the *nonabelian Noether-Lefschetz locus* $NL(X/S, GL(n))$. If $X \to S$ is a family then this is essentially the locus of $s \in S$ where $X_s$ supports an integral variation of Hodge structure. It is the nonabelian analogue of the classical Noether-Lefschetz locus of Hodge cycles. If $S$ is projective then we can see that $NL(X/S, GL(n))$ is algebraic (as would be a consequence of the Hodge-type conjecture that one could formulate, that integral variations of Hodge structure are motivic). We conjecture that this is true even for $S$ quasiprojective, which would be a nonabelian version of the result of [**10**].

In the last section we present an open problem, the problem of understanding the degeneration of all of our structures (Hodge filtration, Gauss-Manin connection, quaternionic structure, etc.) near degenerations of a family $X/S$. This problem is motivated by the problem of proving that $NL(X/S)$ is algebraic when $S$ is quasiprojective—the nonabelian analogue of the work of Cattani, Deligne and Kaplan for the classical Noether-Lefschetz locus of Hodge cycles [**10**].

I would like to thank P. Deligne for sharing with me his ideas on how to construct Hitchin's twistor space. This construction provided the starting point for everything (new) done below.

Everything is over the field $\mathbf{C}$ of complex numbers.

## 2. The nonabelian Hodge theorem

Suppose $X$ is a Riemannian manifold with basepoint $x \in X$, and suppose $G$ is a reductive algebraic group. A representation $\rho : \pi_1(X, x) \to G$ corresponds to a flat principal left $G$-bundle over $X$ (in other words a locally constant sheaf of principal homogeneous spaces for $G$ over $X$), or equally well to a $\mathcal{C}^\infty$ principal $G$-bundle $P$ with an integrable connection $D$. We think of a connection on a principal bundle as a $G$-invariant operator $\nabla$ from functions on $P$ to sections of $T^*(X)|_P$ (satisfying a Leibniz rule with respect to functions pulled back from the base). Such an operator then has a square $\nabla^2$ from functions on $P$ to sections of $\bigwedge^2 T^*(X)|_P$, and the integrability condition is $\nabla^2 = 0$. The flat principal bundle is the sheaf of $\nabla$-horizontal sections of $P$.

Fix a maximal compact subgroup $K \subset G$. A *$K$-reduction* for a principal bundle $P$ is a $\mathcal{C}^\infty$ principal $K$-subbundle $P_K \subset P$ giving $P = P_K \times^K G$ (the notation of a product with superscript means extension of structure group).

A flat principal $G$-bundle $(P, \nabla)$ gives rise to a flat family of homogeneous spaces over $X$ which we can write as $P \times^G (G/K)$. If $P_K \subset P$ is a $K$-reduction for $P$ then the image of $P_K \times (eK)$ in $P_K \times^K G$ is a smooth section of the bundle $P \times^G (G/K)$. This smooth section can also be thought of as a $\rho$-equivariant map $\phi : \tilde{X} \to G/K$. We define the *energy* of the $K$-reduction or of its associated equivariant map by

$$\mathcal{E}(\phi) := \int_X |d\phi|^2,$$

where the integral is taken with respect to the volume form on $X$, and the norm of the differential $d\phi$ is measured with respect to an invariant metric on $G/K$ (note that one has to fix a $K$-invariant metric on the complement $\mathbf{p}$ to $\mathbf{k} \subset \mathbf{g}$ when making this discussion—if $G$ is semisimple then we can fix the Killing form as a canonical choice).

An equivariant map or $K$-reduction $P_K$ is called *harmonic* if it is a critical point of $\mathcal{E}(\phi)$. The Euler-Lagrange equation for for a harmonic equivariant map is $d^* d\phi = 0$. This is a nonlinear equation with the Laplacian as its principal term.

The main theorem in this subject is the following generalisation of the theorem of Eells and Sampson [**20**]:

THEOREM 2.1. (Corlette [**11**]) *If $\rho$ is a representation such that $\rho(\pi_1(X,x))$ is Zariski-dense in $G$ (or such that the Zariski closure is itself reductive), then there exists a harmonic equivariant map $\phi$.*

*Proof:* We indicate here a variant of Corlette's proof which might be useful for people with an algebraic geometry background. Assume that the Zariski closure $G$ is semisimple (the general reductive case may then be obtained by using the linear theory of harmonic forms for $\mathbf{C}^*$ representations).

Eells and Sampson [**20**] prove the existence of harmonic maps from $X$ to a compact negatively curved manifold $M$ by a heat-equation minimization technique. We can start off with an equivariant map $\phi_0$ and apply the same heat equation to obtain a family of maps $\phi_t$. We get the same local estimates. In particular, for $x, y \in \tilde{X}$ the distance from $\phi_t(x)$ to $\phi_t(y)$ is a bounded function of $t$. Furthermore, if for any one point $x$ we can show the "$C^0$-estimate" that $\phi_t(x)$ stays in a compact subset of $G/K$ then the estimates of [**20**] will allow us to show that the $\phi_t$ converge to a smooth harmonic map $\phi$. Note, for example, that the $C^0$-estimate is not true if the Zariski closure of $\rho(\pi_1(X,x))$ is not reductive (in fact, the existence of an equivariant harmonic map implies reductivity of the Zariski closure). In the original case of [**20**] the target was a compact manifold so this problem was avoided.

Now apply a little bit of geometric invariant invariant theory to get the $C^0$-estimate. Choose elements $g_t \in G$ bringing us back to the basepoint: $g_t\phi_t(x) = eK$. Fix a set of generators $\gamma_i$ for $\pi_1(X,x)$. From [**20**] we know that the distance from $\phi_t(x)$ to $\phi_t(\gamma_i x)$ remains bounded. As the distance on $G/K$ is $G$-invariant, we have

$$d(g_t\phi_t(x), g_t\phi_t(\gamma_i x)) \leq C.$$

On the other hand the equivariance of $\phi_t$ gives $\phi_t(\gamma x) = \rho(\gamma_i)\phi_t(x)$ so (also plugging in $g_t\phi_t(x) = eK$) we get

$$d(eK, g_t\rho(\gamma_i)g_t^{-1}eK) \leq C.$$

Since $K$ is compact the map $G \to G/K$ is proper so the $g_t\rho(\gamma_i)g_t^{-1}$ remain in a compact subset of $G$.

The representation variety $R := Hom(\pi_1(X,x), G)$ embeds in a product $G \times \ldots \times G$ by $\rho \mapsto (\ldots, \rho(\gamma_i), \ldots)$. The group $G$ acts on $R$ by the *adjoint action* $Ad(g)(\rho)(\gamma) := g\rho(\gamma)g^{-1}$, and this is compatible with the above embedding via the adjoint action in each variable of $G \times \ldots \times G$. The previous paragraph tells us that, in our situation, $Ad(g_t)\rho$ remain in a compact subset of $R$.

The basic information from the geometric invariant theory of spaces of representations of finitely generated groups, is that the hypothesis that $\rho(\pi_1(X,x))$ is Zariski-dense in $G$ implies that the $Ad(G)$-orbit of $\rho$ is closed in $R$. This is well known [**55**] but we discuss it anyway in the next two paragraphs.

If $G = GL(n, \mathbf{C})$ then $\rho$ corresponds to an irreducible $n$-dimensional representation which we denote $V_\rho$. If $V_0$ is a representation in the closure of the orbit of $\rho$ then we have a family of representations $\{V_t\}$ parametrized by $t$ in a smooth curve with $V_0$ being the value at a point 0 and $V_t \cong V_\rho$ for $t \neq 0$. By semicontinuity there is a nontrivial morphism of representations from $V_\rho$ to $V_0$, but since $V_\rho$ is irreducible this must be an isomorphism and we get that $V_0$ is in the orbit of $V_\rho$.

For a semisimple group $G$ note that $G$ admits a faithful representation $V$ which we can decompose as $V = \bigoplus V^i$ with $V^i$ irreducible. By composition with $\rho$ we get a linear representation $V_\rho = \bigoplus V^i_\rho$ of $\pi_1(X,x)$. Since $\rho$ is Zariski-dense, the $V^i_\rho$ are irreducible. Suppose we have a family $\rho_t$ of representations (parametrized by an affine curve) with $\rho_t \sim \rho$ for $t \neq 0$. This gives a family of linear representations $V_t = \bigoplus V^i_t$ of $\pi_1(X,x)$; as above, semicontinuity and irreducibility of $V^i_\rho$ imply that $V^i_0 \cong V^i_\rho$ and hence $V_0 \cong V_\rho$. This then implies that $\rho$ and $\rho_0$ are conjugate by an automorphism of $G$ where furthermore this automorphism is a limit of inner automorphisms. The group of outer automorphisms being finite (hence discrete) we conclude that $\rho_0$ and $\rho$ are conjugate by an inner automorphism, that is $\rho_0$ is in the $Ad(G)$-orbit of $\rho$.

Now we complete the proof. The $Ad(g_t)\rho$ remain in a compact subset of the orbit $Ad(G)\rho$ because of the fact that the orbit is closed. Since $\rho(\pi_1(X,x))$ is Zariski-dense in $G$, the stabilizer of $\rho$ is just the center of $G$, which is finite (since we have assumed that $G$ is reductive). In particular the map $G \to Ad(G)\rho$ given by the action on $\rho$ is proper, so the $g_t$ themselves remain in a compact subset of $G$. Finally this implies that the $\phi(x) = g_t^{-1}eK$ remain in a compact subset of $G/K$, which is the $C^0$-estimate we need. □

*Remark:* Donaldson proved this theorem for rank 2 representations independently of Corlette [**19**]. It was also proved by Diederich and Ohsawa for representations into $SL(2,\mathbf{R})$ [**18**]. On the other hand there have since been several generalizations to the noncompact case, for example by Corlette [**13**] and Jost and Zuo [**42**] [**43**].

*The Kähler case*

Assume now that $X$ is a compact Kähler manifold. Let $\omega$ denote the Kähler form (of a Kähler metric which we choose); let $\Lambda$ denote the adjoint of wedging with $\omega$; and let $\partial$ and $\overline{\partial}$ denote the operators coming from the complex structure. A holomorphic principal bundle $P$ may be considered as a $\mathcal{C}^\infty$ principal bundle together with a $G$-invariant operator $\overline{\partial}$ from functions on $P$ to sections of $\Omega^{0,1}_X|_P$ satisfying the appropriate Leibniz rule and $\overline{\partial}^2 = 0$.

We say that a section, map or whatever is *pluriharmonic* if it is harmonic when restricted to any locally defined smooth complex subvariety. This condition is independant of the choice of metric. The classical Bochner formula states that harmonic forms are pluriharmonic. The fundamental result about equivariant harmonic maps on Kähler manifolds is just the analogue:

PROPOSITION 2.2. *If $\phi$ is a harmonic equivariant map from $\tilde{X}$ to $G/K$ then $\phi$ is pluriharmonic.*

*Proof:* See [**79**] [**41**] and [**11**]. □

Suppose $P$ is a flat principal $G$-bundle with flat connection denoted by $d = d' + d''$. Suppose $P_K$ is a $K$-reduction corresponding to equivariant harmonic map $\phi$. We can decompose the connection into a component $d^+$ preserving $P_K$ and a component $a$ orthogonal to $P_K$. Then decompose according to type, $d^+ = \partial + \overline{\partial}$ and $a = \theta' + \theta''$. We obtain the decompositions

$$d' = \partial + \theta'$$

and

$$d'' = \overline{\partial} + \theta''.$$

The components orthogonal to $P_K$ operate on functions only via the restrictions of the functions to the fiber, which is to say that they operate on functions via the Lie algebra $ad(P) = P \times^G \mathbf{g}$ of $G$-invariant vector fields on $P$. Thus these components are sections $\theta'$ of $ad(P) \otimes \Omega^{1,0}_X$ and $\theta''$ of $ad(P) \otimes \Omega^{0,1}_X$. The pluriharmonic map equations translate into:

$$\overline{\partial}^2 = 0;$$

$$\overline{\partial}\theta' + \theta'\overline{\partial} = 0 \text{ (which we write } \overline{\partial}(\theta') = 0);$$

and

$$[\theta', \theta'] = 0.$$

In the last equation the form coefficients are wedged and the Lie algebra coefficients bracketed.

We also obtain of course the complex-conjugate equations for $\partial$ and $\theta''$. These are complex conjugates in view of the hermitian or antihermitian properties of $\partial + \overline{\partial}$ and $\theta' + \theta''$ respectively.

The first equation says that $(P, \overline{\partial})$ has a structure of holomorphic principal bundle. This is in general different from the structure of holomorphic principal bundle $(P, d'')$ which comes from the flat structure. The second equation says that $\theta'$ corresponds to a holomorphic section which we now denote simply by $\theta \in H^0(X, ad(P) \otimes \Omega^1_X)$; and the third equation says $[\theta, \theta] = 0$.

We define a *principal Higgs bundle* to be a holomorphic principal $G$-bundle $P$ together with $\theta \in H^0(X, ad(P) \otimes \Omega^1_X)$ such that $[\theta, \theta] = 0$. From the previous results, a flat principal $G$-bundle $P$ with harmonic $K$-reduction $P_K$ gives a principal Higgs bundle $(P, \theta)$.

If $(P, \theta)$ is a principal Higgs bundle and $V$ is a linear representation of $G$ then we obtain an associated Higgs bundle $(E, \theta_E)$ where $E = P \times^G V$ and $\theta_E \in H^0(X, End(E) \otimes \Omega^1_X)$ is the associated form associated to $\theta$.

Recall that a Higgs bundle $(E, \theta)$ is *stable* if for any subsheaf $F \subset E$ preserved by $\theta$ we have $deg(F)/r(F) < deg(E)/r(E)$ (the notion of degree depends on choice of Kähler class). Say that $E$ is *polystable* if it is a direct sum of stable Higgs bundles of the same slope (degree over rank). Say that a principal Higgs bundle $P$ is *polystable* if for every representation $V$, the associated Higgs bundle is polystable. If the generalized first Chern classes of $P$ (corresponding to all degree one invariant polynomials) vanish then it is enough to check this for one faithful representation $V$ (cf [**71**] p. 86).

THEOREM 2.3. *Suppose $\rho : \pi_1(X) \to G$ with $G$ reductive, and suppose that the Zariski closure of the image of $\rho$ is reductive. Let $\mathcal{P}$ be the associated flat bundle and $\mathcal{P}_K$ be a pluriharmonic reduction. The structure of principal Higgs bundle $(P, \theta)$ obtained above doesn't depend on choice of pluriharmonic reduction $P_K$. The principal Higgs bundle has vanishing rational Chern classes and is polystable. Furthermore, any polystable principal Higgs bundle with vanishing rational Chern classes arises from a unique representation $\rho$ in this way.*

*Proof:* See [**36**] [**69**] [**71**]. □

*Moduli spaces*

Fix a reductive complex algebraic group $G$ and a smooth projective variety $X$.

Let $R_{DR}(X,x,G)$ denote the moduli scheme of principal $G$-bundles with integrable connection and frame at $x \in X$ constructed in [**75**]; similarly let $R_{Dol}(X,x,G)$ denote the moduli scheme of semistable principal Higgs bundles with vanishing rational Chern classes with a frame at $x \in X$; and finally let $R_B(X,x,G) := Hom(\pi_1(X,x),G)$ denote the space of representations of the fundamental group in $G$. In all three cases these schemes represent the appropriate functors. We call these spaces the de Rham, Dolbeault and Betti *representation spaces.*

The group $G$ acts on each of the representation spaces. In all three cases, all points are semistable for an appropriate linearized line bundle, so by [**59**] the universal categorical quotients

$$M_{DR}(X,G) := R_{DR}(X,x,G)//G$$

$$M_{Dol}(X,G) := R_{Dol}(X,x,G)//G$$

$$M_B(X,G) := R_B(X,x,G)//G$$

exist [**55**] [**75**] [**62**]. They are independent of the choice of basepoint. The points of these quotients parametrize the closed orbits in the representation spaces. The closed orbit in the closure of an orbit corresponding to a given representation is the *semisimplification* of the representation. Two points in a representation space map to the same point in the moduli space if and only if their semisimplifications coincide.

For some purposes it is useful to think about the *moduli stacks* instead. These are the stack-theoretic quotients

$$\mathcal{M}_{DR}(X,G) := R_{DR}(X,x,G)/G$$

$$\mathcal{M}_{Dol}(X,G) := R_{Dol}(X,x,G)/G$$

$$\mathcal{M}_B(X,G) := R_B(X,x,G)/G.$$

Properly speaking, it is the moduli stacks which should be thought of as the first nonabelian cohomology stacks. The moduli spaces are the hausdorffifications or associated coarse moduli spaces for the stacks (they universally co-represent the functors $\pi_0$ of the stacks).

We have a complex analytic isomorphism $R_{DR}(X,x,G)^{\mathrm{an}} \cong R_B(X,x,G)^{\mathrm{an}}$ compatible with the action of $G$ coming from the Riemann-Hilbert correspondence between holomorphic systems of ODE's and their monodromy representations. This projects to the universal categorical quotients ([**75**] §5) giving $M_{DR}(X,G)^{\mathrm{an}} \cong M_B(X,G)^{\mathrm{an}}$ as well as to the stack quotients giving $\mathcal{M}_{DR}(X,G)^{\mathrm{an}} \cong \mathcal{M}_B(X,G)^{\mathrm{an}}$.

The correspondence of Theorem 2.3 gives an isomorphism between the underlying sets of points of $M_{DR}(X,G)$ and $M_{Dol}(X,G)$, because the points of these spaces correspond exactly to representations which have reductive Zariski closure (really the corresponding de Rham or Dolbeault analogues of this notion defined using the Tannakian formalism). This isomorphism is a homeomorphism of underlying topological spaces [**75**] which we thus write

$$M_{DR}(X,G)^{\mathrm{top}} \cong M_{Dol}(X,G)^{\mathrm{top}}.$$

Hitchin's original point of view [**36**] was slightly different, in that he constructed a single moduli space for all objects, and noted that it had several different complex structures. This amounts to the same thing if one ignores the algebraic structures (and in fact it is difficult to say anything concrete about the relationship between the algebraic structures).

The homeomorphism between the moduli spaces does not lift to a homeomorphism between the representation spaces (cf the counterexample of [**75**] II, pp 38-39). I don't know what happens when we look at the stacks, so I'll give that as a question for future research.

*Question:* Does there exist a natural homeomorphism

$$\mathcal{M}_{DR}(X,G)^{\text{top}} \cong \mathcal{M}_{Dol}(X,G)^{\text{top}}$$

inducing the previous one on moduli spaces?

In order to attack this question one must first define the notion of the "underlying topological space" of a stack.

## 3. The quaternionic structure on the moduli space

Let $M^{\text{sm}}_{DR}(X,G)$ (resp. $M^{\text{sm}}_{Dol}(X,G)$, $M^{\text{sm}}_{B}(X,G)$) denote the open subset of smooth points of $M_{DR}(X,G)$ (resp. $M_{Dol}(X,G)$, $M_B(X,G)$) parametrizing Zariski-dense representations (the notion of Zariski denseness makes sense for the de Rham or Dolbeault spaces using the Tannakian point of view). Then the isomorphism

$$M^{\text{sm}}_{DR}(X,G)^{\text{top}} \cong M^{\text{sm}}_{Dol}(X,G)^{\text{top}}$$

is $\mathcal{C}^\infty$ (and even real analytic) [**36**] [**22**]. Denoting by $M^{\text{sm}}(X,G)$ the differentiable manifold underlying these isomorphic spaces, we obtain two complex structures $I$ and $J$ on $M^{\text{sm}}(X,G)$ coming respectively from $M^{\text{sm}}_{Dol}(X,G)$ and $M^{\text{sm}}_{DR}(X,G)$.

The tangent space to $M^{\text{sm}}_{DR}(X,G)$ (resp. $M^{\text{sm}}_{Dol}(X,G)$ or $M^{\text{sm}}_{B}(X,G)$) at a point corresponding to a principal bundle $P$ is $H^1_{DR}(X,ad(P))$ (resp. $H^1_{Dol}(X,ad(P))$ or $H^1_B(X,ad(P))$). These tangent spaces have natural $L^2$ metrics coming from the interpretation of classes as harmonic forms.

THEOREM 3.1. (Hitchin [**36**]) *Put $K = IJ$. Then the triple $(I,J,K)$ is a quaternionic structure for the manifold $M^{\text{sm}}(X,G)$. Furthermore if $g$ denotes the natural Riemannian metric on $M^{\text{sm}}(X,G)$ obtained from the $L^2$ metric on the tangent space induced by the harmonic metric (these are the same up to a constant for all structures) then $g$ is a Kähler metric for each of the structures $(I,J,K)$, in other words $M^{\text{sm}}(X,G)$ becomes a hyperkähler manifold.*

*Proof:* See [**36**] for the result when $X$ is a curve. The theorem for any $X$ follows from the corresponding theorem for a curve, using the embedding $M(X,G) \subset M(C,G)$ for a curve $C$ which is a complete intersection of hyperplane sections in $X$. On the other hand Fujiki proves this for a general Kähler manifold $X$ [**22**] (where the method of taking hyperplane sections is no longer available).

We will see how to calculate that $I,J,K$ form a quaternionic structure in the proof of Theorem 4.2 below. □

The embedding $M(X,G) \subset M(C,G)$ obtained from a hyperplane section is complex analytic for all structures, along any naturally defined subvariety such as the Whitney strata of the singular locus. Consequently the smooth points of the underlying reduced scheme structure of these strata inherit quaternionic (and even hyperkähler) structures.

### *The twistor space*

The notion of *twistor space* of a quaternionic manifold $N$ is explained, for example, in [**37**]. Suppose $N$ is a manifold with three integrable complex structures

$(I, J, K)$ defining a quaternionic structure on each tangent space. We identify $\mathbf{P}^1$ with the sphere $x^2 + y^2 + z^2 = 1$ by the stereographic projection; this gives

$$\lambda = u + iv \leftrightarrow (x = \frac{1 - |\lambda|^2}{1 + |\lambda|^2}, y = \frac{2u}{1 + |\lambda|^2}, z = \frac{2v}{1 + |\lambda|^2}).$$

The *twistor space* $TW(N)$ is a complex manifold with a $\mathcal{C}^\infty$ trivialisation

$$TW(N) \cong N \times \mathbf{P}^1$$

such that the projection $TW(N) \to \mathbf{P}^1$ is holomorphic; such that for any $n \in N$ the section $\{n\} \times \mathbf{P}^1 \subset TW(N)$ is holomorphic; and such that for any $\lambda \in \mathbf{P}^1$ the complex structure on $M \times \{\lambda\}$ is $xI + yJ + zK$ where $(x, y, z)$ corresponds to $t$ via the stereographic projection defined above (note that $(xI + yJ + zK)^2 = -1$).

If we denote by $I_{u+iv}$ the complex structure corresponding to $\lambda = u + iv \in \mathbf{A}^1$ then one can see using the above definitions that the formula

$$I_{u+iv} = (1 - uK + vJ)^{-1} I (1 - uK + vJ)$$

holds (we'll need this in the proof of Theorem 4.2 below).

This definition serves to determine an almost complex structure on $TW(N) = N \times \mathbf{P}^1$. The almost complex structure is integrable [**37**] [**2**] [**38**] [**66**]; in our case we will indicate below an explicit construction which is integrable so we can avoid using the general integrability result.

The twistor space has various other structures, notably an antilinear involution $\sigma$ covering the antipodal involution $\sigma_{\mathbf{P}^1}$ of $\mathbf{P}^1$. Define $\sigma(n, t) := (n, \sigma_{\mathbf{P}^1}(t))$. This is antilinear because it is antilinear in the horizontal directions (along prefered sections) and in vertical directions because $-xI - yJ - zK$ is the complex conjugate complex structure to $xI + yJ + zK$.

The prefered sections are by definition $\sigma$-invariant.

If $N$ is a quaternionic vector space of quaternionic rank $r$ then the twistor space may be constructed by hand. It the direct sum of $2r$ copies of $\mathcal{O}_{\mathbf{P}^1}(1)$ over $\mathbf{P}^1$. In this case one can check directly that the prefered sections are the only $\sigma$-invariant sections.

*Application: subvarieties of $M(X, \mathbf{G}_m)$ defined by cohomological conditions*

The quaternionic structure gives a nice way of looking at the results of Green-Lazarsfeld [**24**] on subvarieties defined by cohomological conditions. In [**24**] they look at the subvarietes $\Sigma^i_k(Pic^0) \subset Pic^0(X)$ of line bundles $\mathcal{L}$ with $h^i(\mathcal{L}) \geq k$. They show that these are unions of translates of subtori of $Pic^0(X)$. We look at the subvarieties $\Sigma^i_k(M)$ of $M(X, GL(n))$ consisting of those local systems $V$ such that $h^i(X, V) \geq k$. We have not specified whether we look at $M_B$, $M_{DR}$ or $M_{Dol}$ because the same locus is defined in all three cases, and they correspond under the homeomorphisms $M_B \cong M_{DR} \cong M_{Dol}$. This is due to the fact that if $\rho$ is a representation corresponding to vector bundle $V$ with integrable connection and corresponding to Higgs bundle $E$ then the interpretation of cohomology classes as harmonic forms and the Kähler identities between the laplacians [**71**] gives isomorphisms

$$H^i(X, \rho) \cong H^i_{DR}(X, V) \cong H^i_{Dol}(X, E)$$

(see [**71**] Lemma 2.2). Now $\Sigma^i_k(M_{Dol})$ is a complex analytic subvariety of $M_{Dol}$ whereas $\Sigma^i_k(M_{DR})$ is a complex analytic subvariety of $M_{DR}$. At any smooth point of the reduced subvariety, the tangent space of $\Sigma^i_k(M)$ is preserved by both complex structures $I$ and $J$ of the quaternionic structure; thus at smooth points $\Sigma^i_k$

is a quaternionic submanifold of $M^{\rm sm}$. This puts a big restriction on the possibilities for $\Sigma^i_k$ (for example, it must have real dimension divisible by 4 i.e. even complex dimension; the same is true for any stratum in its Whitney stratification, for intersections of various $\Sigma^i_k$, etc.).

Deligne pointed out [**14**] that we can use this compatibility with the quaternionic structure to recover the results of [**24**]. This method (which I described as an alternative in [**72**]) adds to the many various points of view on [**24**] that are now available ([**1**] [**3**] [**9**] [**72**]). It seems worthwhile to mention the quaternionic point of view here, since similar considerations may come into play for local systems of higher rank.

Deligne makes use of the following observation which is probably classical.

LEMMA 3.2. *Any locally defined smooth quaternionic subvariety of a quaternionic vector space is flat (i.e. a linear subspace).*

*Proof:* If it were not flat, the second fundamental form would be a quaternionic quadratic form, but an easy calculation shows that this cannot exist. □

COROLLARY 3.3. *If $G = \mathbf{G}_m$ then the $\Sigma^i_k(M) \subset M(X, \mathbf{G}_m)$ are unions of translates of subtori.*

*Proof:* The universal covering of $M(X, G)$ is just $H^1(X, \mathbf{C})$ and the quaternionic structure here is linear. Thus the previous lemma (which is a local statement) applies to show that $\Sigma^i_k(M)$ is flat at smooth points of reduced irreducible components. A standard argument (such as in [**24**]) gives the conclusion. □

This property and its generalizations (one can show that the translates of subtori are translations by torsion points [**3**] [**9**] [**1**] [**72**]) have implications for the topology of $X$. One can easily fabricate examples of homotopy types such that the corresponding jump loci are not translates of subtori.

We give a crude version here involving additions of 2- and 3-cells to a torus (one can analyze in a similar way examples made by adding cells of any dimensions). Put $\Gamma := \mathbf{Z}^a$ with $a$ even, and put $U_1 = K(\Gamma, 1)$ (which we can take as a real torus). Note that

$$\mathbf{C}\Gamma \cong \mathbf{C}[t_1, t_1^{-1}, \ldots, t_a, t_a^{-1}]$$

is the Laurent polynomial ring in $a$ variables, and

$$M_B(U_1, \mathbf{G}_m) = Hom(\Gamma, \mathbf{G}_m) \cong \mathbf{G}_m^a.$$

In fact one can canonically identify $M_B(U_1, \mathbf{G}_m) \cong Spec(\mathbf{C}\Gamma)$ (be careful that this reasoning only works well for $G = \mathbf{G}_m$). Let $u \in U_1$ be the basepoint and let $U_2$ be obtained from $U_1$ by attaching $m$ 2-spheres at $u$. Finally let $U_3$ be obtained by attaching $\ell$ 3-cells to $U_2$ with attaching maps $\alpha_i \in \pi_2(U_2, u)$ for $i = 1, \ldots, \ell$. We calculate the cohomology jump loci $\Sigma^i_k(U_3) \subset M_B(U_3, \mathbf{G}_m)$ for $i = 2, 3$. Note that $U_1 \hookrightarrow U_3$ induces an isomorphism on $\pi_1$ so $M_B(U_3, \mathbf{G}_m) = Spec(\mathbf{C}\Gamma)$ too. If $L$ is a rank one local system corresponding to a nontrivial representation $\rho : \Gamma \to \mathbf{G}_m$ then $H^2(U_2, L) \cong L_x^m$ and Mayer-Vietoris gives an exact sequence

$$0 \to H^2(U_3, L) \to L_x^m \stackrel{A(\rho)}{\to} L_x^\ell \to H^3(U_3, L) \to 0.$$

The matrix $A(\rho)$ comes from the attaching maps: we have $\pi_2(U_2) \cong (\mathbf{C}\Gamma)^m$ so the collection $\{\alpha_i\}$ can be considered as an $\ell \times m$ matrix $A$ with coefficients in $\mathbf{C}\Gamma$. The matrix $A(\rho)$ is obtained by evaluating $A$ at the algebra homomorphism $\mathbf{C}\Gamma \stackrel{\rho}{\to} \mathbf{C}$. In this case the jump loci $\Sigma^2_k(U_3) = \Sigma^3_{k+\ell-m}(U_3)$ are the sets of $\rho \in M_B(U_3, \Gamma)$

where $A(\rho)$ has rank $\leq m-k$. In particular they are defined by the ideals of $m-k$ by $m-k$ minors of $A$. Since our choice of $\alpha_i$ and hence of $A$ is arbitrary (except that the matrix must actually have coefficients in $\mathbf{Z}\Gamma$), we can get our jump loci to be any subscheme of $\mathbf{G}_m^a$ defined by equations in $\mathbf{Z}\Gamma$, that is any subscheme defined over $\mathbf{Z}$.

We can, for example, get the jump loci to be subschemes which are not of even complex dimension and in any case not unions of translates of subtori—this gives constructions of many homotopy types which cannot be the homotopy types of compact Kähler manifolds. We can arrange that the jump loci do not go through the identity representation (or even any torsion point), in particular this characteristic of the homotopy type will not be seen by rational homotopy theory (we can insure that the cohomology with constant coefficients and even the rational homotopy type are those of the real torus e.g. an abelian variety of dimension $a/2$, or equally well those of any complex subvariety with the same $\pi_1$).

See **[34]** for an analysis, in a similar spirit, of the fundamental group using properties of the jump loci.

Getting back to the result of Corollary 3.3, we can recover the results of Green and Lazarsfeld by looking at the Dolbeault realization. There is a natural embedding $Pic^0(X) \subset M_{Dol}(X, \mathbf{G}_m)$ sending a line bundle $\mathcal{L}$ to the Higgs bundle $(\mathcal{L}, 0)$. The Dolbeault cohomology of $(\mathcal{L}, 0)$ is just the direct sum of the $H^{i-k}(X, \mathcal{L} \otimes \Omega_X^k)$. It is a consequence of semicontinuity that the jump loci for a direct sum must contain as irreducible components the jump loci for each of the factors. Thus the irreducible components of $\Sigma_k^i(Pic^0)$ are among the irreducible components of $\Sigma_k^i(M_{Dol})$, and the conclusion of the corollary implies the result of **[24]**.

## 4. Deligne's construction of $TW(M^{\mathrm{sm}})$

In **[14]** Deligne indicated a complex analytic construction of the complex manifold $TW(M^{\mathrm{sm}})$ (the idea is based on some properties that Hitchin established). This is interesting because the construction given above of the quaternionic structure on $M^{\mathrm{sm}}(X, G)$ comes from the homeomorphism $M_{DR}(X, G) \cong M_{Dol}(X, G)$ which itself comes from the non -complex analytic harmonic metric construction. Of course the trivialization $TW(M^{\mathrm{sm}}(X, G)) \cong M^{\mathrm{sm}}(X, G) \times \mathbf{P}^1$ depends on the harmonic metric construction.

It turns out that Deligne's construction of $TW(M^{\mathrm{sm}}(X, G))$ is useful for two other things that were not mentioned in **[14]**: (1) it gives an approach to defining the nonabelian analogue of the Hodge filtration on $M_{DR}(X, G)$; and (2) it gives a way of compactifying $M_{DR}(X, G)$. Both of these were announced in **[70]** but with only brief sketches of proofs. In this paper we will fill in the details about these two things and give some natural extensions of these ideas. Before doing that, we review Deligne's construction, since it has not otherwise appeared in print (to my knowledge).

An *antilinear morphism* $T \to T'$ between two complex analytic spaces is a morphism of ringed spaces such that the composition $\mathbf{C} \to \mathcal{O}_{T'} \to \mathcal{O}_T$ is the complex conjugate of the structural morphism $\mathbf{C} \to \mathcal{O}_T$. An *antilinear involution* of $T$ is an antilinear morphism $\sigma : T \to T$ with $\sigma^2 = 1$.

Let $\sigma_{\mathbf{P}^1}$ denote the antilinear antipodal involution of $\mathbf{P}^1$. If $z$ is the standard linear coordinate on $\mathbf{A}^1$ then $\sigma(z) = -\overline{z}^{-1}$. Let $\sigma_{\mathbf{G}_m}$ denote the restriction of $\sigma$ to $\mathbf{G}_m \subset \mathbf{P}^1$.

The data of a morphism of complex analytic spaces $T \to \mathbf{P}^1$ together with an antilinear involution $\sigma$ covering $\sigma_{\mathbf{P}^1}$ is equivalent to the data of a morphism $T' \to \mathbf{A}^1$ and an antilinear involution $\sigma'$ of $T'_{\mathbf{G}_m} := T' \times_{\mathbf{A}^1} \mathbf{G}_m$. Given $T'$ and $\sigma'$, let $\overline{T}'$ denote the complex conjugate analytic space (that is the same ringed space but with structural morphism $\mathbf{C} \to \mathcal{O}_{\overline{T}'}$ the complex conjugate of the structural morphism for $T'$). The involution $\sigma'$ becomes a complex linear isomorphism

$$T'_{\mathbf{G}_m} \cong \overline{T}'_{\mathbf{G}_m},$$

which we can use to glue $T'$ to $\overline{T}'$ to obtain $T$. By construction $T$ comes with an antilinear involution $\sigma$ (it comes from the tautological antilinear morphism $T' \to \overline{T}'$ and its inverse). One can see that $\mathbf{P}^1$ is obtained from $\mathbf{A}^1$ and the involution $\sigma_{\mathbf{G}_m}$ by the same construction, so we obtain our map $T \to \mathbf{P}^1$ compatible with involutions.

To give a holomorphic $\sigma$-invariant section $\eta : \mathbf{P}^1 \to T$ it suffices to give a holomorphic section $\eta' : \mathbf{A}^1 \to T'$ such that $\eta'|_{\mathbf{G}_m}$ is $\sigma'$-invariant.

Hitchin noticed that the twistor space $TW(M^{\mathrm{sm}})$ comes equipped with an action of $\mathbf{G}_m$ identifying the fibers over all different $\lambda \in \mathbf{G}_m \subset \mathbf{P}^1$ [**36**] (note however that the twistor space of a general hyperkähler manifold doesn't come equipped with such an action). Deligne's idea is to use this and the remark of the previous paragraph to obtain a direct construction of $TW(M^{\mathrm{sm}})$.

For simplicity we treat the case $G = GL(n, \mathbf{C})$ but the case of a general reductive group can be treated directly by working with principal bundles, or indirectly using the Tannakian formalism such as in [**75**] (note that for the constructions of moduli spaces the indirect Tannakian method is the only one I know of).

Deligne makes the following definition. Suppose $\lambda : S \to \mathbf{A}^1$ is a morphism. A *$\lambda$-connection* on a vector bundle $E$ over $X \times S$ consists of an operator

$$\nabla : E \to E \otimes_{\mathcal{O}} \Omega^1_{X \times S/S}$$

such that $\nabla(ae) = \lambda e \otimes d(a) + a\nabla(e)$ (Leibniz rule multiplied by $\lambda$) and such that $\nabla^2 = 0$ as defined in the usual way (integrability).

Note that if $\lambda = 1$ then this is the same as the usual notion of connection, whereas if $\lambda = 0$ then this is the same as the notion of Higgs field making $(E, \nabla)$ into a Higgs bundle [**36**] [**71**].

PROPOSITION 4.1. *Fix $x \in X$. The functor which to $\lambda : S \to \mathbf{A}^1$ associates the set of triples $(E, \nabla, \beta)$ where $E$ is a vector bundle on $X \times S$, $\nabla$ is a $\lambda$-connection on $E$ (such that the resulting Higgs bundles over $\lambda = 0$ are semistable with vanishing rational Chern classes), and $\beta : E|_{\{x\}\times S} \cong \mathcal{O}_S^n$ is a frame, is representable by a scheme $R_{Hod}(X, x, GL(n)) \to \mathbf{A}^1$. The group $GL(n)$ acts on $R_{Hod}(X, x, GL(n))$ by change of frame and all points are semistable for this action (with respect to the an appropriate linearized bundle). The geometric-invariant theory quotient $M_{Hod}(X, GL(n)) \to \mathbf{A}^1$ is a universal categorical quotient. In particular the fibers of $M_{Hod}$ over $\lambda = 0$ and $\lambda = 1$ are $M_{Dol}(X, GL(n))$ and $M_{DR}(X, GL(n))$ respectively.*

*Proof:* This follows by applying the results of [**75**] to the ring $\Lambda^R$ defined in [**75**] p. 87. □

*Remark:* We can define the notion of $\lambda$-connection on a principal bundle, and obtain the corresponding statement for principal $G$-bundles. We obtain schemes $R_{Hod}(X, x, G)$ and $M_{Hod}(X, G)$. The construction is done by applying the Tannakian considerations of [**75**] §9.

Concerning the terminology $R_{Hod}$ and $M_{Hod}$: this reflects the fact that, as we shall see below, these spaces incarnate the Hodge filtrations on $R_{DR}$ and $M_{DR}$.

Let $\mathcal{M}_{Hod}(X, GL(n))$ (or $M_{Hod}(X, G)$) denote the stack-theoretic quotient of $R_{Hod}(X, x, GL(n))$ by $GL(n)$ (or $R_{Hod}(X, x, G)$ by $G$). The group $\mathbf{G}_m$ acts on the functor $\{(E, \nabla, \beta)\}$ over its action on $\mathbf{A}^1$: if $t \in \mathbf{G}_m(S)$ and $(E, \nabla, \beta)$ is a $\lambda$-connection then $(E, t\nabla, \beta)$ is a $t\lambda$ connection. Since $R_{Hod}(X, x, GL(n))$ represents the functor, we get an action of $\mathbf{G}_m$ on $R_{Hod}(X, x, GL(n))$ covering its action on $\mathbf{A}^1$. Since $M_{Hod}(X, GL(n))$ is a universal categorical quotient, this descends to an action on $M_{Hod}(X, GL(n))$. This action serves to identify the fibers over any $\lambda, \lambda' \neq 0$ in $\mathbf{A}^1$—they are all isomorphic to $M_{DR}(X, GL(n))$.

The space $M_{Hod}(X, GL(n))$ will play the role of the space $T'$ in constructing the twistor space $T$. According to our general discussion, in order to obtain $T$ by glueing, it suffices to have an antilinear involution $\sigma'$ of $M_{Hod}(X, GL(n)) \times_{\mathbf{A}^1} \mathbf{G}_m$. As we have seen above, the action of $\mathbf{G}_m$ gives an isomorphism

$$M_{Hod}(X, GL(n)) \times_{\mathbf{A}^1} \mathbf{G}_m \cong M_{DR}(X, GL(n)) \times \mathbf{G}_m.$$

On the other hand we have an antilinear involution $\tau$ of $M_B(X, GL(n))$ obtained by setting $\tau(\rho)$ equal to the dual of the complex conjugate representation (where complex conjugation is taken with respect to the real structure $GL(n, \mathbf{R})$; the dual of the complex conjugate is also the complex conjugate with respect to the compact real form). To be totally explicit, for $\gamma \in \pi_1(X, x)$ we set $\tau(\rho)(\gamma) := {}^t\overline{\rho(\gamma)}^{-1}$. The complex analytic isomorphism $M_B(X, GL(n))^{\mathrm{an}} \cong M_{DR}(X, GL(n))^{\mathrm{an}}$ given by the Riemann-Hilbert correspondence allows us to interpret $\tau$ as an antilinear involution of $M_{DR}(X, GL(n))$. Finally we define the involution $\sigma'$ of $M_{DR}(X, GL(n)) \times \mathbf{G}_m$ by the formula $\sigma'(u, \lambda) = (\tau(u), -\overline{\lambda}^{-1})$. Using $\sigma'$ and $T' = M_{Hod}(X, GL(n))$ in the general recipe given above, we obtain a space $T$ which we denote $M_{Del}(X, GL(n)) \to \mathbf{P}^1$.

The complex conjugate scheme $\overline{T'} = \overline{M_{Hod}(X, GL(n))}$ which appears above can be identified with $M_{Hod}(\overline{X}, GL(n))$.

*Exercise:* write down the glueing isomorphism between $M_{Hod}(X, GL(n))$ and $M_{Hod}(\overline{X}, GL(n))$ over $\mathbf{G}_m \subset \mathbf{A}^1$. Note that it will be analytic but not algebraic (depending on the Riemann-Hilbert correspondence).

We note rapidly some properties of $M_{Del}(X, GL(n))$ which are immediate consequences of the construction. The fiber of $M_{Del}(X, GL(n)) \to \mathbf{P}^1$ over a point $\lambda \in \mathbf{P}^1$ (which is denoted using the coordinate system of the first embedding of $\mathbf{A}^1$ which corresponds to the part concerning $X$) is equal to $M_{Dol}(X, GL(n))^{\mathrm{an}}$ if $\lambda = 0$; the fiber is isomorphic to

$$M_{DR}(X, GL(n))^{\mathrm{an}} \cong M_B(X, GL(n))^{\mathrm{an}} \cong M_{DR}(\overline{X}, GL(n)))^{\mathrm{an}}$$

if $\lambda \neq 0, \infty$; and the fiber is equal to $M_{Dol}(\overline{X}, GL(n))^{\mathrm{an}}$ if $\lambda = \infty$. There is an analytic action of $\mathbf{G}_m$ covering the standard action on $\mathbf{P}^1$ (this action is constructed by glueing the natural action over the first open set $M_{Hod}(X, GL(n))^{\mathrm{an}}$ with the composition with $i$ of the natural action on the second open set $M_{Hod}(\overline{X}, GL(n))^{\mathrm{an}}$). The antilinear involution $\sigma$ of $M_{Del}(X, GL(n))$ comes from the first version of the construction discussed above.

*Remark:* Let $R_{Hod}(X, x, GL(n))$ denote the representation space constructed with reference to the basepoint $x \in X$. We can construct, exactly as above, an involution $\sigma$ which can be thought of as an isomorphism between the inverse images of $\mathbf{G}_m \subset \mathbf{A}^1$ in $R_{Hod}(X, x, GL(n))$ and $R_{Hod}(\overline{X}, \overline{x}, GL(n))$. We obtain

$R_{Del}(X,x,GL(n)) \to \mathbf{P}^1$ by glueing $R_{Hod}(X,x,GL(n))$ to $R_{Hod}(\overline{X},\overline{x},GL(n))$ using this isomorphism. The group $GL(n,\mathbf{C})$ acts analytically and on each open subset the associated moduli space $M_{Hod}(X,GL(n))$ is a universal categorical quotient in the analytic category ([**75**] §5). The glueing (along invariant open sets which are pullbacks of open sets in the quotients) preserves this property, so

$$R_{Del}(X,x,GL(n)) \to M_{Del}(X,GL(n))$$

is a universal categorical quotient by the action of $GL(n,\mathbf{C})$ in the analytical category. There is again an action of $\mathbf{G}_m$ on $R_{Del}(X,x,GL(n))$ and the fibers are again respectively

$$R_{Dol}(X,x,GL(n))^{\mathrm{an}}, \quad R_B(X,x,GL(n))^{\mathrm{an}}, \quad \text{and} \quad R_{Dol}(\overline{X},\overline{x},GL(n))$$

over $\lambda=0$, $\lambda\neq 0,\infty$, and $\lambda=\infty$ in $\mathbf{P}^1$.

Denote by $\mathcal{M}_{Del}(X,GL(n))$ the stack-theoretic quotient of $R_{Del}(X,x,GL(n))$ by the action of $GL(n)$; it is an analytic stack with a morphism to $\mathbf{P}^1$.

We now show how a *harmonic bundle* defines a section $\mathbf{P}^1 \to M_{Del}(X,GL(n))$ which we refer to as a *prefered section.* As mentioned before, it suffices to obtain a $\sigma'$-invariant section $\mathbf{A}^1 \to M_{Hod}(X,GL(n))$.

Suppose $P$ is a flat principal $GL(n)$-bundle. Choose a pluriharmonic $K$-reduction $P_K$ and consider the decomposition defined previously

$$d' = \partial + \theta',$$

$$d'' = \overline{\partial} + \theta''.$$

For $\lambda \in \mathbf{A}^1$ we define a holomorphic structure

$$\overline{\partial}_\lambda := \overline{\partial} + \lambda\theta'',$$

and an operator

$$\nabla_\lambda := \lambda\partial + \theta'.$$

We claim that $\overline{\partial}_\lambda$ is an integrable holomorphic structure and $\nabla_\lambda$ an integrable holomorphic $\lambda$-connection on $(P,\overline{\partial}_\lambda)$. The equations $\overline{\partial}^2 = 0$, $[\theta'',\theta''] = 0$ and $(d'')^2 = 0$ imply that $\overline{\partial}(\theta'') = 0$ and hence $\overline{\partial}_\lambda^2 = 0$. Similarly, $\overline{\partial}(\theta') = 0$ and $\partial(\theta'') = 0$ and furthermore we get

$$\overline{\partial}\partial + \partial\overline{\partial} + \theta'\theta'' + \theta''\theta' = 0$$

which gives $[\overline{\partial}_\lambda, \nabla_\lambda] = 0$; finally by the same argument as previously $\partial(\theta') = 0$ so $\nabla_\lambda^2 = 0$. This gives the claim. Note that at $\lambda = 0$ we recover the Higgs bundle structure $(\overline{\partial},\theta')$ which we know to be polystable with vanishing Chern classes. This construction thus gives a section $\mathbf{A}^1 \to M_{Hod}(X,GL(n))$. It is holomorphic in $\lambda$ (since $\lambda$ appears linearly in the equations).

We have to check that our section is $\sigma'$-invariant over $\mathbf{G}_m \subset \mathbf{A}^1$. This is a bit technical so feel free to skip it! A point of $M_{Hod}(X,GL(n))$ can be represented as a quadruple $(E,\delta',\delta'',\lambda)$ where $E$ is a $\mathcal{C}^\infty$ bundle, $\lambda\in\mathbf{C}$, $\delta$ is an operator satisfying Leibniz' rule for $\lambda\partial$, and $\delta''$ is an operator satisfying Leibniz' rule for $\overline{\partial}$, such that $(\delta')^2 = 0$, $(\delta'')^2 = 0$, and $\delta'\delta'' + \delta''\delta' = 0$. If $\lambda \neq 0$ this corresponds to a flat bundle $(E, \lambda^{-1}\delta' + \delta'')$. The dual complex conjugate flat bundle (corresponding to the dual of the complex conjugate representation on $X$) is $(\overline{E}^*, \overline{\delta''}^* + \overline{\lambda}^{-1}\overline{\delta'}^*)$ (the superscript $^*$ on the operators means the induced operators on the dual). If we take the point obtained by multiplying this complex conjugate flat bundle by $-\overline{\lambda}^{-1}$, we obtain the point

$$\sigma(E,\delta',\delta'',\lambda) = (\overline{E}^*, -\overline{\lambda}^{-1}\overline{\delta''}^*, \overline{\lambda}^{-1}\overline{\delta'}^*, -\overline{\lambda}^{-1}).$$

We have to check that this operation preserves our preserved section, which is the collection of points of the form $(E, \lambda\partial+\theta', \overline{\partial}+\lambda\theta'', \lambda)$. We have an isomorphism $\overline{E}^* \cong E$ given by the harmonic metric, and via this isomorphism the dual complex conjugation operation has the following effect on operators:

$$\partial \leftrightarrow \overline{\partial},$$

$$\theta' \leftrightarrow -\theta'$$

(this is from the definition of $\overline{\partial}+\partial$ and $\theta'+\theta''$ as the components parallel to and perpendicular to the unitary structure). In view of these formulae, when we apply the operation $\sigma$ to such a point we get

$$\sigma(E,\lambda\partial+\theta',\overline{\partial}+\lambda\theta'',\lambda) = (\overline{E}^* \cong E, -\overline{\lambda}^{-1}\partial+\theta', \overline{\partial}-\overline{\lambda}^{-1}\theta'', -\overline{\lambda}^{-1})$$

which is indeed a point on our prefered section.

It is clear from the definition that through any point of $M_{Del}(X, GL(n))$ passes exactly one prefered section.

The set of prefered sections gives a set-theoretic trivialization

$$M_{Del}(X, GL(n)) \cong M_B(X, GL(n)) \times \mathbf{P}^1.$$

This trivialisation is a homeomorphism, as can be seen by using the techniques of [**75**] which are used in proving that $M_{DR}(X, GL(n))^{\mathrm{top}} \cong M_{Dol}(X, GL(n))^{\mathrm{top}}$. This is also verified in [**22**].

Let $M^{\mathrm{sm}}_{Del}(X, GL(n))$ denote the open subset of $M_{Del}(X, GL(n))$ where the projection to $\mathbf{P}^1$ is smooth. By the etale local triviality of $M_{Hod}$ (explained in §9 below) a point in $M_{Del}(X, GL(n))$ lies in $M^{\mathrm{sm}}_{Del}(X, GL(n))$ if and only if it is a smooth point of the fiber $M_{Dol}(X, GL(n))$, $M_{DR}(X, GL(n))$ or $M_{Dol}(\overline{X}, GL(n))$.

The trivialisation via prefered sections gives

$$M^{\mathrm{sm}}_{Del}(X, GL(n))^{\mathrm{top}} \cong M^{\mathrm{sm}}(X, GL(n))^{\mathrm{top}} \times \mathbf{P}^1.$$

This is in fact a $\mathcal{C}^\infty$ isomorphism, as follows from the construction of the prefered sections and the fact that the harmonic maps or metrics vary smoothly with parameters (since they are solutions of the appropriate kind of nonlinear elliptic equation).

THEOREM 4.2. (Deligne) *The space $M^{\mathrm{sm}}_{Del}(X, GL(n))$ with all of its structures is analytically isomorphic to the twistor space $TW(M^{\mathrm{sm}})$; via this isomorphism, the prefered section trivialisations of $M^{\mathrm{sm}}_{Del}(X, GL(n))$ and $TW(M^{\mathrm{sm}})$ coincide.*

*Proof:* This is actually a consequence of the properties obtained by Hitchin for his twistor space in [**36**]. For intrepid readers, we indicate a self-contained calculation—partly because this also serves to show that $(I, J, K)$ defined a quaternionic structure in the first place. Both the twistor space and $M^{\mathrm{sm}}_{Del}(X, GL(n))$ are $\mathcal{C}^\infty$ isomorphic to the product $M^{\mathrm{sm}} \times \mathbf{P}^1$. This gives the isomorphism between the two. Furthermore we know in both cases that the horizontal sections $\{x\} \times \mathbf{P}^1$ are holomorphic, so the isomorphism is analytic in the horizontal direction. We have to check that this isomorphism is compatible with the complex structures in the vertical direction. Choose a tangent direction to $M^{\mathrm{sm}}$ which we will look at first in the Dolbeault realization. The tangent direction can be thought of as a change of operator $\overline{\partial}+\theta' \mapsto \overline{\partial}+\theta'+\alpha$ where $\alpha$ is an endomorphism-valued form representing the cohomology class of the tangent vector. We may (by gauging back if necessary) assume that the associated harmonic metric remains fixed; the infinitesimal change $\alpha$ then induces a change of operator $\partial+\theta'' \mapsto \partial+\theta''+\beta$. Write $\alpha=\alpha'+\alpha''$ and $\beta=\beta'+\beta''$ according to type. In terms of the isomorphism $E\cong \overline{E}^*$ the condition that the fixed metric still relates our new operators is

$$\beta' = \overline{\alpha''}^*$$
$$\beta'' = -\overline{\alpha'}^*.$$

One can see that if $\alpha$ is harmonic then the form $\beta$ defined by these formulas is also harmonic. If we denote by $B(\alpha)$ the form $\beta$ defined by these formulas then $B$ becomes an endomorphism of the space of harmonic forms. It is antilinear (that is $Bi=-iB$), and $B^2=-1$. Thus $B$ is another complex structure which forms part of a quaternionic triple with $i$.

The complex structure $I$ on $M_{Dol}(X,GL(n))$ corresponds to multiplication of $\alpha$ by $i$ (because $\alpha$ is the representative of our tangent vector in the Dolbeault realization). The complex structure on $M_{DR}(X,GL(n))$ is the operator on $\alpha$ which causes $\alpha+\beta=\alpha+B(\alpha)$ to be multiplied by $i$. Thus we have the formula

$$I(1+B)\alpha=(1+B)J\alpha.$$

From whence $J=IB$. This now shows that the pair $(I,J)$ form a part of a quaternionic triple, for which $B=-K$ (Theorem 3.1).

For $\lambda\in\mathbf{A}^1$ the change of associated $\lambda$-connection is

$$\lambda\partial+\theta'+\overline{\partial}+\lambda\theta'' \mapsto \lambda\partial+\theta'+\overline{\partial}+\lambda\theta''+\lambda\beta'+\alpha'+\alpha''+\lambda\beta''.$$

Thus if $I_\lambda$ denotes the complex structure on the fiber of $M_{Del}(X,GL(n))$ over $\lambda\in\mathbf{A}^1$ then we get the formula

$$I(1+\lambda B)=(1+\lambda B)I_\lambda.$$

One has to be careful about what $\lambda B$ means: if $\lambda=u+iv$ then $\lambda B=uB+vIB$. We obtain (replacing $B$ by $-K$):

$$I_{u+iv}=(1-uK+vJ)^{-1}I(1-uK+vJ).$$

This coincides with the formula given in §3. □

**Question:** Are the prefered sections the only sections which are preserved by the involution $\sigma$?

This is certainly locally true, since the normal bundle to a prefered section is a direct sum of $\mathcal{O}_{\mathbf{P}^1}(1)$. In fact, locally the morphism from the space of all sections to the product of any two distinct fibers is an isomorphism. If we take two antipodal fibers then $\sigma$ gives an antilinear involution of the product of the two fibers, and the prefered sections correspond to the fixed points.

Hitchin's discussion in **[37]** (Theorem 1) is a bit unclear on this point: as written the converse in Hitchin's Theorem 1 would imply that the answer is yes in general, but one can easily imagine that he meant only to look at the real sections in the given family of sections. A glance at **[38]** didn't resolve the problem, so I think that the answer to the above question is not known.

An affirmative answer would mean that $(M_{Del}(X,GL(n)),\sigma)$ determines the twistor space structure and in particular the isomorphism $M_{DR}\cong M_{Dol}$, an interesting point since the construction of $(M_{Del}(X,GL(n)),\sigma)$ is entirely complex analytic, so we could bypass the nonlinear elliptic theory necessary to define the harmonic metrics—conceptually speaking at least.

The answer to this question is 'yes' for the twistor space of a quaternionic vector space. As a consequence we obtain this property for the moduli space of rank one representations:

THEOREM 4.3. *Suppose $G = \mathbf{G}_m$. Then the prefered sections are the only $\sigma$-invariant sections of $TW(M^{\mathrm{sm}}) \to \mathbf{P}^1$.*

*Proof:* The moduli space is a quotient $M = H^1(X,\mathbf{C})/H^1(X,\mathbf{Z})$ (as can be seen by a flat version of the exponential exact sequence). The quaternionic structure is the quotient by the lattice of a linear quaternionic structure on $H^1(X,\mathbf{C})$. Thus the twistor space is the quotient

$$TW(M) = TW(H^1(X,\mathbf{C}))/H^1(X,\mathbf{Z}).$$

Since $\mathbf{P}^1$ is simply connected, the sections from $\mathbf{P}^1$ to $TW(M)$ are just projections of sections from $\mathbf{P}^1$ to $TW(H^1(X,\mathbf{C})$. The involution $\sigma$ acts compatibly on everything. From the theory of the twistor space for quaternionic vector spaces (which is just a bundle which is a direct sum of $\mathcal{O}_{\mathbf{P}^1}(1)$) we see that through any point of $TW(H^1(X,\mathbf{C}))$ there is a unique $\sigma$-invariant section; this gives the same result on $TW(M)$ which implies the theorem. □

## 5. The Hodge filtration

In [**36**] Hitchin introduced an $S^1$ action on the moduli space of representations. This was taken up again in [**71**] as a $\mathbf{C}^*$-action. This action is defined via the isomorphism $M_B^{\mathrm{top}} \cong M_{Dol}^{\mathrm{top}}$ by saying that $t \in \mathbf{C}^*$ sends the Higgs bundle $(E,\theta)$ to $(E, t\theta)$.

The $\mathbf{C}^*$ or $S^1$ actions are the analogue in nonabelian Hodge theory of the Hodge decomposition of cohomology coming from harmonic forms. In the usual case, the Hodge decomposition does not vary holomorphically with parameters, because it includes complex conjugate information. Similarly, if the variety is defined over a small field, there is no particular reason for the Hodge decomposition to be defined over a small field. In order to obtain something which comes from algebraic geometry and thus has the properties of holomorphic variation, and compatibility with fields of definition, one looks at the *Hodge filtration* of the algebraic de Rham cohomology. Our purpose here is to define and investigate the analogue for nonabelian cohomology (these ideas were sketched in [**70**]).

### *The Rees bundle of a filtered vector space*

We begin by looking at filtered vector spaces from a slightly unorthodox point of view. Suppose $V$ is a finite dimensional vector space with complete decreasing filtration $F^{\cdot}$ (complete means that the filtration starts with $V$ and ends with $\{0\}$). Define a locally free sheaf $\xi(V,F)$ over $\mathbf{A}^1$ (which we call the *Rees bundle* of $(V,F^{\cdot})$) with action of $\mathbf{G}_m$ as follows. Let $j:\mathbf{G}_m \to \mathbf{A}^1$ denote the inclusion. Then $\xi(V,F)$ is the subsheaf of $j_*(V\otimes_{\mathbf{C}} \mathcal{O}_{\mathbf{G}_m})$ generated by the sections of the form $z^{-p}v_p$ for $v_p \in F^pV$ where $z$ denotes the coordinate on $\mathbf{A}^1$. Historically the standard approach is to define the *Rees module* of $(V,F)$ as the $\mathbf{C}[z]$-submodule

$$B \subset \mathbf{C}[z,z^{-1}] \otimes_{\mathbf{C}} V$$

generated (as a $\mathbf{C}[z]$-module) by elements of the form $z^{-p}v_p$ for $v_p \in F^pV$. Our locally free sheaf $\xi(V,F)$ is then the coherent sheaf associated to the $\mathbf{C}[z]$-module

$B$. The action of $\mathbf{G}_m$ on $\xi(V,F)$ is induced by the standard action of translation on $j_*(V\otimes_{\mathbf{C}}\mathcal{O}_{\mathbf{G}_m})$. In terms of modules, this means that $\mathbf{G}_m$ acts on $\mathbf{C}[z,z^{-1}]\otimes_{\mathbf{C}}V$ with $\lambda\in\mathbf{G}_m$ sending $z^iv$ to $\lambda^{-i}z^iv$ ($v\in V$). This induces an action on $\mathbf{C}[z,z^{-1}]\otimes_{\mathbf{C}} V$ which preserves the submodule $B$.

If one chooses a basis $\{v_i\}$ for $V$ adapted to the filtration with $v_i\in F^{p(i)}-F^{p(i)+1}$ then the $z^{-p(i)}v_i$ form a basis for the free $\mathbf{C}[z]$-module $B$.

If $(V,F)$ is a filtered vector space then the fiber $\xi(V,F)_0$ over $0\in\mathbf{A}^1$ is naturally identified with the associated-graded $Gr_F^{\cdot}(V):=\bigoplus F^p/F^{p+1}$. In terms of modules and an adapted basis (following the notations of the previous paragraph) the classes $[v_i]$ with $p(i)=p$ form a basis for $Gr_F^p(V)$. The $z^{-p(i)}v_i$ (mod $zB$) form a basis for the fiber $\xi(V,F)_0$ which is equal to $B/zB$. Comparing these bases we get an isomorphism $Gr_F^{\cdot}(V)\cong B/zB$ and it is easy to see that this isomorphism is independant of the choice of adapted basis.

The construction $(V,F^{\cdot})\mapsto\xi(V,F)$ has an inverse. If $W$ is a locally free coherent sheaf on $\mathbf{A}^1$ with action of $\mathbf{G}_m$ then we obtain a decreasing filtration $F$ on the fiber $V:=W_1$ of $W$ over $1\in\mathbf{A}^1$ by looking at orders of zeros of $\mathbf{G}_m$-invariant sections. More precisely, $W$ corresponds to the free $\mathbf{C}[z]$-module $B:=W(\mathbf{A}^1)$ (it is free because all vector bundles on $\mathbf{A}^1$ are trivial). The action of $\mathbf{G}_m$ gives a trivialization of $W$ over $\mathbf{G}_m\subset\mathbf{A}^1$ which, in terms of modules, becomes an isomorphism

$$B\otimes_{\mathbf{C}[z]}\mathbf{C}[z,z^{-1}]\cong V\otimes_{\mathbf{C}}\mathbf{C}[z,z^{-1}].$$

Taking the quotient on the right by the ideal $(z-1)$ yields $V$, whereas on the left we get $B/(z-1)B$. Thus we can say that the vector space $V$ is canonically identified with $B/(z-1)B$ (which is the fiber $W_1$). We obtain an inclusion of $\mathbf{C}[z]$-modules

$$B\subset V\otimes_{\mathbf{C}}\mathbf{C}[z,z^{-1}],$$

so we define a filtration $F^{\cdot}$ on the vector space $V$ by saying that $F^pV$ is equal to the subspace of vectors $v\in V$ such that $z^{-p}v\in B$. This filtration is decreasing and complete.

The construction of the previous paragraph is inverse to the Rees-bundle construction. If we apply it to the Rees-bundle of a filtered vector space, we get back the same filtration (on the same vector space!) as can easily be seen by taking an adapted basis. In the other direction, starting with $W$ and a $\mathbf{G}_m$-action, we obtain $(V,F^{\cdot})$ as above, via the module $B=W(\mathbf{A}^1)$. Let $B'$ denote the Rees module of $(V,F^{\cdot})$. It is clear that $B'\subset B$ since $F^pV$ was defined as the set of sections $v$ with $z^{-p}v\in B$. Note that there is an $n$ such that

$$V\otimes_{\mathbf{C}}z^n\mathbf{C}[z]\subset B'\subset B\subset V\otimes_{\mathbf{C}}z^{-n}\mathbf{C}[z].$$

We have decompositions

$$V\otimes_{\mathbf{C}}z^n\mathbf{C}[z]=\bigoplus_{m\geq n}z^mV.$$

$$V\otimes_{\mathbf{C}}z^{-n}\mathbf{C}[z]=\bigoplus_{m\geq -n}z^mV.$$

and

$$\frac{V\otimes_{\mathbf{C}}z^{-n}\mathbf{C}[z]}{V\otimes_{\mathbf{C}}z^n\mathbf{C}[z]}=\bigoplus_{-n\leq m<n}z^mV.$$

In particular the exact sequence

$$0 \to V \otimes_{\mathbf{C}} z^n\mathbf{C}[z] \to V \otimes_{\mathbf{C}} z^{-n}\mathbf{C}[z] \to \frac{V \otimes_{\mathbf{C}} z^n\mathbf{C}[z]}{V \otimes_{\mathbf{C}} z^{-n}\mathbf{C}[z]} \to 0$$

splits in a way compatible with the $\mathbf{G}_m$-action. The quotient is finite-dimensional and contains $B/V \otimes_{\mathbf{C}} z^n\mathbf{C}[z]$ as a $\mathbf{G}_m$-invariant subspace. The distinct pieces $z^mV$ correspond to distinct characters of $\mathbf{G}_m$, so

$$B/V \otimes_{\mathbf{C}} z^n\mathbf{C}[z] = \bigoplus_{-n \leq m < n} ((B/V \otimes_{\mathbf{C}} z^n\mathbf{C}[z]) \cap z^mV)$$

is a direct sum decomposition. From the fact that $V \otimes_{\mathbf{C}} z^n\mathbf{C}[z] \subset B$ we can lift this decomposition to $B$ and combine it with the decomposition of $V \otimes_{\mathbf{C}} z^n\mathbf{C}[z]$ to obtain the direct sum decomposition

$$B = \bigoplus_{m \geq -n} (B \cap z^mV).$$

We can now show that $B' = B$. Any element $u \in B$ decomposes as a sum of terms of the form $z^mv$ which are also in $B$. But if $z^mv \in B$ then by definition of the filtration $F^{\cdot}$ we have that $v \in F^{-m}V$ and then, by the definition of $B'$, that $z^mv \in B'$. Thus the sum of these pieces $u$ lies in $B'$. This completes the proof that the Rees bundle of $(V, F^{\cdot})$ is the same as the original bundle $W$.

The above pair of constructions establishes an equivalence between finite dimensional vector spaces with complete decreasing filtrations, and vector bundles over $\mathbf{A}^1$ with action of $\mathbf{G}_m$ covering the standard action.

One can perform the same constructions in somewhat greater generality: starting with a possibly infinite dimensional vector space, with a decreasing filtration we can again form the sheaf $\xi(V, F)$ with action of $\mathbf{A}^1$. Under mild conditions on the filtration, this will be a (torsion-free) quasicoherent sheaf. Conversely given a torsion-free quasicoherent sheaf on $\mathbf{A}^1$ with action of $\mathbf{G}_m$ we obtain a filtration on the fiber over 1, by looking at order of zeros of invariant sections. These constructions are still basically inverses but various hypotheses are needed in order to obtain a correct statement (and there seem to be many slightly different versions); we leave that to the reader.

*Filtrations of schemes, sheaves of sets, etc.*

In view of the interpretation of a filtration in terms of the Rees bundle given above, we can define a notion of filtration on pretty much any type of object: a filtration is a family of objects over $\mathbf{A}^1$ together with an action of $\mathbf{G}_m$, with the original object being the fiber over $1 \in \mathbf{A}^1$.

For example, let $\Sigma$ be a sheaf of sets on the big etale site $\mathcal{X}$. We define a *filtration* $\mathcal{F}$ *of* $\Sigma$ to be a sheaf of sets $\Sigma_{\mathcal{F}}$ on $\mathcal{X}$ with morphism $\Sigma_{\mathcal{F}} \to \mathbf{A}^1$ together with action of $\mathbf{G}_m$ (which here means a morphism of sheaves $\Sigma_{\mathcal{F}} \times \mathbf{G}_m \to \Sigma_{\mathcal{F}}$ satisfying the usual axioms) such that the action covers the usual action on $\mathbf{A}^1$; and an isomorphism $\Sigma_{\mathcal{F}} \times_{\mathbf{A}^1} \{1\} \cong \Sigma$. Note that $\Sigma_{\mathcal{F}}$ may be interpreted as a sheaf on $\mathcal{X}/\mathbf{A}^1$, and using this interpretation we can make a similar definition for sheaves of objects of any appropriate category $\mathcal{C}$ (where the necessary limits exist for defining the sheaf condition) by considering functors $(\mathcal{X}/\mathbf{A}^1)^o \to \mathcal{C}$ with action of $\mathbf{G}_m$.

A more general example is obtained by replacing the notion of sheaf of sets by the notion of *n-stack* (the reader may restrict to the case $n = 1$, that of usual stacks—cf [**54**]; or else refer to [**27**] [**7**] [**40**] [**78**] [**76**] [**82**] for preliminary versions of the notion of $n$-stack). If $\Sigma$ is an $n$-stack on $\mathcal{X}$ then a *filtration* $\mathcal{F}$ of $\Sigma$ is an $n$-stack $\Sigma_{\mathcal{F}}$ on $Xx$ with morphism $\Sigma_{\mathcal{F}} \rightarrow \mathbf{A}^1$ together with action of $\mathbf{G}_m$ (which now means a morphism of $n$-stacks $\Sigma_{\mathcal{F}} \times \mathbf{G}_m \rightarrow \Sigma_{\mathcal{F}}$ together with the data of homotopies of coherence and higher coherence, replacing the usual axioms for an action on a sheaf) such that the action covers the usual action on $\mathbf{A}^1$; and as before an isomorphism $\Sigma_{\mathcal{F}} \times_{\mathbf{A}^1} \{1\} \cong \Sigma$.

*Caution:* As we will see in one of our main examples in the section on formal categories, the notion of filtration of a sheaf of sets in the context of stacks is different from the notion of filtration in the context of sheaves of sets, in other words we might have $\Sigma_{\mathcal{F}}$ a stack whereas $\Sigma$ is a set.

Normally we will be interested in the case where $\Sigma$ is represented by a scheme. We obtain the notion of a filtration of a scheme $\Sigma$ by requiring $\Sigma_{\mathcal{F}}$ to be a scheme: the filtration is thus a scheme $\Sigma_{\mathcal{F}}$ with morphism to $\mathbf{A}^1$ and action of $\mathbf{G}_m$ covering the standard action on $\mathbf{A}^1$, and an isomorphism between the original scheme $\Sigma$ and the fiber of $\Sigma_{\mathcal{F}}$ over $1 \in \mathbf{A}^1$.

A remark similar to the above caution is also needed here. It would be possible to have a filtration $\Sigma_{\mathcal{F}}$ of a scheme where $\Sigma_{\mathcal{F}}$ is only a sheaf of sets (but not representable) or even where $\Sigma_{\mathcal{F}}$ is an $n$-stack. In the other direction, if $\Sigma_{\mathcal{F}}$ is to be a scheme, it might be interesting to require that the morphism $\Sigma_{\mathcal{F}} \rightarrow \mathbf{A}^1$ have certain additional properties such as flatness, or perhaps properties extending those of $\Sigma$ (smoothness, properness ... ); and in general, such additional hypotheses will not be satisfied by all filtrations. Actually this situation arises even in the case of filtrations on vector spaces: one obtains an equivalence of categories between filtered finite dimensional vector spaces and locally free coherent sheaves on $\mathbf{A}^1$ with $\mathbf{G}_m$-action. However if one wants to include coherent sheaves which are no longer locally free then the constructions of the previous section are no longer equivalences. (This reflects the fact that filtered vector spaces do not form an abelian category—the category of coherent sheaves on $\mathbf{A}^1$ with $\mathbf{G}_m$-action is abelian, but the subcategory of locally free sheaves is not abelian).

*The Hodge filtration on nonabelian cohomology*

Now, getting back to our main discussion, in terms of the above definition the scheme $M_{Hod}$ defined in Proposition 4.1 together with its morphism $M_{Hod} \rightarrow \mathbf{A}^1$ and the action of $\mathbf{G}_m$, is a filtration on $M_{DR}$ (since the fiber of $M_{Hod}$ over $1 \in \mathbf{A}^1$ is naturally isomorphic to $M_{DR}$). We call this the *Hodge filtration on* $M_{DR}$. Similarly $R_{Hod}$ is the *Hodge filtration on* $R_{DR}$. And most properly speaking, it is the stack $\mathcal{M}_{Hod}$ which provides the *Hodge filtration on the nonabelian cohomology stack* $\mathcal{M}_{DR}$.

In the abelian case the idea of interpreting the Hodge filtration using Rees modules was due to Deninger [**17**]. He looked at a derivation expressing the infinitesimal action instead of the full action of $\mathbf{G}_m$. In turn he refers to Fontaine [**21**] (1979) for a reworking of Hodge theory from this point of view (which is what led Fontaine to all of his rings such as $B_+^{\text{cris}}$ ... I guess ... ).

*Compatibility with the usual Hodge filtration*

In the next section we shall see how this filtration is compatible with the usual Hodge filtration on the nilpotent completion of the fundamental group. For now consider the case of additive abelian coefficients, where we show that the Hodge filtration as defined above on the first cohomology $H^1(X, \mathbf{C})$ (i.e. the filtration obtained from the inverse Rees construction applied to $M_{Hod}(X, \mathbf{G}_a)$) is the same as the standard Hodge filtration.

PROPOSITION 5.1. *Suppose the group scheme $G$ in question is the additive group $\mathbf{G}_a$. Then there is a natural isomorphism between $M_{DR}$ and the vector space $V = \mathbf{H}^1(X, (\Omega^{\cdot}_X, d))$ (the algebraic de Rham cohomology of $X$, isomorphic to $H^1(X^{\text{top}}, \mathbf{C})$). Furthermore if $F^{\cdot}$ is the standard Hodge filtration of $V$ (induced by the "stupid" filtration of $(\Omega^{\cdot}_X, d)$) then there is a natural isomorphism between $M_{Hod}$ and the Rees bundle $\xi(V, F)$ relative to $\mathbf{A}^1$ and preserving the actions of $\mathbf{G}_m$.*

*Proof:* The group $G = \mathbf{G}_a$ may be considered as the group of upper-triangular two by two matrices with the identity on the diagonal. Thus a principal $G$-bundle may be thought of as a rank two vector bundle $V$ with integrable connection $\nabla$ and $\nabla$-invariant sections $a$ of $V$ and $b$ of $V^*$ giving an exact sequence

$$0 \to \mathcal{O}_X \xrightarrow{a} V \xrightarrow{b} \mathcal{O}_X \to 0$$

of vector bundles with integrable connection. Write $X = \bigcup U_\alpha$ as a union of affine open subsets. Then the above extension of bundles splits on each $U_\alpha$. If we choose splittings $\sigma_\alpha : V|_{U_\alpha} \cong \mathcal{O}^2_{U_\alpha}$ compatible with $(a, b)$ then we can write the connection as

$$\sigma_{\alpha,*}(\nabla) = d + \Gamma_\alpha.$$

The $2 \times 2$ matrix $\Gamma_\alpha$ is upper triangular with zeros along the diagonal, and its only nonzero entry is a section $A_\alpha \in H^0(U_\alpha, \Omega^1_{U_\alpha})$. The difference between two splittings $\sigma_\alpha$ and $\sigma_\beta$ is an upper-triangular matrix $G_{\alpha\beta}$ with the identity on the diagonal and a section $g_{\alpha\beta} \in H^0(U_{\alpha\beta}, \mathcal{O}_{U_\alpha})$. These satisfy the relations $d(g_{\alpha\beta}) = A_\alpha - A_\beta$, $dA_\alpha = 0$ and $g_{\alpha\beta} + g_{\beta\gamma} = g_{\alpha\gamma}$. In other words they correspond exactly to a class in $\mathbf{H}^1(X, (\Omega^{\cdot}_X, d))$. Changing the choice of trivializations $\sigma_\alpha$ changes $(A_\alpha, g_{\alpha\beta})$ by a coboundary (and conversely any coboundary comes from a change of the $\sigma_\alpha$), so we obtain an isomorphism between our moduli space $M_{DR}$ and the hypercohomology group (which is the first algebraic de Rham cohomology).

The same calculation, done for $\lambda$-connections, shows that the relative moduli space $M_{Hod}(X, G) \to \mathbf{A}^1$ is the vector bundle associated to the locally free sheaf

$$\mathbf{R}^1 p_*(\Omega^{\cdot}_{X \times \mathbf{A}^1/\mathbf{A}^1}, td)$$

where $p : X \times \mathbf{A}^1 \to \mathbf{A}^1$ is the projection and $t$ is the coordinate on $\mathbf{A}^1$ (pulled back to the product $X \times \mathbf{A}^1$). Note that this higher direct image is locally free by semicontinuity theory, because of the Hodge-theory fact that the Dolbeault cohomology $\mathbf{H}^1(X, (\Omega^{\cdot}_X, 0))$—which is the cohomology of the fiber over $0 \in \mathbf{A}^1$—has the same dimension as the de Rham cohomology (which is the fiber over any $t \neq 0$).

We have to show that

$$\mathbf{R}^1 p_*(\Omega^{\cdot}_{X \times \mathbf{A}^1/\mathbf{A}^1}, td) \text{ equals } \xi(H^1(X, (\Omega^{\cdot}_X, d)).$$

Let $F^{\cdot}\Omega^{\cdot}_X$ be the "stupid filtration", that is

$$F^p\Omega^q_X = \Omega^q_X \ \text{ if } q \geq p, \quad F^p\Omega^q_X = 0 \ \text{ if } q < p.$$

We obtain a complex of locally free sheaves $\xi((\Omega^{\cdot}_X, d), F^{\cdot}\Omega^{\cdot}_X)$ on $X \times \mathbf{A}^1$. In view of the easy description of $F^{\cdot}\Omega^{\cdot}_X$ we have

$$\xi((\Omega^q_X, d), F^{\cdot}\Omega^q_X) = \bigoplus_{q \geq 0} t^{-q}\Omega^q_{X \times \mathbf{A}^1/\mathbf{A}^1}.$$

Multiplication by $t^q$ gives an isomorphism

$$t^{-q}\Omega^q_{X\times \mathbf{A}^1/\mathbf{A}^1} \xrightarrow{t^q} \Omega^q_{X\times \mathbf{A}^1/\mathbf{A}^1},$$

fitting together with the differential $d$ into diagrams of the form

$$\begin{array}{ccc} t^{-q}\Omega^q_{X\times \mathbf{A}^1/\mathbf{A}^1} & \xrightarrow{t^q} & \Omega^q_{X\times \mathbf{A}^1/\mathbf{A}^1} \\ \downarrow d & & \downarrow td \\ t^{-q-1}\Omega^{q+1}_{X\times \mathbf{A}^1/\mathbf{A}^1} & \xrightarrow{t^{q+1}} & \Omega^{q+1}_{X\times \mathbf{A}^1/\mathbf{A}^1}. \end{array}$$

Taking the direct sum over $q$ we obtain an isomorphism of complexes

$$\left(\bigoplus_{q\geq 0} t^{-q}\Omega^q_{X\times \mathbf{A}^1/\mathbf{A}^1}, d\right) \cong (\Omega^{\cdot}_{X\times \mathbf{A}^1/\mathbf{A}^1}, td),$$

in other words

$$\xi((\Omega^{\cdot}_X, d), F^{\cdot}\Omega^{\cdot}_X) \cong (\Omega^{\cdot}_{X\times \mathbf{A}^1/\mathbf{A}^1}, td).$$

We are now reduced to showing that

$$\mathbf{R}^1 p_*\xi((\Omega^{\cdot}_X, d), F^{\cdot}\Omega^{\cdot}_X) \text{ equals } \xi(\mathbf{H}^1(X, (\Omega^{\cdot}_X, d)), F^{\cdot})$$

where the filtration $F^{\cdot}$ on the right is that induced by $F^{\cdot}\Omega^{\cdot}_X$ on the hypercohomology. Since both sides are coherent sheaves on $\mathbf{A}^1$—which is affine—it suffices to give an isomorphism between the global sections, in other words to show that

$$\mathbf{H}^1(X\times \mathbf{A}^1, \xi((\Omega^{\cdot}_X, d), F^{\cdot}\Omega^{\cdot}_X))$$

is equal to the Rees module $\xi(\mathbf{H}^1(X, (\Omega^{\cdot}_X, d)), F^{\cdot})(\mathbf{A}^1)$.

Note that $\mathbf{H}^1(X\times \mathbf{A}^1, \xi((\Omega^{\cdot}_X, d), F^{\cdot}\Omega^{\cdot}_X))$ is a torsion-free $\mathcal{O}^1_{\mathbf{A}}$-module because as pointed out above, the higher direct image sheaf is locally free.

Let $X = \bigcup U_\alpha$ be an affine open covering and let $\mathcal{C}^{\cdot}$ be the Čech complex for $\Omega^{\cdot}_X$ with respect to this covering. It is filtered by subcomplexes $F^p\mathcal{C}^{\cdot}$. Furthermore, $\xi(\mathcal{C}^{\cdot}, F^{\cdot}\mathcal{C}^{\cdot})(\mathbf{A}^1)$ is the Čech complex for $\xi((\Omega^{\cdot}_X, d), F^{\cdot}\Omega^{\cdot}_X)$ on $X\times \mathbf{A}^1$ with respect to the affine covering

$$X\times \mathbf{A}^1 = \bigcup U_\alpha \times \mathbf{A}^1.$$

We can use this covering to calculate the hypercohomology of any complex of coherent sheaves, thus

$$\mathbf{H}^1(X\times \mathbf{A}^1, \xi((\Omega^{\cdot}_X, d), F^{\cdot}\Omega^{\cdot}_X)) = \mathbf{H}^1(\mathbf{A}^1, \xi(\mathcal{C}^{\cdot}, F^{\cdot}\mathcal{C}^{\cdot})).$$

On the other hand, we can calculate hypercohomology on $X$ using the affine covering $\{U\}_\alpha$, so

$$H^1(\mathcal{C}^{\cdot}) = \mathbf{H}^1(X, (\Omega^{\cdot}_X, d))$$

and the filtration $F^{\cdot}$ on the right hand side is that induced by $F^{\cdot}\mathcal{C}^{\cdot}$ on the left hand side.

The result we want follows from the following observations about the construction $\xi$. If $a : U \to V$ is a morphism of filtered vector spaces then (denoting the filtrations by $F_U$ and $F_V$) we have

$$\xi(\ker(a), F_{\ker(a)}) = \ker(\xi(U, F_U) \to \xi(V, F_V));$$

and

$$\xi(\operatorname{coker}(a), F_{\operatorname{coker}(a)}) = \operatorname{coker}(\xi(U, F_U) \to \xi(V, F_V))/T$$

where $T$ is the subsheaf of torsion elements of $\mathrm{coker}(\xi(U, F_U) \to \xi(V, F_V))$. If $a$ is *strict* then

$$\xi(\mathrm{coker}(a), F_{\mathrm{coker}(a)}) = \mathrm{coker}(\xi(U, F_U) \to \xi(V, F_V))$$

and conversely if $\mathrm{coker}(\xi(U, F_U) \to \xi(V, F_V))$ has no torsion then $a$ is strict.

Applying these to the cohomology of the filtered complex $(\mathcal{C}^{\cdot}, F^{\cdot}\mathcal{C}^{\cdot})$ we find that

$$\xi(H^1(\mathcal{C}^{\cdot}), F^{\cdot}) = \mathbf{H}^1(\mathbf{A}^1, \xi(\mathcal{C}^{\cdot}, F^{\cdot}\mathcal{C}^{\cdot}))/\mathrm{torsion}.$$

However, as remarked above, $\mathbf{H}^1(\mathbf{A}^1, \xi(\mathcal{C}^{\cdot}, F^{\cdot}\mathcal{C}^{\cdot}))$ has no torsion. Thus

$$\xi(H^1(\mathcal{C}^{\cdot}), F^{\cdot}) = \mathbf{H}^1(\mathbf{A}^1, \xi(\mathcal{C}^{\cdot}, F^{\cdot}\mathcal{C}^{\cdot})),$$

and putting this together with the previous identifications we get

$$\mathbf{H}^1(X \times \mathbf{A}^1, \xi((\Omega^{\cdot}_X, d), F^{\cdot}\Omega^{\cdot}_X)) = \xi(\mathbf{H}^1(X, (\Omega^{\cdot}_X, d)), F^{\cdot})(\mathbf{A}^1)$$

as desired. □

*Two filtrations*

A word about pairs of filtrations and purity. If $(V, F, \overline{F})$ is a vector space with two filtrations (which could be complex conjugates with respect to a real structure, for example, or otherwise could be independent of each other) then we can define $\xi(V, F, \overline{F})$ to be the vector bundle over $\mathbf{P}^1$ obtained by glueing $\xi(V, F)$ to $\xi(V, \overline{F})$. To be more explicit, the two standard coordinate charts $\mathbf{A}^1$ are glued together to form $\mathbf{P}^1$ by glueing over their intersection $\mathbf{G}_m \subset \mathbf{A}^1$ using the automorphism $z \mapsto z^{-1}$. The restriction of the Rees bundle $\xi(V, F)|_{\mathbf{G}_m}$ is trivial, isomorphic to $\mathbf{G}_m \times V$ (this isomorphism can be thought of as coming from the $\mathbf{G}_m$-action)—here we speak in terms of vector bundles rather than the associated locally free sheaves in order to keep a more geometric language—and similarly $\xi(V, \overline{F})|_{\mathbf{G}_m} \cong \mathbf{G}_m \times V$. Now glue $\xi(V, F)$ to $\xi(V, \overline{F})$ by glueing together their restrictions over $\mathbf{G}_m$ using the automorphism $(z, v) \mapsto (z^{-1}, v)$ of $\mathbf{G}_m \times V$.

The two filtrations $F$ and $\overline{F}$ define a decomposition of the vector space $V$ which is pure of weight $w$, if and only if the vector bundle $\xi(V, F, \overline{F})$ is a direct sum of copies of $\mathcal{O}_{\mathbf{P}^1}(w)$. To see this, suppose first that the two filtrations define a pure decomposition $V = \bigoplus_{p+q=w} V^{p,q}$. Then the triple $(V, F, \overline{F})$ is the direct sum of the pieces $(V^{p,q}, F|_{V^{p,q}}, \overline{F}|_{V^{p,q}})$. On the piece $V^{p,q}$ we have $F^p = V^{p,q}$ and $F^{p+1} = 0$, similarly $\overline{F}^q = V^{p,q}$ and $\overline{F}^{q+1} = 0$. The Rees module of $(V^{p,q}, F|_{V^{p,q}})$ is the $\mathbf{C}[z]$-submodule of $V^{p,q} \otimes_{\mathbf{C}} \mathbf{C}[z, z^{-1}]$ generated by $z^{-p}V^{p,q}$. Thus the sheaf $\xi(V^{p,q}, F|_{V^{p,q}})$ is just the subsheaf of sections of $V^{p,q} \otimes_{\mathbf{C}} \mathcal{O}_{\mathbf{G}_m}$ which have poles of order $\leq p$ at the origin. Similarly the sheaf $\xi(V^{p,q}, \overline{F}|_{V^{p,q}})$ (considered as being on the coordinate chart $\mathbf{A}^1$ which is a neighborhood of infinity) is the subsheaf of sections of $V^{p,q} \otimes_{\mathbf{C}} \mathcal{O}_{\mathbf{G}_m}$ which have poles of order $\leq q$ at infinity. Glueing together these two descriptions we find that $\xi(V^{p,q}, F|_{V^{p,q}}, \overline{F}|_{V^{p,q}})$ is the subsheaf of sections of $V^{p,q} \otimes_{\mathbf{C}} \mathcal{O}_{\mathbf{G}_m}$ which have poles of order $\leq p$ at the origin and poles of order $\leq q$ at infinity, in other words

$$\xi(V^{p,q}, F|_{V^{p,q}}, \overline{F}|_{V^{p,q}}) = V^{p,q} \otimes_{\mathbf{C}} \mathcal{O}_{\mathbf{P}^1}(p \cdot 0 + q \cdot \infty) \cong V^{p,q} \otimes \mathcal{O}_{\mathbf{P}^1}(w)$$

since $p+q = w$. Taking the direct sum over the $(p, q)$ gives the result that $\xi(V, F, \overline{F})$ is semistable of slope $w$, i.e. a direct sum of copies of $\mathcal{O}_{\mathbf{P}^1}(w)$.

Suppose on the other hand that we start off knowing that

$$\xi(V, F, \overline{F}) \cong U \otimes_{\mathbf{C}} \mathcal{O}_{\mathbf{P}^1}(w)$$

for some vector space $U$. Comparing the fibers over $1 \in \mathbf{A}^1$ we obtain a natural isomorphism $U \cong V$.

Let $W$ be a rank one vector space with filtrations $F_W, \overline{F}_W$ with $F_W^w = W$, $F_W^{w+1} = 0$, $\overline{F}_W^0 = W$ and $\overline{F}_W^1 = 0$. In the notation of the above paragraph we have $W = W^{w,0}$. Let $\mathcal{L} = \xi(W, F_W, \overline{F}_W)$ (of course in this case the filtrations $F_W$ and $\overline{F}_W$ are not complex conjugates). It is isomorphic to $\mathcal{O}_{\mathbf{P}^1}(w)$ by the previous part of the argument, and we have chosen a certain lifting of the action of $\mathbf{G}_m$. (It is interesting to note that the choices of lifts of the action of $\mathbf{G}_m$ to actions on $\mathcal{O}_{\mathbf{P}^1}(w)$ correspond exactly to the choices of $(p, q)$ with $p + q = w$.)

Now $Hom(\mathcal{L}, \xi(V, F, \overline{F}))$ is a trivial bundle on $\mathbf{P}^1$, isomorphic in fact to $U \otimes_{\mathbf{C}} \mathcal{O}_{\mathbf{P}^1}$. It has an action of $\mathbf{G}_m$ which yields an action on the space of global sections $U$. In this way we see that there is an action of $\mathbf{G}_m$ on $U$ such that the isomorphism

$$\xi(V, F, \overline{F}) \cong U \otimes_{\mathbf{C}} \mathcal{L}$$

preserves the action of $\mathbf{G}_m$. Diagonalizing the action of $\mathbf{G}_m$ on $U$ we obtain a decomposition $U = \bigoplus U^q$ and if we set $V^{w-q,q} = U^q$ via the isomorphism $V \cong U$ then $V = \bigoplus_{p+q=w} V^{p,q}$ and one can check that this decomposition splits the pair of filtrations $(F, \overline{F})$ (use the description of $F^p$ as being the vectors whose corresponding $\mathbf{G}_m$-invariant sections have poles of order $\leq p$ at the origin, etc.).

We obtain the interpretation that purity of weight $w$ for a pair of filtrations is equivalent to the condition that $\xi(V, F, \overline{F})$ be semistable of slope $w$.

The construction of $\xi(V, F, \overline{F})$ by glueing is practically the same as the construction of $M_{Del}$ in §4. The analogy becomes closer if we include the condition that $\overline{F}$ should be the complex conjugate of $F$ with respect to a real structure, for then the part $\xi(V, \overline{F})$ is just the complex conjugate of $\xi(V, F)$. Thus we think of $M_{Del}$ as the nonabelian analogue of $\xi(V, F, \overline{F})$ where $F$ is replaced by the "filtration" $M_{Hod}$. The fact that $M_{Del}$ is the twistor space for a quaternionic structure is equivalent to the statement that the normal bundle along any prefered section is semistable of slope 1, i.e. a direct sum of $\mathcal{O}_{\mathbf{P}^1}(1)$ (cf [**37**], [**38**]). By analogy with the previous discussion of purity for pairs of filtrations on a vector space, this condition for the normal bundle can be interpreted as purity of weight one (expected since we are looking at cohomology in degree one). I don't know how far one can go toward making this analogy more precise than it is.

*Analogues for higher cohomology*

In this subsection I will indicate how to do an analogous thing to obtain a Hodge filtration on higher nonabelian cohomology. This materiel (added in the course of the revision of the paper) is discussed in [**76**], [**78**] and, since it gets out of our current perspective, the discussion will be brief. The reader may of course skip this section which is not at all required for the remainder of the paper. (On the other hand, among the various notational deficiencies of the present subsection, we will refer here to the discussion of formal categories which takes place in §7 later). For those who decide to stay in, our goal will be to state the analogue of Proposition 5.1 for higher degree cohomology.

We can define an *n-prestack on* $\mathcal{X}$ to be a presheaf of topological spaces $T : \mathcal{X}^o \to Top$ such that for each $X \in \mathcal{X}$, $T(X)$ is $n$-truncated (i.e. has homotopy groups vanishing in degrees $> n$). An *n-stack* is an $n$-prestack satisfying a certain homotopical descent condition. There is an internal $\underline{Hom}(T, T')$ between $n$-stacks (itself an $n$-stack), and an external $Hom(T, T')$ (just the value of the internal $\underline{Hom}$

on the final object $Spec(\mathbf{C}) \in \mathcal{X}$) which is a topological space. This point of view will be related to an eventual purely category-theoretic point of view by equating the topological spaces $T(X)$ with their Poincaré $n$-groupoids. In particular for $n = 1$ one equates the $T(X)$ with their Poincaré groupoids, and $T$ becomes equivalent to a presheaf of groupoids. The descent condition refered to above is, in this case, the usual condition for being a stack.

An $n$-stack $T$ has a sheaf of sets $\pi_0(T)$ and for any basepoint $t \in T(X)$, sheaves of groups $\pi_i(T|_{\mathcal{X}/X}, t)$ on the site $\mathcal{X}/X$ of schemes over $X$.

One can define various good types of $n$-stacks (for example, presentable or very presentable [**77**] or geometric [**78**]—this last condition generalizes the notion of algebraic stack for $n = 1$). The $n$-stacks we would like to consider here are the $T$ which are *connected* (i.e. $\pi_0(T) = *$) and such that the $\pi_i(T, t)$—for the essentially unique basepoint $t \in T(Spec(\mathbf{C}))$—are representable by algebraic group schemes. This condition is the condition of being *connected and presentable* [**77**] (or equivalently *connected and geometric* [**78**]). Often we ask additionally that the group schemes $\pi_i(T, t)$ are affine for $i = 1$ and vector spaces for $i \geq 2$—this is the condition of being *connected and very presentable* [**77**].

The basic examples are $T = K(G, 1) = BG$ for a group scheme $G$ (which should be affine in the "very presentable" case); and $T = K(A, n)$, $n \geq 2$, for an abelian group scheme $A$ (which should be a finite dimensional vector space in the "very presentable" case, in which case we might as well assume $A = \mathbf{C}$). In fact a connected presentable $T$ will always have a Postnikov tower where the stages look like these examples, so in some sense everything can be reduced to looking at these cases.

Suppose $X$ is a smooth projective variety. Let $X_B$ denote the $n$-stack associated to the constant $n$-prestack whose values are always the $n$-truncation (i.e. $n$-th coskeleton) of the topological space $X^{\text{top}}$. Let $X_{DR}$ denote the sheaf of sets $X_{DR}(Y) := X(Y^{red})$. It is the 1-stack (which is actually 0-truncated thus being a sheaf of sets) obtained from the formal groupoid which we also denote $X_{DR}$ below. Suppose $T$ is a connected presentable $n$-stack as defined in the previous paragraph. We define the *Betti cohomology of $X$ with coefficients in $T$* to be the $n$-stack $\mathcal{H}_B(X, T) := \underline{Hom}(X_B, T)$. We define the *de Rham cohomology of $X$ with coefficients in $T$* to be the $n$-stack $\mathcal{H}_{DR}(X, T) := \underline{Hom}(X_{DR}, T)$. If $T$ is very presentable then there is an equivalence of $n$-stacks

$$\mathcal{H}_B(X, T)^{\text{an}} \cong \mathcal{H}_{DR}(X, T)^{\text{an}}$$

where the superscript denotes the corresponding $n$-stack on the analytic site $\mathcal{X}^{\text{an}}$ of complex analytic spaces (with the etale or usual topology).

If $T = K(G, 1)$ for an affine group scheme $G$ then $\mathcal{H}_B(X, T)$ is just the moduli stack $\mathcal{M}_B(X, G)$ and $\mathcal{H}_{DR}(X, T)$ is the moduli stack $\mathcal{M}_{DR}(X, G)$ ([**76**] Lemma 2.11 and Theorem 6.4).

If $T = K(\mathbf{C}, n)$ then $\mathcal{H}_B(X, T)$ is an $n$-stack with

$$\pi_0(\mathcal{H}_B(X, T)) = H^n(X^{\text{top}}, \mathbf{C})$$

([**76**] Lemma 2.11) and $\mathcal{H}_{DR}(X, T)$ is an $n$-stack with

$$\pi_0(\mathcal{H}_B(X, T)) = \mathbf{H}^n(X, (\Omega^{\cdot}_X, d))$$

([**76**] Lemma 6.5) Thus in this case we recover the higher Betti and algebraic de Rham cohomology groups of $X$.

Now we can define the Hodge filtration on $\mathcal{H}_{DR}(X,T)$. For this we need to refer to the formal-category interpretation of the sheaf $X_{DR}$—cf §7 (below, unfortunately ... ). We will define in §7 a formal category $X_{Hod}$ with a morphism $X_{Hod} \to \mathbf{A}^1$ such that the fiber over $1 \in \mathbf{A}^1$ is the formal category which we are denoting by $X_{DR}$, and the fiber over 0 is the formal category which will be denoted $X_{Dol}$. This formal category represents a 1-stack which we again denote $X_{Hod}$ with map to $\mathbf{A}^1$ and again the fiber over 1 is the sheaf of sets $X_{DR}$. There is an action of $\mathbf{G}_m$ on the formal category and hence on the 1-stack $X_{Hod}$, covering the action on $\mathbf{A}^1$. In our previous terminology, $X_{Hod}$ is thus a filtration (which we might as well call the *Hodge filtration*) of $X_{DR}$. Note that, while $X_{Hod}$ is a 1-stack, the fiber $X_{DR}$ is a 0-stack or sheaf of sets. This is the example refered to above where a sheaf of sets can have a stack-theoretic filtration which is not a filtration in terms of sheaves of sets.

The internal $\underline{Hom}$ has a relative version: if $T \to R$ and $T' \to R$ are two $n$-stacks mapping to a third, then we obtain $\underline{Hom}(T/R, T'/R)$ again mapping to $R$, this construction being compabible with change of base $R$ and equal to the usual internal $\underline{Hom}$ when $R$ is the single-point sheaf of sets. If $T$ is a connected very presentable $n$-stack then we obtain the $n$-stack

$$\mathcal{H}_{Hod}(X,T) := \underline{Hom}(X_{Hod}/\mathbf{A}^1, T \times \mathbf{A}^1/\mathbf{A}^1)$$

with a map $\mathcal{H}_{Hod}(X,T) \to \mathbf{A}^1$, an action of $\mathbf{G}_m$ covering the standard action on $\mathbf{A}^1$, and an equivalence of $n$-stacks

$$\mathcal{H}_{Hod}(X,T) \times_{\mathbf{A}^1} \{1\} \cong \mathcal{H}_{DR}(X,T).$$

In other words, following the point of view described at the start of this section, the $n$-stack $\mathcal{H}_{Hod}(X,T)$ is a filtration on $\mathcal{H}_{DR}(X,T)$. It is the *Hodge filtration on the nonabelian cohomology* $\mathcal{H}_{DR}(X,T)$.

In the case $T = K(G,1)$ for an affine group scheme $G$, we have $\mathcal{H}_{Hod}(X, K(G,1)) = \mathcal{M}_{Hod}(X,G)$ and we recover the Hodge filtration on the moduli stack $\mathcal{M}_{DR}(X,G)$ as defined above. In the case $T = K(\mathbf{C}, n)$ a variant of Proposition 5.1 is the following.

PROPOSITION 5.2. *For the case $T = K(\mathbf{C}, n)$ the Hodge filtration as defined in the previous paragraph is, on the level of $\pi_0$, isomorphic*

$$\pi_0(\mathcal{H}_{Hod}(X, K(\mathbf{C}, n))) \cong \xi(H^n_{DR}(X, \mathbf{C}), F)$$

*with the Rees bundle of the usual Hodge filtration on the algebraic de Rham cohomology $H^n_{DR}(X,\mathbf{C})$.*

*Comment about the proof:* After the abstract nonsense it comes down to the statement that the locally free sheaf $\mathbf{R}^n p_*(\Omega^{\cdot}_{X\times\mathbf{A}^1/\mathbf{A}^1}, td)$ on $\mathbf{A}^1$ is equal to the Rees bundle $\xi(H^n_{DR}(X,\mathbf{C}), F)$ with $F$ the usual Hodge filtration—and this statement is proved exactly as in the case $n = 1$. □

*Remark about the analytic site:* It is unreasonable to look at presentable but not very presentable coefficient stacks $T$ when working in the algebraic site (basically because $H^i(X, \mathcal{O}^*_X)$ doesn't have good properties when $i \geq 2$). However this becomes reasonable if everything is done directly over the analytic site $\mathcal{X}^{\text{an}}$ and in this case one can define a Hodge filtration for $\underline{Hom}(X^{\text{an}}_{DR}, T^{\text{an}})$ for any connected presentable $T$. In particular when $T = K(\mathbf{G}_m, n)$ we get a filtration of $H^n(X_{DR}, \mathbf{G}_m)$ which is probably closely related to the Deligne cohomology groups (I haven't investigated this relationship, though).

## 6. The nilpotent completion of $\pi_1$ and representations near the identity

We will justify our definition of $M_{Hod}$ as the Hodge filtration by making the connection with the usual Hodge filtration on the nilpotent completion of the fundamental group [**58**] [**30**]. For simplicity we work with the group algebra $\mathbf{C}\pi_1^{\wedge}$ (completed at the augmentation ideal). The mixed Hodge structure on $\pi_1$ is usually defined via the mixed Hodge structure on $\mathbf{C}\pi_1^{\wedge}$ and the inclusion of the Lie algebra corresponding to $\pi_1$ into this group algebra.

We first consider the relationship between the completed group algebra and the completions of the spaces of representations at the identity.

Suppose $A$ is an augmented $\mathbf{C}$-algebra which is complete with respect to the augmentation ideal $J_A$. Let $R(A,n)$ denote the functor of artinian local $\mathbf{C}$-algebras $B$ defined by setting

$$R(A,n)(B) := Hom^{\mathrm{aug}}(A, M_n(B))$$

where $Hom^{\mathrm{aug}}$ denotes the set of algebra homomorphisms sending $J_A$ to the ideal $M_n(\mathbf{m}_B)$ ($\mathbf{m}_B$ denotes the maximal ideal of $B$).

If $A = \mathbf{C}\Gamma^{\wedge}$ is the completion of the group algebra of a finitely presented group $\Gamma$ then $R(A,n)$ is pro-representable by the completion at the identity representation $R(\Gamma, GL(n))^{\wedge}$ of the space of representations of $\Gamma$ in $GL(n)$ (there are probably more abstract conditions on $A$ which could be used to insure representability but we don't need those here).

Conversely let $\mathcal{C}$ denote the category of algebras which are direct products of algebras of the form $M_n(\mathbf{C})$. Suppose $\Upsilon : \mathcal{C} \to ForSch$ is a functor from $\mathcal{C}$ to the category of formal schemes, compatible with products (so we can think of $\Upsilon$ as a collection of formal schemes $\Upsilon_n$ together with morphisms of functoriality corresponding to morphisms of products of algebras in $\mathcal{C}$). Then we define $\mathcal{A}(\Upsilon)$ to be the algebra of natural transformations $\Upsilon \to 1_{\mathcal{C}}$. The elements of $\mathcal{A}(\Upsilon)$ are functions $a$ which for each $n$ associate a section $a_n : \Upsilon_n \to M_n(\mathbf{C})$ with the $a_n$ compatible with morphisms of products of objects of $\mathcal{C}$.

LEMMA 6.1. *Suppose $A = \mathbf{C}\Gamma^{\wedge}$ is the completion of the group algebra of a finitely presented group. Then the $R(A,n)$ give a functor $R(A,\cdot) : \mathcal{C} \to ForSch$ and we can recover $A$ by the construction of the previous paragraph:*

$$A = \mathcal{A}(R(A,\cdot)).$$

□

*Remark:* The morphisms of functoriality defining $R(A,\cdot)$ can be obtained from the morphisms of functoriality of $R(\Gamma, GL(n))$ for morphisms between products of groups $GL(n)$.

Now we investigate the Hodge filtrations. Suppose $X$ is a smooth projective variety. We obtain a family of formal completions $R_{Hod}(X,x,GL(n))^{\wedge} \to \mathbf{A}^1$, with an action of $\mathbf{G}_m$. The technique of Goldman and Millson used in ([**75**] §10) to give an isomorphism $R_{Dol}(X,x,GL(n))^{\wedge} \cong R_{DR}(X,x,GL(n))^{\wedge}$ actually gives a trivialization

$$R_{Hod}(X,x,GL(n))^{\wedge} \cong R_{Dol}(X,x,GL(n))^{\wedge} \hat{\times} \mathbf{A}^1,$$

with the action of $\mathbf{G}_m$ coming from the action defined in [**71**] on $R_{Dol}(X,x,GL(n))^{\wedge}$ and the standard action on $\mathbf{A}^1$. We discuss this further in §9 below.

Deligne's original geometric interpretation of this trivialization in terms of $R_{Del}$ was to say that we look (formally near the identity section) at all of the sections

$\mathbf{P}^1 \to R_{Del}$ which send $\infty$ to the identity object $1 \in R_{Del} \times_{\mathbf{P}^1} \{\infty\} = R_{Dol}(\overline{X})$. The space of these sections maps isomorphically to any fiber over a point different from $\infty$ in $\mathbf{P}^1$ (again this statement is meant formally in a neighborhood of the identity). A sketch of proof is obtained by noting that sufficiently many such sections are given by Goldman-Millson theory (§9 below) and on the other hand, formally nearby the identity section there are no other sections than these, because the normal bundle is a direct sum of $\mathcal{O}_{\mathbf{P}^1}(1)$'s. We don't actually need this nice interpretation for the argument of the previous paragraph, where it suffices to have the compatibility of the Goldman-Millson trivialization with the $\mathbf{G}_m$-action (easy to check by looking carefully at the discussion of §9).

We obtain a functor $R_{Hod}(X, x, \cdot) : \mathcal{C} \to \text{ForSch}/\mathbf{A}^1$ which we think of as a family of functors parametrized by $\mathbf{A}^1$ (with action of $\mathbf{G}_m$). Because of the trivialization we can apply the previous lemma. This family of functors gives rise to a completed algebra $\mathcal{A}$ over $\mathbf{A}^1$, by a relative version of the construction of Lemma 6.1 (which poses no problem since everything is a product). Conversely starting from $\mathcal{A}$ we get back the $R_{Hod}(X, x, GL(n))^\wedge$. Finally, the fiber $\mathcal{A}_1$ over $1 \in \mathbf{A}^1$ is isomorphic to the completed group algebra $\mathbf{C}\pi_1(X, x)^\wedge$ again by the above discussion. This family of algebras together with $\mathbf{G}_m$-action (which as we have seen is equivalent to the data of the $R_{Hod}$ functorially in $n$) corresponds to a filtration of $\mathbf{C}\pi_1(X, x)^\wedge$. We claim that this filtration is the Hodge filtration of Morgan-Hain [**58**] [**30**].

To see this, note as a consequence of the trivializations and $\mathbf{G}_m$-actions in the above discussion, that the filtration on $\mathbf{C}\pi_1(X, x)^\wedge$ corresponding to $\mathcal{A}$ is just the filtration associated to the grading given by the $\mathbf{G}_m$-action on $\mathbf{C}\pi_1(X, x)^\wedge$. One has to check that this $\mathbf{G}_m$-action is equal to that which was defined in [**71**]. Finally, in §§5-6 of [**71**] it was verified that this $\mathbf{G}_m$-action gives rise to the Hodge filtration of Morgan-Hain.

To sum up, starting with the completed group algebra $\mathbf{C}\pi_1(X, x)^\wedge$ and the Morgan-Hain Hodge filtration $F$ we can form $\mathcal{A} = \xi(\mathbf{C}\pi_1(X, x)^\wedge, F)$, a family of algebras over $\mathbf{A}^1$ with $\mathbf{G}_m$-action; then the family of completed representation spaces associated to this family of algebras is isomorphic (together with $\mathbf{G}_m$-action) to the completion $R_{Hod}(X, x, GL(n))^\wedge$ along the identity section. In the other direction, the data of the completions $R_{Hod}(X, x, GL(n))^\wedge$ functorially in $GL(n)$ serve to define (via Lemma 6.1) a family of completed algebras $\mathcal{A}$ over $\mathbf{A}^1$, again with $\mathbf{G}_m$-action and isomorphism $\mathcal{A}_1 \cong \mathbf{C}\pi_1(X, x)^\wedge$, and this family yields the Morgan-Hain Hodge filtration on $\mathbf{C}\pi_1(X, x)^\wedge$ by reversing the construction $\xi$.

Thus the completion of our Hodge filtration at the identity representation corresponds to the Hodge filtration on the nilpotent completion of the fundamental group. The whole Hodge filtration $R_{Hod}(X, x, G)$ or $M_{Hod}(X, G)$ should be thought of as an analytic continuation of the Hodge filtration on the nilpotent completion.

It might be interesting to try to express the existence of this analytic continuation in terms of estimates on the mixed Hodge structure on $\mathbf{C}\pi_1(X, x)^\wedge$ in the spirit of Hadamard's technique [**29**].

*Writing formulas*

We can combine what we know so far to sketch a method which should allow, in principle, to write down the local equations for the correspondence $M_{Dol} \cong M_{DR}$ (and hence for the quaternionic structure) near a complex variation of Hodge structure in $M^{\text{sm}}$.

The method sketched above should work equally well along any prefered section coming from a complex variation of Hodge structure $\rho$. The Hodge filtration on the relative Malcev completion [**31**] should give a Hodge filtration $F$ on the complete local ring $\widehat{\mathcal{O}}_{M_{DR},\rho}$; then taking $\xi(\widehat{\mathcal{O}}_{M_{DR},\rho}, F)$ we get a family of complete local rings over $\mathbf{A}^1$, and taking a formal spectrum we get a formal scheme along $\mathbf{A}^1$. The real structure or at least the invariant indefinite hermitian form underlying the variation $\rho$ should give an involution allowing us to glue the formal scheme with itself to get a formal scheme along $\mathbf{P}^1$. We should get back in this way the formal completion of $M_{Del}$ along the prefered section.

The normal bundle of $M_{Del}$ along a prefered section is a direct sum of $\mathcal{O}_{\mathbf{P}^1}(1)$'s, so the space of sections near the given section (which has a structure of formal scheme in this case) will map isomorphically to the product of any two fibers. Taking two antipodal fibers (for example the fibers over 0 and $\infty$) we obtain explicitly the involution on the space of sections and the prefered sections are those which are invariant. Finally looking at the isomorphism from the space of sections to the product of the formal completions of $M_{Dol}$ and $M_{DR}$, the space of invariant sections gives the graph of the real analytic isomorphism $M_{Dol} \cong M_{DR}$.

One can imagine following out this entire construction explicitly to obtain the Taylor series for the isomorphism $M_{Dol} \cong M_{DR}$ near the point $\rho$. The only ingredients are the Hodge filtration and the real structure (and of course an analysis of the space of sections of our formal scheme, but this is an algebraic question).

One can see just from the existence of this method that the algebraically closed field generated by the coefficients of the Hodge filtration on the relative Malcev completion (and their complex conjugates) will contain the coefficients of the power series for the isomorphism $M_{Dol} \cong M_{DR}$ and hence for the power series of the quaternionic structure.

I have not checked the details of this construction any more than what is written above.

## 7. Formal categories

One of the main properties of the Hodge filtration on usual abelian cohomology is Griffiths transversality. This is a property of the variation of the Hodge filtration with respect to the Gauss-Manin connection which arises from a smooth family of varieties. We would like to obtain a similar property for our nonabelian Hodge filtration. Let's first look at how to interpret the usual Griffiths transversality in terms of the construction $\xi$.

*A digression: the Rees bundle construction for filtered vector bundles*

Suppose $S$ is a smooth variety and $V$ is a vector bundle on $S$. If $F^{\cdot}$ is a complete, decreasing filtration of $V$ by strict subbundles, then a relative version of the Rees bundle construction gives a vector bundle $\xi(V,F)$ on $S \times \mathbf{A}^1$ together with action of $\mathbf{G}_m$ which covers the product action (trivial on $S$ and the standard action on $\mathbf{A}^1$). The fiber $\xi(V,F)|_{S\times\{0\}}$ is naturally isomorphic to the bundle $Gr^{\cdot}_F(V)$. Conversely, given a vector bundle $W$ on $S \times \mathbf{A}^1$ with action of $\mathbf{G}_m$ then the inverse construction yields a filtration $F^{\cdot}$ (decreasing, complete) of $V = W|_{S\times\{1\}}$ by strict subbundles.

LEMMA 7.1. *Suppose that $V$ is a vector bundle on $S$ with integrable connection $\nabla$. If $F^{\cdot}$ is a decreasing filtration of $V$ by strict subbundles, then $(F^{\cdot}, \nabla)$ satisfies the Griffiths transversality condition $\nabla F^p \subset F^{p-1} \otimes \Omega^1_S$ if and only if the action of*

*$T(S)$ on $V$ extends to a $\mathbf{G}_m$-invariant action of the sheaf $T(S\times \mathbf{A}^1/\mathbf{A}^1)(-S\times\{0\})$ of relative tangent vector fields vanishing to order one along $S\times\{0\}$ on $\xi(V,F^{\cdot})$.*

*Proof:* The question is local on $S$ (even in the etale or analytic topology) so we may assume that $V = \bigoplus V^p$ is a direct sum of trivial bundles with the filtration given by $F^p = \bigoplus_{q\geq p} V^q$. The connection $\nabla$ is given in terms of the trivial connection $d$ by

$$\nabla = d + A$$

where $A$ is an $End(V)$-valued 1-form. We can decompose $A = \oplus A_{pq}$ where $A_{pq}$ is a $Hom(V^p, V^q)$-valued 1-form. The Griffiths transversality condition becomes

$$A_{pq} = 0 \text{ if } q < p-1.(GT)$$

On the other hand we can write

$$\xi(V,F^{\cdot}) = \bigoplus_p t^{-p}\mathcal{O}_{\mathbf{A}^1}\otimes V^p.$$

As usual there is an isomorphism

$$\varphi : \xi(V,F^{\cdot}) \cong \bigoplus_p \mathcal{O}_{\mathbf{A}^1}\otimes V^p$$

decomposing into components $\varphi_p$ equal to multiplication by $t^p$. The connection $\nabla$ extends to a $\mathbf{G}_m$-invariant action of the tangent vector fields in the $S$-direction, defined over $S\times \mathbf{G}_m \subset S\times \mathbf{A}^1$. This is equivalent to an action of the tangent vector fields in the $S$-direction on $\bigoplus_p \mathcal{O}_{\mathbf{A}^1}\otimes V^p$ via the isomorphism $\varphi$.

The tangent fields in question are written as sums of the form $\sum_i t^i\eta_i$ for $\eta_i$ tangent vector fields on $S$ (this sum has to be interpreted in an appropriate sense depending on the open set of definition: as is, for an open set of the form $U\times\mathbf{A}^1$—in which case $i\geq 0$—or $U\times \mathbf{G}_m$—in which case $i$ can be negative; or as a convergent series, for an analytic open set of $S\times\mathbf{A}^1$).

A tangent vector field $\sum_i t^i\eta_i$ acts on

$$\xi(V,F^{\cdot}) = \bigoplus_p t^{-p}\mathcal{O}_{\mathbf{A}^1}\otimes V^p$$

via the matrix $\sum_i t^i A(\eta_i)$. If we translate via the isomorphism $\varphi$ then it acts on $\bigoplus_p \mathcal{O}_{\mathbf{A}^1}\otimes V^p$ via the matrix

$$\sum_{i,p,q} t^{i+q-p}A_{pq}(\eta_i).$$

This is now the matrix for the action in terms of a basis of $\xi(V,F^{\cdot})$ over a neighborhood of a point in $S\times\{0\}$ (whereas the previous expression was not). The action of the vector field extends across $S\times\{0\}$ if and only if the terms in negative powers of $t$ vanish. The condition that this extension exists for any vector field of the form $\sum_{i\geq 1} t^i\eta_i$ is thus equivalent to the condition that $A_{pq} = 0$ for $q-p < -1$. The former is the condition of extension of actions of vector fields in the statement of the Lemma; the latter is exactly the Griffiths transversality condition $(GT)$. □

Note that if $(F^{\cdot},\nabla)$ satisfies Griffiths transversality then the action of $T(S\times\mathbf{A}^1/\mathbf{A}^1)(-S\times\{0\})$ given by the lemma may be interpreted as a $\lambda$-connection on the bundle $\xi(V,F^{\cdot})$ over $S\times\mathbf{A}^1$. In particular, for $\lambda = 0$ we obtain a Higgs bundle structure with underlying vector bundle $\xi(V,F)|_{S\times\{0\}} \cong Gr_F^{\cdot}(V)$. This Higgs bundle is actually a *system of Hodge bundles* (notation of [**68**], cf also [**71**]), and the Higgs field is the residual part of the connection (often called the *Kodaira-Spencer map*). See Lemma 7.3 below for a further statement taking into account semistability.

*Nonabelian Griffiths-transversality*

To continue the motivation for introducing formal groupoids, we consider in a preliminary way the nonabelian Gauss-Manin connection (this discussion is continued in §8). If $X \to S$ is a smooth projective morphism, which we think of as a family of varieties parametrized by $S$, then we obtain the relative moduli space $M_{DR}(X/S,G)$. This in turn should be thought of as the family of moduli spaces of flat $G$-bundles on the fibers $X_s$ (for example, $M_{DR}(X/S,G)_s = M_{DR}(X_s,G)$). But analytically and locally in the usual topology of $S$, the family of topological spaces $X_s^{\rm top}$ is trivial so the family of moduli spaces of flat $G$-bundles $M_B(X_s,G)$ is locally constant as $s$ varies in $S$. Our analytic isomorphisms $M_{DR}(X_s,G)^{\rm an} \cong M_B(X_s,G)^{\rm an}$—which vary analytically in $s$—induce an analytic trivialization of the family $M_{DR}(X/S,G)^{\rm an} \to S$ locally over the analytic topology of $S$. In other words, we get a *flat connection* on $M_{DR}(X/S,G)^{\rm an}$ relative to $S$. If $M_{DR}(X/S,G) \to S$ were smooth this connection could be simply interpreted as a way of lifting vector fields on the base $S$ to vector fields on the total space $M_{DR}(X/S,G)$ (and in fact this interpretation is valid over the open subset of smooth points). If the $M_{DR}(X_s,G)$ have singularities then we are forced to adopt the formal category or "crystalline" point of view ([**28**]) which will be presented shortly.

An interesting point about this situation is that the connection, which we first see in an analytic way, actually comes from an algebraically defined connection (it was S. Mochizuki who first asked me the question of whether the connection could be defined algebraically). For example, over the smooth points, the lifts of algebraic vector fields on $S$ will be algebraic vector fields on $M_{DR}(X/S,G)$.

We can now imagine defining a notion of Griffiths transversality in the nonabelian context: proceed by analogy with the previous interpretation of Griffiths transversality in terms of the Rees bundle of a filtered vector bundle. Recall that for the nonabelian Hodge filtration, the analogue of the Rees bundle is $M_{Hod}$. Thus we would like our "Griffiths transversality" to say that the lifting of vector fields on $S$ to vector fields on $M_{DR}(X/S,G)$ given by the Gauss-Manin connection, extends to a $\mathbf{G}_m$-invariant lifting of sections of $T(S\times \mathbf{A}^1/\mathbf{A}^1)(-S\times\{0\})$ to vector fields on $M_{Hod}(X/S,G)$. To have an interpretation which makes sense on non-smooth points of the moduli space we need the point of view of formal groupoids.

*Formal groupoids*

Start with the definition of the Gauss-Manin connection which works even over singular points in the moduli space (cf [**70**] [**75**], following [**28**] for the abelian version). The Gauss-Manin connection on $M_{DR}(X/S,G)$ is an isomorphism

$$p_1^*M_{DR}(X/S,G)|_{(S\times S)^\wedge} \cong p_2^*M_{DR}(X/S,G)|_{(S\times S)^\wedge}$$

where $p_1,p_2 : S\times S \to S$ are the projections and $(S\times S)^\wedge$ is the formal completion of the diagonal. If we set $N := (S\times S)^\wedge$ then the pair $(S,N)$ has a structure of category in the category of formal schemes. The notion of formal category codifies this example. It provides a general framework for operations on families of things over $S$.

What follows comes from Berthelot [**4**] and Illusie [**39**], inspired by Grothendieck [**28**].

A *formal category* is a pair $(X, N)$ consisting of a scheme $X$ and a formal scheme $N$ mapping to $X \times X$, together with a structure of category, that is morphisms $N \times_X N \to N$ giving composition and $X \to N$ giving the identity, subject to the usual axioms for a category. A formal category gives in a natural way a presheaf of categories on $Sch/\mathbf{C}$. A *formal groupoid* is a formal category such that the values of the associated presheaf are groupoids. We say that a formal groupoid is *of smooth type* if the underlying scheme of $N$ is the scheme $X$ (via the identity morphism), and the morphism $N \to X$ is formally smooth.

Let $X_N$ denote the stack over $Sch/\mathbf{C}$ associated to the presheaf of groupoids given by $(X, N)$. We have a morphism $p : X \to X_N$. Note that $N$ represents the functor $X \times_{X_N} X$, so we can recover $(X, N)$ from the stack $X_N$ with its morphism $X \to X_N$. In practice we confuse the notions (and notations) of formal groupoid $(X, N)$ and associated stack $X_N$. If $(X, N)$ is of smooth type then the morphism of stacks $X \to X_N$ is formally smooth, because for any morphism from a scheme $S \to X_N$ there exists (locally on $S$) a lifting $S \to X$ and then

$$X \times_{X_N} S = (X \times_{X_N} X) \times_X S = N \times_X S$$

and $N \times_X S \to S$ is formally smooth since $N \to X$ is.

Suppose $(X, N)$ is a formal groupoid of smooth type. Note that the structure sheaf $\mathcal{O}$ of $Sch/\mathbf{C}$ restricts to a sheaf of rings which we also denote by $\mathcal{O}$ on $Sch/X_N$. There is a complex of $\mathcal{O}$-modules over $X_N$ which we denote by $p_*\Omega^{\cdot}_{X/X_N}$ with differential denoted $d$, giving a resolution

$$\mathcal{O} \to p_*\Omega^0_{X/X_N} \to p_*\Omega^1_{X/X_N} \to \ldots \to p_*\Omega^n_{X/X_N} \to 0$$

(here $n = dim(N/X)$ is the relative (formal) dimension of the formally smooth scheme $N$ over $X$ via either of the projections).

The notation $p_*\Omega^{\cdot}_{X/X_N}$ is justified by the fact that each component of this resolution is actually the direct image of a locally free sheaf $\Omega^i_{X/X_N}$ on $X$. This locally free sheaf comes from $\Omega^i_{N/X}$ by descent from $N = X \times_{X_N} X$ to $X$. Note that the differential is in general a differential operator (of first order) between these component sheaves, so it becomes a morphism only over $X_N$.

A *local system $V$ on $X_N$* is a sheaf on $Sch/X_N$ which locally isomorphic to $\mathcal{O}^a$. This is equivalent to a vector bundle $V_X$ on $X$ together with *$N$-connection* $\varphi : V_X \to V_X \otimes_{\mathcal{O}_X} \Omega^1_{X/X_N}$ satisfying an integrability condition $\varphi^2 = 0$.

We obtain the resolution $p_*\Omega^{\cdot}_{X/X_N} \otimes_{\mathcal{O}} V$ of $V$, which we can use to calculate the cohomology of $V$ over $X_N$. Note that the morphism $p$ is cohomologically trivial for coherent sheaves on $X$. So the cohomology of $V$ may be calculated as the hypercohomology of the complex $\Omega^i_{X/X_N} \otimes_{\mathcal{O}} V$ of Zariski (or etale) sheaves on $X$; in particular there is a spectral sequence

$$H^i(X, \Omega^j_{X/X_N} \otimes_{\mathcal{O}} V) \Rightarrow H^{i+j}(X_N, V).$$

The cohomology groups are finite dimensional vector spaces if $X$ is proper.

To a formal category $(X, N)$ of smooth type we can associate a split almost polynomial sheaf of rings of differential operators $\Lambda_N$ [**39**] [**75**]. It is the sheaf of rings associated to the differentials in the above complex. Note that $p_*(\mathcal{O}_N)$ is naturally a projective limit of locally free sheaves on $X$. We can construct $\Lambda_N$ as the continuous dual, which is a union of locally free sheaves. The ring structure is dual to the cogebra structure

$$p_*(\mathcal{O}_N) \to p_*(\mathcal{O}_N) \otimes_{\mathcal{O}_X} p_*(\mathcal{O}_N)$$

which itself comes from the composition morphism $N \times_X N \to N$. A local system $V$ on $X_N$ is the same thing as a $\Lambda_N$-module; the underlying $\mathcal{O}_X$-module is $V_X$ and the $\Lambda_N$-module structure is given by $\varphi$.

A *principal $G$-bundle* on $X_N$ means a $G$-torsor on the stack $X_N$. This is the same thing as a principal $G$-bundle $P$ on $X$ together with an isomorphism $\varphi : s^*(P) \cong b^*(P)$ on $N$ (where $s, b : N \to X$ are the two tautological morphisms), such that $\varphi$ satisfies the appropriate cocycle condition on $N \times_X N$.

There is a notion of semistability for local systems over $X_N$ which is analogous to the usual notion: we can define a notion of coherent sheaf over $X_N$ (a coherent sheaf on $X$ with descent data to $X_N$) and a bundle over $X_N$ is semistable if for every $X_N$-subsheaf, the normalized Hilbert polynomial is less than or equal to that of the original object. Again as usual we can define the notion of semistability of a principal $G$-bundle on $X_N$. Lacking a direct proof of the conservation of semistability by tensor product in the case of local systems over a general formal category (this is a good question for further research), we put in the definition here that the local systems associated to all representations of $G$ should be semistable.

When considering a relative situation $X_N \to S$, semistability means semistability on each fiber $(X_N)_s$ (it is an open condition on the base $S$).

Because, in all of the examples we interested in in this paper, it is necessary to include a condition of vanishing rational Chern classes when defining the moduli spaces, we put this directly into the definition in the formal category setting. Of course, for formal groupoids different from our examples, this condition may not necessarily be a sensible one; and even in our examples, it may also be interesting to consider other components of the moduli spaces. One could make the same definitions and obtain moduli spaces without this condition, but we include the condition here for simplicity of notation.

Suppose $(X, N) \to S$ is a morphism from a formal groupoid of smooth type to a base scheme $S$, such that $X$ is smooth and proper over $S$ and $N$ is formally smooth over $S$. Suppose $x : S \to X$ is a section. Define the functor $\mathcal{R}(X_N/S, x, G)$ which to an $S$-scheme $S'$ associates $\{(P, \varphi, \beta)\}$ where $(P, \varphi)$ is a principal $G$-bundle over $X_N \times_S S'$, semistable and with vanishing rational Chern classes relative to $S'$; and $\beta : x^*(P) \cong G \times S'$ is a framing along the section $x$.

THEOREM 7.2. *The functor $\mathcal{R}(X_N/S, x, G)$ is representable by a scheme which we denote by $R(X_N/S, x, G) \to S$. Furthermore all points are semistable for the action of $G$ so a universal categorical quotient $M(X_N/S, G) = R(X_N/S, x, G)//G$ exists.*

This theorem follows from the interpretation of local systems over $X_N$ as $\Lambda_N$-modules [**75**]; the construction of the moduli and representation schemes for $\Lambda_N$-modules; and from the tannakian point of view used in [**75**] §9. □

We denote the *stack quotient* by

$$\mathcal{M}(X_N/S, G) := R(X_N/S, x, G)/G.$$

This is the first relative nonabelian cohomology of $X_N/S$ with coefficients in $G$. It is an algebraic stack.

*The basic examples*

The main example (which we used to introduce this section and which you meet in most treatments of the subject—cf [**4**] for example) is the formal groupoid obtained by setting $N := (X \times X)^\wedge$ (completion along the diagonal). We denote this formal groupoid (or the associated stack) by $X_{DR}$. In this stack there is at most one morphism between any pair of objects, so the stack is equivalent to a sheaf of sets which we also denote $X_{DR}$. The sheaf of sets is the quotient of $X$ by the equivalence relation $N$: heuristically we identify any two points which are infinitesimallly close together, and it is this infinitesimal glue which makes it so that $X_{DR}$ actually reflects the topology of the underlying usual space. More precisely if $S$ is any scheme over $\mathbf{C}$ then the $S$-valued points of $X_{DR}$ are the $S$-valued points of $X$ modulo the relation that two points are equivalent if their restrictions to the underlying reduced scheme $S^{\rm red}$ are the same; except that we have to divide by this equivalence relation and then sheafify. Since $X$ is smooth, any $S^{\rm red}$-valued point extends, locally on $S$, to an $S$-valued point. Thus after taking the quotient and sheafifying the result is simply that $X_{DR}(S) = X(S^{\rm red})$.

The sheaf of rings of differential operators associated to the formal groupoid $X_{DR}$ is just the full ring $\mathcal{D}_X$ of differential operators on $X$ (denoted $\Lambda_{DR}$ in [**75**]). A principal bundle on $X_{DR}$ is just a principal bundle on $X$ with integrable connection, and a vector bundle or local system over $X_{DR}$ is just a vector bundle on $X$ with integrable connection. We recover

$$M(X_{DR}, G) = M_{DR}(X, G)$$

and similarly for the representation spaces and stacks

$$R(X_{DR}, x, G) = R_{DR}(X, x, G), \quad \mathcal{M}(X_{DR}, G) = \mathcal{M}_{DR}(X, G).$$

The cohomology of $X_{DR}$ with coefficients in a local system is just the algebraic de Rham cohomology of $X$ with coefficients in the corresponding vector bundle with integrable connection.

We now define a formal groupoid $X_{Dol}$ which gives rise to the Dolbeault theory in the same way as $X_{DR}$ gave rise to the de Rham theory. In this formal groupoid the object object is $X$ and the morphism object is the formal completion of the zero section in the tangent bundle of $X$, lying over the diagonal in $X \times X$. A principal bundle on $X_{Dol}$ is just a principal Higgs bundle; a local system is a Higgs bundle; the cohomology of a local system is the Dolbeault cohomology; and the associated sheaf of rings of differential operators is the ring $\Lambda_{Dol}$ defined in [**75**]. The ring $\Lambda_{Dol}$ is actually the associated-graded of $\Lambda_{DR} = \mathcal{D}_X$ (with respect to the filtration by degrees of operators) which in turn is the symmetric algebra on the tangent bundle; we write

$$\Lambda_{Dol} = Gr^{\cdot}(\mathcal{D}_X) = Sym^{\cdot}(TX).$$

Semistability for $X_{Dol}$-modules as defined on the previous page is the same as the usual condition of semistability for Higgs bundles (this is also the same as semistability for $\Lambda_{Dol}$-modules), and we have also included the condition of vanishing Chern classes for the objects we consider. Thus we recover

$$M(X_{Dol}, G) = M_{Dol}(X, G)$$

and similarly for the representation spaces and stacks

$$R(X_{Dol}, x, G) = R_{Dol}(X, , x, G), \quad \mathcal{M}(X_{Dol}, G) = \mathcal{M}_{Dol}(X, G).$$

Now we define a formal groupoid $X_{Hod} \to \mathbf{A}^1$ which serves as a deformation from $X_{DR}$ to $X_{Dol}$ and from which we can recover the moduli spaces $M_{Hod}(X,G)$. It is the stack associated to the realisation of the nerve of the presheaf of groupoids given by a formal groupoid which we denote by $\widetilde{X}_{Hod} \to \mathbf{A}^1$. The object object is

$$Ob(\widetilde{X}_{Hod}) := X \times \mathbf{A}^1.$$

Let $Y$ be the complement of the strict transform of $X \times X \times \{0\}$ in the blow-up of $X \times X \times \mathbf{A}^1$ along $\Delta(X) \times \{0\}$. (Here $\Delta(X)$ is the diagonal.) There is a unique composition

$$Y \times_{X \times \mathbf{A}^1} Y \to Y$$

compatible with the trivial composition

$$(X \times X \times \mathbf{A}^1) \times_{X \times \mathbf{A}^1} (X \times X \times \mathbf{A}^1)$$

via the morphism $Y \to X \times X \times \mathbf{A}^1$.

There is a morphism $\Delta' : X \times \mathbf{A}^1 \to Y$ covering the inclusion $\Delta : X \to X \times X \times \mathbf{A}^1$. This section provides an identity for the above composition—in particular, setting the morphism object equal to $Y$ would define a category. As with the case of $X_{DR}$ itself, we take the formal completion of this morphism object: the morphism object $Mor(\widetilde{X}_{Hod})$ is defined to be the formal completion of $Y$ along $\Delta'(X \times \mathbf{A}^1)$.

The formal groupoid defined in this way is a groupoid; it maps to $\mathbf{A}^1$; it has fiber over $\{0\}$ equal to the formal groupoid defining $X_{Dol}$; and its fiber over $\{t\}$ for any $t \neq 0$ is equal to the formal groupoid defining $X_{DR}$.

The general construction of Theorem 7.2 gives back for $X_{Hod}$ over $\mathbf{A}^1$ the result of Proposition 4.1:

$$M(X_{Hod}/\mathbf{A}^1, G) = M_{Hod}(X,G)$$

and similarly for the representation space and moduli stack

$$R(X_{Hod}/\mathbf{A}^1, x, G) = R_{Hod}(X,x,G), \quad \mathcal{M}(X_{Hod}/\mathbf{A}^1, G) = \mathcal{M}_{Hod}(X,G).$$

We will now study the $\mathbf{G}_m$-equivariant sections of the moduli stack $\mathcal{M}_{Hod}(X,G) \to \mathbf{A}^1$. Since this latter is a *stack*, the notion of $\mathbf{G}_m$-equivariance of a section $\sigma : \mathbf{A}^1 \to \mathcal{M}_{Hod}(X,G)$ is more subtle than would be the notion of $\mathbf{G}_m$-invariance for a morphism of schemes.

An action of a group ($\mathbf{G}_m$ say) on a stack $\mathcal{U}$ consists of the data of a morphism

$$\mu : \mathbf{G}_m \times \mathcal{U} \to \mathcal{U}$$

together with a homotopy $\alpha$ making commutative the diagram

$$\begin{array}{ccc} \mathbf{G}_m \times \mathbf{G}_m \times \mathcal{U} & \to & \mathbf{G}_m \times \mathcal{U} \\ \downarrow & & \downarrow \\ \mathbf{G}_m \times \mathcal{U} & \to & \mathcal{U} \end{array}$$

such that the composition of the homotopies on the sides of the cube one forms with entries $\mathbf{G}_m^3 \times \mathcal{U}$ (once), $\mathbf{G}_m^2 \times \mathcal{U}$ (three times), $\mathbf{G}_m \times \mathcal{U}$ (three times) and $\mathcal{U}$ (once), is trivial. An equivariant morphism $\varphi : \mathbf{A}^1 \to \mathcal{U}$ is the data of the morphism $\varphi$ plus a homotopy $\beta$ making commutative the diagram

$$\begin{array}{ccc} \mathbf{G}_m \times \mathbf{A}^1 & \to & \mathbf{A}^1 \\ \downarrow & & \downarrow \\ \mathcal{U} & = & \mathcal{U} \end{array}$$

such that the composition of the sides of the natural cube that one can form, is zero.

(In the above two definitions, we start with the usual data of a multiplication or a map, but replace the condition that needs to be satisfied, with a homotopy; but then the homotopy is required to satisfy a condition of higher order.)

In our case, take $\mathcal{U} = \mathcal{M}_{Hod}(X,G)$ and suppose for simplicity that $G = GL(n)$. The action of $\mathbf{G}_m$ on $\mathcal{M}_{Hod}(X,G)$ is obtained via the action on $X_{Hod}$ over $\mathbf{A}^1$. More precisely there is a universal $\lambda$-connection $(\mathcal{E},\nabla)$ on $\mathcal{M}_{Hod}(X,G)\times X$ and pulling this back to $\mathbf{G}_m \times \mathcal{M}_{Hod}(X,G)\times X$ we get a $t_1\lambda$-connection $(p_{23}^*(\mathcal{E}), t_1\nabla)$ (where $t_1$ is the pullback of the coordinate on $\mathbf{G}_m$) which gives precisely a morphism

$$\mathbf{G}_m \times \mathcal{M}_{Hod}(X,G) \to \mathcal{M}_{Hod}(X,G).$$

There is a natural isomorphism between the two pullback $t_1t_2\lambda$-connections on $\mathbf{G}_m\times\mathbf{G}_m\times\mathcal{M}_{Hod}(X,G)\times X$, this gives the desired homotopy (we leave it to the reader to write this down explicitly).

A morphism $\varphi : \mathbf{A}^1 \to \mathcal{M}_{Hod}(X,G)$ corresponds to a $\lambda$-connection $(E,\nabla)$ on $\mathbf{A}^1\times X$ over $\mathbf{A}^1$. Note that $\varphi$ is a section of the usual projection $\mathcal{M}_{Hod}(X,G)$ if and only if the $\lambda$ here is the coordinate on $\mathbf{A}^1$.

A homotopy of $\mathbf{G}_m$-equivariance for the section corresponding to a $\lambda$-connection $(E,\nabla)$ on $\mathbf{A}^1\times X$ is an isomorphism

$$\beta : (p_{23}^*(E), t_1\nabla) \cong ((\mu\times 1)^*(E),\nabla)$$

of $t_1\lambda$-connections on $\mathbf{G}_m\times\mathbf{A}^1\times X$, where: $t_1$ is the pullback of the coordinate of $\mathbf{G}_m$; $p_{23}$ is the projection to $\mathbf{A}^1\times X$; and $\mu : \mathbf{G}_m\times\mathbf{A}^1\to\mathbf{A}^1$ is the standard action whence the morphism $\mu\times 1 : \mathbf{G}_m\times\mathbf{A}^1\times X\to\mathbf{A}^1\times X$. The coherence condition on the homotopy $\beta$ amounts to saying that we get an action of $\mathbf{G}_m$ on the bundle $E$ lying over the standard action on $\mathbf{A}^1$ (and respecting $\nabla$ with the twist of multiplying by $t$).

LEMMA 7.3. *Keep the notations of the previous paragraphs.*
*(1) A structure of $\mathbf{G}_m$-equivariant section in the above sense is the same thing as an isomorphism between $(E,\nabla)$ and the Rees bundle over $X\times\mathbf{A}^1$ of a vector bundle $V = E|_{X\times\{1\}}$ on $X$ with integrable connection $\nabla_1$ together with a filtration $F^{\cdot}$ satisfying Griffiths transversality and such that the associated-graded $Gr_F^{\cdot}(V)$ is semistable as a Higgs bundle. Note that such a Rees bundle has a natural $\lambda$-connection in the $X$-direction and our isomorphism is required to be compatible with the $\lambda$-connections.*
*(2) The relative cohomology of our vector bundle with $\lambda$-connection (i.e. the higher direct image, under the projection from $X_{Hod}$ to $\mathbf{A}^1$, of the bundle $\mathcal{W}$ on $X_{Hod}$ associated to the $\lambda$-connection $(E,\nabla)$) is just the Rees bundle on $\mathbf{A}^1$ corresponding to the filtration induced by $F^{\cdot}$ and the stupid filtration on $\Omega^{\cdot}_X$ on the de Rham cohomology of $(V,\nabla)$.*

*Proof:* From the preceeding discussion, a $\mathbf{G}_m$-equivariant section of the moduli stack is just a $\lambda$-connection $(E,\nabla)$ on $X\times\mathbf{A}^1$ (which is the same thing as a bundle $\mathcal{W}$ on $X_{Hod}$) together with action of $\mathbf{G}_m$ covering the standard action on $\mathbf{A}^1$ and compatible with $\nabla$ (i.e. an action of $\mathbf{G}_m$ on the bundle over $X_{Hod}$ covering the action of $\mathbf{G}_m$ on $X_{Hod}$). In particular our vector bundle $E$ on the underlying scheme $X\times\mathbf{A}^1$ has an action of $\mathbf{G}_m$; by the relative version of the inverse of $\xi$ this corresponds to a bundle $V$ on $X$ with filtration by strict subbundles $F^p$. The

$\lambda$-connection (i.e. descent data down to $X_{Hod}$) is determined by the restriction of $\nabla$ to $X \times \mathbf{G}_m \subset X \times \mathbf{A}^1$, and by the $\mathbf{G}_m$-equivariance of our section, this is determined by the connection $\nabla|_{X\times\{1\}}$, which is an integrable connection $\nabla_1$ on $V := E|_{X\times\{1\}}$. By Lemma 7.1 the connection on $X \times \mathbf{G}_m$ induced from $\nabla_1$ by equivariance, extends to a $\lambda$-connection $\nabla$ on $E$ if and only if the filtration $F^\cdot$ on $V$ satisfies Griffiths transversality with respect to $\nabla_1$. The $\lambda$-connection $(E, \nabla)$ corresponds to a section of the moduli stack $\mathcal{M}_{Hod}(X, G)$ if and only if the fibers are semistable with vanishing Chern classes. The fibers over points different from 0 work automatically, in particular the Chern classes of the fiber over 0 vanish by flatness. Thus the only condition is that the fiber over 0—which is $Gr_F^\cdot(V)$ with its induced Higgs field—be a semistable Higgs bundle. This shows (1).

For (2) note that the higher direct image of $\mathcal{W}$ down to $\mathbf{A}^1$ is the same as the hyper-higher direct image of the de Rham complex associated to $(E, \nabla)$

$$\mathbf{R}^i p_{2,*}(E \otimes \Omega^\cdot_{X\times\mathbf{A}^1/\mathbf{A}^1}, \nabla)$$

(this comes from the resolution of $\mathcal{O}$ on $X_{Hod}$ mentionned above in the context of a general formal category). We have the formula

$$(E \otimes \Omega^\cdot_{X\times\mathbf{A}^1/\mathbf{A}^1}, \nabla) = \xi((V \otimes \Omega^\cdot_X, \nabla_1), F^\cdot)$$

where the filtration on the right is obtained by combining the filtration of $V$ with the stupid filtration of $\Omega^\cdot_X$ (only then do we get a filtered complex because $\nabla_1$ doesn't actually respect the filtration on $V$ alone). As we have seen in Proposition 5.1 the construction $\xi$ commutes with higher direct images, giving (2). □

*Remark:* If $V$ is a vector bundle with integrable connection, let $[V]$ denote the corresponding point in $\mathcal{M}_{DR}(X, GL(n)) \subset \mathcal{M}_{Hod}(X, GL(n))$ (this inclusion being the fiber over the point $1 \in \mathbf{A}^1$). The orbit $\mathbf{G}_m \cdot [V]$ in $\mathcal{M}_{Hod}(X, GL(n))$ extends to a section $\mathbf{A}^1 \to \mathcal{M}_{Hod}(X, GL(n))$—this is the result of Theorem 10.1 below. However, this extension might not be unique, even up to translation of the filtration, even if $V$ is irreducible. This non-uniqueness occurs when the limiting Higgs bundle is not irreducible (or more precisely, when it is not simple). An example may be obtained as follows: let $E$ be the trivial bundle of rank $r$ on a Riemann surface $X$, with the trivial connection $d_0$. Let $\alpha$ be some matrix of holomorphic one-forms and set $\nabla = d_0 + \alpha$. For any filtration $F^\cdot$ of $E$ by trivial subbundles (corresponding to a filtration of $H^0(X, E)$) such that $\alpha F^p \subset F^{p-1} \otimes \Omega^1_X$ (the Griffiths transversality condition) we have that $Gr_F(E)$ is an extension of Higgs bundles of the form $\mathcal{O}_X$, in particular it is semistable. Thus any such choice of $F$ is possible (and note, for example, that a two-step $F$ automatically satisfies the transversality condition). On the other hand, for generic $\alpha$ (in genus $\geq 2$) the flat bundle $(E, \nabla)$ will be irreducible. Thus there exist irreducible connections where the extension as in the previous lemma is not at all unique.

On the other hand if the limiting Higgs bundle is stable then the filtration is unique up to translation of indices, cf Proposition 10.4 below.

## 8. The Gauss-Manin connection

Suppose $X \to S$ is a smooth projective morphism. Then we obtain

$$M_{DR}(X/S, G) \to S,$$

a family whose fiber over $s \in S$ is $M_{DR}(X_s, G)$. This family has an algebraic integrable connection [**70**] [**75**]. We can interpret the connection in terms of formal categories in the following way (this is a simple variant of the crystalline interpretation of [**75**]). We have a morphism $X_{DR} \to S_{DR}$ and the fiber product $X_{DR} \times_{S_{DR}} S$ has a structure of smooth formal groupoid over $S$, which we call $X_{DR/S}$. The morphism space is the formal completion of the diagonal in $X \times_S X$. The moduli stack $\mathcal{M}_{DR}(X/S)$ is just the nonabelian cohomology of $X_{DR/S}$ relative to $S$ with coefficients in $G$, or in our previous notations

$$\mathcal{M}_{DR}(X/S, G) = \mathcal{M}(X_{DR/S}/S, G).$$

The same holds for the moduli spaces and representation spaces. But we could equally well take the nonabelian cohomology of $X_{DR}$ relative to $S_{DR}$. We obtain a stack over $S_{DR}$ which, when pulled back to $S$, gives $\mathcal{M}_{DR}(X/S, G)$. To put this another way, we get descent data for $\mathcal{M}_{DR}(X/S, G)$ from $S$ down to $S_{DR}$. This is exactly the data of an integrable connection which is the nonabelian version of the Gauss-Manin connection.

One can see that the associated analytic connection on the analytic family is the same as that induced by the fact that (locally over the base) all of the fibers of $\mathcal{M}_{DR}(X/S)$ are of the form $\mathcal{M}_B(\Gamma)$ for $\Gamma$ the fundamental group of the fiber ([**75**], Theorem 8.6).

In abelian Hodge theory there are two principal results about the Gauss-Manin connection: Griffiths transversality with respect to the Hodge filtration, and regular singularities at the singular points of a family. We obtain their analogues for nonabelian cohomology by using the theory of formal categories and following the above description of the connection.

These properties are easily obtained by using a variant of our construction of the connection. Suppose $X_N \to S_K$ is a morphism of formal categories of smooth type such that the fiber product $X_N \times_{S_K} S$ is a formal groupoid of smooth type on $X/S$. Then the schemes $M(X_N \times_{S_K} S/S, G)$ and stacks $\mathcal{M}(X_N \times_{S_K} S/S, G)$ have descent data down to $S_K$, that is they are pullbacks of sheaves or stacks on $Sch/S_K$. The same is true for the $R(X_N \times_{S_K} S/S, x, G)$ if $x : S_K \to X_N$ is a section.

*Remark:* The same type of discussion as given below in the case of nonabelian first cohomology, works also to give Griffiths transversality and regularity of the Gauss-Manin connection for the higher nonabelian cohomology that we briefly discussed in §5 above (see [**76**], [**78**] for details).

*Griffiths transversality*

Suppose $X \to S$ is a smooth family of projective varieties (and we now ask that the base be smooth, although we don't need it to be projective). Then we obtain a morphism of formal categories $X_{Hod} \to S_{Hod}$ over $\mathbf{A}^1$. Put

$$X_{Hod/S} := X_{Hod} \times_{S_{Hod}} (S \times \mathbf{A}^1).$$

It is given by a smooth formal groupoid on $X \times \mathbf{A}^1$ relative to $S \times \mathbf{A}^1$. The relative nonabelian cohomology is

$$\mathcal{M}_{Hod}(X/S, G) := \mathcal{M}(X_{Hod/S}/S, G) \to S \times \mathbf{A}^1.$$

This morphism is provided with an action of the formal groupoid $S_{Hod}$ (i.e. $\mathcal{M}_{Hod}(X/S, G)$ is the pullback of a stack over $S_{Hod}$). In particular over $\lambda \neq 0$ we get back the action of $S_{DR}$, that is to say the Gauss-Manin connection. There is a $\mathbf{G}_m$-action compatible with everything. This whole situation is the nonabelian analogue of Griffiths transversality, an interpretation which, comparing with the abelian case, is justified by Lemma 7.3.

After we discuss the compactification below we will give another interpretation in terms of poles of the Gauss-Manin connection at infinity.

*Regularity of the Gauss-Manin connection*

Suppose $S' \subset S$ is an open subset whose complement is a divisor $D$ with normal crossings. Define $S_{DR}(\log D)$ to be the formal groupoid which is associated to the bundle of vector fields tangent to $D$. To construct it explicitly we first treat the case where $S$ is a smooth curve and $D$ a point. Then $S \times S$ is a smooth surface. Blow up at the point $(D, D)$, and take the formal completion along strict transform of the diagonal, as morphism scheme. The maps of a smooth scheme into the blow up minus the transform of $D \times S$ are the same as maps into $S \times S$ which send the intersection with the divisor $D \times S$ to the point $(D, D)$. From this description we obtain the composition. The formal completion then becomes a formal groupoid. Now, for any $(S, D)$, glue this construction in along each component of the divisor (or more precisely along an $i$-fold intersection of the divisor in an $n$-dimensional $S$, glue the product of $i$ copies of this construction with $n-i$ copies of a smooth curve with the usual de Rham construction.

A vector bundle or local system over $S_{DR}(\log D)$ is just a vector bundle on $S$ with integrable connection with logarithmic singularities along $D$. In particular, the condition that a vector bundle on $S'_{DR}$ (corresponding to a vector bundle with integrable connection on $S'$) extends to a bundle on $S_{DR}(\log D)$ is equivalent to the condition that the connection have regular singularities.

Now suppose that we have a projective morphism of smooth varieties $f : S \rightarrow S$ and a divisor $D \subset S$ which has normal crossings, such that $Y := f^{-1}(D)$ has normal crossings, and such that $f : X' \rightarrow S'$ is smooth where $S' = S - D$ and $X' = f^{-1}(S') = X - Y$. There is a morphism of formal groupoids $S_{DR}(\log D) \rightarrow S_{DR}$. Let

$$X_{DR}(\log D) := X_{DR} \times_{S_{DR}} S_{DR}(\log D).$$

Again we have a morphism of formal groupoids

$$X_{DR}(\log D) \rightarrow S_{DR}(\log D),$$

and the fiber product

$$X_{DR/S}(\log D) := X_{DR}(\log D) \times_{S_{DR}(\log D)} S$$

is a formal groupoid on $X$ over $S$. Even though $X/S$ is not smooth, this formal groupoid corresponds to a split almost polynomial sheaf of rings of differential operators $\Lambda_{DR/S(\log D)}$ on $X$ over $S$. Thus the moduli problems are solved by **[75]** so the first relative nonabelian cohomology stack $\mathcal{M}(X_{DR/S}(\log D)/S, G)$ is an algebraic stack; the associated representation functor for objects provided with a frame along a fixed section (not passing through the singular points) is representable by a scheme, and the universal categorical quotient scheme $M(X_{DR/S}(\log D)/S, G)$ exists.

Finally, since $X_{DR/S}(\log D)$ comes from fiber product as in our general situation, these moduli spaces, moduli stacks and representation spaces descend to $S_{DR}(\log D)$. This statement is the regularity of the Gauss-Manin connection.

In the case when $G = GL(n)$ we can do slightly better. Let us recall that $M(X_{DR/S}(\log D)/S, GL(n))$ parametrizes bundles with descent data to the formal groupoid $X_{DR/S}(\log D)$, which are semistable with vanishing rational Chern classes on the fibers $X_s$. But over the singular fibers, where the extra structure is a connection away from the singularities but simply a logarithmic connection near the singularities, it is also natural to consider objects which are no longer bundles but torsion-free sheaves. This moduli space, which we can denote by $M^{\mathrm{tf}}(X_{DR/S}(\log D)/S, GL(n))$, will have the advantage that, when combined with the techniques used below for compactifying $M_{DR}(X_s, GL(n))$, will give a compact total space over projective $S$.

*Caution:* One might be tempted to try the above argument with $X_{DR}/S_{DR}$. In this case the fiber product $X_{DR}\times_{S_{DR}} S$ no longer has the required smoothness properties, and the relative moduli stack is no longer an algebraic stack—otherwise we would have an extension of the Gauss-Manin connection over the singularities! In the case $G = \mathbf{G}_a$ for example this would give an extension of the Gauss-Manin connection for ordinary first cohomology, easily seen as impossible in most examples.

*Exercise:* Explain what goes wrong in an example (such as a family of curves aquiring a node) if we try to do the previous construction with $X_{DR}/S_{DR}$. This can be done in the context of abelian cohomology.

## 9. Etale local triviality of $M_{Hod}$

*Goldman-Millson theory*

A deformation problem is often controlled by a differential graded Lie algebra (dgla) $L^{\cdot}$ over the base field which we are assuming is $\mathbf{C}$. As explained in [**23**] this means that the deformations with values in an artin local ring $A$ (with maximal ideal $\mathbf{m}$) correspond to elements $\eta \in L^1 \otimes_{\mathbf{C}} \mathbf{m}$ such that

$$d(\eta) + \frac{1}{2}[\eta, \eta] = 0.$$

The isomorphisms between deformations $\eta$ and $\eta'$ correspond to elements $s \in L^0 \otimes_{\mathbf{C}} \mathbf{m}$ with

$$\eta' = d(s) + e^{-s}\eta e^{s}.$$

If $P$ is a principal $G$-bundle, let $A^{\cdot}(ad\, P)$ denote the graded Lie algebra of $\mathcal{C}^{\infty}$ forms on $X$ with coefficients in the adjoint bundle $P \times^G \mathbf{g}$ (with Lie bracket combining the wedge of forms and the Lie bracket of $\mathbf{g}$). If $(P, \theta)$ is a principal Higgs bundle structure then we obtain a dgla $(A^{\cdot}(ad\, P), \overline{\partial}+\theta)$ which gives the deformation theory of $(P, \theta)$. If $(P, \nabla)$ is a principal bundle with integrable connection then we obtain a dgla $(A^{\cdot}(ad\, P), \nabla)$ which gives the deformation theory of $(P, \nabla)$ [**23**].

More generally, suppose $(X, N)$ is a formal groupoid structure for $X$; then we have the algebra of differentials $\Omega^{\cdot}_{X/X_N}$ (cf §7 above). Let

$$A^i_N := \bigoplus_{p+q=i} A^{0,q}(\Omega^p_{X/X_N})$$

This is a differential graded algebra with differential equal to $\overline{\partial} + \delta$ where $\delta$ is the first order differential operator corresponding to the differential of $p_*\Omega^{\cdot}_{X/X_N}$. If $P$ is a principal $G$-bundle over $X_N$ then we obtain a dgla

$$A^{\cdot}_N(ad\, P) := A^{\cdot}_N \otimes_{\mathcal{O}} (P \times^G \mathbf{g}).$$

The differential comes from that of $A^{\cdot}_N$. This dgla controls the deformation theory of $P$ as a principal bundle on $X_N$ (we leave the proof as an exercise following [**23**] and [**75**] §10).

If $X_N = X_{DR}$ then we get $A^{\cdot}_{DR}(ad\,P) := (A^{\cdot}(ad\,P), \nabla)$, whereas if $X_N = X_{Dol}$ then we get $A^{\cdot}_{Dol}(ad\,P) := (A^{\cdot}(ad\,P), \overline{\partial} + \theta)$.

If $L^{\cdot}$ is a dgla then let $H^{\cdot}(L^{\cdot})$ denote the dgla of cohomology with differential equal to zero. Recall that we say that $L^{\cdot}$ is *formal* if there is a quasiisomorphism between $L^{\cdot}$ and $H^{\cdot}(L^{\cdot})$. According to the theory of Goldman and Millson [**23**], a quasiisomorphism induces an equivalence of deformation theories, and on the other hand the deformation theory of a dgla with zero differential is *quadratic*, in other words the universal deformation space is the quadratic cone in $H^1$ which is defined as the zero scheme of the map $H^1 \to H^2$, $\eta \mapsto [\eta, \eta]$. It follows that the deformation theory of a formal dgla is quadratic.

We need a relative version of this theory for $X \times \mathbf{A}^1/\mathbf{A}^1$ near a prefered section. It is actually an interesting question (which I don't think has yet been addressed) to develop a relative version of Goldman-Millson theory in all generality. In our case we are helped by the fact that the total space is a product. Also we will restrict to deformations over artinian base (whereas in a better version one should consider arbitrary base with nilpotent ideal).

Suppose $P$ is a flat principal $G$-bundle with harmonic $K$-reduction $P_K$ and associated operators $d' = \partial + \theta'$ and $d'' = \overline{\partial} + \theta''$. We define the differential graded Lie algebra over $\mathbf{C}[\lambda]$

$$A^{\cdot}_{Hod} := (A^{\cdot}(ad\,P) \otimes_{\mathbf{C}} \mathbf{C}[\lambda], \lambda\partial + \theta' + \overline{\partial} + \lambda\theta'').$$

Note that the differential has square zero and, as varying with parameters, gives a deformation from the de Rham to the Dolbeault differentials. Notice also that the components of $A^{\cdot}_{Hod}$ are flat $\mathbf{C}[\lambda]$-modules. For any dgla $L^{\cdot}$ over $\mathbf{C}[\lambda]$ where the components are flat, we define a stack of groupoids over the category of artinian local $\mathbf{C}[\lambda]$-algebras $(B, \mathbf{m})$ in the same way as above (except that tensor products are taken over $\mathbf{C}[\lambda]$): the objects are elements $\eta \in L^1 \otimes_{\mathbf{C}[\lambda]} \mathbf{m}$ such that

$$d(\eta) + \frac{1}{2}[\eta, \eta] = 0.$$

The isomorphisms between deformations $\eta$ and $\eta'$ correspond to elements $s \in L^0 \otimes_{\mathbf{C}[\lambda]} \mathbf{m}$ with

$$\eta' = d(s) + e^{-s}\eta e^{s}.$$

Again we have the same theorem that quasiisomorphisms between flat dgla's induce equivalences of deformation groupoids. And, on the other hand, the dgla $A^{\cdot}_{Hod}$ gives the deformation theory of $\mathcal{M}_{Hod}$ near the prefered section corresponding to $P$. More precisely the groupoid defined above for $(B, \mathbf{m})$ is equivalent to the groupoid of morphisms $Spec(B) \to \mathcal{M}_{Hod}$ with isomorphism between the morphism $Spec(B/\mathbf{m}) \to \mathcal{M}_{Hod}$ and the composed morphism $Spec(B/\mathbf{m}) \to \mathbf{A}^1 \to \mathcal{M}_{Hod}$ (where the first morphism is the projection to $\mathbf{A}^1 = Spec(\mathbf{C}[\lambda])$ and the second morphism is the prefered section corresponding to $P$). The proofs are left as exercises.

Finally, the dgla $A^{\cdot}_{Hod}$ defined above is formal over $\mathbf{C}[\lambda]$. Let $\ker(\partial + \theta'')[\lambda]$ denote the subcomplex of $A^{\cdot}_{Hod}$ consisting of forms $\alpha$ with $\partial\alpha + \theta''\alpha = 0$. Since this operator doesn't depend on $\lambda$, the subcomplex is just a tensor product of the usual complex $\ker(\partial + \theta'')$ with $\mathbf{C}[\lambda]$. Furthermore the subcomplex is a sub-dgla; and finally note that the differential of the sub-complex is $\overline{\partial} + \theta'$ which also doesn't depend on $\lambda$. The "principle of two types" (cf Lemma 2.2 of [**71**]) implies that the morphisms

$$H^{\cdot}(A^{\cdot}_{Hod}) \leftarrow \ker(\partial + \theta'')[\lambda] \rightarrow A^{\cdot}_{Hod}$$

are quasiisomorphisms. Finally, let $\mathcal{H}^{\cdot}$ denote the $\mathbf{C}$-vector space of harmonic forms in $A^{\cdot}(ad\, P)$ (which is the same for the flat connection or the Dolbeault operator $\overline{\partial} + \theta'$ or, for that matter, anything in between). Then the morphism

$$\mathcal{H}^{\cdot} \otimes_{\mathbf{C}} \mathbf{C}[\lambda] \rightarrow H^{\cdot}(A^{\cdot}_{Hod})$$

is an isomorphism of graded vector spaces (the space of harmonic forms does not *a priori* have a product structure). This shows that $H^{\cdot}(A^{\cdot}_{Hod})$ is flat over $\mathbf{C}[\lambda]$ so we can apply the quasiisomorphism result to conclude that the deformation theory of $A^{\cdot}_{Hod}$ is the same as that of the graded Lie algebra $H^{\cdot}(A^{\cdot}_{Hod})$. Finally, in order to obtain local triviality we have to show that there is a product structure on $\mathcal{H}^{\cdot}$ such that the isomorphism $H^{\cdot}(A^{\cdot}_{Hod}) \cong \mathcal{H}^{\cdot} \otimes_{\mathbf{C}} \mathbf{C}[\lambda]$ becomes an isomorphism with product structure. This can be seen by identifying $\mathcal{H}^{\cdot}$ with the cohomology of the complex $\ker(\partial + \theta'')$ and noting that this latter has a product structure. With this result, the deformation theory along the prefered section becomes a product. We have shown this result for deformations over artinian base, so we get the result on the level of formal completions. In terms of representation spaces we can state this as

$$R_{Hod}(X, x, G)^{\wedge} \cong R_{Dol}(X, x, G)^{\wedge} \hat{\times} \mathbf{A}^1$$

where the completions are along the prefered section or at the point in $R_{Dol}(X, x, G)$. Passing to the stack-theoretic quotient we obtain the same statement for the moduli stacks. Artin approximation then gives that locally in the etale topology at any point of a prefered section, $\mathcal{M}_{Hod}$ is a product. Any point (even non-semisimple) is isomorphic to a point in a neighborhood of a semisimple point, so we obtain local triviality at any point in the union $\mathcal{M}_{Hod}(X, G)$ of components corresponding to bundles with vanishing rational Chern classes.

The formal trivialization along the prefered section is the total-space version of the *isosingularity principle* stated in the introduction and §10 of [**75**].

If the preferred section is preserved by the $\mathbf{G}_m$-action (or more precisely if we are given a lifting of the $\mathbf{G}_m$-action to the bundle) then the above construction is compatible with the $\mathbf{G}_m$-action and our formal trivialization is compatible with the action. Note that the condition that the prefered section be preserved by the $\mathbf{G}_m$ action is equivalent to asking that it come from a complex variation of Hodge structure. In passing, note that for the completion around the identity section in the case $G = GL(n)$, the above is the trivialization that was needed in §6.

We can sum up in the following theorem.

THEOREM 9.1. *Suppose $X$ is a smooth projective variety. Let $M_{Hod}(X, G) \rightarrow \mathbf{A}^1$ denote the union of components corresponding to objects with vanishing rational Chern classes. Then etale locally upstairs $M_{Hod}(X, G)$ is a product, in other words any point $P \in M_{Hod}(X, G)$ over $\lambda \in \mathbf{A}^1$ admits an etale neighborhood $P \in U \rightarrow M_{Hod}(X, G)$ with an etale morphism*

$$U \rightarrow M_{Hod}(X, G)_{\lambda} \times \mathbf{A}^1.$$

*The same holds for $R_{Hod}(X, x, G)$ and the moduli stack $\mathcal{M}_{Hod}(X, G)$.*

□

COROLLARY 9.2. *The morphism $M_{Hod}(X, G, 0) \rightarrow \mathbf{A}^1$ is flat (similarly for $R_{Hod}(X, x, G)$ and the moduli stack $\mathcal{M}_{Hod}(X, G)$).*

□

On the other hand, if we have a family of varieties $X \to S$ then the Gauss-Manin connection guarantees that the family $M_{DR}(X/S, G) \to S$ is etale locally a product. It seems reasonable to guess that these two results combine into the following.

CONJECTURE 9.3. *Let us suppose that $X \to S$ is a smooth projective family. Let $M_{Hod}(X/S, G) \to S \times \mathbf{A}^1$ denote the relative $M_{Hod}$ space. Then etale locally (above) $M_{Hod}(X/S, G)$ is a product, in other words any point $P \in M_{Hod}(X/S, G)$ over $(s, \lambda) \in S \times \mathbf{A}^1$ admits an etale neighborhood $P \in U \to M_{Hod}(X/S, G)$ with an etale morphism $U \to M_{Hod}(X/S, G)_{(s,\lambda)} \times S \times \mathbf{A}^1$. The same holds for $R_{Hod}(X/S, x, G)$ and the moduli stack $\mathcal{M}_{Hod}(X/S, G)$.*

To prove this one would have to analyze much more closely the deformation theory. In particular, the fact that the total space is a product, which helped a lot in the previous argument, is no longer there to help us here.

We do give an argument showing this conjecture over smooth points (i.e. showing that the morphism to $S$ is smooth at smooth points of the fibers) in the subsection "Griffiths transversality revisited" of §11 below.

A consequence of this conjecture would be that the deformation class of $M_{Hod}$ over a point $(s, 0) \in X \times \mathbf{A}^1$ is determined by a class $\zeta \in H^1(M_{Dol}(X_s), \Theta_{M_{Dol}}) \otimes (TS_s \oplus \mathbf{C})$, where $\Theta_{M_{Dol}}$ is the etale sheaf of infinitesimal automorphisms of $M_{Dol}$. Note that in any case the deformation theory of $M_{Hod}$ over $(s, \lambda)$ with $\lambda \neq 0$ is trivial due to the Gauss-Manin connection in the $S$-direction and the $\mathbf{G}_m$-action in the $\mathbf{A}^1$-direction. This is why we took the point $(s, 0)$ in the first sentence.

## 10. A weight property for the Hodge filtration

From its very definition, the usual Hodge filtration on cohomology has the property that $F^0V = V$. In terms of the Rees bundle $\xi(V, F)$ this translates to the statement that if $v \in V$ then the $\mathbf{G}_m$-orbit $\mathbf{G}_m v \subset \xi(V, F)$ has a limit point, i.e. it extends to a section $\mathbf{A}^1 \to \xi(V, F)$. In effect we can choose a splitting $V \cong \bigoplus_{p \geq 0} V^p$ of the filtration $F^{\cdot}$ and work on each piece separately. Recall that $\xi(V^p, F)$ is the sheaf of sections of $V \otimes \mathcal{O}_{\mathbf{G}_m}$ which have a pole of order $\leq p$ at the origin; in particular the constant section (which is $\mathbf{G}_m$-invariant) having no poles, is a section over $\mathbf{A}^1$. Note that if $V$ had had a component $V^p$ with $p < 0$ then the constant section, for a nonzero vector in this component, wouldn't have had the required zero at the origin so it wouldn't have been a section of $\xi(V, F)(\mathbf{A}^1)$.

Among other things this property allows us to write

$$V = H^0(\mathbf{A}^1, \xi(V, F))^{\mathbf{G}_m}.$$

We will establish a similar property for $M_{Hod}$ and $\mathcal{M}_{Hod}$—this will be the extension property alluded to in the remarks following Lemma 7.3.

In this section we will use the notion of almost-polynomial sheaf of rings of differential operators $\Lambda$ on $X/S$, for $X \to S$ a projective morphism, from [**75**]. Such a thing is a sheaf of rings $\Lambda$ with increasing filtration $\Lambda^i \subset \Lambda$, with $\Lambda^0 = \mathcal{O}_X$, such that the associated-graded is a polynomial ring, and such that $\Lambda^1$ acts (by commutation) on $\Lambda^0 = \mathcal{O}_X$ by differential operators of order 1.

Recall that for any formal groupoid $(X, N)$ of smooth type we obtain an almost-polynomial sheaf of rings of differential operators $\Lambda_N$ as described in [**39**] and in

§7 above. The $\Lambda_N$-modules are just the coherent sheaves on $X$ with descent data down to $X_N$.

From our basic examples of formal groupoids, we get the basic examples of sheaves of rings $\Lambda$. For example the formal groupoid $X_{DR}$ gives rise to $\Lambda_{DR} = \mathcal{D}_X$, the full sheaf of rings of differential operators on $X$. A $\Lambda_{DR}$-module (i.e. $\mathcal{D}_X$-module—which, it should be noted, we are here assuming always to be $\mathcal{O}_X$-coherent) is just a vector bundle with a flat connection, interpreted also as a vector bundle on $X$ with descent data down to $X_{DR}$ (i.e. a crystal). The formal groupoid $X_{Dol}$ gives rise to $\Lambda_{Dol} = Sym^{\cdot}(TX)$, the polynomial ring obtained from the tangent bundle. It is the associated-graded of $\mathcal{D}_X$. A $\Lambda_{Dol}$-module is a Higgs sheaf (this interpretation is given in Lemma 6.5 of [**75**], for example). Again, a Higgs sheaf may be interpreted as a coherent sheaf on $X$ with descent data down to $X_{Dol}$.

The case which interests us for this section is the formal groupoid $X_{Hod}$ on $X \times \mathbf{A}^1$ over $\mathbf{A}^1$. We can explicitly describe the sheaf of rings $\Lambda_{Hod}$ (it is the sheaf of rings denoted $\Lambda^R$ in [**75**]). Recall that $\Lambda_{DR}$ has a filtration with $\Lambda^i_{DR}$ being the differential operators of order $\leq i$. Define a decreasing filtration by indexing negatively, $F^{-i} = \Lambda^i_{DR}$. This filtration is compatible with the ring structure so the construction $\xi$ gives a sheaf of rings on $X \times \mathbf{A}^1$ over $\mathbf{A}^1$,

$$\Lambda_{Hod} := \xi(\Lambda_{DR}, F).$$

One checks that it is the sheaf of rings associated to the formal groupoid $X_{Hod}$. The relative moduli theory for $\Lambda_{Hod}$-modules on $X \times \mathbf{A}^1/\mathbf{A}^1$ yields the moduli space $M_{Hod}$ and representation space $R_{Hod}$. The stack theoretic quotient gives the moduli stack $\mathcal{M}_{Hod}$. Note that, as usual for the Rees bundle construction, the fiber of $\Lambda_{Hod}$ over $0 \in \mathbf{A}^1$ is the associated-graded of $\Lambda_{DR}$, i.e. it is $\Lambda_{Dol}$.

We would like to investigate how to take the closure of $\mathbf{G}_m$-orbits in $M_{Hod}$ or $\mathcal{M}_{Hod}$. For this, we simply generalize the technique of S. Langton to the case of modules over a split almost polynomial sheaf of rings of differential operators. We state this in generality rather than for $\Lambda_{Hod}$ because, first of all, the generality will actually simplify notation for the upcoming subsection, and second of all, the result might be useful in other contexts.

*Langton theory*

For this subsection, we assume familiarity with the notations and terminology of [**75**] (although the reader should be able to get by with the previous listing of the examples of $\Lambda$ which interest us, plus a little bit of imagination). In particular, $p$-semistablity and $p$-stability refer to Gieseker's definition involving Hilbert polynomials. We will work with $G = GL(n)$ at the start.

The following theorem is the generalisation of Langton's theory [**51**] of properness of moduli spaces, in the case of sheaves of $\Lambda$-modules (M. Maruyama pointed out to me that Langton's theory carries over in this type of general context).

THEOREM 10.1. *Suppose $S = Spec(A)$ where $A$ is a discrete valuation ring with fraction field $K$ and residue field $A/\mathbf{m} = \mathbf{C}$. Let $\eta$ denote the generic point and $s$ the closed point of $S$. Suppose $X \to S$ is a projective morphism of schemes with relatively very ample $\mathcal{O}(1)$ on $X$. Suppose $\Lambda$ is a split almost polynomial sheaf of rings of differential operators on $X/S$ as in [**75**]. Suppose $\mathcal{F}$ is a sheaf of $\Lambda$-modules on $X$ which is relatively of pure dimension $d$, flat over $S$, and such that the generic fiber $\mathcal{F}_\eta$ is p-semistable. Then there exists a sheaf of $\Lambda$-modules $\mathcal{F}'$ on*

*$X$ which is relatively of pure dimension $d$, flat over $S$, and such that $\mathcal{F}'_\eta \cong \mathcal{F}_\eta$ and also $\mathcal{F}'_s$ is p-semistable.*

*Proof:* Langton's proof carries over into our situation. We give a brief sketch for compatibility with our notations. Let $p_X(\cdot)$ denote the absolute normalized Hilbert polynomial of a sheaf on $X$ with proper support, and let $p_{X/S}(\cdot)$ denote the relative normalized Hilbert polynomial of a sheaf flat over $S$. Let $\mathcal{F}_n$ denote the sheaf of $\Lambda$-modules

$$\mathcal{F}_n := \mathcal{F} \otimes_A A/\mathbf{m}^{n+1}.$$

It is of pure dimension $d$ on $X$. Let

$$\mathcal{F}_n \to \mathcal{G}_n$$

denote the destabilizing quotient, that is the quotient with the minimal normalized Hilbert polynomial. (Note that if $\mathcal{G}_0 = \mathcal{F}_0$ then $\mathcal{F}_0$ is $p$-semistable and we're done—so we assume that this is not the case). We make the following claims:

1. Let $p_0 = p_X(\mathcal{G}_0)$. Then for all $n$, $p_X(\mathcal{G}_n) = p_0$.
2. There are morphisms $\mathcal{G}_n \to \mathcal{G}_{n-1}$ compatible with the morphisms $\mathcal{F}_n \to \mathcal{F}_{n-1}$.
3. There there is $q$ such that for $n \geq q$ we have $\mathcal{G}_n \stackrel{\cong}{\to} \mathcal{G}_q$.

*Proof of 1:* Assume it is known for $n-1$. We have an exact sequence of $\Lambda$-modules

$$0 \to \mathcal{F}_0 \to \mathcal{F}_n \to \mathcal{F}_{n-1} \to 0.$$

From this, we obtain a quotient $\mathcal{F}_n \to \mathcal{G}_{n-1}$, with $p_X(\mathcal{G}_{n-1}) = p_0$. This shows that the normalized Hilbert polynomial of the destabilizing quotient $\mathcal{G}_n$ is $\leq p_0$. On the other hand,

$$p_X(\mathrm{im}(\mathcal{F}_0 \to \mathcal{G}_n)) \leq p_X(\mathcal{G}_n)$$

(since $\mathcal{G}_n$ is $p$-semistable)—unless this morphism is zero in which case $\mathcal{G}_n = \mathcal{G}_{n-1}$ and we're done anyway. Therefore (by the definition of $p_0$) we have that $p_0 \leq p_X(\mathcal{G}_n)$. This proves claim (1).

*Proof of 2:* Since $\mathcal{G}_{n-1}$ is a quotient of $\mathcal{F}_n$ with $p_X(\mathcal{G}_{n-1}) = p_0$, it factors through the destabilizing quotient giving the morphism $\mathcal{G}_n \to \mathcal{G}_{n-1}$.

*Proof of 3:* Suppose not. We may assume that $A$ is complete. Let $\mathcal{G} := \lim_{\leftarrow} \mathcal{G}_n$. This gives a quotient of $\mathcal{F}$ destabilizing $\mathcal{F}$ over the generic point.

Now we proceed with the construction of $\mathcal{F}'$. Starting with $\mathcal{F}$ (and assuming that $\mathcal{F}_0$ is not $p$-semistable), we construct the quotient $\mathcal{G}_q$ as above. By statement (3), $\mathcal{G}_q$ is the maximal quotient of $\mathcal{F}$ which has normalized Hilbert polynomial $\leq p_0$ and which is supported over some $Spec(A/\mathbf{m}^n)$.

Let $\mathcal{F}^{(1)}$ be the kernel of the map $\mathcal{F} \to \mathcal{G}_q$. Let $p_1$ be the normalized Hilbert polynomial of the destabilizing quotient $\mathcal{Q}$ of $\mathcal{F}^{(1)} \otimes_A A/\mathbf{m}$. We claim that $p_1 > p_0$. To see this, suppose to the contrary that $p_1 \leq p_0$. Let $\mathcal{K}$ be the kernel of the map $\mathcal{F}^{(1)} \to \mathcal{Q}$, and let $\mathcal{G}' = \mathcal{F}/\mathcal{K}$. Then $\mathcal{G}'$ is a quotient of $\mathcal{F}$ which is an extension of $\mathcal{G}_q$ by $\mathcal{Q}$; in particular its normalized Hilbert polynomial is $\leq p_0$. Furthermore $\mathcal{G}'$ is supported over $Spec(A/\mathbf{m}^{q+2})$. This contradicts maximality of $\mathcal{G}_q$, showing that $p_1 > p_0$.

Now start with $\mathcal{F}^{(1)}$ and repeat the same construction to obtain $\mathcal{F}^{(2)}$ etc.; and for each $i$ let $p_i$ be the normalized Hilbert polynomial of the destabilizing quotient of $\mathcal{F}^{(i)} \otimes_A A/\mathbf{m}$ (we stop if $\mathcal{F}^{(i)} \otimes_A A/\mathbf{m}$ is $p$-semistable). We have $p_0 < p_1 < p_2 < \ldots$. Since all of these sheaves are flat over $S$ (they are subsheaves of $\mathcal{F}$ and hence without $A$-torsion), the Hilbert polynomials of $\mathcal{F}^{(i)} \otimes_A A/\mathbf{m}$ are all equal to the Hilbert polynomial of $\mathcal{F} \otimes_A K$. Finally, the set of $\Lambda$-modules with a given Hilbert polynomial and with destabilizing quotient having normalized Hilbert polynomial $\geq p_0$, is bounded. Thus the set of possible normalized Hilbert polynomials of the destabilizing quotients is finite. This shows that the process must stop. At the stopping point $\mathcal{F}^{(i)} \otimes_A A/\mathbf{m}$ is $p$-semistable, and we take $\mathcal{F}' := \mathcal{F}^{(i)}$. □

*Application to $M_{Hod}$*

We can apply Langton theory to limits of $\mathbf{G}_m$-orbits in $M_{Hod}$. In fact this applies equally well to the moduli stack $\mathcal{M}_{Hod}$. Suppose $p \in \mathcal{M}_{Hod}(X_s, GL(n))$. It corresponds to a $\lambda(1)$-connection $(\mathcal{E}, \varphi)$ on $X_s$. The $\mathbf{G}_m$-orbit of $p$ is a morphism $\mathbf{G}_m \to \mathcal{M}_{Hod}$ which corresponds to a $\lambda$-connection

$$(\mathcal{F}', \nabla') = (p_1^*(\mathcal{E}), t\varphi)$$

on $X_s \times \mathbf{G}_m$ (where $t$ is the coordinate on $\mathbf{G}_m$ and $\lambda = t\lambda(1) : \mathbf{G}_m \to \mathbf{A}^1$ is the projection of the orbit).

COROLLARY 10.2. *With the notations of the above paragraph, there is an extension $(\mathcal{F}, \nabla)$ of $(\mathcal{F}', \nabla')$ to a $\lambda$-connection on $X_s \times \mathbf{A}^1$, such that $\mathcal{F}|_{X_s \times \{0\}}$ is a bundle, is semistable and has vanishing rational Chern classes. (Note by way of notation that the extension in question is from $\mathbf{G}_m$ to $\mathbf{A}^1$ via the standard inclusion $\mathbf{G}_m \subset \mathbf{A}^1$ and not via the map $\lambda$, which could be identically zero.)*

*Proof:* First of all note that there exists an extension $(\mathcal{F}_1, \nabla_1)$. We can simply put $\mathcal{F}_1 = p_1^*(\mathcal{E})$ and $\nabla_1 = t\varphi$ on $X \times \mathbf{A}^1$.

Theorem 10.1 implies that there exists an extension $(\mathcal{F}, \nabla)$ which is semistable over $X_s \times \{0\}$. Note that $\lambda(0) = 0$ so the restriction to $X_s \times \{0\}$ is a Higgs sheaf. By flatness of $\mathcal{F}$ over $\mathbf{A}^1$, the restriction has vanishing rational Chern classes. By ([**71**] Theorem 2 p. 39), our restriction is actually a bundle. □

COROLLARY 10.3. *If $P \in M_{Hod}(X/S, GL(n))$ then the $\mathbf{G}_m$-orbit $\mathbf{G}_m \cdot P$ extends to a unique morphism $\mathbf{A}^1 \to M_{Hod}(X/S, GL(n))$.*

*Proof:* The existence is given above. Uniqueness comes from the fact that the moduli scheme $M_{Hod}(X/S, GL(n))$ is separated. □

PROPOSITION 10.4. *Suppose that $\lambda(1) = 1$ so that $(\mathcal{E}, \varphi) \in M_{DR}$ is a vector bundle with integrable connection. Suppose that the limit $(\mathcal{F}_0, \nabla_0)$ is a stable Higgs bundle. (Note that this condition may be expressed solely in terms of the corresponding point in $M_{Hod}$ so it is* a priori *independent of the choice of extension $(\mathcal{F}, \nabla)$.) Then the extension $(\mathcal{F}, \nabla)$ is unique up to isomorphism; and it may be provided with a structure of action of $\mathbf{G}_m$, in other words (by 7.3) there exists a complete decreasing filtration $F^{\cdot}$ of $\mathcal{E}$ satisfying the Griffiths transversality condition with respect to the connection $\varphi$ and such that the system of Hodge bundles $Gr_F^{\cdot}(V)$ with Higgs field $\kappa : Gr_V^p(F) \to Gr_F^{p-1}(V)$ induced by $\varphi$, is semistable. The limiting Higgs bundle $(\mathcal{F}_0, \nabla_0)$ is $(Gr_F^{\cdot}(V), \kappa)$. Finally, under the current hypotheses the filtration is unique up to translation of indices.*

*Proof:* Suppose $(\mathcal{F}, \nabla)$ and $(\mathcal{F}', \nabla')$ are two extensions as given by the previous corollary. Note that by definition they are isomorphic over $\lambda \neq 0$. By hypothesis $(\mathcal{F}_0, \nabla_0)$ and $(\mathcal{F}_0, \nabla_0)$ are stable, so the space of morphisms between them has dimension at most one. By semicontinuity it is at least one, so there is exactly one nonzero morphism up to scalars—and this in either direction. The bundle over $\mathbf{A}^1$ of isomorphisms between $(\mathcal{F}, \nabla)$ and $(\mathcal{F}', \nabla')$ is locally free of rank one, so it is trivial and we can choose a nonvanishing section. This section provides an isomorphism. Thus the conclusion is that, under our hypotheses of stability, there is only one possible extension up to isomorphism.

For any point $t \in \mathbf{G}_m$ note that, from the previous paragraph, the translate $t^*(\mathcal{F}, \nabla)$ is isomorphic to $(\mathcal{F}, \nabla)$. Any two such isomorphisms differ by multiplication by a section of $\mathcal{O}^*_{\mathbf{A}^1}(\mathbf{A}^1) = \mathbf{C}^*$. Thus, over $\mathbf{G}_m$ we obtain a principal $\mathbf{G}_m$-bundle of such isomorphisms, in other words a line bundle. But all line bundles on $\mathbf{G}_m$ are trivial so we can choose a family of isomorphisms

$$\varphi_t : (\mathcal{F}, \nabla) \cong t^*(\mathcal{F}, \nabla)$$

varying algebraically with $t \in \mathbf{G}_m$. Normalize so that $\varphi_1$ is the identity. Finally the cocycle

$$\alpha(t, t') := \varphi_{tt'}^{-1}\varphi_t\varphi_{t'}$$

is a morphism $\mathbf{G}_m \times \mathbf{G}_m \to \mathbf{G}_m$. Any such morphism of varieties is very close to being a morphism of groups: we can write

$$\alpha(t, t') = ct^a(t')^b.$$

The condition $\alpha(t, 1) = 1$ (consequence of the normalization $\varphi_1 = Id$) means that $c = 1$ and $a = 0$; and the condition $\alpha(1, t') = 1$ means that $b = 0$. Thus $\alpha(t, t') = 1$ and $\varphi_t$ forms an action of $\mathbf{G}_m$ on $(\mathcal{F}, \nabla)$ covering the action on $\mathbf{A}^1$. (What we have actually shown in this paragraph is that $H^2(B\mathbf{G}_m, \mathbf{G}_m) = \{1\}$.)

The action of $\mathbf{G}_m$ gives rise to the desired filtration by 7.3. The action $t \mapsto \varphi_t$ is unique up to multiplying by a character $\mathbf{G}_m \to \mathbf{G}_m$, and multiplication by a character corresponds to translation of the filtration index. We obtain the uniqueness of the filtration up to translation (because conversely any choice of filtration with the given properties corresponds by 7.3 to an action of $\mathbf{G}_m$). □

The subset of points of $M_{DR}$ such that the limit of the $\mathbf{G}_m$-orbit is a stable Higgs bundle, is a locally closed subset. It is a union of locally closed subsets on which the morphism to the Hilbert scheme of filtrations is well defined and a regular morphism of schemes (normalize the translations of indices of the filtrations so that the first nonzero graded piece is $F^0/F^1$).

If $(V, \nabla)$ is the vector bundle with integrable connection underlying an irreducible complex variation of Hodge structure then the limiting Higgs bundle is the system of Hodge bundles $Gr_F^{\cdot}(V)$ where $F$ is the Hodge filtration; this is stable as a Higgs bundle so the above discussion applies, i.e. $(V, \nabla)$ lies in the above-defined locally closed subset of $M_{DR}$. Of course, in this case the Hodge filtration $F^{\cdot}$ is equal to the filtration provided by Proposition 10.4. In particular the above discussion gives a recipe for recovering the Hodge filtration from the data of $(V, \nabla)$.

We return to some direct corollaries of Langton theory.

COROLLARY 10.5. *Suppose $G$ is any reductive group. If $p \in M_{Hod}(X/S, G)$ then the limit* $\lim_{t\to 0} t \cdot p$ *exists in $M_{Hod}(X/S, G)$.*

*Proof:* We can choose an injection $G \hookrightarrow GL(n)$. By a variant of [**75**] Corollary 9.15 concerning $M_{Hod}$ (we can get this by using the topological trivialization $M_{Hod} \cong M_{DR} \times \mathbf{A}^1$ which is functorial in $G$) the induced map $M_{Hod}(X_s, G) \to M_{Hod}(X_s, GL(n))$ is finite. Since the limit exists in $M_{Hod}(X_s, GL(n))$ by the previous corollary, it exists in $M_{Hod}(X_s, G)$. □

**Question:** What happens for non-reductive groups? If $G = \mathbf{G}_a$ then the limits again exist (this is exactly the weight property of the Hodge filtration refered to at the start of the section), so it seems likely that this will be true in general.

LEMMA 10.6. *Suppose $G$ is a reductive group. Let $\mathbf{V} \subset M_{Hod}(X, G)$ be the fixed point set of the $\mathbf{G}_m$-action (note that $\mathbf{V}$ is concentrated over the origin so in fact $V \subset M_{Dol}(X, G)$). Then $\mathbf{V}$ is proper over $S$.*

*Proof:* The fixed point set lies over the origin in $\mathbf{A}^1$, so it is just the fixed point set of the $\mathbf{G}_m$-action on the moduli space of Higgs bundles. For $G = GL(n)$ this fixed point set is proper by ([**36**], [**62**], [**75**] Theorem 6.11). For any $G$, argue as in the previous corollary. Alternatively one can obtain properness using Langton theory as above. □

## 11. Compactification of $M_{DR}$

The space $M_{Hod}(X/S, G) \rightarrow \mathbf{A}^1$ together with the action of $\mathbf{G}_m$ and the isomorphism between $M_{DR}(X/S, G)$ and the fiber over $\lambda = 1$, allow us to compactify $M_{DR}(X/S, G)$ relative to $S$. This depends on the properness results of the previous section.

*Structure theory for $\mathbf{G}_m$-orbits and construction of some quotients*

This subsection is entirely independant of the rest of the paper (although it will be used later). In particular the notations will mean different things: for example, what we denote by $X$ in this subsection will typically be a moduli space $M_{Hod}(X, G)$ elsewhere.

Suppose $X \rightarrow S$ is a projective morphism with an action of $\mathbf{G}_m$ covering the trivial action on $S$. Choose a relatively very ample line bundle $\mathcal{L}$ and a compatible action of $\mathbf{G}_m$. Let $V_i$ denote the connected components of the fixed point set $V$. For each $i$ there is an integer $\alpha_i$ such that $t \in \mathbf{G}_m$ acts by $t^{\alpha_i}$ on $\mathcal{L}|_{V_i}$.

Define a partial ordering $\leq$ on $V$, by saying that $u \leq v$ if there is a sequence of points $x_1, \ldots, x_m \in X$ with

$$\lim_{t \rightarrow 0} x_1 = u$$

$$\lim_{t \rightarrow 0} x_k = \lim_{t \rightarrow \infty} x_{k+1}$$

$$\lim_{t \rightarrow \infty} x_m = v.$$

Notice that if $u \leq v$ and $u \in V_i$, $v \in V_j$ then $\alpha_i \geq \alpha_j$ (and if $\alpha_i = \alpha_j$ then $u = v$).

Suppose $V = V_+ \cup V_-$ is a decomposition of the fixed point set into two disjoint closed subsets (which are consequently unions of connected components), with the properties that

$$v \in V_+, \ u \in V, \ u \leq v \Rightarrow u \in V_+.$$

and

$$v \in V_-, \ u \in V, \ u \geq v \Rightarrow u \in V_-.$$

Put

$$Y_+ = \{y \in X \ : \ \lim_{\lambda \rightarrow \infty} \lambda \cdot y \in V_+\}$$

and

$$Y_- = \{y \in X \ : \ \lim_{\lambda \rightarrow 0} \lambda \cdot y \in V_-\}.$$

These are disjoint closed subsets. They are closed by an argument similar to the proof of properness in Theorem 11.1 below. They are disjoint because if there existed $y \in Y_+ \cap Y_-$ then

$$\lim_{\lambda \rightarrow 0} \lambda \cdot y \leq \lim_{\lambda \rightarrow \infty} \lambda \cdot y,$$

so we would obtain two points $u, v$ with $u \in V_-$ and $v \in V_+$ but $u \leq v$; whence $u \in V_+$ and $v \in V_-$ (by the conditions on $V_-$ and $V_+$) contradicting the disjointness of $V_+$ and $V_-$. Finally, note that $Y_+$ and $Y_-$ are, by the nature of their definitions, $\mathbf{G}_m$-invariant.

Let $U := X - Y_+ - Y_-$. This is a $\mathbf{G}_m$-invariant open set in $X$.

THEOREM 11.1. *With the above notations, a universal geometric quotient* $U/\mathbf{G}_m$ *exists. It is separated and proper over* $S$.

*Remark:* When the subsets $V_+$ and $V_-$ are defined by choosing $a \in \mathbf{Q} - \mathbf{Z}$ and setting $V_+ = \bigcup_{\alpha_i > a} V_i$ and $V_- = \bigcup_{\alpha_i < a} V_i$ then the quotient defined above is just the geometric invariant theory quotient of the set of semistable points (for the linearized action obtained when the linearization is translated by $a$). In particular, in this case the quotient is projective. I don't know if the quotient given by this theorem will be projective in general, nor if it is projective in our example (the compactification of $M_{DR}$) below.

*Proof:* Let $X^{(pre)}$ denote the set of pre-stable points [**59**]. In our case it is easy to see that $X^{(pre)} = X - V$ is just the complement of the fixed point set. Mumford constructs a universal geometric quotient $\phi : X^{(pre)} \to X^{(pre)}/\mathbf{G}_m$. This morphism is submersive so $\phi(U)$ is open, and by the universality we obtain a universal geometic quotient $\phi : U \to U/\mathbf{G}_m$. The only problem is to prove that $U/\mathbf{G}_m$ is separated and proper over $S$.

Suppose $R$ the henselian local ring of $\mathbf{C}[z]$ at the origin $P$, with maximal ideal $\mathbf{m}$ and residue field $R/\mathbf{m} = \mathbf{C}$. Let $K$ be the fraction field of $R$, and let $z \in R$ denote a uniformizing parameter. Let $\tilde{K}$ denote the algebraic closure of $K$ and let $\tilde{R}$ denote the normalization of $R$ in $\tilde{K}$. The extension $\tilde{K}$ is obtained from $K$, as $\tilde{R}$ is obtained from $R$, by adjoining the elements $z^{1/n}$. Let $\tilde{\mathbf{m}}$ denote the maximal ideal of $\tilde{R}$. Note that $\tilde{R}$ is a valuation ring with $\mathbf{Q}$ as value group, $\tilde{\mathbf{m}}$ is the valuation ideal, and $\tilde{R}/\tilde{m} = \mathbf{C}$.

Any finite extension of $K$ is isomorphic to $K$ (by changing the parameter).

Suppose $\eta : Spec(K) \to U$ is a point. We have to show that there is $\varphi \in \mathbf{G}_m(\tilde{K})$ such that $\varphi\eta$ extends to a point $Spec(\tilde{R}) \to U$, and furthermore that $\varphi$ is unique up to $\mathbf{G}_m(\tilde{R})$.

The action of $\mathbf{G}_m$ on the point $\eta$ gives a morphism $Spec(K) \times \mathbf{G}_m \to X$ which completes to $Spec(K) \times \mathbf{P}^1 \to X$. Let

$$\eta_0 := \lim_{t \to 0} t \cdot \eta$$

$$\eta_\infty := \lim_{t \to \infty} t \cdot \eta$$

as points $Spec(K) \to X$. These complete to points $Spec(R) \to X$. There is a scheme $W$ (the closure of the graph of the previous morphism in $Spec(R) \times X$) with a diagram

$$\begin{array}{ccccc} Spec(K) \times \mathbf{P}^1 & \hookrightarrow & W & \to & X \\ \downarrow & & \downarrow & & \downarrow \\ Spec(K) & \hookrightarrow & Spec(R) & \to & S \end{array}$$

where the vertical arrows are proper, and where $\mathbf{G}_m$ acts compatibly on everything in the top row. The fiber of $W$ over the closed point of $Spec(R)$ decomposes as a string of $\mathbf{P}^1$'s meeting at fixed points for the action. Let $y_1, \ldots, y_r$ denote the images in $X$ of the fixed points in the string of $\mathbf{P}^1$'s over the origin. Since $W$ was taken as the closure of the graph, these points are distinct. We can order them so that $y_1 = \eta_0(P)$, $y_r = \eta_\infty(P)$, and $y_i$ is joined to $y_{i+1}$ by a $\mathbf{P}^1$ in the fiber. In this case for a general point $x$ on the $\mathbf{P}^1$ joining $y_i$ to $y_{i+1}$ we have (*)

$$\lim_{t\to 0} t\cdot x = y_i, \quad \lim_{t\to\infty} t\cdot x = y_{i+1}.$$

In particular, refering to our partial ordering above we have

$$\eta_0(P) = y_1 < y_2 < \ldots < y_r = \eta_\infty(P).$$

Note that since $\eta \in U(K)$ we have $\eta_0 \in V_+(K)$ and $\eta_\infty \in V_-(K)$. Thus $y_1 \in V_+$ and $y_r \in V_-$ (as $V_+$ and $V_-$ are closed). By the conditions on $V_+$ and $V_-$ there is $k$ such that $y_1, \ldots, y_k \in V_+$ and $y_{k+1}, \ldots, y_r \in V_-$. ¿From the definitions of $Y_+$ and $Y_-$ as well as the the property (*) we find that the $\mathbf{P}^1$ joining $y_i$ to $y_{i+1}$ lies in $Y_+$ if $i < k$ and in $Y_-$ if $i > k$, whereas it meets $U$ if $i = k$. The uniqueness of the $\mathbf{G}_m$-orbit meeting $U$ in the closed fiber gives the separatedness. Choose $\varphi \in \mathbf{G}_m(\tilde{K})$ so that $\varphi\eta : Spec(\tilde{K}) \to W$ completes to a point $Spec(\tilde{R}) \to W$ with $P$ mapping to a general point on the $\mathbf{P}^1$ joining $y_k$ to $y_{k+1}$. This gives the desired $\varphi$ for properness. □

We obtain the following theorem as a corollary.

THEOREM 11.2. *Suppose $Z \to S$ is an $S$-scheme on which $\mathbf{G}_m$ acts (acting trivially on $S$). Suppose that the fixed point set $W \subset Z$ is proper over $S$, and that for any $z \in Z$ the limit $\lim_{t\to 0} t\cdot z$ exists in $W$. Let $U \subset Z$ be the subset of points $z$ such that the limit $\lim_{t\to\infty} t\cdot z$ does not exist in $Z$. Then $U$ is open and there exists a geometric quotient $Q = U/\mathbf{G}_m$ by the action of $\mathbf{G}_m$. This geometric quotient is separated and proper over $S$.*

*Proof:* Chose a $\mathbf{G}_m$-linearized very ample line bundle $\mathcal{L}$ on $Z$ (this exists by [**59**]). Then $\mathbf{G}_m$ acts in a locally finite way on $H^0(Z,\mathcal{L})$ so we may choose a fixed subspace which gives a projective embedding of $Z$. Thus we may assume that $Z \subset \mathbf{P}^N$ as a locally closed subscheme, and that $\mathbf{G}_m$ acts linearly on $\mathbf{P}^N$ preserving $Z$ and inducing the given action there. Taking the graph we can consider this as an embedding $Z \subset \mathbf{P}^N \times S$. Let $X$ be the subscheme closure of $Z$ in $\mathbf{P}^N \times S$ (that is, the subscheme defined by the homogeneous ideal of forms which vanish on $Z$). Note that $X$ is projective over $S$ and that $\mathbf{G}_m$ acts on $X$ preserving the open set $Z$. Let $V$ denote the fixed point set in $X$, and let $V_+ := W$ be the fixed point set in $Z$. Let $V_- := V \cap (X - Z)$ denote the fixed point set in the complement. Note that the complement $X - Z$ is closed, hence proper over $S$, so $V_-$ is proper over $S$. By hypothesis $V_+$ is proper over $S$. We obtain a decomposition $V = V_+ \cup V_-$ as a disjoint union of two closed subsets.

Suppose $u, v \in V$ with $v \leq u$. This means that there is a sequence of points $v_0 = v, \ldots, v_n = u$ such that $v_i$ is joined to $v_{i+1}$ by a $\mathbf{G}_m$-orbit (i.e. there is an orbit whose limits are $v_i$ at $\lambda \to 0$ and $v_{i+1}$ at $\lambda \to \infty$). Suppose $v_i \in V_-$. Then the orbit corresponds to a point $x \in X$ with $\lim_{\lambda\to 0} \lambda\cdot x = v_i$. But if $x \in Z$ then our hypothesis would give $v_i \in V_+$, so $x$ must be in $X - Z$. Since $X - Z$ is closed, the other limit $v_{i+1}$ must be in $X - Z$ also. We thus show by induction that if $v = v_0$ is in $V_-$ then so is $u = v_n$. The contrapositive says that if $u$ is in $V_+$ then so is $v$. We have shown on the one hand that if $v \in V_-$ and $u \in V$ with $v \leq u$ then $u \in V_-$; and on the other hand that if $u \in V_+$ and $v \in V$ with $v \leq u$ then $v \in V_+$.

We are now ready to apply the general construction above. Define the subsets $Y_+$ and $Y_-$ as before, and let $U'$ be the complement $U' = X - Y_+ - Y_-$. We claim that $Y_+$ is the set of points $x \in Z$ such that $\lim_{\lambda\to\infty} \lambda\cdot x \in Z$. Recall that $Y_+ := \{x \in X \ : \ \lim_{\lambda\to\infty} \lambda\cdot x \in V_+\}$. But if $x \in Z$ with $\lim_{\lambda\to\infty} \lambda\cdot x \in Z$ then this limit is in $V \cap Z = V_+$. On the other hand, if $x \in X$ with $\lim_{\lambda \to \infty} \lambda \cdot y \in V_+$ then $x \notin (X - Z)$ because $X - Z$ is closed and $V_+ \cap (X - Z) = \emptyset$. This shows the claim. We next show that $Y_- = X - Z$. To see this recall that $Y_- := \{x \in X : \lim_{\lambda \to 0} \lambda \cdot x \in V_-\}$. If $x \in Z$ then by hypothesis $\lim_{\lambda \to 0} \lambda \cdot x \in Z$ and $V_- \cap Z = \emptyset$, so this shows that $Y_- \subset X - Z$. On the other hand if $x \in X - Z$ then since $X - Z$ is closed, $\lim_{\lambda \to 0} \lambda \cdot x \in X - Z$ and hence this limit is in $Y_-$, which shows that $Y_- = X - Z$.

With the two statements of the previous paragraph we obtain that the complement $U'$ of $Y_+$ and $Y_-$ is equal to the subset of points of $Z$ whose limits at $\lambda \to \infty$ do not exist in $Z$, that is to say that $U'$ is the same as ths subset $U$ described in the statement of the theorem. The result of Theorem 11.1 now gives the universal geometric quotient $U/\mathbf{G}_m$ which is separated and proper over $S$. □

*Relative compactification of $M_{DR}(X/S, G)$*

Suppose $G$ is a reductive group and $X \to S$ a smooth projective morphism. Apply Theorem 11.2 to $Z = M_{Hod}(X/S, G)$. The hypotheses on $Z$ are given by Corollary 10.5 and Lemma 10.6. Note that the open set $U$ certainly contains the open set

$$M_{Hod}(X/S, G) \times_{\mathbf{A}}^{1} \mathbf{G}_m \cong M_{DR}(X/S, G) \times \mathbf{G}_m$$

as a $\mathbf{G}_m$-invariant open set. Since the quotient $U \to Q$ is a geometric quotient, the image of the open set $M_{DR}(X/S, G) \times \mathbf{G}_m$ is a geometric quotient of $M_{DR}(X/S, G) \times \mathbf{G}_m$, but we already know the geometric quotient here, it is just $M_{DR}(X/S, G)$. Thus our quotient $Q$ contains $M_{DR}(X/S, G)$ as an open set, and $Q$ is proper over $S$. We have proved the following theorem.

THEOREM 11.3. *If $G$ is a reductive group and $X \to S$ a smooth projective morphism, then there exists a natural relative compactification $\overline{M}_{DR}(X/S, G)$ proper over $S$ and containing $M_{DR}(X/S, G)$ as an open subset.*

*Proof:* Take $\overline{M}_{DR}(X/S, G)$ to be the quotient $Q$ of the previous paragraph. □

*Remark:* There is a natural stack-theoretic compactification $\overline{\mathcal{M}}_{DR}(X/S, GL(n))$ containing $\mathcal{M}_{DR}(X/S, GL(n))$ as an open subset and satisfying the valuative criterion of properness over $S$. The valuative criterion comes from Corollary 10.2. We state this only in the case $G = GL(n)$ because the finiteness result ([**75**] Corollary 9.15) used to pass to any group $G$ in the proof of Corollary 10.5 is only available for the moduli spaces, not for the moduli stacks.

There is a natural Cartier divisor on $\mathcal{M}_{Hod}(X/S, GL(n))$ given as the pullback of the divisor $\{0\} \subset \mathbf{A}^1$. This divisor is $\mathbf{G}_m$-invariant, so it projects to a Cartier divisor in the stack-theoretic quotient $\overline{\mathcal{M}}_{DR}(X/S, GL(n))$; and the open set $\mathcal{M}_{DR}(X/S, GL(n))$ is just the complement of this divisor. Thus the "divisor at infinity" exists as a natural Cartier divisor.

In the moduli space compactification $\overline{M}_{DR}(X/S, G)$ the divisor at infinity is only a Weil divisor. We can define an intermediate *orbifold compactification* by taking the quotient $M_{Hod}(X/S, G)/\mathbf{G}_m$ in the sense of stacks. Here the divisor at infinity is again a Cartier divisor. In the orbifold compactification there may be orbifold points corresponding to fixed points of finite subgroups of $\mathbf{G}_m$. In the scheme-theoretic compactification these project to certain quotient singularities. They correspond to objects which are like systems of Hodge bundles (or variations of Hodge structure) except that the Hodge bundles are only indexed by a cyclic group instead of $\mathbf{Z}$ so the Kodaira-Spencer map $\theta$ can go "around and around" to no longer be nilpotent.

The orbifold structure of the divisor at infinity is that of the quotient $U \times_{\mathbf{A}^1} \{0\}/\mathbf{G}_m$. But this is the quotient of the open subset of $M_{Dol}(X/S, G)$ corresponding to Higgs bundles with non-nilpotent Higgs field, by the action of $\mathbf{G}_m$.

*Interpretation of the Hodge filtration in terms of the compactification*

Let $\overline{M}'_{DR}$ denote the orbifold compactification described above. There is a principal $\mathbf{G}_m$-bundle over this space, it is just the total space from before taking the quotient. This principal bundle corresponds to a line bundle $\mathcal{L}$. One can see that $\mathcal{L} = \mathcal{O}_{\overline{M}'_{DR}}(-D)$ where $D$ is the divisor at infinity. Conversely, from the data of $\overline{M}'_{DR}$ and the divisor at infinity $D$ (which has multiplicity one) we obtain a line bundle and hence a principal $\mathbf{G}_m$-bundle with a section defined over $M_{DR}$. There is only one function on this total space which is constant on the multiples of our copy of $M_{DR}$. This function is the morphism to $\mathbf{A}^1$. The total space is $M^*_{Hod}$, the open subset which is the complement of the locus of Higgs bundles with nilpotent Higgs field. Thus we recover most but not all of the Hodge filtration $M_{Hod}$ from our compactification with its divisor at infinity.

*Griffiths transversality revisited*

In the relative case we have obtained a family of compactifications

$$\overline{M_{DR}}(X/S) \to S.$$

On the other hand, recall that $M_{DR}(X/S)$ has the Gauss-Manin connection which, analytically, translates the fact that $M_{DR}(X/S)^{\mathrm{an}}$ is locally over $S^{\mathrm{an}}$ a product of the form $S^{\mathrm{an}} \times M_B$ where $M_B$ is the moduli space of representations of the fundamental group of the fiber $X_s$. The Griffiths transversality condition basically says that the Gauss-Manin connection has poles of order 1 along the divisor at infinity.

In order to make this precise we restrict ourselves to a case where the moduli space is smooth. Fix a family $X \to S$ over a base $S$, and suppose $S$ is a smooth curve. Suppose for simplicity that $G$ and Chern class data $c$ are fixed so that the corresponding unions of components $M_{DR}(X_s, G)_c$ is smooth and equidimensional (for example $X/S$ a family of curves, $G = PGL(n)$ and $c$ means we look at bundles of degree $d$ prime to $n$). We obtain

$$M_{Hod}(X/S, G)_c \to S \times \mathbf{A}^1.$$

We claim that this map is smooth. This is a special case of Conjecture 9.3, and requires some care. Apply the criterion of ([**33**] Chapter III Lemma 10.3.A—for which Hartshorne refers to Bourbaki and Altman and Kleiman) where $t$ is the coordinate on $\mathbf{A}^1$. We have to show that $t$ is not a zero divisor upstairs, and that $M_{Dol}(X/S, G)_c \to S$ is flat. Since all of the fibers of our map are smooth, the only way $t$ could be a zero divisor is if there were an irreducible component lying over $0 \in \mathbf{A}^1$. But Theorem 9.1 shows that this is not the case. To show that $M_{Dol}(X/S, G)_c \to S$ is flat it suffices (again in view of the smoothness of the fibers) to show that no irreducible component of $M_{Dol}(X/S, G)_c$ lies over a point in the curve $S$. Any component of $M_{Dol}(X_s, G)_c$ is contained in the closure of $M_{DR}(X_s, G)_c \times \mathbf{G}_m$, so any component of $M_{Dol}(X/S, G)_c$ is contained in the closure of $M_{DR}(X/S, G)_c \times \mathbf{G}_m$. If $n$ denotes the dimension of any components of $M_{DR}(X_s, G)_c$ (by hypothesis these dimensions are all the same) then the dimension of any component of $M_{Hod}$ is at least $n+2$ and (since $M_{Dol}$ is defined by one equation $t=0$ inside $M_{Hod}$) the dimension of any component of $M_{Dol}(X/S,G)_c$ must be at least $n+1$. But the fibers $M_{Dol}(X_s,G)_c$ are all of dimension $n$, so they cannot contain irreducible components of $M_{Dol}(X/S,G)_c$. This proves that our map is flat. As the fibers are smooth, the map is smooth.

Now we can get back to the thread of our discussion. Let $U \subset M_{Hod}(X/S,G)_c$ denote the open set used in defining the compactification. Let $\overline{M}'_{DR}(X/S,G)$ denote the orbifold compactification of $M_{DR}(X/S,G)$ obtained by taking the quotient $U/\mathbf{G}_m$ in the sense of stacks. This can introduce orbifold points at places where the stabilizer is a nontrivial finite subgroup of $\mathbf{G}_m$ (it has to be the $m$-th roots of unity). These orbifold points would be replaced by the corresponding cyclic quotient singularities in the usual compactification defined previously. The advantage here is that $\overline{M}'_{DR}(X/S,G)$ is smooth over $S$.

Let $D \subset \overline{M}'_{DR}(X/S,G)$ denote the divisor at infinity. It is reduced (since $M_{Hod}$ is smooth over $\mathbf{A}^1$). The Gauss-Manin connection can be interpreted as a lifting of vector fields on $S$ to vector fields on $M_{DR}(X/S,G)$. Fix a vector field $\nu$ on $S$. The Gauss-Manin connection give a vector field $\eta$ on $M_{DR}(X/S,G)$, more precisely

$$\eta \in H^0(M_{DR}(X/S,G), T(M_{DR}(X/S,G))).$$

Note that it projects to the vector field $\nu$, which means that flowing along $\eta$ takes us from one fiber of $M_{DR}(X/S,G)/S$ to another. The Gauss-Manin connection is equivalent to the data of these liftings $\eta(\nu)$ for all vector fields $\nu$ on $S$.

In the following statement, $\overline{M}'_{DR/S}$ means $\overline{M}'_{DR}(X/S,G)$.

THEOREM 11.4. *The Griffiths transversality property says that the vector field $\eta$ has simple poles along $D$, and the residue is tangent to $D$. More precisely $\eta$ extends to a section*

$$\overline{\eta} \in H^0(\overline{M}'_{DR/S}, T(\overline{M}'_{DR/S}) \otimes \mathcal{O}(D))$$

*and $\overline{\eta}|_D$ is a section of*

$$T(D) \otimes \mathcal{O}_D(D) \subset T(\overline{M}'_{DR/S}) \otimes \mathcal{O}_D(D).$$

*Proof:* Let $U \subset M_{Hod}(X/S,G)$ be the open subset of points whose $\mathbf{G}_m$-orbits don't have limits at infinity. Thus

$$\overline{M}'_{DR/S} = U/\mathbf{G}_m$$

as an orbifold quotient (i.e. a stack quotient which is an orbifold). Let $t$ denote the coordinate on $\mathbf{A}^1$ (possibly pulled back to $S \times \mathbf{A}^1$ or to $U$ which lies over $S \times \mathbf{A}^1$). Let $U' \subset U$ be the complement of the divisor $t=0$, thus

$$U' \cong \mathbf{G}_m \times M_{DR}(X/S,G).$$

Let $\zeta$ denote the vector field on $U'$ corresponding to $\eta$ in the $M_{DR}(X/S,G)$-direction. It lifts the vector field $(0,\nu)$ on $\mathbf{G}_m \times S \subset \mathbf{A}^1 \times S$. The Griffiths transversality condition as we have defined it above in terms of formal categories means that the vector fields on $\mathbf{A}^1 \times S$ parallel to the $S$-direction and vanishing on $\{0\} \times S$, lift to vector fields on $M_{Hod}(X/S,G)$. Thus $t\zeta$ extends to a vector field on $U$.

If we let $E \subset U$ denote the divisor defined by $t=0$ then the previous sentence says that $\zeta$ has a pole of order one along $E$, so can consider it as a section $\overline{\zeta}$ of $T(U) \otimes \mathcal{O}_U(E)$. Furthermore the residue is tangent to $E$ (i.e. the image of $\overline{\zeta}$ in

$T(U)\otimes\mathcal{O}_E(E)$ lies in $T(E)\otimes\mathcal{O}_E(E)$). In fact we have the stronger statement that $\overline{\zeta}$ is tangent to the fiber over any point in $\mathbf{A}^1$, simply because $\zeta$ was a lifting of the vector field $(0,\nu)$ which has no component in the $\mathbf{A}^1$-direction.

Finally note that $\zeta$ is a lifting of the vector field $\eta$ under the morphism $U'\rightarrow M_{DR}(X/S,G)$, and on the other hand $E\subset U$ is the inverse image of the divisor $D\subset\overline{M}'_{DR/S}$ under the morphism $U\rightarrow\overline{M}'_{DR/S}$. The fact that $\zeta$ descends to a vector field $\eta$ implies, since $U$ is integral, that $\overline{\zeta}$ descends to a section $\overline{\eta}\in H^0(\overline{M}'_{DR/S},T(\overline{M}'_{DR/S})\otimes\mathcal{O}_{\overline{M}'_{DR/S}}(D))$. The restriction $\overline{\zeta}|_E$ descends to the restriction $\overline{\eta}|_D$. In this way we obtain that $\eta$ has a pole of order $\leq 1$ along $D$ and furthermore that the residue is tangent to $D$. □

It should be interesting to study the behavior of the dynamical system given by this vector field with poles. The transport between fibers $M_{DR}(X_s,G)$ and $M_{DR}(X_t,G)$ has the effect of composing the analytic isomorphisms

$$M_{DR}(X_s,G)^{\mathrm{an}}\cong M_B(X_s,G)^{\mathrm{an}}$$
$$=M_B(X_t,G)^{\mathrm{an}}\cong M_{DR}(X_t,G)^{\mathrm{an}}$$

where the left and right isomorphisms are the Riemann-Hilbert correspondence and the middle equality comes from the isomorphism of fundamental groups (which depends on the path we choose from $s$ to $t$).

*Exercise:* Interpret the regularity of the Gauss-Manin connection in way similar to the above interpretation of Griffiths transversality. The lifts of vector fields to the moduli space $M(X_{DR/S}(\log D)/S,G)$ given by the Gauss-Manin connection will have simple poles along inverse image of the singular set in $S$.

### *Compactifications of spaces of Λ-modules*

By a technique similar to our construction of the relative compactification of $M_{DR}$ we have the following general theorem.

THEOREM 11.5. *Suppose $X\rightarrow S$ is a projective flat morphism, and $\Lambda$ is a split almost polynomial sheaf of rings of differential operators on $X/S$. Let $M(\Lambda,P)\rightarrow S$ denote the moduli space of semistable $\Lambda$-modules with Hilbert polynomial $P$ on $X/S$. Then there exists a relative compactification, a scheme $\overline{M(\Lambda)}\rightarrow S$ containing $M(\Lambda)$ as an open set and which is proper over $S$.*

*Proof:* The proof is the same as the previous one, with the following changes. We replace the ring $\Lambda_{Hod}$ by the ring $\xi(\Lambda,F)$ on $X\times\mathbf{A}^1$ for the filtration $F$ of $\Lambda$ by degree. Even if not admitted in the definition of $M(\Lambda,P)$, we must now admit torsion-free objects in the space $M(\xi(\Lambda,F),P)$ used to get the compactification. □

We could even obtain the same result with $\Lambda$-modules which are of pure dimension $d<dim(X/S)$—this is an interpretation of the statement for $deg(P)=d$. The proof is again exactly the same.

### *A total space compactification of $M_{DR}(X/S,GL(n))$*

Suppose $S$ is smooth, projective, with $S'=S-D$ the complement of a divisor with normal crossings. Suppose $X\rightarrow S$ is smooth over $S'$ (we denote $X':=X\times_S S'$) and has inverse image of $D$ being a divisor with normal crossings. In this situation We can get a compactification for the total space $M_{DR}(X'/S',GL(n))$. Let $\Lambda=\Lambda_{DR/S(\log D)}$ be the split almost polynomial sheaf of rings of differential operators corresponding to the formal groupoid $X_{DR/S}(\log D)$ defined in §8. Apply the construction of Theorem 11.5 to obtain a compactification. We can describe this more precisely. There is a formal groupoid $X_{Hod/S}(\log D)$ combining all of the constructions of §8, with underlying scheme $X \times \mathbf{A}^1$. Let $M^{\rm tf}(X_{Hod/S}(\log D), GL(n))$ denote the moduli space for torsion free semistable objects on $X_{Hod/S}(\log D)$ (with vanishing rational Chern classes). Our *total space compactification* is

$$\overline{M}^{\rm tf}(X_{DR/S}(\log D), GL(n)) := U/\mathbf{G}_m$$

where $U \subset M^{\rm tf}(X_{Hod/S}(\log D), GL(n))$ is the open set of points $p$ such that $\lim_{t\to\infty} tp$ does not exist. It is proper over $S$ and since $S$ itself is proper, it is compact.

The necessity to include torsion-free sheaves over the singular fibers is why we must assume here that the structure group is $GL(n)$.

Combining the interpretations of Griffiths transversality and regularity of Gauss-Manin (exercise), we get that the lifts of vector fields on $S$ given by the Gauss-Manin connection on $M_{DR}(X/S, GL(n))$, have simple poles at all components of infinity in the total space compactification of $M_{DR}(X/S, GL(n))$.

## 12. The nonabelian Noether-Lefschetz locus

A. Beilinson made a comment to the effect that one could get the moduli for $\mathbf{Z}$-variations of Hodge structure as an intersection between the integral representations, and the filtered local systems. Of course for $X$ fixed the moduli space is just a finite set of points, but this becomes interesting when we let $X$ vary in a family.

For this section we suppose $G = GL(n)$.

Suppose $X \to S$ is a smooth projective morphism. Let $V \subset M_{Dol}(X/S, GL(n))$ denote the fixed point set of the $\mathbf{G}_m$-action; it is also the moduli space for systems of Hodge bundles or equivalently for complex variations of Hodge structure [**69**]. Let $V_{DR}$ denote the image in $M_{DR}(X/S, GL(n))$ (note that this is not a complex analytic subset). On the other hand, let $M_B(X_s, GL(n, \mathbf{Z})) \subset M_B(X_s, GL(n))$ denote the image of $Hom(\pi_1(X_s, x), GL(n, \mathbf{Z}))$ (it is the subset of representations which admit an integral structure), and let $M_{DR}(X/S, GL(n, \mathbf{Z}))$ denote the subset of points of $M_{DR}(X/S, GL(n))$ which for each fiber $X_s$ correspond to elements of $M_B(X_s, GL(n, \mathbf{Z}))$. Note that $M_{DR}(X/S, GL(n, \mathbf{Z}))$ is a complex analytic subset of $M_{DR}(X/S, GL(n))$. Finally put

$$NL(X/S, GL(n)) := V_{DR} \cap M_{DR}(X/S, GL(n, \mathbf{Z})).$$

There is a morphism $NL(X/S, GL(n)) \to S$.

THEOREM 12.1. *For each $s \in S$, the fiber $NL(X/S, GL(n))_s$ is the set of isomorphism classes (up to conjugacy by $GL(n, \mathbf{C})$) of integral representations $\rho$ such that $\rho \oplus \rho$ underlies an integral variation of Hodge structure. The morphism $NL(X/S, GL(n)) \to S$ is proper, and $NL(X/S, GL(n))$ has a unique structure of reduced analytic variety and the inclusion $NL(X/S, GL(n)) \to M_{DR}(X/S, GL(n))$ is a closed immersion of complex analytic spaces. The morphism from the normalization $\widetilde{NL}(X/S, GL(n))$ of $NL(X/S, GL(n))$ to $M_{Dol}(X/S, GL(n))$ is complex analytic.*

*Remark:* In this first treatment of the subject, we are ignoring a possibly more natural non-reduced structure of complex analytic space on $NL(X/S, GL(n))$. Also it seems likely that the morphism $NL(X/S, GL(n)) \to M_{Dol}(X/S, GL(n))$ is analytic but I don't see at the moment how to prove this.

*Proof:* If $\rho \in NL(X/S, GL(n))_s$ then $\rho$ is a complex variation of Hodge structure on $X_s$, and $\rho$ is integral. It is easy to see that $\rho \oplus \rho$ has a structure of integral variation of Hodge structure (see for example the arguments in [**15**], [**16**], [**71**]). Conversely if $\rho \oplus \rho$ has a structure of variation of Hodge structure then $\rho$ is fixed by the $\mathbf{G}_m$ action, so $\rho$ lies in $V$.

To see that $NL(X/S, GL(n)) \to S$ is proper, it suffices to work locally near a point $s_0 \in S$. By the main result of [**15**] (done with small variations in the parameters, which still works) there is a finite subset of points of $M_{DR}(X_{s_0}, GL(2n, \mathbf{Z}))$ which correspond to representations which are doubles of $n$-dimensional representations and which could possibly be $\mathbf{Z}$-variations of Hodge structure on $X_s$ with $s$ in a given relatively compact neighborhood of $s_0$. Thus, over this neighborhood, there is a finite set of sections of $M_B(X/S, GL(2n, \mathbf{Z}))$ corresponding to representations whose double could possibly have a structure of variation of Hodge structure. Over our neighborhood of $s_0$, $NL(X/S, GL(n))$ is the intersection of $V_{DR}$ (which is proper over $S$ [**75**]) with this finite set of sections; thus $NL(X/S, GL(n))$ is proper over $S$.

We show the complex analyticity of $NL(X/S, GL(n))$ inside $M_{DR}(X/S, GL(n))$. The question is local over $S$, so—by the discussion of the previous paragraph—we can fix a neighborhood $U$ of $s \in S$ and a section $\sigma : U \to M_{DR}(X/S, GL(n, \mathbf{Z}))$ corresponding to an integral representation of $\pi_1(X_s)$; and we have to show that $\sigma^{-1}(V_{DR})$ is an analytic subset of $U$. Our representation $\rho$ corresponds to a local system $W$ on $X_s$. Let $W_i$ denote the complex irreducible factors of $W$. Then $\sigma^{-1}(V_{DR})$ is the intersection of the subsets $N_i$ of points in $u$ where $W_i$ admits a complex variation of Hodge structure (one knows that the irreducible factors of the representation underlying a CVHS themselves admit structures of CVHS—cf [**68**]). It suffices to show that $N_i$ are complex analytic. If $W_i$ does not admit a flat hermitian form then $N_i$ is empty, so we can assume that it does admit such a form. This form $\langle\ ,\ \rangle$ is uniquely determined up to a scalar. Fix an integer $w$, and data of ranks $r_i$ and degrees $d_i$. Let

$$HF(X/S, W_i, r_i, d_i) \to U$$

denote the parameter scheme for filtrations $F^{\cdot}$ of the local systems $W_i(s)$ (considered as vector bundles with integrable connection on the fibers $X_s$) with $r(F^i) = r_i$ and $deg(F^i) = d_i$, and satisfying the Griffiths transversality condition. It is an analytic variety over $U$, in fact the pullback of a quasiprojective variety over $M_{DR}(X/S, GL(n))$ by the section $\sigma$ (the parameter variety is the closed subscheme of the Hilbert scheme of filtrations defined by the Griffiths transversality condition). Let

$$HF(X/S, W_i, \langle\ ,\ \rangle, w, r_i, d_i) \subset HF(X/S, W_i, r_i, d_i)$$

denote the open subset of filtrations such that $F^{\cdot}$ and $F^{\cdot,\perp}$ together determine a Hodge structure of weight $w$ on the fiber over every $x \in X$. The morphism

$$HF(X/S, W_i, \langle\ ,\ \rangle, w, r_i, d_i) \to U$$

is injective (because there is at most one structure of complex variation of Hodge structure on the irreducible representation $W_i$ up to translations, but the translations are fixed by specifying the ranks $r_i$), so the image is an analytic subset. There are only a finite number of possible sets of degrees and ranks which can occur, and it suffices to consider $w = 0$ for example, so the union of this finite number of analytic subsets is $N_i$. This shows that $N_i$ is analytic, hence that the intersection is analytic. Thus, taking the union over the finite number of sections we have to consider, $NL(X/S, GL(n))$ is an analytic subset of $M_{DR}(X/S, GL(n))$.

Give $NL(X/S, GL(n))$ the structure of complex analytic space which is a closed subspace of $M_{DR}(X/S, GL(n))$. Note that the inclusion is then by definition a closed immersion, and conversely the condition that the inclusion be a closed immersion fixes uniquely the reduced complex analytic structure of $NL(X/S, GL(n))$.

What remains to be proven is the fact that the morphism from the normalization of $NL(X/S, GL(n))$ to $M_{Dol}(X/S, GL(n))$ is analytic. Recall from §10 the process of taking the limit of the $\mathbf{G}_m$-orbit in $M_{Hod}(X/S, GL(n))$. Let $K \subset M_{DR}(X/S, GL(n))$ be the subset of points where the reducibility type of the Higgs bundle at the limit of the $\mathbf{G}_m$-orbit is the same as the reducibility type of the original vector bundle with integrable connection. Note that $K$ is a constructible subset, and the process of taking the limit of the $\mathbf{G}_m$-orbit yields a constructible morphism $\psi$ (i.e. a morphism which is regular on each piece of a decomposition of $K$ into locally closed subsets) from $K$ to $M_{Dol}(X/S, GL(n))$.

The morphism $\psi$ not generically related to the correspondence between Higgs bundles and local systems. However, on points $p \in K$ corresponding to complex variations of Hodge structure, $\psi(p)$ is the associated-graded of the Hodge filtration and hence is equal to the corresponding Higgs bundle. To prove this, recall from Proposition 10.4 that an irreducible local system whose limit point is an irreducible Higgs bundle has a unique decreasing filtration satisfying Griffiths transversality and such that the associated-graded is stable; the limit point is then this associated-graded. But in the case of a complex variation of Hodge structure the already-existing Hodge filtration satisfies the required conditions so it must be equal to the unique filtration and in particular $\psi(p)$ is the associated-graded of the Hodge filtration. By the way note that this same argument shows that the subset of complex variations of Hodge structure is contained in the subset $K$ of points where the limiting Higgs bundle has the same irreducibility type as the original point.

The conclusion of the previous paragraph is that $V_{DR} \subset K$ and that the standard morphism $V_{DR} \to M_{Dol}(X/S, GL(n))$ is equal to the constructible morphism $\psi$. In particular the restriction of the standard morphism to $NL(X/S, GL(n))$ is equal to $\psi|_{NL(X/S,GL(n))}$. Thus our standard morphism $NL(X/S, GL(n)) \to M_{Dol}(X/S, GL(n))$ is regular on each piece of a decomposition of $NL(X/S, GL(n))$ into analytic constructible subsets—note that

$$NL(X/S, GL(n)) \subset K \subset M_{DR}(X/S, GL(n))$$

is analytic which implies that the restriction of the constructible decomposition of $K$ to $NL(X/S, GL(n))$ gives a decomposition of the latter into analytic constructible subsets.

We obtain that the morphism $NL(X/S, GL(n)) \to M_{Dol}(X/S, GL(n))$, which is continuous, is analytic on the smooth strata of any stratification of $NL(X/S, GL(n))$ into smooth pieces. It follows that the map from the normalization $\widetilde{NL}(X/S, GL(n))$ to $M_{Dol}(X/S, GL(n))$ is analytic. Unfortunately I don't see how to conclude that the morphism from the originally defined $NL(X/S, GL(n))$ to the Dolbeault moduli space $M_{Dol}(X/S, GL(n))$ is analytic (this probably requires a much closer analysis of what happens near points where an irreducible representation degenerates into a reducible one). $\square$

COROLLARY 12.2. *If $S$ is projective, then $NL(X/S, GL(n))$ has a structure of projective variety such that the morphisms*

$$NL(X/S, GL(n)) \to M_{DR}(X/S, GL(n))$$

*and*

$$\widetilde{NL}(X/S, GL(n)) \to M_{Dol}(X/S, GL(n))$$

*are algebraic morphisms.*

*Proof:* Think of $NL(X/S, GL(n))$ as an analytic subvariety of $M_{DR}(X/S, GL(n))$ which is proper over $S$. In particular it is a closed subvariety of $\overline{M}_{DR}(X/S, GL(n))$ so we can apply GAGA to say that it is algebraic. Its normalization $\widetilde{NL}(X/S, GL(n))$ is algebraic so again by GAGA the morphism to $M_{Dol}(X/S, GL(n))$ is algebraic. □

The results of the theorem and its corollary are relatively surprising because the homeomorphism $M_{DR} \cong M_{Dol}$ is very far from being holomorphic (on each fiber over $s \in S$ the two complex structures form a part of a quaternionic triple). The reader might then well ask if $NL(X/S, GL(n))$ isn't just a finite collection of points. In fact, $NL(X/S, GL(n))$ can have positive dimensional components (but any such component must be finite over the base $S$). For example if $Z$ is a surface with a $\mathbf{Z}$-variation of Hodge structure $V$ and if

$$\begin{array}{ccc} X & \to & Z \\ \downarrow & & \\ S & & \end{array}$$

is a pencil of curves on $Z$, the family of restrictions $\{V|_{X_s}\}$ gives a component of $NL(X/S, GL(n))$ dominating $S$. More generally if $X/S$ contains a pencil of curves on $Z$ as a subfamily, then we get a component of $NL(X/S, GL(n))$ dominating this subfamily. One might ask whether all positive dimensional components of $NL(X/S, GL(n))$ come from this type of construction.

We can also ask for an extension of the result of Corollary 12.2 to the quasiprojective case.

CONJECTURE 12.3. *If $S$ is a quasiprojective variety then $NL(X/S, GL(n))$ is an algebraic variety and the morphisms to $M_{DR}(X/S, GL(n))$ and $M_{Dol}(X/S, GL(n))$ are algebraic.*

This conjecture would be the nonabelian analogue of the result of [**10**]. It is a globalized version of a problem Deligne posed to me some time ago, of obtaining a generalization of the finiteness results of [**15**] uniformly near singularities. In [**10**] it is explained how their result would be a consequence of the Hodge conjecture. In a similar way, Conjecture 12.3 would be a consequence of the following nonabelian version of the Hodge conjecture.

CONJECTURE 12.4. *The points of $NL(X_s, GL(n))$ (i.e. the $\mathbf{Z}$-variations of Hodge structure on $X_s$) are motivic representations on $X_s$.*

I don't know who first thought of this conjecture but that must have been a long time ago. Of course there is little chance of making any progress on this—I have presented it only for the light it sheds on Conjecture 12.3.

The main problem in proving Conjecture 12.3 is to make a local analysis around the singularities of a good completion of the family $X/S$ to $\overline{X}/\overline{S}$. We hope that the total space compactification constructed above will be useful for studying this algebraicity question locally at the singularities of the family.

## 13. An open problem: degeneration of nonabelian Hodge structure

I would like to end this paper by proposing an open problem for further research. This is the problem (motivated at the end of the previous section) of studying the degeneration of nonabelian Hodge structure which arises from a degenerating family of varieties $X \to S$. It is already complicated enough to study the case of a degenerating family of curves, so we can suppose that $X \to S$ is a family of curves. Let $0 \in S$ denote a point where the fiber $X_0$ has a simple singularity (node). Let $U = S - \{0\}$ and suppose $X_U$ is smooth over $U$. We assume $G = GL(n)$ for reasons we will see in a minute. Then the relative moduli space $M_{DR}(X_U/U, GL(n))$ over $U$ is provided with the Gauss-Manin connection; with a family of hyperkähler structures; with a family of Hodge filtrations satisfying Griffiths transversality; and with a Noether-Lefschetz locus $NL(X_U/U, GL(n)) \subset M_{DR}(X_U/U, GL(n))$ for which we would like to prove algebraicity. The problem, then, is to study the degeneration of all of these structures as one approaches $0 \in S$.

The first step in studying degenerations of variations of Hodge structure in the abelian case was to have a canonical extension of the underlying holomorphic bundle, and a regular singularity theorem for the flat connection. We have obtained the analogues of these things for the nonabelian case. Note that in defining a canonical extension, it is important to get something which is proper over $S$. This was perhaps not apparent in the abelian case, where nobody cares about the fact that a vector space is noncompact! But in the nonabelian case, a noncompact extension would leave room for some asymptotic behavior as one goes off to infinity, difficult to see. To get a compactification of $M_{DR}(X_U/U, GL(n))$ we take the total space compactification $\overline{M}^{\mathrm{tf}}(X_{DR/S}(\log 0)/S, GL(n))$ which comes from looking at the moduli of torsion free sheaves with $\lambda$-connection logarithmic at the singularities, and applying the construction of §§10-11. After doing all of this we obtain a compactification of $M_{DR}(X_U/U, GL(n))$ which is the analogue of the canonical extension (together with the Hodge filtration which corresponds to the fiberwise compactification). The regularity of the Gauss-Manin connection, coupled with Griffiths transversality, give that the Gauss-Manin connection on $M_{DR}(X_U/U, GL(n))$ over $U$ has poles of order one at infinity (both along the fiberwise infinity and at the singular point $0 \in S$).

We need a good local description of this compactified moduli space near all points at infinity, specially torsion free sheaves on the singular fibers but also at infinity in the fiberwise direction.

One approach to the analysis of the degeneration might be to directly analyze the lifted vector field giving the Gauss-Manin connection with its simple poles, and try to deduce asymptotic properties of the transport along this vector field.

Another possibility would be to try to analyze the degeneration of the hyperkähler structure, thinking of it as a family of quaternionic structures on the fixed $M_B(X_s, GL(n))$.

Among the goals of this study should be: to prove algebraicity of the Noether-Lefschetz locus, in a nonabelian version of the work of Cattani-Deligne-Kaplan [**10**]; to obtain a nonabelian version of the Clemens-Schmid exact sequence; to obtain estimates and asymptotic expansions for everything in the spirit of the $SL_2$ and nilpotent orbit theorems; and finally to be able to apply all of this to obtain a devissage principle for nonabelian cohomology (even with coefficients in higher homotopy types [**76**]), i.e. a version of the Leray spectral sequence.

## References

[1] D. Arapura. Higgs line bundles, Green-Lazarsfeld sets and maps of Kähler manifolds to curves. *Bull. A.M.S.* **26** (1992), 310-314.

[2] M. Atiyah, N. Hitchin, I. Singer. Self duality in four dimensional Riemannian geometry. *Proc. Royal Soc. Lond.* **A362** (1978), 425-461.

[3] A. Beauville. Annulation du $H^1$ et systèmes paracanoniques sur les surfaces. *Crelle Journ.* **388** (1988), 149-157.

[4] P. Berthelot. *Cohomologie Cristalline des Schémas de Caractéristique* $p > 0$, Springer L.N.M. **407** (1974).

[5] O. Biquard. Fibrés de Higgs et connexions intégrables: le cas logarithmique (diviseur lisse). Preprint.

[6] O. Biquard. Fibrés paraboliques stables et connexions singulières plates. *Bull. S.M.F.* **119** (1991), 231-257.

[7] L. Breen. On the classification of 2-gerbs and 2-stacks. *Astérisque* **225**, Soc. Math. de France (1994).

[8] J. Carlson and D. Toledo. Harmonic mappings of Kähler manifolds to locally symmetric spaces. *Publ. Math. I.H.E.S.* **69** (1989), 173-201.

[9] F. Catanese. Moduli and classification of irregular Kähler manifolds (and algebraic varieties) with Albanese general type fibrations. *Invent. Math.* **104** (1991), 263-289.

[10] E. Cattani, P. Deligne, A. Kaplan. On the locus of Hodge classes. *Jour. A.M.S.* **8** (1995), 483-506.

[11] K. Corlette. Flat $G$-bundles with canonical metrics. *J. Diff. Geom.* **28** (1988), 361-382.

[12] K. Corlette. Rigid representations of Kählerian fundamental groups. *J. Diff. Geom.* **33** (1991), 239-252.

[13] K. Corlette. Nonabelian Hodge theory and square integrability. *Actes du colloque "Géométrie et Topologie des Variétés Projectives" Toulouse, 1992* (preprint).

[14] P. Deligne. Various letters to the author.

[15] P. Deligne. Un théorème de finitude pour la monodromie. *Discrete Groups in Geometry and Analysis*, Birkhauser (1987), 1-19.

[16] P. Deligne. Théorie de Hodge II. *Publ. Math. I.H.E.S.* **40** (1971), 5-58.

[17] C. Deninger. On the $\Gamma$-factors attached to motives. *Invent. Math.* **104** (1991), 245-263.

[18] K. Diederich and T. Ohsawa. Harmonic mappings and disc bundles over compact Kähler manifolds. *Publ. R.I.M.S.* **21** (1985), 819-833.

[19] S. Donaldson. Twisted harmonic maps and self-duality equations. *Proc. London Math. Soc.* **55** (1987), 127-131.

[20] J. Eells and J. Sampson. Harmonic mappings of Riemannian manifolds. *Amer. J. of Math.* **86** (1964), 109-160.

[21] J.M. Fontaine. Modules galoisiens, modules filtrés et anneaux de Barsotti-Tate, *Journée de Géometrie Algébrique de Rennes, Astérisque* **65** (1979), 3-80.

[22] A. Fujiki. Hyperkähler structure on the moduli space of flat bundles. Preprint (Kyoto University).

[23] W. Goldman and J. Millson. The deformation theory of representations of fundamental groups of compact Kähler manifolds. *Publ. Math. I.H.E.S.* **67** (1988), 43-96.

[24] M. Green and R. Lazarsfeld. Deformation theory, generic vanishing theorems, and some conjectures of Enriques, Catanese and Beauville. *Invent. Math.* **90** (1987), 389-407.

[25] M. Gromov. Sur le groupe fondamentale d'une variété kählerienne. *C.R. Acad. Sci. Paris* Ser. I Math. **308** (1989), 67-70.

[26] M. Gromov and R. Schoen. Harmonic maps into singular spaces and $p$-adic superrigidity for lattices in groups of rank one. *Publ. Math. I.H.E.S.* **76** (1992), 165-246.

[27] A. Grothendieck. *Pursuing Stacks*, manuscript available from Université de Montpellier 2.

[28] A. Grothendieck. *Dix Exposés sur la Cohomologie des Schémas* North-Holland, Amsterdam (1968).

[29] J. Hadamard. Essai sur l'étude des fonctions données par leurs developpements de Taylor. *J. Math. Pures Appl.* **8** (1892), 101-186.

[30] R. Hain. The de Rham homotopy theory of complex algebraic varieties I. *K-theory* **1** (1987), 271-324.

[31] R. Hain. Completions of mapping class groups and the cycle $C - C^-$. Preprint.

[32] R. Hain. Torelli groups and geometry of moduli spaces of curves. Preprint alg-geom 9403015 (plus later versions).
[33] R. Hartshorne. *Algebraic Geometry.* Springer, New York (1977).
[34] E. Hironaka. Alexander stratifications of character varieties. Preprint alg-geom 9602004.
[35] N. Hitchin. Stable bundles and integrable systems. *Duke Math. J.* **54** (1987), 91-114.
[36] N. Hitchin. The self-duality equations on a Riemann surface. *Proc. London Math. Soc.* **55** (1987), 59-126.
[37] N. Hitchin. Hyperkähler manifolds. Exposé 748, *Seminaire Bourbaki, Astérisque* **206** (1992), 137-166.
[38] N. Hitchin, A. Karlhede, U. Lindstrom, M. Roček. Hyperkäher metrics and supersymmetry. *Comm. Math. Phys.* **108** (1987), 535-559.
[39] L. Illusie. *Complexe cotangent et déformations I*, Springer L.N.M. **239** (1971); *II*, Springer L.N.M. **283**.
[40] J.F. Jardine. Simplicial presheaves, *J. Pure and Appl. Algebra* **47** (1987), 35-87.
[41] J. Jost and S.T. Yau. Harmonic maps and Kähler manifolds. *Math. Ann.* **262** (1983), 145-166.
[42] J. Jost and K. Zuo. Harmonic maps into Tits buildings, factorization of nonintegral $p$-adic and nonrigid representations of $\pi_1$ of algebraic varieties. Preprint.
[43] J. Jost and K. Zuo. Harmonic maps and $Sl(r, \mathbf{C})$-representations of fundamental groups of quasiprojective manifolds. Preprint.
[44] L. Katzarkov. Factorization theorems for the representations of the fundamental groups of quasiprojective varieties and some applications. Preprint alg-geom 9402012.
[45] L. Katzarkov. Nilpotent groups and universal coverings of smooth projective varieties. Preprint alg-geom 9510009.
[46] L. Katzarkov. On the Shafarevich maps. Preprint.
[47] L. Katzarkov and T. Pantev. Representations of fundamental groups whose Higgs bundles are pullbacks. *J. Diff. Geo.* **39** (1994), 103-121.
[48] L. Katzarkov and T. Pantev. Stable $G_2$ bundles and algebraically completely integrable systems. *Compositio Math.* **92** (1994), 43-60.
[49] L. Katzarkov and M. Ramachandran. On the Shafarevich conjecture for surfaces. Preprint (1994).
[50] J. Kollar. Shafarevich maps and plurigenera of algebraic varieties. *Invent. Math.* **113** (1993), 177-215.
[51] S. Langton. Valuative criteria for families of vector bundles on algebraic varieties. *Ann. of Math.* **101** (1975), 88-110.
[52] B. Lasell. Complex local systems and morphisms of varieties. *Compositio Math.* **98** (1995), 141-166.
[53] B. Lasell and M. Ramachandran. Local systems on Kähler manifolds and harmonic maps. Preprint (1994).
[54] G. Laumon, L. Moret-Bailly. Champs algébriques. Preprint, Orsay **42** (1992).
[55] A. Lubotzky and A. Magid. Varieties of representations of finitely generated groups. *Memoirs A.M.S.* **336** (1985).
[56] M. Maruyama and K. Yokogawa. Moduli of parabolic stable sheaves. *Math. Ann.* **293** (1992), 77-99.
[57] N. Mok. Factorization of semisimple discrete representations of Kähler groups. *Invent. Math.* **110** (1992), 557-614.
[58] J. Morgan. The algebraic topology of smooth algebraic varieties. *Publ. Math. I.H.E.S.* **48** (1978), 137-204.
[59] D. Mumford. *Geometric Invariant Theory*, Springer, New York (1965).
[60] T. Napier. Convexity properties of coverings of smooth projective varieties. *Math. Ann.* **286** (1990), 433-480.
[61] T. Napier and M. Ramachandran. Structure theorems for complete Kähler manifolds and applications to a Lefschetz type of theorem. To appear.
[62] N. Nitsure. Moduli spaces of semistable pairs on a curve. *Proc. London Math. Soc.* **62** (1991), 275-300.
[63] N. Nitsure. Moduli of semistable logarithmic connections. *Jour. A. M. S.* **6** (1993), 597-610.
[64] A. Reznikov. All regulators of flat bundles are torsion. *Ann. of Math.* **141** (1995), 373-386.

[65] A. Reznikov. Harmonic maps, hyperbolic cohomology and higher Milnor inequalities. *Topology* **32** (1993), 899-907.
[66] S. Salamon. Differential geometry of quaternionic manifolds. *Ann. Sci. E.N.S.* **19** (1986), 31-55.
[67] J. Sampson. Applications of harmonic maps to Kähler geometry. *Contemp. Math.* **49** (1986), 125-133.
[68] C. Simpson. Yang-Mills theory and uniformization. *Lett. Math. Phys.* **14** (1987), 371-377.
[69] C. Simpson. Constructing variations of Hodge structure using Yang-Mills theory and applications to uniformization. *Journ. A. M. S.* **1** (1988), 867-918.
[70] C. Simpson. Nonabelian Hodge theory. *Proceedings, ICM-90, Kyoto* Springer, Tokyo (1991), 198-230.
[71] C. Simpson. Higgs bundles and local systems. *Publ. Math. I.H.E.S.* **75** (1992), 5-95.
[72] C. Simpson. Subspaces of moduli spaces of rank one local systems. *Ann. Sci. Ec. Norm. Sup.* **26** (1993), 361-401.
[73] C. Simpson. Lefschetz theorems for the integral leaves of a holomorphic one form. *Compositio Math.* **87** (1993), 99-113.
[74] C. Simpson. Some families of local systems over smooth projective varieties. *Annals of Math.* **138** (1993), 337-425.
[75] C. Simpson. Moduli of representations of the fundamental group of a smooth projective variety, I: *Publ. Math. I.H.E.S.* **79** (1994), 47-129; II: *Publ. Math. I.H.E.S.* **80** (1994), 5-79.
[76] C. Simpson. Homotopy over the complex numbers and de Rham cohomology. *Moduli of Vector Bundles, Lecture Notes in Pure and Applied Mathematics* **179** M. Maruyama, Ed., Marcel Dekker (1996), 229-263.
[77] C. Simpson. A relative notion of algebraic Lie group and application to $n$-stacks. Preprint, available on alg-geom 96-07.
[78] C. Simpson. Algebraic (geometric) $n$-stacks. Preprint, available on alg-geom 96-09.
[79] Y.T. Siu. Complex analyticity of harmonic maps and strong rigidity of complex Kähler manifolds. *Ann. of Math.* **112** (1980), 73-110.
[80] Y. T. Siu. Complex analyticity of harmonic maps, vanishing and Lefschetz theorems. *J. Diff. Geo.* **17** (1982), 55-138.
[81] Y.T. Siu. Strong rigidity for Kähler manifolds and the construction of bounded holomorphic functions, in *Discrete Groups in Geometry and Analysis* (the Mostow Volume). Ed. R. Howe, Birkhäuser, Boston (1987).
[82] Z. Tamsamani. Sur des notions de $n$-catégorie et $n$-groupoide non strictes via des ensembles multi-simpliciaux. Thesis, Université Paul Sabatier (1996), available on alg-geom 95-12 and 96-07.
[83] D. Toledo. Projective varieties with non-residually finite fundamental group. *Publ. Math. I.H.E.S.* **77** (1993), 103-119.
[84] K. Yokogawa. Compactification of moduli of parabolic sheaves and moduli of parabolic Higgs sheaves. *J. Math. Kyoto Univ.* **33** (1993), 451-504.
[85] K. Yokogawa. Infinitesimal deformations of parabolic Higgs sheaves. *Internat. J. Math.* **6** (1995), 125-148.
[86] K. Zuo. Some structure theorems for semisimple representations of $\pi_1$ of algebraic manifolds. *Math. Ann.* **295** (1993), 365-382.
[87] K. Zuo. *Representations of Fundamental Groups of Algebraic Varieties*, Habilitation, Universität Kaiserslautern (1995), to appear as a book.

LABORATORE EMILE PICARD
UMR 5580, CNRS UNIVERSITÉ PAUL SABATIER
31062 TOULOUSE CEDEX, FRANCE
*E-mail address*: carlos@picard.ups-tlse.fr

# VII. Complex Geometry

Proceedings of Symposia in Pure Mathematics
Volume **62.2**, 1997

# Algebraic Criteria for Kobayashi Hyperbolic Projective Varieties and Jet Differentials

Jean-Pierre Demailly

ABSTRACT. These notes are an expanded version of five lectures delivered at the Santa Cruz AMS Summer School on Algebraic Geometry. The main goal of the notes is to study complex varieties (mostly compact or projective algebraic ones), through a few geometric questions related to hyperbolicity in the sense of Kobayashi. A convenient framework for this is the category of "directed manifolds", that is, the category of pairs $(X, V)$ where $X$ is a complex manifold and $V$ a holomorphic subbundle of $T_X$. If $X$ is compact, the pair $(X, V)$ is hyperbolic if and only if there are no nonconstant entire holomorphic curves $f : \mathbb{C} \to X$ tangent to $V$ (Brody's criterion). We describe a construction of projectivized $k$-jet bundles $P_k V$, which generalizes a construction made by Semple in 1954 and allows to analyze hyperbolicity in terms of negativity properties of the curvature. More precisely, $\pi_k : P_k V \to X$ is a tower of projective bundles over $X$ and carries a canonical line bundle $\mathcal{O}_{P_k V}(1)$; the hyperbolicity of $X$ is then conjecturally equivalent to the existence of suitable singular hermitian metrics of negative curvature on $\mathcal{O}_{P_k V}(-1)$ for $k$ large enough. The direct images $(\pi_k)_\star \mathcal{O}_{P_k V}(m)$ can be viewed as bundles of algebraic differential operators of order $k$ and degree $m$, acting on germs of curves and invariant under reparametrization. Following an approach initiated by Green and Griffiths, we establish a basic Ahlfors-Schwarz lemma in the situation when $\mathcal{O}_{P_k V}(-1)$ has a (possibly singular) metric of negative curvature, and we infer that every nonconstant entire curve $f : \mathbb{C} \to V$ tangent to $V$ must be contained in the base locus of the metric. This basic result is then used to obtain a proof of the Bloch theorem, according to which the Zariski closure of an entire curve in a complex torus is a translate of a subtorus. Our hope, supported by explicit Riemann-Roch calculations and other geometric considerations, is that the Semple bundle construction should be an efficient tool to calculate the base locus. Necessary or sufficient algebraic criteria for hyperbolicity are then obtained in terms of inequalities relating genera of algebraic curves drawn on the variety, and singularities of such curves. We finally describe some techniques introduced by Siu in value distribution theory, based on a use of meromorphic connections. These techniques have been developped later by Nadel to produce elegant examples of hyperbolic surfaces of low degree in projective 3-space; thanks to a suitable concept of "partial projective projection" and the associated Wronskian operators, substantial improvements on Nadel's degree estimate will be achieved here.

1991 *Mathematics Subject Classification.* 32H20, 32L10, 53C55, 14J40.

*Key words and phrases.* Kobayashi hyperbolic variety, directed manifold, genus of curves, jet bundle, jet differential, jet metric, Chern connection and curvature, negativity of jet curvature, variety of general type.

## Contents



## 0. Introduction

In these notes, we investigate some geometric questions related to the concept of hyperbolic variety in the sense of Kobayashi [Kob70]. Hyperbolic algebraic varieties have attracted considerable attention, in part because of their conjectured diophantine properties. For instance, [Lang86] has conjectured (among other things) that any hyperbolic complex projective variety over a number field $K$ can contain only finitely many rational points over $K$; this conjecture, which seems at present far beyond reach, may be regarded as a higher dimensional analogue of the Mordell conjecture. The reader can consult P. Vojta [Voj87] for aspects connected to diophantine problems.

We will be concerned here only with the geometric aspects of the theory which, although a priori more tractable than the diophantine aspects, are still conjectural for a major part; in fact very few satisfactory general purpose theorems are available. We hope that some of the ideas presented here will prove useful to achieve substantial progress. The reader is referred to S. Lang's survey [Lang86] and book [Lang87] for an overview of the theory until the mid 80's, and to J. Noguchi-T. Ochiai [NoOc90], P.M. Wong [Wong93] and M. Zaidenberg [Zai93] for a good exposition of more recent problems. Our goal here is not to provide an exhaustive compilation of known results, but rather to emphasize two or three important ideas around the concepts of jet bundles and jet metrics. Similar ideas have been applied successfully in a somewhat special situation in the recent work [SiYe96a] by Siu and Yeung, where the authors prove the hyperbolicity of the complement of an irreducible generic curve of high degree $d \geqslant 10^{13}$ in $\mathbb{P}^2$. Let us fix here our terminology: the word "generic" will refer to a property which holds true in the complement of a global algebraic or analytic subset in the parameter space, and the expression "very generic" will be used when the exceptional set of parameters is a countable union of algebraic or analytic subsets. As we will see in several instances, the geometry of jets conveys many natural interesting problems concerning the relationship between hyperbolicity and jet curvature negativity.

We now give a short outline of the contents. Recall that a complex variety is *hyperbolic in the sense of Kobayashi* if the family of holomorphic maps $f : \Delta \to X$

from the unit disk into $X$ is a normal family. If $X$ is compact (e.g. projective algebraic), it is well known that $X$ is Kobayashi hyperbolic if and only if it is *Brody hyperbolic*, that is, if there are no nonconstant entire holomorphic curve $f : \mathbb{C} \to X$. In particular $X$ has no rational or elliptic curves, and more generally every holomorphic map $f : Z \to X$ from an abelian variety (or complex torus) to $X$ must be constant. Conversely, it has been suggested by Kobayashi [Kob70] and [Lang86] that these algebraic properties could be equivalent to hyperbolicity. To prove this, one would have to construct a torus $Z$ and a nontrivial holomorphic map $f : Z \to X$ whenever $X$ is non hyperbolic. A hint that this should be true is given by the following observation: if $X$ is hyperbolic, there is an absolute constant $\varepsilon > 0$ such that the genus of any compact curve of $X$ is bounded below by $\varepsilon$ times the degree; conversely, this property fails to be true in many examples of nonhyperbolic projective varieties. Our belief, supported by some heuristic arguments, is that any sequence of compact curves $(C_\ell)$ with $\mathrm{genus}(C_\ell)/\mathrm{degree}(C_\ell) \to 0$ should have a cluster set swept out by the image of a map $f : Z \to X$ from a complex torus $Z$, such that the limit of some subsequence of the sequence of universal covering maps $\Delta \to C_\ell \to X$ (suitably reparametrized) coincides with the image of a (non necessarily compact) straight line of $Z$ into $X$. A related conjecture of [Lang86] states that a projective variety is hyperbolic if and only if all its irreducible algebraic varieties are of general type. The most elementary step would be to exclude the case of manifolds with $c_1 \equiv 0$ (say, Calabi-Yau manifolds and symplectic manifolds), by showing for instance that they do admit a sequence of compact curves $(C_\ell)$ with $\mathrm{genus}(C_\ell)/\mathrm{degree}(C_\ell) \to 0$.

We next introduce jet bundles and jet differentials, extending some ideas of Green and Griffiths [GrGr80] (actually, the idea of using jet differentials can be traced back to the work of A. Bloch [Blo26, 26′], H. Cartan [Car28], L. Ahlfors [Ahl41] and T. Ochiai [Och77]). The basic idea is to introduce a bundle $E_{k,m}$ of algebraic differential operators $Q(f', f'', \ldots, f^{(k)})$ of weighted degree $m$, acting on germs of holomorphic curves ("*jet differentials of order $k$ and degree $m$*"). When no other restrictions are made on $Q$, one obtains a "huge" bundle which will be denoted $E^{\mathrm{GG}}_{k,m}$ in reference to Green-Griffiths' work. In our case, the main goal is to determine the conformal type of entire curves drawn on the variety. Therefore, the way curves are parametrized is irrelevant. For this reason, one is led to introduce a subbundle $E_{k,m} \subset E^{\mathrm{GG}}_{k,m}$ of "special jet differential operators" $Q(f', f'', \ldots, f^{(k)})$, namely operators which have the property of being invariant by reparametrization, i.e.

$$Q((f \circ \varphi)', (f \circ \varphi)'', \ldots, (f \circ \varphi)^{(k)}) = (\varphi')^m Q(f', f'', \ldots, f^{(k)})$$

for any local reparametrization $\varphi : (\mathbb{C}, 0) \to (\mathbb{C}, 0)$ of the curve. For instance, all Wronskian determinants appearing in the wedge products $f'$, $f' \wedge f''$, $f' \wedge f'' \wedge f'''$, $\ldots$ , $f' \wedge f'' \wedge \cdots \wedge f^{(\ell)}$ (computed in any coordinate system), and all polynomial combinations of these give rise to local sections of $E_{k,m}$. The subbundle $E_{k,m}$ turns out to have better positivity properties than $E^{\mathrm{GG}}_{k,m}$ (see section § 12). The bundles $E^{\mathrm{GG}}_{k,m}$ and $E_{k,m}$ have natural filtrations for which the graded pieces split in irreducible tensor product representations of $T^\star_X$ (Schur fonctor representations). However, in the case of $E_{k,m}$, it seems to be a highly non trivial question to decide which representations are actually involved, except for the simpler cases of jet differentials of order $k \leqslant 2$.

A basic observation is that any entire curve $f : \mathbb{C} \to X$ must automatically satisfy all algebraic differential equations $Q(f', f'', \ldots, f^{(k)}) = 0$ arising from global jet differential operators $Q \in H^0(X, E_{k,m} \otimes \mathcal{O}(-A))$ which vanish on some ample divisor $A$. Our proof is based on a strong pointwise version of the Ahlfors-Schwarz lemma (Lemma 3.2 and Theorem 7.8), and actually completes the scheme of proof suggested in [GrGr80] in the case of invariant jet differentials; the general pointwise case of the Ahlfors-Schwarz lemma for non necessarily invariant jet differentials, however, seems to be still unsettled. Let us mention, although we will not need it here, that the above vanishing theorem is still true with sections of $E^{\mathrm{GG}}_{k,m} \otimes \mathcal{O}(-A)$ in place of $E_{k,m} \otimes \mathcal{O}(-A)$ (see Remark 7.11 and [SiYe96c]). The vanishing theorem provides a way of investigating the hyperbolicity of $X$ by trying to compute the base locus of sections in $H^0(X, E_{k,m} \otimes \mathcal{O}(-A))$ for large $k$ and $m$ (we will call this set the "*Green-Griffiths locus*" of $X$, although [GrGr80] deals rather with sections of $E^{\mathrm{GG}}_{k,m} \otimes \mathcal{O}(-A)$). We believe that the use of $E_{k,m}$ in place of $E^{\mathrm{GG}}_{k,m}$ should make easier to understand the structure of the base locus (especially in the case of higher values of $k$ and $n = \dim X$), since the dimension of the projectivized jet bundles under consideration is smaller and $E_{k,m}$ is "more positive" than $E^{\mathrm{GG}}_{k,m}$. At least in the case of surfaces of general type, the existence of sections in $E_{k,m}$ and $E^{\mathrm{GG}}_{k,m}$ can be obtained by Riemann-Roch computations and suitable vanishing theorems proved by Bogomolov [Bog79] (here again, the conditions involved for the existence of sections in $E_{k,m}$ are better than those for $E^{\mathrm{GG}}_{k,m}$). It is reasonable to hope that suitable refinements of these ideas could lead in the future to a complete proof of the conjecture that every surface of general type only has finitely many rational and elliptic curves. Such a result is indeed obtained in [Bog77] for the class of surfaces $X$ satisfying $c_1(X)^2 > c_2(X)$, thanks to a finiteness theorem for integral curves of foliations on surfaces (cf. Jouanolou [Jou78]). When $c_1(X)^2 > 2\,c_2(X)$, Lu-Yau [LuYa90] proved the additional result that the transcendental curves are algebraically degenerate; the argument is based on a result of Miyaoka asserting that the base locus of the first order jet differentials in $H^0(X, S^kT^\star_X \otimes \mathcal{O}(-A))$ is at most 1-dimensional for $c_1(X)^2 > 2\,c_2(X)$ and $k \gg 0$. Recently, S. Lu [Lu96] also obtained a proof for the limit case $c_1(X)^2 = 2\,c_2(X)$. By a different method, Lu-Miyaoka [LuMi95] investigate the case of arbitrary surfaces of general type, and get the desired finiteness results under suitable restrictions on the curve singularities.

In a similar manner, for varieties of general type of arbitrary dimension, it is conjectured that all entire curves $f : \mathbb{C} \to X$ are contained in a proper algebraic subset. One of the most celebrated result in this direction is the proof of the "Bloch theorem", i.e. the special case of the above conjecture when $X$ has irregularity $q = h^0(X, \Omega^1_X) > \dim X$ ; the subject was revived by Ochiai [0ch77], who considerably clarified the ideas introduced in [Blo26], and formulated a technical result that would yield what he termed the "Bloch conjecture". The Bloch theorem was finally settled in the affirmative, by means of various techniques, in fundamental papers by Noguchi [Nog77, 81, 84], Kawamata [Kaw80] and Green-Griffiths [GrGr80] (to avoid the slight technical difficulty with the proof of the Ahlfors-Schwarz lemma in [GrGr80], we will prove here Bloch's theorem by replacing the Green-Griffiths jet differentials $E^{\mathrm{GG}}_{k,m}$ with the invariant jet differentials $E_{k,m}$). A very interesting related result is the proof by Siu-Yeung [SiYe96b] of the hyperbolicity of complements of ample divisors in abelian varieties (see also Noguchi [Nog96a] for an extension to the semi-abelian case). In § 10, we give a brief overview of these results, following an alternative approach suggested by Dethloff and Lu [DLu96], which makes use of a "logarithmic version" of our jet bundles. An earlier interesting result is the construction by Nadel [Nad89] of explicit hyperbolic algebraic surfaces in $\mathbb{P}^3$. Nadel's method is based on a use of meromorphic connections with low pole order, according to ideas introduced by Siu [Siu87]. In our setting, Nadel's technique is just the very special case when the jet differential operators under consideration are Wronskian operators associated with a meromorphic connection. Unfortunately, the method seems to work only for a restricted class of hypersurfaces or complete intersections defined by polynomials with few monomials. Thanks to a new flexible concept of *partial projective connection*, we have been able to improve Nadel's bounds and show the existence of hyperbolic algebraic surfaces in $\mathbb{P}^3$ of any degree $\geqslant 11$ (the conjectured optimal bound should be 5); see §11 and [EG96, 97], [DeEG97]*.

Contrary to most prior methods, including J. Noguchi's "jet projection method" (see [NoOc90], [Nog96]), our method does not use any Nevanlinna theory at all. Our approach is to construct suitable jet metrics of negative curvature, following original ideas of Grauert-Reckziegel [GRec65], [Gra89], Kobayashi [Kob75], Cowen-Griffiths [CoGr76] and Green-Griffiths [GrGr80]. In fact, the sheaves $\mathcal{O}(E_{k,m})$ of jet differentials defined above are the direct image sheaves of some canonical invertibles sheaves $\mathcal{O}_{X_k}(m)$ defined over suitable "projectivized $k$-jet bundles" $X_k \to X$. The $k$-jet bundle $X_k$ is a tower of projective bundles, and can be obtained by iterating a natural fonctorial construction $(X, V) \rightsquigarrow (X_1, V_1)$ in the category of "*directed manifolds*". By definition, objects of this category are pairs $(X, V)$, where $X$ is a complex manifold and $V$ a holomorphic subbundle of $T_X$, and the arrows are holomorphic morphisms preserving the $V$ subbundles. We show in § 6 that the bundle $X_k$ is a canonical smooth compactification of the bundle of "geometric" $k$-jets of regular curves (by "geometric jets", we mean that one does not pay attention to the way curves are parametrized). Such bundles $X_k$ appear to be a natural generalization of a construction introduced by Semple [Sem54] in 1954, which has been used recently as a tool for establishing enumerative formulas dealing with the order of contact of plane curves (see [Coll88], [ASS92], [CoKe94]).

In fact, almost all concepts pertaining to hyperbolicity can be extended in the general framework of directed manifolds $(X, V)$, which we may think of as a "relative" situation (in fact, it is not necessary to assume that $V$ is an integrable subbundle of $T_X$, but the case when $V = T_{X/S}$ is the relative tangent bundle of a smooth map $X \to S$ is of special interest). For instance, $(X, V)$ is said to be Brody hyperbolic if there are no global holomorphic curve $f : \mathbb{C} \to X$ tangent to $V$. In the inductive definition of the $k$-jet spaces $(X_k, V_k)$, $X_k$ is simply the projectivization $P(V_{k-1})$ of $V_{k-1} \to X_{k-1}$, and $\mathcal{O}_{X_{k-1}}(-1)$ is the tautological line subbundle of the inverse image of $V_{k-1}$ over $X_k$. Now, we say that $X$ has $k$-jet negative curvature if $\mathcal{O}_{X_k}(-1)$ can be equipped with a hermitian metric (for which some type of singularities have to be allowed), such that the $(1,1)$ curvature form of the metric is negative along $V_k$. The negativity property of the $k$-jet curvature is closely related to the existence of sections of large degree in $H^0(X, E_{k,m})$, exactly in the same way positivity and ampleness are related. A variant of the Ahlfors-Schwarz lemma shows that the negativity of $k$-jet curvature implies hyperbolicity.

* After these notes were completed, similar results have been announced independently in [SiYe96c].

Conversely, Kobayashi [Kob70] and [Lang86] raised the question whether hyperbolicity is equivalent to 1-jet negativity (in our terminology). We show that this optimistic picture is unfortunately wrong. In fact the $k$-jet negativity property yields the following necessary algebraic condition: there exists a constant $\varepsilon > 0$ such that every algebraic curve $C \subset X$ satisfies

$$2g(\overline{C}) - 2 \geqslant \varepsilon \deg(C) + \sum_{x \in C_{k-1}} (m_{C_{k-1}}(x) - 1),$$

where $\overline{C}$ is the normalization of $C$ and $m_{C_{k-1}}(x)$ are the multiplicities of the singular points in the $(k-1)$-st jet lifting of $C$. Using this criterion, we construct for every integer $k_0$ an hyperbolic algebraic surface which cannot have any $k$-jet metric of negative curvature when $k \leqslant k_0$. It is nevertheless reasonable to expect that hyperbolicity is equivalent to the existence of a sufficiently large integer $k_1$ such that $X$ has $k$-jet negative curvature for $k \geqslant k_1$.

We want to stress that many important questions have been left out in these notes, especially Nevanlinna theory and its applications to hyperbolicity theory [CaGr72], [Nog83], [Siu87], [Wong89], [RuSt91], [Nog91]. Especially noticeable in this respect is the work of Dethloff-Wong-Schumacher [DSW92, 94] on the hyperbolicity of complements of 3 or more generic curves in the projective plane, and the construction by Masuda-Noguchi [MaNo93] of hyperbolic hypersurfaces of large degree in $\mathbb{P}^n$. Also, in a more algebraic setting, there is an extensive literature dealing with the question of computing genus of curves in algebraic surfaces, bearing an intimate connection with hyperbolicity ([Bog77], [Cle86], [CKM88], [LuYa90], [Lu91], [LuMi95], [Lu96] [Xu94]). Last but not least, there are several important questions of Number Theory which either depend on Nevanlinna theory or suggest new tools for the study of differential geometric problems. The reader may profitably consult McQuillan's paper [McQu96], in which the method of Vojta-Faltings is adapted to give a completely new proof of the Bloch theorem.

I wish to express my gratitude to the organizers of the AMS Summer Institute held at Santa Cruz in July 1995 for giving me the opportunity of making a series of lectures on hyperbolicity theory. I warmly thank Gerd Dethloff, Siegmund Kosarew, Steven Lu, Bernard Shiffman, Yum-Tong Siu and Mikhail Zaidenberg for formal or informal discussions which got me started in the subject and helped me to improve these notes.

## 1. Hyperbolicity Concepts and Directed Manifolds

We first recall a few basic facts concerning the concept of hyperbolicity, according to S. Kobayashi [Kob70, Kob76]. Let $X$ be a complex $n$-dimensional manifold. We denote by $f : \Delta \to X$ an arbitrary holomorphic map from the unit disk $\Delta \subset \mathbb{C}$ to $X$. The *Kobayashi-Royden infinitesimal pseudometric* on $X$ is the Finsler pseudometric on the tangent bundle $T_X$ defined by

$$\mathbf{k}_X(\xi) = \inf\{\lambda > 0\,;\, \exists f : \Delta \to X,\, f(0) = x,\, \lambda f'(0) = \xi\}, \qquad x \in X,\ \xi \in T_{X,x}$$

(see H. Royden [Roy71], [Roy74]). In the terminology of Kobayashi [Kob75], a *Finsler metric* (resp. *pseudometric*) on a vector bundle $E$ is a homogeneous positive (resp. nonnegative) positive function $N$ on the total space $E$, that is,

$$N(\lambda\xi) = |\lambda|\, N(\xi) \qquad \text{for all } \lambda \in \mathbb{C} \text{ and } \xi \in E.$$

A Finsler (pseudo-)metric on $E$ is thus nothing but a hermitian (semi-)norm on the tautological line bundle $\mathcal{O}_{P(E)}(-1)$ of lines of $E$ over the projectivized bundle $Y = P(E)$. The *Kobayashi pseudodistance* $d_K(x,y)$ is the geodesic pseudodistance obtained by integrating the Kobayashi-Royden infinitesimal metric. The manifold $X$ is said to be *hyperbolic* (in the sense of Kobayashi) if $d_K$ is actually a distance, namely if $d_K(x,y) > 0$ for all pairs of distinct points $(x,y)$ in $X$. In this context, we have the following well-known results of Brody [Bro78].

1.1. BRODY REPARAMETRIZATION LEMMA. *Let $\omega$ be a hermitian metric on $X$ and let $f : \Delta \to X$ be a holomorphic map. For every $\varepsilon > 0$, there exists a radius $R \geqslant (1-\varepsilon)\|f'(0)\|_\omega$ and a homographic transformation $\psi$ of the disk $D(0,R)$ onto $(1-\varepsilon)\Delta$ such that*

$$\|(f\circ\psi)'(0)\|_\omega = 1, \qquad \|(f\circ\psi)'(t)\|_\omega \leqslant \frac{1}{1-|t|^2/R^2} \quad \text{for every } t\in D(0,R).$$

*In particular, if $X$ is compact, given any sequence of holomorphic mappings $f_\nu : \Delta \to X$ such that $\lim \|f'_\nu(0)\|_\omega = +\infty$, one can find a sequence of homographic transformations $\psi_\nu : D(0,R_\nu) \to (1-1/\nu)\Delta$ with $\lim R_\nu = +\infty$, such that, after passing possibly to a subsequence, $(f_\nu\circ\psi_\nu)$ converges uniformly on every compact subset of $\mathbb{C}$ towards a non constant holomorphic map $g : \mathbb{C} \to X$ with $\|g'(0)\|_\omega = 1$ and $\sup_{t\in\mathbb{C}} \|g'(t)\|_\omega \leqslant 1$.*

PROOF. The first assertion of Brody's lemma is obtained by selecting $t_0 \in \Delta$ such that $(1-|t|^2)\|f'((1-\varepsilon)t)\|_\omega$ reaches its maximum for $t = t_0$. The reason for this choice is that $(1-|t|^2)\|f'((1-\varepsilon)t)\|_\omega$ is the norm of the differential $f'((1-\varepsilon)t) : T_\Delta \to T_X$ with respect to the Poincaré metric $|dt|^2/(1-|t|^2)^2$ on $T_\Delta$, which is conformally invariant under $\mathrm{Aut}(\Delta)$. One then adjusts $R$ and $\psi$ so that $\psi(0) = (1-\varepsilon)t_0$ and $|\psi'(0)|\,\|f'(\psi(0))\|_\omega = 1$. As $|\psi'(0)| = \frac{1-\varepsilon}{R}(1-|t_0|^2)$, the only possible choice for $R$ is

$$R = (1-\varepsilon)(1-|t_0|^2)\|f'(\psi(0))\|_\omega \geqslant (1-\varepsilon)\|f'(0)\|_\omega.$$

The inequality for $(f\circ\psi)'$ follows from the fact that the Poincaré norm is maximum at the origin, where it is equal to 1 by the choice of $R$. □

1.2. COROLLARY (Brody's theorem). *A compact complex manifold $X$ is hyperbolic if and only if there are no non constant entire holomorphic maps $g : \mathbb{C} \to X$.*

PROOF. The arguments are rather standard and will be developped in more detail in the proof of Prop. 1.5 below. □

Now, more generally, let $(X,V)$ be a complex manifold equipped with a *holomorphic subbundle* $V \subset T_X$. We will refer to such a pair as being a *complex directed manifold.* A morphism $\Phi : (X,V) \to (Y,W)$ in the category of complex directed manifolds is a holomorphic map such that $\Phi_\star(V) \subset W$. Our philosophy is that directed manifolds are also useful to study the "absolute case", i.e. the case $V = T_X$, because there are fonctorial constructions which work better in the category of directed manifolds (see e.g. § 4, 5, 6). We think of directed manifolds as a kind of "relative situation", covering e.g. the case when $V$ is the relative tangent sheaf to a smooth map $X \to S$. We want to stress here that no assumption need be made on the Lie bracket tensor $[\ ,\ ] : V\times V \to T_X/V$, and the rank $r = \operatorname{rank} V$ may be an arbitrary integer in the range $1 \leqslant r \leqslant n := \dim_{\mathbb{C}} X$. For the sake of generality, one might also wish to allow singularities in the subbundle $V$: for this, one can take $V$ to be given by an arbitrary coherent subsheaf $\mathcal{V} \subset \mathcal{O}(T_X)$ such that $\mathcal{O}(T_X)/\mathcal{V}$ has no torsion; then $V$ is a subbundle outside an analytic subset of codimension at least 2 (it is however somewhat safer to view $\mathcal{V}^\star$ as given by a quotient sheaf morphism $\Omega^1_X \to \mathcal{V}^\star$ and let $V^\star$ be the associated linear space, see Remark 3.10 below). For the sake of simplicity, we will assume most of the time that $V$ is actually a subbundle of $T_X$. In this situation, we generalize the notion of hyperbolicity as follows.

1.3. DEFINITION. *Let $(X, V)$ be a complex directed manifold.*

i) *The Kobayashi-Royden infinitesimal metric of $(X, V)$ is the Finsler metric on $V$ defined for any $x \in X$ and $\xi \in V_x$ by*

$$\mathbf{k}_{(X,V)}(\xi) = \inf\big\{\lambda > 0\,;\, \exists f : \Delta \to X,\ f(0) = x,\ \lambda f'(0) = \xi,\ f'(\Delta) \subset V\big\}.$$

*Here $\Delta \subset \mathbb{C}$ is the unit disk and the map $f$ is an arbitrary holomorphic map which is tangent to $V$, i.e., such that $f'(t) \in V_{f(t)}$ for all $t \in \Delta$. We say that $(X, V)$ is infinitesimally hyperbolic if $\mathbf{k}_{(X,V)}$ is positive definite on every fiber $V_x$ and satisfies a uniform lower bound $\mathbf{k}_{(X,V)}(\xi) \geqslant \varepsilon\|\xi\|_\omega$ in terms of any smooth hermitian metric $\omega$ on $X$, when $x$ describes a compact subset of $X$.*

ii) *More generally, the Kobayashi-Eisenman infinitesimal pseudometric of $(X, V)$ is the pseudometric defined on all decomposable $p$-vectors $\xi = \xi_1 \wedge \cdots \wedge \xi_p \in \Lambda^p V_x$, $1 \leqslant p \leqslant r = \operatorname{rank} V$, by*

$$\mathbf{e}^p_{(X,V)}(\xi) = \inf\big\{\lambda > 0\,;\, \exists f : \mathbb{B}_p \to X,\ f(0) = x,\ \lambda f_\star(\tau_0) = \xi,\ f_\star(T_{\mathbb{B}_p}) \subset V\big\}$$

*where $\mathbb{B}_p$ is the unit ball in $\mathbb{C}^p$ and $\tau_0 = \partial/\partial t_1 \wedge \cdots \wedge \partial/\partial t_p$ is the unit $p$-vector of $\mathbb{C}^p$ at the origin. We say that $(X, V)$ is infinitesimally $p$-measure hyperbolic if $\mathbf{e}^p_{(X,V)}$ is positive definite on every fiber $\Lambda^p V_x$ and satisfies a locally uniform lower bound in terms of any smooth metric.*

If $\Phi : (X, V) \to (Y, W)$ is a morphism of directed manifolds, it is immediate to check that we have the monotonicity property

$$\mathbf{k}_{(Y,W)}(\Phi_\star\xi) \leqslant \mathbf{k}_{(X,V)}(\xi), \qquad \forall \xi \in V, \tag{1.4}$$

$$\mathbf{e}^p_{(Y,W)}(\Phi_\star\xi) \leqslant \mathbf{e}^p_{(X,V)}(\xi), \qquad \forall \xi = \xi_1 \wedge \cdots \wedge \xi_p \in \Lambda^p V. \tag{1.4$^p$}$$

The following proposition shows that virtually all reasonable definitions of the hyperbolicity property are equivalent if $X$ is compact (in particular, the additional assumption that there is locally uniform lower bound for $\mathbf{k}_{(X,V)}$ is not needed). We merely say in that case that $(X, V)$ is *hyperbolic*.

1.5. PROPOSITION. *For an arbitrary directed manifold $(X, V)$, the Kobayashi-Royden infinitesimal metric $\mathbf{k}_{(X,V)}$ is upper semicontinuous on the total space of $V$. If $X$ is compact, $(X, V)$ is infinitesimally hyperbolic if and only if there are no non constant entire curves $g : \mathbb{C} \to X$ tangent to $V$. In that case, $\mathbf{k}_{(X,V)}$ is a continuous (and positive definite) Finsler metric on $V$.*

PROOF. The proof is almost identical to the standard proof for $\mathbf{k}_X$, so we only give a brief outline of the ideas. In order to prove the upper semicontinuity, let $\xi_0 \in V_{x_0}$ and $\varepsilon > 0$ be given. Then there is a curve $f : \Delta \to X$ tangent to $V$ such that $f(0) = x_0$ and $\lambda f'(0) = \xi_0$ with $0 < \lambda < \mathbf{k}_X(\xi_0) + \varepsilon$. Take $\lambda = 1$ for simplicity, and replace $\xi_0$ by $\lambda^{-1}\xi_0$. We may assume that $f$ is a proper embedding, otherwise we replace $(X,V)$ by $(X',V')=(X\times\Delta,\mathrm{pr}_1^\star V\oplus\mathrm{pr}_2^\star T_\Delta)$, $f$ by $f\times\mathrm{Id}_\Delta$, $\xi_0$ by $\xi_0\oplus 1$, and use a monotonicity argument for the projection $\mathrm{pr}_1:X'\to X$. If $f$ is an embedding, then $f(\Delta)$ is a Stein submanifold of $X$, and thus $f(\Delta)$ has a Stein neighborhood $\Omega$. As $\Omega$ is Stein, there exists a section $\theta\in H^0(\Omega,\mathcal{O}(V))$ extending $f'\in H^0(f(\Delta),\mathcal{O}(V))$. The map $f$ can be viewed as the solution of the differential equation $f'=\theta(f)$ with initial value $f(0)=x_0$. Take a small perturbation $g'=\theta_\eta(g)$ with initial value $g(0)=x$, where $\theta_\eta=\theta+\sum\eta_js_j$ and $s_1,\ldots,s_N$ are finitely many sections of $H^0(\Omega,\mathcal{O}(V))$ which generate $V$ in a neighborhood of $x_0$. We can achieve that $g'(0)=\theta_\eta(x)$ is equal to any prescribed vector $\xi\in V_x$ close to $\xi_0=\theta(x_0)$, and the solution $g$ exists on $(1-\varepsilon)\Delta$ if the perturbation is small enough. We conclude that $\mathbf{k}_{(X,V)}$ is upper semicontinuous by considering $t\mapsto g((1-\varepsilon)t)$.

If there exists a non constant entire curve $g:\mathbb{C}\to X$ tangent to $V$, it is clear that $\mathbf{k}_{(X,V)}(g'(t))\equiv 0$, hence $(X,V)$ cannot be hyperbolic. Conversely, if $X$ is compact and if there are no non constant entire curves $g:\mathbb{C}\to X$ tangent to $V$, Brody's lemma implies that there is an absolute bound $\|f'(0)\|_\omega\leqslant C$ for all holomorphic maps $f:\Delta\to X$ tangent to $V$; hence $\mathbf{k}_{(X,V)}(\xi)\geqslant C^{-1}\|\xi\|_\omega$ and $(X,V)$ is infinitesimally hyperbolic. By reparametrizing $f$ with an arbitrary automorphism of $\Delta$, we find $\|f'(t)\|_\omega\leqslant C/(1-|t|^2)$. The space of maps $f:\Delta\to X$ tangent to $V$ is therefore compact for the topology of uniform convergence on compact subsets of $\Delta$, thanks to Ascoli's theorem. We easily infer from this that $\mathbf{k}_{(X,V)}$ is lower semicontinuous on $V$. □

We conclude this section by showing that hyperbolicity is an open property.

1.6. PROPOSITION. *Let $(\mathcal{X},\mathcal{V})\to S$ be a holomorphic family of compact directed manifolds (by this, we mean a proper holomorphic map $\mathcal{X}\to S$ together with a holomorphic subbundle $\mathcal{V}\subset T_{\mathcal{X}/S}$ of the relative tangent bundle, defining a deformation $(X_t,V_t)_{t\in S}$ of the fibers). Then the set of $t\in S$ such that the fiber $(X_t,V_t)$ is hyperbolic is open in $S$ with respect to the euclidean topology.*

PROOF. Take a sequence of non hyperbolic fibers $(X_{t_\nu},V_{t_\nu})$ with $t_\nu\to t$ and fix a hermitian metric $\omega$ on $\mathcal{X}$. By Brody's lemma, there is a sequence of entire holomorphic maps $g_\nu:\mathbb{C}\to X_{t_\nu}$ tangent to $V_{t_\nu}$, such that $\|g'_\nu(0)\|_\omega=1$ and $\|g'_\nu\|\leqslant 1$. Ascoli's theorem shows that there is a subsequence of $(g_\nu)$ converging uniformly to a limit $g:\mathbb{C}\to X_t$, tangent to $V_t$, with $\|g'(0)\|_\omega=1$. Hence $(X_t,V_t)$ is not hyperbolic, and the collection of non hyperbolic fibers is closed in $S$. □

## 2. Hyperbolicity and Bounds for the Genus of Curves

In the case of projective algebraic varieties, hyperbolicity is expected to be related to other properties of a more algebraic nature. Theorem 2.1 below is a first step in this direction.

2.1. THEOREM. *Let $(X,V)$ be a compact complex directed manifold and let $\sum\omega_{jk}dz_j\otimes d\overline{z}_k$ be a hermitian metric on $X$, with associated positive $(1,1)$-form $\omega=\frac{i}{2}\sum\omega_{jk}dz_j\wedge d\overline{z}_k$. Consider the following three properties, which may or not be satisfied by $(X,V)$ :*

i) *$(X,V)$ is hyperbolic.*

ii) *There exists* $\varepsilon > 0$ *such that every compact irreducible curve* $C \subset X$ *tangent to* $V$ *satisfies*

$$-\chi(\overline{C}) = 2g(\overline{C}) - 2 \geqslant \varepsilon \deg_\omega(C)$$

*where* $g(\overline{C})$ *is the genus of the normalization* $\overline{C}$ *of* $C$, $\chi(\overline{C})$ *its Euler characteristic and* $\deg_\omega(C) = \int_C \omega$. *(This property is of course independent of* $\omega$.*)*

iii) *There does not exist any non constant holomorphic map* $\Phi : Z \to X$ *from an abelian variety* $Z$ *to* $X$ *such that* $\Phi_\star(T_Z) \subset V$.

*Then* i) $\Rightarrow$ ii) $\Rightarrow$ iii).

PROOF. i) $\Rightarrow$ ii). If $(X, V)$ is hyperbolic, there is a constant $\varepsilon_0 > 0$ such that $\mathbf{k}_{(X,V)}(\xi) \geqslant \varepsilon_0 \|\xi\|_\omega$ for all $\xi \in V$. Now, let $C \subset X$ be a compact irreducible curve tangent to $V$ and let $\nu : \overline{C} \to C$ be its normalization. As $(X, V)$ is hyperbolic, $\overline{C}$ cannot be a rational or elliptic curve, hence $\overline{C}$ admits the disk as its universal covering $\rho : \Delta \to \overline{C}$.

The Kobayashi-Royden metric $\mathbf{k}_\Delta$ is the Finsler metric $|dz|/(1-|z|^2)$ associated with the Poincaré metric $|dz|^2/(1-|z|^2)^2$ on $\Delta$, and $\mathbf{k}_{\overline{C}}$ is such that $\rho^\star \mathbf{k}_{\overline{C}} = \mathbf{k}_\Delta$. In other words, the metric $\mathbf{k}_{\overline{C}}$ is induced by the unique hermitian metric on $\overline{C}$ of constant Gaussian curvature $-4$. If $\sigma_\Delta = \frac{i}{2} dz \wedge d\overline{z}/(1-|z|^2)^2$ and $\sigma_{\overline{C}}$ are the corresponding area measures, the Gauss-Bonnet formula (integral of the curvature $= 2\pi\,\chi(\overline{C})$) yields

$$\int_{\overline{C}} d\sigma_{\overline{C}} = -\frac{1}{4}\int_{\overline{C}} \operatorname{curv}(\mathbf{k}_{\overline{C}}) = -\frac{\pi}{2}\chi(\overline{C})$$

On the other hand, if $j : C \to X$ is the inclusion, the monotonicity property (1.4) applied to the holomorphic map $j \circ \nu : \overline{C} \to X$ shows that

$$\mathbf{k}_{\overline{C}}(t) \geqslant \mathbf{k}_{(X,V)}\big((j\circ\nu)_\star t\big) \geqslant \varepsilon_0 \big\|(j\circ\nu)_\star t\big\|_\omega, \qquad \forall t \in T_{\overline{C}}.$$

From this, we infer $d\sigma_{\overline{C}} \geqslant \varepsilon_0^2 (j\circ\nu)^\star\omega$, thus

$$-\frac{\pi}{2}\chi(\overline{C}) = \int_{\overline{C}} d\sigma_{\overline{C}} \geqslant \varepsilon_0^2 \int_{\overline{C}} (j\circ\nu)^\star\omega = \varepsilon_0^2 \int_C \omega.$$

Property ii) follows with $\varepsilon = 2\varepsilon_0^2/\pi$.

ii) $\Rightarrow$ iii). First observe that ii) excludes the existence of elliptic and rational curves tangent to $V$. Assume that there is a non constant holomorphic map $\Phi : Z \to X$ from an abelian variety $Z$ to $X$ such that $\Phi_\star(T_Z) \subset V$. We must have $\dim \Phi(Z) \geqslant 2$, otherwise $\Phi(Z)$ would be a curve covered by images of holomorphic maps $\mathbb{C} \to \Phi(Z)$, and so $\Phi(Z)$ would be elliptic or rational, contradiction. Select a sufficiently general curve $\Gamma$ in $Z$ (e.g., a curve obtained as an intersection of very generic divisors in a given very ample linear system $|L|$ in $Z$). Then all isogenies $u_m : Z \to Z$, $s \mapsto ms$ map $\Gamma$ in a $1:1$ way to curves $u_m(\Gamma) \subset Z$, except maybe for finitely many double points of $u_m(\Gamma)$ (if $\dim Z = 2$). It follows that the normalization of $u_m(\Gamma)$ is isomorphic to $\Gamma$. If $\Gamma$ is general enough, similar arguments show that the images

$$C_m := \Phi(u_m(\Gamma)) \subset X$$

are also generically $1:1$ images of $\Gamma$, thus $\overline{C}_m \simeq \Gamma$ and $g(\overline{C}_m) = g(\Gamma)$. We would like to show that $C_m$ has degree $\geqslant \operatorname{Const} m^2$. This is indeed rather easy to check if

$\omega$ is Kähler, but the general case is slightly more involved. We write

$$\int_{C_m} \omega = \int_\Gamma (\Phi \circ u_m)^\star \omega = \int_Z [\Gamma] \wedge u_m^\star(\Phi^\star \omega),$$

where $\Gamma$ denotes the current of integration over $\Gamma$. Let us replace $\Gamma$ by an arbitrary translate $\Gamma + s$, $s \in Z$, and accordingly, replace $C_m$ by $C_{m,s} = \Phi \circ u_m(\Gamma + s)$. For $s \in Z$ in a Zariski open set, $C_{m,s}$ is again a generically $1:1$ image of $\Gamma + s$. Let us take the average of the last integral identity with respect to the unitary Haar measure $d\mu$ on $Z$. We find

$$\int_{s\in Z}\left(\int_{C_{m,s}} \omega\right) d\mu(s) = \int_Z \left(\int_{s\in Z} [\Gamma + s]\, d\mu(s)\right) \wedge u_m^\star(\Phi^\star \omega).$$

Now, $\gamma := \int_{s\in Z} [\Gamma + s]\, d\mu(s)$ is a translation invariant positive definite form of type $(p-1, p-1)$ on $Z$, where $p = \dim Z$, and $\gamma$ represents the same cohomology class as $[\Gamma]$, i.e. $\gamma \equiv c_1(L)^{p-1}$. Because of the invariance by translation, $\gamma$ has constant coefficients and so $(u_m)_\star \gamma = m^2 \gamma$. Therefore we get

$$\int_{s\in Z} d\mu(s) \int_{C_{m,s}} \omega = m^2 \int_Z \gamma \wedge \Phi^\star \omega.$$

In the integral, we can exclude the algebraic set of values $z$ such that $C_{m,s}$ is not a generically $1:1$ image of $\Gamma + s$, since this set has measure zero. For each $m$, our integral identity implies that there exists an element $s_m \in Z$ such that $g(\overline{C}_{m,s_m}) = g(\Gamma)$ and

$$\deg_\omega(C_{m,s_m}) = \int_{C_{m,s_m}} \omega \geqslant m^2 \int_Z \gamma \wedge \Phi^\star \omega.$$

As $\int_Z \gamma \wedge \Phi^\star \omega > 0$, the curves $C_{m,s_m}$ have bounded genus and their degree is growing quadratically with $m$, contradiction to property ii). □

2.2. DEFINITION. *We say that a projective directed manifold $(X, V)$ is "algebraically hyperbolic" if it satisfies property* 2.1 ii), *namely, if there exists $\varepsilon > 0$ such that every algebraic curve $C \subset X$ tangent to $V$ satisfies*

$$2g(\overline{C}) - 2 \geqslant \varepsilon \deg_\omega(C).$$

A nice feature of algebraic hyperbolicity is that it satisfies an algebraic analogue of the openness property.

2.3. PROPOSITION. *Let $(\mathcal{X}, \mathcal{V}) \to S$ be an algebraic family of projective algebraic directed manifolds (given by a projective morphism $\mathcal{X} \to S$). Then the set of $t \in S$ such that the fiber $(X_t, V_t)$ is algebraically hyperbolic is open with respect to the "countable Zariski topology" of $S$ (by definition, this is the topology for which closed sets are countable unions of algebraic sets).*

PROOF. After replacing $S$ by a Zariski open subset, we may assume that the total space $\mathcal{X}$ itself is quasi-projective. Let $\omega$ be the Kähler metric on $\mathcal{X}$ obtained by pulling back the Fubini-Study metric via an embedding in a projective space. If integers $d > 0$, $g \geqslant 0$ are fixed, the set $A_{d,g}$ of $t \in S$ such that $X_t$ contains an algebraic 1-cycle $C = \sum m_j C_j$ tangent to $V_t$ with $\deg_\omega(C) = d$ and $g(\overline{C}) = \sum m_j\, g(\overline{C}_j) \leqslant g$ is a closed algebraic subset of $S$ (this follows from the existence of a relative cycle space of curves of given degree, and from the fact that the geometric genus is Zariski lower semicontinuous). Now, the set of non algebraically hyperbolic fibers is by definition

$$\bigcap_{k>0} \bigcup_{2g-2<d/k} A_{d,g}.$$

This concludes the proof (of course, one has to know that the countable Zariski topology is actually a topology, namely that the class of countable unions of algebraic sets is stable under arbitrary intersections; this can be easily checked by an induction on dimension). □

2.4. REMARK. More explicit versions of the openness property have been dealt with in the literature. H. Clemens ([Cle86] and [CKL88]) has shown that on a very generic surface of degree $d \geqslant 5$ in $\mathbb{P}^3$, the curves of type $(d,k)$ are of genus $g > kd(d-5)/2$ (recall that a very generic surface $X \subset \mathbb{P}^3$ of degree $\geqslant 4$ has Picard group generated by $\mathcal{O}_X(1)$ thanks to the Noether-Lefschetz theorem, thus any curve on the surface is a complete intersection with another hypersurface of degree $k$; such a curve is said to be of type $(d,k)$; genericity is taken here in the sense of the countable Zariski topology). Improving on this result of Clemens, Geng Xu [Xu94] has shown that every curve contained in a very generic surface of degree $d \geqslant 5$ satisfies the sharp bound $g \geqslant d(d-3)/2 - 2$. This actually shows that a very generic surface of degree $d \geqslant 6$ is algebraically hyperbolic. Although a very generic quintic surface has no rational or elliptic curves, it seems to be unknown whether a (very) generic quintic surface is algebraically hyperbolic in the sense of Definition 2.2.

2.5. REMARK. It would be interesting to know whether algebraic hyperbolicity is open with respect to the euclidean topology; still more interesting would be to know whether Kobayashi hyperbolicity is open for the countable Zariski topology (of course, both properties would follow immediately if one knew that Zariski and Kobayashi hyperbolicity coincide, but they seem otherwise highly non trivial to establish). The latter openness property has raised an important amount of work around the following more particular question: is a (very) generic hypersurface $X \subset \mathbb{P}^{n+1}$ of degree $d$ large enough (say $d \geqslant 2n+1$) Kobayashi hyperbolic? Again, "very generic" is to be taken here in the sense of the countable Zariski topology. Brody-Green [BrGr77] and Nadel [Nad89] produced examples of hyperbolic surfaces in $\mathbb{P}^3$ for all degrees $d \geqslant 50$, and Masuda-Noguchi [MaNo93] recently gave examples of such hypersurfaces in $\mathbb{P}^n$ for arbitrary $n \geqslant 2$, of degree $d \geqslant d_0(n)$ large enough. The question of studying the hyperbolicity of complements $\mathbb{P}^n \smallsetminus D$ of generic divisors is in principle closely related to this; in fact if $D = \{P(z_0, \ldots, z_n) = 0\}$ is a smooth generic divisor of degree $d$, one may look at the hypersurface

$$X = \{z_{n+1}^d = P(z_0, \ldots, z_n)\} \subset \mathbb{P}^{n+1}$$

which is a cyclic $d:1$ covering of $\mathbb{P}^n$. Since any holomorphic map $f : \mathbb{C} \to \mathbb{P}^n \smallsetminus D$ can be lifted to $X$, it is clear that the hyperbolicity of $X$ would imply the hyperbolicity of $\mathbb{P}^n \smallsetminus D$. The hyperbolicity of complements of divisors in $\mathbb{P}^n$ has been investigated by many authors. M. Green [Green77] proved the hyperbolicity of the complement of $(2n+1)$ generic hyperplanes in $\mathbb{P}^n$. Zaidenberg [Zai89] showed the existence of curves with hyperbolic complement for every degree $d \geqslant 5$. In response to a conjecture of [Zai89], [DSW92, 94] showed that the complement of the union of at least 3 generic curves is hyperbolic, when the sum of degrees is at least 5. More recently, Siu and Yeung [SiYe96a] proved the harder fact that the complement of a generic irreducible curve of high degree in $\mathbb{P}^2$ is hyperbolic. Their approach uses jet bundle techniques, and it is one of our goals to explain some of the underlying geometric ideas. □

In the "absolute case" $V = T_X$, it seems reasonable to expect that all three properties 2.1 i), ii), iii) are equivalent, in particular that Kobayashi and algebraic hyperbolicity coincide. In fact, S. Lang made the following conjecture.

2.6. CONJECTURE ([Lang 86, 87]). *A compact complex manifold $X$ is hyperbolic if and only if there are no nontrivial holomorphic maps $Z \to X$ where $Z = \mathbb{C}^p/\Lambda$ is a compact complex torus.*

The "only if" part of the conjecture is of course clearly true. On the other hand, if $X$ is projective algebraic, every holomorphic map $Z \to X$ of a compact complex torus $Z$ to $X$ admits a factorization $Z \to Z' \to X$ where $Z'$ is an abelian variety (see e.g. A. Weil [We57]). Thus, for $X$ projective algebraic and $V = T_X$, a positive solution to Conjecture 2.5 would imply the equivalence of properties i), ii), iii) in Theorem 2.1.

2.7. HINT OF HEURISTIC PROOF OF LANG'S CONJECTURE. Although Lang's conjecture seems at present far beyond reach, we would like to present here a heuristic argument showing how things could possibly work. These ideas arose through discussions with S. Kosarew. Let $\mathcal{B}_X$ be the space of all Brody curves, that is, the set of all entire holomorphic curves $f : \mathbb{C} \to X$ with $\|f'\|_\omega \leqslant 1$ for some given hermitian metric $\omega$ on $X$. For each $\varepsilon \geqslant 0$, we define a distance $\delta_\varepsilon$ on $\mathcal{B}_X$ by putting

$$\delta_\varepsilon(f, g) = \sup_{t \in \mathbb{C}} d_\omega(f(t), g(t)) e^{-\varepsilon|t|}$$

where $d_\omega$ is the geodesic distance on $X$. For all $\varepsilon > 0$, $d_\varepsilon$ defines the topology of uniform convergence of compact sets, whilst $d_0$ is the topology of uniform convergence up to infinity. Hence $(\mathcal{B}_X, d_\varepsilon)$ is a compact metric space for $\varepsilon > 0$. Assume that this is still true for $\varepsilon = 0$ and assume moreover that $X$ is not hyperbolic, i.e. $\mathcal{B}_X \neq \emptyset$. We then consider the compact topological group $G_\mathcal{B}$ of isometries of $(\mathcal{B}_X, d_0)$ and look at the group homomorphism

$$\Psi : (\mathbb{C}, +) \to (G_\mathcal{B}, \circ), \qquad a \mapsto f_a \quad \text{where } f_a(t) = f(t - a).$$

We claim that $G_\mathcal{B}$ should be a finite dimensional Lie group and $Z = \overline{\Psi(\mathbb{C})} \subset G_\mathcal{B}$ a compact commutative complex subgroup, thus a complex torus. In fact, a compact Banach Lie group is finite dimensional, and the "Lie algebra" of $G_\mathcal{B}$ seems to be interpretable as a closed subspace of the Banach space of bounded holomorphic sections in $H^0(\mathcal{B}_X \times \mathbb{C}, \mathrm{ev}^\star T_X)$ where $\mathrm{ev} : \mathcal{B}_X \times \mathbb{C} \to X$ is the evaluation map $(f, t) \mapsto f(t)$. The complex structure on $Z$ should arise from the complex structure on that Banach space. Now, we obtain a non trivial holomorphic map $\Phi : Z \to X$ by selecting an $f \in \mathcal{B}_X$ which is not a fixed point of $Z$ and putting $\Phi(\gamma) = \gamma(f)(0)$, $\gamma \in Z$. □

In the general context of directed manifolds, algebraic hyperbolicity can be strictly weaker than Kobayashi hyperbolicity. The simplest example is provided by a 2-dimensional abelian variety $X = \mathbb{C}^2/\Lambda$ equipped with a constant subbundle $V \subset T_X$ given by a complex line $V_0 \subset \mathbb{C}^2$ such that $V_0 \cap \Lambda = \{0\}$. Then all leaves of $V$ are isomorphic to complex lines, in particular there are no compact curves tangent to $V$, and thus 2.1 ii) and 2.1 iii) are satisfied (as void conditions). This pathology can somehow be corrected by observing that the hyperbolicity of $(X, V)$ implies a statement analogue to 2.1 ii) but substantially stronger, namely a lower bound of $-\chi(\overline{C})$ for curves $C$ which are almost tangent to $V$, in the sense that their "deviation with respect to $V$" is small.

2.8. DEFINITION. *Let $X$ be equipped with a hermitian $(1,1)$-form $\omega$, and let $C \subset X$ be a compact curve in $X$. We define the $L^2$-deviation of $C$ with respect to $V$ by*

$$\mathrm{dev}^2_\omega(C/V) = \int_C \omega_{V^\perp}$$

*where $\omega = \omega_V \oplus \omega_{V^\perp}$ is the orthogonal decomposition of $\omega$ on $V \oplus V^\perp$. Similarly, if $\nu : \overline{C} \to X$ is the normalization map and $\overline{C} \not\simeq \mathbb{P}^1$, we define the $L^\infty$-deviation (resp. the $L^p$-deviation) to be*

$$\mathrm{dev}^\infty_\omega(C/V) = \sup_{t\in\overline{C}} \frac{\nu^\star\omega_{V^\perp}(t)}{d\widetilde{\sigma}(t)} = \sup_{t\in\overline{C}} \|\nu'(t)^\perp\|^2_{\widetilde{\sigma},\omega_{V^\perp}},$$

$$\mathrm{dev}^p_\omega(C/V) = \Big[\int_{t\in\overline{C}} \Big(\frac{\nu^\star\omega_{V^\perp}(t)}{d\widetilde{\sigma}(t)}\Big)^{p/2} d\widetilde{\sigma}(t)\Big]^{2/p} = \Big[\int_{t\in\overline{C}} \|\nu'(t)^\perp\|^p_{\widetilde{\sigma},\omega_{V^\perp}} d\widetilde{\sigma}(t)\Big]^{2/p},$$

*where $d\widetilde{\sigma}$ is the normalized Poincaré metric on $\overline{C}$ (hermitian metric of constant curvature with $\int_{\overline{C}} d\widetilde{\sigma} = 1$), and $\nu'(t)^\perp$ is the projection of the tangent vector $\nu'(t)$ on $V^\perp$. If $\overline{C} \simeq \mathbb{P}^1$, we set instead*

$$\mathrm{dev}^\infty_\omega(C/V) = \inf_{\gamma\in\mathrm{PGL}_2(\mathbb{C})} \sup_{t\in\overline{C}} \frac{\nu^\star\omega_{V^\perp}(t)}{\gamma^\star d\widetilde{\sigma}(t)}$$

$$\mathrm{dev}^p_\omega(C/V) = \inf_{\gamma\in\mathrm{PGL}_2(\mathbb{C})} \Big[\int_{t\in\overline{C}} \|\nu'(t)^\perp\|^p_{\gamma^\star\widetilde{\sigma},\omega_{V^\perp}} d\gamma^\star\widetilde{\sigma}(t)\Big]^{2/p}.$$

2.9. PROPOSITION. *Let $(X, V)$ be a compact directed manifold equipped with a hermitian metric $\omega$. If $(X, V)$ is hyperbolic, there exists a constant $\varepsilon > 0$ such that*

$$\max\big(-\chi(\overline{C}), \mathrm{dev}^\infty_\omega(C/V)\big) \geqslant \varepsilon \deg_\omega(C),$$

*for every compact curve $C \subset X$.*

PROOF. Otherwise, there would exist a sequence of curves $(C_\ell)$ and a sequence of positive numbers $\varepsilon_\ell$ converging to 0, such that

$$-\chi(\overline{C}_\ell) \leqslant \varepsilon_\ell \deg_\omega(C_\ell), \qquad \mathrm{dev}^\infty_\omega(C_\ell/V)) \leqslant \varepsilon_\ell \deg_\omega(C_\ell).$$

First assume that all curves $C_\ell$ have geometric genus $g(\overline{C}_\ell) \geqslant 2$. Let $\nu_\ell : \overline{C}_\ell \to X$ be the normalization map of $C_\ell$, and let $d\sigma_\ell$ be the area measure associated with the Poincaré metric on $\overline{C}_\ell$ and $d\widetilde{\sigma}_\ell = \lambda_\ell^{-1} d\sigma_\ell$ the normalized Poincaré metric with $\lambda_\ell = \int_{\overline{C}_\ell} d\sigma_\ell = \frac{\pi}{2}(-\chi(C_\ell))$. Select a point $x_\ell \in \overline{C}_\ell$ where the ratio $\nu_\ell^\star\omega/d\sigma_\ell$ is maximum. Since $\int_{\overline{C}_\ell} \nu_\ell^\star\omega = \deg_\omega(C_\ell)$, we have

$$R_\ell^2 := \frac{\nu_\ell^\star\omega(x_\ell)}{d\sigma_\ell(x_\ell)} \geqslant \frac{\int_{\overline{C}_\ell} \nu_\ell^\star\omega}{\int_{\overline{C}_\ell} d\sigma_\ell} = \frac{\deg_\omega(C_\ell)}{\lambda_\ell} \geqslant \frac{2}{\pi\varepsilon_\ell} \to +\infty.$$

Let $\rho_\ell : \Delta \to \overline{C}_\ell$ be the universal covering map of $\overline{C}_\ell$, chosen such that $\rho_\ell(0) = x_\ell$. We get a holomorphic map $f_\ell = \nu_\ell \circ \rho_\ell : \Delta \to X$ such that $\|f_\ell'(0)\|_\omega = R_\ell \to +\infty$. By Brody's reparametrization lemma, we can reparametrize $f_\ell$ as $g_\ell(t) = f_\ell(t/R_\ell)$ so that some subsequence of $(g_\ell)$ converges uniformly on every compact set to a limit $g : \mathbb{C} \to X$ with $\|g'(0)\|_\omega = 1$. We claim that $g$ must be tangent to $V$. In fact, by definition of the $L^\infty$ deviation, we get $\nu_\ell^\star \omega_{V^\perp} \leqslant \varepsilon_\ell \deg_\omega(C_\ell) d\widetilde{\sigma}_\ell$, thus $f_\ell^\star \omega_{V^\perp} = \rho_\ell^\star(\nu_\ell^\star \omega_{V^\perp})$ and $g_\ell^\star \omega_{V^\perp}$ satisfy

$$f_\ell^\star \omega_{V^\perp} \leqslant \frac{1}{\lambda_\ell} \varepsilon_\ell \deg_\omega(C_\ell) \frac{|dt|^2}{(1-|t|^2)^2},$$

$$g_\ell^\star \omega_{V^\perp} \leqslant \frac{\deg_\omega(C_\ell)}{\lambda_\ell} \varepsilon_\ell \frac{R_\ell^{-2}|dt|^2}{(1-|t|^2/R_\ell^2)^2} \leqslant \varepsilon_\ell \frac{|dt|^2}{(1-|t|^2/R_\ell^2)^2}.$$

From this we conclude that $g^\star \omega_{V^\perp} = 0$, hence $g$ is tangent to $V$, contradiction. If the curves $C_\ell$ are of genus 0 or 1, the arguments are similar and will be left to the reader. □

## 3. The Ahlfors-Schwarz Lemma for Metrics of Negative Curvature

One of the most basic ideas is that hyperbolicity should somehow be related with suitable negativity properties of the curvature. For instance, it is a standard fact already observed in Kobayashi [Kob70] that the negativity of $T_X$ (or the ampleness of $T_X^\star$) implies the hyperbolicity of $X$. There are many ways of improving or generalizing this result. We present here a few simple examples of such generalizations. If $(V, h)$ is a holomorphic vector bundle equipped with a smooth hermitian metric, we denote by $\nabla_h = \nabla_h' + \nabla_h''$ the associated Chern connection and by $\Theta_h(V) = \frac{i}{2\pi}\nabla_h^2$ its Chern curvature tensor.

3.1. PROPOSITION. *Let $(X, V)$ be a compact directed manifold. Assume that $V^\star$ is ample. Then $(X, V)$ is hyperbolic.*

PROOF (from an original idea of [Kob75]). Recall that a vector bundle $E$ is said to be ample if $S^m E$ has enough global sections $\sigma_1, \ldots, \sigma_N$ so as to generate 1-jets of sections at any point, when $m$ is large. One obtains a Finsler metric $N$ on $E^\star$ by putting

$$N(\xi) = \Big(\sum_{1 \leqslant j \leqslant N} |\sigma_j(x) \cdot \xi^m|^2\Big)^{1/2m}, \qquad \xi \in E_x^\star,$$

and $N$ is then a strictly plurisubharmonic function on the total space of $E^\star$ minus the zero section (in other words, the line bundle $\mathcal{O}_{P(E^\star)}(1)$ has a metric of positive curvature). By the ampleness assumption on $V^\star$, we thus have a Finsler metric $N$ on $V$ which is strictly plurisubharmonic outside the zero section. By Brody's lemma, if $(X, V)$ is not hyperbolic, there is a non constant entire curve $g : \mathbb{C} \to X$ tangent to $V$ such that $\sup_\mathbb{C} \|g'\|_\omega \leqslant 1$ for some given hermitian metric $\omega$ on $X$. Then $N(g')$ is a bounded subharmonic function on $\mathbb{C}$ which is strictly subharmonic on $\{g' \neq 0\}$. This is a contradiction, for any bounded subharmonic function on $\mathbb{C}$ must be constant. □

This result can be generalized a little bit further by means of the Ahlfors-Schwarz lemma (see e.g. [Lang87]).

3.2. AHLFORS-SCHWARZ LEMMA. *Let $\gamma(t) = \gamma_0(t)\, i\, dt \wedge d\bar{t}$ be a hermitian metric on $\Delta_R$ where $\log \gamma_0$ is a subharmonic function such that $i\,\partial\bar\partial \log \gamma_0(t) \geqslant A\,\gamma(t)$ in the sense of currents, for some positive constant $A$. Then $\gamma$ can be compared with the Poincaré metric of $\Delta_R$ as follows:*

$$\gamma(t) \leqslant \frac{2}{A} \frac{R^{-2}|dt|^2}{(1-|t|^2/R^2)^2}.$$

*More generally, let $\gamma = i \sum \gamma_{jk} dt_j \wedge d\bar{t}_k$ be an almost everywhere positive hermitian form on the ball $B(0,R) \subset \mathbb{C}^p$, such that $-\operatorname{Ricci}(\gamma) := i\,\partial\bar\partial \log\det\gamma \geqslant A\gamma$ in the sense of currents, for some constant $A > 0$ (this means in particular that $\det\gamma = \det(\gamma_{jk})$ is such that $\log\det\gamma$ is plurisubharmonic). Then*

$$\det(\gamma) \leqslant \Big(\frac{p+1}{AR^2}\Big)^p \frac{1}{(1-|t|^2/R^2)^{p+1}}.$$

PROOF. It is of course sufficient to deal with the more general case of a ball in $\mathbb{C}^p$. First assume that $\gamma$ is smooth and positive definite on $\overline{B}(0,R)$. Take a point $t_0 \in B(0,R)$ at which $(1-|t|^2/R^2)^{p+1}\det(\gamma(t))$ is maximum. The logarithmic $i\,\partial\bar\partial$-derivative of this function at $t_0$ must be $\leqslant 0$, hence

$$i\,\partial\bar\partial \log\det\gamma(t)_{t=t_0} - (p+1)\, i\,\partial\bar\partial \log(1-|t|^2/R^2)^{-1}_{t=t_0} \leqslant 0.$$

The hypothesis on the Ricci curvature implies

$$A^p\,\gamma(t_0)^p \leqslant \big(i\,\partial\bar\partial \log\det\gamma(t)_{t=t_0}\big)^p \leqslant (p+1)^p \big(i\,\partial\bar\partial \log(1-|t|^2/R^2)^{-1}_{t=t_0}\big)^p.$$

An easy computation shows that the determinant of $i\,\partial\bar\partial \log(1-|t|^2/R^2)^{-1}$ is equal to $R^{-2p}(1-|t|^2/R^2)^{-p-1}$. From this, we conclude that

$$(1-|t|^2/R^2)^{p+1}\det\gamma(t) \leqslant (1-|t_0|^2/R^2)^{p+1}\det\gamma(t_0) \leqslant \Big(\frac{p+1}{AR^2}\Big)^p.$$

If $\gamma$ is not smooth, we use a regularization argument. Namely, we shrink $R$ a little bit and look at the maximum of the function

$$u(t) = (1-|t|^2/R^2)^{p+1}\exp\big(\rho_\varepsilon \star \log\det\gamma(t)\big)$$

where $(\rho_\varepsilon)$ is a family of regularizing kernels. The argument goes through because

$$i\,\partial\bar\partial(\rho_\varepsilon \star \log\det\gamma) \geqslant A\,\rho_\varepsilon \star \gamma$$

and $\log\det(\rho_\varepsilon \star \gamma) \geqslant \rho_\varepsilon \star \log\det\gamma$ by concavity of the $\log\det$ function. □

3.3. PROPOSITION. *Let $(X,V)$ be a compact directed manifold. Assume that $V^\star$ is "very big" in the following sense: there exists an ample line bundle $L$ and a sufficiently large integer $m$ such that the global sections in $H^0(X, S^mV^\star \otimes L^{-1})$ generate all fibers over $X \smallsetminus Y$, for some analytic subset $Y \subsetneq X$. Then all entire curves $f : \mathbb{C} \to X$ tangent to $V$ satisfy $f(\mathbb{C}) \subset Y$ [under our assumptions, $X$ is a projective algebraic manifold and $Y$ is an algebraic subvariety, thus it is legitimate to say that the entire curves are "algebraically degenerate"].*

PROOF. Let $\sigma_1, \dots, \sigma_N \in H^0(X, S^m V^\star \otimes L^{-1})$ be a basis of sections generating $S^m V^\star \otimes L^{-1}$ over $X \smallsetminus Y$. If $f : \mathbb{C} \to X$ is tangent to $V$, we define a semipositive hermitian form $\gamma(t) = \gamma_0(t)\,|dt|^2$ on $\mathbb{C}$ by putting

$$\gamma_0(t) = \sum \|\sigma_j(f(t)) \cdot f'(t)^m\|_{L^{-1}}^{2/m}$$

where $\| \; \|_L$ denotes a hermitian metric with positive curvature on $L$. If $f(\mathbb{C}) \not\subset Y$, the form $\gamma$ is not identically 0 and we then find

$$i\,\partial\overline{\partial}\log\gamma_0 \geqslant \frac{2\pi}{m} f^\star\Theta(L)$$

where $\Theta(L)$ is the curvature form. The positivity assumption combined with an obvious homogeneity argument yield

$$\frac{2\pi}{m} f^\star\Theta(L) \geqslant \varepsilon\|f'(t)\|_\omega^2\,|dt|^2 \geqslant \varepsilon'\,\gamma(t)$$

for any given hermitian metric $\omega$ on $X$. Now, for any $t_0$ with $\gamma_0(t_0) > 0$, the Ahlfors-Schwarz lemma shows that $f$ can only exist on a disk $D(t_0, R)$ such that $\gamma_0(t_0) \leqslant \frac{2}{\varepsilon'} R^{-2}$, contradiction. □

There are similar results for $p$-measure hyperbolicity, e.g.

3.4. PROPOSITION. *Let $(X, V)$ be a compact directed manifold. Assume that $\Lambda^p V^\star$ is ample. Then $(X, V)$ is infinitesimally $p$-measure hyperbolic. More generally, assume that $\Lambda^p V^\star$ is very big with base locus contained in $Y \subsetneq X$ (see 3.3). Then $\mathbf{e}^p$ is non degenerate over $X \smallsetminus Y$.*

PROOF. By the ampleness assumption, there is a smooth Finsler metric $N$ on $\Lambda^p V$ which is strictly plurisubharmonic outside the zero section. We select also a hermitian metric $\omega$ on $X$. For any holomorphic map $f : \mathbb{B}_p \to X$ we define a semipositive hermitian metric $\widetilde{\gamma}$ on $\mathbb{B}_p$ by putting $\widetilde{\gamma} = f^\star\omega$. Since $\omega$ need not have any good curvature estimate, we introduce the function $\delta(t) = N_{f(t)}(\Lambda^p f'(t) \cdot \tau_0)$, where $\tau_0 = \partial/\partial t_1 \wedge \cdots \wedge \partial/\partial t_p$, and select a metric $\gamma = \lambda\widetilde{\gamma}$ conformal to $\widetilde{\gamma}$ such that $\det\gamma = \delta$. Then $\lambda^p$ is equal to the ratio $N/\Lambda^p\omega$ on the element $\Lambda^p f'(t)\cdot\tau_0 \in \Lambda^p V_{f(t)}$. Since $X$ is compact, it is clear that the conformal factor $\lambda$ is bounded by an absolute constant independent of $f$. From the curvature assumption we then get

$$i\,\partial\overline{\partial}\log\det\gamma = i\,\partial\overline{\partial}\log\delta \geqslant (f, \Lambda^p f')^\star(i\,\partial\overline{\partial}\log N) \geqslant \varepsilon f^\star\omega \geqslant \varepsilon'\,\gamma.$$

By the Ahlfors-Schwarz lemma we infer that $\det\gamma(0) \leqslant C$ for some constant $C$, i.e., $N_{f(0)}(\Lambda^p f'(0) \cdot \tau_0) \leqslant C'$. This means that the Kobayashi-Eisenman pseudometric $\mathbf{e}^p_{(X,V)}$ is positive definite everywhere and uniformly bounded from below. In the case $\Lambda^p V^\star$ is very big with base locus $Y$, we use essentially the same arguments, but we then only have $N$ being positive definite on $X \smallsetminus Y$. □

3.5. COROLLARY ([Gri71], KobO71]). *If $X$ is a projective variety of general type, the Kobayashi-Eisenmann volume form $\mathbf{e}^n$, $n = \dim X$, can degenerate only along a proper algebraic set $Y \subsetneq X$.*

The converse of Corollary 3.5 is expected to be true, namely, the generic non degeneracy of $\mathbf{e}^n$ should imply that $X$ is of general type, but this is only known for surfaces (see [GrGr80] and [MoMu82]):

3.6. CONJECTURE (Green-Griffiths [GrGr80]). *A projective algebraic variety $X$ is almost measure hyperbolic (i.e. $\mathbf{e}^n$ degenerates only along a proper algebraic subvariety) if and only if $X$ is of general type.*

In the same vein, Green-Griffiths and Lang proposed the following conjectures.

3.7. CONJECTURE (Green-Griffiths [GrGr80]). *If $X$ is a variety of general type, there exists a proper algebraic set $Y \subsetneq X$ such that every entire holomorphic curve $f : \mathbb{C} \to X$ is contained in $Y$.*

The most outstanding result in the direction of Conjecture 3.7 is the proof of the Bloch theorem, as proposed by Bloch [Blo26] and Ochiai [Och77]. The Bloch theorem is the special case of 3.7 when the irregularity of $X$ satisfies $q = h^0(X, \Omega^1_X) > \dim X$. Various solutions have then been obtained in fundamental papers of Noguchi [Nog77, 81, 84], Kawamata [Kaw80] and Green-Griffiths [GrGr80], by means of different techniques. See section § 9 for a proof based on jet bundle techniques.

3.8. CONJECTURE ([Lang86, 87]). *A projective algebraic variety $X$ is hyperbolic if and only if all its algebraic subvarieties (including $X$ itself) are of general type.*

An essential step in the proof of the necessity of having general type subvarieties would be to show that manifolds of Kodaira dimension 0 (say, Calabi-Yau manifolds and symplectic manifolds, all of which have $c_1(X) = 0$) are not hyperbolic, e.g. by exhibiting a sequence of curves $C_\ell$ such that $(2g(\overline{C}_\ell) - 2)/\deg(C_\ell) \to 0$. In fact, it is even expected that there are covering families of such curves, whereby proving that such manifolds are not measure hyperbolic. An analogous conjecture in the relative situation might stand as follows.

3.9. CONJECTURE. *Let $(X, V)$ be a projective directed manifold. Assume that $\det V^\star$ is big and that $V$ is semistable in some sense (e.g. with respect to $\det V^\star$, if $\det V^\star$ is ample). Then there is a proper algebraic subset $Y \subsetneq X$ such that every entire curve $f : \mathbb{C} \to X$ tangent to $V$ satisfies $f(\mathbb{C}) \subset Y$.*

Recall that a line bundle $L$ is said to be *big* if it has maximal Kodaira dimension, in other words, if $H^0(X, L^{\otimes m}) \geqslant c\, m^{\dim X}$ for some constant $c > 0$, when $m \geqslant m_0$ is sufficiently large. Some sort of semistability condition is clearly required, otherwise one might take $X$ to be an abelian variety of dimension $\geqslant 3$ and $V = L + L'$ where $L \subset T_X$ is a constant line subbundle with dense trajectories, and $L' \subset T_X$ a "generic" sufficiently negative line bundle (also take a blow-up $\widetilde{X} \to X$ to resolve the singularities of $V$, so as to obtain a subbundle $\widetilde{V} \subset T_{\widetilde{X}}$). In the absolute case $V = T_X$, the semistability condition is probably not needed since $T_X$ tends to be always semistable in some sense (if $K_X$ is ample, there is always a Kähler-Einstein metric, hence $T_X$ is $K_X$-semistable).

3.10. REMARK. One should take care of the fact that Propositions 3.1, 3.3 and 3.4 cannot be extended without modifications to the case when $V$ admits singularities. For instance, take $X = \mathbb{P}^n$ and let $\ell = \mathbb{P}^1 \subset X$ be a line. Take a section of $T_\ell \otimes \mathcal{O}_\ell(d) \simeq \mathcal{O}_\ell(d+2)$ admitting only one zero $z_0$ of multiplicity $d+2$, and extend it as a section $\sigma$ of $T_{\mathbb{P}^n} \otimes \mathcal{O}(d)$ admitting only isolated zeroes (this is always possible for arbitrary $d > 0$). Then $\sigma$ defines a sheaf injection $\sigma : \mathcal{O}(-d) \hookrightarrow \mathcal{O}(T_{\mathbb{P}^n})$ with $\mathcal{V} = \sigma(\mathcal{O}(-d)) \simeq \mathcal{O}(-d)$ negative, nevertheless $\mathcal{V}$

admits a complex line $\ell \smallsetminus \{z_0\} \simeq \mathbb{C}$ as one of its integral curves. The correct assumption guaranteeing the hyperbolicity of $(X, V)$ is that $\mathcal{W} = \mathrm{Im}(\Omega^1_X \to \mathcal{V}^\star)$ should be ample.

## 4. Projectivization of a Directed Manifold

The basic idea is to introduce a fonctorial process which produces a new complex directed manifold $(\widetilde{X}, \widetilde{V})$ from a given one $(X, V)$. The new structure $(\widetilde{X}, \widetilde{V})$ plays the role of a space of 1-jets over $X$. We let

$$\widetilde{X} = P(V), \qquad \widetilde{V} \subset T_{\widetilde{X}}$$

be the projectivized bundle of lines of $V$, together with a subbundle $\widetilde{V}$ of $T_{\widetilde{X}}$ defined as follows: for every point $(x, [v]) \in \widetilde{X}$ associated with a vector $v \in V_x \smallsetminus \{0\}$,

$$\widetilde{V}_{(x,[v])} = \big\{\xi \in T_{\widetilde{X},\,(x,[v])}\,;\ \pi_\star\xi \in \mathbb{C}\,v\big\}, \qquad \mathbb{C}\,v \subset V_x \subset T_{X,x}, \tag{4.1}$$

where $\pi : \widetilde{X} = P(V) \to X$ is the natural projection and $\pi_\star : T_{\widetilde{X}} \to \pi^\star T_X$ is its differential. On $\widetilde{X} = P(V)$ we have a tautological line bundle $\mathcal{O}_{\widetilde{X}}(-1) \subset \pi^\star V$ such that $\mathcal{O}_{\widetilde{X}}(-1)_{(x,[v])} = \mathbb{C}\,v$. The bundle $\widetilde{V}$ is characterized by the two exact sequences

$$0 \longrightarrow T_{\widetilde{X}/X} \longrightarrow \widetilde{V} \xrightarrow{\ \pi_\star\ } \mathcal{O}_{\widetilde{X}}(-1) \longrightarrow 0, \tag{4.2}$$

$$0 \longrightarrow \mathcal{O}_{\widetilde{X}} \longrightarrow \pi^\star V \otimes \mathcal{O}_{\widetilde{X}}(1) \longrightarrow T_{\widetilde{X}/X} \longrightarrow 0, \tag{4.2$'$}$$

where $T_{\widetilde{X}/X}$ denotes the relative tangent bundle of the fibration $\pi : \widetilde{X} \to X$. The first sequence is a direct consequence of the definition of $\widetilde{V}$, whereas the second is a relative version of the Euler exact sequence describing the tangent bundle of the fibers $P(V_x)$. From these exact sequences we infer

$$\dim \widetilde{X} = n + r - 1, \qquad \mathrm{rank}\,\widetilde{V} = \mathrm{rank}\,V = r, \tag{4.3}$$

and by taking determinants we find $\det(T_{\widetilde{X}/X}) = \pi^\star \det V \otimes \mathcal{O}_{\widetilde{X}}(r)$, thus

$$\det \widetilde{V} = \pi^\star \det V \otimes \mathcal{O}_{\widetilde{X}}(r-1). \tag{4.4}$$

By definition, $\pi : (\widetilde{X}, \widetilde{V}) \to (X, V)$ is a morphism of complex directed manifolds. Clearly, our construction is fonctorial, i.e., for every morphism of directed manifolds $\Phi : (X, V) \to (Y, W)$, there is a commutative diagram

$$\begin{array}{ccc} (\widetilde{X}, \widetilde{V}) & \xrightarrow{\ \pi\ } & (X, V) \\ {\scriptstyle\widetilde{\Phi}}\ \vdots\downarrow & & \downarrow\ {\scriptstyle\Phi} \\ (\widetilde{Y}, \widetilde{W}) & \xrightarrow{\ \pi\ } & (Y, W) \end{array} \tag{4.5}$$

where the left vertical arrow is the meromorphic map $P(V) \dashrightarrow P(W)$ induced by the differential $\Phi_\star : V \to \Phi^\star W$ ($\widetilde{\Phi}$ is actually holomorphic if $\Phi_\star : V \to \Phi^\star W$ is injective).

Now, suppose that we are given a holomorphic curve $f : \Delta_R \to X$ parametrized by the disk $\Delta_R$ of centre 0 and radius $R$ in the complex plane, and that $f$ is a *tangent trajectory* of the directed manifold, i.e., $f'(t) \in V_{f(t)}$ for every $t \in \Delta_R$. If $f$ is non constant, there is a well defined and unique tangent line $[f'(t)]$ for every $t$, even at stationary points, and the map

$$\widetilde{f} : \Delta_R \to \widetilde{X}, \qquad t \mapsto \widetilde{f}(t) := (f(t), [f'(t)]) \tag{4.6}$$

is holomorphic (at a stationary point $t_0$, we just write $f'(t) = (t - t_0)^s u(t)$ with $s \in \mathbb{N}^\star$ and $u(t_0) \neq 0$, and we define the tangent line at $t_0$ to be $[u(t_0)]$, hence $\widetilde{f}(t) = (f(t), [u(t)])$ near $t_0$; even for $t = t_0$, we still denote $[f'(t_0)] = [u(t_0)]$ for simplicity of notation). By definition $f'(t) \in \mathcal{O}_{\widetilde{X}}(-1)_{\widetilde{f}(t)} = \mathbb{C}\, u(t)$, hence the derivative $f'$ defines a section

$$f' : T_{\Delta_R} \to \widetilde{f}^\star \mathcal{O}_{\widetilde{X}}(-1). \tag{4.7}$$

Moreover $\pi \circ \widetilde{f} = f$, therefore

$$\pi_\star \widetilde{f}'(t) = f'(t) \in \mathbb{C}\, u(t) \Longrightarrow \widetilde{f}'(t) \in \widetilde{V}_{(f(t), u(t))} = \widetilde{V}_{\widetilde{f}(t)}$$

and we see that $\widetilde{f}$ is a tangent trajectory of $(\widetilde{X}, \widetilde{V})$. We say that $\widetilde{f}$ is the *canonical lifting* of $f$ to $\widetilde{X}$. Conversely, if $g : \Delta_R \to \widetilde{X}$ is a tangent trajectory of $(\widetilde{X}, \widetilde{V})$, then by definition of $\widetilde{V}$ we see that $f = \pi \circ g$ is a tangent trajectory of $(X, V)$ and that $g = \widetilde{f}$ (unless $g$ is contained in a vertical fiber $P(V_x)$, in which case $f$ is constant).

For any point $x_0 \in X$, there are local coordinates $(z_1, \ldots, z_n)$ on a neighborhood $\Omega$ of $x_0$ such that the fibers $(V_z)_{z \in \Omega}$ can be defined by linear equations

$$V_z = \Big\{ \xi = \sum_{1 \leqslant j \leqslant n} \xi_j \frac{\partial}{\partial z_j} \,;\, \xi_j = \sum_{1 \leqslant k \leqslant r} a_{jk}(z) \xi_k \text{ for } j = r+1, \ldots, n \Big\}, \tag{4.8}$$

where $(a_{jk})$ is a holomorphic $(n - r) \times r$ matrix. It follows that a vector $\xi \in V_z$ is completely determined by its first $r$ components $(\xi_1, \ldots, \xi_r)$, and the affine chart $\xi_j \neq 0$ of $P(V)_{\restriction \Omega}$ can be described by the coordinate system

$$\Big( z_1, \ldots, z_n; \frac{\xi_1}{\xi_j}, \ldots, \frac{\xi_{j-1}}{\xi_j}, \frac{\xi_{j+1}}{\xi_j}, \ldots, \frac{\xi_r}{\xi_j} \Big). \tag{4.9}$$

Let $f \simeq (f_1, \ldots, f_n)$ be the components of $f$ in the coordinates $(z_1, \ldots, z_n)$ (we suppose here $R$ so small that $f(\Delta_R) \subset \Omega$). It should be observed that $f$ is uniquely determined by its initial value $x$ and by the first $r$ components $(f_1, \ldots, f_r)$. Indeed, as $f'(t) \in V_{f(t)}$, we can recover the other components by integrating the system of ordinary differential equations

$$f_j'(t) = \sum_{1 \leqslant k \leqslant r} a_{jk}(f(t)) f_k'(t), \qquad j > r, \tag{4.10}$$

on a neighborhood of 0, with initial data $f(0) = x$. We denote by $m = m(f, t_0)$ the *multiplicity* of $f$ at any point $t_0 \in \Delta_R$, that is, $m(f, t_0)$ is the smallest integer $m \in \mathbb{N}^\star$ such that $f_j^{(m)}(t_0) \neq 0$ for some $j$. By (4.10), we can always suppose $j \in \{1, \ldots, r\}$, for example $f_r^{(m)}(t_0) \neq 0$. Then $f'(t) = (t - t_0)^{m-1} u(t)$ with $u_r(t_0) \neq 0$, and the lifting $\widetilde{f}$ is described in the coordinates of the affine chart $\xi_r \neq 0$ of $P(V)_{\restriction \Omega}$ by

$$\widetilde{f} \simeq \Big( f_1, \ldots, f_n; \frac{f_1'}{f_r'}, \ldots, \frac{f_{r-1}'}{f_r'} \Big). \tag{4.11}$$

We end this section with a few curvature computations. Assume that $V$ is equipped with a smooth hermitian metric $h$. Denote by $\nabla_h = \nabla_h' + \nabla_h''$ the associated Chern connection and by $\Theta_h(V) = \frac{i}{2\pi}\nabla_h^2$ its Chern curvature tensor. For every point $x_0 \in X$, there exists a "normalized" holomorphic frame $(e_\lambda)_{1\leqslant\lambda\leqslant r}$ on a neighborhood of $x_0$, such that

$$\langle e_\lambda, e_\mu\rangle_h = \delta_{\lambda\mu} - \sum_{1\leqslant j,k\leqslant n} c_{jk\lambda\mu} z_j \overline{z}_k + O(|z|^3), \tag{4.12}$$

with respect to any holomorphic coordinate system $(z_1, \ldots, z_n)$ centered at $x_0$. A computation of $d'\langle e_\lambda, e_\mu\rangle_h = \langle \nabla'_h e_\lambda, e_\mu\rangle_h$ and $\nabla_h^2 e_\lambda = d''\nabla'_h e_\lambda$ then gives

$$\nabla'_h e_\lambda = -\sum_{j,k,\mu} c_{jk\lambda\mu}\overline{z}_k\, dz_j \otimes e_\mu + O(|z|^2),$$

$$\Theta_h(V)_{x_0} = \frac{i}{2\pi}\sum_{j,k,\lambda,\mu} c_{jk\lambda\mu} dz_j \wedge d\overline{z}_k \otimes e_\lambda^\star \otimes e_\mu. \tag{4.13}$$

The above curvature tensor can also be viewed as a hermitian form on $T_X \otimes V$. In fact, one associates with $\Theta_h(V)$ the hermitian form $\langle\Theta_h(V)\rangle$ on $T_X \otimes V$ defined for all $(\zeta, v) \in T_X \times_X V$ by

$$\langle\Theta_h(V)\rangle(\zeta \otimes v) = \sum_{1\leqslant j,k\leqslant n,\ 1\leqslant\lambda,\mu\leqslant r} c_{jk\lambda\mu}\zeta_j\overline{\zeta}_k v_\lambda \overline{v}_\mu. \tag{4.14}$$

Let $h_1$ be the hermitian metric on the tautological line bundle $\mathcal{O}_{P(V)}(-1) \subset \pi^\star V$ induced by the metric $h$ of $V$. We compute the curvature $(1,1)$-form $\Theta_{h_1}(\mathcal{O}_{P(V)}(-1))$ at an arbitrary point $(x_0, [v_0]) \in P(V)$, in terms of $\Theta_h(V)$. For simplicity, we suppose that the frame $(e_\lambda)_{1\leqslant\lambda\leqslant r}$ has been chosen in such a way that $[e_r(x_0)] = [v_0]$ and $|v_0|_h = 1$. We get holomorphic local coordinates $(z_1, \ldots, z_n\,;\, \xi_1, \ldots, \xi_{r-1})$ on a neighborhood of $(x_0, [v_0])$ in $P(V)$ by assigning

$$(z_1, \ldots, z_n\,;\, \xi_1, \ldots, \xi_{r-1}) \longmapsto (z, [\xi_1 e_1(z) + \cdots + \xi_{r-1}e_{r-1}(z) + e_r(z)]) \in P(V).$$

Then the function

$$\eta(z, \xi) = \xi_1 e_1(z) + \cdots + \xi_{r-1}e_{r-1}(z) + e_r(z)$$

defines a holomorphic section of $\mathcal{O}_{P(V)}(-1)$ in a neighborhood of $(x_0, [v_0])$. By using the expansion (4.12) for $h$, we find

$$|\eta|_{h_1}^2 = |\eta|_h^2 = 1 + |\xi|^2 - \sum_{1\leqslant j,k\leqslant n} c_{jkrr} z_j\overline{z}_k + O((|z| + |\xi|)^3),$$

$$\Theta_{h_1}(\mathcal{O}_{P(V)}(-1))_{(x_0,[v_0])} = -\frac{i}{2\pi} d'd'' \log |\eta|_{h_1}^2$$

$$= \frac{i}{2\pi}\Big(\sum_{1\leqslant j,k\leqslant n} c_{jkrr} dz_j \wedge d\overline{z}_k - \sum_{1\leqslant\lambda\leqslant r-1} d\xi_\lambda \wedge d\overline{\xi}_\lambda\Big). \tag{4.15}$$

Now, the connection $\nabla_h$ on $V$ defines on $\widetilde{X} = P(V)$ a $C^\infty$ decomposition

$$T_{\widetilde{X}} = {}^H T_{\widetilde{X}} \oplus {}^V T_{\widetilde{X}}, \qquad {}^H T_{\widetilde{X},(x,[v])} \simeq T_{X,x}, \quad {}^V T_{\widetilde{X},(x,[v])} \simeq T_{P(V_x),[v]},$$

in horizontal and vertical components. With respect to this decomposition, (4.15) can be rewritten as

$$\langle\Theta_{h_1}(\mathcal{O}_{P(V)}(-1))\rangle_{(x_0,[v_0])}(\tau) = \langle\Theta_h(V)\rangle_{x_0}({}^H\tau \otimes v_0) - |{}^V\tau|_{\mathrm{FS}}^2 \tag{4.16}$$

where $|\ |_{\rm FS}$ is the Fubini-Study metric along the fibers $T_{P(V_x)}$. By definition of $\widetilde V$, we have $\widetilde V_{(x,[v])} \subset V_x \oplus T_{P(V_x),[v]}$ with respect to the decomposition. By this observation, if we equip $P(V)$ with the Fubini-Study metric rescaled by $\rho^2 > 0$, the metric $h$ on $V$ induces a canonical hermitian metric $\widetilde h_\rho$ on $\widetilde V$ such that

$$|w|^2_{\widetilde h_\rho} = |{}^H w|^2_h + \rho^2 |{}^V w|^2_h \qquad \text{for } w \in \widetilde V_{(x_0,[v_0])},$$

where ${}^H w \in \mathbb{C}\, v_0 \subset V_{x_0}$ and ${}^V w \in T_{P(V_{x_0}),[v_0]}$ is viewed as an element of $v_0^\perp \subset V_{x_0}$. A computation (left to the reader) gives the formula

$$\begin{aligned}\langle \Theta_{\widetilde h_\rho}(\widetilde V)\rangle_{(x_0,[v_0])}(\tau\otimes w) &= \langle \Theta_h(V)\rangle_{x_0}({}^H\tau\otimes v_0)\,\big(|{}^H w|^2_h - \rho^2|{}^V w|^2_h\big)\\ &\quad + \rho^2 \langle \Theta_h(V)\rangle_{x_0}({}^H\tau\otimes {}^V w)\\ &\quad + \rho^2\big(|\langle {}^V\tau, {}^V w\rangle_h|^2 + |{}^V\tau|^2_h|{}^V w|^2_h\big) - |{}^V\tau|^2_h |{}^H w|^2_h\\ &\quad + O(\rho)|\tau|^2_\omega |w|^2_{\widetilde h_\rho}, \qquad \tau \in T_{\widetilde X},\ w \in \widetilde V,\end{aligned} \tag{4.17}$$

where $|\tau|^2_\omega$ is computed from a fixed hermitian metric $\omega$ on $T_X$.

## 5. Jets of Curves and Semple Jet Bundles

Let $X$ be a complex $n$-dimensional manifold. Following ideas of Green-Griffiths [GrGr80], we let $J_k \to X$ be the bundle of $k$-jets of germs of parametrized curves in $X$, that is, the set of equivalence classes of holomorphic maps $f : (\mathbb{C}, 0) \to (X, x)$, with the equivalence relation $f \sim g$ if and only if all derivatives $f^{(j)}(0) = g^{(j)}(0)$ coincide for $0 \leqslant j \leqslant k$, when computed in some local coordinate system of $X$ near $x$. The projection map $J_k \to X$ is simply $f \mapsto f(0)$. If $(z_1, \ldots, z_n)$ are local holomorphic coordinates on an open set $\Omega \subset X$, the elements $f$ of any fiber $J_{k,x}$, $x \in \Omega$, can be seen as $\mathbb{C}^n$-valued maps

$$f = (f_1, \ldots, f_n) : (\mathbb{C}, 0) \to \Omega \subset \mathbb{C}^n,$$

and they are completetely determined by their Taylor expansion of order $k$ at $t = 0$

$$f(t) = x + t\, f'(0) + \frac{t^2}{2!} f''(0) + \cdots + \frac{t^k}{k!} f^{(k)}(0) + O(t^{k+1}).$$

In these coordinates, the fiber $J_{k,x}$ can thus be identified with the set of $k$-tuples of vectors $(f'(0), \ldots, f^{(k)}(0)) \in (\mathbb{C}^n)^k$. It follows that $J_k$ is a holomorphic fiber bundle with typical fiber $(\mathbb{C}^n)^k$ over $X$ (however, $J_k$ is not a vector bundle for $k \geqslant 2$, because of the nonlinearity of coordinate changes; see formula (6.2) in § 6).

According to the philosophy developed throughout this paper, we describe the concept of jet bundle in the general situation of complex directed manifolds. If $X$ is equipped with a holomorphic subbundle $V \subset T_X$, we associate to $V$ a $k$-jet bundle $J_k V$ as follows.

5.1. Definition. *Let $(X, V)$ be a complex directed manifold. We define the bundle $J_k V \to X$ to be the set of all $k$-jets of curves $f : (\mathbb{C}, 0) \to X$ which are tangent to $V$, i.e., such that $f'(t) \in V_{f(t)}$ for all $t$ in a neighborhood of $0$, together with the projection map $f \mapsto f(0)$ onto $X$.*

It is easy to check that $J_k V$ is actually a subbundle of $J_k$. In fact, by using (4.8) and (4.10), we see that the fibers $J_k V_x$ are parametrized by

$$\big((f_1'(0), \ldots, f_r'(0)); (f_1''(0), \ldots, f_r''(0)); \ldots ; (f_1^{(k)}(0), \ldots, f_r^{(k)}(0))\big) \in (\mathbb{C}^r)^k$$

for all $x \in \Omega$, hence $J_k V$ is a locally trivial $(\mathbb{C}^r)^k$-subbundle of $J_k$. □

We now describe a convenient process for constructing "projectivized jet bundles", which will later appear as natural quotients of our jet bundles $J_k V$ (or rather, as suitable desingularized compactifications of the quotients). Such spaces have already been considered since a long time, at least in the special case $X = \mathbb{P}^2$, $V = T_{\mathbb{P}^2}$ (see Gherardelli [Ghe41], Semple [Sem54]), and they have been mostly used as a tool for establishing enumerative formulas dealing with the order of contact of plane curves (see [Coll88], [CoKe94]); the article [ASS92] is also concerned with such generalizations of jet bundles*.

We define inductively the *projectivized $k$-jet bundle* $P_k V = X_k$ (or *Semple $k$-jet bundle*) and the associated subbundle $V_k \subset T_{X_k}$ by

$$(5.2) \qquad (X_0, V_0) = (X, V), \qquad (X_k, V_k) = (\widetilde{X}_{k-1}, \widetilde{V}_{k-1}).$$

In other words, $(P_k V, V_k) = (X_k, V_k)$ is obtained from $(X, V)$ by iterating $k$-times the lifting construction $(X, V) \mapsto (\widetilde{X}, \widetilde{V})$ described in § 4. By (4.2–4.7), we find

$$(5.3) \qquad \dim P_k V = n + k(r-1), \qquad \operatorname{rank} V_k = r,$$

together with exact sequences

$$(5.4) \qquad 0 \longrightarrow T_{P_k V / P_{k-1} V} \longrightarrow V_k \xrightarrow{(\pi_k)_\star} \mathcal{O}_{P_k V}(-1) \longrightarrow 0,$$

$$(5.4') \qquad 0 \longrightarrow \mathcal{O}_{P_k V} \longrightarrow \pi_k^\star V_{k-1} \otimes \mathcal{O}_{P_k V}(1) \longrightarrow T_{P_k V / P_{k-1} V} \longrightarrow 0.$$

where $\pi_k$ is the natural projection $\pi_k : P_k V \to P_{k-1} V$ and $(\pi_k)_\star$ its differential. Formula (4.4) yields

$$(5.5) \qquad \det V_k = \pi_k^\star \det V_{k-1} \otimes \mathcal{O}_{P_k V}(r-1).$$

Every non constant tangent trajectory $f : \Delta_R \to X$ of $(X, V)$ lifts to a well defined and unique tangent trajectory $f_{[k]} : \Delta_R \to P_k V$ of $(P_k V, V_k)$. Moreover, the derivative $f'_{[k-1]}$ gives rise to a section

$$(5.6) \qquad f'_{[k-1]} : T_{\Delta_R} \to f_{[k]}^\star \mathcal{O}_{P_k V}(-1).$$

In coordinates, one can compute $f_{[k]}$ in terms of its components in the various affine charts (4.9) occurring at each step: we get inductively

$$(5.7) \qquad f_{[k]} = (F_1, \ldots, F_N), \qquad f_{[k+1]} = \Big(F_1, \ldots, F_N, \frac{F'_{s_1}}{F'_{s_r}}, \ldots, \frac{F'_{s_{r-1}}}{F'_{s_r}}\Big)$$

where $N = n + k(r-1)$ and $\{s_1, \ldots, s_r\} \subset \{1, \ldots, N\}$. If $k \geqslant 1$, $\{s_1, \ldots, s_r\}$ contains the last $r-1$ indices of $\{1, \ldots, N\}$ corresponding to the "vertical" components of the projection $P_k V \to P_{k-1} V$, and in general, $s_r$ is an index such that $m(F_{s_r}, 0) = m(f_{[k]}, 0)$, that is, $F_{s_r}$ has the smallest vanishing order among all components $F_s$ ($s_r$ may be vertical or not, and the choice of $\{s_1, \ldots, s_r\}$ need not be unique).

---

* Very recently, a preprint [LaTh96] by Laksov and Thorup has also appeared, dealing in depth with algebraic-theoretic properties of jet differentials. The formalism of "higher order" differentials has been part of the mathematical folklore during the 18th and 19th centuries (without too much concern, in those times, on the existence of precise definitions!). During the 20th century, this formalism almost disappeared, before getting revived in several ways. See e.g. the interested article by P.A. Meyer [Mey89], which was originally motivated by applications to probability theory.

By definition, there is a canonical injection $\mathcal{O}_{P_kV}(-1) \hookrightarrow \pi_k^\star V_{k-1}$, and a composition with the projection $(\pi_{k-1})_\star$ (analogue for order $k-1$ of the arrow $(\pi_k)_\star$ in sequence (5.4)) yields for all $k \geqslant 2$ a canonical line bundle morphism

$$\mathcal{O}_{P_kV}(-1) \hookrightarrow \pi_k^\star V_{k-1} \xrightarrow{(\pi_k)^\star(\pi_{k-1})_\star} \pi_k^\star \mathcal{O}_{P_{k-1}V}(-1), \tag{5.8}$$

which admits precisely $D_k = P(T_{P_{k-1}V/P_{k-2}V}) \subset P(V_{k-1}) = P_kV$ as its zero divisor (clearly, $D_k$ is a hyperplane subbundle of $P_kV$). Hence we find

$$\mathcal{O}_{P_kV}(1) = \pi_k^\star \mathcal{O}_{P_{k-1}V}(1) \otimes \mathcal{O}(D_k). \tag{5.9}$$

Now, we consider the composition of projections

$$\pi_{j,k} = \pi_{j+1} \circ \cdots \circ \pi_{k-1} \circ \pi_k : P_kV \longrightarrow P_jV. \tag{5.10}$$

Then $\pi_{0,k} : P_kV \to X = P_0V$ is a locally trivial holomorphic fiber bundle over $X$, and the fibers $P_kV_x = \pi_{0,k}^{-1}(x)$ are $k$-stage towers of $\mathbb{P}^{r-1}$-bundles. Since we have (in both directions) morphisms $(\mathbb{C}^r, T_{\mathbb{C}^r}) \leftrightarrow (X,V)$ of directed manifolds which are bijective on the level of bundle morphisms, the fibers are all isomorphic to a "universal" nonsingular projective algebraic variety of dimension $k(r-1)$ which we will denote by $\mathbb{R}_{r,k}$ ; it is not hard to see that $\mathbb{R}_{r,k}$ is rational (as will indeed follow from the proof of Theorem 6.8 below). The following Proposition will help us to understand a little bit more about the geometric structure of $P_kV$. As usual, we define the *multiplicity* $m(f,t_0)$ of a curve $f : \Delta_R \to X$ at a point $t \in \Delta_R$ to be the smallest integer $s \in \mathbb{N}^\star$ such that $f^{(s)}(t_0) \neq 0$, i.e., the largest $s$ such that $\delta(f(t), f(t_0)) = O(|t-t_0|^s)$ for any hermitian or riemannian geodesic distance $\delta$ on $X$. As $f_{[k-1]} = \pi_k \circ f_{[k]}$, it is clear that the sequence $m(f_{[k]}, t)$ is non increasing with $k$.

5.11. PROPOSITION. *Let $f : (\mathbb{C},0) \to X$ be a non constant germ of curve tangent to $V$. Then for all $j \geqslant 2$ we have $m(f_{[j-2]},0) \geqslant m(f_{[j-1]},0)$ and the inequality is strict if and only if $f_{[j]}(0) \in D_j$. Conversely, if $w \in P_kV$ is an arbitrary element and $m_0 \geqslant m_1 \geqslant \cdots \geqslant m_{k-1} \geqslant 1$ is a sequence of integers with the property that*

$$\forall j \in \{2,\ldots,k\}, \qquad m_{j-2} > m_{j-1} \quad \textit{if and only if } \pi_{j,k}(w) \in D_j,$$

*there exists a germ of curve $f : (\mathbb{C},0) \to X$ tangent to $V$ such that $f_{[k]}(0) = w$ and $m(f_{[j]},0) = m_j$ for all $j \in \{0,\ldots,k-1\}$.*

PROOF. i) Suppose first that $f$ is given and put $m_j = m(f_{[j]},0)$. By definition, we have $f_{[j]} = (f_{[j-1]}, [u_{j-1}])$ where $f'_{[j-1]}(t) = t^{m_{j-1}-1}u_{j-1}(t) \in V_{j-1}$, $u_{j-1}(0) \neq 0$. By taking the composition with the differential of the projection $\pi_{j-1} : P_{j-1}V \to P_{j-2}V$, we find $f'_{[j-2]}(t) = t^{m_{j-1}-1}(\pi_{j-1})_\star u_{j-1}(t)$. Therefore

$$m_{j-2} = m_{j-1} + \mathrm{ord}_{t=0}(\pi_{j-1})_\star u_{j-1}(t),$$

and so $m_{j-2} > m_{j-1}$ if and only if $(\pi_{j-1})_\star u_{j-1}(0) = 0$, that is, if and only if $u_{j-1}(0) \in T_{P_{j-1}V/P_{j-2}V}$, or equivalently $f_{[j]}(0) = (f_{[j-1]}(0), [u_{j-1}(0)]) \in D_j$.

ii) Suppose now that $w \in P_kV$ and $m_0, \ldots, m_{k-1}$ are given. We denote by $w_{j+1} = (w_j, [\eta_j])$, $w_j \in P_jV$, $\eta_j \in V_j$, the projection of $w$ to $P_{j+1}V$. Fix coordinates $(z_1, \ldots, z_n)$ on $X$ centered at $w_0$ such that the $r$-th component $\eta_{0,r}$ of $\eta_0$ is non zero. We prove the existence of the germ $f$ by induction on $k$, in the form of a Taylor expansion

$$f(t) = a_0 + t\,a_1 + \cdots + t^{d_k} a_{d_k} + O(t^{d_k+1}), \qquad d_k = m_0 + m_1 + \cdots + m_{k-1}.$$

If $k = 1$ and $w = (w_0, [\eta_0]) \in P_1 V_x$, we simply take $f(t) = w_0 + t^{m_0}\eta_0 + O(t^{m_0+1})$. In general, the induction hypothesis applied to $P_k V = P_{k-1}(V_1)$ over $X_1 = P_1 V$ yields a curve $g : (\mathbb{C}, 0) \to X_1$ such that $g_{[k-1]} = w$ and $m(g_{[j]}, 0) = m_{j+1}$ for $0 \leqslant j \leqslant k-2$. If $w_2 \notin D_2$, then $[g'_{[1]}(0)] = [\eta_1]$ is not vertical, thus $f = \pi_1 \circ g$ satisfies $m(f, 0) = m(g, 0) = m_1 = m_0$ and we are done.

If $w_2 \in D_2$, we express $g = (G_1, \ldots, G_n; G_{n+1}, \ldots, G_{n+r-1})$ as a Taylor expansion of order $m_1 + \cdots + m_{k-1}$ in the coordinates (4.9) of the affine chart $\xi_r \neq 0$. As $\eta_1 = \lim_{t\to 0} g'(t)/t^{m_1-1}$ is vertical, we must have $m(G_s, 0) > m_1$ for $1 \leqslant j \leqslant n$. It follows from (5.7) that $G_1, \ldots, G_n$ are never involved in the calculation of the liftings $g_{[j]}$. We can therefore replace $g$ by $f \simeq (f_1, \ldots, f_n)$ where $f_r(t) = t^{m_0}$ and $f_1, \ldots, f_{r-1}$ are obtained by integrating the equations $f'_j(t)/f'_r(t) = G_{n+j}(t)$, i.e., $f'_j(t) = m_0 t^{m_0-1} G_{n+j}(t)$, while $f_{r+1}, \ldots, f_n$ are obtained by integrating (4.10). We then get the desired Taylor expansion of order $d_k$ for $f$. □

Since we can always take $m_{k-1} = 1$ without restriction, we get in particular:

5.12. COROLLARY. *Let $w \in P_k V$ be an arbitrary element. Then there is a germ of curve $f : (\mathbb{C}, 0) \to X$ such that $f_{[k]}(0) = w$ and $f'_{[k-1]}(0) \neq 0$ (thus the liftings $f_{[k-1]}$ and $f_{[k]}$ are regular germs of curve). Moreover, if $w_0 \in P_k V$ and $w$ is taken in a sufficiently small neighborhood of $w_0$, then the germ $f = f_w$ can be taken to depend holomorphically on $w$.*

PROOF. Only the holomorphic dependence of $f_w$ with respect to $w$ has to be guaranteed. If $f_{w_0}$ is a solution for $w = w_0$, we observe that $(f_{w_0})'_{[k]}$ is a non vanishing section of $V_k$ along the regular curve defined by $(f_{w_0})_{[k]}$ in $P_k V$. We can thus find a non vanishing section $\xi$ of $V_k$ on a neighborhood of $w_0$ in $P_k V$ such that $\xi = (f_{w_0})'_{[k]}$ along that curve. We define $t \mapsto F_w(t)$ to be the trajectory of $\xi$ with initial point $w$, and we put $f_w = \pi_{0,k} \circ F_w$. Then $f_w$ is the required family of germs. □

Now, we can take $f : (\mathbb{C}, 0) \to X$ to be regular at the origin (by this, we mean $f'(0) \neq 0$) if and only if $m_0 = m_1 = \cdots = m_{k-1} = 1$, which is possible by Proposition 5.11 if and only if $w \in P_k V$ is such that $\pi_{j,k}(w) \notin D_j$ for all $j \in \{2, \ldots, k\}$. For this reason, we define

$$\tag{5.13} \begin{aligned} P_k V^{\mathrm{reg}} &= \bigcap_{2 \leqslant j \leqslant k} \pi_{j,k}^{-1}(P_j V \smallsetminus D_j), \\ P_k V^{\mathrm{sing}} &= \bigcup_{2 \leqslant j \leqslant k} \pi_{j,k}^{-1}(D_j) = P_k V \smallsetminus P_k V^{\mathrm{reg}}, \end{aligned}$$

in other words, $P_k V^{\mathrm{reg}}$ is the set of values $f_{[k]}(0)$ reached by all regular germs of curves $f$. One should take care however that there are singular germs which reach the same points $f_{[k]}(0) \in P_k V^{\mathrm{reg}}$, e.g., any $s$-sheeted covering $t \mapsto f(t^s)$. On the other hand, if $w \in P_k V^{\mathrm{sing}}$, we can reach $w$ by a germ $f$ with $m_0 = m(f, 0)$ as large as we want.

5.14. COROLLARY. *Let $w \in P_k V^{\mathrm{sing}}$ be given, and let $m_0 \in \mathbb{N}$ be an arbitrary integer larger than the number of components $D_j$ such that $\pi_{j,k}(w) \in D_j$. Then there is a germ of curve $f : (\mathbb{C},0) \to X$ with multiplicity $m(f,0) = m_0$ at the origin, such that $f_{[k]}(0) = w$ and $f'_{[k-1]}(0) \neq 0$.*

## 6. Jet Differentials

Following Green-Griffiths [GrGr80], we now introduce the concept of jet differential. This concept gives an intrinsic way of describing holomorphic differential equations that a germ of curve $f : (\mathbb{C},0) \to X$ may satisfy. In the sequel, we fix a directed manifold $(X, V)$ and suppose implicitly that all germs $f$ are tangent to $V$.

Let $\mathbb{G}_k$ be the group of germs of $k$-jets of biholomorphisms of $(\mathbb{C},0)$, that is, the group of germs of biholomorphic maps

$$t \mapsto \varphi(t) = a_1 t + a_2 t^2 + \cdots + a_k t^k, \qquad a_1 \in \mathbb{C}^\star,\ a_j \in \mathbb{C},\ j \geqslant 2,$$

in which the composition law is taken modulo terms $t^j$ of degree $j > k$. Then $\mathbb{G}_k$ is a $k$-dimensional nilpotent complex Lie group, which admits a natural fiberwise right action on $J_k V$. The action consists of reparametrizing $k$-jets of maps $f : (\mathbb{C},0) \to X$ by a biholomorphic change of parameter $\varphi : (\mathbb{C},0) \to (\mathbb{C},0)$, that is, $(f, \varphi) \mapsto f \circ \varphi$. There is an exact sequence of groups

$$1 \to \mathbb{G}'_k \to \mathbb{G}_k \to \mathbb{C}^\star \to 1$$

where $\mathbb{G}_k \to \mathbb{C}^\star$ is the obvious morphism $\varphi \mapsto \varphi'(0)$, and $\mathbb{G}'_k = [\mathbb{G}_k, \mathbb{G}_k]$ is the group of $k$-jets of biholomorphisms tangent to the identity. Moreover, the subgroup $\mathbb{H} \simeq \mathbb{C}^\star$ of homotheties $\varphi(t) = \lambda t$ is a (non normal) subgroup of $\mathbb{G}_k$, and we have a semidirect decomposition $\mathbb{G}_k = \mathbb{G}'_k \ltimes \mathbb{H}$. The corresponding action on $k$-jets is described in coordinates by

$$\lambda \cdot (f', f'', \ldots, f^{(k)}) = (\lambda f', \lambda^2 f'', \ldots, \lambda^k f^{(k)}).$$

Following [GrGr80], we introduce the vector bundle $E^{\mathrm{GG}}_{k,m} V^\star \to X$ whose fibers are complex valued polynomials $Q(f', f'', \ldots, f^{(k)})$ on the fibers of $J_k V$, of weighted degree $m$ with respect to the $\mathbb{C}^\star$ action defined by $H$, that is, such that

$$Q(\lambda f', \lambda^2 f'', \ldots, \lambda^k f^{(k)}) = \lambda^m Q(f', f'', \ldots, f^{(k)}) \tag{6.1}$$

for all $\lambda \in \mathbb{C}^\star$ and $(f', f'', \ldots, f^{(k)}) \in J_k V$. Here we view $(f', f'', \ldots, f^{(k)})$ as indeterminates with components

$$\big((f'_1, \ldots, f'_r); (f''_1, \ldots, f''_r); \ldots; (f^{(k)}_1, \ldots, f^{(k)}_r)\big) \in (\mathbb{C}^r)^k.$$

Notice that the concept of polynomial on the fibers of $J_k V$ makes sense, for all coordinate changes $z \mapsto w = \Psi(z)$ on $X$ induce polynomial transition automorphisms on the fibers of $J_k V$, given by a formula

$$(\Psi \circ f)^{(j)} = \Psi'(f) \cdot f^{(j)} + \sum_{s=2}^{s=j} \sum_{j_1 + j_2 + \cdots + j_s = j} c_{j_1 \ldots j_s} \Psi^{(s)}(f) \cdot (f^{(j_1)}, \ldots, f^{(j_s)}) \tag{6.2}$$

with suitable integer constants $c_{j_1 \ldots j_s}$ (this is easily checked by induction on $s$). In the "absolute case" $V = T_X$, we simply write $E^{\mathrm{GG}}_{k,m} T^\star_X = E^{\mathrm{GG}}_{k,m}$. If $V \subset W \subset T_X$ are holomorphic subbundles, there are natural inclusions

$$J_k V \subset J_k W \subset J_k, \qquad P_k V \subset P_k W \subset P_k.$$

The restriction morphisms induce surjective arrows

$$E_{k,m}^{\mathrm{GG}} \to E_{k,m}^{\mathrm{GG}} W^\star \to E_{k,m}^{\mathrm{GG}} V^\star,$$

in particular $E_{k,m}^{\mathrm{GG}} V^\star$ can be seen as a quotient of $E_{k,m}^{\mathrm{GG}}$. (The notation $V^\star$ is used here to make the contravariance property implicit from the notation).

If $Q \in E_{k,m}^{\mathrm{GG}} V^\star$ is decomposed into multihomogeneous components of multidegree $(\ell_1, \ell_2, \ldots, \ell_k)$ in $f', f'', \ldots, f^{(k)}$ (the decomposition is of course coordinate dependent), these multidegrees must satisfy the relation

$$\ell_1 + 2\ell_2 + \cdots + k\ell_k = m.$$

The bundle $E_{k,m}^{\mathrm{GG}} V^\star$ will be called the *bundle of jet differentials of order $k$ and weighted degree $m$*. It is clear from (6.2) that a coordinate change $f \mapsto \Psi \circ f$ transforms every monomial $(f^{(\bullet)})^\ell = (f')^{\ell_1}(f'')^{\ell_2} \cdots (f^{(k)})^{\ell_k}$ of partial weighted degree $|\ell|_s := \ell_1 + 2\ell_2 + \cdots + s\ell_s$, $1 \leqslant s \leqslant k$, into a polynomial $((\Psi \circ f)^{(\bullet)})^\ell$ in $(f', f'', \ldots, f^{(k)})$ which has the same partial weighted degree of order $s$ if $\ell_{s+1} = \cdots = \ell_k = 0$, and a larger or equal partial degree of order $s$ otherwise. Hence, for each $s = 1, \ldots, k$, we get a well defined (i.e., coordinate invariant) decreasing filtration $F_s^\bullet$ on $E_{k,m}^{\mathrm{GG}} V^\star$ as follows:

$$(6.3) \qquad F_s^p(E_{k,m}^{\mathrm{GG}} V^\star) = \left\{ \begin{matrix} Q(f', f'', \ldots, f^{(k)}) \in E_{k,m}^{\mathrm{GG}} V^\star \text{ involving} \\ \text{only monomials } (f^{(\bullet)})^\ell \text{ with } |\ell|_s \geqslant p \end{matrix} \right\}, \qquad \forall p \in \mathbb{N}.$$

The graded terms $\mathrm{Gr}_{k-1}^p(E_{k,m}^{\mathrm{GG}} V^\star)$ associated with the filtration $F_{k-1}^p(E_{k,m}^{\mathrm{GG}} V^\star)$ are precisely the homogeneous polynomials $Q(f', \ldots, f^{(k)})$ whose monomials $(f^\bullet)^\ell$ all have partial weighted degree $|\ell|_{k-1} = p$ (hence their degree $\ell_k$ in $f^{(k)}$ is such that $m - p = k\ell_k$, and $\mathrm{Gr}_{k-1}^p(E_{k,m}^{\mathrm{GG}} V^\star) = 0$ unless $k|m-p$). The transition automorphisms of the graded bundle are induced by coordinate changes $f \mapsto \Psi \circ f$, and they are described by substituting the arguments of $Q(f', \ldots, f^{(k)})$ according to formula (6.2), namely $f^{(j)} \mapsto (\Psi \circ f)^{(j)}$ for $j < k$, and $f^{(k)} \mapsto \Psi'(f) \circ f^{(k)}$ for $j = k$ (when $j = k$, the other terms fall in the next stage $F_{k-1}^{p+1}$ of the filtration). Therefore $f^{(k)}$ behaves as an element of $V \subset T_X$ under coordinate changes. We thus find

$$(6.4) \qquad G_{k-1}^{m-k\ell_k}(E_{k,m}^{\mathrm{GG}} V^\star) = E_{k-1,m-k\ell_k}^{\mathrm{GG}} V^\star \otimes S^{\ell_k} V^\star.$$

Combining all filtrations $F_s^\bullet$ together, we find inductively a filtration $F^\bullet$ on $E_{k,m}^{\mathrm{GG}} V^\star$ such that the graded terms are

$$(6.5) \qquad \mathrm{Gr}^\ell(E_{k,m}^{\mathrm{GG}} V^\star) = S^{\ell_1} V^\star \otimes S^{\ell_2} V^\star \otimes \cdots \otimes S^{\ell_k} V^\star, \qquad \ell \in \mathbb{N}^k, \quad |\ell|_k = m.$$

The bundles $E_{k,m}^{\mathrm{GG}} V^\star$ have other interesting properties. In fact,

$$E_{k,\bullet}^{\mathrm{GG}} V^\star := \bigoplus_{m \geqslant 0} E_{k,m}^{\mathrm{GG}} V^\star$$

is in a natural way a bundle of graded algebras (the product is obtained simply by taking products of polynomials). There are natural inclusions $E_{k,\bullet}^{\mathrm{GG}} V^\star \subset E_{k+1,\bullet}^{\mathrm{GG}} V^\star$ of algebras, hence $E_{\infty,\bullet}^{\mathrm{GG}} V^\star = \bigcup_{k \geqslant 0} E_{k,\bullet}^{\mathrm{GG}} V^\star$ is also an algebra. Moreover, the sheaf of holomorphic sections $\mathcal{O}(E^{\mathrm{GG}}_{\infty,\bullet}V^\star)$ admits a canonical derivation $\nabla$ given by a collection of $\mathbb{C}$-linear maps

$$\nabla : \mathcal{O}(E^{\mathrm{GG}}_{k,m}V^\star) \to \mathcal{O}(E^{\mathrm{GG}}_{k+1,m+1}V^\star), \tag{6.6}$$

constructed in the following way. A holomorphic section of $E^{\mathrm{GG}}_{k,m}V^\star$ on a coordinate open set $\Omega \subset X$ can be seen as a differential operator on the space of germs $f : (\mathbb{C},0) \to \Omega$ of the form

$$Q(f) = \sum_{|\alpha_1|+2|\alpha_2|+\cdots+k|\alpha_k|=m} a_{\alpha_1\ldots\alpha_k}(f)\,(f')^{\alpha_1}(f'')^{\alpha_2}\cdots(f^{(k)})^{\alpha_k}$$

in which the coefficients $a_{\alpha_1\ldots\alpha_k}$ are holomorphic functions on $\Omega$. Then $\nabla Q$ is given by the formal derivative $(\nabla Q)(f)(t) = d(Q(f))/dt$ with respect to the 1-dimensional parameter $t$ in $f(t)$. For example, in dimension 2, if $Q \in H^0(\Omega, \mathcal{O}(E^{\mathrm{GG}}_{2,4}))$ is the section of weighted degree 4

$$Q(f) = a(f_1,f_2)\,f_1'^3 f_2' + b(f_1,f_2)\,f_1''^2,$$

we find that $\nabla Q \in H^0(\Omega, \mathcal{O}(E^{\mathrm{GG}}_{3,5}))$ is given by

$$\begin{aligned}(\nabla Q)(f) &= \frac{\partial a}{\partial z_1}(f_1,f_2)\,f_1'^4 f_2' + \frac{\partial a}{\partial z_2}(f_1,f_2)\,f_1'^3 f_2'^2 + \frac{\partial b}{\partial z_1}(f_1,f_2)\,f_1' f_1''^2 \\ &\quad + \frac{\partial b}{\partial z_2}(f_1,f_2)\,f_2' f_1''^2 + a(f_1,f_2)\,\big(3f_1'^2 f_1'' f_2' + f_1'^3 f_2''\big) + b(f_1,f_2)\,2f_1''f_1'''.\end{aligned}$$

Associated with the graded algebra bundle $E^{\mathrm{GG}}_{k,\bullet}V^\star$, we have an analytic fiber bundle $\mathrm{Proj}(E^{\mathrm{GG}}_{k,\bullet}V^\star) = J_kV^{\mathrm{nc}}/\mathbb{C}^\star$ over $X$, which has weighted projective spaces $\mathbb{P}(r,\ldots,r\,;1,2,\ldots,k)$ as fibers (these weighted projective spaces are singular for $k>1$, but they only have quotient singularities, see [Dol81]; here $J_kV^{\mathrm{nc}}$ denotes the set of non constant jets of order $k$; we refer e.g. to Hartshorne's book [Har77] for a definition of the Proj fonctor). However, we are not really interested in the bundles $J_kV^{\mathrm{nc}}/\mathbb{C}^\star$ themselves, but rather on their quotients $J_kV^{\mathrm{nc}}/\mathbb{G}_k$ (would such nice complex space quotients exist!). We will see that the Semple bundle $P_kV$ constructed in § 5 plays the role of such a quotient. First we introduce a canonical bundle subalgebra of $E^{\mathrm{GG}}_{k,\bullet}V^\star$.

6.7. Definition. *We introduce a subbundle $E_{k,m}V^\star \subset E^{\mathrm{GG}}_{k,m}V^\star$, called the bundle of invariant jet differentials of order $k$ and degree $m$, defined as follows: $E_{k,m}V^\star$ is the set of polynomial differential operators $Q(f',f'',\ldots,f^{(k)})$ which are invariant under arbitrary changes of parametrization, i.e., for every $\varphi \in \mathbb{G}_k$*

$$Q\big((f\circ\varphi)',(f\circ\varphi)'',\ldots,(f\circ\varphi)^{(k)}\big) = \varphi'(0)^m Q(f',f'',\ldots,f^{(k)}).$$

Alternatively, $E_{k,m}V^\star = (E^{\mathrm{GG}}_{k,m}V^\star)^{\mathbb{G}'_k}$ is the set of invariants of $E^{\mathrm{GG}}_{k,m}V^\star$ under the action of $\mathbb{G}'_k$. Clearly, $E_{\infty,\bullet}V^\star = \bigcup_{k\geqslant 0}\bigoplus_{m\geqslant 0} E_{k,m}V^\star$ is a subalgebra of $E^{\mathrm{GG}}_{k,m}V^\star$ (observe however that this algebra is not invariant under the derivation $\nabla$, since e.g. $f_j'' = \nabla f_j$ is not an invariant polynomial). In addition to this, there are natural induced filtrations $F^p_s(E_{k,m}V^\star) = E_{k,m}V^\star \cap F^p_s(E^{\mathrm{GG}}_{k,m}V^\star)$ (all locally trivial over $X$). These induced filtrations will play an important role in Section 12.

6.8. THEOREM. *Suppose that $V$ has rank $r \geqslant 2$. Let $\pi_{0,k} : P_kV \longrightarrow X$ be the Semple jet bundles constructed in section 5, and let $J_kV^{\rm reg}$ be the bundle of regular $k$-jets of maps $f : (\mathbb{C}, 0) \to X$, that is, jets $f$ such that $f'(0) \neq 0$.*

i) *The quotient $J_kV^{\rm reg}/\mathbb{G}_k$ has the structure of a locally trivial bundle over $X$, and there is a holomorphic embedding $J_kV^{\rm reg}/\mathbb{G}_k \hookrightarrow P_kV$ over $X$, which identifies $J_kV^{\rm reg}/\mathbb{G}_k$ with $P_kV^{\rm reg}$ (thus $P_kV$ is a relative compactification of $J_kV^{\rm reg}/\mathbb{G}_k$ over $X$).*

ii) *The direct image sheaf*

$$(\pi_{0,k})_\star \mathcal{O}_{P_kV}(m) \simeq \mathcal{O}(E_{k,m}V^\star)$$

*can be identified with the sheaf of holomorphic sections of $E_{k,m}V^\star$.*

iii) *For every $m > 0$, the relative base locus of the linear system $|\mathcal{O}_{P_kV}(m)|$ is equal to the set $P_kV^{\rm sing}$ of singular $k$-jets. Moreover, $\mathcal{O}_{P_kV}(1)$ is relatively big over $X$.*

PROOF. i) For $f \in J_kV^{\rm reg}$, the lifting $\widetilde{f}$ is obtained by taking the derivative $(f, [f'])$ without any cancellation of zeroes in $f'$, hence we get a uniquely defined $(k-1)$-jet $\widetilde{f} : (\mathbb{C}, 0) \to \widetilde{X}$. Inductively, we get a well defined $(k-j)$-jet $f_{[j]}$ in $P_jV$, and the value $f_{[k]}(0)$ is independent of the choice of the representative $f$ for the $k$-jet. As the lifting process commutes with reparametrization, i.e., $(f \circ \varphi)^\sim = \widetilde{f} \circ \varphi$ and more generally $(f \circ \varphi)_{[k]} = f_{[k]} \circ \varphi$, we conclude that there is a well defined set-theoretic map

$$J_kV^{\rm reg}/\mathbb{G}_k \to P_kV^{\rm reg}, \qquad f \bmod \mathbb{G}_k \mapsto f_{[k]}(0).$$

This map is better understood in coordinates as follows. Fix coordinates $(z_1, \ldots, z_n)$ near a point $x_0 \in X$, such that $V_{x_0} = \mathrm{Vect}(\partial/\partial z_1, \ldots, \partial/\partial z_r)$. Let $f = (f_1, \ldots, f_n)$ be a regular $k$-jet tangent to $V$. Then there exists $i \in \{1, 2, \ldots, r\}$ such that $f_i'(0) \neq 0$, and there is a unique reparametrization $t = \varphi(\tau)$ such that $f \circ \varphi = g = (g_1, g_2, \ldots, g_n)$ with $g_i(\tau) = \tau$ (we just express the curve as a graph over the $z_i$-axis, by means of a change of parameter $\tau = f_i(t)$, i.e. $t = \varphi(\tau) = f_i^{-1}(\tau)$). Suppose $i = r$ for the simplicity of notation. The space $P_kV$ is a $k$-stage tower of $\mathbb{P}^{r-1}$-bundles. In the corresponding inhomogeneous coordinates on these $\mathbb{P}^{r-1}$'s, the point $f_{[k]}(0)$ is given by the collection of derivatives

$$\big((g_1'(0), \ldots, g_{r-1}'(0)); (g_1''(0), \ldots, g_{r-1}''(0)); \ldots ; (g_1^{(k)}(0), \ldots, g_{r-1}^{(k)}(0))\big).$$

[Recall that the other components $(g_{r+1}, \ldots, g_n)$ can be recovered from $(g_1, \ldots, g_r)$ by integrating the differential system (4.10)]. Thus the map $J_kV^{\rm reg}/\mathbb{G}_k \to P_kV$ is a bijection onto $P_kV^{\rm reg}$, and the fibers of these isomorphic bundles can be seen as unions of $r$ affine charts $\simeq (\mathbb{C}^{r-1})^k$, associated with each choice of the axis $z_i$ used to describe the curve as a graph. The change of parameter formula $\frac{d}{d\tau} = \frac{1}{f_r'(t)} \frac{d}{dt}$ expresses all derivatives $g_i^{(j)}(\tau) = d^j g_i/d\tau^j$ in terms of the derivatives $f_i^{(j)}(t) =$

$d^j f_i/dt^j$

$$(6.9)\quad \begin{aligned}(g'_1,\ldots,g'_{r-1}) &= \Big(\frac{f'_1}{f'_r},\ldots,\frac{f'_{r-1}}{f'_r}\Big);\\ (g''_1,\ldots,g''_{r-1}) &= \Big(\frac{f''_1 f'_r - f''_r f'_1}{f'^3_r},\ldots,\frac{f''_{r-1}f'_r - f''_r f'_{r-1}}{f'^3_r}\Big);\ \ldots;\\ (g^{(k)}_1,\ldots,g^{(k)}_{r-1}) &= \Big(\frac{f^{(k)}_1 f'_r - f^{(k)}_r f'_1}{f'^{k+1}_r},\ldots,\frac{f^{(k)}_{r-1}f'_r - f^{(k)}_r f'_{r-1}}{f'^{k+1}_r}\Big) + (\text{order} < k).\end{aligned}$$

Also, it is easy to check that $f'^{2k-1}_r g^{(k)}_i$ is an invariant polynomial in $f', f'', \ldots, f^{(k)}$ of total degree $2k-1$, i.e., a section of $E_{k,2k-1}$.
ii) Since the bundles $P_kV$ and $E_{k,m}V^\star$ are both locally trivial over $X$, it is sufficient to identify sections $\sigma$ of $\mathcal{O}_{P_kV}(m)$ over a fiber $P_kV_x = \pi^{-1}_{0,k}(x)$ with the fiber $E_{k,m}V^\star_x$, at any point $x \in X$. Let $f \in J_kV^{\rm reg}_x$ be a regular $k$-jet at $x$. By (5.6), the derivative $f'_{[k-1]}(0)$ defines an element of the fiber of $\mathcal{O}_{P_kV}(-1)$ at $f_{[k]}(0) \in P_kV$. Hence we get a well defined complex valued operator

$$(6.10)\qquad Q(f', f'', \ldots, f^{(k)}) = \sigma(f_{[k]}(0)) \cdot (f'_{[k-1]}(0))^m.$$

Clearly, $Q$ is holomorphic on $J_kV^{\rm reg}_x$ (by the holomorphicity of $\sigma$), and the $\mathbb{G}_k$-invariance condition of Def. 6.7 is satisfied since $f_{[k]}(0)$ does not depend on reparametrization and $(f\circ\varphi)'_{[k-1]}(0) = f'_{[k-1]}(0)\varphi'(0)$. Now, $J_kV^{\rm reg}_x$ is the complement of a linear subspace of codimension $n$ in $J_kV_x$, hence $Q$ extends holomorphically to all of $J_kV_x \simeq (\mathbb{C}^r)^k$ by Riemann's extension theorem (here we use the hypothesis $r \geqslant 2$; if $r = 1$, the situation is anyway not interesting since $P_kV = X$ for all $k$). Thus $Q$ admits an everywhere convergent power series

$$Q(f', f'', \ldots, f^{(k)}) = \sum_{\alpha_1,\alpha_2,\ldots,\alpha_k \in \mathbb{N}^r} a_{\alpha_1\ldots\alpha_k} (f')^{\alpha_1}(f'')^{\alpha_2}\cdots(f^{(k)})^{\alpha_k}.$$

The $\mathbb{G}_k$-invariance (6.7) implies in particular that $Q$ must be multihomogeneous in the sense of (6.1), and thus $Q$ must be a polynomial. We conclude that $Q \in E_{k,m}V^\star_x$, as desired.

Conversely, Corollary 5.12 implies that there is a holomorphic family of germs $f_w : (\mathbb{C},0) \to X$ such that $(f_w)_{[k]}(0) = w$ and $(f_w)'_{[k-1]}(0) \neq 0$, for all $w$ in a neighborhood of any given point $w_0 \in P_kV_x$. Then every $Q \in E_{k,m}V^\star_x$ yields a holomorphic section $\sigma$ of $\mathcal{O}_{P_kV}(m)$ over the fiber $P_kV_x$ by putting

$$(6.11)\qquad \sigma(w) = Q(f'_w, f''_w, \ldots, f^{(k)}_w)(0)\,\big((f_w)'_{[k-1]}(0)\big)^{-m}.$$

iii) By what we saw in i-ii), every section $\sigma$ of $\mathcal{O}_{P_kV}(m)$ over the fiber $P_kV_x$ is given by a polynomial $Q \in E_{k,m}V^\star_x$, and this polynomial can be expressed on the Zariski open chart $f'_r \neq 0$ of $P_kV^{\rm reg}_x$ as

$$(6.12)\qquad Q(f', f'', \ldots, f^{(k)}) = f'^m_r \widehat{Q}(g', g'', \ldots, g^{(k)}),$$

where $\widehat{Q}$ is a polynomial and $g$ is the reparametrization of $f$ such that $g_r(\tau) = \tau$. In fact $\widehat{Q}$ is obtained from $Q$ by substituting $f'_r = 1$ and $f^{(j)}_r = 0$ for $j \geqslant 2$, and conversely $Q$ can be recovered easily from $\widehat{Q}$ by using the substitutions (6.9).

In this context, the jet differentials $f \mapsto f'_1, \ldots, f \mapsto f'_r$ can be viewed as sections of $\mathcal{O}_{P_kV}(1)$ on a neighborhood of the fiber $P_kV_x$. Since these sections vanish exactly on $P_kV^{\rm sing}$, the relative base locus of $\mathcal{O}_{P_kV}(m)$ is contained in $P_kV^{\rm sing}$ for every $m>0$. We see that $\mathcal{O}_{P_kV}(1)$ is big by considering the sections of $\mathcal{O}_{P_kV}(2k-1)$ associated with the polynomials $Q(f',\ldots,f^{(k)})=f_r'^{2k-1}g_i^{(j)}$, $1\leqslant i\leqslant r-1$, $1\leqslant j\leqslant k$; indeed, these sections separate all points in the open chart $f_r'\neq 0$ of $P_kV_x^{\rm reg}$.

Now, we check that every section $\sigma$ of $\mathcal{O}_{P_kV}(m)$ over $P_kV_x$ must vanish on $P_kV_x^{\rm sing}$. Pick an arbitrary element $w\in P_kV^{\rm sing}$ and a germ of curve $f:(\mathbb{C},0)\to X$ such that $f_{[k]}(0)=w$, $f'_{[k-1]}(0)\neq 0$ and $s=m(f,0)\gg 0$ (such an $f$ exists by Corollary 5.14). There are local coordinates $(z_1,\ldots,z_n)$ on $X$ such that $f(t)=(f_1(t),\ldots,f_n(t))$ where $f_r(t)=t^s$. Let $Q$, $\widehat{Q}$ be the polynomials associated with $\sigma$ in these coordinates and let $(f')^{\alpha_1}(f'')^{\alpha_2}\cdots(f^{(k)})^{\alpha_k}$ be a monomial occurring in $Q$, with $\alpha_j\in\mathbb{N}^r$, $|\alpha_j|=\ell_j$, $\ell_1+2\ell_2+\cdots+k\ell_k=m$. Putting $\tau=t^s$, the curve $t\mapsto f(t)$ becomes a Puiseux expansion $\tau\mapsto g(\tau)=(g_1(\tau),\ldots,g_{r-1}(\tau),\tau)$ in which $g_i$ is a power series in $\tau^{1/s}$, starting with exponents of $\tau$ at least equal to 1. The derivative $g^{(j)}(\tau)$ may involve negative powers of $\tau$, but the exponent is always $\geqslant 1+\frac{1}{s}-j$. Hence the Puiseux expansion of $\widehat{Q}(g',g'',\ldots,g^{(k)})$ can only involve powers of $\tau$ of exponent $\geqslant\ -\max_\ell((1-\frac{1}{s})\ell_2+\cdots+(k-1-\frac{1}{s})\ell_k)$. Finally $f_r'(t)=st^{s-1}=s\tau^{1-1/s}$, thus the lowest exponent of $\tau$ in $Q(f',\ldots,f^{(k)})$ is at least equal to

$$\begin{aligned}\Big(1-\frac{1}{s}\Big)m-\max_\ell\Big(\Big(1-\frac{1}{s}\Big)\ell_2+\cdots+\Big(k-1-\frac{1}{s}\Big)\ell_k\Big)\\ \geqslant\min_\ell\Big(1-\frac{1}{s}\Big)\ell_1+\Big(1-\frac{1}{s}\Big)\ell_2+\cdots+\Big(1-\frac{k-1}{s}\Big)\ell_k\end{aligned}$$

where the minimum is taken over all monomials $(f')^{\alpha_1}(f'')^{\alpha_2}\cdots(f^{(k)})^{\alpha_k}$, $|\alpha_j|=\ell_j$, occurring in $Q$. Choosing $s\geqslant k$, we already find that the minimal exponent is positive, hence $Q(f',\ldots,f^{(k)})(0)=0$ and $\sigma(w)=0$ by (6.11). □

Theorem (6.8 iii) shows that $\mathcal{O}_{P_kV}(1)$ is never relatively ample over $X$ for $k\geqslant 2$. In order to overcome this difficulty, we define for every $\boldsymbol{a}=(a_1,\ldots,a_k)\in\mathbb{Z}^k$ a line bundle $\mathcal{O}_{P_kV}(\boldsymbol{a})$ on $P_kV$ such that

(6.13)
$$\mathcal{O}_{P_kV}(\boldsymbol{a})=\pi_{1,k}^\star\mathcal{O}_{P_1V}(a_1)\otimes\pi_{2,k}^\star\mathcal{O}_{P_2V}(a_2)\otimes\cdots\otimes\mathcal{O}_{P_kV}(a_k).$$

By (5.9), we have $\pi_{j,k}^\star\mathcal{O}_{P_jV}(1)=\mathcal{O}_{P_kV}(1)\otimes\mathcal{O}_{P_kV}(-\pi_{j+1,k}^\star D_{j+1}-\cdots-D_k)$, thus by putting $D_j^\star=\pi_{j+1,k}^\star D_{j+1}$ for $1\leqslant j\leqslant k-1$ and $D_k^\star=0$, we find an identity

(6.14)
$$\begin{aligned}&\mathcal{O}_{P_kV}(\boldsymbol{a})=\mathcal{O}_{P_kV}(b_k)\otimes\mathcal{O}_{P_kV}(-\boldsymbol{b}\cdot D^\star),\qquad\text{where}\\ &\boldsymbol{b}=(b_1,\ldots,b_k)\in\mathbb{Z}^k,\quad b_j=a_1+\cdots+a_j,\\ &\boldsymbol{b}\cdot D^\star=\sum_{1\leqslant j\leqslant k-1}b_j\,\pi_{j+1,k}^\star D_{j+1}.\end{aligned}$$

In particular, if $\boldsymbol{b}\in\mathbb{N}^k$, i.e., $a_1+\cdots+a_j\geqslant 0$, we get a morphism

(6.15)
$$\mathcal{O}_{P_kV}(\boldsymbol{a})=\mathcal{O}_{P_kV}(b_k)\otimes\mathcal{O}_{P_kV}(-\boldsymbol{b}\cdot D^\star)\to\mathcal{O}_{P_kV}(b_k).$$

6.16. Proposition. *Let $\boldsymbol{a}=(a_1,\ldots,a_k)\in\mathbb{Z}^k$ and $m=a_1+\cdots+a_k$.*

i) *We have the direct image formula*

$$(\pi_{0,k})_\star\mathcal{O}_{P_kV}(\boldsymbol{a})\simeq\mathcal{O}(\overline{F}^{\boldsymbol{a}}E_{k,m}V^\star)\subset\mathcal{O}(E_{k,m}V^\star)$$

*where* $\overline{F}^{\boldsymbol{a}} E_{k,m}V^\star$ *is the subbundle of polynomials* $Q(f', f'', \ldots, f^{(k)}) \in E_{k,m}V^\star$ *involving only monomials* $(f^{(\bullet)})^\ell$ *such that*

$$\ell_{s+1} + 2\ell_{s+2} + \cdots + (k-s)\ell_k \leqslant a_{s+1} + \cdots + a_k$$

*for all* $s = 0, \ldots, k-1$.

ii) *If* $a_1 \geqslant 3a_2, \ldots, a_{k-2} \geqslant 3a_{k-1}$ *and* $a_{k-1} \geqslant 2a_k \geqslant 0$, *the line bundle* $\mathcal{O}_{P_kV}(\boldsymbol{a})$ *is relatively nef over* $X$.

iii) *If* $a_1 \geqslant 3a_2, \ldots, a_{k-2} \geqslant 3a_{k-1}$ *and* $a_{k-1} > 2a_k > 0$, *the line bundle* $\mathcal{O}_{P_kV}(\boldsymbol{a})$ *is relatively ample over* $X$.

PROOF. i) By (6.15), we find a sheaf injection

$$(\pi_{0,k})_\star \mathcal{O}_{P_kV}(\boldsymbol{a}) \hookrightarrow (\pi_{0,k})_\star \mathcal{O}_{P_kV}(m) = \mathcal{O}(E_{k,m}V^\star).$$

Given a section $\sigma$ of $\mathcal{O}_{P_kV}(\boldsymbol{a})$ over a fiber $P_kV_x$, the associated polynomial

$$Q(f', f'', \ldots, f^{(k)}) \in E_{k,m}V_x^\star$$

is given by the identity

$$Q(f', f'', \ldots, f^{(k)}) = \sigma(f_{[k]}(0)) \cdot (f'(0))^{a_1} \cdot (f'_{[1]}(0))^{a_2} \cdots (f'_{[k-1]}(0))^{a_k}.$$

Indeed, we see this from (6.10) and from the fact that $f'_{[k-1]}(0)$ is mapped to $f'_{[j-1]}(0)$ by the projection morphism

$$(\pi_{j-1,k-1})_\star : \mathcal{O}_{P_kV}(-1) \to \pi_{j,k}^\star \mathcal{O}_{P_jV}(-1)$$

(cf. (5.8)), which is dual to the corresponding morphism (6.15). Now, we prove the inclusion $(\pi_{0,k})_\star \mathcal{O}_{P_kV}(\boldsymbol{a}) \subset \mathcal{O}(\overline{F}^{\boldsymbol{a}} E_{k,m}V^\star)$ by induction on $k$. For $s = 0$, the desired inequality comes from the weighted homogeneity condition, hence we may assume $s \geqslant 1$. Let $f$ run over all regular germs having their first derivative $f'(0)$ fixed. This means that $\sigma$ is viewed as a section of $\pi_{2,k}^\star \mathcal{O}_{P_2V}(a_2) \otimes \cdots \otimes \mathcal{O}_{P_kV}(a_k)$ on the fibers of the projection $P_kV = P_{k-1}V_1 \to X_1 = P_1V$. Then we get a polynomial $Q_1 \in E_{k-1,m-a_1}V_1^\star$ such that

$$Q_1(f'_{[1]}, f''_{[1]}, \ldots, f_{[1]}^{(k-1)}) = Q(f', f'', \ldots, f^{(k)}).$$

In the affine chart $f'_r \neq 0$, the map $f_{[1]}$ is defined in coordinates by

$$f_{[1]} \simeq (f_1, \ldots, f_n\,;\, f'_1/f'_r, \ldots, f'_{r-1}/f'_r).$$

Its derivative $f'_{[1]} \in V_1$ can thus be described by $f'_{[1]} \simeq \big((f'_1/f'_r)', \ldots, (f'_{r-1}/f'_r)', f'_r\big)$, by taking $r-1$ vertical components and a horizontal one. All this becomes much simpler if we replace $f$ by $g = f \circ f_r^{-1}$, since $g_r(t) = t$ and $g'_r(t) = 1$. Then we get

$$(g', g'', \ldots, g^{(k)}) \simeq \big((g'_1, \ldots, g'_{r-1}, 1), (g''_1, \ldots, g''_{r-1}, 0), \ldots, (g_1^{(k)}, \ldots, g_{r-1}^{(k)}, 0)\big),$$
$$(g'_{[1]}, g''_{[1]}, \ldots, g_{[1]}^{(k)}) \simeq \big((g''_1, \ldots, g''_{r-1}, 1), (g'''_1, \ldots, g'''_{r-1}, 0), \ldots, (g_1^{(k)}, \ldots, g_{r-1}^{(k)}, 0)\big)$$

in the corresponding charts of $J_kV$ and $J_{k-1}V_1$. The inequality (6.16 i) for the monomials $(g^{(\bullet)})^\ell$ of $Q(g', g'', \ldots, g^{(k)})$ follows clearly from the corresponding inequality on the monomials $(g_{[1]}^{(\bullet)})^\ell$ of $Q_1$, when $(k, s)$ is replaced by $(k-1, s-1)$. Now, thanks to (6.9), we get $Q(f', f'', \ldots, f^{(k)}) = (f'_r)^m Q(g', g'', \ldots, g^{(k)})$, and the desired inequality (6.16 i) for the monomials $(f^{(\bullet)})^\ell$ follows easily. In the opposite direction, if we are given a section $Q(f', f'', \ldots, f^{(k)}) \in \mathcal{O}(\overline{F}^{\boldsymbol{a}} E_{k,m} V^\star)$, we see by induction on $k$ that $Q$ defines a section of

$$\mathcal{O}_{P_1 V}(a_1) \otimes (\pi_{1,k})_\star \big(\pi_{2,k}^\star \mathcal{O}_{P_2 V}(a_2) \otimes \cdots \otimes \mathcal{O}_{P_k V}(a_k)\big)$$

on $P_1 V$, and we conclude that we get a section of $(\pi_{0,k})_\star \mathcal{O}_{P_k V}(\boldsymbol{a})$ by taking the direct image by $(\pi_1)_\star$.

ii-iii) By induction on $k$, we construct a relatively ample line bundle $L_{k-1}$ on $P_{k-1}V$ such that $\mathcal{O}_{P_k V}(1) \otimes \pi_k^\star L_{k-1}$ is relatively nef; by definition, this is equivalent to saying that the vector bundle $V_{k-1}^\star \otimes L_{k-1}$ is relatively nef (for the notion of a nef vector bundle, see e.g. [DPS94]). Since $\mathcal{O}_{P_1 V}(1)$ is relatively ample, we can start with $L_0 = \mathcal{O}_X$. Suppose that $L_{k-1}$ has been constructed. The dual of (5.4) yields an exact sequence

$$0 \longrightarrow \mathcal{O}_{P_k V}(1) \longrightarrow V_k^\star \longrightarrow T^\star_{P_k V / P_{k-1} V} \longrightarrow 0.$$

As an extension of nef vector bundles is nef, it is enough to select $L_k$ in such a way that

(6.17) $\quad \mathcal{O}_{P_k V}(1) \otimes L_k \quad$ and $\quad T^\star_{P_k V/P_{k-1}V} \otimes L_k \quad$ are relatively nef.

By taking the second wedge power of the central term in (5.4′), we get an injection

$$0 \longrightarrow T_{P_k V/P_{k-1}V} \longrightarrow \Lambda^2\big(\pi_k^\star V_{k-1} \otimes \mathcal{O}_{P_k V}(1)\big).$$

By dualizing and twisting with $\mathcal{O}_{P_{k-1}V}(2) \otimes \pi_k^\star L_{k-1}^{\otimes 2}$, we find a surjection

$$\pi_k^\star \Lambda^2 (V_{k-1}^\star \otimes L_{k-1}) \longrightarrow T^\star_{P_k V/P_{k-1}V} \otimes \mathcal{O}_{P_k V}(2) \otimes \pi_k^\star L_{k-1}^{\otimes 2} \longrightarrow 0.$$

As $V_{k-1}^\star \otimes L_{k-1}$ is relatively nef by the induction hypothesis, we obtain that its quotient $T^\star_{P_k V/P_{k-1}V} \otimes \mathcal{O}_{P_k V}(2) \otimes \pi_k^\star L_{k-1}^{\otimes 2}$ is also relatively nef. Hence Condition (6.17) is achieved if we take $L_k \geqslant \pi_k^\star L_{k-1}$ and $L_k \geqslant \mathcal{O}_{P_k V}(2) \otimes \pi_k^\star L_{k-1}^{\otimes 2}$ (the ordering relation $\geqslant$ is the one given by the cone of relatively nef line bundles). We need only define $L_k$ inductively by

$$L_k = \mathcal{O}_{P_k V}(2) \otimes \pi_k^\star L_{k-1}^{\otimes 3}.$$

The relative ampleness of $L_k$ is then clear by induction, since $\mathcal{O}_{P_k V}(1) \otimes \pi_k^\star L_{k-1}$ is relatively nef over $X$ and relatively ample over $P_{k-1}V$. The resulting formula for $L_k$ is

$$L_k = \mathcal{O}_{P_k V}\big((2 \cdot 3^{k-1}, 2 \cdot 3^{k-2}, \ldots, 6, 2)\big).$$

By definition, we then find

$$\mathcal{O}_{P_k V}(1) \otimes \pi_k^\star L_{k-1} = \mathcal{O}_{P_k V}\big((2 \cdot 3^{k-2}, 2 \cdot 3^{k-3}, \ldots, 6, 2, 1)\big) \qquad \text{relatively nef.}$$

These properties imply ii) and iii) by taking suitable convex combinations. □

6.18. REMARK. As in Green-Griffiths [GrGr80], Riemann's extension theorem shows that for every meromorphic map $\Phi : X \dashrightarrow Y$ there are well-defined pullback morphisms

$$\Phi^\star : H^0(Y, E_{k,m}^{\mathrm{GG}}) \to H^0(X, E_{k,m}^{\mathrm{GG}}), \qquad \Phi^\star : H^0(Y, E_{k,m}) \to H^0(X, E_{k,m}).$$

In particular the dimensions $h^0(X, E_{k,m}^{\mathrm{GG}})$ and $h^0(X, E_{k,m}^{\mathrm{GG}})$ are bimeromorphic invariants of $X$. The same is true for spaces of sections of any subbundle of $E_{k,m}^{\mathrm{GG}}$ or $E_{k,m}$ constructed by means of the canonical filtrations $F_s^\bullet$.

## 7. $k$-Jet Metrics with Negative Curvature

The goal of this section is to show that hyperbolicity is closely related to the existence of $k$-jet metrics with suitable negativity properties of the curvature. The connection between these properties is in fact a simple consequence of the Ahlfors-Schwarz lemma. Such ideas have been already developed long ago by Grauert-Reckziegel [GRec65], Kobayashi [Kob75] for 1-jet metrics (i.e., Finsler metrics on $T_X$) and by Cowen-Griffiths [CoGr76], Green-Griffiths [GrGr80] and Grauert [Gra89] for higher order jet metrics. However, even in the standard case $V = T_X$, the definition given below differs from that of [GrGr80], in which the $k$-jet metrics are not supposed to be $\mathbb{G}_k'$-invariant. We prefer to deal here with $\mathbb{G}_k'$-invariant objects, because they reflect better the intrinsic geometry. Grauert [Gra89] actually deals with $\mathbb{G}_k'$-invariant metrics, but he apparently does not take care of the way the quotient space $J_k^{\mathrm{reg}}V/\mathbb{G}_k$ can be compactified; also, his metrics are always induced by the Poincaré metric, and it is not at all clear whether these metrics have the expected curvature properties (see 7.14 below). In the present situation, it is important to allow also hermitian metrics possessing some singularities ("singular hermitian metrics" in the sense of [Dem90]).

7.1. DEFINITION. *Let $L \to X$ be a holomorphic line bundle over a complex manifold $X$. We say that $h$ is a singular metric on $L$ if for any trivialization $L_{\restriction U} \simeq U \times \mathbb{C}$ of $L$, the metric is given by $|\xi|_h^2 = |\xi|^2 e^{-\varphi}$ for some real valued weight function $\varphi \in L^1_{\mathrm{loc}}(U)$. The curvature current of $L$ is then defined to be the closed $(1,1)$-current $\Theta_h(L) = \frac{i}{2\pi}\partial\overline{\partial}\varphi$, computed in the sense of distributions. We say that $h$ admits a closed subset $\Sigma \subset X$ as its degeneration set if $\varphi$ is locally bounded on $X \smallsetminus \Sigma$ and is unbounded on a neighborhood of any point of $\Sigma$.*

An especially useful situation is the case when the curvature of $h$ is positive definite. By this, we mean that there exists a smooth positive definite hermitian metric $\omega$ and a continuous positive function $\varepsilon$ on $X$ such that $\Theta_h(L) \geqslant \varepsilon\omega$ in the sense of currents, and we write in this case $\Theta_h(L) \gg 0$. We need the following basic fact (quite standard when $X$ is projective algebraic; however we want to avoid any algebraicity assumption here, so as to be able the case of general complex tori in § 9).

7.2. PROPOSITION. *Let $L$ be a holomorphic line bundle on a compact complex manifold $X$.*

i) *$L$ admits a singular hermitian metric $h$ with positive definite curvature current $\Theta_h(L) \gg 0$ if and only if $L$ is big.*

*Now, define $B_m$ to be the base locus of the linear system $|H^0(X, L^{\otimes m})|$ and let*

$$\Phi_m : X \smallsetminus B_m \to \mathbb{P}^N$$

*be the corresponding meromorphic map. Let $\Sigma_m$ be the closed analytic set equal to the union of $B_m$ and of the set of points $x \in X \smallsetminus B_m$ such that the fiber $\Phi_m^{-1}(\Phi_m(x))$ is positive dimensional.*

ii) *If $\Sigma_m \neq X$ and $G$ is any line bundle, the base locus of $L^{\otimes k} \otimes G^{-1}$ is contained in $\Sigma_m$ for $k$ large. As a consequence, $L$ admits a singular hermitian metric $h$ with degeneration set $\Sigma_m$ and with $\Theta_h(L)$ positive definite on $X$.*

iii) *Conversely, if $L$ admits a hermitian metric $h$ with degeneration set $\Sigma$ and positive definite curvature current $\Theta_h(L)$, there exists an integer $m > 0$ such that the base locus $B_m$ is contained in $\Sigma$ and $\Phi_m : X \smallsetminus \Sigma \to \mathbb{P}_m$ is an embedding.*

iv) *Assume that $L$ admits a singular hermitian metric $h$ with positive definite curvature current, such that the degeneration set $\Sigma$ is an analytic subset of $X$. Assume moreover that for each irreducible component $\Sigma_j$ of $\Sigma$, $L_{|\Sigma_j}$ admits a singular hermitian metric $h_j$ with positive definite curvature current on $\Sigma_j$ and degeneration set $\Sigma_{j,k} \subset \Sigma_j$. Then $L$ admits a singular hermitian metric $\widetilde{h}$ of positive curvature current on $X$, with degeneration set $\widetilde{\Sigma} = \bigcup_{j,k} \Sigma_{j,k}$.*

PROOF. i) is proved e.g. in [Dem90, 92], so we will only briefly sketch the details. If $L$ is big, then $X$ is Moishezon and we can even assume that $X$ is projective algebraic after taking a suitable modification $\widetilde{X}$ (apply Hironaka [Hir64]; observe moreover that the direct image of a strictly positive current is strictly positive). So, assume that $X$ is projective algebraic. Then it is well-known that some large multiple of $L$ can be written as $L^{\otimes m} = \mathcal{O}_X(D + A)$ with divisors $D$, $A$ such that $D$ is effective and $A$ ample. The invertible sheaf $\mathcal{O}_X(D)$ can be viewed as a subsheaf of the sheaf of meromorphic functions. We get a singular metric $|s|^2$ on sections of $\mathcal{O}_X(D)$ by just taking the square of the modulus of $s$ viewed as a complex valued (meromorphic) function. By the Lelong-Poincaré equation, the curvature current of that metric is equal to the current of integration $[D] \geqslant 0$ over the divisor $D$. We thus get $\Theta(L) = \frac{1}{m}([D] + \Theta(A)) \geqslant \frac{1}{m}\Theta(A) \gg 0$ for a suitable choice of the metric on $\mathcal{O}_X(A)$. In the other direction, if $\Theta_h(L)$ is positive, one can construct a "lot of" sections in $H^0(X, L^{\otimes m})$, $m \gg 0$, by using Hörmander's $L^2$ estimates; the Hörmander-Bombieri-Skoda technique implies that these sections can be taken to have arbitrary jets at all points in a given finite subset of $X \smallsetminus \Sigma$, if $\Sigma$ is the degeneration set of $h$. This also proves property iii).

ii) The assumption $\Sigma_m \neq X$ shows that there is a generically finite meromorphic map from $X$ to an algebraic variety, and this implies again that $X$ is Moishezon. By blowing-up the ideal

$$\mathcal{I}_m = \operatorname{Im}\left(H^0(X, L^{\otimes m}) \otimes \mathcal{O}_X(L^{\otimes -m}) \to \mathcal{O}_X\right) \subset \mathcal{O}_X$$

and resolving the singularities, we obtain a smooth modification $\mu : \widetilde{X} \to X$ and a line bundle $\widetilde{L} = \mu^\star(L^{\otimes m}) \otimes \mathcal{O}_{\widetilde{X}}(-E)$ (where $E$ is a $\mu$-exceptional divisor with support in $\mu^{-1}(\Sigma_m)$, such that $\widetilde{L}$ is base point free; after possibly blowing-up again, we may assume furthermore that $\widetilde{X}$ is projective algebraic. Clearly, it is enough to prove the result for $\widetilde{L}$, and we are thus reduced to the case when $L$ is base point free and $X$ is projective algebraic. We may finally assume that $G$ is very ample (other we add a large ample divisor to $G$ to make it very ample). In this situation, we have a holomorphic map $\Phi_m : X \to \mathbb{P}^N$ such that $L^{\otimes m} = \Phi_m^{-1}(\mathcal{O}(1))$, and $\Phi_m$ is finite-to-one outside $\Sigma_m$. Hence, if $x \in X \smallsetminus \Sigma_m$, the set $\Phi_m^{-1}(\Phi_m(x))$ is finite, and we can take a smooth divisor $D \in |G|$ such that $D \cap \Phi_m^{-1}(\Phi_m(x)) = \emptyset$. Thus $\Phi_m(D) \not\ni \varphi_m(x)$ in $\mathbb{P}^N$. It follows that there exists a hypersurface

$$H = \sigma^{-1}(0) \in |\mathcal{O}_{\mathbb{P}^N}(k)|$$

of sufficiently large degree $k$, such that $H$ contains $\Phi_m(D)$ but does not pass through $\Phi_m(x)$. Then $\Phi_m^\star\sigma$ can be viewed as a section of $\Phi_m^\star\mathcal{O}_{\mathbb{P}^N}(k) \otimes \mathcal{O}_X(-D) = L^{\otimes km} \otimes G^{-1}$, and $\Phi_m^\star\sigma$ does not vanish at $x$. By the Noetherian property, there exists $k_0$ such that the base locus of $L^{\otimes km} \otimes G^{-1}$ is contained in $\Sigma_m$ for $k \geqslant k_0$ large. Claim ii) follows.

iv) is obtained by extending the metric $h_j$ to a metric $\widetilde{h}_j$ on a neighborhood of $\Sigma_j$ (it is maybe necessary to modify $\widetilde{h}_j$ slightly by adding some "transversally convex terms" in the weight, so as to obtain positive curvature in all directions of $T_X$, on a suitable neighborhood of $\Sigma_j$), and then taking $\widetilde{h} = \min(h, \varepsilon\widetilde{h}_j)$ with $\varepsilon > 0$ small enough. □

We now come to the main definitions. By (5.6), every regular $k$-jet $f \in J_kV$ gives rise to an element $f'_{[k-1]}(0) \in \mathcal{O}_{P_kV}(-1)$. Thus, measuring the "norm of $k$-jets" is the same as taking a hermitian metric on $\mathcal{O}_{P_kV}(-1)$.

7.3. Definition. *A smooth, (resp. continuous, resp. singular) $k$-jet metric on a complex directed manifold $(X, V)$ is a hermitian metric $h_k$ on the line bundle $\mathcal{O}_{P_kV}(-1)$ over $P_kV$ (i.e. a Finsler metric on the vector bundle $V_{k-1}$ over $P_{k-1}V$), such that the weight functions $\varphi$ representing the metric are smooth (resp. continuous, $L^1_{\mathrm{loc}}$). We let $\Sigma_{h_k} \subset P_kV$ be the singularity set of the metric, i.e., the closed subset of points in a neighborhood of which the weight $\varphi$ is not locally bounded.*

We will always assume here that the weight function $\varphi$ is quasi psh. Recall that a function $\varphi$ is said to be quasi psh if $\varphi$ is locally the sum of a plurisubharmonic function and of a smooth function (so that in particular $\varphi \in L^1_{\mathrm{loc}}$). Then the curvature current

$$\Theta_{h_k^{-1}}(\mathcal{O}_{P_kV}(1)) = \frac{i}{2\pi}\partial\overline{\partial}\varphi.$$

is well defined as a current and is locally bounded from below by a negative $(1,1)$-form with constant coefficients.

7.4. Definition. *Let $h_k$ be a $k$-jet metric on $(X, V)$. We say that $h_k$ has negative jet curvature (resp. negative total jet curvature) if $\Theta_{h_k}(\mathcal{O}_{P_kV}(-1))$ is negative definite along the subbundle $V_k \subset T_{P_kV}$ (resp. on all of $T_{P_kV}$), i.e., if there is $\varepsilon > 0$ and a smooth hermitian metric $\omega_k$ on $T_{P_kV}$ such that*

$$\langle\Theta_{h_k^{-1}}(\mathcal{O}_{P_kV}(1))\rangle(\xi) \geqslant \varepsilon|\xi|^2_{\omega_k}, \qquad \forall\xi \in V_k \subset T_{P_kV} \quad (\textit{resp.} \quad \forall\xi \in T_{P_kV}).$$

*(If the metric $h_k$ is not smooth, we suppose that its weights $\varphi$ are quasi psh, and the curvature inequality is taken in the sense of distributions.)*

It is important to observe that for $k \geqslant 2$ there cannot exist any smooth hermitian metric $h_k$ on $\mathcal{O}_{P_kV}(1)$ with positive definite curvature along $T_{X_k/X}$, since $\mathcal{O}_{P_kV}(1)$ is not relatively ample over $X$. However, it is relatively big, and Prop. 7.2 i) shows that $\mathcal{O}_{P_kV}(-1)$ admits a singular hermitian metric with negative total jet curvature (whatever the singularities of the metric are) if and only if $\mathcal{O}_{P_kV}(1)$ is big over $P_kV$. It is therefore crucial to allow singularities in the metrics in Def. 7.4.

7.5. Special case of 1-jet metrics. A 1-jet metric $h_1$ on $\mathcal{O}_{P_1V}(-1)$ is the same as a Finsler metric $N = \sqrt{h_1}$ on $V \subset T_X$. Assume until the end of this paragraph that $h_1$ is smooth. By the well known Kodaira embedding theorem, the existence of a smooth metric $h_1$ such that $\Theta_{h_1^{-1}}(\mathcal{O}_{P_1V}(1))$ is positive on all of $T_{P_1V}$ is equivalent to $\mathcal{O}_{P_1V}(1)$ being ample, that is, $V^\star$ ample. In the absolute case $V = T_X$, there are only few examples of varieties $X$ such that $T_X^\star$ is ample, mainly quotients of the ball $\mathbb{B}_n \subset \mathbb{C}^n$ by a discrete cocompact group of automorphisms. The 1-jet negativity condition considered in Definition 7.4 is much weaker. For example, if the hermitian metric $h_1$ comes from a (smooth) hermitian metric $h$ on $V$, then formula (4.16) implies that $h_1$ has negative total jet curvature (i.e. $\Theta_{h_1^{-1}}(\mathcal{O}_{P_1V}(1))$ is positive) if and only if $\langle\Theta_h(V)\rangle(\zeta \otimes v) < 0$ for all $\zeta \in T_X \smallsetminus \{0\}$, $v \in V \smallsetminus \{0\}$, that is, if $(V, h)$ is *negative in the sense of Griffiths*. On the other hand, $V_1 \subset T_{P_1V}$ consists by definition of tangent vectors $\tau \in T_{P_1V,(x,[v])}$ whose horizontal projection ${}^H\tau$ is proportional to $v$, thus $\Theta_{h_1}(\mathcal{O}_{P_1V}(-1))$ is negative definite on $V_1$ if and only if $\Theta_h(V)$ satisfies the much weaker condition that the *holomorphic sectional curvature* $\langle\Theta_h(V)\rangle(v \otimes v)$ is negative on every complex line. □

We now come back to the general situation of jets of arbitrary order $k$. Our first observation is the fact that the $k$-jet negativity property of the curvature becomes actually weaker and weaker as $k$ increases.

7.6. Lemma. *Let $(X, V)$ be a compact complex directed manifold. If $(X, V)$ has a $(k-1)$-jet metric $h_{k-1}$ with negative jet curvature, then there is a $k$-jet metric $h_k$ with negative jet curvature such that $\Sigma_{h_k} \subset \pi_k^{-1}(\Sigma_{h_{k-1}}) \cup D_k$. (The same holds true for negative total jet curvature).*

Proof. Let $\omega_{k-1}$, $\omega_k$ be given smooth hermitian metrics on $T_{P_{k-1}V}$ and $T_{P_kV}$. The hypothesis implies

$$\langle\Theta_{h_{k-1}^{-1}}(\mathcal{O}_{P_{k-1}V}(1))\rangle(\xi) \geqslant \varepsilon|\xi|^2_{\omega_{k-1}}, \qquad \forall\xi \in V_{k-1}$$

for some constant $\varepsilon > 0$. On the other hand, as $\mathcal{O}_{P_kV}(D_k)$ is relatively ample over $P_{k-1}V$ ($D_k$ is a hyperplane section bundle), there exists a smooth metric $\widetilde{h}$ on $\mathcal{O}_{P_kV}(D_k)$ such that

$$\langle\Theta_{\widetilde{h}}(\mathcal{O}_{P_kV}(D_k))\rangle(\xi) \geqslant \delta|\xi|^2_{\omega_k} - C|(\pi_k)_\star\xi|^2_{\omega_{k-1}}, \qquad \forall\xi \in T_{P_kV}$$

for some constants $\delta, C > 0$. Combining both inequalities (the second one being applied to $\xi \in V_k$ and the first one to $(\pi_k)_\star\xi \in V_{k-1}$), we get

$$\begin{aligned}\langle\Theta_{(\pi_k^\star h_{k-1})^{-p}\widetilde{h}}(\pi_k^\star\mathcal{O}_{P_{k-1}V}(p) \otimes \mathcal{O}_{P_kV}(D_k))\rangle(\xi) \geqslant&\\ \geqslant \delta|\xi|^2_{\omega_k} + (p\varepsilon - C)|(\pi_k)_\star\xi|^2_{\omega_{k-1}}, &\qquad \forall\xi \in V_k.\end{aligned}$$

Hence, for $p$ large enough, $(\pi_k^\star h_{k-1})^{-p}\widetilde{h}$ has positive definite curvature along $V_k$. Now, by (5.9), there is a sheaf injection

$$\mathcal{O}_{P_kV}(-p) = \pi_k^\star\mathcal{O}_{P_{k-1}V}(-p) \otimes \mathcal{O}_{P_kV}(-pD_k) \hookrightarrow \big(\pi_k^\star\mathcal{O}_{P_{k-1}V}(p) \otimes \mathcal{O}_{P_kV}(D_k)\big)^{-1}$$

obtained by twisting with $\mathcal{O}_{P_kV}((p-1)D_k)$. Therefore $h_k := ((\pi_k^\star h_{k-1})^{-p}\widetilde{h})^{-1/p} = (\pi_k^\star h_{k-1})\widetilde{h}^{-1/p}$ induces a singular metric on $\mathcal{O}_{P_kV}(1)$ in which an additional degeneration divisor $p^{-1}(p-1)D_k$ appears. Hence we get $\Sigma_{h_k} = \pi_k^{-1}\Sigma_{h_{k-1}} \cup D_k$

and

$$\Theta_{h_k^{-1}}(\mathcal{O}_{P_kV}(1)) = \frac{1}{p}\Theta_{(\pi_k^\star h_{k-1})^{-p}\widetilde{h}} + \frac{p-1}{p}[D_k]$$

is positive definite along $V_k$. The same proof works in the case of negative total jet curvature. □

One of the main motivations for the introduction of $k$-jets metrics is the following list of algebraic sufficient conditions.

7.7. Algebraic sufficient conditions. We suppose here that $X$ is projective algebraic, and we make one of the additional assumptions i), ii) or iii) below.
i) Assume that there exist integers $k, m > 0$ and $\boldsymbol{b} \in \mathbb{N}^k$ such that the line bundle $\mathcal{O}_{P_kV}(m) \otimes \mathcal{O}_{P_kV}(-\boldsymbol{b}\cdot D^\star)$ is ample over $P_kV$. Set $A = \mathcal{O}_{P_kV}(m) \otimes \mathcal{O}_{P_kV}(-\boldsymbol{b}\cdot D^\star)$. Then there is a smooth hermitian metric $h_A$ on $A$ with positive definite curvature on $P_kV$. By means of the morphism $\mu : \mathcal{O}_{P_kV}(-m) \to A^{-1}$, we get an induced metric $h_k = (\mu^\star h_A^{-1})^{1/m}$ on $\mathcal{O}_{P_kV}(-1)$ which is degenerate on the support of the zero divisor $\operatorname{div}(\mu) = \boldsymbol{b}\cdot D^\star$. Hence $\Sigma_{h_k} = \operatorname{Supp}(\boldsymbol{b}\cdot D^\star) \subset P_kV^{\rm sing}$ and

$$\Theta_{h_k^{-1}}(\mathcal{O}_{P_kV}(1)) = \frac{1}{m}\Theta_{h_A}(A) + \frac{1}{m}[\boldsymbol{b}\cdot D^\star] \geqslant \frac{1}{m}\Theta_{h_A}(A) > 0.$$

In particular $h_k$ has negative total jet curvature.
ii) Assume more generally that there exist integers $k, m > 0$ and an ample line bundle $L$ on $X$ such that $H^0(P_kV, \mathcal{O}_{P_kV}(m) \otimes \pi_{0,k}^\star L^{-1})$ has non zero sections $\sigma_1, \ldots, \sigma_N$. Let $Z \subset P_kV$ be the base locus of these sections; necessarily $Z \supset P_kV^{\rm sing}$ by 6.8 iii). By taking a smooth metric $h_L$ with positive curvature on $L$, we get a singular metric $h'_k$ on $\mathcal{O}_{P_kV}(-1)$ such that

$$h'_k(\xi) = \Big(\sum_{1\leqslant j\leqslant N} |\sigma_j(w)\cdot\xi^m|^2_{h_L^{-1}}\Big)^{1/m}, \qquad w \in P_kV, \quad \xi \in \mathcal{O}_{P_kV}(-1)_w.$$

Then $\Sigma_{h'_k} = Z$, and by computing $\frac{i}{2\pi}\partial\overline{\partial}\log h'_k(\xi)$ we obtain

$$\Theta_{h_k'^{-1}}(\mathcal{O}_{P_kV}(1)) \geqslant \frac{1}{m}\pi_{0,k}^\star\Theta(L).$$

By (6.15) and 6.16 iii), there exists $\boldsymbol{b} \in \mathbb{Q}_+^k$ such that $\mathcal{O}_{P_kV}(1) \otimes \mathcal{O}_{P_kV}(-\boldsymbol{b}\cdot D^\star)$ is relatively ample over $X$. Hence $A = \mathcal{O}_{P_kV}(1) \otimes \mathcal{O}_{P_kV}(-\boldsymbol{b}\cdot D^\star) \otimes \pi_{0,k}^\star L^{\otimes p}$ is ample on $X$ for $p \gg 0$. The arguments used in i) show that there is a $k$-jet metric $h''_k$ on $\mathcal{O}_{P_kV}(-1)$ with $\Sigma_{h''_k} = \operatorname{Supp}(\boldsymbol{b}\cdot D^\star) = P_kV^{\rm sing}$ and

$$\Theta_{h_k''^{-1}}(\mathcal{O}_{P_kV}(1)) = \Theta(A) + [\boldsymbol{b}\cdot D^\star] - p\,\pi_{0,k}^\star\Theta(L),$$

where $\Theta(A)$ is positive definite on $P_kV$. The metric $h_k = (h_k'^{mp} h''_k)^{1/(mp+1)}$ then satisfies $\Sigma_{h_k} = \Sigma_{h'_k} = Z$ and

$$\Theta_{h_k^{-1}}(\mathcal{O}_{P_kV}(1)) \geqslant \frac{1}{mp+1}\Theta(A) > 0.$$

iii) If $E_{k,m}V^\star$ is ample, there is an ample line bundle $L$ and a sufficiently high symmetric power such that $S^p(E_{k,m}V^\star) \otimes L^{-1}$ is generated by sections. These sections can be viewed as sections of $\mathcal{O}_{P_kV}(mp) \otimes \pi_{0,k}^\star L^{-1}$ over $P_kV$, and their base locus is exactly $Z = P_k V^{\rm sing}$ by 6.8 iii). Hence the $k$-jet metric $h_k$ constructed in ii) has negative total jet curvature and satisfies $\Sigma_{h_k} = P_k V^{\rm sing}$. □

An important fact, first observed by [GRe65] for 1-jet metrics and by [GrGr80] in the higher order case, is that $k$-jet negativity implies hyperbolicity. In particular, the existence of enough global jet differentials implies hyperbolicity.

7.8. THEOREM. *Let $(X, V)$ be a compact complex directed manifold. If $(X, V)$ has a $k$-jet metric $h_k$ with negative jet curvature, then every entire curve $f : \mathbb{C} \to X$ tangent to $V$ is such that $f_{[k]}(\mathbb{C}) \subset \Sigma_{h_k}$. In particular, if $\Sigma_{h_k} \subset P_k V^{\rm sing}$, then $(X, V)$ is hyperbolic.*

PROOF. The main idea is to use the Ahlfors-Schwarz lemma, following the approach of [GrGr80]. However we will give here all necessary details because our setting is slightly different. Assume that there is a $k$-jet metric $h_k$ as in the hypotheses of Theorem 7.8. Let $\omega_k$ be a smooth hermitian metric on $T_{P_k V}$. By hypothesis, there exists $\varepsilon > 0$ such that

$$\langle \Theta_{h_k^{-1}}(\mathcal{O}_{P_k V}(1)) \rangle(\xi) \geqslant \varepsilon |\xi|^2_{\omega_k} \qquad \forall \xi \in V_k.$$

Moreover, by (5.4), $(\pi_k)_\star$ maps $V_k$ continuously to $\mathcal{O}_{P_k V}(-1)$ and the weight $e^\varphi$ of $h_k$ is locally bounded from above. Hence there is a constant $C > 0$ such that

$$|(\pi_k)_\star \xi|^2_{h_k} \leqslant C |\xi|^2_{\omega_k}, \qquad \forall \xi \in V_k.$$

Combining these inequalities, we find

$$\langle \Theta_{h_k^{-1}}(\mathcal{O}_{P_k V}(1)) \rangle(\xi) \geqslant \frac{\varepsilon}{C} |(\pi_k)_\star \xi|^2_{h_k}, \qquad \forall \xi \in V_k.$$

Now, let $f : \Delta_R \to X$ be a non constant holomorphic map tangent to $V$ on the disk $\Delta_R$. We use the line bundle morphism (5.6)

$$F = f'_{[k-1]} : T_{\Delta_R} \to f^\star_{[k]} \mathcal{O}_{P_k V}(-1)$$

to obtain a pullback metric

$$\gamma = \gamma_0(t)\, dt \otimes d\bar{t} = F^\star h_k \qquad \text{on } T_{\Delta_R}.$$

If $f_{[k]}(\Delta_R) \subset \Sigma_{h_k}$ then $\gamma \equiv 0$. Otherwise, $F(t)$ has isolated zeroes at all singular points of $f_{[k-1]}$ and so $\gamma(t)$ vanishes only at these points and at points of the degeneration set $(f_{[k]})^{-1}(\Sigma_{h_k})$ which is a polar set in $\Delta_R$. At other points, the Gaussian curvature of $\gamma$ satisfies

$$\frac{i\,\partial\bar\partial \log \gamma_0(t)}{\gamma(t)} = \frac{-2\pi\, (f_{[k]})^\star \Theta_{h_k}(\mathcal{O}_{P_k V}(-1))}{F^\star h_k} = \frac{\langle \Theta_{h_k^{-1}}(\mathcal{O}_{P_k V}(1)) \rangle (f'_{[k]}(t))}{|f'_{[k-1]}(t)|^2_{h_k}} \geqslant \frac{\varepsilon}{C},$$

since $f'_{[k-1]}(t) = (\pi_k)_\star f'_{[k]}(t)$. The Ahlfors-Schwarz lemma 3.2 implies that $\gamma$ can be compared with the Poincaré metric as follows:

$$\gamma(t) \leqslant \frac{2C}{\varepsilon} \frac{R^{-2}|dt|^2}{(1-|t|^2/R^2)^2} \qquad \Longrightarrow \qquad |f'_{[k-1]}(t)|^2_{h_k} \leqslant \frac{2C}{\varepsilon} \frac{R^{-2}}{(1-|t|^2/R^2)^2}.$$

If $f : \mathbb{C} \to X$ is an entire curve tangent to $V$ such that $f_{[k]}(\mathbb{C}) \not\subset \Sigma_{h_k}$, the above estimate implies as $R \to +\infty$ that $f_{[k-1]}$ must be a constant, hence also $f$. Now, if $\Sigma_{h_k} \subset P_k V^{\rm sing}$, the inclusion $f_{[k]}(\mathbb{C}) \subset \Sigma_{h_k}$ implies $f'(t) = 0$ at every point, hence $f$ is a constant and $(X, V)$ is hyperbolic. □

Combining Theorem 7.8 with 7.7 ii) and iii), we get the following consequences.

7.9. COROLLARY. *Assume that there exist integers $k, m > 0$ and an ample line bundle $L$ on $X$ such that $H^0(P_kV, \mathcal{O}_{P_kV}(m) \otimes \pi_{0,k}^\star L^{-1}) \simeq H^0(X, E_{k,m}(V^\star) \otimes L^{-1})$ has non zero sections $\sigma_1, \ldots, \sigma_N$. Let $Z \subset P_kV$ be the base locus of these sections. Then every entire curve $f : \mathbb{C} \to X$ tangent to $V$ is such that $f_{[k]}(\mathbb{C}) \subset Z$. In other words, for every global $\mathbb{G}_k$-invariant polynomial differential operator $P$ with values in $L^{-1}$, every entire curve $f$ must satisfy the algebraic differential equation $P(f) = 0$.* □

7.10. COROLLARY. *Let $(X, V)$ be a compact complex directed manifold. If $E_{k,m}V^\star$ is ample for some positive integers $k, m$, then $(X, V)$ is hyperbolic.* □

7.11. REMARK. Green and Griffiths [GrGr80] stated that Corollary 7.9 is even true with sections $\sigma_j \in H^0(X, E_{k,m}^{\mathrm{GG}}(V^\star) \otimes L^{-1})$, in the special case $V = T_X$ they consider. We refer to the recent preprint [SiYe96c] by Siu and Yeung for a detailed proof of this fact, based on a use of the well-known logarithmic derivative lemma in Nevanlinna theory (the original proof given in [GrGr80] does not seem to be complete, as it relies on an unsettled pointwise version of the Ahlfors-Schwarz lemma for general jet differentials); other proofs seem to have been circulating in the literature in the last years. We give here a very short proof for the case when $f$ is supposed to have a bounded derivative (thanks to Brody's theorem, this is enough if one is merely interested in proving hyperbolicity, thus Corollary 7.10 will be valid with $E_{k,m}^{\mathrm{GG}}V^\star$ in place of $E_{k,m}V^\star$). In fact, if $f'$ is bounded, one can apply the Cauchy inequalities to all components $f_j$ of $f$ with respect to a finite collection of coordinate patches covering $X$. As $f'$ is bounded, we can do this on sufficiently small discs $D(t, \delta) \subset \mathbb{C}$ of constant radius $\delta > 0$. Therefore all derivatives $f'$, $f''$, $\ldots$ $f^{(k)}$ are bounded. From this we conclude that $\sigma_j(f)$ is a bounded section of $f^\star L^{-1}$. Its norm $|\sigma_j(f)|_{L^{-1}}$ (with respect to any positively curved metric $|\ |_L$ on $L$) is a bounded subharmonic function, which is moreover strictly subharmonic at all points where $f' \neq 0$ and $\sigma_j(f) \neq 0$. This is a contradiction unless $f$ is constant or $\sigma_j(f) \equiv 0$. □

The above results justify the following definition and problems.

7.12. DEFINITION. *We say that $X$, resp. $(X, V)$, has non degenerate negative $k$-jet curvature if there exists a $k$-jet metric $h_k$ on $\mathcal{O}_{P_kV}(-1)$ with negative jet curvature such that $\Sigma_{h_k} \subset P_kV^{\mathrm{sing}}$.*

7.13. CONJECTURE. *Let $(X, V)$ be a compact directed manifold. Then $(X, V)$ is hyperbolic if and only if $(X, V)$ has nondegenerate negative $k$-jet curvature for $k$ large enough.*

This is probably a hard problem. In fact, we will see in the next section that the smallest admissible integer $k$ must depend on the geometry of $X$ and need not be uniformly bounded as soon as $\dim X \geqslant 2$ (even in the absolute case $V = T_X$). On the other hand, if $(X, V)$ is hyperbolic, we get for each integer $k \geqslant 1$ a generalized Kobayashi-Royden metric $\mathbf{k}_{(P_{k-1}V, V_{k-1})}$ on $V_{k-1}$ (see Definition 1.3), which can be also viewed as a $k$-jet metric $h_k$ on $\mathcal{O}_{P_kV}(-1)$ ; we will call it the *Grauert $k$-jet metric* of $(X, V)$, although it formally differs from the jet metric considered in [Gra89] (see also [DGr91]). By looking at the projection $\pi_k : (P_kV, V_k) \to (P_{k-1}V, V_{k-1})$, we see that the sequence $h_k$ is monotonic, namely $\pi_k^\star h_k \leqslant h_{k+1}$ for every $k$. If $(X, V)$ is hyperbolic, then $h_1$ is nondegenerate and therefore by monotonicity $\Sigma_{h_k} \subset P_kV^{\mathrm{sing}}$ for $k \geqslant 1$. Conversely, if the Grauert metric satisfies $\Sigma_{h_k} \subset P_kV^{\mathrm{sing}}$, it is easy to see that $(X, V)$ is hyperbolic. The following problem is thus especially meaningful.

7.14. PROBLEM. *Estimate the $k$-jet curvature $\Theta_{h_k^{-1}}(\mathcal{O}_{P_kV}(1))$ of the Grauert metric $h_k$ on $(P_kV, V_k)$ as $k$ tends to $+\infty$.*

## 8. Algebraic Criterion for the Negativity of Jet Curvature

Our goal is to show that the negativity of $k$-jet curvature implies strong restrictions of an algebraic nature, similar to property 2.1 ii). Using this we give examples, for any prescribed integer $k$, of hyperbolic projective surfaces which do not admit any $k$-jet metric of negative jet curvature.

8.1. THEOREM. *Let $(X, V)$ be a compact complex directed manifold and let $\omega$ be a hermitian metric on $X$. If $(X, V)$ has negative $k$-jet curvature, there exists a constant $\varepsilon > 0$ such that every closed irreducible curve $C \subset X$ tangent to $V$ satisfies*

$$-\chi(\overline{C}) = 2g(\overline{C}) - 2 \geqslant \varepsilon \deg_\omega(C) + \sum_{t\in\overline{C}} (m_{k-1}(t) - 1) > 0$$

*where $g(\overline{C})$ is the genus of the normalization $\nu : \overline{C} \to C \subset X$, and $m_k(t)$ is the multiplicity at point $t$ of the $k$-th lifting $\nu_{[k]} : \overline{C} \to P_kV$ of $\nu$.*

PROOF. By (5.6), we get a lifting $\nu_{[k]} : \overline{C} \to P_kV$ of the normalization map $\nu$, and there is a canonical map

$$\nu'_{[k-1]} : T_{\overline{C}} \to \nu^\star_{[k]}\mathcal{O}_{P_kV}(-1).$$

Let $t_j \in \overline{C}$ be the singular points of $\nu_{[k-1]}$, and let $m_j = m_{k-1}(t_j)$ be the corresponding multiplicity. Then $\nu'_{[k-1]}$ vanishes at order $m_j - 1$ at $t_j$ and thus we find

$$T_{\overline{C}} \simeq \nu^\star_{[k]}\mathcal{O}_{P_kV}(-1) \otimes \mathcal{O}_{\overline{C}}\Big(-\sum (m_j - 1)p_j\Big).$$

Taking any $k$-jet metric $h_k$ with negative jet curvature on $\mathcal{O}_{P_kV}(-1)$, the Gauss-Bonnet formula yields

$$2g(\overline{C}) - 2 = \int_{\overline{C}} \Theta(T^\star_{\overline{C}}) = \sum (m_j - 1) + \int_{\overline{C}} \nu^\star_{[k]}\Theta_{h_k^{-1}}(\mathcal{O}_{P_kV}(1)).$$

Now, the curvature hypothesis implies

$$\langle \Theta_{h_k^{-1}}(\mathcal{O}_{P_kV}(1))\rangle(\xi) \geqslant \varepsilon'|\xi|^2_{\omega_k} \geqslant \varepsilon''|(\pi_{0,k})_\star\xi|^2_\omega \qquad \forall \xi \in V_k,$$

for some $\varepsilon', \varepsilon'' > 0$ and some smooth hermitian metric $\omega_k$ on $P_kV$. As $\pi_{0,k} \circ \nu_{[k]} = \nu$, we infer from this $\nu^\star_{[k]}\Theta_{h_k^{-1}}(\mathcal{O}_{P_kV}(1)) \geqslant \nu^\star\omega$, hence

$$\int_{\overline{C}} \nu^\star_{[k]}\Theta_{h_k^{-1}}(\mathcal{O}_{P_kV}(1)) \geqslant \frac{\varepsilon''}{2\pi}\int_{\overline{C}} \nu^\star\omega = \varepsilon \deg_\omega(C)$$

with $\varepsilon = \varepsilon''/2\pi$. Theorem 8.1 follows. □

8.2. THEOREM. *Let $k \geqslant 1$ be any positive integer. Then there is a nonsingular algebraic surface $X$ (depending on $k$) which is hyperbolic, but does not carry any nondegenerate $k$-jet metric with negative jet curvature. In fact, given any two curves $\Gamma, \Gamma'$ of genus at least 2, the surface $X$ may be constructed as a fibration $X \to \Gamma$ in which one of the fibers $C_0$ is singular and has $\Gamma'$ as its normalization.*

PROOF. The idea is to construct $X$ in such a way that the singular fiber $C$ which is normalized by $\Gamma'$ violates the inequality obtained in Theorem 8.1. For this we need only having a singular point $t_0$ such that $m_{k-1}(t_0) - 1 > 2g(\overline{C}) - 2$, i.e., $m_{k-1}(t_0) \geqslant 2g(\Gamma')$. Moreover, as $\Gamma$ is hyperbolic, $X$ will be hyperbolic if and only if all fibers of $X \to \Gamma$ have geometric genus at least 2.

We first construct from $\Gamma'$ a singular curve $\Gamma''$ with normalization $\overline{\Gamma}'' = \Gamma'$, simply by modifying the structure sheaf $\mathcal{O}_{\Gamma'}$ at one given point $w_0 \in \Gamma'$. Let $t$ be a holomorphic coordinate on $\Gamma'$ at $w_0$. We replace $\mathcal{O}_{\Gamma',w_0} = \mathbb{C}\{t\}$ by $\mathcal{O}_{\Gamma'',w_0} = \mathbb{C}\{t^a, t^b\}$, where $a < b$ are relatively prime integers. The corresponding singularity is described by the germ of embedding $t \mapsto f(t) = (t^a, t^b)$ in $(\mathbb{C}^2, 0)$. Now, $f'(t) = (at^{a-1}, bt^{b-1})$, thus $[f'(t)] \in \mathbb{P}^1 \simeq \mathbb{C} \cup \{\infty\}$ is given by $[f'(t)] = \frac{b}{a}t^{b-a}$. By induction, we see that the singularity of the $j$-th lifting $f_{[j]}$ is described by the embedding

$$t \mapsto (t^a, t^b, c_1 t^{b-a}, \ldots, c_j t^{b-ja}) \in \mathbb{C}^{j+2}, \qquad c_j = a^{-j} b(b-a) \cdots (b-(j-1)a)$$

if $b > ja$. Then we have $m(f_{[j]}, 0) = \min(a, b - ja)$. If we take for instance $a = 2g(\Gamma')$ and $b = ka + 1$, then $m(f_{[k-1]}, 0) = a$. We embed $\Gamma''$ in some projective space $\mathbb{P}^n$ and let $C = p(\Gamma'')$ to be a generic projection to a plane $\mathbb{P}^2 \subset \mathbb{P}^n$ in such a way that $C$ has only $x_0 = p(w_0)$ and some nodes (ordinary double points) as its singular points. By construction, the Zariski tangent space to $\Gamma''$ at $w_0$ is 2-dimensional, so we may assume that $p$ projects that plane injectively into $T_{\mathbb{P}^2}$. Then we get a curve $C \subset \mathbb{P}^2$ with $\overline{C} = \Gamma'$, such that $m(\nu_{[k-1]}, w_0) = a = 2g(\overline{C})$, if $\nu : \overline{C} \to \mathbb{P}^2$ is the normalization.

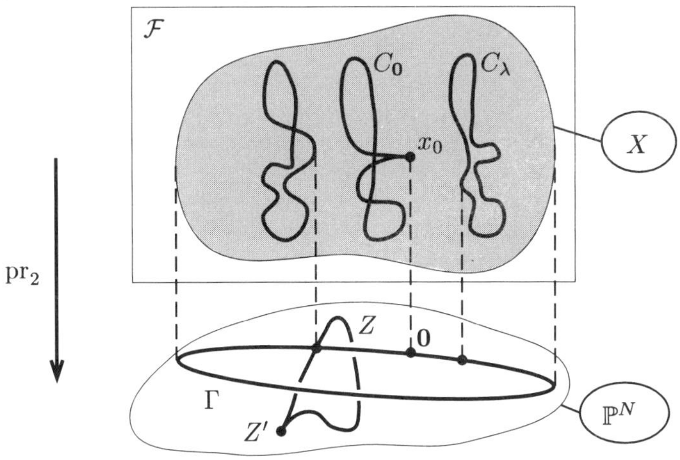


**Figure 1.** Construction of the surface $X$

Let $P_0(z_0, z_1, z_2) = 0$ be an equation of $C$ in $\mathbb{P}^2$. Since $C$ has geometric genus at least 2, we have $d = \deg P_0 \geqslant 4$. We complete $P_0$ into a basis $(P_0, \dots, P_N)$ of the space of homogeneous polynomials of degree $d$, and consider the universal family

$$\mathcal{F} = \left\{([z_0 : z_1 : z_2], [\lambda_0, \lambda_1, \dots, \lambda_N]) \in \mathbb{P}^2 \times \mathbb{P}^N \,;\, \sum \lambda_j P_j(z) = 0\right\} \subset \mathbb{P}^2 \times \mathbb{P}^N$$

of curves $C_\lambda = \{\sum \lambda_j P_j(z) = 0\}$ of degree $d$ in $\mathbb{P}^2$. As is well known, the set $Z$ of points $\lambda \in \mathbb{P}^N$ such that $C_\lambda$ is a singular curve is an algebraic hypersurface, and the set $Z' \subset Z$ of points $\lambda$ such that $C_\lambda$ has not just a node in its singularity set satisfies $\operatorname{codim} Z' \geqslant 2$. The curve $C = C_{\mathbf{0}}$ itself corresponds to the point $\mathbf{0} = [1 : 0 : \cdots : 0] \in Z'$. Since $\operatorname{codim} Z' \geqslant 2$, we can embed $\Gamma$ in $\mathbb{P}^N$ in such a way that $\Gamma \cap Z' = \{\mathbf{0}\}$. We then take $X \to \Gamma$ to be the family of curves $(C_\lambda)_{\lambda \in \Gamma}$. If $X$ is singular, we move $\Gamma$ by a generic automorphism of $\mathbb{P}^N$ leaving $\mathbf{0}$ fixed. Then, since $\mathcal{F}$ is smooth (it is a smooth $\mathbb{P}^{N-1}$ subbundle of $\mathbb{P}^2 \times \mathbb{P}^N$ over $\mathbb{P}^2$), Bertini's theorem implies that $X \smallsetminus C_{\mathbf{0}}$ will become nonsingular. That $X$ will be also nonsingular near $C_{\mathbf{0}}$ depends only on the following first order condition: if $[1 : \alpha\lambda_1^0 : \cdots : \alpha\lambda_N^0]$, $\alpha \in \mathbb{C}$, is the tangent line to $\Gamma$ at $\mathbf{0}$, then $\sum_{j \geqslant 1} \lambda_j^0 P_j(z)$ does not vanish at any of the singular points of $C_{\mathbf{0}}$. Now, all nonsingular fibers $C_\lambda$ of the fibration $X \to \Gamma$ have genus $(d-1)(d-2)/2 \geqslant 3$, and the singular ones other than $C_{\mathbf{0}}$ only have one node, so their genus is $(d-1)(d-2)/2 - 1 \geqslant 2$. □

If we make an assumption on the total jet curvature (as is the case with the algebraic sufficient conditions 7.7), Theorem 8.1 can be strengthened to curves which are not necessarily tangent to $V$, again by introducing the concept of deviation. We start with a general purpose statement.

8.3. Proposition. *Let $(X, V)$ be a compact complex directed manifold and let $L$ be a holomorphic line bundle over $X$. Assume that $L$ is equipped with a singular* *hermitian metric $h$ of degeneration set $\Sigma_h$, such that the curvature (computed in the sense of distributions) satisfies*

$$\Theta_h(L) \geqslant \alpha, \qquad \alpha_{\restriction V} \geqslant \delta\omega_{\restriction V}$$

*where $\delta$ is a positive constant, $\omega$ a smooth hermitian metric and $\alpha$ is a continuous real $(1,1)$-form on $X$. Then for every compact irreducible curve $C \subset X$ not contained in $\Sigma_h$, there exists a constant $\varepsilon > 0$ such that the following a priori inequality holds*

$$\max\big(L\cdot C, \mathrm{dev}^2_\omega(C/V)\big) \geqslant \varepsilon\ \deg_\omega(C).$$

PROOF. By the continuity of $\alpha$ and the compactness of $X$, our assumption $\alpha_{\restriction V} \geqslant \delta\omega$ implies that there is a constant $M > 0$ such that

$$\alpha + M\,\omega^{V^\perp} \geqslant \frac{\delta}{2}\omega$$

(to get this, one merely needs to apply the Cauchy-Schwarz inequality to mixed terms $V^\star \otimes (V^\perp)^\star$ in a hermitian form on $V$). In particular, we find

$$\Theta_h(L) + M\,\omega^{V^\perp} \geqslant \frac{\delta}{2}\omega$$

This inequality gives rise to a corresponding numerical inequality on every irreducible curve $C \not\subset \Sigma_h$, for the difference has a well defined and nonnegative restriction to $C$ (we use here the fact that the weight of $h$ is quasi-psh and locally bounded at some point of $C$, hence locally integrable along $C$). From this we infer

$$L\cdot C + M\,\mathrm{dev}^2_\omega(C/V) \geqslant \frac{\delta}{2}\deg_\omega(C),$$

and the left hand side is at most equal to $(M+1)\max\big(L\cdot C, \mathrm{dev}^2_\omega(C/V)\big)$. □

8.4. PROPOSITION. *Let $(X,V)$ be a compact complex directed manifold. Assume that there are integers $k,m > 0$, $\boldsymbol{b} \in \mathbb{N}^k$ such that $\mathcal{O}_{P_kV}(m) \otimes \mathcal{O}_{P_kV}(-\boldsymbol{b}\cdot D^\star)$ is an ample line bundle over $P_kV$. Then $(X,V)$ is hyperbolic and there exists $\varepsilon > 0$ such that every closed curve $C \subset X$ satisfies*

$$\max\Big(-\chi(\overline{C}) - \sum_{t\in\overline{C}}(m_{k-1}(t)-1), \mathrm{dev}^\infty_\omega(C/V)\Big) \geqslant \varepsilon\ \deg_\omega(C).$$

Proposition 8.4 is likely to be true also if we assume more generally that $(X,V)$ has non degenerate total $k$-jet curvature but, in this case, some technical difficulties appear in the construction of the required singular hermitian metric $h_k$ on $\mathcal{O}_{P_kT_X}(1)$ (see the proof below).

PROOF. The hyperbolicity of $(X,V)$ follows from 7.7 i) and Theorem 7.8. Now, the identity map defines a natural monomorphism $(X,V) \to (X,T_X)$ of directed manifolds and therefore induces an embedding $P_kV \hookrightarrow P_kT_X$ for each $k$. With respect to this embedding, we have

$$\mathcal{O}_{P_kT_X}(1)_{\restriction P_kV} = \mathcal{O}_{P_kV}(1),$$
$$\mathcal{O}_{P_kT_X}(m)\otimes\mathcal{O}_{P_kT_X}(-\boldsymbol{b}\cdot D^\star)_{\restriction P_kV} = \mathcal{O}_{P_kV}(m)\otimes\mathcal{O}_{P_kV}(-\boldsymbol{b}\cdot D^\star)$$

By our assumptions, $\mathcal{O}_{P_kT_X}(m)\otimes\mathcal{O}_{P_kT_X}(-\boldsymbol{b}\cdot D^\star)$ is ample over $P_kV$ and over the fibers of the projection $P_kT_X \to X$. Hence, we can find a smooth hermitian metric

$h_{k,m,\boldsymbol{b}}$ on $\mathcal{O}_{P_kT_X}(m)\otimes\mathcal{O}_{P_kT_X}(-\boldsymbol{b}\cdot D^\star)$ such that the curvature form is positive definite on a neighborhood $U$ of $P_kV$ and satisfies

$$\Theta\big(\mathcal{O}_{P_kT_X}(m)\otimes\mathcal{O}_{P_kT_X}(-\boldsymbol{b}\cdot D^\star)\big)\geqslant -C\pi_{k,0}^\star\omega$$

for some Kähler metric $\omega$ over $X$. This metric $h_{k,m,\boldsymbol{b}}$ gives rise to a hermitian metric $h_k$ on $\mathcal{O}_{P_kT_X}(1)$ with singularity set $\Sigma_{h_k}\subset P_k^{\rm sing}T_X$ and similar curvature properties, that is

$$\text{(8.5)}\qquad \begin{cases}\Theta_{h_k}(\mathcal{O}_{P_kT_X}(1))\geqslant -C\pi_{k,0}^\star\omega & \text{on } P_kT_X,\\ \Theta_{h_k}(\mathcal{O}_{P_kT_X}(1))\geqslant \delta\omega_k\geqslant\delta'\pi_{k,0}^\star\omega & \text{on } U\supset P_kV,\end{cases}$$

where $\omega_k$ is a hermitian metric on $P_kT_X$ and $\delta,\,\delta'>0$. Now, assume that the conclusion of Prop. 8.4 is wrong. Then there would exist a sequence of curves $(C_\ell)$ and a sequence of positive numbers $\varepsilon_\ell$ converging to 0, such that

$$\mathcal{O}_{P_kT_X}(1)\cdot C_{\ell,[k]}\leqslant\varepsilon_\ell\deg_\omega(C_\ell),\qquad \mathrm{dev}_\omega^\infty(C_\ell/V))\leqslant\varepsilon_\ell\deg_\omega(C_\ell)$$

where $C_{\ell,[k]}$ is the lifting of $C_\ell$ to $P_kT_X$ [indeed, we have $\mathcal{O}_{P_kT_X}(1)\cdot C_{\ell,[k]}=-\chi(\overline{C}_\ell)-\sum(m_{k-1}(t)-1)$]. Let $\nu_\ell:\overline{C}_\ell\to X$ be the normalization map. As $\mathrm{dev}_\omega^\infty(C_\ell/V))=\sup\nu_\ell^\star(\omega_{V^\perp})/d\widetilde{\sigma}$ where $d\sigma$ is the Poincaré metric and $d\widetilde{\sigma}$ the associated normalized metric, the second condition means

$$\sup\|\,\mathrm{pr}_{V^\perp}\nu_\ell'\|_{\sigma,\omega}^2=\sup\frac{\nu_\ell^\star(\omega_{V^\perp})}{d\sigma}\leqslant\frac{\varepsilon_\ell\deg_\omega(C_\ell)}{\int_{\overline{C}\ell}d\sigma}=\varepsilon_\ell\frac{\int_{\overline{C}\ell}\nu_\ell^\star\omega}{\int_{\overline{C}\ell}d\sigma}.$$

In addition to this, we have

$$\frac{\int_{\overline{C}\ell}\nu_\ell^\star\omega}{\int_{\overline{C}\ell}d\sigma}\leqslant R_\ell^2:=\sup\|\nu_\ell'\|_{\sigma,\omega}^2$$

and $R=\sup R_\ell<+\infty$, otherwise the proof of Prop. 2.9 would produce a non constant entire curve $g:\mathbb{C}\to X$ tangent to $V$, contradicting the hyperbolicity of $(X,V)$. An application of the Cauchy inequalities to the components of $\mathrm{pr}_{V^\perp}$ on sufficiently small disks in the universal covering of $\overline{C}_\ell$ and in suitable trivializations of $T_X/V$ shows that there is a constant $M_k\geqslant 0$ such that

$$\sup_{1\leqslant j\leqslant k}\|\,\mathrm{pr}_{V^\perp}\nu_\ell^{(j)}\|_{\sigma,\omega}^2\leqslant M_k\sup\|\,\mathrm{pr}_{V^\perp}\nu_\ell'\|_{\sigma,\omega}^2\leqslant M_k\varepsilon_\ell\frac{\int_{\overline{C}\ell}\nu_\ell^\star\omega}{\int_{\overline{C}\ell}d\sigma}.$$

As $\int_{\overline{C}_\ell}\|\nu_\ell'\|_{\sigma,\omega}^{-2}\nu_\ell^\star\omega=\int_{\overline{C}_\ell}d\sigma$, we infer

$$\text{(8.6)}\qquad \int_{\overline{C}_\ell}\frac{\sup_{1\leqslant j\leqslant k}\|\,\mathrm{pr}_{V^\perp}\nu_\ell^{(j)}\|_{\sigma,\omega}^2}{\|\nu_\ell'\|_{\sigma,\omega}^2}\nu_\ell^\star\omega\leqslant M_k\varepsilon_\ell\int_{\overline{C}\ell}\nu_\ell^\star\omega.$$

Since $U$ is a neighborhood of $P_kV$, there exists a constant $\eta>0$ such that

$$\frac{\sup_{1\leqslant j\leqslant k}\|\,\mathrm{pr}_{V^\perp}\nu_\ell^{(j)}(t)\|_{\sigma,\omega}^2}{\|\nu_\ell'(t)\|_{\sigma,\omega}^2}<\eta\implies\nu_{\ell,[k]}(t)\in U$$

for any $t \in \overline{C}_\ell$. By the integral estimate (8.6), the set $S_\eta$ of "bad points" $t \in \overline{C}_\ell$ at which the left hand inequality does not hold has area $< M_k\varepsilon_\ell \deg_\omega(C_\ell)/\eta$ with respect to $\nu_\ell^\star\omega$. By (8.5), we then get

$$\begin{aligned}\mathcal{O}_{P_kT_X}(1)\cdot C_{\ell,[k]} &= \int_{\overline{C}_\ell\smallsetminus S_\eta} \nu_{\ell,[k]}^\star\Theta(\mathcal{O}_{P_kT_X}(1)) + \int_{S_\eta}\nu_{\ell,[k]}^\star\Theta(\mathcal{O}_{P_kT_X}(1))\\ &\geqslant \delta'\int_{\overline{C}_\ell\smallsetminus S_\eta}\nu_\ell^\star\omega - C\int_{S_\eta}\nu_\ell^\star\omega\\ &= \big(\delta'(1-M_k\varepsilon_\ell/\eta) - CM_k\varepsilon_\ell/\eta\big)\deg_\omega(C_\ell).\end{aligned}$$

This contradicts our initial hypothesis that $\mathcal{O}_{P_kT_X}(1)\cdot C_{\ell,[k]} \leqslant \varepsilon_\ell\deg_\omega(C_\ell)$ when $\varepsilon_\ell$ is small enough. □

The above results lead in a natural way to the following questions, dealing with the "directed manifold case" of Kleiman's criterion (Kleiman's criterion states that a line bundle $L$ on $X$ is ample if and only if there exists $\varepsilon > 0$ such that $L\cdot C \geqslant \varepsilon\deg_\omega C$ for every curve $C \subset X$).

8.7. QUESTIONS. *Let $(X, V)$ be a compact directed manifold and let $L$ be a line bundle over $X$. Fix $p \in [2, +\infty]$.*

i) *Assume that*
$$\max\big(L\cdot C, \mathrm{dev}_\omega^p(C/V)\big) \geqslant \varepsilon\deg_\omega(C)$$
*for every algebraic curve $C \subset X$ (and some $\varepsilon > 0$). Does $L$ admit a smooth hermitian metric $h$ with $\Theta_h(L)_{\restriction V}$ positive definite?*

ii) *Assume more generally that there is an analytic subset $Y \subsetneq X$ such that* i) *holds for all curves $C \not\subset Y$. Does $L$ admit a singular hermitian metric $h$ with $\Theta_h(L)_{\restriction V}$ positive definite, and with degeneration set $\Sigma_h \subset Y$?*

iii) *Assume that there exists $\varepsilon > 0$ such that every closed curve $C \subset X$ satisfies*
$$\max\Big(-\chi(\overline{C}) - \sum_{t\in\overline{C}}(m_{k-1}(t)-1), \mathrm{dev}_\omega^p(C/V)\Big) \geqslant \varepsilon\ \deg_\omega(C).$$
*Does it follow that $(X, V)$ admits non degenerate negative $k$-jet (total) curvature?*

The answer to 8.7 i) is positive if $V$ is the vertical tangent sheaf of a smooth map $X \to S$, and in that case one can even restrict oneself to curves that are tangent to $V$ (i.e. vertical curves): this is just the relative version of Kleiman's criterion. However, in general, it is not sufficient to deal only with curves tangent to $V$ (if $X$ is an abelian variety and $V$ is a constant line subbundle of $T_X$ with non closed leaves, the condition required for algebraic curves $C$ is void, hence $L$ can be taken negative on $X$; then, of course, the curvature cannot be made positive along $V$.)

## 9. Proof of the Bloch Theorem

The core of the result can be expressed as a characterization of the Zariski closure of an entire curve drawn on a complex torus. The proof will be obtained as a simple consequence of the Ahlfors-Schwarz lemma (more specifically Theorem 7.8), combined with a jet bundle argument. Our argument works in fact without any algebraicity assumption on the complex tori under consideration (only the case of abelian or semi-abelian varieties seems to have been treated earlier).

9.1. THEOREM. *Let $Z$ be a complex torus and let $f : \mathbb{C} \to Z$ be a holomorphic map. Then the (analytic) Zariski closure $\overline{f(\mathbb{C})}^{\mathrm{Zar}}$ is a translate of a subtorus, i.e. of the form $a + Z'$, $a \in Z$, where $Z' \subset Z$ is a subtorus.*

The converse is of course also true: for any subtorus $Z' \subset Z$, we can choose a dense line $L \subset Z'$, and the corresponding map $f : \mathbb{C} \simeq a + L \hookrightarrow Z$ has Zariski closure $\overline{f(\mathbb{C})}^{\mathrm{Zar}} = a + Z'$.

PROOF (based on the ideas of [GrGr80]). Let $f : \mathbb{C} \to Z$ be an entire curve and let $X$ be the Zariski closure of $f(\mathbb{C})$. We denote by $Z_k = P_k(T_Z)$ the $k$-jet bundle of $Z$ and by $X_k$ the closure of $X_k^{\mathrm{reg}} = P_k(T_{X^{\mathrm{reg}}})$ in $Z_k$. As $T_Z$ is trivial, we have $Z_k = Z \times \mathbb{R}_{n,k}$ where $\mathbb{R}_{n,k}$ is the rational variety introduced in § 5. By Proposition 6.16 iii), there is a weight $\boldsymbol{a} \in \mathbb{N}^k$ such that $\mathcal{O}_{Z_k}(\boldsymbol{a})$ is relatively very ample. This means that there is a very ample line bundle $\mathcal{O}_{\mathbb{R}_{n,k}}(\boldsymbol{a})$ over $\mathbb{R}_{n,k}$ such that $\mathcal{O}_{Z_k}(\boldsymbol{a}) = \mathrm{pr}_2^\star \mathcal{O}_{\mathbb{R}_{n,k}}(\boldsymbol{a})$. Consider the map $\Phi_k : X_k \to \mathbb{R}_{n,k}$ which is the restriction to $X_k$ of the second projection $Z_k \to \mathbb{R}_{n,k}$. By fonctoriality, we have $\mathcal{O}_{X_k}(\boldsymbol{a}) = \Phi_k^\star \mathcal{O}_{\mathbb{R}_{n,k}}(\boldsymbol{a})$.

Define $B_k \subset X_k$ to be the set of points $x \in X_k$ such that the fiber of $\Phi_k$ through $x$ is positive dimensional. Assume that $B_k \neq X_k$. By Proposition 7.2 ii), $\mathcal{O}_{X_k}(\boldsymbol{a})$ carries a hermitian metric with degeneration set $B_k$ and with strictly positive definite curvature on $X_k$ (if necessary, blow-up $X_k$ along the singularities and push the metric forward). Theorem 7.8 shows that $f_{[k]}(\mathbb{C}) \subset B_k$, and this is of course also true if $B_k = X_k$. The inclusion $f_{[k]}(\mathbb{C}) \subset B_k$ means that through every point $f_{[k]}(t_0)$ there is a germ of positive dimensional variety in the fiber $\Phi_k^{-1}(\Phi_k(f_{[k]}(t_0)))$, say a germ of curve $t' \mapsto u(t') = (z(t'), j_k) \in X_k \subset Z \times \mathbb{R}_{n,k}$ with $u(0) = f_{[k]}(t_0) = (z_0, j_k)$ and $z_0 = f(t_0)$. Then $(z(t'), j_k)$ is the image of $f_{[k]}(t_0)$ by the $k$-th lifting of the translation $\tau_s : z \mapsto z + s$ defined by $s = z(t') - z_0$. Now, we have $f(\mathbb{C}) \not\subset X^{\mathrm{sing}}$ since $X$ is the Zariski closure of $f(\mathbb{C})$, and we may therefore choose $t_0$ so that $f(t_0) \in X^{\mathrm{reg}}$ and $f(t_0)$ is a regular point. Let us define

$$A_k(f) = \big\{s \in Z \ :\ f_{[k]}(t_0) \in P_k(X) \cap P_k(\tau_{-s}(X))\big\}.$$

Clearly $A_k(f)$ is an analytic subset of $Z$ containing the curve $t' \mapsto s(t') = z(t') - z_0$ through 0. Since

$$A_1(f) \supset A_2(f) \supset \cdots \supset A_k(f) \supset \cdots,$$

the Noetherian property shows that the sequence stabilizes at some $A_k(f)$. Therefore, there is a curve $D(0, r) \to Z$, $t' \mapsto s(t')$ such that the infinite jet $j_\infty$ defined by $f$ at $t_0$ is $s(t')$-translation invariant for all $t'$. By uniqueness of analytic continuation, we conclude that $s(t') + f(t) \in X$ for all $t \in \mathbb{C}$ and $t' \in D(0, r)$. As $X$ is the Zariski closure of $f(\mathbb{C})$, we must have $s(t') + X \subset X$ for all $t' \in D(0, r)$; also, $X$ is irreducible, thus we have in fact $s(t') + X = X$. Define

$$W = \big\{s \in Z\,;\ s + X = X\big\}.$$

Then $W$ is a closed positive dimensional subgroup of $Z$. Let $p : Z \to Z/W$ be the quotient map. As $Z/W$ is a complex torus with $\dim Z/W < \dim Z$, we conclude by induction on dimension that the curve $\widehat{f} = p \circ f : \mathbb{C} \to Z/W$ has its Zariski closure $\widehat{X} := \overline{\widehat{f}(\mathbb{C})}^{\mathrm{Zar}} = p(X)$ equal to a translate $\widehat{s} + \widehat{T}$ of some subtorus $\widehat{T} \subset Z/W$. Since

$X$ is $W$-invariant, we get $X = s + p^{-1}(\widehat{T})$, where $p^{-1}(\widehat{T})$ is a closed subgroup of $Z$. This implies that $X$ is a translate of a subtorus, as expected. □

We now state two simple corollaries, and then the "Bloch theorem" itself (see also [Och77], [Nog77, 81, 84], [Kaw80] for other approaches in the algebraic case).

9.2. COROLLARY. *Let $X$ be a complex analytic subvariety in a complex torus $Z$. Then $X$ is hyperbolic if and only if $X$ does not contain any translate of a subtorus.*

9.3. COROLLARY. *Let $X$ be a complex analytic subvariety of a complex torus $Z$. Assume that $X$ is not a translate of a subtorus. Then every entire curve drawn in $X$ is analytically degenerate.*

9.4. BLOCH THEOREM. *Let $X$ be a compact complex Kähler variety such that the irregularity $q = h^0(X, \Omega_X^1)$ is larger than the dimension $n = \dim X$. Then every entire curve drawn in $X$ is analytically degenerate.*

Here $X$ may be singular and $\Omega_X^1$ can be defined in any reasonable way (direct image of the $\Omega_{\widehat{X}}^1$ of a desingularization $\widehat{X}$ or direct image of $\Omega_U^1$ where $U$ is the set of regular points in the normalization of $X$).

PROOF. By blowing-up, we may assume that $X$ is smooth. Then the Albanese map $\alpha : X \to \mathrm{Alb}(X)$ sends $X$ onto a proper subvariety $Y \subset \mathrm{Alb}(X)$ (as $\dim Y \leqslant \dim X < \dim \mathrm{Alb}(X)$), and $Y$ is not a translate of a subtorus by the universal property of the Albanese map. Hence, for every entire curve $f : \mathbb{C} \to X$ we infer that $\alpha \circ f : \mathbb{C} \to Y$ is analytically degenerate; it follows that $f$ itself is analytically degenerate. □

## 10. Logarithmic Jet Bundles and a Conjecture of Lang

We want to report here briefly about an important question raised by S. Lang, namely *whether the complement of an ample divisor in an Abelian variety is Kobayashi hyperbolic*? This statement has been first settled in the affirmative by Siu and Yeung [SiYe96b], using an extension of some of the methods used to prove Bloch's theorem. We will adopt here a slightly different approach of G. Dethloff and S. Lu [DLu96], who followed a suggestion made during our Santa Cruz lectures in July 1995. Namely, there should exist a theory of *logarithmic jet bundles* extending Semple's construction, which would allow to study the hyperbolicity properties of open varieties of the form $X \smallsetminus D$ ($D$ being a divisor in a projective variety $X$). We give here a short account of Dethloff and Lu's technique, referring to [DLu96] for details, and to [SiYe96b], [Nog96a] for alternative approaches.

Let $(X, V)$ be a compact directed manifold and $D$ a reduced divisor in $X$. Recall that the sheaf $\Omega_X^1\langle D\rangle$ of *holomorphic* 1*-forms with logarithmic poles along* $D$ is defined to be the coherent sheaf generated by $\Omega_X^1$ and $ds_j/s_j$, where $s_j = 0$ are local equations for the irreducible components of $D$. It is locally free as soon as $D$ is a normal crossing divisor (we may always suppose that this is the case after blowing up $X$ along smooth centers contained in $D$). Similarly, one introduces the sheaf $\mathcal{O}(V^\star\langle D\rangle)$ to be the sheaf of holomorphic 1-forms along $V$ with logarithmic poles along $D$ (this is just the quotient of $\Omega_X^1\langle D\rangle$ by the conormal sheaf $V^o \subset V^\star$ of $V$). It is locally free as soon as $D$ has normal crossings and its components $D_{(j)}$ are everywhere tranversal to $V$ (by this we mean that $T_{D_{(j)}} + V = T_X$ along $D_{(j)}$).

Under this assumption, we consider the dual (locally free) sheaves

$$(10.1)\qquad \mathcal{O}(T_X\langle D\rangle) := (\Omega^1_X\langle D\rangle)^\star, \qquad \mathcal{O}(V\langle D\rangle) := (V^\star\langle D\rangle)^\star.$$

One easily checks that $\mathcal{O}(T_X\langle D\rangle)$ (resp. $\mathcal{O}(V\langle D\rangle)$) is the sheaf of germs of vector fields in $\mathcal{O}(T_X)$ (resp. $\mathcal{O}(V)$) which are tangent to each component of $D$. Now, one defines a sequence

$$(10.2)\qquad (X_k, D_k, V_k)$$

of logarithmic $k$-jet bundles exactly in the same way as we proceeded in section § 4 and § 5: if $X_0 = X$, $D_0 = D$ and $V_0 = V\langle D\rangle$, one sets inductively $X_k = P(V_{k-1})$, $D_k = (\pi_{k,0})^{-1}(D)$, and $V_k$ is the set of tangent vectors in $T_{X_k}\langle D_k\rangle$ which project into the line defined by the tautological line bundle $\mathcal{O}_{X_k}(-1) \subset \pi_k^\star V_{k-1}$. In this case, the direct image formula given in Theorem 6.8 reads

$$(10.3)\qquad (\pi_{k,0})_\star\mathcal{O}_{X_k}(m) = \mathcal{O}(E_{k,m}V^\star\langle D\rangle),$$

where $\mathcal{O}(E_{k,m}V^\star\langle D\rangle)$ is the sheaf generated by all polynomial differential operators in the derivatives of order $1, 2, \ldots, k$ of the components $f_1, \ldots, f_n$, together with the extra function $\log s_j(f)$ along the $j$-th component of $D$.

Just as before, a *logarithmic k-jet metric* is just a singular hermitian metric on $\mathcal{O}_{X_k}(-1)$. Dethloff and Lu [DLu96] state the following results 10.4–10.9, which extend our results of sections § 7 and § 9 (most of these results can already be derived from [SiYe96b] as well).

10.4. THEOREM. *Let $(X, D, V)$ be as above. Let $\Sigma_{k,m}$ be the union of the base locus of $\mathcal{O}_{X_k}(m)$ and of the positive dimensional fibers of the canonical map defined by the corresponding linear system. Then*

i) *If $\Sigma_{k,m} \neq X_k$, there exists a logarithmic k-jet metric $h_k$ with strictly negative jet curvature and $\Sigma_{h_k} = \Sigma_{k,m}$.*
ii) *For every entire map $f : \mathbb{C} \to X \smallsetminus D$ tangent to $V$, one has $f_{[k]}(\mathbb{C}) \subset \Sigma_{k,m}$.*
iii) *For every holomorphic map $f : \Delta^\star \to X \smallsetminus D$ tangent to $V$ (where $\Delta^\star$ is the punctured disk), one has: either $f$ extends to a holomorphic map $\overline{f} : \Delta \to X$ or $f_{[k]}(\Delta^\star) \subset \Sigma_{k,m}$.*

Consider now a semi-abelian variety $Z$ (that is, a commutative algebraic group $\mathbb{C}^n/\Gamma$), and let $D \subset Z$ be a reduced algebraic divisor.

10.5. THEOREM. *Let $(Z, D)$ be as above.*

i) *For every entire curve $f : \mathbb{C} \to Z$, the Zariski closure $\overline{f(\mathbb{C})}^{\mathrm{Zar}}$ is a translate of an algebraic subgroup of $Z$.*
ii) *For every entire curve $f : \mathbb{C} \to Z \smallsetminus D$, we have $\overline{f(\mathbb{C})}^{\mathrm{Zar}} \cap D = \emptyset$.*

10.6. COROLLARY. *If $D$ has non empty intersection with any translate of an algebraic subgroup of $Z$ of positive dimension, then $Z \smallsetminus D$ is Brody hyperbolic. This is true e.g. if $Z$ is abelian and $D$ is ample.*

10.7. REMARK. Theorem 10.5 and its corollary have been obtained independently by Noguchi [Nog96a], and also by Siu-Yeung [SiYe96b] in the case of abelian varieties. Both of their proofs use value distribution theory, whilst the present approach uses only negative curvature arguments. It is likely that Theorem 10.5 can be extended to arbitrary commutative (non necessarily algebraic) Lie groups $\mathbb{C}^n/\Gamma$.

10.8. THEOREM. *The following properties hold true.*

A) *Let* $f : \Delta^\star \to Z$ *be a holomorphic map. Then either it extends to a holomorphic map* $\overline{f} : \Delta \to \overline{Z}$ *or there exists a maximal algebraic subgroup* $Z'$ *of* $Z$ *of positive dimension such that* $\overline{f(\Delta^\star)}^{\mathrm{Zar}}$ *is foliated by translates of* $Z'$.

B) *Let* $f : \Delta^\star \to Z \smallsetminus D$ *be a holomorphic map. Then one of the following holds:*

   i) $f$ *extends to a holomorphic map* $\overline{f} : \Delta \to Z$.

   ii) $\overline{f(\Delta^\star)}^{\mathrm{Zar}} \cap D = \emptyset$.

   iii) *There exists an algebraic subgroup* $Z''$ *of* $Z'$ *of positive dimension such that* $\overline{f(\Delta^\star)}^{\mathrm{Zar}} \cap D$ *is foliated by translates of* $Z''$.

C) *Assume here that* $Z$ *is an abelian variety and let* $f : \Delta^\star \to Z \smallsetminus D$ *be a holomorphic map. Then one of the following holds:*

   i) $f$ *extends to a holomorphic map* $\overline{f} : \Delta \to Z$.

   ii) *There exists an algebraic subgroup* $Z''$ *of* $Z'$ *of positive dimension such that* $D$ *is foliated by translates of* $Z''$.

Part A) of Theorem 10.8 is due to Noguchi [Nog96a] (again with a proof based on Nevanlinna theory).

10.9. COROLLARY. *If* $Z$ *is abelian and* $D$ *is ample, then every holomorphic map* $f : \Delta^\star \to Z \smallsetminus D$ *extends to a holomorphic map* $\overline{f} : \Delta \to Z$.

## 11. Projective Meromorphic Connections and Wronskians

We describe here an important method introduced by Siu [Siu87] and later developped by Nadel [Nad89], which is powerful enough to provide explicit examples of algebraic hyperbolic surfaces. It yields likewise interesting results about the algebraic degeneration of entire curves in higher dimensions. The main idea is to use meromorphic connections with low pole orders, and the associated Wronskian operators. In this way, Nadel produced examples of hyperbolic surfaces in $\mathbb{P}^3$ for any degree of the form $p = 6k + 3 \geqslant 21$. We present here a variation of Nadel's method, based on the more general concept of *partial projective connection*, which allows us to extend his result to all degrees $p \geqslant 11$. This approach is inspired from a recent work of J. El Goul [EG96], and is in some sense a formalization of his strategy.

Let $X$ be a complex $n$-dimensional manifold. A *meromorphic connection* $\nabla$ on $T_X$ is a $\mathbb{C}$-linear sheaf morphism

$$\mathcal{M}(U, T_X) \longrightarrow \mathcal{M}(U, \Omega^1_X \otimes T_X)$$

(where $\mathcal{M}(U, \bullet)$ stands for meromorphic sections over $U$), satisfying the Leibnitz rule

$$\nabla(fs) = df \otimes s + f\nabla s$$

whenever $f \in \mathcal{M}(U)$ (resp. $s \in \mathcal{M}(U, T_X)$) is a meromorphic function (resp. section of $T_X$). Let $(z_1, \ldots, z_n)$ be holomorphic local coordinates on an open set $U \subset X$. The *Christoffel symbols* of $\nabla$ with respect to these coordinates are the coefficients $\Gamma^\lambda_{j\mu}$ such that

$$\Gamma^\lambda_\mu = \sum_{1 \leqslant j \leqslant n} \Gamma^\lambda_{j\mu} dz_j = \lambda\text{-th component of } \nabla\Big(\frac{\partial}{\partial z_\mu}\Big).$$

The associated *connection form* on $U$ is the tensor

$$\Gamma = \sum_{1\leqslant j,\lambda,\mu\leqslant n} \Gamma_{j\mu}^{\lambda}\, dz_j \otimes dz_\mu \otimes \frac{\partial}{\partial z_\lambda} \in \mathcal{M}(U, T_X^\star \otimes T_X^\star \otimes T_X).$$

Then, for all local sections $v = \sum_{1\leqslant\lambda\leqslant n} v_\lambda \frac{\partial}{\partial z_\lambda}$, $w = \sum_{1\leqslant\lambda\leqslant n} w_\lambda \frac{\partial}{\partial z_\lambda}$ of $\mathcal{M}(U, T_X)$, we get

$$\nabla v = \sum_{1\leqslant\lambda\leqslant n} \Big(dv_\lambda + \sum_{1\leqslant\mu\leqslant n} \Gamma_\mu^\lambda v_\mu\Big)\frac{\partial}{\partial z_\lambda} = dv + \Gamma\cdot v,$$

$$\nabla_w v = \sum_{1\leqslant j,\lambda\leqslant n} \Big(w_j \frac{\partial v_\lambda}{\partial z_j} + \sum_{1\leqslant\mu\leqslant n} \Gamma_{j\mu}^\lambda w_j v_\mu\Big)\frac{\partial}{\partial z_\lambda} = d_w v + \Gamma\cdot(w,v).$$

The connection $\nabla$ is said to be *symmetric* if it satisfies $\nabla_v w - \nabla_w v = [v,w]$, or equivalently, if the Christoffel symbols $\Gamma_{j\mu}^\lambda = \Gamma_{\mu j}^\lambda$ are symmetric in $j$, $\mu$.

We now turn ourselves to the important concept of *Wronskian operator.* Let $B$ be the divisor of poles of $\nabla$, that is, the divisor of the least common multiple of all denominators occuring in the meromorphic functions $\Gamma_{j\mu}^\lambda$. If $\beta \in H^0(X, \mathcal{O}(B))$ is the canonical section of divisor $B$, then the operator $\beta\nabla$ has holomorphic coefficients. Given a holomorphic curve $f : D(0,r) \to X$ whose image does not lie in the support $|B|$ of $B$, one can define inductively a sequence of covariant derivatives

$$f', \quad f''_\nabla = \nabla_{f'}(f'), \ \ldots\ , \ f_\nabla^{(k+1)} := \nabla_{f'}(f_\nabla^{(k)}).$$

These derivatives are given in local coordinates by the explicit inductive formula

$$f_\nabla^{(k+1)}(t)_\lambda = \frac{d}{dt}\big(f_\nabla^{(k)}(t)_\lambda\big) + \sum_{1\leqslant\mu\leqslant n} (\Gamma_{j\mu}^\lambda \circ f)\, f_j'\, f_\nabla^{(k)}(t)_\mu. \tag{11.1}$$

Therefore, if $\operatorname{Im} f \not\subset |B|$, one can define the *Wronskian* of $f$ relative to $\nabla$ as

$$W_\nabla(f) = f' \wedge f''_\nabla \wedge \cdots \wedge f_\nabla^{(n)}. \tag{11.2}$$

Clearly, $W_\nabla(f)$ is a meromorphic section of $f^\star(\Lambda^n T_X)$. By induction $\beta(f)^{k-1} f_\nabla^{(k)}$ is holomorphic for all $k \geqslant 1$. We infer that $\beta(f)^{n(n-1)/2} W_\nabla(f)$ is holomorphic and can be seen as a holomorphic section of the line bundle $f^\star(\Lambda^n T_X \otimes \mathcal{O}_X(\frac{1}{2}n(n-1)B)$. From (11.1) and (11.2) we see that $P = \beta^{n(n-1)/2} W_\nabla$ is a global holomorphic polynomial operator $f \mapsto P(f', f'', \ldots, f^{(n)})$ of order $n$ and total degree $n(n+1)/2$, with values in $\Lambda^n T_X \otimes \mathcal{O}_X(\frac{1}{2}n(n-1)B)$. Moreover, if we take a biholomorphic reparametrization $\varphi$, we get inductively

$$(f\circ\varphi)_\nabla^{(k)} = (\varphi')^k f_\nabla^{(k)} \circ \varphi + \text{linear combination of } f_\nabla^{(j)} \circ \varphi,\ j < k.$$

Therefore

$$W_\nabla(f\circ\varphi) = (\varphi')^{n(n+1)} W_\nabla(f)$$

and $\beta^{n(n-1)/2} W_\nabla$ can be viewed as a section

$$\beta^{n(n-1)/2} W_\nabla \in H^0(X, E_{n,n(n+1)/2} T_X^\star \otimes L^{-1}), \tag{11.3}$$

where $L$ is the line bundle

$$L = K_X \otimes \mathcal{O}_X\Big(-\frac{1}{2}n(n-1)B\Big).$$

From this, we get the following theorem, which is essentially due to [Siu87] (with a more involved proof based on suitable generalizations of Nevanlinna's second main theorem).

11.4. THEOREM (Y.T. Siu). *Let $X$ be a compact complex manifold equipped with a meromorphic connection $\nabla$ of pole divisor $B$. If $K_X \otimes \mathcal{O}_X(-\frac{1}{2}n(n-1)B)$ is ample, then for every non constant entire curve $f : \mathbb{C} \to X$, one has either $f(\mathbb{C}) \subset |B|$ or $W_\nabla(f) \equiv 0$.*

PROOF. By Corollary 7.9 applied with $P = \beta^{n(n-1)/2}W_\nabla$, we conclude that $\beta^{n(n-1)/2}(f)W_\nabla(f) \equiv 0$, whence the result. □

11.5. BASIC OBSERVATION. It is not necessary to know all Christoffel coefficients of the meromorphic connection $\nabla$ in order to be able to compute its Wronskian $W_\nabla$. In fact, assume that $\widetilde{\nabla}$ is another connection such that there are meromorphic 1-forms $\alpha$, $\beta$ with

$$\widetilde{\nabla} = \nabla + \alpha \otimes \mathrm{Id}_{T_X} + (\beta \otimes \mathrm{Id}_{T_X})_{\tau_{12}}, \qquad \text{i.e.,}$$
$$\widetilde{\nabla}_w v = \nabla_w v + \alpha(w)v + \beta(v)w,$$

where $\tau_{12}$ means transposition of first and second arguments in the tensors of $T^\star_X \otimes T^\star_X \otimes T_X$. Then $W_\nabla = W_{\widetilde{\nabla}}$. Indeed, the defining formula $f^{(k+1)}_{\widetilde{\nabla}} = \widetilde{\nabla}_{f'}(f^{(k)}_{\widetilde{\nabla}})$ implies that $f^{(k+1)}_{\widetilde{\nabla}} = \nabla_{f'}(f^{(k)}_{\widetilde{\nabla}}) + \alpha(f')f^{(k)}_{\widetilde{\nabla}} + \beta(f^{(k)}_{\widetilde{\nabla}})f'$, and an easy induction then shows that the $\widetilde{\nabla}$ derivatives can be expressed as linear combinations with meromorphic coefficients

$$f^{(k)}_{\widetilde{\nabla}}(t) = f^{(k)}_{\nabla}(t) + \sum_{1\leqslant j<k} \gamma_j(t)\, f^{(j)}_{\nabla}(t). \qquad \square$$

The essential consequence of Remark 11.5 is that we need only have a "partial projective connection" $\nabla$ on $X$, in the following sense.

11.6. DEFINITION. *A (meromorphic) partial projective connection $\nabla$ on $X$ is a section of the quotient sheaf of meromorphic connections modulo addition of meromorphic tensors in $(\Omega^1_X \otimes \mathrm{Id}_{T_X}) \oplus (\Omega^1_X \otimes \mathrm{Id}_{T_X})_{\tau_{12}}$. In other words, it can be defined as a collection of meromorphic connections $\nabla_j$ relative to an open covering $(U_j)$ of $X$, satisfying the compatibility conditions*

$$\nabla_k - \nabla_j = \alpha_{jk} \otimes \mathrm{Id}_{T_X} + (\beta_{jk} \otimes \mathrm{Id}_{T_X})_{\tau_{12}}$$

*for suitable meromorphic 1-forms $\alpha_{jk}$, $\beta_{jk}$ on $U_j \cap U_k$.*

If we have similar more restrictive compatibility relations with $\beta_{jk} = 0$, the connection form $\Gamma$ is just defined modulo $\Omega^1_X \otimes \mathrm{Id}_{T_X}$ and can thus be seen as a 1-form with values in the Lie algebra $\mathfrak{pgl}(n,\mathbb{C}) = \mathfrak{sl}(n,\mathbb{C})$ rather than in $\mathfrak{gl}(n,\mathbb{C})$. Such objects are sometimes referred to as "projective connections", although this terminology has been also employed in a completely different meaning. In any event, Proposition 11.4 extends (with a completely identical proof) to the more general case where $\nabla$ is just a partial projective connection. Accordingly, the pole divisor $B$ can be taken to be the pole divisor of the trace free part

$$\Gamma^0 = \Gamma \mod (\Omega^1_X \otimes \mathrm{Id}_{T_X}) \oplus (\Omega^1_X \otimes \mathrm{Id}_{T_X})_{\tau_{12}}.$$

Such partial projective connections occur in a natural way when one considers quotient varieties under the action of a Lie group. Indeed, let $W$ be a complex manifold in which a connected complex Lie group $G$ acts freely and properly (on the left, say), and let $X = W/G$ be the quotient complex manifold. We denote by $\pi : W \to X$ the projection. Given a connection $\widetilde{\nabla}$ on $W$ and a local section $\sigma : U \to W$ of $\pi$, one gets an induced connection on $T_{X|U}$ by putting

$$\nabla = \pi_\star \circ (\sigma^\star \widetilde{\nabla}), \tag{11.7}$$

where $\sigma^\star\widetilde{\nabla}$ is the induced connection on $\sigma^\star T_W$ and $\pi_\star : T_W \to \pi^\star T_X$ is the projection. Of course, the connection $\nabla$ may depend on the choice of $\sigma$, but we nevertheless have the following simple criterion ensuring that it yields an intrinsic partial projective connection.

11.8. LEMMA. *Let $\widetilde{\nabla} = d + \widetilde{\Gamma}$ be a meromorphic connection on $W$. Assume that $\widetilde{\nabla}$ satisfies the following conditions:*

i) *$\widetilde{\nabla}$ is $G$-invariant;*

ii) *there are meromorphic 1-forms $\alpha$, $\beta \in \mathcal{M}(W, T_{W/X})$ along the relative tangent bundle of $X \to W$, such that for all $G$-invariant holomorphic vector fields $v$, $\tau$ on $W$ (possibly only defined locally over $X$) such that $\tau$ is tangent to the $G$-orbits, the vector fields*

$$\widetilde{\nabla}_\tau v - \alpha(\tau)v, \qquad \widetilde{\nabla}_v \tau - \beta(\tau)v$$

*are again tangent to the $G$-orbits ($\alpha$ and $\beta$ are thus necessarily $G$-invariant, and $\alpha = \beta$ if $\widetilde{\nabla}$ is symmetric).*

*Then Formula (11.7) yields a partial projective connection $\nabla$ which is globally defined on $X$ and independent of the choice of the local sections $\sigma$.*

PROOF. Since the expected conclusions are local with respect to $X$, it is enough to treat the case when $W = X \times G$ and $G$ acts on the left on the second factor. Then $W/G \simeq X$ and $\pi : W \to X$ is the first projection. If $d_G$ is the canonical left-invariant connection on $G$, we can write $\widetilde{\nabla}$ as

$$\widetilde{\nabla} = d_X + d_G + \widetilde{\Gamma}, \qquad \widetilde{\Gamma} = \widetilde{\Gamma}(x, g), \quad x \in X, \ g \in G,$$

where $d_X$ is some connection on $X$, e.g. the "coordinate derivative" taken with respect to given local coordinates $(z_1, \ldots, z_n)$ on $X$. Then $\widetilde{\nabla}$ is left invariant on $W = X \times G$ if and only if $\widetilde{\Gamma}(x, g) = \Gamma(x)$ is independent of $g \in G$ (this is meaningful since the tangent bundle to $G$ is trivial), and condition ii) means that

$$\Gamma(x) \cdot (\tau, v) - \alpha(\tau)v \quad \text{and} \quad \Gamma(x) \cdot (v, \tau) - \beta(\tau)v$$

are tangent to the $G$-orbits. A local section $\sigma : U \to W$ of $\pi$ can be written $\sigma(x) = (x, h(x))$ for some holomorphic function $h : U \to G$. Formula (11.7) says more explicitly that

$$\nabla_w v = \pi_\star\big((\sigma^\star\widetilde{\nabla})_w v\big) = \pi_\star\big(d_{\sigma_\star w}\sigma_\star v + (\widetilde{\Gamma} \circ \sigma) \cdot (\sigma_\star w, \sigma_\star v)\big).$$

Let $v = \sum v_j(z)\partial/\partial z_j$, $w = \sum w_j(z)\partial/\partial z_j$ be local vector fields on $U \subset X$. Since $\sigma_\star v = v + dh(v)$, we get

$$\begin{aligned}(\sigma^\star\widetilde{\nabla})_w v &= d_{w+dh(w)}(v + dh(v)) + \widetilde{\Gamma}(x, h(x)) \cdot \big(w + dh(w), v + dh(v)\big)\\ &= d_w v + d^2h(w, v) + \Gamma(x) \cdot (w + dh(w), v + dh(v)).\end{aligned}$$

As $v$, $w$, $dh(v)$, $dh(w)$ depend only on $X$, they can be seen as $G$-invariant vector fields over $W$, and $dh(v)$, $dh(w)$ are tangent to the $G$-orbits. Hence

$$\Gamma(x)\cdot(dh(w),v)-\alpha(dh(w))v,\quad \Gamma(x)\cdot(w,dh(v))-\beta(dh(v))w,\quad \Gamma(x)\cdot(dh(w),dh(v))$$

are tangent to the $G$-orbits, i.e., in the kernel of $\pi_\star$. We thus obtain

$$\nabla_w v=\pi_\star\big((\sigma^\star\widetilde{\nabla})_w v\big)=d_w v+\Gamma(x)\cdot(w,v)+\alpha(dh(w))v+\beta(dh(v))w.$$

From this it follows by definition that the local connections $\nabla_{\restriction U_j}$ defined by various sections $\sigma_j : U_j\to W$ can be glued together to define a global partial projective connection $\nabla$ on $X$. □

11.9. Remark. Lemma 11.8 is also valid when $\widetilde{\nabla}$ is a partial projective connection. Hypothesis 11.8 ii) must then hold with local meromorphic 1-forms $\alpha_j$, $\beta_j\in\mathcal{M}(\widetilde{U}_j,T_{W/X})$ relatively to some open covering $\widetilde{U}_j$ of $W$. □

In the special case $\mathbb{P}^n=(\mathbb{C}^{n+1}\smallsetminus\{0\})/\mathbb{C}^\star$, we get

11.10. Corollary. *Let $\widetilde{\nabla}=d+\widetilde{\Gamma}$ be a meromorphic connection on $\mathbb{C}^{n+1}$. Let $\varepsilon=\sum z_j\partial/\partial z_j$ be the Euler vector field on $\mathbb{C}^{n+1}$ and $\pi:\mathbb{C}^{n+1}\smallsetminus\{0\}\to\mathbb{P}^n$ be the canonical projection. Then $\widetilde{\nabla}$ induces a meromorphic partial projective connection on $\mathbb{P}^n$ provided that*

i) *the Christoffel symbols $\Gamma^\lambda_{j\mu}$ are homogeneous rational functions of degree $-1$ (homothety invariance of the connection $\widetilde{\nabla}$);*

ii) *there are meromorphic functions $\alpha$, $\beta$ and meromorphic 1-forms $\gamma$, $\eta$ such that*

$$\widetilde{\Gamma}\cdot(\varepsilon,v)=\alpha v+\gamma(v)\varepsilon,\qquad \widetilde{\Gamma}\cdot(w,\varepsilon)=\beta w+\eta(w)\varepsilon$$

*for all vector fields $v$, $w$.* □

Now, our goal is to study certain hypersurfaces $Y$ of sufficiently high degree in $\mathbb{P}^n$. Assume for the moment that $Y$ is an hypersurface in some $n$-dimensional manifold $X$, and that $Y$ is defined locally by a holomorphic equation $s=0$. We say that $Y$ is *totally geodesic* with respect to a meromorphic connection $\nabla$ on $X$ if $Y$ is not contained in the pole divisor $|B|$ of $\nabla$, and for all pairs $(v,w)$ of (local) vector fields tangent to $Y$ the covariant derivative $\nabla_w v$ is again tangent to $Y$. (Notice that this concept also makes sense when $\nabla$ is a partial projective connection.) If $Y$ is totally geodesic, the ambient connection $\nabla$ on $T_X$ induces by restriction a connection $\nabla_{\restriction Y}$ on $T_Y$.

We now want to derive explicitly a condition for the hypersurface $Y=\{s=0\}$ to be totally geodesic in $(X,\nabla)$. A vector field $v$ is tangent to $Y$ if and only if $ds\cdot v=0$ along $s=0$. By taking the differential of this identity along another vector field $w$ tangent to $Y$, we find

$$d^2s\cdot(w,v)+ds\cdot(d_w v)=0 \tag{11.11}$$

along $s=0$ (this is meaningful only with respect to some local coordinates). On the other hand, the condition that $\nabla_w v=d_w v+\Gamma\cdot(w,v)$ is tangent to $Y$ is

$$ds\cdot\nabla_w v=ds\cdot(d_w v)+ds\circ\Gamma\cdot(w,v)=0.$$

By subtracting the above from (11.11), we get the following equivalent condition: $(d^2s-ds\circ\Gamma)\cdot(w,v)=0$ for all vector fields $v,w$ in the kernel of $ds$ along $s=0$. Therefore we obtain the

11.12. CHARACTERIZATION OF TOTALLY GEODESIC HYPERSURFACES. *The hypersurface $Y = \{s = 0\}$ is totally geodesic with respect to $\nabla$ if and only if there are holomorphic 1-forms $a = \sum a_j dz_j$, $b = \sum b_j dz_j$ and a 2-form $c = \sum c_{j\mu} dz_j \otimes dz_\mu$ such that*

$$\nabla^\star(ds) = d^2 s - ds \circ \Gamma = a \otimes ds + ds \otimes b + s\, c$$

*in a neighborhood of every point of $Y$ (here $\nabla^\star$ is the induced connection on $T_Y^\star$).*

From this, we derive the following useful lemma.

11.13. LEMMA. *Let $Y \subset X$ be an analytic hypersurface which is totally geodesic with respect to a meromorphic connection $\nabla$, and let $n = \dim X = \dim Y + 1$. Let $f : D(0, R) \to X$ be a holomorphic curve such that $W_\nabla(f) \equiv 0$. Assume that there is a point $t_0 \in D(0, R)$ such that*

i) *$f(t_0)$ is not contained in the poles of $\nabla$ ;*

ii) *the system of vectors $(f'(t), f''_\nabla(t), \ldots, f_\nabla^{(n-1)}(t))$ achieves its generic rank (i.e. its maximal rank) at $t = t_0$ ;*

iii) *$f(t_0) \in Y$ and $f'(t_0)$, $f''_\nabla(t_0), \ldots, f_\nabla^{(n-1)}(t_0) \in T_{Y, f(t_0)}$.*

*Then $f(D(0, R)) \subset Y$.*

PROOF. Since $W_\nabla(f) \equiv 0$, the vector fields $f'$, $f''_\nabla, \ldots, f_\nabla^{(n)}$ are linearly dependent and satisfy a non trivial relation

$$u_1(t) f'(t) + u_2(t) f''_\nabla(t) + \cdots + u_n(t) f_\nabla^{(n)}(t) = 0$$

with suitable meromorphic coefficients $u_j(t)$ on $D(0, R)$. If $u_n$ happens to be $\equiv 0$, we take $\nabla$-derivatives in the above relation so as to reach another relation with $u_n \not\equiv 0$. Hence we can always write

$$f_\nabla^{(n)} = v_1 f' + v_2 f''_\nabla + \cdots + v_{n-1} f_\nabla^{(n-1)}$$

for some meromorphic functions $v_1, \ldots, v_{n-1}$. We can even prescribe the $v_j$ to be 0 eXcept for indices $j = j_k \in \{1, \ldots, n-1\}$ such that $(f_\nabla^{(j_k)}(t))$ is a minimal set of generators at $t = t_0$. Then the coefficients $v_j$ are uniquely defined and are holomorphic near $t_0$. By taking further derivatives, we conclude that $f_\nabla^{(k)}(t_0) \in T_{X, f(t_0)}$ for all $k$. We now use the assumption that $X$ is totally geodesic to prove the following claim: if $s = 0$ is a local equation of $Y$, the $k$-th derivative $\frac{d^k}{dt^k}(s \circ f(t))$ can be expressed as a holomorphic linear combination

$$\frac{d^k}{dt^k}\big(s \circ f(t)\big) = \gamma_{0k}(t)\, s \circ f(t) + \sum_{1 \leqslant j \leqslant k} \gamma_{jk}(t)\, ds_{f(t)} \cdot f_\nabla^{(j)}(t)$$

on a neighborhood of $t_0$. This will imply $\frac{d^k}{dt^k}(s \circ f)(t_0) = 0$ for all $k \geqslant 0$, hence $s \circ f \equiv 0$. Now, the above claim is clearly true for $k = 0, 1$. By taking the derivative and arguing inductively, we need only show that

$$\frac{d}{dt}\big(ds_{f(t)} \cdot f_\nabla^{(j)}(t)\big)$$

is again a linear combination of the same type. However, Leibnitz's rule for covariant differentiations together with 11.12 yield

$$\begin{aligned}\frac{d}{dt}\big(ds_{f(t)}\cdot f_\nabla^{(j)}(t)\big) &= ds_{f(t)}\cdot\Big(\frac{\nabla}{dt}f_\nabla^{(j)}(t)\Big)+\nabla^\star(ds)_{f(t)}\cdot\big(f'(t),f_\nabla^{(j)}(t)\big)\\ &= ds\cdot f_\nabla^{(j+1)}(t)+\big(a\cdot f'(t)\big)\big(ds\cdot f_\nabla^{(j)}(t)\big)\\ &\quad+\big(ds\cdot f'(t)\big)\big(b\cdot f_\nabla^{(j)}(t)\big)+\big(s\circ f(t)\big)\big(c\cdot(f'(t),f_\nabla^{(j)}(t))\big),\end{aligned}$$

as desired. □

If $Y=\{s=0\}\subset X$ is given and a connection $\nabla$ on $X$ is to be found so that $Y$ is totally geodesic, condition 11.12 amounts to solving a highly underdetermined linear system of equations

$$\frac{\partial^2 s}{\partial z_j\partial z_\mu}-\sum_{1\leqslant\lambda\leqslant n}\Gamma_{j\mu}^\lambda\frac{\partial s}{\partial z_\lambda}=a_j\frac{\partial s}{\partial z_\mu}+b_\mu\frac{\partial s}{\partial z_j}+s\,c_{j\mu},\qquad 1\leqslant j,\mu\leqslant n,$$

in terms of the unknowns $\Gamma_{j\mu}^\lambda$, $a_j$, $b_\mu$ and $c_{j\mu}$. Nadel's idea is to take advantage of this indeterminacy to achieve that all members in a large linear system $(Y_\alpha)$ of hypersurfaces are totally geodesic with respect to $\nabla$. The following definition is convenient.

11.14. DEFINITION. *For any $(n+2)$-tuple of integers $(p,k_0,k_1,\ldots,k_n)$ with $0<k_j<p/2$, let $\mathcal{S}_{p;\,k_0,\ldots,k_n}$ be the space of homogeneous polynomials*

$$s\in\mathbb{C}[z_0,z_1,\ldots,z_n]$$

*of degree $p$ on $\mathbb{C}^{n+1}$ such that every monomial of $s$ is a product of a power $z_j^{p-k_j}$ of one of the variables with a lower degree monomial of degree $k_j$. Any polynomial $s\in\mathcal{S}_{p;\,k_0,\ldots,k_n}$ admits a unique decomposition*

$$s=s_0+s_1+\cdots+s_n,\qquad s_j\in\mathcal{S}_{p;\,k_0,\ldots,k_n}$$

*where $s_j$ is divisible by $z_j^{p-k_j}$.*

Given a homogeneous polynomial $s=s_0+s_1+\cdots+s_n\in\mathcal{S}_{p;\,k_0,\ldots,k_n}$, we consider the linear system

$$Y_\alpha=\big\{\alpha_0s_0+\alpha_1s_1+\cdots+\alpha_ns_n=0\big\},\quad \alpha=(\alpha_0,\ldots,\alpha_n)\in\mathbb{C}^n. \tag{11.15}$$

Our goal is to study smooth varieties $Z$ which arise as complete intersections $Z=Y_{\alpha^1}\cap\cdots\cap Y_{\alpha^q}$ of members in the linear system (the $\alpha^j$ being linearly independent elements in $\mathbb{C}^{n+1}$). For this, we want to construct a (partial projective) meromorphic connection $\nabla$ on $\mathbb{P}^n$ such that all $Y_\alpha$ are totally geodesic. Corollary 11.10 shows that it is enough to construct a meromorphic connection $\widetilde\nabla=d+\widetilde\Gamma$ on $\mathbb{C}^{n+1}$ satisfying 11.10 i) and ii), such that the conic affine varieties $\widetilde Y_\alpha\subset\mathbb{C}^{n+1}$ lying over the $Y_\alpha$ are totally geodesic with respect to $\widetilde\nabla$. Now, Characterization 11.12 yields a sufficient condition in terms of the linear system of equations

$$\sum_{0\leqslant\lambda\leqslant n}\widetilde\Gamma_{j\mu}^\lambda\frac{\partial s_\kappa}{\partial z_\lambda}=\frac{\partial^2 s_\kappa}{\partial z_j\partial z_\mu},\qquad 0\leqslant j,\kappa,\mu\leqslant n. \tag{11.16}$$

(We just fix the choice of $a_j$, $b_\mu$ and $c_{j\mu}$ to be 0). This linear system can be considered as a collection of decoupled linear systems in the unknowns $(\widetilde\Gamma_{j\mu}^\lambda)_\lambda$, when $j$ and $\mu$ are fixed. Each of these has format $(n+1)\times(n+1)$ and can be solved by Cramer's rule if the principal determinant

$$\delta := \det\left(\frac{\partial s_\kappa}{\partial z_\lambda}\right)_{0\leqslant\kappa,\lambda\leqslant n} \not\equiv 0 \tag{11.17}$$

is not identically zero. We always assume in the sequel that this non degeneracy assumption is satisfied. As $\partial s_\kappa/\partial z_\lambda$ is homogeneous of degree $p-1$ and $\partial^2 s_\kappa/\partial z_j\partial z_\mu$ is homogeneous of degree $p-2$, the solutions $\widetilde{\Gamma}^\lambda_{j\mu}(z)$ are homogeneous rational functions of degree $-1$ (condition 11.10 i)). Moreover, $\widetilde{\nabla}$ is symmetric, for $\partial^2 s/\partial z_j\partial z_\mu$ is symmetric in $j,\mu$. Finally, if we multiply (11.16) by $z_j$ and take the sum, Euler's identity yields

$$\sum_{0\leqslant j,\lambda\leqslant n} z_j\widetilde{\Gamma}^\lambda_{j\mu}\frac{\partial s_\kappa}{\partial z_\lambda} = \sum_{0\leqslant j\leqslant n} z_j\frac{\partial^2 s_\kappa}{\partial z_j\partial z_\mu} = (p-1)\frac{\partial s_\kappa}{\partial z_\mu}, \qquad 0\leqslant\kappa,\mu\leqslant n.$$

The non degeneracy assumption implies $(\sum_j z_j\widetilde{\Gamma}^\lambda_{j\mu})_{\lambda\mu} = (p-1)\,\mathrm{Id}$, hence

$$\widetilde{\Gamma}(\varepsilon,v) = \widetilde{\Gamma}(v,\varepsilon) = (p-1)v$$

and condition 11.10 ii) is satisfied. From this we infer

11.18. PROPOSITION. *Let $s = s_0+\cdots+s_n \in \mathcal{S}_{p;\,k_0,\ldots,k_n}$ be satisfying the non degeneracy condition $\delta := \det(\partial s_\kappa/\partial z_\lambda)_{0\leqslant\kappa,\lambda\leqslant n}\not\equiv 0$. Then the solution $\widetilde{\Gamma}$ of the linear system* (11.16) *provides a partial projective meromorphic connection on $\mathbb{P}^n$ such that all hypersurfaces*

$$Y_\alpha = \{\alpha_0 s_0+\cdots+\alpha_n s_n = 0\}$$

*are totally geodesic. Moreover, the divisor of poles $B$ of $\nabla$ has degree at most equal to $n+1+\sum k_j$.*

PROOF. Only the final degree estimate on poles has to be checked. By Cramer's rule, the solutions are expressed in terms of ratios

$$\widetilde{\Gamma}^\lambda_{j\mu} = \frac{\delta^\lambda_{j\mu}}{\delta},$$

where $\delta^\lambda_{j\mu}$ is the determinant obtained by replacing the column of index $\lambda$ in $\det(\partial s_\kappa/\partial z_\lambda)_{0\leqslant\kappa,\lambda\leqslant n}$ by the column $(\partial^2 s_\kappa/\partial z_j\partial z_\mu)_{0\leqslant\kappa\leqslant n}$. Now, $\partial s_\kappa/\partial z_\lambda$ is a homogeneous polynomial of degree $p-1$ which is divisible by $z_\kappa^{p-k_\kappa-1}$, hence $\delta$ is a homogeneous polynomial of degree $(n+2)(p-1)$ which is divisible by $\prod z_j^{p-k_j-1}$. Similarly, $\partial^2 s_\kappa/\partial z_j\partial z_\mu$ has degree $p-2$ and is divisible by $z_\kappa^{p-k_\kappa-2}$. This implies that $\delta^\lambda_{j\mu}$ is divisible by $\prod z_j^{p-k_j-2}$. After removing this common factor in the numerator and denominator, we are left with a denominator of degree

$$\sum_{0\leqslant j\leqslant n}\big((p-1)-(p-k_j-2)\big) = \sum(k_j+1) = n+1+\sum k_j,$$

as stated. □

An application of Theorem 11.4 then yields the following theorem on certain complete intersections in projective spaces.

11.19. THEOREM. *Let $s \in \mathcal{S}_{p;\,k_0,\ldots,k_{n+q}} \subset \mathbb{C}[z_0, z_1, \ldots, z_{n+q}]$ be a homogeneous polynomial satisfying the non degeneracy assumption* $\det(\partial s_\kappa/\partial z_\lambda) \not\equiv 0$ *in* $\mathbb{C}^{n+q+1}$. *Let*

$$Y_\alpha = \{\alpha_0 s_0 + \alpha_1 s_1 + \cdots + \alpha_{n+q} s_{n+q} = 0\} \subset \mathbb{P}^{n+q}$$

*be the corresponding linear system, and let*

$$Z = Y_{\alpha^1} \cap \cdots \cap Y_{\alpha^q} \subset \mathbb{P}^{n+q}$$

*be a smooth $n$-dimensional complete intersection, for some linearly independent elements $\alpha^j \in \mathbb{C}^{n+q+1}$ such that $ds_{\alpha^1} \wedge \cdots \wedge ds_{\alpha^q}$ does not vanish along $Z$. Assume that $Z$ is not contained in the set of poles $|B|$ of the meromorphic connection $\nabla$ defined by* (11.16), *nor in any of the coordinate hyperplanes $z_j = 0$, and that*

$$pq > n + q + 1 + \frac{1}{2}n(n-1)\Big(n + q + 1 + \sum k_j\Big).$$

*Then every nonconstant entire curve $f : \mathbb{C} \to Z$ is algebraically degenerate and satisfies either*

i) $f(\mathbb{C}) \subset Z \cap |B|$ *or*

ii) $f(\mathbb{C}) \subset Z \cap Y_\alpha$ *for some member $Y_\alpha$ which does not contain $Z$.*

PROOF. By Proposition 11.18, the pole divisor of $\nabla$ has degree at most equal to $n+q+1+\sum k_j$, hence, if we let $B = \mathcal{O}(n+q+1+\sum k_j)$, we can find a section $\beta \in H^0(\mathbb{P}^{n+q}, B)$ such that the operator $f \mapsto \beta^{n(n+1)/2}(f)\, W_{Z,\nabla}(f)$ is holomorphic. Moreover, as $Z$ is smooth, the adjunction formula yields

$$K_Z = \big(K_{\mathbb{P}^{n+q}} \otimes \mathcal{O}(pq)\big)_{\restriction Z} = \mathcal{O}_Z(pq - n - q - 1).$$

By (11.3), the differential operator $\beta^{n(n-1)/2}(f)\, W_{Z,\nabla}(f)$ defines a section in $H^0(Z, E_{n,n(n+1)/2} T^\star_Z \otimes L^{-1})$ with

$$\begin{aligned} L &= K_Z \otimes \mathcal{O}_Z\Big(-\frac{1}{2}n(n-1)B\Big) \\ &= \mathcal{O}_Z\Big(pq - n - q - 1 - \frac{1}{2}n(n-1)\Big(n+q+1+\sum k_j\Big)\Big). \end{aligned}$$

Hence, if $f(\mathbb{C}) \not\subset |B|$, we know by Theorem 11.4 that $W_{Z,\nabla}(f) \equiv 0$. Fix a point $t_0 \in \mathbb{C}$ such that $f(t_0) \notin |B|$ and $(f'(t_0), f''_\nabla(t_0), \ldots, f^{(n)}_\nabla(t_0))$ is of maximal rank $r < n$. There must exist an hypersurface $Y_\alpha \not\supset Z$ such that

$$f(t_0) \in Y_\alpha, \quad f'(t_0),\ f''_\nabla(t_0), \ldots,\ f^{(n)}_\nabla(t_0) \in T_{Y_\alpha, f(t_0)}.$$

In fact, these conditions amount to solve a linear system of equations

$$\sum_{0 \leqslant j \leqslant n+q} \alpha_j s_j(f(t_0)) = 0, \qquad \sum_{0 \leqslant j \leqslant n+q} \alpha_j ds_j(f^{(j)}_\nabla(t_0)) = 0$$

in the unknowns $(\alpha_0, \alpha_1, \ldots, \alpha_{n+q}) = \alpha$, which has rank $\leqslant r+1 \leqslant n$. Hence the solutions form a vector space Sol of dimension at least $q+1$, and we can find a solution $\alpha$ which is linearly independent from $\alpha^1, \ldots, \alpha^q$. We complete $(\alpha, \alpha^1, \ldots, \alpha^q)$ into a basis of $\mathbb{C}^{n+q+1}$ and use the fact that the determinant $\delta = \det(\partial s_\kappa/\partial s_\lambda)$ does not vanish identically on $Z$, since

$$Z \cap \{\delta = 0\} \subset Z \cap \Big(|B| \cup \Big\{\prod z_j = 0\Big\}\Big) \subsetneq Z.$$

From this we see that $\sum \alpha_j ds_j$ does not vanish identically on $Z$, in particular $Z \not\subset Y_\alpha$. By taking a generic element $\alpha \in \mathrm{Sol}$, we get a smooth $n$-dimensional hypersurface $Z_\alpha = Y_\alpha \cap Y_{\alpha^2} \cap \cdots \cap Y_{\alpha^q}$ in $W = Y_{\alpha^2} \cap \cdots \cap Y_{\alpha^q}$. Lemma 11.13 applied to the pair $(Z_\alpha, W)$ shows that $f(\mathbb{C}) \subset Z_\alpha$, hence $f(\mathbb{C}) \subset Z \cap Z_\alpha = Z \cap Y_\alpha$, as desired. □

If we want to decide whether $Z$ is hyperbolic, we are thus reduced to decide whether the hypersurfaces $Z \cap |B|$ and $Z \cap Y_\alpha$ are hyperbolic. This may be a very hard problem, especially if $Z \cap |B|$ and $Z \cap Y_\alpha$ are singular. (In the case of a smooth intersection $Z \cap Y_\alpha$, we can of course apply the theorem again to $Z' = Z \cap Y_\alpha$ and try to argue by induction). However, when $Z$ is a surface, $Z \cap |B|$ and $Z \cap Y_\alpha$ are curves and the problem can in principle be solved directly through explicit genus calculations.

11.20. EXAMPLES.

i) Consider the Fermat hypersurface of degree $p$

$$Z = \{z_0^p + z_1^p + \cdots + z_{n+1}^p = 0\}$$

in $\mathbb{P}^{n+1}$, which is defined by an element in $\mathcal{S}_{p;\,0,\dots,0}$. A simple calculation shows that $\delta = p^{n+2} \prod z_j^{p-1} \not\equiv 0$ and that the Christoffel symbols are given by $\widetilde{\Gamma}^j_{jj} = (p-1)/z_j$ (with all other coefficients being equal to 0). Theorem 11.19 shows that all nonconstant entire curves $f : \mathbb{C} \to Y$ are algebraically degenerate when

$$p > n + 2 + \frac{1}{2} n(n-1)(n+2).$$

In fact the term $\frac{1}{2}n(n-1)(n+2)$ coming from the pole order estimate of the Wronskian is by far too pessimistic. A more precise calculation shows in that case that $(z_0 \cdots z_{n+1})^{n-1}$ can be taken as a denominator for the Wronskian. Hence the algebraic degeneracy occurs for $p > n+2+(n+2)(n-1)$, i.e., $p \geqslant (n+1)^2$. However, the Fermat hypersurfaces are not hyperbolic. For instance, when $n = 2$, they contain rational lines $z_1 = \omega z_0$, $z_3 = \omega' z_2$ associated with any pair $(\omega, \omega')$ of $p$-th roots of $-1$.

ii) Following J. El Goul ([EG96, 97]), let us consider surfaces $Z \subset \mathbb{P}^3$ of the form

$$Z = \{z_0^p + z_1^p + z_2^p + z_3^{p-2}(\varepsilon_0 z_0^2 + \varepsilon_1 z_1^2 + \varepsilon_2 z_2^2 + z_3^2) = 0\},$$

defined by the element in $\mathcal{S}_{p;\,0,0,0,2}$ such that $s_3 = z_3^{p-2}(\varepsilon_0 z_0^2 + \varepsilon_1 z_1^2 + \varepsilon_2 z_2^2 + z_3^2)$ and $s_j = z_j^p$ for $0 \leqslant j \leqslant 2$. One can check that $Z$ is smooth provided that

$$\sum_{j \in J} \varepsilon_j^{\frac{p}{p-2}} \neq \frac{2}{p-2}\Big(-\frac{p}{2}\Big)^{\frac{p}{p-2}}, \qquad \forall J \subset \{0,1,2\}, \tag{11.21}$$

for any choice of complex roots of order $p-2$. The connection $\widetilde{\nabla} = d + \widetilde{\Gamma}$ is computed by solving linear systems with principal determinant $\delta = \det(\partial s_\kappa / \partial z_\lambda)$

equal to

$$\begin{vmatrix} pz_0^{p-1} & 0 & 0 & 0 \\ 0 & pz_1^{p-1} & 0 & 0 \\ 0 & 0 & pz_2^{p-1} & 0 \\ 2\varepsilon_0 z_0 z_3^{p-2} & 2\varepsilon_1 z_1 z_3^{p-2} & 2\varepsilon_2 z_2 z_3^{p-2} & (p-2)z_3^{p-3}\big(\varepsilon_0 z_0^2+\varepsilon_1 z_1^2+\varepsilon_2 z_2^2+\frac{p}{p-2}z_3^2\big) \end{vmatrix}$$
$$= p^3(p-2)z_0^{p-1}z_1^{p-1}z_2^{p-1}z_3^{p-3}\Big(\varepsilon_0 z_0^2+\varepsilon_1 z_1^2+\varepsilon_2 z_2^2+\frac{p}{p-2}z_3^2\Big)\not\equiv 0.$$

The numerator of $\widetilde{\Gamma}^{\lambda}_{j\mu}$ is obtained by replacing the column of index $\lambda$ of $\delta$ by $(\partial^2 s_\kappa/\partial z_j \partial z_\mu)_{0\leqslant\kappa\leqslant 3}$, and $z_0^{p-2}z_1^{p-2}z_2^{p-2}z_3^{p-4}$ cancels in all terms. Hence the pole order of $\widetilde{\nabla}$ and of $W_{\widetilde{\nabla}}$ is 6 (as given by Proposition 11.18), with

$$z_0 z_1 z_2 z_3\Big(\varepsilon_0 z_0^2+\varepsilon_1 z_1^2+\varepsilon_2 z_2^2+\frac{p}{p-2}z_3^2\Big)$$

as the denominator, and its zero divisor as the divisor $B$. The condition on $p$ we get is $p > n+2+6 = 10$. An explicit calculation shows that all curves $Z\cap|B|$ and $Z\cap Y_\alpha$ have geometric genus $\geqslant 2$ under the additional hypothesis

$$\text{(11.22)}\qquad \begin{cases} \text{none of the pairs } (\varepsilon_i,\varepsilon_j) \text{ is equal to } (0,0), \\ \varepsilon_i/\varepsilon_j \neq -\theta^2 \text{ whenever } \theta \text{ is a root of } \theta^p=-1. \end{cases}$$

[(11.22) excludes the existence of lines in the intersections $Z\cap Y_\alpha$.]

11.23. Corollary. *Under conditions* (11.21) *and* (11.22), *the algebraic surface*

$$Z=\big\{z_0^p+z_1^p+z_2^p+z_3^{p-2}(\varepsilon_0 z_0^2+\varepsilon_1 z_1^2+\varepsilon_2 z_2^2+z_3^2)=0\big\}\subset\mathbb{P}^3$$

*is smooth and hyperbolic for all* $p\geqslant 11$. □

Another question which has raised considerable interest is to decide when the complement $\mathbb{P}^2\smallsetminus C$ of a plane curve $C$ is hyperbolic. If $C=\{\sigma=0\}$ is defined by a polynomial $\sigma(z_0,z_1,z_2)$ of degree $p$, we can consider the surface $X$ in $\mathbb{P}^3$ defined by $z_3^p=\sigma(z_0,z_1,z_2)$. The projection

$$\rho : X\to\mathbb{P}^2,\qquad (z_0,z_1,z_2,z_3)\mapsto(z_0,z_1,z_2)$$

is a finite $p:1$ morphism, ramified along $C$. It follows that $\mathbb{P}^2\smallsetminus C$ is hyperbolic if and only if its unramified covering $X\smallsetminus\rho^{-1}(C)$ is hyperbolic; hence a sufficient condition is that $X$ itself is hyperbolic. If we take $\varepsilon_2=0$ in Cor. 11.23 and exchange the roles of $z_2$, $z_3$, we get the following

11.24. Corollary. *Consider the plane curve*

$$C=\big\{z_0^p+z_1^p+z_2^{p-2}(\varepsilon_0 z_0^2+\varepsilon_1 z_1^2+z_2^2)=0\big\}\subset\mathbb{P}^2,\qquad \varepsilon_0,\,\varepsilon_1\in\mathbb{C}^\star.$$

*Assume that neither of the numbers* $\varepsilon_0$, $\varepsilon_1$, $\varepsilon_0+\varepsilon_1$ *is equal to* $\frac{2}{p-2}\big(-\frac{p}{2}\big)^{\frac{p}{p-2}}$ *and that* $\varepsilon_1/\varepsilon_0\neq-\theta^2$ *whenever* $\theta^p=-1$. *Then* $\mathbb{P}^2\smallsetminus C$ *is hyperbolic.* □

## 12. Decomposition of Jets in Irreducible Representations

Let us first briefly recall the definition of the Schur fonctors $\Gamma^\bullet$ (they are frequently denoted $S_\bullet$ in the literature, but we want to avoid any confusion with ordinary symmetric powers). Let $V$ be a complex vector space of dimension $r$. To the set of nonincreasing $r$-tuples $(a_1, a_2, \ldots, a_r) \in \mathbb{Z}^r$, $a_1 \geqslant a_2 \geqslant \cdots \geqslant a_r$, one associates in a fonctorial way a collection of vector spaces $\Gamma^{(a_1,a_2,\ldots,a_r)}V$ which provide the list of all irreducible representations of the linear group $\mathrm{GL}(V)$, up to isomorphism (here, $(a_1, \ldots, a_r)$ is the highest weight of the action of a maximal torus $(\mathbb{C}^\star)^r \subset \mathrm{GL}(V)$). The Schur fonctors can be defined in an elementary way as follows. Let $\mathbb{U}_r = \left\{ \begin{pmatrix} 1 & 0 \\ \star & 1 \end{pmatrix} \right\}$ be the group of lower triangular unipotent $r \times r$ matrices. If all $a_j$ are nonnegative, one defines

$$\Gamma^{(a_1,a_2,\ldots,a_r)}V \subset S^{a_1}V \otimes \cdots \otimes S^{a_r}V$$

to be the set of polynomials $P(\xi_1, \ldots, \xi_r)$ on $(V^\star)^r$ which are homogeneous of degree $a_j$ with respect to $\xi_j$ and which are invariant under the left action of $\mathbb{U}_r$ on $(V^\star)^r = \mathrm{Hom}(V, \mathbb{C}^r)$, namely such that

$$P(\xi_1, \ldots, \xi_{j-1}, \xi_j + \xi_k, \xi_{j+1}, \ldots, \xi_r) = P(\xi_1, \ldots, \xi_r) \qquad \forall k < j.$$

We agree that $\Gamma^{(a_1,a_2,\ldots,a_r)}V = 0$ unless $(a_1, a_2, \ldots, a_r)$ is nonincreasing. As a special case, we recover symmetric and exterior powers

$$\begin{aligned} S^kV &= \Gamma^{(k,0,\ldots,0)}V, \\ \Lambda^kV &= \Gamma^{(1,\ldots,1,0,\ldots,0)}V, \qquad \text{(with } k \text{ indices 1)} \\ \det V &= \Gamma^{(1,\ldots,1)}V. \end{aligned} \tag{12.1}$$

The Schur fonctors satisfy the well-known formula

$$\Gamma^{(a_1+\ell,\ldots,a_r+\ell)}V = \Gamma^{(a_1,\ldots,a_r)}V \otimes (\det V)^\ell. \tag{12.2}$$

This formula can of course be used to define $\Gamma^{(a_1,\ldots,a_r)}V$ if any of the $a_j$'s happens to be negative.

Now, by what we saw in section §6, the group $\mathbb{G}'_k$ of germs of reparametrizations $\varphi(t) = t + b_2t^2 + \cdots + b_kt^k + O(t^{k+1})$ tangent to identity acts on $k$-tuples $(f', f'', \ldots, f^{(k)})$ of derivatives of $f$ at 0 by the formulas

$$(f \circ \varphi)' = f', \quad (f \circ \varphi)'' = f'' + 2b_2f', \quad (f \circ \varphi)''' = f''' + 3b_2f'' + 3b_3f', \ldots.$$

This is clearly a unipotent action, induced by the action of $\mathbb{U}_k$ through an embedding

$$\mathbb{G}'_k \hookrightarrow \mathbb{U}_k, \qquad \varphi \longmapsto \begin{pmatrix} 1 & 0 & 0 & \cdots & 0 & 0 \\ 2b_2 & 1 & 0 & \cdots & 0 & 0 \\ 3b_3 & 3b_2 & 1 & \cdots & 0 & 0 \\ \vdots & & & & \vdots & \vdots \\ & & & \cdots & 1 & 0 \\ kb_k & & & \cdots & kb_2 & 1 \end{pmatrix}.$$

By formula (6.5), we find that the graded bundle of $E_{k,m}V^\star$ is

$$\mathrm{Gr}^\bullet E_{k,m}V^\star = \left(\bigoplus_{\ell\in\mathbb{N}^k,\,\ell_1+2\ell_2+\cdots+k\ell_k=m} S^{\ell_1}V^\star\otimes S^{\ell_2}V^\star\otimes\cdots\otimes S^{\ell_k}V^\star\right)^{\mathbb{G}'_k}.$$

Since the action of $\mathbb{G}'_k$ does not preserve each individual component in the summation, the computation of the invariants is quite difficult in general. We will see however that everything is easy if $k \leqslant 2$. In fact, if $k=1$, then

$$E_{1,m}V^\star = E^{\mathrm{GG}}_{1,m}V^\star = S^mV^\star. \tag{12.3}$$

If $k=2$, the effect of a parameter change $(f', f'') \mapsto (f', f''+\lambda f')$ on a weighted homogeneous polynomial $Q(f',f'') = \sum_{|\alpha_1|+2|\alpha_2|=m} a_{\alpha_1\alpha_2}(f')^{\alpha_1}(f'')^{\alpha_2}$ is to replace each monomial $(f')^{\alpha_1}(f'')^{\alpha_2}$ by a sum

$$\sum_\beta C_\beta\lambda^{|\beta|}(f')^{\alpha_1+\beta}(f'')^{\alpha_2-\beta}.$$

It follows that terms $(f')^{\alpha_1}(f'')^{\alpha_2}$ corresponding to different values of the pair $(|\alpha_1|,|\alpha_2|) =: (\ell_1,\ell_2)$ cannot produce monomials with the same multidegree and the same exponent $|\beta|$ of $\lambda$. Hence the various components $S^{\ell_1}V^\star\otimes S^{\ell_2}V^\star$ do not mix up and we get

$$\mathrm{Gr}^\bullet E_{2,m}V^\star = \bigoplus_{\ell_1+2\ell_2=m}(S^{\ell_1}V^\star\otimes S^{\ell_2}V^\star)^{\mathbb{G}'_k} = \bigoplus_{\ell_1+2\ell_2=m}\Gamma^{(\ell_1,\ell_2,0,\ldots,0)}V^\star. \tag{12.4}$$

In the special case when $r=\operatorname{rank}V=2$, (12.1) and (12.2) yield $\Gamma^{(\ell_1,\ell_2)}V^\star = S^{\ell_1-\ell_2}V^\star\otimes(\det V^\star)^{\ell_2}$. Hence we get the simpler formula

$$\mathrm{Gr}^\bullet E_{2,m}V^\star = \bigoplus_{0\leqslant j\leqslant m/3} S^{m-3j}V^\star\otimes(\det V^\star)^j \qquad (k=r=2). \tag{12.5}$$

Similar calculations can be done for low values of $k$ and $m$, but it is a major unsolved problem to compute the decomposition formula of $\mathrm{Gr}^\bullet E_{k,m}V^\star$ for arbitrary $k$ and $m$.

12.6. SPECIAL CASE. Assume that $X$ is a surface and consider the absolute case $V=T_X$. We find

$$\mathrm{Gr}^\bullet E_{2,m}T^\star_X = \bigoplus_{0\leqslant j\leqslant m/3} S^{m-3j}T^\star_X\otimes K^j_X,$$

where $E_{1,m}T^\star_X = S^mT^\star_X$ is a subbundle of $E_{2,m}T^\star_X$. We thus get an exact sequence

$$0\to S^mT^\star_X\to E_{2,m}T^\star_X\to Q_m\to 0,$$

and $Q_m$ admits a filtration with

$$\mathrm{Gr}^\bullet Q_m = \bigoplus_{1\leqslant j\leqslant m/3} S^{m-3j}T^\star_X\otimes K^j_X.$$

The simplest case is $m=3$, which yields the interesting exact sequence

$$0\to S^3T^\star_X\to E_{2,3}T^\star_X\to K_X\to 0.$$

12.7. COMPLEMENT. Assume that $X$ is a surface of degree $d$ in $\mathbb{P}^3$. Then $K_X = \mathcal{O}_X(d-4)$. As $T^\star_X$ is a quotient bundle of $T^\star_{\mathbb{P}^3|X}$ and as $T^\star_{\mathbb{P}^n|X} \otimes \mathcal{O}(2)$ is generated by sections, we conclude that $S^{m-3j}T^\star_X \otimes K^j_X$ is (very) ample whenever $j(d-4) > 2(m-3j)$. This condition is most restrictive when $j=1$. In particular, $Q_m$ is ample for $d > 2m-2$, and we see that there is at most a "very small part" of $E_{2,m}T^\star_X$, namely $S^m T^\star_X$, which need not be ample when the degree $d$ is large. By contrast, the Green-Griffiths graded bundle

$$G^\bullet E^{\mathrm{GG}}_{2,m}T^\star_X = \sum_{\ell_1+2\ell_2=m} S^{\ell_1}T^\star_X \otimes S^{\ell_2}T^\star_X$$

does not such exhibit such strong positivity properties. This is one of the main reasons for which we believe that the invariant jet bundles $E_{k,m}T^\star_X$ are more appropriate to the study of hyperbolicity questions.

## 13. Riemann-Roch Calculations and Study of the Base Locus

In view of the Green-Griffiths conjecture 3.7 concerning algebraic degeneration of entire curves, the main point is to compute the base loci

$$(13.1) \qquad B_k = \bigcap_{m>0} \mathrm{Bs}\left(H^0(X_k, \mathcal{O}_{X_k}(m) \otimes \pi^\star_{k,0}\mathcal{O}(-A))\right) \subset X_k$$

where $X_k = P_k T_X$ and $A$ is an ample divisor over $X$. By corollary 7.9, every nonconstant entire curve $f : \mathbb{C} \to X$ must satisfy $f_{[k]}(\mathbb{C}) \subset B_k$. If the set $Y = \bigcap_{k>0} \pi_{k,0}(B_k)$ is distinct from $X$, then $f(\mathbb{C}) \subset Y \subsetneq X$ and every entire curve is thus algebraically degenerate. We will call $Y$ the *Green-Griffiths locus of* $X$, although Green and Griffiths did use ordinary jet bundles in place of the Semple jet bundles. Unfortunately, it turns out that $Y$ is extremely hard to compute, especially in the case when $X$ is an hypersurface or complete intersection in projective space. (However, an important breakthrough has been achieved in [SiYe96a] for the case of complements of curves in $\mathbb{P}^2$ ; noticeably, the authors obtain an explicit construction of global jet differentials of order 1 and 2, which allows them to show that the base locus is small enough.) Here, we will derive a few sufficient conditions for the existence of sections, mostly based on Riemann-Roch computations and a use of (semi-)stability inequalities. From now on, we restrict ourselves to the case when $X$ is an algebraic surface of general type.

The easiest case is the case of order 1 jets $E_{1,m}T^\star_X = S^m T^\star_X$, namely symmetric differentials. The Riemann-Roch formula then gives

$$(13.2) \qquad \chi(X, S^m T^\star_X \otimes \mathcal{O}(-A)) = \frac{m^3}{6}(c_1^2 - c_2) + O(m^2),$$

where $c_1$ and $c_2$ are the Chern classes of $X$. This can be seen e.g. by computing $h^3$ for the hyperplane bundle first Chern class $h = c_1(\mathcal{O}_{PT_X}(1))^3$ and using the identity $h^2 + c_1 h + c_2 = 0$. By the Bogomolov vanishing theorem 14.1 of the Appendix, we get $h^2(X, S^m T^\star_X \otimes \mathcal{O}(-A)) = 0$ for $m$ large, thus

$$(13.3) \qquad h^0(X, S^m T^\star_X \otimes \mathcal{O}(-A)) \geqslant \frac{m^3}{6}(c_1^2 - c_2) - O(m^2).$$

As a consequence, if $c_1^2 > c_2$, there are non trivial symmetric differentials $\sigma$ with values in $\mathcal{O}(-A)$, and every entire curve must satisfy the corresponding order 1 differential equation $\sigma(f')=0$. This is especially interesting in connection with the following result of Jouanolou [Jou78].

13.4. THEOREM (Jouanolou). *Let $Z$ be a compact complex manifold such that the Hodge spectral sequence degenerates in $E_2$, and let $\mathcal{L}\subset\Omega^1_Z$ be a rank 1 coherent subsheaf such that $\Omega^1_Z/\mathcal{L}$ has no torsion. Let $\mathcal{V}\subset\mathcal{O}(T_X)$ be the dual distribution of hyperplanes in $T_Z$. Then either $\mathcal{V}$ is the relative tangent sheaf of a meromorphic fibration from $Z$ to a curve, or there are only finitely many compact hypersurfaces tangent to $\mathcal{V}$.*

(Jouanolou [Jou78] even obtains a precise upper bound for the number of hypersurfaces which may occur in terms of $h^0(X,\Omega^2_X\otimes\mathcal{L}^{-1})$ and of the Picard number of $X$). As a consequence, one recovers the following result due to Bogomolov [Bog77].

13.5. THEOREM (Bogomolov). *On a surface $X$ of general type such that $c_1^2>c_2$, there are only finitely many rational or elliptic curves.*

PROOF. By the results of § 7, these curves must be integral curves of some multivalued distribution of lines in $X$, associated with the zero divisor $Z\subset P(T_X)$ of any nonzero section in

$$H^0(P(T_X),\mathcal{O}_{P(T_X)}(m)\otimes\pi^\star_{1,0}\mathcal{O}(-A)).$$

At a generic point of $Z$ over a point $x\in X$, this distribution defines a unique line in $T_{X,x}$, and we thus get a rank 1 subsheaf of $T_{\widetilde{Z}}$ (or $\Omega^1_{\widetilde{Z}}$) on any desingularization $\widetilde{Z}$ of $Z$. By Jouanolou's result applied to $\widetilde{Z}$, either these integral curves form a family or there are only a finite number of them. If they form a family, not all of them can be rational or elliptic, otherwise $X$ would be a ruled or elliptic surface; hence the general fiber has genus at least 2. In both cases, there are only finitely many rational or elliptic curves. □

The above result of Bogomolov does not give information on transcendental curves, essentially because very little is known on transcendental leaves of a randomly chosen meromorphic foliation (e.g., one does not know how to decide whether there are only finitely many integral curves of parabolic type). As observed by Lu and Yau [LuYa90], one can say more if the *topological index* $c_1^2-2c_2$ is positive, using the following result of Schneider-Tancredi [ScTa85] (the special case when $E=T^\star_X$ is due to Miyaoka [Miy82]).

13.6. THEOREM (Schneider-Tancredi). *Let $E$ be rank 2 vector bundle over a projective algebraic surface $X$. Assume that* $\det E$ *is nef and big (i.e. $c_1(E)$ is numerically nonnegative and $c_1(E)^2>0$), that $E$ is* $(\det E)$*-semistable and that $c_1(E)^2-2\,c_2(E)>0$. Then $E$ is almost ample in the sense that $S^mE$ generates all 1-jets of sections outside a finite union of curves in $X$, when $m$ is large enough.*

PROOF (SKETCH). Let $P=P(E^\star)$ be the hyperplane bundle and $H=\mathcal{O}_P(1)$. Then $P$ is a ruled 3-fold and the hypotheses imply $c_1(H)^3=c_1(E)^2-c_2(E)>0$. Hence by Riemann-Roch and Serre duality, either $h^0(X,S^mE)$ or $h^0(X,S^mE^\star)$ grow fast. The latter case is impossible by the assumption on semistability and the assumption $\det E$ nef. Therefore $H$ is big. Fix an ample divisor $A$ on $P$. We have to show that the base locus of $mH-A$ in $P$ projects to a curve in $X$ when $m$ is large. Otherwise, let $D$ be an irreducible component of a divisor in the linear system $|mH-A|$. In the Picard group $\mathrm{Pic}(P)=\mathrm{Pic}(X)\oplus\mathbb{Z}[H]$ we then have

$D = kH - \pi^\star F$ for some integer $k > 0$ and some divisor $F$ on $X$. Observing that the multiplication by the canonical section of $H^0(P, \mathcal{O}(D))$ yields an injection of sheaves

$$\mathcal{O}(F) \hookrightarrow \pi_\star \mathcal{O}(kH) = \mathcal{O}(S^k E),$$

we find by semistability

$$c_1(F) \cdot c_1(E) \leqslant \frac{1}{k+1} c_1(S^k E) \cdot c_1(E) = \frac{k}{2} c_1(E)^2.$$

From this, we infer

$$\begin{aligned} H^2 \cdot D = H^2 \cdot (kH - \pi^\star F) &= k(c_1(E)^2 - c_2(E)) - c_1(E) \cdot c_1(F) \\ &\geqslant \frac{k}{2}\big(c_1(E)^2 - 2\, c_2(E)\big) > 0, \end{aligned}$$

therefore $(mH - A)^2 \cdot D > 0$ for $m$ large. By Riemann-Roch, either

$$h^0(D, p(mH - A)_{|D}) \quad \text{or} \quad h^2(D, p(mH - A)_{|D})$$

grows fast as $p$ goes to infinity. By stability again, the latter case cannot occur, as we see by looking at the exact sequence

$$0 \to \mathcal{O}(-D) \otimes \mathcal{O}(p(mH - A)) \to \mathcal{O}_P \otimes \mathcal{O}(p(mH - A)) \to \mathcal{O}_D \otimes \mathcal{O}(p(mH - A)) \to 0,$$

and descending everything at the $h^2$ and $h^3$ level down to $X$ by the Leray spectral sequence. Hence $H_{|D}$ is big and the claim follows by 7.2 iv). □

13.7. THEOREM ([LuYa90]). *Let $X$ be a smooth algebraic surface of general type such that $c_1^2 - 2c_2 > 0$. Then there are only finitely many rational or elliptic curves in $X$, and every non constant entire curve $f : \mathbb{C} \to X$ maps to one of these.*

PROOF. One may assume that $X$ is minimal, i.e. that $K_X$ is nef (and big). By the work of Bogomolov [Bog79], $T_X^\star$ is semi-stable. The result of Schneider and Tancredi now implies that $T_X^\star$ is almost ample. Theorem 7.8 concludes the proof. □

We now turn ourselves to the case of jet differentials of degree 2. A simple Riemann-Roch computation based on Formula 12.6 shows that

$$\chi(X, E_{2,m}T_X^\star) = \chi(X, \mathrm{Gr}^\bullet E_{2,m}T_X^\star) = \frac{m^4}{648}(13\, c_1^2 - 9\, c_2) + O(m^3) \tag{13.8}$$

where $c_1$, $c_2$ are the Chern classes of $X$ (only the terms of bidegree $(2,2)$ in $\mathrm{Ch}(\mathrm{Gr}^\bullet E_{2,m}T_X^\star)$ play a role). This formula should be put in perspective with the one obtained by Green and Griffiths [GrGr80] for the jet bundles $E_{k,m}^{\mathrm{GG}}T_X^\star$. In the case of surfaces, they obtain

$$\chi(X, E_{k,m}^{\mathrm{GG}}T_X^\star) = \frac{m^{2k+1}}{(k!)^2(2k+1)!}(\alpha_k\, c_1^2 - \beta_k\, c_2) + O(m^{2k}),$$

where $\alpha_k \sim \frac{1}{2}(\log k)^2$ and $\beta_k = O(\log k)$ (especially $\lim \beta_k/\alpha_k = 0$). In the special case $n = k = 2$, their formula yields

$$\chi(X, E_{2,m}^{\mathrm{GG}}T_X^\star) = \frac{m^5}{384}(7\, c_1^2 - 5\, c_2) + O(m^4).$$

This is weaker than formula (13.8) in the sense that the ratio 5/7 is larger than 9/13. In general, we expect analogous estimates of the form

$$\chi(X, E_{k,m}T_X^\star) \sim m^{k+2}(\gamma_k\, c_1^2 - \delta_k\, c_2) + O(m^{k+1})$$

with $\lim \delta_k/\gamma_k = 0$ (and even similar higher dimensional estimates with a leading term of the form $c_{n,k}m^{(n-1)k+n}(-c_1)^n$ when $m \gg k \gg 1$). Unfortunately, our lack of knowledge of the combinatorics of the Schur representations involved makes the computation hard to achieve.

In the special case when $X$ is a surface of degree $d$ in $\mathbb{P}^3$, we have $c_1 = (4-d)h$ and $c_2 = (d^2 - 4d + 6)h^2$ where $h = c_1(\mathcal{O}(1)_{|X})$, $h^2 = d$, thus

$$\chi(X, E_{2,m}T_X^\star) = \frac{m^4}{648}\, d(4\,d^2 - 68\,d + 154) + O(m^3).$$

This estimate is especially useful in combination with vanishing theorems for holomorphic tensor fields (see Theorem 14.1 in the Appendix).

13.9. COROLLARY. *If $X$ is an algebraic surface of general type and $A$ an ample line bundle over $X$, then*

$$h^0(X, E_{2,m}T_X^\star \otimes \mathcal{O}(-A)) \geqslant \frac{m^4}{648}(13\,c_1^2 - 9\,c_2) - O(m^3).$$

*In particular, every smooth surface $X \subset \mathbb{P}^3$ of degree $d \geqslant 15$ admits non trivial sections of $E_{2,m}T_X^\star \otimes \mathcal{O}(-A)$ for $m$ large, and every entire curve $f : \mathbb{C} \to X$ must satisfy the corresponding algebraic differential equations.*

PROOF. First note that the leading term in the Riemann-Roch estimate does not depend on taking a tensor product by a line bundle $\mathcal{O}(-A)$. The claim will follow from the computation of the Euler characteristic made in (13.8) if we check that $h^2(X, E_{2,m}T_X^\star \otimes \mathcal{O}(-A)) = 0$ for $m$ large. However

$$H^2(X, E_{2,m}T_X^\star \otimes \mathcal{O}(-A)) = H^0(X, K_X \otimes (E_{2,m}T_X^\star)^\star \otimes \mathcal{O}(A))$$

by Serre duality. Since $K_X \otimes (E_{2,m}T_X^\star)^\star \otimes \mathcal{O}(A)$ admits a filtration with graded pieces

$$S^{m-3j}T_X \otimes K_X^{\otimes 1-j} \otimes \mathcal{O}(A),$$

we easily deduce the vanishing of global sections from Bogomolov's result 14.1, using the fact that $K_X^{\otimes \nu} \otimes \mathcal{O}(-A)$ is big for $\nu \geqslant \nu_0$ large enough. □

OTHER APPROACH USING WEIGHTED LINE BUNDLES $\mathcal{O}_{X_k}(\boldsymbol{a})$. We show here how a use of the weighted line bundles $\mathcal{O}_{X_k}(\boldsymbol{a})$ may yield further information on the base locus. Consider a directed manifold $(X, V)$ with $\dim X = n$ and $\operatorname{rank} V = r$. We set $u_k = c_1(\mathcal{O}_{X_k}(1))$ and let

$$c_\bullet^{[k]} = 1 + c_1^{[k]} + \cdots + c_r^{[k]} := c_\bullet(V_k)$$

be the total chern class of $V_k$. Then the cohomology ring of $X_k = P(V_{k-1})$ is defined in terms of generators and relations as the polynomial algebra $H^\bullet(X)[u_1, \ldots, u_k]$ with relations

$$u_j^r + c_1^{[j-1]}u_j^{r-1} + \cdots + c_{r-1}^{[j-1]}u_j + c_r^{[j-1]} = 0, \qquad 1 \leqslant j \leqslant k \tag{13.10}$$

(we omit all pull-backs $\pi_j^\star$ for simplicity of notation). Moreover, the exact sequences (5.4) and (5.4′) yield the inductive formula

$$c_\bullet^{[k]} = c_\bullet(\mathcal{O}_{X_k}(-1))\, c_\bullet(T_{X_k/X_{k-1}}) = (1-u_k)\, c_\bullet(T_{X_k/X_{k-1}}),$$
$$c_\bullet(T_{X_k/X_{k-1}}) = c_\bullet(\pi_k^\star V_{k-1}\otimes \mathcal{O}_{X_k}(1)) = \sum_{0\leqslant j\leqslant r} c_j^{[k-1]}(1+u_k)^{r-j},$$

in other words

$$c_\bullet^{[k]} = (1-u_k)\sum_{0\leqslant j\leqslant r} c_j^{[k-1]}(1+u_k)^{r-j}. \tag{13.11}$$

In particular, if $r = \operatorname{rank} V = 2$, we find

$$\begin{aligned} &u_k^2 + c_1^{[k-1]}u_1 + c_2^{[k-1]} = 0,\\ &c_1^{[k]} = c_1^{[k-1]} + u_k, \qquad c_2^{[k]} = c_2^{[k-1]} - u_k^2, \end{aligned} \tag{13.12}$$

hence

$$c_1^{[k]} = c_1^{[0]} + u_1 + \cdots + u_k, \qquad c_2^{[k]} = c_2^{[0]} - u_1^2 - \cdots - u_k^2. \tag{13.13}$$

From now on, we concentrate again on the surface case. The 2-jet bundle

$$X_2 \to X_1 \to X$$

is a 2-step tower of $\mathbb{P}^1$-bundles over $X$ and therefore has dimension 4. The exact sequence (5.4) shows that $V_1$ has splitting type $V_{1\restriction F_1} = \mathcal{O}(2)\oplus\mathcal{O}(-1)$ along the fibers $F_1$ of $X_1 \to X$, since $T_{X_1/X\restriction F_1} = \mathcal{O}(2)$. Hence the fibers $F_2$ of $X_2 \to X$ are Hirzebruch surfaces $P(\mathcal{O}(2)\oplus\mathcal{O}(-1)) \simeq P(\mathcal{O}\oplus\mathcal{O}(-3))$ and

$$\mathcal{O}_{X_2}(1)_{\restriction F_2} = \mathcal{O}_{P(\mathcal{O}(2)\oplus\mathcal{O}(-1))}(1).$$

The weighted line bundle $\mathcal{O}_{X_2}(2,1)$ is relatively nef over $X$, as follows from our general result (6.16 ii) or from the equality

$$\mathcal{O}_{X_2}(2,1)_{\restriction F_2} = \mathcal{O}_{P(\mathcal{O}(2)\oplus\mathcal{O}(-1))}(1)\otimes \pi^\star\mathcal{O}_{\mathbb{P}^1}(2) = \mathcal{O}_{P(\mathcal{O}\oplus\mathcal{O}(-3))}(1).$$

Its multiples have zero higher order direct images $R^q(\pi_{2,0})_\star\mathcal{O}_{X_2}(2m,m)$, $q\geqslant 1$, and lower order direct images

$$(\pi_{2,0})_\star\mathcal{O}_{X_2}(2m,m) = (\pi_{2,0})_\star\mathcal{O}_{X_2}(3m) = E_{2,3m}T_X^\star$$

[either apply (6.16 i) or observe that

$$(\pi_{1,0})_\star\big(\mathcal{O}_{P(\mathcal{O}\oplus\mathcal{O}(-3))}(m)\big) = S^m\big(\mathcal{O}\oplus\mathcal{O}(3)\big),$$
$$(\pi_{1,0})_\star\big(\mathcal{O}_{P(\mathcal{O}(2)\oplus\mathcal{O}(-1))}(3m)\big) = S^{3m}\big(\mathcal{O}(-2)\oplus\mathcal{O}(1)\big)$$

have the same sections over $\mathbb{P}^1$]. By the Leray spectral sequence, we conclude that

$$h^q(X_2, \mathcal{O}_{X_2}(2m,m)) = h^q(X, E_{2,3m}T_X^\star), \qquad 0\leqslant q\leqslant 2,$$

in particular the Euler characteristics are equal and grow as $\frac{1}{8}m^4(13\,c_1^2 - 9c_2)$ when $m\to+\infty$. This can also be checked directly by computing $\frac{1}{4!}(2u_1+u_2)^4$. In fact, (13.12) and (13.13) easily provide

$$u_1^4 = 0,\quad u_1^3u_2 = c_1^2 - c_2,\quad u_1^2u_2^2 = c_2,\quad u_1u_2^3 = c_1^2 - 3c_2,\quad u_1^4 = 5c_2 - c_1^2.$$

The main difficulty when trying to check the hyperbolicity of $X$ is to show that the base locus of $\mathcal{O}_{X_2}(2,1)$ is small enough. Proving that the base locus is one dimensional would imply that $X$ only admits a finite number of rational and elliptic curves, and that every entire curve $f:\mathbb{C}\to X$ maps into one of these. A possibility for this would be to check that $(2u_1+u_2)^3\cdot Y>0$ for every 3-fold $Y\subset X_2$ and $(2u_1+u_2)^2\cdot S>0$ for every surface $S\subset X_2$. Unfortunately, such estimates are rather hard to check, since we would need to evaluate the numerical cones of effective codimension 1 and codimension 2 cycles in the 4-fold $X_2$. The codimension 1 case, however, can be treated by using semi-stability arguments (although possibly the conditions obtained in this way are far from being optimal). The following computation is due to J. El Goul [EG97].

13.14. PROPOSITION ([EG97]). *Let $X$ be a minimal algebraic surface of general type. If $c_1^2-\frac{9}{7}c_2>0$, then $\mathcal{O}_{X_2}(2,1)$ is almost ample on $X_2$ with a base locus of dimension 2 at most.*

PROOF (SKETCH). We proceed as in the proof of the result by Miyaoka and Schneider-Tancredi. Let $Y$ be a 3-dimensional irreducible component of the base locus, if any. In $\mathrm{Pic}(X_2)=\mathrm{Pic}(X)\oplus\mathbb{Z}u_1\oplus\mathbb{Z}u_2$, we then find an equality $Y=a_1u_1+a_2u_2-\pi^\star F$ for some integers $a_1$, $a_2\in\mathbb{Z}$ and some divisor $F$ on $X$. As $Y$ is effective, we must have $a_1\geqslant 0$, $a_2\geqslant 0$. Moreover, $\mathcal{O}(F)$ can be viewed as a subsheaf of $\pi_\star(\mathcal{O}_{X_2}(a_1,a_2))\subset E_{2,m}T^\star_X$ where $m=a_1+a_2$. Thus there is a non trivial morphism $\mathcal{O}(F)\hookrightarrow S^{m-3j}T^\star_X\otimes K^j_X$ for some $j$, and the semistability inequality yields

$$F\cdot K_X\leqslant\frac{m-j}{2}K_X^2\leqslant\frac{m}{2}c_1^2.$$

A short computation now yields

$$(2u_1+u_2)^2\cdot Y=(a_1+a_2)(13\,c_1^2-9\,c_2)-12\,c_1\cdot F\geqslant m(7\,c_1^2-9\,c_2). \qquad \square$$

LOGARITHMIC CASE. Similar computations can be made in this situation. In fact, if $X$ is a surface and $D$ is a smooth effective divisor in $X$, the bundle $E_{2,m}T^\star_X\langle D\rangle$ of logarithmic jet differentials of order 2 and degree $m$ admits a filtration with

$$\mathrm{Gr}^\bullet E_{2,m}T^\star_X\langle D\rangle=\bigoplus_{0\leqslant j\leqslant m/3}S^{m-3j}(T^\star_X\langle D\rangle)\otimes\det(T^\star_X\langle D\rangle)^j. \tag{13.15}$$

We thus get

$$\begin{aligned}h^0(X,E_{2,m}T^\star_X\langle D\rangle)&\geqslant\chi(X,E_{2,m}T^\star_X\langle D\rangle)\\&\geqslant\frac{m^4}{648}\big(13\,c_1^2(T_X\langle D\rangle)-9\,c_2(T_X\langle D\rangle)\big)-O(m^3).\end{aligned} \tag{13.16}$$

The exact sequence $0\to T_X\langle D\rangle\to T_X\to(i_D)_\star N_{X/D}\to 0$ yields

$$c_\bullet(T_X\langle D\rangle)=c_\bullet(T_X)c_\bullet((i_D)_\star N_{X/D})^{-1}=(1+c_1+c_2)(1+\delta)^{-1},$$

where $\delta=c_1(\mathcal{O}_X(D))$ and $c_\bullet((i_D)_\star N_{X/D})=c_\bullet(\mathcal{O}_X(D))=1+\delta$, thus

$$c_1(T_X\langle D\rangle)=c_1-\delta,\qquad c_2(T_X\langle D\rangle)=c_2-c_1\cdot\delta+\delta^2.$$

Moreover, the expected vanishing theorem for $h^2(X, E_{2,m}T^\star_X\langle D\rangle)$ still holds since $T_X\langle D\rangle$ is a subbundle of $T_X$. In particular, if $X = \mathbb{P}^2$ and $D$ is a smooth curve of degree $d$, we find $c_1(T_X\langle D\rangle) = 3 - d$, $c_2(T_X\langle D\rangle) = 3 - 3d + d^2$ and

$$h^0(X, E_{2,m}T^\star_{\mathbb{P}^2}\langle D\rangle) \geqslant \frac{m^4}{648}(4\,d^2 - 51\,d + 90) - O(m^3).$$

From this, one infers that every entire curve $f : \mathbb{C} \to \mathbb{P}^2 \smallsetminus D$ must satisfy a non trivial algebraic differential equation of order 2 if $d \geqslant 11$.

## 14. Appendix: a Vanishing Theorem for Holomorphic Tensor Fields

In this appendix, we prove a basic vanishing theorem for holomorphic tensor fields on minimal varieties of general type. It has been observed since a long time that the existence of holomorphic tensor fields on a compact Kähler manifold is governed in a rather precise way by the sign of the Ricci curvature (in case the Ricci curvature does admit some definite sign, semipositive or seminegative). See for instance the papers [Li67, 71] by Lichnerowicz for the case of sections of $\Lambda^k T_X$ or $\Lambda^k T^\star_X$, and S. Kobayashi's articles [Kob80, 81] for the more general case of tensors in $\Gamma^a T_X$. However, we want here to consider the situation of varieties of general type (i.e. with $K_X$ big), and it is unknown whether $K_X$ should be semipositive even if $K_X$ is assumed to be big and nef. On the other hand, it is a consequence of Bogomolov's work [Bog79] (dealing with the so-called "Bogomolov stability" concept), that such vanishing theorems hold when $T_X$ is semistable; this is the case for instance if $X$ is a minimal surface of general type. Tsuji [Tsu88] has proved more generally that the tangent bundle $T_X$ is semistable for any minimal nonsingular projective variety of general type (here, $X$ "minimal" means that $K_X$ is nef) . Thus, the following theorem 14.1 below can be obtained as a combination of the above mentioned results of Bogomolov and Tsuji. For the convenience of the reader, we give instead a direct proof based on a use of approximate Kähler-Einstein metrics in combination with the Bochner formula. Our hope is that similar a priori estimates could produce as well vanishing theorems for higher degree cohomology groups $H^q$.

14.1. THEOREM. *Let $X$ be a projective algebraic manifold, $n = \dim X$, and let $L$ be a holomorphic line bundle over $X$. Assume that $X$ is of general type and minimal (i.e. $K_X$ is big and nef), and let $a = (a_1, \ldots, a_n) \in \mathbb{Z}^n$, $a_1 \geqslant \cdots \geqslant a_n$, be a weight. If either $L$ is pseudoeffective and $|a| = \sum a_j > 0$, or $L$ is big and $|a| \geqslant 0$, then*

$$H^0(X, \Gamma^a T_X \otimes L^\star) = 0.$$

Recall that a line bundle $L$ is said to be pseudoeffective if $c_1(L)$ belongs to the closure of the cone of effective divisors, or equivalently, if $L$ carries a singular hermitian metric $h$ with curvature current $\Theta_h(L) \geqslant 0$. Also notice that the result is invariant by modifications, hence it extends to the case when $X$ is of general type and possesses a smooth minimal model $\widetilde{X}$ ; this is always the case when $X$ is a surface. On the other hand, it is likely that the result holds for all varieties $X$ of general type, in view of Mori's minimal model conjecture (however, the differential geometric proof given below might be difficult to extend to the case when the minimal model is singular).

PROOF OF THEOREM 14.1. We will use the following notation: if all $a_j$ are nonnegative integers, $\Gamma^a T_X$ is viewed as a subbundle of $(T_X)^{\otimes p}$ with $p = |a|$. In particular, given coordinates $(z_1, \ldots, z_n)$ on $X$, any tensor of $\Gamma^a T_X$ can be expressed as a linear combination of the elements

$$(\partial/\partial z)^I := \frac{\partial}{\partial z_{i_1}} \otimes \cdots \otimes \frac{\partial}{\partial z_{i_p}}, \qquad I = (i_1, \ldots, i_p), \quad 1 \leqslant i_k \leqslant n,$$

which form a basis of $(T_X)^{\otimes p}$. If some $a_j$ is negative, we use instead the identity

$$\Gamma^{(a_1, \ldots, a_r)} T_X = \Gamma^{(a_1+\ell, \ldots, a_r+\ell)} T_X \otimes (\det T_X)^{-\ell}$$

with $\ell = \max(-a_j)$, and consider the basis elements $(\partial/\partial z)^J \otimes (dz_1 \wedge \cdots \wedge dz_n)^\ell$ with $|J| = p + n\ell$. Same notation with the elements of $\Gamma^a T_X^\star$ in terms of the basis $(dz)^I = dz_{i_1} \otimes \cdots \otimes dz_{i_p}$, resp. $(dz)^J \otimes (dz_1 \wedge \cdots \wedge dz_n)^{-\ell}$.

14.2. LEMMA. *Let $L$ be a holomorphic line bundle over $X$ equipped with a smooth hermitian metric $h$, and let $\omega$ be a Kähler metric over $X$. We denote by $\#$ the conjugate linear $C^\infty$-isomorphism $T_X \to T_X^\star$, $v \mapsto i\overline{v} \lrcorner \omega$, defined by contracting $(0,1)$-vectors with the Kähler metric $\omega$. Denote also by $\# : \Gamma^a T_X \otimes L \to \Gamma^a T_X^\star \otimes L^\star$ the induced $C^\infty$ isomorphism on the Schur tensor powers of $T_X$ and $T_X^\star$, combined with the conjugate linear (metric) isomorphism $L \to L^\star$. Then for an arbitrary smooth section $v$ of $\Gamma^a T_X \otimes L$ we have*

$$\int_X \|\overline{\partial}(\# v)\|^2 dV_\omega = \int_X \|\overline{\partial} v\|^2 dV_\omega + \int_X \langle \mathcal{R}_a(v), v\rangle + \gamma |v|^2 \, dV_\omega$$

*where $dV_\omega$ is the Kähler element of volume, $\gamma$ the trace ( = sum of eigenvalues) of $\Theta_h(L)$ with respect to $\omega$, and $\mathcal{R}_a$ is the hermitian operator*

$$v = \sum_{|I|=p} v_I (\partial/\partial z)^I \otimes s \longmapsto \mathcal{R}_a(v) = \sum_{|I|=p} \Big( \sum_{1 \leqslant k \leqslant p} \rho_{i_k} \Big) v_I (\partial/\partial z)^I \otimes s,$$

$$(\textit{resp. } \mathcal{R}_a v = \mathcal{R}_{a+\ell(1,\ldots,1)} v - \ell(\textstyle\sum_j \rho_j) v \textit{ with } \ell = \max(-a_j), \textit{ if } a \notin \mathbb{N}^n)$$

*associated with the Ricci curvature form: $\rho_k$ denotes the eigenvalues of* Ricci$(\omega)$ *and $(\partial/\partial z_k)$, $s$ are supposed to be orthonormal frames of $(T_X, \omega)$ and $(L, h)$.*

PROOF. We first make a pointwise calculation of $\overline{\partial}^\star \overline{\partial} v$ and $\overline{\partial}^\star \overline{\partial}(\# v)$ in a normal coordinate system for the Kähler metric $\omega$ and in a normalized holomorphic frame $(s)$ for $(L, h)$. In suitable such coordinates we can write

$$\omega = i \sum_{1 \leqslant m \leqslant n} dz_m \wedge d\overline{z}_m - i \sum_{1 \leqslant j,k,\ell,m \leqslant n} c_{jk\ell m} z_j \overline{z}_k dz_\ell \wedge d\overline{z}_m + O(|z|^3),$$

$$|s|^2 = 1 - \sum_{1 \leqslant j \leqslant n} \gamma_{jk} z_j \overline{z}_k + O(|z|^3)$$

where $(c_{jk\ell m})$ is the curvature tensor of $T_X$ with respect to $\omega$, and the $\gamma_{jk}$'s are the coefficients of $\Theta_h(L)$. The Kähler property shows that we have the symmetry relations $c_{jk\ell m} = c_{\ell kjm} = c_{jm\ell k}$, and the Ricci tensor $R = \sum R_{\ell m} dz_\ell \wedge d\overline{z}_m$ is obtained as the trace: $R_{\ell m} = \sum_j c_{jj\ell m}$. Since $\omega$ is tangent of order 2 to a flat metric at the center $x_0$ of the chart, we easily see that the first order operator $\overline{\partial}^\star$ has the same formal expression at $x_0$ as in the case of the flat metric on $\mathbb{C}^n$ : if $w$ if a smooth $(0,q)$-form with values in a holomorphic vector bundle $E$ trivialized locally by a holomorphic frame $(e_\lambda)$ such that $(e_\lambda(x_0))$ is orthonormal and $De_\lambda(x_0) = 0$, we have at $x_0$ the formula

$$w = \sum_{|J|=q,\, 1\leqslant\lambda\leqslant r} w_{J,\lambda}\, d\overline{z}_J \otimes e_\lambda, \qquad \overline{\partial}^\star w = -\sum_{|J|=q,\, \lambda,\, k} \frac{\partial w_{J,\lambda}}{\partial z_k}\Big(\frac{\partial}{\partial \overline{z}_k} \lrcorner\, d\overline{z}_J\Big) \otimes e_\lambda.$$

We apply this to smooth sections $v = \sum v_I\,(\partial/\partial z)^I \otimes s$ of $\Gamma^a T_X \otimes L$ and $w = \sum w_I\,(dz)^I \otimes s^\star$ of $\Gamma^a T^\star_X \otimes L^\star$ where $s^\star$ denotes the holomorphic section of $L^\star$ such that $s^\star(s) = 1$. We get

$$\overline{\partial}^\star\overline{\partial} v = -\sum_{I,k} \frac{\partial^2 v_I}{\partial z_k \partial \overline{z}_k}\,(\partial/\partial z)^I \otimes s, \qquad \overline{\partial}^\star\overline{\partial} w = -\sum_{I,k} \frac{\partial^2 w_I}{\partial z_k \partial \overline{z}_k}\,(dz)^I \otimes s^\star$$

at $x_0$. Now, we find

$$\begin{aligned}
&\#\,\frac{\partial}{\partial z_m} = i\frac{\partial}{\partial \overline{z}_m} \lrcorner\, \omega = dz_m - \sum_{j,k,\ell} c_{jk\ell m} z_j \overline{z}_k dz_\ell + O(|z|^3),\\
&\#\, s = \Big(1 - \sum_{1\leqslant j,k\leqslant n} \gamma_{jk} z_j \overline{z}_k + O(|z|^3)\Big) s^\star\\
&\#\, v = \sum_I \overline{v}_I (dz)^I \otimes s^\star - \sum_{I,j,k,\ell,m} \overline{v}_I c_{jk\ell m} z_j \overline{z}_k \Big(dz_\ell \otimes \frac{\partial}{\partial z_m}\Big) \lrcorner\,(dz)^I \otimes s^\star,\\
&\qquad\qquad - \sum_{I,j,k} \overline{v}_I\, \gamma_{jk} z_j \overline{z}_k (dz)^I \otimes s^\star + O(|z|^3)
\end{aligned}$$

where (by definition)

$$\Big(dz_\ell \otimes \frac{\partial}{\partial z_m}\Big) \lrcorner\,(dz)^I := \sum_{1\leqslant k\leqslant p,\, i_k = m} dz_{i_1} \otimes \cdots \otimes dz_{i_{k-1}} \otimes dz_\ell \otimes dz_{i_{k+1}} \otimes \cdots \otimes dz_{i_p}.$$

Computing $\overline{\partial}^\star\overline{\partial}(\#\, v)$ at $x_0$ we obtain

$$\begin{aligned}
\overline{\partial}^\star\overline{\partial}(\#\, v) &= -\sum_{I,k} \frac{\partial^2 \overline{v}_I}{\partial z_k \partial \overline{z}_k} (dz)^I \otimes s^\star + \sum_{I,k,\ell,m} \overline{v}_I c_{kk\ell m} \Big(dz_\ell \otimes \frac{\partial}{\partial z_m}\Big) \lrcorner\,(dz)^I \otimes s^\star\\
&\qquad + \sum_j \gamma_{jj} \sum_I \overline{v}_I (dz)^I \otimes s^\star\\
&= \#\,(\overline{\partial}^\star\overline{\partial} v) + \sum_{I,\ell,m} \overline{v}_I R_{\ell m} \Big(dz_\ell \otimes \frac{\partial}{\partial z_m}\Big) \lrcorner\,(dz)^I \otimes s^\star + \gamma(\#\, v)\\
&= \#\,(\overline{\partial}^\star\overline{\partial} v) + \#\,\mathcal{R}_a(v) + \gamma(\#\, v)
\end{aligned}$$

where $\gamma = \sum_j \gamma_{jj}$. Lemma 14.2 then follows from this identity by writing

$$\int_X \|\overline{\partial}(\#\, v)\|^2 dV_\omega = \int_X \langle \overline{\partial}^\star\overline{\partial}(\#\, v), \#\, v\rangle\, dV_\omega = \int_X \langle \overline{\partial}^\star\overline{\partial} v + \mathcal{R}_a(v) + \gamma v, v\rangle\, dV_\omega. \quad \square$$

Proof of Theorem 14.1 (end). Our goal is to apply the Bochner formula of Lemma 14.2 to show that every section $v$ of $H^0(X, \Gamma^a T_X \otimes L^\star)$ must vanish. We first make a reduction to the case when $L$ is ample. In fact, by raising $v$ to some tensor power, we get a section $v^m \in H^0(X, \Gamma^{ma} T_X \otimes L^{\otimes -m})$. If $L$ is big, some power $L^{\otimes m}$ can be written as $\mathcal{O}(A+D)$ where $A$ is an ample divisor and $D$ an effective divisor. It is then enough to prove the vanishing of $H^0(\Gamma^{ma} T_X \otimes \mathcal{O}(-A))$. If $L$ is just pseudoeffective, then $|a| > 0$ by hypothesis and we write

$$\Gamma^{ma} T_X \otimes L^{\otimes -m} = \Gamma^{ma-(1,\ldots,1)} T_X \otimes (L^{\otimes m} \otimes K_X)^{-1}$$

where $|ma - (1,\ldots,1)| = m|a| - n > 0$ for $m > n$ and $L^{\otimes m} \otimes K_X$ is big. We can now proceed as before to reduce the situation to the case of an ample $L$.

If, in addition to this, $K_X$ is also ample, the statement is a straightforward consequence of the Aubin-Calabi-Yau theorem ([Aub77], [Yau77]). In fact, we can choose $\omega$ to be Kähler-Einstein, i.e., $\mathrm{Ricci}(\omega) = -\omega$. Then, for any holomorphic section $v \in H^0(X, \Gamma^a T_X \otimes L^\star$, Lemma 14.2 yields

$$\begin{aligned} 0 &= \int_X \|\overline{\partial} v\|^2 dV_\omega = \int_X \|\overline{\partial}(\# v)\|^2 dV_\omega \int_X -\langle \mathcal{R}_a(v), v\rangle + \gamma |v|^2 \, dV_\omega \\ &\geqslant \int_X (|a| + \gamma)|v|^2 \, dV_\omega \end{aligned}$$

($\gamma$ becomes $-\gamma$ since we changed $L$ into $L^\star$, and all Ricci eigenvalues are equal to $-1$ in that case). As $|a| \geqslant 0$ and $\gamma > 0$ by the ampleness of $L$, we get the desired conclusion.

If $K_X$ is only big and nef, we take $\omega$ to be a Kähler form in the positive class $c_1(K_X) + \varepsilon c_1(L) = -c_1(X) + \varepsilon c_1(L)$, such that

$$\mathrm{Ricci}(\omega) = -\omega + \varepsilon\theta$$

where $\theta = \Theta_h(L) > 0$ (the existence of such $\omega$ is is a well-known consequence of the theory of Monge-Ampère equations). Then the Ricci curvature eigenvalues satisfy $\rho_j = -1 + \varepsilon\gamma_j \leqslant -1 + \varepsilon\gamma$ and we get

$$\langle -\mathcal{R}_a(v), v\rangle + \gamma|v|^2 \geqslant (|a| + \gamma - N\varepsilon\gamma)|v|^2$$

where $N$ is an integer depending only on the weight $a = (a_1, \ldots, a_n)$ ; for instance, the choice $N = |a|$ works if all $a_j$ are nonnegative, otherwise we can take $N = |a| + n \max(-a_j)$. □

## References


[Ahl41] L. Ahlfors, *The theory of meromorphic curves*, Acta Soc. Sci. Finn. N.S., A **3** (1941), 1–31.

[ASS92] E. Arrondo, I. Sols, R. Speiser, *Global moduli for contacts*, Preprint 1992, to appear.

[Blo26] A. Bloch, *Sur les systèmes de fonctions uniformes satisfaisant à l'équation d'une variété algébrique dont l'irrégularité dépasse la dimension*, Jour. de Math. **5** (1926), 19–66.

[Blo26′] A. Bloch, *Sur les systèmes de fonctions holomorphes à variétés linéaires lacunaires*, Ann. Ecole Normale, **43** (1926), 309–362.

[Bog77] F.A. Bogomolov, *Families of curves on a surface of general type*, Soviet Math. Dokl. **18** (1977), 1294–1297.

[Bog79] F.A. Bogomolov, *Holomorphic tensors and vector bundles on projective varieties*, Math. USSR Izvestija **13/3** (1979), 499–555.

[Bro78] R. Brody, *Compact manifolds and hyperbolicity.*, Trans. Amer. Math. Soc. **235** (1978), 213–219.

[BrGr77] R. Brody, M. Green, *A family of smooth hyperbolic surfaces in* $\mathbb{P}^3$, Duke Math. J. **44** (1977), 873–874.

[CaGr72] J. Carlson, P. Griffiths, *A defect relation for equidimensional holomorphic mappings between algebraic varieties*, Ann. Math. **95** (1972), 557–584.

[Car28] H. Cartan, *Sur les systèmes de fonctions holomorphes à variétés linéaires lacunaires et leurs applications, Thèse, Paris*, Ann. Ecole Normale, **45** (1928), 255–346.

[CKM88] H. Clemens, J. Kollár, S. Mori, *Higher dimensional complex geometry*, Astérisque **166** (1988).

[Cle86] H. Clemens, *Curves on generic hypersurfaces*, Ann. Sci. Ec. Norm. Sup. **19** (1986), 629–636.

[CoKe94] S.J. Colley, G. Kennedy, *The enumeration of simultaneous higher order contacts between plane curves*, Compositio Math. **93** (1994), 171–209.

[Coll88] A. Collino, *Evidence for a conjecture of Ellingsrud and Strømme on the Chow ring of* $\mathrm{Hilb}_d(\mathbb{P}^2)$, Illinois J. Math. **32** (1988), 171–210.

[CoGr76] M. Cowen, P. Griffiths, *Holomorphic curves and metrics of negative curvature*, J. Analyse Math. **29** (1976), 93–153.

[Dem90] J.-P. Demailly, *Singular hermitian metrics on positive line bundles*, Proceedings of the Bayreuth conference "Complex algebraic varieties", April 2–6, 1990, edited by K. Hulek, T. Peternell, M. Schneider, F. Schreyer, Lecture Notes in Math. n° 1507, Springer-Verlag (1992), 87–104.

[Dem92] J.-P. Demailly, *Regularization of closed positive currents and Intersection Theory*, J. Alg. Geom. **1** (1992), 361–409.

[DeEG97] J.-P. Demailly, J. El Goul, *Connexions méromorphes projectives et variétés algébriques hyperboliques*, C. R. Acad. Sci. Paris, Série I, Math. **324** (janvier 1997).

[DPS94] J.-P. Demailly, Th. Peternell, M. Schneider, *Compact complex manifolds with numerically effective tangent bundles*, J. Algebraic Geometry **3** (1994), 295–345.

[DGr91] G. Dethloff, H. Grauert, *On the infinitesimal deformation of simply connected domains in one complex variable.*, International Symposium in Memory of Hua Loo Keng, Vol. II, Beijing, 1988 (1991), Springer, Berlin, 57–88.

[DLu96] G. Dethloff, S. Lu, *Logarithmic jet bundles and a conjecture of Lang*, Seminar conference at MSRI, Univ. of California, Berkeley, March 1996, preprint in preparation.

[DSW92] G. Dethloff, G. Schumacher, P.M. Wong, *Hyperbolicity of the complement of plane algebraic curves*, Amer. J. Math. **117** (1995), 573–599.

[DSW94] G. Dethloff, G. Schumacher, P.M. Wong, *On the hyperbolicity of the complements of curves in Algebraic surfaces: the three component case*, Duke Math. Math. **78** (1995), 193–212.

[Dol81] I. Dolgachev, *Weighted projective varieties*, Proceedings Polish-North Amer. Sem. on Group Actions and Vector Fields, Vancouver, 1981, J.B. Carrels editor, Lecture Notes in Math. **956** (1982), Springer-Verlag, New-York, 34–71.

[EG96] J. El Goul, *Algebraic families of smooth hyperbolic surfaces of low degree in* $\mathbb{P}^3_{\mathbb{C}}$, Manuscripta Math. **90** (1996), 521–532.

[EG97] J. El Goul, *Propriétés de négativité de courbure des variétés algébriques hyperboliques.*, Thèse de Doctorat, Univ. de Grenoble I (1997).

[Ghe41] G. Gherardelli, *Sul modello minimo della varieta degli elementi differenziali del* 2° *ordine del piano projettivo*, Atti Accad. Italia. Rend., Cl. Sci. Fis. Mat. Nat. (7) **2** (1941), 821–828.

[Gra89] H. Grauert, *Jetmetriken und hyperbolische Geometrie*, Math. Zeitschrift **200** (1989), 149–168.

[GRec65] H. Grauert, H. Reckziegel, *Hermitesche Metriken und normale Familien holomorpher Abbildungen*, Math. Zeitschrift **89** (1965), 108–125.

[Green75] M. Green, *Some Picard theorems for holomorphic maps to algebraic varieties*, Amer. J. Math. **97** (1975), 43–75.

[Green78] M. Green, *Holomorphic maps to complex tori*, Amer. J. Math. **100** (1978), 615–620.

[GrGr80] M. Green, P. Griffiths, *Two applications of algebraic geometry to entire holomorphic mappings*, The Chern Symposium 1979, Proc. Internal. Sympos. Berkeley, CA, 1979, (1980), Springer-Verlag, New York, 41–74.

[Gri71] P. Griffiths, *Holomorphic mappings into canonical algebraic varieties*, Ann. of Math. **98** (1971), 439–458.

[Har77] R. Hartshorne, *Algebraic geometry*, Springer-Verlag, Berlin, 1977.

[Hir64] H. Hironaka, *Resolution of singularities of an algebraic variety over a field of characteristic zero*, Ann. of Math. **79** (1964), 109–326.

[Jou79] J.-P. Jouanolou, *Hypersurfaces solutions d'une équation de Pfaff analytique*, Math. Ann. **232** (1978), 239–245.

[Kaw80] Y. Kawamata, *On Bloch's conjecture*, Invent. Math. **57** (1980), 97–100.

[Kob70] S. Kobayashi, *Hyperbolic manifolds and holomorphic mappings*, Marcel Dekker, New York, 1970.

[Kob75] S. Kobayashi, *Negative vector bundles and complex Finsler structures*, Nagoya Math. J. **57** (1975), 153–166.

[Kob76] S. Kobayashi, *Intrinsic distances, measures and geometric function theory*, Bull. Amer. Math. Soc. **82** (1976), 357–416.

[Kob80] S. Kobayashi, *The first Chern class and holomorphic tensor fields*, Nagoya Math. J. **77** (1980), 5–11.

[Kob81] S. Kobayashi, *Recent results in complex differential geometry*, Jber. dt. Math.-Verein. **83** (1981), 147–158.

[KobO71] S. Kobayashi, T. Ochiai, *Mappings into compact complex manifolds with negative first Chern class*, J. Math. Soc. Japan **23** (1971), 137–148.

[KobO75] S. Kobayashi, T. Ochiai, *Meromorphic mappings into compact complex spaces of general type*, Invent. Math. **31** (1975), 7–16.

[KobR91] R. Kobayashi, *Holomorphic curves into algebraic subvarieties of an abelian variety*, Internat. J. Math. **2** (1991), 711–724.

[LaTh96] D. Laksov, A. Thorup, *These are the differentials of order n*, Preprint Royal Inst. Technology, Stockholm, (June 1996), 65 p.

[Lang86] S. Lang, *Hyperbolic and Diophantine analysis*, Bull. Amer. Math. Soc. **14** (1986), 159–205.

[Lang87] S. Lang, *Introduction to complex hyperbolic spaces*, Springer-Verlag, New York, 1987.

[Li67] A. Lichnerowicz, *Variétés kähleriennes et première classe de Chern*, J. Diff. Geom. **1** (1967), 195–224.

[Li71] A. Lichnerowicz, *Variétés Kählériennes à première classe de Chern non négative et variétés riemanniennes à courbure de Ricci généralisée non négative.*, J. Diff. Geom. **6** (1971), 47–94.

[Lu91] S.S.Y. Lu, *On meromorphic maps into varieties of log-general type*, Proc. Symposia Pure Math. **52,** Part 2 (1991), Amer. Math. Soc., Providence, RI, 305–333.

[Lu96] S.S.Y. Lu, *On hyperbolicity and the Green-Griffiths conjecture for surfaces*, Geometric Complex Analysis, ed. by J. Noguchi et al., World Scientific Publishing Co. (1996), 401–408.

[LuMi95] S.S.Y. Lu, Y. Miyaoka, *Bounding curves in algebraic surfaces by genus and Chern numbers*, Math. Research Letters **2** (1995), 663-676.

[LuMi96] S.S.Y. Lu, Y. Miyaoka, *Bounding codimension one subvarieties and a general inequality between Chern numbers*, submitted to the Amer. J. of Math.

[LuYa90] S.S.Y. Lu, S.T. Yau, *Holomorphic curves in surfaces of general type*, Proc. Nat. Acad. Sci. USA **87** (January 1990), 80–82.

[MaNo93] K. Masuda, J. Noguchi, *A construction of hyperbolic hypersurface of* $\mathbb{P}^n(\mathbb{C})$, Preprint Tokyo Inst. Technology, Ohokayama, Tokyo (1993), 27 p.

[McQu96] M. McQuillan, *A new proof of the Bloch conjecture*, J. Alg. Geom. **5** (1996), 107–117.

[Mey89] P.-A. Meyer, *Qu'est ce qu'une différentielle d'ordre n ?*, Expositiones Math. **7** (1989), 249–264.

[Miy82] Y. Miyaoka, *Algebraic surfaces with positive indices. Classification of algebraic and analytic manifolds*, Katata Symp. Proc. 1982, Progress in Math., vol. 39, Birkhäuser (1983), 281–301.

[MoMu82] S. Mori, S. Mukai, *The uniruledness of the moduli space of curves of genus* 11, In: Algebraic Geometry Conference Tokyo-Kyoto 1982, Lecture Notes in Math. **1016** (1982), 334–353.

[Nad89] A. Nadel, *Hyperbolic surfaces in* $\mathbb{P}^3$, Duke Math. J. **58** (1989), 749–771.

[Nog77] J. Noguchi, *Holomorphic curves in algebraic varieties*, Hiroshima Math. J. **7** (1977), 833–853.

[Nog81] J. Noguchi, *A higher-dimensional analogue of Mordell's conjecture over function fields*, Math. Ann. **258** (1981), 207-212.

[Nog83] J. Noguchi, *Lemma on logarithmic derivatives and holomorphic curves in algebraic varieties.*, Nagoya Math. J. **83** (1981), 213–233.

[Nog86] J. Noguchi, *Logarithmic jet spaces and extensions of de Franchis' theorem.*, Contributions to Several Complex Variables, Aspects of Math., E9, Vieweg, Braunschweig (1986), 227–249.

[Nog91] J. Noguchi, *Hyperbolic manifolds and diophantine geometry*, Sugaku Expositiones **4** (1991), Amer. Math. Soc., Providence, RI, 63–81.

[Nog96a] J. Noguchi, *On holomorphic curves in semi-abelian varieties*, MSRI preprint, Univ. of California, Berkeley (1996).

[Nog96b] J. Noguchi, *Chronicle of Bloch's conjecture*, Private communication (1996).

[NoOc90] J. Noguchi, T. Ochiai, *Geometric function theory in several complex variables*, Japanese edition, Iwanami, Tokyo, 1984; English translation, xi + 282 p., Transl. Math. Monographs, vol. 80, Amer. Math. Soc., Providence, Rhode Island, 1990.

[Och77] T. Ochiai, *On holomorphic curves in algebraic varieties with ample irregularity*, Invent. Math. **43** (1977), 83-96.

[Roy71] H. Royden, *Remarks on the Kobayashi metric*, Proc. Maryland Conference on Several Complex Variables, Lecture Notes **185** (1971), Springer-Verlag, Berlin.

[Roy74] H. Royden, *The extension of regular holomorphic maps*, Amer. Math. Soc. **43** (1974), 306–310.

[RuSt91] M. Ru, W. Stoll, *The second main theorem for moving targets*, J. of Geom. Analysis **1** (1991), 99–138.

[ScTa85] M. Schneider, A. Tancredi, *Positive vector bundles on complex surfaces*, Manuscripta Math. **50** (1985), 133–144.

[Sem54] J.L. Semple, *Some investigations in the geometry of curves and surface elements*, Proc. London Math. Soc. (3) **4** (1954), 24–49.

[Siu87] Y.T. Siu, *Defect relations for holomorphic maps between spaces of different dimensions*, Duke Math. J. **55** (1987), 213–251.

[SiYe96a] Y.T. Siu, S.K. Yeung, *Hyperbolicity of the complement of a generic smooth curve of high degree in the complex projective plane*, Invent. Math. **124** (1996), 573–618.

[SiYe96b] Y.T. Siu, S.K. Yeung, *A generalized Bloch's theorem and the hyperbolicity of the complement of an ample divisor in an abelian variety*, To appear in Math. Annalen (1996).

[SiYe96c] Y.T. Siu, S.K. Yeung, *Defects for ample divisors of Abelian varieties, Schwarz lemma and hyperbolic surfaces of low degree*, Preprint (Fall 1996).

[Tsu88] H. Tsuji, *Stability of tangent bundles of minimal algebraic varieties*, Topology **27** (1988), 429–442.

[Voj87] P. Vojta, *Diophantine approximations and value distribution theory*, Lecture Notes in Math. **1239** (1987), Springer-Verlag, Berlin.

[Won89] P.M. Wong, *On the second main theorem of Nevanlinna theory*, Am. J. Math. **111** (1989), 549–583.

[Won93] P.M. Wong, *Recent results in hyperbolic geometry and Diophantine geometry*, International Symposium on Holomorphic mappings, Diophantine Geometry and Related topics, R.I.M.S. Lecture Notes ser., vol. 819, R.I.M.S., Kyoto University, 1993, pp. 120–135.

[Xu94] G. Xu, *Subvarieties of general hypersurfaces in projective space.*, J. Differential Geometry **39** (1994), 139–172.

[Zai86] M. Zaidenberg, *On hyperbolic embedding of complements of divisors and the limiting behaviour of the Kobayashi-Royden metric*, Math. USSR Sbornik **55** (1986), 55–70.

[Zai87] M. Zaidenberg, *The complement of a generic hypersurface of degree $2n$ in $\mathbb{C}\mathbb{P}^n$ is not hyperbolic*, Siberian Math. J. **28** (1987), 425–432.

[Zai89] M. Zaidenberg, *Stability of hyperbolic embeddedness and construction of examples*, Math. USSR Sbornik **63** (1989), 351–361.

[Zai93] M. Zaidenberg, *Hyperbolicity in projective spaces*, International Symposium on Holomorphic mappings, Diophantine Geometry and Related topics, R.I.M.S. Lecture Notes ser. **819** (1993), R.I.M.S., Kyoto University, 136–156.

UMR 5582 du C.N.R.S., BP 74, 38402 Saint-Martin d'Hères, France
*E-mail address*: demailly@fourier.ujf-grenoble.fr

Proceedings of Symposia in Pure Mathematics
Volume **62.2**, 1997

# Twistors for Tourists: A Pocket Guide for Algebraic Geometers

Claude LeBrun

## 1. Introduction

Think of your favorite compact complex manifold. Chances are, you're thinking about a smooth complex projective variety. This is completely understandable, not only because projective varieties provide such a rich source of examples, but also because Serre's GAGA principle [**63**] allows one to analyze such examples quite completely by algebraic means. On the other hand, since small deformations of projective varieties need not be projective — e.g. most complex tori are not Abelian varieties — algebraic geometers often broaden their circle of acquaintance to include the manifolds of Kähler type. In the minds of most researchers, though, mainstream mathematics stops there.

Kodaira [**34**], however, discovered that a great deal can be said about compact complex surfaces without imposing the Kählerian condition. Much of his approach relied on the use of curves to explore the nature of the ambient surface, and this turns out to be particularly efficient because a curve in a surface also happens to be a divisor. How should this be generalized to higher dimensions? The traditional approach is to stick with divisors, but little will then be revealed about spaces with few meromorphic functions. But even on projective varieties, with divisors galore, Mori [**50**] has shown us that it is often more fruitful to study curves — specifically, $\mathbb{CP}_1$'s — rather than hypersurfaces. In this article, we will see that some remarkable things can also be learned about non-Kählerian manifolds by exploring them with families of rational curves.

Moduli spaces of curves in a complex manifold carry a natural holomorphic differential geometry. Indeed, this is even true of moduli spaces of compact complex submanifolds of *any* dimension. The key discovery of Penrose [**53**] is that these differential geometries often characterize the original complex manifold. On the other hand, the differential geometries that arise in this way are typically constrained by non-trivial curvature conditions, and so may therefore be thought of as solutions of interesting partial differential equations. The resulting dictionary, translating

1991 *Mathematics Subject Classification.* Primary 32J25;Secondary 14J30,32L25,53C21.
Supported in part by NSF grant DMS-9505744.

between the unrelated languages of complex manifold theory and differential geometry, sometimes turns even the worn-out truisms of one mathematical culture into complete surprises for the other.

The purpose of this article is therefore two-fold. First, I will describe some natural differential-geometric constructions which yield compact complex manifolds quite unlike anything in the familiar algebro-geometric zoo. Second, I will point out some tractable-looking problems in algebraic geometry that are disguised versions of harder-looking problems in differential geometry.

I have seen fit to preface the hard facts of twistor geometry with some glittering generalities [**28, 37, 47**] concerning the geometry of moduli spaces. This prologue is far from essential, however, and I recommend that it be viewed as a mere backdrop for the ensuing action. Like a theater-set, it is supposed to provide a context for the dialogue, but should not itself become a source of distraction.

## 2. Glittering Generalities

Let $Y$ be a complex manifold — not necessarily compact — and let $X \subset Y$ be a compact complex submanifold. The space of compact complex submanifolds of $Y$ forms a complex analytic space [**17**]; let $\overline{\mathcal{M}}$ denote the irreducible component which contains $X$. Now assume that the normal bundle $N = TY/TX$ of $X$ satisfies $H^1(X_p, \mathcal{I}_{x_1x_2}(N)) = 0$, where $\mathcal{I}_{x_1x_2}$ is the ideal sheaf of any pair of points $x_1, x_2 \in X$. Let $\mathcal{M} \subset \overline{\mathcal{M}}$ denote the set of points $p$ of the moduli space for which the analogous vanishing assertions

$$(1) \qquad H^1(X_p, \mathcal{I}_{x_1x_2}(N_p)) = 0 \ \forall x_1, x_2 \in X_p$$

hold for the corresponding deformations $X_p \subset Y$ of $X$. By semi-continuity [**24**], $\mathcal{M}$ is connected, open, and dense in $\overline{\mathcal{M}}$. More importantly, (1) guarantees that $H^1(X_p, \mathcal{O}(N_p)) = 0$ for all $p \in \mathcal{M}$, so the pioneering work of Kodaira [**35**] tells us that $\mathcal{M}$ is a complex manifold, with tangent space at $p$ given by $T_p\mathcal{M} = H^0(X_p, \mathcal{O}(N_p))$. Hypothesis (1) also guarantees that the normal bundle $N_p$ of $X_p$ is very ample, and so in particular generated by its global holomorphic sections. The submanifolds $X_p$ therefore sweep out an open subset $\mathcal{Y}$ of $Y$, and the situation is succinctly described by a *double fibration*

$$\begin{array}{ccccc} & & \mathcal{F} & & \\ & {\scriptstyle\mu}\swarrow & & \searrow{\scriptstyle\nu} & \\ \mathcal{Y} & & & & \mathcal{M} \end{array}$$

meaning that the derivatives $\mu_*$ and $\nu_*$ are surjective and that $\ker \mu_* \cap \ker \nu_* = 0$. Here the underlying point set of $\mathcal{F} \overset{\mu\times\nu}{\hookrightarrow} \mathcal{Y}\times\mathcal{M}$ is the disjoint union of the $X_p$, $p \in \mathcal{M}$. The right-hand projection $\nu$ is proper, but the fibers of the left-hand projection $\mu$ will typically be non-compact. The moduli space $\mathcal{M}$ comes equipped with a system of preferred submanifolds $\nu(\mu^{-1}(y))$, $y \in \mathcal{Y}$, and one could in principle reconstruct $\mathcal{Y} \subset Y$ as the parameter space of these submanifolds.

Broadly speaking, this way of reconstructing the complex manifold $Y$ is called twistor theory. In order for such an approach to succeed, we of course need some non-tautological way of characterizing the submanifolds $\nu(\mu^{-1}(y)) \subset \mathcal{M}$, to which purpose we would like to give our moduli space $\mathcal{M}$ a differential geometry which determines these submanifolds. This may be done by observing that (1) guarantees that each $X_p$ naturally embeds in the Grassmannian $Gr_k(\Omega^1_p\mathcal{M})$, where $k$ is the codimension of $X \subset Y$. The designation of a subfamily $\mathcal{F}$ of the Grassmann bundle $Gr_k(\Omega^1_{\mathcal{M}})$ is the first part of the desired geometry. In addition, we need a holomorphic sub-bundle $D \subset T\mathcal{F}$ to play the rôle of $\ker \mu_*$. This sub-bundle is required to sit inside the tautological codimension-$k$ sub-bundle on $Gr_k(\Omega^1_p\mathcal{M})$, satisfy $D \cap \ker \nu_* = 0$, and be *Frobenius integrable*:

$$[\mathcal{O}(D), \mathcal{O}(D)] \subset \mathcal{O}(D).$$

Given any such differential geometry on $\mathcal{M}$, one can then define a new space $\mathcal{Y}$ as the leaf-space of the foliation tangent to $D$. Provided we are willing to perhaps replace $\mathcal{M}$ with some neighborhood of any point $p \in \mathcal{M}$, this new space $\mathcal{Y}$ will then be a complex manifold. In particular, deformations of the given complex manifold $Y$ will correspond to deformations of a differential-geometric structure on the moduli space $\mathcal{M}$. In a closely related way, the cohomology of $Y$ with coefficients in any holomorphic vector bundle has an interpretation on $M$ in terms of holomorphic differential equations; this arises by a pull-back/push-down procedure known as the *Penrose transform* [**4, 26**].

The picture becomes much more concrete and useful if we impose the additional hypotheses that

$$H^1(X, TX) = H^1(X, N^* \otimes N) = H^1(X, N^* \otimes TX) = 0.$$

The first two of these vanishing hypotheses tell us that $X$ and $N \to X$ are infinitesimally rigid, so that, perhaps by replacing $\mathcal{M}$ with an open dense subset, we may assume that all the manifolds $X_p$ are copies of $X$ and have normal bundle $N$. Invoking the last vanishing hypothesis then allows one to show that every point $p \in \mathcal{M}$ has a neighborhood which admits a holomorphic connection $\nabla : \mathcal{O}(T\mathcal{M}) \to \Omega^1(T\mathcal{M})$ such that $\mathcal{F} \subset Gr_k(\Omega^1)$ is invariant under the corresponding notion of parallel transport and such that $D$ is contained in the horizontal sub-bundle of $\nabla$. The Frobenius integrability constraint on $D$ then amounts to the requirement that a certain piece of the curvature of $\nabla$ must vanish: namely, if $P \subset T_p\mathcal{M}$ is the codimension-$k$ plane corresponding to a point of $\mathcal{F} \subset Gr_k(\Omega^1\mathcal{M})$, then

$$\mathcal{R}(u, v, w, \varphi) = 0 \ \forall u, v, w \in P \subset T_p\mathcal{M}, \varphi \in P^\perp \subset \Omega^1_p$$

where $\mathcal{R}$ denotes the curvature tensor of the connection $\nabla$.

We are now almost in the realm of mainstream differential geometry. One peculiar feature remains, though: our geometries are complex-analytic. In the cases of greatest interest, however, this peculiarity can be eliminated. Namely, suppose that $Y$ has a *real structure*, meaning an anti-holomorphic map $\sigma : Y \to Y$ such that $\sigma^2$ is the identity. Then we can define a real structure $\rho : \mathcal{M} \to \mathcal{M}$ by

$$\rho(p) = q \ \Leftrightarrow \ \sigma(X_p) = X_q.$$

The fixed-point set $M \subset \mathcal{M}$ of $\rho$, if non-empty, is then a real-analytic manifold whose real dimension equals the complex dimension of $\mathcal{M}$. Moreover, the relevant geometry of $\mathcal{M}$ is invariant under $\rho$, and so restricts to a 'real' geometry on $M$, in the sense that, for example, the local connections $\nabla$ can be taken to to induce real (real-analytic) connections on $TM$. By the principle of analytic continuation, the geometry of $\mathcal{M}$ is completely determined by the geometry of $M$, so that the complex manifold $Y$ is largely determined by the real geometry.

This last statement can be made much more precise if we also assume that $\sigma$ acts without fixed points, and that the 'real slice' $M$ is a compact manifold of real dimension $\geq 2k$. These harmless-looking assumptions imply that $\mu(\nu^{-1}(M))$ is an open and compact subset of $\mathcal{Y} \subset Y$, so it follows that $Y = \mathcal{Y} = \mu(\nu^{-1}(M))$ and is compact. If $\dim M = 2k$, moreover, the smooth fiber bundle $\nu^{-1}(M)$ over $M$ is a finite cover of $Y$, and the complex structure of $Y$, pulled back to this cover, is completely determined by the geometric structure of $M$.

## 3. The Penrose Correspondence

Let us now get down to specifics. Suppose that $Z$ is a complex 3-manifold, and suppose that $Z$ contains a rational curve $C \cong \mathbb{P}_1$ with normal bundle $\mathcal{O}(1) \oplus \mathcal{O}(1)$. We may then consider the connected component $\mathcal{M}$ of $C$ in the space of smooth rational curves in $Z$ which have normal bundle $\mathcal{O}(1) \oplus \mathcal{O}(1)$. Let $\mathcal{Z} \subset Z$ be the open set swept out by these curves. As predicted by our glittering generalities, $\mathcal{M}$ is a complex 4-manifold, and comes equipped with a family of 2-dimensional submanifolds parametrized by the points of $Z$. For example, if $Z = \mathbb{P}_3$, $\mathcal{M} = Gr_2(\mathbb{C}^4) \cong Q_4$, and each of the submanifolds in question is a linearly embedded $\mathbb{P}_2 \subset Q_4 \subset \mathbb{P}_5$.

The geometry of such moduli spaces $\mathcal{M}$ was first understood by Penrose [**53**]. In the double fibration

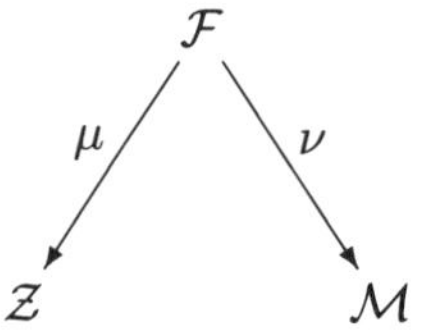

the right-hand projection $\nu$ is a holomorphic $\mathbb{P}_1$-bundle, and the fiber $\mathbb{P}_1$ over $p \in \mathcal{M}$ comes equipped with a natural embedding in the Grassmannian $Gr_2(\Omega^1_p)$ of 2-planes in the cotangent space $\Omega^1_p$. Let us now embed the Grassmann bundle $Gr_2(\Omega^1)$ in the $\mathbb{P}_5$-bundle $\mathbb{P}(\Omega^2)$ by

$$\text{span}(\varphi, \psi) \mapsto [\varphi \wedge \psi].$$

The image of this embedding is the 4-quadric bundle

$$\mathcal{Q} = \{[\omega] \in \mathbb{P}(\Omega^2) \mid \omega \wedge \omega = 0\},$$

and, for reasons of degree, each fiber of $\mathcal{F} \subset \mathcal{Q}$ is a smooth plane conic. Let $\Omega^+ \subset \Omega^2$ be the rank-3 holomorphic vector bundle spanned by $\mathcal{F}$, and let $\Omega^-$ denote its $\wedge$-orthogonal complement in the bundle of holomorphic 2-forms. We then have an orthogonal direct sum decomposition

$$\Omega^2 = \Omega^+ \oplus \Omega^- \tag{2}$$

with respect to the wedge product $\wedge$. Relative to this decomposition, a 2-form $\omega \in \Omega^2$ is called *self-dual* if $\omega \in \Omega^+$ or *anti-self-dual* if $\omega \in \Omega^-$.

The fiber $\nu^{-1}(p) \subset Gr_2(\Omega^1_p)$ may also be thought of more directly as a pencil of lines in the projective space $\mathbb{P}(T_p\mathcal{M}) \cong \mathbb{P}_3$. The union of these lines gives an embedding of the projectivized normal bundle $\mathbb{P}_1 \times \mathbb{P}_1$ of $\mu(\nu^{-1}(p))$ in $\mathbb{P}(T_p\mathcal{M})$, as a quadric hypersurface. Each point of $\mathcal{M}$ therefore has a neighborhood on which one can find a non-degenerate quadratic differential $\mathbf{g}$ whose zero locus in $\mathbb{P}(T\mathcal{M})$ is precisely this bundle of quadrics, and this requirement specifies $\mathbf{g}$ up to multiplication by a non-zero holomorphic function. Thus $\mathcal{M}$ naturally comes equipped with the holomorphic analogue of a conformal structure — a 'metric' $\mathbf{g}$ modulo functional rescaling. The decomposition (2) completely determines the 'conformal class' $[\mathbf{g}]$ because

$$SO(6,\mathbb{C})/[SO(3,\mathbb{C}) \times SO(3,\mathbb{C})] = SL(4,\mathbb{C})/SO(4,\mathbb{C});$$

conversely, $[\mathbf{g}]$ determines (2) up to the interchange of $\Omega^+$ and $\Omega^-$ because a quadric hypersurface in $\mathbb{P}_3$ contains exactly 2 pencils of lines, corresponding to the two factors of $\mathbb{P}_1 \times \mathbb{P}_1$. More explicitly, $[\mathbf{g}]$ and (2) are related by the requirement that two linearly independent vectors $v, w \in T_p\mathcal{M}$ satisfy $\mathbf{g}(v,v) = \mathbf{g}(w,w) = \mathbf{g}(v,w) = 0$ iff $\mathbf{g}(v,\cdot) \wedge \mathbf{g}(w,\cdot) \in \Omega^{\pm}$. As we will see more clearly in a moment, singling out a factor of the decomposition (2) is the complex analogue of choosing an orientation on a real 4-manifold with conformal structure.

Suppose we now choose a holomorphic 'metric' $\mathbf{g}$ in our conformal class $[\mathbf{g}]$. Then there is a unique torsion-free holomorphic connection $\nabla$ such that $\nabla\mathbf{g} = 0$, analoguous to the usual Levi-Cività connection of Riemannian geometry. If $[\mathbf{g}]$ arises by the above construction, one can show that $D = \ker \mu_*$ is horizontal with respect to the Levi-Cività connection of any $\mathbf{g} \in [\mathbf{g}]$, and our glittering generalities therefore tell us that the curvature $\mathcal{R}$ of $\nabla$ must satisfy the constraint

$$\mathcal{R}(v,w,v,w) = 0 \text{ whenever } v \wedge w \in \Omega^+, \tag{3}$$

where $\mathbf{g}$ has been freely used to identify $T$ with $\Omega^1$. Conversely, this constraint suffices to guarantee that $[\mathbf{g}]$ locally arises from this construction.

The constraint (3) can be better understood by observing that the curvature $\mathcal{R}$ of $\mathbf{g}$ may be identified, via $\mathbf{g}$, with a linear map $\Omega^2 \to \Omega^2$, and we may define a holomorphic tensor $\mathcal{W}_+$ as the trace-free part of the induced map $\Omega^+ \to \Omega^+$ relative to the decomposition (2). The constraint (3) is satisfied iff $\mathcal{W}_+ = 0$, and a complex conformal class $[\mathbf{g}]$ is called *anti-self-dual* when this occurs.

Now suppose that $Z$ is endowed with a real structure $\sigma : Z \to Z$ which has no fixed points. Let $M \subset \mathcal{M}$ denote the fixed point set of the induced real structure on $\mathcal{M}$, and assume that $M$ is non-empty. Then the real-analytic 4-manifold $M$ inherits a geometrical structure from $\mathcal{M}$ consisting precisely of an orientation and a conformal class $[g]$ of real-analytic Riemannian metrics satisfying $W_+ = 0$, where $W_+$ is a piece of the Riemann curvature tensor of $g \in [g]$ which is known as the *self-dual Weyl tensor.* A conformal class satisfying this condition is said to be *anti-self-dual.* The situation is thus summed up by the following result:

THEOREM 1 (Penrose). *Let $Z$ be a complex 3-manifold with a fixed-point-free anti-holomorphic involution $\sigma : Z \to Z$. Let $M$ denote the set of smooth $\sigma$-invariant rational curves $C \subset Z$ with normal bundle $\mathcal{O}(1) \oplus \mathcal{O}(1)$. Then $M$ is a real-analytic 4-manifold, and carries a naturally induced anti-self-dual conformal class $[g]$ of real-analytic Riemannian metrics.*

If a component $M_0$ of $M$ happens to be compact, then so is $Z$, and a finite cover of the underlying real manifold of $Z$ may be identified with the $S^2$-bundle $\nu^{-1}(M_0)$ over $M_0$. Turning this observation in its head, we will next see that every oriented Riemannian 4-manifold with $W_+ = 0$ gives rise to an associated complex 3-manifold $Z$, called its *twistor space.*

## 4. Four-Dimensional Geometry

Just as the theory of Riemann surfaces provides a powerful equivalence between conformal Riemannian geometry and complex analysis in real dimension 2, twistor theory provides a dictionary between special conformal Riemannian 4-manifolds and a class of complex 3-folds. The route from complex manifolds to conformal geometry was explained in the last section. We will now describe a direct route from 4-dimensional Riemannian geometry back to complex analysis.

Let $M$ be an oriented 4-manifold with smooth Riemannian metric $g$. The metric and orientation determine a volume 4-form $\text{vol}_g$, and the metric also induces a point-wise inner product $\langle\cdot,\cdot\rangle_g$ on 2-forms. Using these, one can in turn define the Hodge star operator $\star = \star_g : \Lambda^2 \to \Lambda^2$ by requiring that

$$\phi \wedge \star\psi = \langle \phi, \psi \rangle \text{vol}_g$$

for all 2-forms $\phi$ and $\psi$. One can check that that $\star^2 = 1$, so the eigenvalues of $\star$ are $\pm 1$. This gives rise to a decomposition

$$\Lambda^2 = \Lambda^+ \oplus \Lambda^- \tag{4}$$

of the rank 6 bundle of 2-forms into two rank-3 bundles, the $(\pm 1)$-eigenspaces of $\star$. Sections of $\Lambda^+$ (respectively, $\Lambda^-$) are called *self-dual* (respectively, *anti-self-dual*) 2-forms. The decomposition (4) is *conformally invariant*, in the sense it is unchanged if $g$ is replaced by $ug$ for any positive function $u$; but reversing the orientation of $M$ interchanges $\Lambda^+$ and $\Lambda^-$.

The central importance of (4) stems from the fact that curvature tensors are bundle-valued 2-forms and thus, on a Riemannian 4-manifold, can be broken up into self-dual and anti-self-dual parts. For the Riemannian curvature of our metric $g$, however, one can go even further; using the metric to reinterpret the curvature tensor as the *curvature operator* endomorphism $R$ of the bundle of 2-forms, it is apparent that $R : \Lambda^+ \oplus \Lambda^- \to \Lambda^+ \oplus \Lambda^-$ may be considered as consisting of more primitive pieces

$$R = \left( \begin{array}{c|c} W_+ + \frac{s}{12} & \overset{\circ}{r} \\ \hline \overset{\circ}{r} & W_- + \frac{s}{12} \end{array} \right), \tag{5}$$

where trace $W_\pm = 0$. The number $s$ is a famous object called the scalar curvature, while $\overset{\circ}{r}$ is a disguised form of the the trace-free part of the Ricci curvature tensor. The remaining two tensors $W_\pm$ are respectively called the *self-dual* and *anti-self-dual Weyl curvatures*, and are conformally invariant; they both vanish iff $g$ is locally conformal to a flat metric. When $W_+ = 0$, the metric $g$ is said to be *anti-self-dual*. Notice that this is a conformally invariant condition.

As was already indicated in the previous section, anti-self-duality is a crucial concept for understanding the Penrose correspondence. This can be beautifully re-explained from the Riemannian point of view, as follows. Let $(M, g)$ be an oriented Riemannian 4-manifold, and let $Z$ be total space of the $S^2$-bundle $\wp : S(\Lambda^+) \to M$, which we define to be the set of self-dual 2-forms of length $\sqrt{2}$. Then $Z$ naturally carries an almost-complex structure $J : TZ \to TZ$, $J^2 = -1$, which preserves the decomposition of $TZ$ into horizontal and vertical components with respect to the Levi-Cività connection:

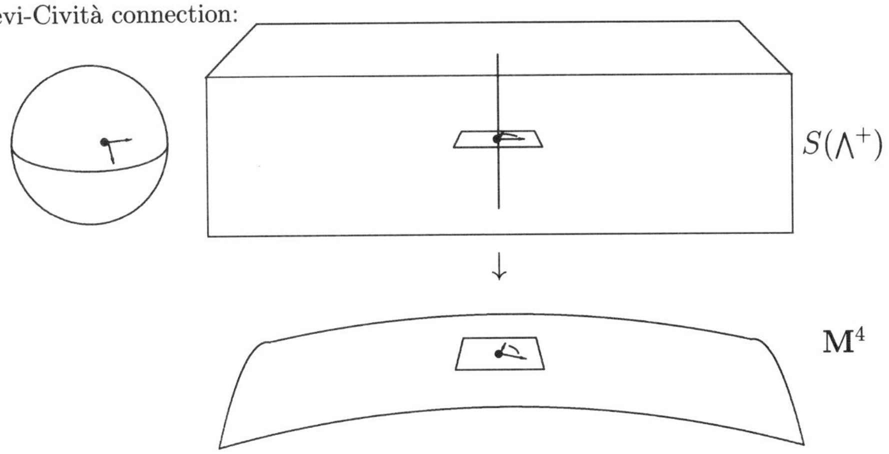


On the tangent spaces of the fiber $S^2$'s, $J$ simply acts by rotation by $-90°$. Meanwhile, in the horizontal subspace at $\phi \in S(\Lambda^+)$, which we identify with $T_{\wp(\phi)}M$ via the derivative of the canonical projection $\wp$, $J$ acts as $g^{-1} \circ \phi$. Although it is hardly obvious, this almost-complex structure $J$ is actually conformally invariant—despite the fact that replacing $g$ with $ug$ changes the horizontal subspaces on $Z = (\Lambda^+ - 0)/\mathbb{R}^+$.

However, the almost-complex structure $J$ is not generally integrable — one cannot generally find $\mathbb{C}^3$-valued coordinate charts on $Z$ which identify $J$ with the standard constant-coefficient tensor on $\mathbb{C}^3 = \mathbb{R}^6$. The obstruction **[51]** to finding such complex charts on an almost-complex manifold is in general encoded by a vector-valued 2-form called the Nijenhuis tensor; but in our case **[2]** the Nijenhuis tensor of $(Z, J)$ is determined by the self-dual part $W_+$ of the Weyl curvature of $(M, g)$. When this vanishes, the adapted complex charts are then interrelated by biholomorphisms, and $Z$ acquires the structure of a complex manifold:

THEOREM 2 (Atiyah-Hitchin-Singer). *The almost-complex manifold $(Z, J)$ is a complex 3-manifold iff $W_+ = 0$. Moreover, a complex 3-manifold arises by this construction iff it admits a fixed-point-free anti-holomorphic involution $\sigma : Z \to Z$ and a foliation by $\sigma$-invariant rational curves $\mathbb{CP}_1$, each of which has normal bundle $\mathcal{O}(1) \oplus \mathcal{O}(1)$. Finally, the complex manifold $(Z, J)$ and the real structure $\sigma$ suffice to determine the metric $g$ on $M$ up to conformal rescaling.*

DEFINITION 1. The complex 3-manifold $(Z, J)$ associated with an anti-self-dual 4-manifold $(M, g)$ by Theorem 2 is called the *twistor space* of $M$. The map $\wp : Z \to M$ is called the *twistor projection*, and its fibers are called *real twistor lines*.

Notice that Theorem 2 is not an exact converse of Theorem 1. In particular, putting the two results together gives us a new result for free:

COROLLARY 1. *Let $M$ be an oriented 4-manifold, and suppose that $[g]$ is a conformal class of twice-differentiable anti-self-dual Riemannian metrics on $M$. Then $M$ admits an atlas of coordinate charts relative to which $[g]$ is represented by real-analytic metrics.*

Indeed, the Atiyah-Hitchin-Singer construction embeds $M$ as an open set in the space of real rational curves in the twistor space $Z$. It therefore suffices to check that the conformal class of Theorem 1 coincides with the given one.

While it might seem intriguing that compact anti-self-dual 4-manifolds give rise to compact complex 3-manifolds via this twistor construction, it is only fair to demand some examples before giving the matter further thought. Fortunately there are some excellent examples [**2**] which can be explained without difficulty.

EXAMPLE 1. Let $M = S^4$ and let $g$ be the standard 'round' metric induced on the unit sphere in $\mathbb{R}^5$ by the Euclidean metric. Because stereographic projection shows that $g$ is locally conformally flat, both $W_+$ and $W_-$ vanish, and $(S^4, g)$ is therefore anti-self-dual for either choice of orientation. Because the isometry group $SO(5)$ acts transitively on $S(\Lambda^2_+) \to S^4$, the twistor space $Z$ of $S^4$ is a homogeneous complex manifold. This line of reasoning in fact allows one to show without difficulty that $Z$ is just complex projective 3-space $\mathbb{CP}_3$, and that the twistor projection $\wp : Z \to M$ is just the Hopf map

$$[z^1 : z^2 : z^3 : z^4] \mapsto [z^1 + jz^2 : z^3 + jz^4]$$

from $\mathbb{CP}_3 = (\mathbb{C}^4 - 0)/\mathbb{C}^\times$ to the quaternionic projective line $\mathbb{HP}_1 = (\mathbb{H}^2 - 0)/\mathbb{H}^\times$.

EXAMPLE 2. More generally, let $(M, [g])$ be any *conformally flat* oriented 4-manifold. The choice of such a flat conformal structure $[g]$ on $M$ is equivalent to a coordinate atlas for which the transition functions are elements of the conformal automorphism group $SO(5,1) \cong PSL(4,\mathbb{C})$ of the 5-sphere.

For instance, let $(M, g)$ be a compact quotient of hyperbolic 4-space. Thus $M$ is obtained from the upper hemisphere of $S^4$ by dividing by the action of an appropriate discrete sub-group of $SO(4,1) \subset SO(5,1)$. The corresponding twistor space $Z$ is then obtained by dividing the region $|z^1|^2 + |z^2|^2 > |z^3|^2 + |z^4|^2$ of $\mathbb{CP}_3$ by the action of $\pi_1(M) \subset PSL(4,\mathbb{C})$.

Notice that this shows that twistor spaces can be highly non-Kählerian; indeed, the fundamental group of a compact hyperbolic 4-manifold can never be the fundamental group of any Kähler manifold [**11**].

The conformally flat 4-manifolds play a special rôle in the theory of anti-self-dual spaces. If $g$ is any Riemannian metric on a compact oriented 4-manifold $M$, the Hirzebruch index theorem gives us Gauss-Bonnet-like formula

$$\tau(M) = \frac{1}{3}\int_M p_1(M) = \frac{1}{12\pi^2}\int_M (\|W_+(g)\|^2 - \|W_-(g)\|^2) d\mathrm{vol}_g$$

for the signature $\tau(M) = b_+(M) - b_-(M)$ of the manifold. Here $b_\pm$ is defined to be the maximal dimension of a linear subspace of $H^2(M, \mathbb{R})$ on which the cup product $\cup : H^2 \times H^2 \to H^4$ is positive (respectively, negative) definite. If $g$ is an anti-self-dual metric on $M$, we therefore have

$$\tau(M) \leq 0,$$

with equality iff $g$ is conformally flat.

EXAMPLE 3. Let $g$ be the usual Fubini-Study metric on $\mathbb{CP}_2$, and let $\overline{\mathbb{CP}}_2$ be $\mathbb{CP}_2$ with the non-standard orientation. Then $(\overline{\mathbb{CP}}_2, g)$ is anti-self-dual, since any point $p \in \overline{\mathbb{CP}}_2$ is the fixed point of an isometric $U(2)$-action which acts irreducibly on $\Lambda^+_p$, and so forces the trace-free quadratic form $W_+|_p$ on $\Lambda^+_p$ to vanish. Again the isometry group acts transitively on the twistor space, so that $Z$ is again homogeneous. Indeed, the twistor space $Z$ turns out to be the full flag variety

$$F_{1,2}(\mathbb{C}^3) = \{([z],[w]) \in \mathbb{CP}_2 \times \mathbb{CP}_2 \mid z \cdot w = 0\},$$

and the twistor projection $\wp : F_{1,2}(\mathbb{C}^3) \to \overline{\mathbb{CP}}_2$ is given by $([z],[w]) \mapsto [z \times \bar{w}]$, where $\times$ is the usual 3-dimensional cross product.

EXAMPLE 4. Let $X$ be a compact complex surface with $c_1^2 = 3c_2 > 0$, and let $M = \overline{X}$ be the oriented 4-manifold obtained by equipping $X$ with the non-standard orientation. Then $X$ admits **[69]** a Kähler-Einstein metric $g$ such that the universal cover of $(X,g)$ is the symmetric space $\mathbb{C}\mathcal{H}_2 = SU(2,1)/U(2)$, the usual model for which is the unit ball in $\mathbb{C}^2$ equipped with the Bergman metric. By the same isotropy argument as above, $\overline{\mathbb{C}\mathcal{H}_2}$ is anti-self-dual, and the same is therefore true of $(M,g)$. The twistor space of $\overline{\mathbb{C}\mathcal{H}_2}$ is the open set

$$\{([z],[w]) \in \mathbb{CP}_2 \times \mathbb{CP}_2 \mid z \cdot w = 0, z \cdot z^\dagger > 0, w \cdot w^\dagger > 0\}$$

of $F_{1,2}(\mathbb{C}^3)$, where $(z_1,z_2,z_3)^\dagger := (-\bar{z}_1, \bar{z}_2, \bar{z}_3)$, and the twistor space of $(M,g)$ is therefore this set modulo an action of $\pi_1(X)$.

EXAMPLE 5. Let $M$ be a complex surface, oriented in the usual way, and let $g$ be a Kähler metric on $M$. Then the bundle of complex-valued self-dual 2-forms on $(M,g)$ is just

$$\Lambda^+ \otimes \mathbb{C} = \mathbb{C}\omega \oplus \Lambda^{2,0} \oplus \Lambda^{0,2}$$

where $\omega$ is the Kähler form of $g$. Now the curvature of any Kähler metric is an element of $\Lambda^{1,1} \otimes \Lambda^{1,1}$ because the complex structure is parallel with respect to the Levi-Cività connection of such a metric. Thus the block

$$W_+ + \frac{s}{12} \in \mathrm{End}(\Lambda^+)$$

of the curvature operator must act by zero on $\Lambda^{2,0} \oplus \Lambda^{0,2}$ and be a scalar multiple of the identity on $\mathbb{C}\omega$; and this scalar must be the trace of $W_+ + s/12$, namely $s/4$. Thus the eigenvalues of $W_+$ are therefore $-s/12$, with multiplicity 2, and $s/6$, with multiplicity 1, and so **[22]** *a Kähler surface is anti-self-dual iff the scalar curvature $s$ is identically zero.*

In particular, let $M$ be a $K3$ surface. Then **[69]** there is a Ricci-flat Kähler metric in each Kähler class on $M$. Since the scalar curvature is the trace of the Ricci curvature, these Calabi-Yau metrics are all anti-self-dual. Varying the Kähler class and complex structure yields a 57-parameter family of anti-self-dual conformal classes on $M$.

Since reversing the orientation of $M$ interchanges $W_+$ and $W_-$, there is but a looking-glass difference between anti-self-dual manifolds and *self-dual* manifolds, which by definition instead satisfy $W_- = 0$. Our preference for anti-self-duality is largely motivated by the importance of the scalar-flat Kähler surfaces.

## 5. Kählerian Twistor Spaces

Among the the numerous examples of anti-self-dual 4-manifolds discussed in the last section, there were two, $S^4$ and $\overline{\mathbb{CP}}_2$, whose twistor spaces were immediately recognizable as projective algebraic varieties. Don't let this whet your appetite too much, though — there are **[27]** no other examples of this kind!

THEOREM 3 (Hitchin). *Up to conformal isometry, the only compact anti-self-dual 4-manifolds with Kählerian twistor spaces are the spaces $S^4$ and $\overline{\mathbb{CP}}_2$ described above.*

Indeed, suppose that the twistor space $Z$ of a compact anti-self-dual 4-manifold $(M, g)$ admits a Kähler form $\omega$. If $\sigma : Z \to Z$ is the real structure (anti-holomorphic involution) given by the twistor construction, then $\hat{\omega} = \omega - \sigma^*\omega$ is also a Kähler form, and is anti-invaraint with respect to $\sigma$. If $C \subset Z$ is a twistor line, we may arrange that $\int_C \hat{\omega} = 4$ by multiplying it by a suitable positive constant. However, $c_1(Z)$ is the only element of $H^2(Z, \mathbb{R})$ which is anti-invariant under $\sigma$ and equals 4 on a twistor line, since $H^2(Z, \mathbb{R}) = H^2(M, \mathbb{R}) \oplus \mathbb{R}c_1(Z)$, and the factors in this decompositions are exactly the $(\pm 1)$-eigenspaces of $\sigma^*$. Thus any Kählerian twistor space is a Fano manifold — i.e. has $c_1 > 0$. In particular, a 4-manifold with Kählerian twistor space must be simply connected.

Now if $b_2(Z) = 1$, it follows that $b_2(M) = b_+(M) + b_-(M) = 0$, and hence $\tau = b_+ - b_- = 0$, too. But our Gauss-Bonnet formula for $\tau$ then tells us that any anti-self-dual metric $g$ on $M$ must be conformally flat. But if $Z$ is Kählerian, we have just seen that $M$ is simply connected, and it therefore follows that $(M, [g])$ is $S^4$ with its standard conformal metric.

It remains to consider the case in which $b_2(Z) > 1$. Notice that the Chern class of the vertical line bundle of the twistor projection $\wp : Z \to M$ is anti-invariant under $\sigma$ and equals 2 on a twistor line. Since $H^2(Z, \mathbb{R}) = H^2(M, \mathbb{R}) \oplus \mathbb{R}c_1(Z)$, this shows that $c_1(Z)$ is divisible by 2. We can therefore finish our proof [**41, 70**] by quoting the following beautiful consequence [**67**] of Mori theory:

THEOREM 4 (Wiśniewski). *Let $X$ be a Fano manifold of dimension $2r - 1$ for which $r | c_1$. If $b_2(X) > 1$, then $X$ is one of the following:*

- $\mathbb{CP}_{r-1} \times Q_r$*;*
- $F_{1,r}(\mathbb{C}^{r+1})$*; or*
- $\mathbb{CP}_{2r-1}$ *blown up along* $\mathbb{CP}_{r-2}$.

*Here $Q_r$ denotes the $r$-quadric.*

Because $\mathbb{CP}_1 \times \mathbb{CP}_1 \times \mathbb{CP}_1$ does not contain any $\mathbb{CP}_1$'s with normal bundle $\mathcal{O}(1) \oplus \mathcal{O}(1)$, and the blow-up of $\mathbb{CP}_3$ along a line is not covered by such curves, the only viable candidate is the flag manifold $F_{1,2}(\mathbb{C}^3)$, and this is exactly the twistor space of $\overline{\mathbb{CP}}_2$.

## 6. Moishezon Twistor Spaces

We have just seen that there are only two twistor spaces which are projective algebraic manifolds. There is, however, a weaker notion of algebraicity [**33**]; a complex manifold is said to be *Moishezon* [**49**] if it is bimeromorphic to a projective variety. This weaker notion has wildly different ramifications in the present context; in fact, there are infinitely many deformation types of Moishezon twistor spaces!

Conic bundles over $\mathbb{CP}_1 \times \mathbb{CP}_1$ provide many interesting examples of such spaces; the geometric explanation for this is closely related to an essentially *linear* construction of anti-self-dual 4-manifolds, which will now be explained. Suppose for a moment that $U$ is a neighborhood of the diagonal in $\mathbb{CP}_1 \times \mathbb{CP}_1$, and let $L \to U$ be a holomorphic line bundle with $c_1(L) = 0$; such line bundles are parameterized by the vector space $H^1(U, \mathcal{O})$. Now consider the smooth rational curves $P \subset U$ which are homologous to the diagonal; these are parameterized by an open set in the space $PSL(2,\mathbb{C})$ of degree 1 holomorphic maps $\mathbb{CP}_1 \to \mathbb{CP}_1$, since each such $P$ is the graph of a function. Now the restriction of $L$ to any such $P$ is trivial, since $H^1(\mathbb{CP}_1, \mathcal{O}) = 0$, and we can therefore always choose a non-zero holomorphic section of $L|_P$, the image of which will be a rational curve $C$ in the total space of $L$. Now the normal bundle $N$ of $C$ projects onto the normal bundle $\mathcal{O}(2)$ of $P$, so we have an exact sequence

$$0 \to \mathcal{O} \to N \to \mathcal{O}(2) \to 0$$

which does not necessarily split, since $H^1(\mathbb{CP}_1, \mathcal{O}(-2)) = \mathbb{C} \neq 0$, and this extension class exactly measures whether or not the restriction of $L$ to the first infinitesimal neighborhood of $P$ is non-zero. When the obstruction is non-zero on some $P$, the associated curve $C \subset L$ has normal bundle $\mathcal{O}(1) \oplus \mathcal{O}(1)$, and one can invoke §3 in order to associate a (complex) self-dual 4-manifold with some open subset of $L$.

Now since $H^1(U, \mathcal{O})$ is a vector space, there must be some way of 'super-posing' anti-self-dual metrics which arise in this manner — despite the fact that the anti-self-duality condition is a very non-linear PDE! In fact, this superposition can be carried out completely explicitly. Notice that each element of $H^1(U, \mathcal{O})$ gives rise to a holomorphic function on an open subset of $PSL(2,\mathbb{C})$ by associating to each curve $P$ the image of the class in $H^1(P, \mathcal{O}_U/\mathcal{I}_P^2) = \mathbb{C}$. Now $PSL(2,\mathbb{C})$ may be thought of as a complexification of hyperbolic 3-space $\mathcal{H}^3$, which corresponds to the *Hermitian* $2 \times 2$ unimodular matrices. Thus, by restriction, any element of $H^1(U, \mathcal{O})$ gives rise to a complex-valued function $V$ on some open subset of $\mathcal{H}^3$. Rather than being arbitrary, this function is actually *harmonic*, in the sense that it satisfies

$$\Delta V = 0,$$

where $\Delta$ is the Laplace-Beltrami operator of hyperbolic 3-space $\mathcal{H}^3$; conversely, an application of the Penrose Transform [**4**] shows that every harmonic function on any convex domain $\hat{U} \subset \mathcal{H}^3$ arises from a unique element of $H^1(U, \mathcal{O})$, where $U \subset \mathbb{CP}_1 \times \mathbb{CP}_1$ represents the set of oriented hyperbolic geodesics meeting $U$. When the function $V$ is real and non-zero, so is the corresponding anti-self-dual conformal metric. Moreover, if $V$ is a non-zero harmonic function on a convex subset $U \subset \mathcal{H}^3$, the corresponding conformal class is explicitly represented by

$$g = h + V^{-2}(dt + A)^2$$

on $\hat{U} \times S^1$; here $h$ is the hyperbolic metric on $\mathcal{H}^3$, $t$ is the usual angular coordinate on $S^1 = \mathbb{R}/2\pi\mathbb{Z}$, and $A$ is a 1-form such that $dA = \star dV$, where $\star$ is the Hodge star operator of $\mathcal{H}^3$.

This recipe for cooking up anti-self-dual metrics from harmonic functions can be generalized to situations in which the domain $\hat{U} \subset \mathcal{H}^3$ of $V$ is an arbitrary open subset. This may be done by replacing $\hat{U} \times S^1$ with an $S^1$ bundle and replacing $dt + A$ with a connection form with curvature $\star dV$, and where $\frac{1}{2\pi} \star dV$ is now required to represent an integral class in de Rham cohomology. For example, if $V$ is the harmonic function $1 + (e^{2r} - 1)^{-1}$, where $r$ is the hyperbolic distance from a point in $\mathcal{H}^3$, then the corresponding metric is conformally equivalent to the Fubini-Study metric on an open dense subset of $\overline{\mathbb{CP}}_2$; by contrast, $V = 1$ gives rise to a conformal model of an open dense set of the standard 4-sphere $S^4$. It is therefore quite natural to choose $m$ distinct points $p_1, \dots, p_m \in \mathcal{H}^3$,

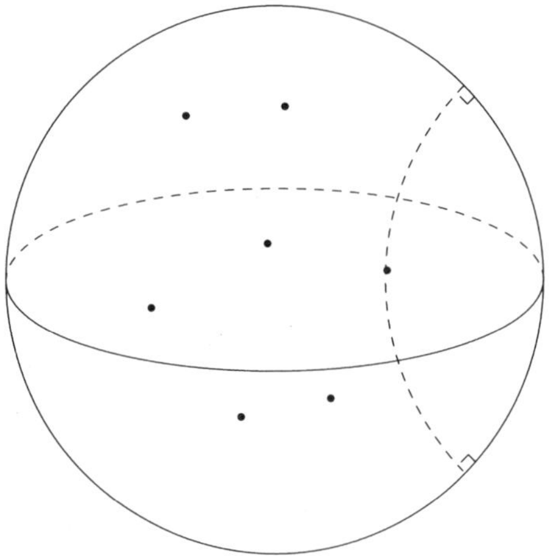

let $r_1, \dots, r_m$ denote the corresponding hyperbolic distance functions, and consider the harmonic function

$$V = 1 + \sum_{j=1}^{m} \frac{1}{e^{2r_j} - 1},$$

which can be shown to give rise to an anti-self-dual conformal class which extends to the iterated connected sum

$$m\overline{\mathbb{C}P}_2 := \underbrace{\overline{\mathbb{C}P}_2 \# \cdots \# \overline{\mathbb{C}P}_2}_{m}.$$

Here the connected sum $M_1 \# M_2$ of two compact connected oriented $n$-manifolds $M_1$ and $M_2$ is obtained by removing a small ball from each and identifying the resulting $S^{n-1}$ boundaries via a reflection. Since $M \# S^4 \approx M$ for any 4-manifold $M$, we will adopt the convenient convention that $0\overline{\mathbb{CP}}_2 = S^4$.

Because we are now considering potentials $V$ whose domains are non-convex, the corresponding twistor spaces are no longer just line bundles over open sets in $\mathbb{CP}_1 \times \mathbb{CP}_1$; however, they still admit a meromorphic map to $\mathbb{CP}_1 \times \mathbb{CP}_1$ and a holomorphic $\mathbb{C}^\times$-action which commutes with the map to $\mathbb{CP}_1 \times \mathbb{CP}_1$. Let $\mathcal{O}(k, \ell)$ denote the unique holomorphic line-bundle over $\mathbb{CP}_1 \times \mathbb{CP}_1$ with degree $k$ on the first factor and degree $\ell$ on the second, and identify hyperbolic 3-space $\mathcal{H}^3$ with one component of the Hermitian unimodular $2 \times 2$ matrices. The data points $p_1, \dots, p_m \in \mathcal{H}^3 \subset \mathbb{R}^{3,1}$ then determine $m$ sections $\mathcal{P}_1, \dots, \mathcal{P}_m \in \Gamma(\mathbb{CP}_1 \times \mathbb{CP}_1, \mathcal{O}(1,1))$. In the total space of the $\mathbb{CP}_2$-bundle

$$\mathbb{P}(\mathcal{O}(m-1,1) \oplus \mathcal{O}(1,m-1) \oplus \mathcal{O}) \longrightarrow \mathbb{CP}_1 \times \mathbb{CP}_1 ,$$

now define an algebraic variety $\tilde{Z}$ by the equation

$$xy = t^2 \prod_{j=1}^{m} \mathcal{P}_j ,$$

where $x \in \mathcal{O}(m-1, 1)$, $y \in \mathcal{O}(1, m-1)$, and $t \in \mathcal{O} = \mathcal{O}(0,0)$. The twistor space $Z$ of the relevant metric anti-self-dual metric on $m\overline{\mathbb{CP}}_2$ is then obtained from $\tilde{Z}$ by making small resolutions of the singular points and blowing down the surfaces $x = t = 0$ and $y = t = 0$ to $\mathbb{CP}_1$'s. In particular, these twistor spaces are all Moishezon, as promised. While the $m = 0$ and 1 cases just correspond to the most obvious metrics on $S^4$ and $\overline{\mathbb{CP}}_2$, the $m = 2$ metrics are sufficiently non-obvious that their discovery by Poon [**56**] was greeted with an excited flurry of new research. For $m \geq 3$, these twistor spaces and the corresponding metrics were discovered by the present author [**38**], while related results concerning the $m = 3$ case were discovered independently by Kurke and his school [**36**].

THEOREM 5. *For every* $m \geq 0$, *there exist anti-self-dual metrics on*

$$m\overline{\mathbb{C}P}_2 := \underbrace{\overline{\mathbb{C}P}_2 \# \cdots \# \overline{\mathbb{C}P}_2}_{m}$$

*for which the corresponding twistor spaces are Moishezon.*

A near-converse of this result was discovered by Campana [**12**]:

THEOREM 6 (Campana). *Suppose that* $M$ *is a compact 4-manifold which admits an anti-self-dual conformal class for which the the corresponding twistor space* $Z$ *is bimeromorphic to a Kähler manifold. Then* $Z$ *is Moishezon, and* $M$ *is homeomorphic to* $m\overline{\mathbb{CP}}_2$ *for some* $m \geq 0$.

Indeed, Poon [**57**] had already shown that $Z$ Moishezon implies that $b_+(M) = 0$ and $b_1(M) = 0$. Campana was able to simultaneously weaken the hypothesis and strengthen the latter conclusion to $\pi_1(M) = 0$ by studying curves rather than divisors in $Z$. Donaldson's thesis [**15**] then tells one that the intersection form $\cup : H^2(M, \mathbb{Z}) \times H^2(M, \mathbb{Z}) \to H^4(M, \mathbb{Z})$ is diagonalizable. But Freedman's work [**19**] tells us that simply connected smooth 4-manifolds are determined up to homeomorphism by their intersection forms, so the result follows.

It is still unknown whether a 4-manifold with Moishezon twistor space must in general be *diffeomorphic* to a connected sum $m\overline{\mathbb{CP}}_2$. If $m \leq 5$, however, this is indeed the case [**52**].

## 7. Connected Sums

The 4-manifolds discussed in the previous section were connected sums of $\overline{\mathbb{CP}}_2$'s, and their geometry essentially degenerates into that of $n$ copies of the Fubini-Study metric if the 'data points' in hyperbolic 3-space are widely separated. This is indicative of the fact that there is a general machine [**16, 18, 29**] for producing anti-self-dual metrics $g$ on appropriate connected sums $M = M_1 \# M_2$ of anti-self-dual 4-manifolds. As we will see, however, the relationship between this general point of view and the previous construction has a few surprises in store for us.

Suppose for a moment that $(M_1, [g_1])$ and $(M_2, [g_2])$ are two compact conformally flat 4-manifolds, and suppose that we wish to produce a conformal class $[g]$ of conformally flat metrics on $M_1 \# M_2$. Well, there is an obvious procedure for accomplishing this. Choose coordinate systems near arbitrarily chosen points $x_1 \in M_1$ and $x_2 \in M_2$ in which the given conformal classes become the conformal class of the Euclidean metric on $\mathbb{R}^4$, delete coordinate balls about the $x_j$, and then glue the complements of these balls by an inversion $\vec{v} \mapsto c\vec{v}/|\vec{v}|^2$ of $\mathbb{R}^4$. Because any such inversion is a conformal map of $\mathbb{R}^4 - 0$, the conformal classes $[g_1]$ and $[g_2]$ fit together to yield a conformal class of conformally flat metrics on the connected sum $M_1 \# M_2$. Notice that this procedure depends not only on a choice of the points $x_j \in M_j$, but also on the choice of an orientation-reversing isometry between the $T_{x_j} M_j$, implicit in the above choice of coordinates. Even having fixed this information, however, we still have a family of such metrics on $M$, since we are free to change the radii of the deleted ball about the points $x_j$. By letting these radii tend to zero, we may think of this family $[g_t]$ of conformal geometries on $M_1\#M_2$ as being smoothed out versions of the object obtained by letting $(M_1, [g_1])$ and $(M_2, [g_2])$ touch tangentially at a single point.

This construction has a surprising and beautiful reinterpretation on the level of twistor spaces. Let $Z_j$ denote the twistor spaces of $(M_j, [g_j])$, and let us blow up each of these twistor spaces along the twistor line of $x_j$ so as to produce a pair of complex 3-folds $\tilde{Z}_j$. Since a twistor line is a $\mathbb{CP}_1$ with normal bundle $\mathcal{O}(1) \oplus \mathcal{O}(1)$, the corresponding exceptional divisors in $\tilde{Z}_j$ are each $\mathbb{CP}_1 \times \mathbb{CP}_1$'s with normal bundle $\mathcal{O}(1, -1)$. An orientation-reversing isometry between $T_{x_1}M_1$ and $T_{x_2}M_2$ induces an isomorphism between the these two $\mathbb{CP}_1 \times \mathbb{CP}_1$'s which interchanges the $\mathbb{CP}_1$ factors. Now let $Z_0$ be obtained by identifying $Z_1$ and $Z_2$ along these $\mathbb{CP}_1 \times \mathbb{CP}_1$ exceptional divisors as specified by the above map. Thus $Z_0$ is a 3-dimensional complex space whose only singularity is of *normal crossing* type. Moreover, the normal bundles of the singular locus $Q \cong \mathbb{CP}_1 \times \mathbb{CP}_1$ relative to $Z_1$ and $Z_2$ are dual as line bundles, and so the singularity satisfies the so-called *d-semi-stable condition.* Now quite remarkably, as was first noticed by Donaldson, the twistor spaces $Z_t$ of the conformal metrics $[g_t]$ on $M_1\#M_2$ are small deformations of $Z_0$, with the added distinguishing feature that they admit anti-holomorphic involutions $\rho_t : Z_t \to Z_t$ which are small deformation of the real structure of $Z_0$.

With these paradigmatic examples in mind, Donaldson and Friedman **[16]** tackled the problem of finding anti-self-dual metrics on connected sums. Given compact anti-self-dual 4-manifolds $(M_1, [g_1])$ and $(M_2, [g_2])$, one again produces a space $Z_0$ with normal crossing singularity $\mathbb{CP}_1 \times \mathbb{CP}_1$ by blowing up the corresponding twistor spaces $Z_1$ and $Z_2$ along a pair of twistor lines and then identifying the two spaces along their exceptional divisors. The real structures of $Z_1$ and $Z_2$ then fit together to form an anti-holomorphic involution of $Z_0$ provided that one once again stipulated that the gluing procedure is specified by an orientation-reversing isometry between the appropriate tangent spaces of $M_1$ and $M_2$. One then applies the deformation theory of spaces with normal crossing **[19]** in order to look for smooth deformations

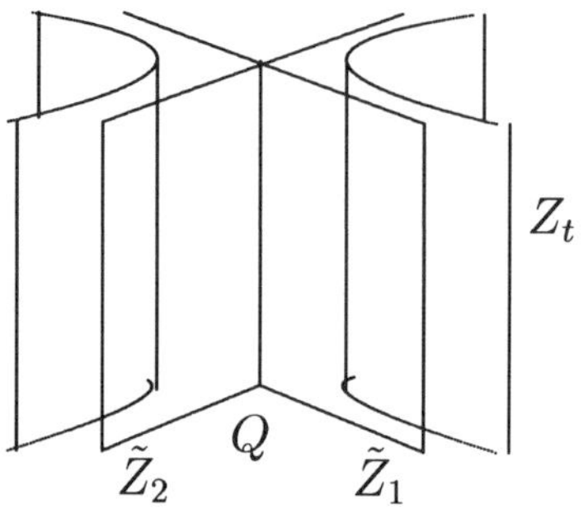


of $Z_0$ with real structure. This deformation theory is then unobstructed provided that Kodaira-Spencer theory is unobstructed for $Z_1$ and $Z_2$ separately.

THEOREM 7 (Donaldson-Friedman). *Suppose that the oriented 4-manifolds $M_1$ and $M_2$ admit anti-self-dual metrics for which the corresponding twistor spaces $Z_1$ and $Z_2$ satisfy $H^2(Z_j, \mathcal{O}(TZ_j)) = 0$. Then the connected sum $M_1\#M_2$ also admits anti-self-dual metrics with this same property.*

Because the above vanishing hypothesis holds for the standard metric on $\overline{\mathbb{CP}}_2$, it follows by induction that such metrics exist on all the connected sums $m\overline{\mathbb{CP}}_2$.

Inspection of the Friedman-Donaldson proof in fact yields more: for $m \geq 3$, there is a $(7m-15)$-dimensional moduli space of anti-self-dual conformal structures on this space; and for $m \geq 5$ the generic point in this moduli space represents a conformal structure for which the twistor space has *algebraic dimension zero* — i.e. does not admit any non-constant meromorphic functions. On the other hand, the explicit $(3m-5)$-parameter family of metrics encountered in the previous section all had twistor spaces of *algebraic dimension three.* Reconciling these results leads to the conclusion that there are Moishezon twistor spaces that have small deformations which are non-Moishezon twistor spaces. However, a beautiful result of Campana [**12**], proved by studying the space of rational curves through a general point, tells us that a twistor spaces can be bimeromorphic to a Kähler manifold only if it is Moishezon. This leads one to the following surprising conclusion [**13, 43**]:

THEOREM 8 (Campana/LeBrun-Poon). *The class $\mathcal{C}$, consisting of complex manifolds which are bimeromorphic to Kähler manifolds, is not stable under small deformations of complex structure.*

There is now a considerable literature concerning the algebraic dimension of twistor spaces and its relationship to the conformal geometry of the corresponding 4-manifold.

It needs to be pointed out that the above order of presentation is idiosyncratic rather than historical. The first complete proof of the existence of anti-self-dual metrics on $m\overline{\mathbb{CP}}_2$, $m \geq 3$ was actually that of Donaldson-Friedman, whereas the hyperbolic ansatz came only later. It is also worth pointing out that the Donaldson-Friedman method can also be applied even without the stated vanishing hypotheses, but the results are considerably more delicate. For example, the Calabi-Yau metrics on $K3$ are anti-self-dual, but their twistor spaces satisfy $h^2(Z, \mathcal{O}(TZ)) = 5$. The blow-up $K3\#m\overline{\mathbb{CP}}_2$ of $K3$ at $m \geq 3$ points can nonetheless be shown to admit anti-self-dual metrics by a careful application of this technique [**16**].

## 8. Anti-Self-Duality as a PDE

Thus far, we have analyzed the equation $W_+ = 0$ by just thinking about it as the integrability for the complex structure on twistor space. However, treating it directly as a partial differential equation for the unknown conformal metric $[g]$ leads to some remarkable results which are at present inaccessible by other techniques.

The first important attempt in this direction was made by Floer [**19**], who explored much the same ground treated by Donaldson-Friedman [**16**], but without the use of twistor theory. By imitating the cut-and-paste construction for conformally flat metrics on a connected sum of conformally flat manifolds, Floer observes that connected sums of anti-self-dual manifolds admit metrics for which $W_+$ is quite small. Given such a metric $g$, one then considers perturbations $g+h$, where $h$ is a uniformly small trace-free symmetric tensor field, and looks for such perturbations which are anti-self-dual. We thus need to consider the equation

$$0 = W_+(g+h) = W_+(g) + L(h) + O(|h|^2),$$

where the second order linear differential operator $L$ is the linearization of $W_+$ at $g$. In order to stand a decent chance of solving this equation, one should demand that $L$ be surjective (so the equation can at least be solved modulo the correction term) and that $W_+(g)$ is small (so that solving the linearized equation for $h$ nearly solves the non-linear equation, too). In this way, Floer discovered what amounts to a statement of Theorem 7. However, his paper is rather heuristic in that it avoids discussing the analytic issues involved in passing from a solution of the linearized equation to a solution of the non-linear equation $W_+ = 0$.

Despite this, Floer was certainly on the right track, as Taubes [**64, 65**] was later able to establish that one can indeed correct for the higher-order terms via an iteration scheme. But in his hands, this approach does much more than reprove Theorem 7; in fact, he is proves

THEOREM 9 (Taubes). *For each smooth compact oriented 4-manifold $M$, there is an integer $m_0(M)$ such that the connected sum $M\#m\overline{\mathbb{CP}}_2$ admits anti-self-dual metrics whenever $m \geq m_0(M)$.*

Notice that the minimal such $m_0(M)$ is a diffeomorphism invariant of $M$. This number is typically non-zero; for instance, the Gauss-Bonnet-type formula

$$\tau(M) = \frac{1}{3}\int_M p_1(M) = \frac{1}{12\pi^2}\int_M (\|W_+\|^2 - \|W_-\|^2)\mathrm{vol}_g$$

for the signature $\tau = b_+ - b_-$ tells us that $M$ cannot admit anti-self-dual metrics if $\tau(M) > 0$. On the other hand, it is extremely difficult to compute $\min m_0(M)$; for example, its value for $M = \mathbb{CP}_2$ still seems to be unknown, though we do know [**46**] that the answer is between 2 and 14.

Taubes' proof begins by arguing that, for $m$ sufficiently large, $M\#m\overline{\mathbb{CP}}_2$ admits Riemannian metrics for which $W_+$ is small; the point is that the standard metric on $\overline{\mathbb{CP}}_2$ is anti-self-dual, and one is free to cut out pieces of $M$ where $W_+$ is large and replace them with $\overline{\mathbb{CP}}_2$'s. He then uses the contraction mapping theorem to show that any metric with sufficiently small $W_+$ (as measured by a cunningly constructed norm) is actually near an anti-self-dual metric.

Theorem 9 has a truly remarkable folk-corollary, perhaps first noticed by Jim Carlson:

COROLLARY 2. *Let $\Gamma$ be any finitely-presented group. Then there exists a compact complex 3-manifold $Z$ with $\pi_1(Z) \cong \Gamma$.*

PROOF. By performing elementary surgeries on $S^4$, one can construct a compact oriented 4-manifold $M$ with $\pi_1(M) \cong \Gamma$; indeed, one just needs to replace an arbitrary $S^0 \times D^4$ with a $D^1 \times S^3$ for each generator, and then replace a suitable $S^1 \times D^3$ with a $D^2 \times S^2$ for each relation. Now Theorem 9 tells us that that there is an $m$ such that $\tilde{M} = M\#m\overline{\mathbb{CP}}_2$ admits an anti-self-dual conformal metric $[g]$, and $\pi_1(\tilde{M}) = \pi_1(M) = \Gamma$ by the Seifert-van Kampen Theorem. By the Theorem 2, the twistor space $(Z, J)$ of $(\tilde{M}, [g])$ is a complex 3-manifold, and $Z$ is an $S^2$-bundle over $\tilde{M}$. Since the long exact homotopy sequence of this fibration reads

$$\cdots \to \pi_1(S^2) \to \pi_1(Z) \to \pi_1(\tilde{M}) \to 0,$$

the compact complex manifold $Z$ therefore has fundamental group $\Gamma$. □

By contrast, it is known [**11**] that the fundamental groups of Kähler manifolds satisfy highly non-trivial constraints. This shows that most compact complex manifolds are topologically quite unlike the algebraic varieties.

## 9. Special Anti-Self-Dual Manifolds

Since the condition $W_+ = 0$ is conformally invariant, it should be considered as an equation for a conformal class rather than for a metric. Now on a *compact*

4-manifold, one can largely ignore this, because Schoen's solution of the Yamabe problem [**62**] allows us to eliminate most of the conformal freedom $g \mapsto ug$ by requiring that $g$ have *constant scalar curvature* $s$ and some fixed volume. The solution to this problem is not generally unique, however; for example, the round metric on $S^4$ has lots of conformal automorphisms which are not isometries. Nonetheless, one can certainly define many global conformal invariants of an anti-self-dual 4-manifold in this way, and in particular *sign* of the constant scalar curvature — positive, negative, or zero — of the scalar curvature is an invariant of the conformal class. The relationship between this invariant and twistor theory [**23, 30, 54**] is interesting, but rather complicated; for example [**57**], an anti-self-dual 4-manifold with Moishezon twistor space of algebraic dimension $\geq 2$ must have positive scalar curvature, but the converse is false.

By imposing a stronger condition, one may instead attempt to single out a preferred metric $g \in [g]$ even if $M$ is non-compact. For example, one could try to require that $g$ be *Einstein*, meaning [**8**] that $r = \lambda g$, where $r$ is the Ricci curvature tensor of $g$ and $\lambda$ is a constant. When the conformal class is fixed, this is an overdetermined condition, and there are usually no solutions. Nonetheless, there is a beautiful twistor interpretation [**53, 26, 66**] of this condition.

PROPOSITION 1 (Penrose/Hitchin/Ward). *Let $(M, [g])$ be an anti-self-dual 4-manifold with twistor space $Z$. Then there is a one-to-one correspondence between Einstein metrics $\tilde{g}$ in the conformal class $[g]$ and twisted holomorphic 1-forms $\theta \in \Gamma(Z, \Omega^1(K^{-1/2}))$ which are non-zero on the tangent space of each real twistor line and satisfy $\sigma^*\theta = \bar{\theta}$.*

Here $K$ denotes the canonical line bundle of $Z$, and its square-root is chosen so that the underlying smooth bundle of $K^{-1/2}$ is isomorphic to the vertical line bundle of $Z \to M$. Notice that if $\Gamma(Z, \Omega^1(K^{-1/2})) \neq 0$, it then follows that $[g]$ contains an Einstein metric defined on an open dense subset of $M$. Given $\theta$ as above, $\theta \wedge d\theta$ is a holomorphic function on $Z$, and so constant; moreover, this constant is real, and turns out to represent the scalar curvature of the corresponding Einstein metric $g$.

While this correspondence was originally used by Penrose to produce large families of non-compact Einstein manifolds, it turns out that *compact* anti-self-dual Einstein manifolds are very hard to come by. Essentially as a consequence [**20**] of Theorem 3, there are only two such manifolds with positive scalar curvature, namely $S^4$ and $\overline{\mathbb{CP}}_2$. When the scalar curvature is zero, the only possibilities [**8**] are covered by flat tori or Calabi-Yau K3's. On the other hand, we have absolutely no theorems regarding the negative-scalar-curvature case; the only known examples are quotients of hyperbolic 4-space $\mathcal{H}^4$ and the reverse-oriented complex-hyperbolic plane $\overline{\mathbb{C}\mathcal{H}}_2$.

Another interesting class of anti-self-dual 4-manifolds consists of the Kähler manifolds of vanishing scalar curvature and complex dimension 2; these will henceforth be called *scalar-flat Kähler surfaces.* Their twistor theory is as follows [**55**]:

PROPOSITION 2 (Pontecorvo). *Let $(M, [g])$ be an anti-self-dual 4-manifold with twistor space $Z$. Then there is a one-to-one correspondence between pairs $(\tilde{g}, J)$, where $J$ is an orientation-compatible complex structure on $M$ and $\tilde{g} \in [g]$ is a $J$-compatible scalar-flat Kähler metric, and sections $u \in \Gamma(Z, \mathcal{O}(K^{-1/2}))$ which are non-zero on each real twistor line and satisfy $\sigma^*u = \bar{u}$.*

Certainly the Ricci-flat Kähler metrics **[69]** on K3's, complex tori, and their quotients give us some examples of compact scalar-flat Kähler surfaces. On the other hand, because the Ricci form of a Kähler manifold is the curvature of the anti-canonical line bundle, the total scalar curvature of any compact Kähler surface is given by $\int_M s\, d\mu = 4\pi c_1 \cdot [\omega]$, where $[\omega]$ is the Kähler class, so a compact scalar-flat Kähler surface must satisfy $c_1 \cdot [\omega] = 0$. Unless some power of the canonical line bundle is trivial, this tells us that the Kodaira dimension must be $-\infty$. Any non-Ricci-flat example must therefore be a (possibly non-minimal) ruled surface.

The latter possibility actually occurs, and indeed one can construct **[39]** many explicit examples by a modification of the 'hyperbolic ansatz' of §6. This involves replacing $\mathcal{H}^3$ with $X = \mathcal{H}^3/\Gamma \approx \mathbb{R} \times \Sigma$, where $\Gamma$ is the fundamental group of a compact Riemann surface $\Sigma$ of genus $\geq 2$; by uniformization, the choice of a complex structure on $\Sigma$ determines an action of $\Gamma$ on $\mathcal{H}^3$ via the obvious inclusion $SO(1,2) \hookrightarrow SO(1,3)$. Configurations

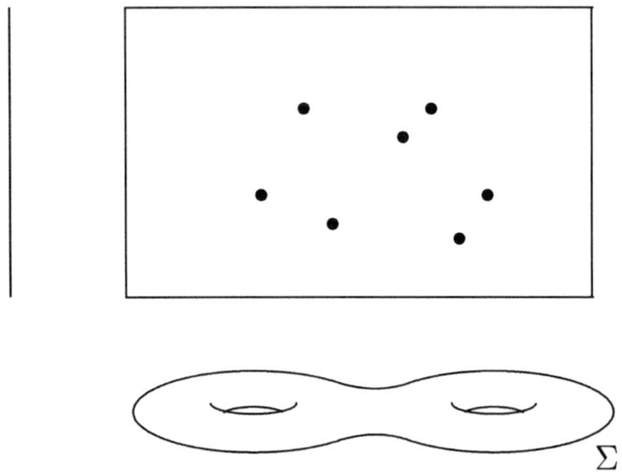


of $m \geq 2$ points in $X$ give rise **[39]** to scalar-flat Kähler metrics over non-minimal ruled surfaces $\approx (\Sigma \times S^2)\#m\overline{\mathbb{CP}}_2$ whose minimal models are $\mathbb{CP}_1$-bundles over $\Sigma$. These surfaces are non-generic because they carry holomorphic vector fields. However, their small deformations also can be shown **[45]** to carry scalar-flat Kähler metrics, so these examples give one to a fairly comfortable understanding of scalar-flat Kähler metrics on ruled surfaces over curves of large genus.

To get at the low genus cases, however, one must work considerably harder. Nonetheless, one may prove the existence of many such surfaces **[32]** by exploiting Proposition 2 and an orbifold version of Theorem 7. This allows one to prove the following result:

THEOREM 10 (Kim-LeBrun-Pontecorvo). *Let $(M, J)$ be a compact complex 2-manifold which admits a Kähler class $[\omega]$ with $c_1 \cdot [\omega] \geq 0$. Then either*

- *$(M, J)$ admits a Ricci-flat Kähler metric; or else*
- *$(M, J)$ has blow-ups $(\tilde{M}, \tilde{J})$ which admit scalar-flat Kähler metrics. Moreover, any blow-up of such an $(\tilde{M}, \tilde{J})$ admits scalar-flat Kähler metrics, too.*

## 10. Higher Dimensions

While these notes have thus far focused on anti-self-dual 4-manifolds, it has been observed that in principle the same techniques apply to many other geometries. We will now see that twistor theory leads to many new insights concerning Riemannian manifolds with parallel tensor fields.

A tensor field $\varphi \neq 0$ on a Riemannian $m$-manifold $(M, g)$ is said to be *parallel* if it satisfies $\nabla\varphi = 0$, where $\nabla$ is the Levi-Cività connection of the Riemannian metric $g$. How many such tensor fields are there? Obviously, any oriented Riemannian manifold admits some such fields: the metric, the volume form, their inverses, and linear combinations of contracted tensor products of these. There will be more only if the geometry of $M$ is special in the sense that the *holonomy group* — that is [**7, 8, 61**], the group of linear transformation of a tangent space induced by parallel transport around loops — is a proper subgroup of $SO(m)$, and a fundamental research topic for present-day Riemannian geometry is the problem of classifying those Riemannian manifolds whose holonomy geometries are special in this sense. In fact, if we exclude the locally symmetric spaces (for which the curvature tensor $\mathcal{R}$ satisfies $\nabla\mathcal{R} = 0$) and local products of manifolds of lower dimension, there are (up to conjugation) only seven possible families of connected holonomy groups: $SO(m)$, $U(m/2)$, $SU(m/2)$, $G_2$ $(m = 7)$, Spin(7) $(m = 8)$, $Sp(m/4)$, and $[Sp(m/4) \times Sp(1)]/\mathbb{Z}_2$ $(m \geq 8)$. Twistor theory turns out to be particularly well adapted to the study of the last two of these holonomy families.

DEFINITION 2. A Riemannian manifold $(M, g)$ of dimension $4n$, $n \geq 2$, is said to be quaternion-Kähler if its holonomy group is (conjugate to) a subgroup of $[Sp(n) \times Sp(1)]/\mathbb{Z}_2 \subset SO(4n)$.

This terminology may be justified by the fact [**8, 61**] that any such manifold carries a 4-form $\varphi$ such that $\nabla\varphi = 0$, and this 4-form may be thought of as a quaternionic analogue of the 2-form which plays such a central rôle in geometry of Kähler manifolds. (Note that the Kähler condition is equivalent to stipulating that the holonomy be contained in $U(m/2)$.)

It should be emphasized that a quaternion-Kähler manifold is typically *not* a Kähler manifold; $[Sp(n) \times Sp(1)]/\mathbb{Z}_2 \not\subset U(2n)$. Nonetheless, questions about such manifolds can, in the positive case, be reduced to problems in algebraic geometry over $\mathbb{C}$ by means of a twistor construction [**59, 6**]:

THEOREM 11 (Salamon/Bérard-Bergery). *Let $(M^{4n}, g)$ be a quaternion-Kähler manifold of scalar curvature $s$. Then there is a complex $(2n+1)$-manifold $Z$, called the* twistor space *of $(M, g)$, such that*

- *(i) the underlying smooth manifold of $Z$ is the total space of an $S^2$-bundle $\wp : Z \to M$;*
- *(ii) each fiber of $\wp$ is a rational curve $\mathbb{CP}_1 \subset Z$, with normal bundle isomorphic to $[\mathcal{O}(1)]^{\oplus 2n}$;*
- *(iii) there is a free anti-holomorphic involution $\sigma : Z \to Z$ which commutes with $\wp$;*
- *(iv) there is an $(n+1)^{\text{st}}$ root $K^{1/(n+1)}$ of the canonical line bundle of $Z$, and an isomorphism $\sigma^* K^{1/(n+1)} \cong \overline{K^{1/(n+1)}}$ induced by an anti-holomorphic involution of the total space of $K^{1/(n+1)}$; and*

- *(v) there is a twisted 1-form $\theta \in \Gamma(Z, \Omega^1(K^{-1/(n+1)}))$ which is non-zero when restricted to the tangent space of each fiber of $\wp$, and such that $\sigma^*\theta = \bar{\theta}$, $\theta \wedge (d\theta)^{\wedge n} = \wp^* s$.*

EXAMPLE 6. Quaternionic projective space $\mathbb{HP}_n$, equipped with its standard $Sp(n)$-invariant metric, is a quaternion-Kähler manifold. Its twistor space $Z$ is just $\mathbb{CP}_{2n+1}$, and the twistor projection $\wp : Z \to M$ is just the Hopf map $\mathbb{CP}_{2n+1} \to \mathbb{HP}_n$ induced by the obvious identification of $\mathbb{C}^{2n+2}$ with $\mathbb{H}^{n+1}$. The real structure $\sigma$ may be expressed in homogeneous coordinates as left multiplication by the quaternion $j$ on $\mathbb{C}^{2n+2} = \mathbb{H}^{n+1}$, and $\theta \in \Gamma(\mathbb{CP}_{2n+1}, \Omega^1(2))$ may be written in homogeneous coordinates as $\theta = \sum_{j=1}^{n+1} z^{2j-1} dz^{2j}$.

To prove *(i)–(iv)* one only needs [**3, 60, 29**] a torsion-free reduction of the structure group of $TM$ to $GL(n, \mathbb{H}) \times Sp(1)/\mathbb{Z}_2$; such a *quaternionic structure* is a higher-dimensional analogue of an anti-self-dual conformal structure. In the statement of *(v)*, we have identified $\Omega^{2n+1}(K^{(-n-1)/(n+1)})$ with $\mathcal{O}$ in the tautological way, and because every holomorphic function on $Z$ is constant (even if $M$ is non-compact) we are, in particular, asserting that the scalar curvature of a quaternion-Kähler manifold is necessarily constant. The latter can be seen more directly because the quaternion-Kähler condition turns out to imply that $(M, g)$ is an *Einstein manifold.* This suggests the correct way of supplementing Definition 2 to allow for the $n = 1$ case while simultaneously ensuring that Theorem 11 remains true:

DEFINITION 3. An oriented Riemannian manifold $(M, g)$ of dimension 4 is said to be quaternion-Kähler if it is Einstein and has anti-self-dual Weyl curvature.

Quaternion-Kähler geometry comes in three distinct flavors, depending on the sign of scalar curvature $s$. For example, if $s = 0$, the sub-bundle $\ker\theta$ of $TZ$ is integrable, and there is a finite cover $\tilde{M}$ of $M$ for which the corresponding twistor space is a holomorphic family over $\mathbb{CP}_1$, with vertical tangent bundle $\ker\theta$. The latter is equivalent to saying that the holonomy group of $\tilde{M}$ is contained in $Sp(n)$, and one then says that $(\tilde{M}, g)$ is *hyper-Kähler.* This name indicates the fact that such a metric is Kähler with respect to an $S^2$'s worth of distinct complex structures; and choosing any such complex structure converts $(\tilde{M}, g)$ into a Kähler-Einstein manifold which carries a holomorphic 2-form $\alpha \in \Gamma(\Omega^2)$ such that $\alpha^{\wedge n} \neq 0$. Yau's solution [**69**] of the Calabi conjecture thus reduces the problem of classifying these manifolds to the problem of classifying complex-symplectic manifolds of Kähler type. While such a classification still eludes us, an important family of examples has been constructed by Beauville [**5**]. The hyper-Kähler metrics one finds in this way are by no means unique; instead, moduli space of such metrics has dimension $3b_2(\tilde{M}) - 8$ in the irreducible case, and this number is typically fairly large.

On the other hand, if the scalar curvature $s$ is non-zero one has $\theta \wedge (d\theta)^{\wedge n} \neq 0$, so $\theta$ is a complex *contact structure* on $Z$. Darboux's theorem [**1**] asserts that all such structures are locally standard, so this amounts to specifying an atlas of coordinate charts on $Z$ for which the coordinate transformations preserve the span of the 1-form $dz^{2n+1} + \sum_{j=1}^{n} z^j dz^{j+n}$ on $\mathbb{C}^{2n+1}$.

The metric $g$ on $M$ can be used to define a canonical Hermitian structure on the $K^{-1/(n+1)}$, because the $n^{\text{th}}$ power of this bundle is exactly $\det\ker\theta$ and $\wp^* g$ is a Hermitian metric on $\ker\theta$. If $-i\omega$ is the curvature form of this Hermitian structure, then $\omega$ corresponds to a symmetric tensor field $h$ which would be a

Kähler metric were it only positive definite. In point of fact, the restriction of $h$ *is* positive definite on $\ker \wp_*$, but equals $\frac{s}{4n(n+2)}\wp^* g$ on $\ker \theta$; moreover, $\ker \wp_*$ and $\ker \theta$ are orthogonal with respect to $h$. This shows that $Z$ is Fano ($c_1(Z) > 0$) iff $s > 0$. Further computation reveals that when $s > 0$, $h$ is actually a Kähler-Einstein metric on $Z$; when $s < 0$, it is instead an *indefinite* Kähler-Einstein metric in the sense of pseudo-Riemannian geometry.

It is now obviously worth introducing the following piece of terminology:

DEFINITION 4. If $\theta \in \Gamma(Z, \Omega^1(K^{-1/(n+1)}))$ is a complex contact structure on a compact complex $(2n+1)$-manifold $Z$ with $c_1(Z) > 0$, we will say that $(Z, \theta)$ is a *Fano contact manifold.*

So far, we have seen that Theorem 11 associates a Fano contact manifold with every compact quaternion-Kähler manifold of $s > 0$, and this Fano manifold admits Kähler-Einstein metrics. Conversely [**42**], every such Fano manifold arises in this way:

THEOREM 12. *A Fano contact manifold $(Z, \theta)$ is the twistor space of some compact quaternion-Kähler manifold $(M, g)$ of positive scalar curvature iff $Z$ admits a Kähler-Einstein metric. Moreover, $(M, g)$ is completely determined up to homothety by the complex manifold $Z$.*

It is unknown whether there are Fano contact manifolds which do not admit Kähler-Einstein metrics; indeed, it is even unknown whether there are Fano contact manifolds which are not homogeneous! However, an infinitesimal rigidity result together with boundedness of degree for Fanos tells one [**41, 44**] that there only finitely many Fano contact manifolds of any fixed dimension. This implies

THEOREM 13. *Up to isometries and rescaling, there are only finitely many compact positive-scalar-curvature quaternion-Kähler manifolds of any fixed dimension.*

This offers a certain plausibility to what is widely considered to be *the* outstanding conjecture in the subject: that every compact positive-scalar-curvature quaternion-Kähler manifold is locally symmetric; this would follow [**68**] if every Fano contact manifold were homogeneous. In a sense, this would be quite a surprising result, because the analogous assertion quaternion-Kähler *orbifolds* is wildly false [**21, 9**]. Nonetheless, it is certainly true in the lowest dimensions [**27, 58**], and further support is offered by the following consequence [**41, 44**] of Theorem 4:

THEOREM 14. *Let $M^{4n}$ be a compact quaternion-Kähler manifold of positive scalar curvature. Then $\pi_1(M) = 0$ and*

$$\pi_2(M) = \begin{cases} 0 & \textit{iff } M \textit{ is } \mathbb{HP}_n \\ \mathbb{Z} & \textit{iff } M \textit{ is } Gr_2(\mathbb{C}^{n+2}) \\ \textit{finite with 2-torsion} & \textit{otherwise.} \end{cases}$$

It therefore seems reasonable to expect that the resolution of the the conjecture in question will hinge on further advances in algebraic geometry.

Finally, a few remarks should be made in regard to the $s < 0$ case. Here the non-compact but complete case is characterized by flexibility rather than rigidity. For example [**40**], twistor methods allow one to show that the moduli space of complete $s < 0$ quaternion-Kähler metrics on $\mathbb{R}^{4n}$ has an infinite-dimensional component. On the other hand, the only known *compact* examples are quotients of symmetric spaces, and the super-rigidity **[14]** of these examples is echoed by an infinitesimal rigidity result that can be be proved **[25]** for any compact quaternion-Kähler manifold of negative scalar curvature.

## 11. Problem List

This article's focus has been narrower than might be desirable; in particular, it was not feasible to include some beautiful recent applications **[10, 48]** of twistor theory to affine holonomy. Instead, the topics have been chosen so as to highlight areas with a strong intrinsic interest for algebraic geometers. In the same spirit, let me now close with a list of problems that illustrate our current level of ignorance and should appeal to algebraic geometers.

1. Is every contact Fano manifold the twistor space of a compact positive-scalar-curvature quaternion-Kähler manifold? (For those admitting Kähler-Einstein metrics, the answer is yes. For a direct algebro-geometric solution, it would be enough to construct a real structure and a suitable family of rational curves.)
2. Is every contact Fano manifold homogeneous? (Equivalently, does the linear system $|K^{-1/(n+1)}|$ gives a projective embedding of any contact Fano $(2n+1)$-manifold? An affirmative answer would imply that every compact positive-scalar-curvature quaternion-Kähler manifold is symmetric.)
3. If a Fano manifold $Z$ admits two distinct complex contact structures, must it be complex projective space? (If $Z$ admits a Kähler-Einstein metric, the answer is affirmative.)
4. Let $Z_{2n+1}$ be the twistor space of a compact $4n$-manifold $M$ with torsion-free $GL(n, \mathbb{H}) \times Sp(1)/\mathbb{Z}_2$-structure, $n \geq 2$. If $Z$ is Kählerian, can the quaternionic structure of $M$ always be represented by a quaternion-Kähler metric of positive scalar curvature? (Note that the stated hypotheses certainly imply that $Z$ is Fano, and the answer is therefore yes if $H_2(M)$ is torsion-free.)
5. Is the the twistor space $Z$ of any compact positive-scalar-curvature quaternion-Kähler manifold automatically rigid? (Since $Z$ is Fano, this is the same as asking whether $H^1(Z, \mathcal{O}(TZ)) = 0$. Because $(Z, \theta)$ is rigid as a contact manifold, this would follow if the previous problem had an affirmative answer.)
6. Classify the complex-symplectic manifolds of Kähler type.
7. Is the moduli space of hyper-Kähler metrics on a fixed manifold always connected? Is there a global Torelli theorem for such metrics?
8. If a Moishezon 3-fold $Z$ is the twistor space of an anti-self-dual 4-manifold $M$, must $M$ be diffeomorphic to a connected sum $m\overline{\mathbb{CP}}_2$, $m \geq 0$? (It is known that $M$ must be *homeomorphic* to such a connected sum.)
9. Let $M$ be a compact anti-self-dual manifold with positive scalar curvature and $b_1 = 0$. Is $M$ simply connected? (Note that this would imply that $M$ is homeomorphic to a connected sum of $\overline{\mathbb{CP}}_2$'s.)
10. Does every anti-self-dual conformal class on $M = \overline{\mathbb{CP}}_2$ have a twistor space of positive algebraic dimension? (If so, the only such class is that of the Fubini-Study metric.)
11. Same question for $M = K3$. (This would imply that every such class is represented by a Calabi-Yau metric.)

12. Find a homeomorphic pair of 4-manifolds such that one admits an anti-self-dual metric but the other does not. (This might be done, for example, by showing that homotopy K3 surfaces of Kodaira dimension 1 never admit anti-self-dual metrics.)
13. The only known compact anti-self-dual Einstein manifolds of negative scalar curvature are quotients of hyperbolic 4-space $\mathcal{H}^4$ or reverse-oriented complex-hyperbolic 2-space $\overline{\mathbb{C}\mathcal{H}_2}$. Are there others?
14. Suppose that the compact 4-manifold $M$ admits an anti-self-dual Einstein metric. Is every other Einstein metric on $M$ also anti-self-dual? (This is true of all known examples with non-positive scalar curvature.)

## References


[1] V.I. Arnold, **Mathematical Methods in Classical Mechanics**, Springer, 1989.

[2] M. Atiyah, N. Hitchin and I. Singer, "Self-Duality in Four Dimensional Riemannian Geometry," **Proc. R. Soc. Lond. A 362** (1978) 425–461.

[3] T.N. Baily and M.G. Eastwood, "Complex Paraconformal Manifolds— Their Differential Geometry and Twistor Theory," **Forum Math. 3** (1991) 61-103.

[4] R. Baston and M.G. Eastwood, **The Penrose Transform: Its Interaction with Representation Theory**,

[5] A. Beauville, "Variétés Kähleriennes dont la Première Classe de Chern est Nulle," **J. Diff. Geom. 18** (1983) 755–782.

[6] L. Bérard-Bergery, "Variétés Quaternioniennes," unpublished lecture notes, Espalion, 1979.

[7] M. Berger, "Sur les Groupes d'Holonomie des Variétés à Connexion Affine et des Variétés Riemanniennes," **Bull. Soc. Math. France 83** (1955) 278-330

[8] A. L. Besse, **Einstein Manifolds**, Springer-Verlag, 1986.

[9] C.P. Boyer, K. Galicki, and B.M. Mann, "Quaternionic Reduction and Einstein Manifolds," **Comm. An. Geom. 1** (1993) 229-279.

[10] R.L. Bryant, "Two Exotic Holonomies in Dimension Four, Path Geometries, and Twistor Theory," **Proc. Symp. Pure Math. 53** (1989) 33-88.

[11] J.A. Carlson and T. Domingo, "Harmonic Mappings of Kähler Manifolds to Locally Symmetric Spaces," **Publ. Math. IHES 69** (1989) 173–201.

[12] F. Campana, "On Twistor Spaces of the Class $\mathcal{C}$," **J. Diff. Geom. 33** (1991) 541-549.

[13] F. Campana, "The Class $\mathcal{C}$ is not Stable by Small Deformations," **Math. Ann. 290** (1991) 19–30.

[14] K. Corlette, "Archimedean Superrigidity and Hyperbolic Geometry," **Ann. Math. 135** (1992) 165–182.

[15] S. Donaldson, "An Application of Gauge Theory to Four-Dimensional Topology," **J. Diff. Geom. 18** (1983) 279-315.

[16] S. Donaldson and R. Friedman, "Connected Sums of Self-Dual Manifolds and Deformations of Singular Spaces," **Nonlinearity 2** (1989) 197–239.

[17] A. Douady, "Le Problème de Modules pour les Sous-Espaces Analytiques d'un Espace Anaytique Donné," **Ann. Inst. Fourier 16** (1966) 1–95.

[18] A. Floer, "Self-Dual Conformal Structures on $\ell\mathbb{CP}^2$," **J. Diff. Geom. 33** (1991) 551–573.

[19] M.H. Freedman, "The Topology of Four-Dimensional Manifolds," **J. Diff. Geom. 17** (1982) 357–454.

[20] T. Friedrich and H. Kurke, "Compact Four-Dimensional Self-Dual Einstein Manifolds with Positive Scalar Curvature," **Math. Nachr. 106** (1982) 271–299.

[21] K. Galicki and H.B. Lawson, "Quaternionic Reduction and Quaternionic Oribifolds," **Math. Ann. 282** (1988) 1–21.

[22] P. Gauduchon, "Surfaces Kählériennes dont la Courbure Vérifie Certaines Conditions de Positivité," in **Géometrie Riemannienne en Dimension 4**, (Bérard-Bergery, Berger, and Houzel, eds.), CEDIC/Fernand Nathan, 1981.

[23] P. Gauduchon, "Structures de Weyl et Théorèmes d'Annulation sur une Variété Conforme Autoduale," **Ann. Sc. Norm. Sup. Pisa Cl. Sci. 18** (1991) 563–629.

[24] H. Grauert and R. Remmert, **Coherent Analytic Sheaves**, Springer-Verlag, 1984.

[25] R. Horan, "A Rigidity Theorem for Quaternionic-Kähler Manifolds," **Diff. Geom. Appl. 6** (1996) 189–196.

[26] N.J. Hitchin, "Linear Field Equations on Self-Dual Spaces," **Proc. R. ASoc. Lond. A 370** (1980) 173–191.

[27] N.J. Hitchin, "Kählerian Twistor Spaces," **Proc. Lond. Math. Soc. 43** (1981) 133–150.

[28] N.J. Hitchin, "Complex Manifolds and Einstein's Equations," **Lec. Notes Math. 970** (1982) 73–99.

[29] Joyce, D. "The Hypercomplex Quotient and the Quaternionic Quotient," **Math. Ann. 290** (1991) 323–340.

[30] J.-S. Kim, "On the Scalar Curvature of Self-Dual Manifolds," **Math. Ann. 297** (1993) 235-251.

[31] J.-S. Kim and M. Pontecorvo, "A New Method of Constructing Scalar-Flat Kähler Metrics," **J. Diff. Geom.**

[32] J.-S. Kim, C. LeBrun, and M. Pontecorvo, "Scalar-Flat Kähler Surfaces of All Genera," **J. reine angew. Math.** *to appear.*

[33] D. Knutson, *Algebraic Spaces*, **Lec. Notes Math. 203**, Springer-Verlag, 1971.

[34] K. Kodaira, "On Compact Complex Analytic Surfaces I," **Ann. Math. 71** (1960) 111–152; "... II" **Ann. Math. 77** (1963) 563–626; "... III" **Ann. Math. 78** (1963) 1–40.

[35] K. Kodaira, "A Theorem of Completeness of Characteristic Systems for Analytic Families of Compact Submanifolds of a Complex Manifold," **Ann. Math. (2)75**(1962)146-162

[36] B. Kreussler and H. Kurke, "Twistor Spaces over the Connect Sum of 3 Projective Planes," **Comp. Math. 82** (1992) 25–55.

[37] C. LeBrun, "Spaces of Complex Null Geodesics in Complex-Riemannian Geometry," **Trans. Am. Math. Soc.278**(1983)209-231

[38] C. LeBrun, "Explicit Self-Dual Metrics on $\mathbb{CP}_2\#\cdots\#\mathbb{CP}_2$," **J. Diff. Geom. 34** (1991) 223–253.

[39] C. LeBrun, "Scalar-Flat Kähler Metrics on Blown-Up Ruled Surfaces," **J. Reine Angew. Math.** 420 (1991) 161–177.

[40] C. LeBrun, "On Complete Quaternionic-Kähler Manifolds," **Duke Math. J. 63** (1991) 723–743.

[41] C. LeBrun, "A Finiteness Theorem for Quaternionic-Kähler Manifolds with Positive Scalar Curvature," in *The Penrose Transform and Analytic Cohomology in Representation Theory*, Eastwood, Wolf, Zierau, *eds.* **Contemp. Maths. 154** (1993) 89-101.

[42] C. LeBrun, "Fano Manifolds, Contact Structures, and Quaternionic Geometry," **Int. J. Math. 3** (1995) 419–437.

[43] C. LeBrun and Y.-S. Poon, "Twistors, Kähler Manifolds, and Bimeromorphic Geometry II," **J. Am. Math. Soc. 5**(1992) 317–325.

[44] C. LeBrun and S. Salamon, "Strong Rigidity of Quaternion-Kähler Manifolds," **Inv. Math. 118** (1994) 109–132.

[45] C. LeBrun and M.A. Singer, "Existence and Deformation Theory for Scalar-Flat Kähler Metrics on Compact Complex Surfaces," **Inv. Math. 112** (1993) 273–313.

[46] C. LeBrun and M.A. Singer, "A Kummer-type Construction of Self-Dual 4-Manifolds," Math. Ann. 300 (1994) 165–180.

[47] Y.I. Manin, **Gauge Field Theory and Complex Geometry**, Springer-Verlag, 1988.

[48] S.A. Merkulov, "Moduli of Compact Legendre Submanifolds of Complex Contact Manifolds," **Math. Res. Lett. 1** (1994) 717–727.

[49] B.G. Moishezon, "On $n$-Dimensional Compact Varieties with $n$ algebraically independent Meromorphic Functions I,II,III," **AMS Translations 63** (1967) 51–177.

[50] S. Mori, "Hartshorne Conjecture and Extremal Ray," **Sugaku Expositions 0** (1988) 15-37.

[51] A. Newlander and L. Nirenberg, "Complex Analytic Coordinates in Almost-Complex Manifolds," **Ann. Math. 65** (1957) 391–404.

[52] H. Pedersen and Y.-S. Poon, "Self-Duality and Differentiable Structures on the Connected Sum of Projective Planes," **Proc. Am. Math. Soc. 121** (1994) 859–864.

[53] R. Penrose, "Non-linear Gravitons and Curved Twistor Theory," **Gen. Rel. Grav. 7** (1976) 31–52.

[54] M. Pontecorvo, "Algebraic Dimension of Twistor Spaces and Scalar Curvature of Anti-Self-Dual Metrics," **Math. Ann. 291** (1991) 113–122.

[55] M. Pontecorvo, "On Twistor Spaces of Anti-Self-Dual Hermitian Surfaces," **Trans. Am. Math. Soc. 331** (1992) 653–661.

[56] Y.-S. Poon, "Compact Self-Dual Manifolds with Positive Scalar Curvature," **J. Diff. Geom. 24** (1986) 97–132.

[57] Y.-S. Poon, "On the Algebraic Dimension of Twistor Spaces," **Math. Ann. 282** (1988) 621-627.

[58] Y.-S. Poon and S.M. Salamon, "Eight-Dimensional Quaternionic-Kähler Manifolds with Positive Scalar Curvature," **J. Diff. Geom. 33** (1991) 363-378.

[59] S.M. Salamon, "Quaternionic-Kähler Manifolds," **Inv. Math. 67** (1982) 143-171.

[60] S.M. Salamon, "Differential Geometry of Quaternionic Manifolds," **Ann. Scient. ENS 19** (1987) 31-55.

[61] S.M. Salamon, **Riemannian Geometry and Holonomy Groups**, Pitman Research Notes in Mathematics 201, Longman Scientific, 1989.

[62] R. Schoen, "Conformal Deformation of a Riemannian Metric to Constant Scalar Curvature," **J. Diff. Geom. 20** (1984) 478-495.

[63] J.P. Serre, "Géométrie Algébrique et Géométrie Analytique," **Ann. Inst. Fourier 6** (1956) 1–42.

[64] C.H. Taubes, "The Existence of Anti-Self-Dual Metrics," **J. Diff. Geom. 36** (1992)163–253.

[65] C.H. Taubes, **Metrics, Connections, and Gluing Theorems**, CBMS Regional Conference Series #89, American Mathematical Society, 1996.

[66] R.S. Ward, "Self-Dual Space-Times with Cosmological Constant," **Comm. Math. Phys. 78** (1980) 1–17.

[67] J.A. Wiśniewski, "On Fano Manifolds of Large Index," **Manus. Math. 70** (1991) 145–152.

[68] J. A. Wolf, "Complex Homogeneous Contact Structures and Quaternionic Symmetric Spaces," **J. Math. Mech. 14** (1965) 1033-1047.

[69] S.-T. Yau, "On the Ricci-Curvature of a Complex Kähler Manifold and the Complex Monge-Ampère Equations," **Comment. Pure Appl. Math. 31** (1978) 339–411.

[70] Y.-G. Ye, "A Note on Complex Threefolds Admitting a Holomorphic Contact Structure," **Inv. Math. 115** (1994) 311-314.

DEPT. OF MATHEMATICS, SUNY AT STONY BROOK, STONY BROOK, NY 11794-3651

# VIII. Toric Geometry

Proceedings of Symposia in Pure Mathematics
Volume **62.2**, 1997

# Recent Developments in Toric Geometry

David A. Cox

ABSTRACT. This paper will survey some recent developments in the theory of toric varieties, including new constructions of toric varieties and relations to symplectic geometry, combinatorics and mirror symmetry.

## Introduction

A toric variety over $\mathbb{C}$ is a $n$-dimensional normal variety $X$ containing $(\mathbb{C}^*)^n$ as a Zariski open set in such a way that the natural action of $(\mathbb{C}^*)^n$ on itself extends to an action of $(\mathbb{C}^*)^n$ on $X$. This seemingly simple definition leads to a fascinating combinatorial structure and some surprisingly rich mathematics. In this article, we will discuss some recent developments in toric varieties, including novel applications and new foundations for the entire theory.

Toric varieties were discovered in the early 1970's independently by several groups of people. From the beginning, the theory of toric varieties has led to some notable applications, including the following:

- The characterization of algebraic subgroups of maximal rank of Cremona groups, in Demazure's 1970 paper [**Dem**].
- The stable reduction theorem, in the 1973 book [**KKMS**] by Knudsen, Kempf, Mumford and Saint-Donat.
- The construction of nice (meaning "toroidal") compactifications of discrete quotients of bounded symmetric domains, begun in the 1973 paper [**Sat**] by Satake and the 1975 book [**AMRT**] by Ash, Mumford, Rapaport and Tai.
- The rich connections between Newton polytopes, toric varieties and singularities, first explored by Kushnirenko [**Kus**] in 1976 and Khovanskii [**Kho**] in 1977.
- The use of Hard Lefschetz for simplicial toric varieties to prove McMullen's conjectures for the number of vertices, edges, faces, etc. of convex simplical polytopes, in Stanley's 1980 paper [**Sta**].

This brief list, of course, does not do justice to the work of many other people who have written about toric varieties.

---

1991 *Mathematics Subject Classification.* Primary 14M25.
The author was supported in part by NSF Grant #DMS–9301161.

Besides these applications of toric varieties, many people have come to the realization, as noted by Fulton in his 1993 book [**Ful**], that "toric varieties have provided a remarkably fertile testing ground for general theories." An example of this philosophy can be found in Reid's 1983 paper [**Rei2**] which studies Mori theory in the context of toric varieties.

The theory of toric varieties has also had the benefit of some superb exposition. We've already mentioned Fulton's book [**Ful**], and two other classic references in the field are Danilov's 1978 survey [**Dan2**] and Oda's 1988 book [**Oda2**]. We warmly recommend these to anyone who wants to learn more about this fascinating part of algebraic geometry.

Although the classical theory seems fairly complete, the last few years have seen an explosion of new ideas and tools for studying toric varieties as well as significant new applications and relations to other areas of mathematics (and physics!). This paper will survey some of these new developments. We begin in §1 with a review of the notation and terminology we will use. Then §2 introduces a new construction of toric varieties, similar to way that $\mathbb{P}^n$ is realized as a quotient of $\mathbb{C}^{n+1} - \{0\}$. It follows that simplicial toric varieties have "homogeneous coordinates" which enable us to define subvarieties by global equations. In §3, we briefly discuss the Kähler cone of a toric variety. Then §4 describes a related construction of smooth simplicial toric varieties coming from symplectic geometry.

In §5, we discuss a yet another method for constructing toric varieties which has gained prominence recently. Here, a toric variety is defined to be the closure of an equivariant map from a torus. Such toric varieties need not be normal, which gives the theory a slightly different flavor. Then §6 shows how classical results of Griffiths on the cohomology of projective hypersurfaces can be generalized to the toric case. The *secondary fan* is the subject of §7. This fan, which has some remarkable applications, begins with a collection of rays and asks how many fans have these rays as their 1-dimensional cones. In §8 we will discuss reflexive polytopes and Calabi-Yau hypersurfaces, and then §9 will touch briefly on the work of Gelfand, Kapranov and Zelevinsky on resultants, discriminants and hypergeometric equations.

Mirror symmetry is the topic of §10. This field represents a fascinating interaction between mathematics and physics. Toric geometry plays a prominent role in many mirror symmetry contructions, and there are even physical theories specially designed for toric varieties. In particular, we will see that mirror symmetry makes use of virtually *everything* in §§2–9. Finally, §11 will discuss some other research being done on toric varieties.

In this survey, we will assume that the reader is familiar with basic theory of toric varieties as presented in [**Dan2**], [**Ful**] or [**Oda2**]. For simplicity, we will usually work over the complex numbers $\mathbb{C}$.

To keep the bibliography from getting too large, references to some topics are not complete—it was often more convenient to refer to later papers rather than the original ones. A fuller picture of recent work on toric varieties can be obtained by checking the references in the papers mentioned in this survey. The reader may also want to consult the 1989 and 1996 toric surveys of Oda [**Oda3**] and [**Oda4**], both of which contain extensive bibliographies.

## §1. Notation and terminology for toric varieties

The basic combinatorial object associated to a toric variety is a fan. One starts with an integer lattice $M \simeq \mathbb{Z}^n$ and its dual lattice $N$. Then a *fan* $\Sigma$ in $N_{\mathbb{R}} = N \otimes \mathbb{R}$

consists of a finite collection of *strongly convex rational polyhedral cones* $\sigma \subset N_{\mathbb{R}}$ which is closed under intersection and taking faces. The 1-dimensional cones $\rho$ play a prominent role in the theory, and it is customary to denote the unique generator of $\rho \cap N$ by the same letter $\rho$. Finally, we say that $\Sigma$ is *simplicial* if the minimal generators of each $\sigma \in \Sigma$ are linearly independent over $\mathbb{R}$. (For details of these and other definitions in this section, consult the references given in the introduction.)

**The classical construction.** Given a fan $\Sigma$, each cone $\sigma \in \Sigma$ has a *dual cone* $\sigma^\vee = \{v \in M_{\mathbb{R}} : \langle v, \sigma\rangle \geq 0\}$, which determines the semigroup algebra $\mathbb{C}[\sigma^\vee \cap M]$. In the classical formulation of the theory, the toric variety $X = X_\Sigma$ is obtained from $\Sigma$ by gluing together the affine toric varieties $X_\sigma = \operatorname{Spec}(\mathbb{C}[\sigma^\vee \cap M])$ for $\sigma \in \Sigma$. We will see in §2 that there are now other ways to construct $X$ from $\Sigma$.

The torus $T = N \otimes \mathbb{C}^*$ sits inside $X$ as described in the introduction. Hence $N$ gives the 1-parameter subgroups of $T$ and $M$ is the character group. So each $m \in M$ gives $\chi^m : T \to \mathbb{C}^*$, which can be regarded as a rational function on $X$.

Although a toric variety $X$ can be singular, it is always Cohen-Macaulay (this is proved in [**Hoc**] and [**Dan2**]), so that duality theory works nicely. In particular, the dualizing sheaf $\omega_X$ coincides with the sheaf $\widehat{\Omega}^n_X$ of Zariski $n$-forms on $X$. An especially nice case is when $X$ is simplicial (meaning that $\Sigma$ is simplicial). In this case, $X$ is a $V$-manifold and for many purposes (including cohomology over $\mathbb{Q}$ and Hodge theory) behaves like a manifold.

**The role of polyhedra.** One way to see the connection with polyhedra is via divisors on $X$. Each 1-dimensional cone $\rho \in \Sigma$ corresponds to a Weil divisor $D_\rho \subset X$. A divisor $D = \sum_\rho a_\rho D_\rho$ gives a (possibly unbounded) convex polyhedron

$$\Delta_D = \{m \in M_{\mathbb{R}} : \langle m, \rho\rangle \geq -a_\rho\} \subset M_{\mathbb{R}}. \tag{1.1}$$

To see how $\Delta_D$ relates to $D$, consider the reflexive sheaf $\mathcal{O}_X(D)$ whose sections over $U \subset X$ are those rational functions $f$ such that $\operatorname{div}(f) + D \geq 0$ on $U$. Then

$$H^0(X, \mathcal{O}_X(D)) = \bigoplus_{m \in \Delta_D \cap M} \mathbb{C} \cdot \chi^m. \tag{1.2}$$

When $D = \sum_\rho a_\rho D_\rho$ is Cartier, there is a *support function* $\psi_D$ with the property that for each $\sigma \in \Sigma$, there is $m_\sigma \in M$ such that $\psi_D(\rho) = \langle m_\sigma, \rho\rangle = -a_\rho$ for all $\rho \subset \sigma$. In particular, $\psi_D$ is linear and integral on each cone in $\Sigma$.

When $X$ is complete, a classic fact is that $D$ is ample if and only if $\psi_D$ is strictly convex. In this case, $\Delta_D$ is a $n$-dimensional integral convex polytope (= bounded polyhedron) which is combinatorially dual to $\Sigma$, i.e., facets of $\Delta_D$ (faces of dimension $n-1$) correspond to 1-dimensional cones $\rho \in \Sigma$ and, more generally, $i$-dimensional faces of $\Delta_D$ correspond to $(n-i)$-dimensional cones of $\Sigma$.

Conversely, given a $n$-dimensional integral convex polytope $\Delta \subset M_{\mathbb{R}}$, there is a unique fan (sometimes called the *normal fan* of $\Delta$) such that corresponding toric variety $X_\Delta$ has a Cartier divisor which gives $\Delta$ exactly (see [**Oda1**, Sect. 2.4]). In §5 we will give a method (due to Batyrev) for obtaining $X_\Delta$ directly from $\Delta$.

**Other remarks.** A minor omission in the classic references for toric varieties is that line bundles are discussed in detail, but not the reflexive sheaves coming from Weil divisors $\sum_\rho a_\rho D_\rho$. Basic facts about reflexive sheaves on normal varieties can be found in [**Rei1**]. We should also mention that there is a standard conflict of notation: some authors use $\Delta$ for the fan (see [**Ful**], [**Oda2**], [**Stu2**]), while others use $\Sigma$ for the fan and $\Delta$ for a polytope (see [**Dan2**], [**Bat1**]).

In [**Ful**, Sect. 3.4], it is stated that $\mathrm{Pic}(X)$ is always free. This is not quite correct, since the smooth toric surface with 1-dimensional cone generators $(1,0),(1,2)$ and no 2-dimensional cones has Picard group $\mathbb{Z}/2\mathbb{Z}$. However, one can prove that $\mathrm{Pic}(X)$ *is* free when $X$ is complete. See [**Kaj**] for the details.

It is also possible to consider infinite fans. Gluing together affine toric varieties for cones in such a fan would lead to a scheme which is only locally of finite type. However, given an appropriate discrete group action, one can then take a quotient to get a variety. This is the approach used in [**KMMS**] and [**AMRT**]. A nice example is the resolution of a 2-dimensional Hilbert cusp singularity [**Oda1**, Sect. 4.1].

## §2. Global coordinates for toric varieties

Projective space is one of the simplest examples of a toric variety. The way $\mathbb{P}^n$ is obtained by gluing together affine spaces is a special case of the classic construction of a toric variety. But $\mathbb{P}^n$ can also be constructed as a quotient

$$\mathbb{P}^n = (\mathbb{C}^{n+1} - \{0\})/\mathbb{C}^*,$$

which is where we get the homogeneous coordinates on projective space. Recently, this construction has been generalized to most toric varieties. We will describe this construction and some of its consequences.

**The construction.** We begin by fixing a fan $\Sigma$ in $N_{\mathbb{R}} \simeq \mathbb{R}^n$ and letting $\Sigma(1)$ be the set of 1-dimensional cones in $\Sigma$. We will assume that

$$\text{the 1-dimensional cones } \rho \in \Sigma(1) \text{ span } N_{\mathbb{R}} \tag{2.1}$$

(where as usual we regard $\rho$ as the integral generator of its cone). Any complete fan satisfies this condition. Then consider the affine space $\mathbb{C}^{\Sigma(1)}$ with variables $x_\rho$ for $\rho \in \Sigma(1)$. We need to remove a certain *exceptional subset* from $\mathbb{C}^{\Sigma(1)}$. This is defined as follows. For each $\sigma \in \Sigma$, consider the monomial $\hat{x}_\sigma = \Pi_{\rho \not\subset \sigma} x_\rho$. These generate the monomial ideal $B(\Sigma) = \langle \hat{x}_\sigma : \sigma \in \Sigma \rangle$, and the exceptional set $Z(\Sigma) \subset \mathbb{C}^{\Sigma(1)}$ is the subvariety defined by $B(\Sigma)$.

The toric variety $X = X_\Sigma$ will be a quotient of $\mathbb{C}^{\Sigma(1)} - Z(\Sigma)$ by the group

$$G = \mathrm{Hom}_{\mathbb{Z}}(A_{n-1}(X), \mathbb{C}^*),$$

where $A_{n-1}(X)$ is the Chow group of Weil divisors modulo rational equivalence. To see how this group acts on $\mathbb{C}^{\Sigma(1)} - Z(\Sigma)$, recall the exact sequence

$$0 \longrightarrow M \stackrel{\alpha}{\longrightarrow} \bigoplus_{\rho \in \Sigma(1)} \mathbb{Z} \cdot D_\rho \stackrel{\beta}{\longrightarrow} A_{n-1}(X) \longrightarrow 0 \tag{2.2}$$

where $\alpha$ sends $m \in M$ to $\mathrm{div}(\chi^m) = \sum_\rho \langle m, \rho \rangle D_\rho$ and $\beta$ is the obvious map from Weil divisors to the Chow group. The injectivity of $\alpha$ is equivalent to (2.1). Applying $\mathrm{Hom}_{\mathbb{Z}}(-, \mathbb{C}^*)$ yields the exact sequence

$$1 \longrightarrow G \longrightarrow (\mathbb{C}^*)^{\Sigma(1)} \longrightarrow N \otimes \mathbb{C}^* \longrightarrow 1. \tag{2.3}$$

This gives a natural action of $G$ on $\mathbb{C}^{\Sigma(1)}$, and since $Z(\Sigma)$ is a union of coordinate subspaces, $G$ preserves $\mathbb{C}^{\Sigma(1)} - Z(\Sigma)$.

A nice example of what this looks like is given by $\mathbb{P}^n$. The fan $\Sigma$ of $\mathbb{P}^n$ is well-known, and we leave it to the reader to check that in this case, the monomial ideal $B(\Sigma)$ is the "irrelevant" ideal $\langle x_0, \dots, x_n \rangle$, so that the exceptional set $Z(\Sigma)$ consists of the origin. Furthermore, the exact sequence (2.2) becomes

$$0 \longrightarrow \mathbb{Z}^n \longrightarrow \mathbb{Z}^{n+1} \xrightarrow{\beta} \mathbb{Z} \longrightarrow 0,$$

where $\beta(a_0, \dots, a_n) = \sum_{i=0}^n a_i$. By (2.3), the action of $G$ on $\mathbb{C}^{\Sigma(1)} - Z(\Sigma)$ is the usual action of $\mathbb{C}^*$ on $\mathbb{C}^{n+1} - \{0\}$. The quotient, of course, is $\mathbb{P}^n$.

Returning to the general case, assume for the moment that $X$ is given by the quotient $\big(\mathbb{C}^{\Sigma(1)} - Z(\Sigma)\big)/G$. Then the natural inclusion $(\mathbb{C}^*)^{\Sigma(1)} \subset \mathbb{C}^{\Sigma(1)} - Z(\Sigma)$, combined with (2.3), gives an inclusion $T = N \otimes \mathbb{C}^* \subset X_\Sigma$ as a dense open set. The action of $T$ on $X$ is also easy to see in this picture: it is inherited from the action of the "big torus" $(\mathbb{C}^*)^{\Sigma(1)}$ on $\mathbb{C}^{\Sigma(1)} - Z(\Sigma)$.

Taking quotients in algebraic geometry can be a bit subtle, and the "quotient" $\big(\mathbb{C}^{\Sigma(1)} - Z(\Sigma)\big)/G$ is no exception. The precise relation between this quotient and the toric variety is as follows.

THEOREM 2.1. *Let $X$ be a toric variety whose fan $\Sigma$ satisfies* (2.1). *Then:*
1. *$X$ is the universal categorical quotient $\big(\mathbb{C}^{\Sigma(1)} - Z(\Sigma)\big)/G$.*
2. *$X$ is a geometric quotient $\big(\mathbb{C}^{\Sigma(1)} - Z(\Sigma)\big)/G$ if and only if $\Sigma$ is simplicial.*

REMARKS ON THEOREM 2.1. A universal categorical quotient is a $G$-equivariant map $\pi : \mathbb{C}^{\Sigma(1)} - Z(\Sigma) \to X$ (where the action on $X$ is trivial) which is universal in the obvious sense. It is a geometric quotient precisely when the fibers of $\pi$ coincide with the $G$-orbits. Hence the simplicial case is the one closest to the way we think about projective space.

To understand the quotient in the general case, observe that $\mathbb{C}^{\Sigma(1)} - Z(\Sigma)$ is the union of affine open sets $U_\sigma = \{\hat{x}_\sigma \neq 0\}$ for $\sigma \in \Sigma$. Then the categorical quotient $U_\sigma/G$ is determined by the ring of invariants of $G$ acting on the coordinate ring of $U_\sigma$. In [**Cox2**], this ring of invariants is identified with the semigroup algebra $\mathbb{C}[\sigma^\vee \cap M]$, which explains why the quotients $U_\sigma/G$ patch together to give $X$.

An interesting aspect of this construction is that it was found by several people independently at approximately the same time. Audin [**Aud**] (following ideas of Delzant [**Del**] and Kirwan [**Kir**]) described the construction in the context of symplectic geometry, while Musson [**Mus**] was studying differential operators on toric varieties, and Cox [**Cox2**] was more interested in the algebraic aspects of the situation. This result was also discovered by Batyrev [**Bat3**] and Fine [**Fin**].

It is possible to develop the entire theory of toric varieties using Theorem 2.1 as the *definition* of toric variety.

**The exceptional set.** In $X = \big(\mathbb{C}^{\Sigma(1)} - Z(\Sigma)\big)/G$, note that the group $G$ depends only on the 1-dimensional cones of $\Sigma$, while the exceptional set $Z(\Sigma)$ depends on the full fan. The combinatorics of $Z(\Sigma)$ are quite interesting. For projective space, this set is very small, but it is usually bigger as the following result of [**BC**, Sect. 2] shows.

PROPOSITION 2.2. *Let $X$ be a $n$-dimensional complete simplicial toric variety with fan $\Sigma$. Then either*
1. $2 \leq \operatorname{codim} Z(\Sigma) \leq \lfloor n/2 \rfloor + 1$, *or*
2. $Z(\Sigma) = \{0\}$ *and $X$ is a finite quotient of a weighted projective space.*

When $X$ is simplicial, the coordinate subspaces making up $Z(\Sigma)$ can be described in terms of Batyrev's notion of a *primitive collection*, which is a subset $\mathcal{P} \subset \Sigma(1)$ with the property that $\mathcal{P}$ is not the set of generators of a cone in $\Sigma$ while every proper subset of $\mathcal{P}$ is. Then the decomposition of $Z(\Sigma)$ into irreducible components is given by

$$Z(\Sigma) = \bigcup_{\mathcal{P}} \mathbf{A}(\mathcal{P}),$$

where $\mathbf{A}(\mathcal{P})$ is the coordinate subspace determined by $x_\rho$ for $\rho \in \mathcal{P}$ and the union is over all primitive collections $\mathcal{P}$. When $X$ is smooth and complete, Batyrev has conjectured [**Bat2**] that the number of irreducible components of $Z(\Sigma)$ (= the number of primitive collections) is bounded by a constant depending only on the the Picard number of $X$.

**The homogeneous coordinate ring.** We next explore the algebraic consequences of $X = \big(\mathbb{C}^{\Sigma(1)} - Z(\Sigma)\big)/G$. The basic idea is that like projective space, homogeneous coordinates allow us to define subvarieties using global equations. When $X$ is a simplicial toric variety, a point of $X$ has "homogeneous coordinates" $(t_\rho) \in \mathbb{C}^{\Sigma(1)} - Z(\Sigma)$, which are well-defined up to the action of $G$. The corresponding polynomial ring is

$$S = \mathbb{C}[x_\rho : \rho \in \Sigma(1)]$$

with a grading induced by the action of $G$ on $\mathbb{C}^{\Sigma(1)}$. The character group of $G$ is $A_{n-1}(X)$ and the grading on $S$ can be viewed as defining the "degree" of a monomial $x^{\mathbf{a}} = \Pi_\rho x_\rho^{a_\rho}$ to be $\deg(x^{\mathbf{a}}) = \big[\sum_\rho a_\rho D_\rho\big] \in A_{n-1}(X)$. With this grading, $S = \mathbb{C}[x_\rho]$ is called the *homogeneous coordinate ring* of $X$.

EXAMPLES. 1. In the case of $\mathbb{P}^n$ or $\mathbb{P}^n \times \mathbb{P}^m$, the coordinate rings are the classical rings of homogeneous or bihomogeneous polynomials.

2. Another good example is the blow-up of $\mathbb{C}^n$ at the origin. We leave it as an exercise for the reader to show that the coordinate ring is $\mathbb{C}[t, x_1, \dots, x_n]$, where $\deg(t) = -1$, $\deg(x_i) = 1$. Furthermore, the exceptional set in $\mathbb{C}^{n+1}$ is $Z = \mathbb{C} \times \{(0, \dots, 0)\}$ and $G = \mathbb{C}^*$ acts on $\mathbb{C}^{n+1} = \mathbb{C} \times \mathbb{C}^n$ by $\mu \cdot (t, \mathbf{x}) = (\mu^{-1}t, \mu\mathbf{x})$. Then, given a point $(t, \mathbf{x}) \in \mathbb{C}^{n+1} - Z$, we have

$$\begin{aligned} (t, \mathbf{x}) &\sim_G (1, t\mathbf{x}) \quad \text{if } t \neq 0 \\ (0, \mathbf{x}) &\sim_G (0, \mu\mathbf{x}) \quad \text{if } \mu \neq 0. \end{aligned}$$

From this, it should be clear that $X = (\mathbb{C}^{n+1} - Z)/G$ is the blow-up of $\mathbb{C}^n$ at the origin. Notice also that the blow-up map $X \to \mathbb{C}^n$ is given by $(t, \mathbf{x}) \mapsto t\mathbf{x}$.

The coordinate ring $S = \mathbb{C}[x_\rho]$ allows us to define subvarieties of $X$ using homogeneous ideals of $S$. The relation between ideals and varieties is similar to what happens in $\mathbb{P}^n$, with the ideal $B(\Sigma) = \langle \hat{x}_\rho : \rho \in \Sigma(1) \rangle$ playing the role of the irrelevant ideal (see [**Cox2**]). One can also define sheaves on $X$ using graded $S$-modules. Here is an example of how this works.

EXAMPLE. Consider the graded $S$-module $\widehat{\Omega}^p_S$ defined as the kernel of the map

$$\gamma : S \otimes \Lambda^p M \longrightarrow \bigoplus\nolimits_\rho S/\langle x_\rho \rangle \otimes \Lambda^{p-1} M,$$

where the $\rho$th component of $\gamma$ is $\gamma_\rho(f \otimes \omega) = f \bmod x_\rho \otimes i_\rho(\omega)$ and $i_\rho(\omega)$ is interior product. Then [**BC**, Sect. 8] shows that the sheaf corresponding to $\widehat{\Omega}^p_S$ is the sheaf of Zariski $p$-forms $\widehat{\Omega}^p_X$ on $X$. Furthermore, given $\alpha \in A_{n-1}(X)$, we can define a shifted module $\widehat{\Omega}^p_S(\alpha)$ in the usual way. This gives a sheaf $\widehat{\Omega}^p_X(\alpha)$ with global sections

$$H^0(X, \widehat{\Omega}^p_X(\alpha)) \simeq (\widehat{\Omega}^p_S)_\alpha$$

(where the subscript refers to the graded piece in degree $\alpha$).

If $\mathcal{L}$ is a line bundle (or, more generally, a rank one torsion-free reflexive sheaf) on $X$, then we get $\alpha = [\mathcal{L}] \in A_{n-1}(X)$, and one can prove that

$$H^0(X, \mathcal{L}) \simeq S_\alpha \tag{2.4}$$

(see [**Cox2**]). When $\mathcal{L} = \mathcal{O}_X(D)$ and $X$ is complete, (1.2) then shows that $\dim S_\alpha$ is the number $l(\Delta)$ of integer points in the polytope $\Delta_D$. From (2.4) we also obtain a ring isomorphism

$$\bigoplus_{k=0}^{\infty} H^0(X, \mathcal{L}^{\otimes k}) \simeq \bigoplus_{k=0}^{\infty} S_{k\alpha} \subset S. \tag{2.5}$$

In particular, if $X$ is projective, the homogeneous coordinate ring $S$ contains the coordinate rings (in the usual sense) of *all* possible projective embeddings of $X$.

For toric varieties with enough invariant effective Cartier divisors, Kajiwara [**Kaj**] defined a generalization $S_\Sigma$ of the homogeneous coordinate $S$. Using this ring, $X$ is a geometric quotient of a Zariski open subset of $\mathrm{Spec}(S_\Sigma)$ under the action of $G = \mathrm{Hom}_{\mathbb{Z}}(\mathrm{Pic}(X), \mathbb{C}^*)$. In contrast, Theorem 2.1 only gives a geometric quotient in the simplicial case. One has $S = S_\Sigma$ when $X$ is smooth and complete.

**Applications.** There have been several recent applications of global coordinates for toric varieties. We've already mentioned [**Mus**], which studies differential operators on toric varieties. In [**Per1**] and [**Per2**], homogeneous coordinates on a smooth toric variety $X$ and the Euler sequence

$$0 \longrightarrow \Omega^1_X \longrightarrow \bigoplus_\rho \mathcal{O}_X(-D_\rho) \longrightarrow A_{n-1}(X) \otimes \mathcal{O}_X \longrightarrow 0$$

(see [**BC**] and [**Jac**]) are used to compute the principal parts of line bundles and to study highly inflected toric varieties in dimensions $\leq 3$.

Using homogeneous coordinates, maps to a toric variety can be studied in much the same way one describes maps to projective space—see [**Cox1**], [**Gue**], [**Jac**] and [**Kaj**]. Homogeneous coordinates were also used in [**Cox2**] to show that Demazure's results on the automorphism group of a smooth complete toric variety (see [**Oda1**, Sect. 3.4] for a description) remain valid in the simplicial case.

Further applications of homogeneous coordinates will be given in §§4 and 6, and in §10, we will also see how homogeneous coordinates are used in mirror symmetry.

## §3. The Kähler cone

The Kähler classes of a smooth projective variety $X = X_\Sigma$ form an open cone in $H^{1,1}(X, \mathbb{R})$ called the *Kähler cone*. This cone can be complicated in general, but it is pleasantly simple when $X$ is a smooth projective toric variety.

In this case, the cohomology classes $[D_\rho]$ span $H^2(X, \mathbb{R}) = H^{1,1}(X, \mathbb{R}) = A_{n-1}(X) \otimes \mathbb{R}$. Since $X$ is smooth, a class $\mathbf{a} = [\sum_\rho a_\rho D_\rho]$ (with $a_\rho \in \mathbb{R}$) has a *support function* $\psi : N_\mathbb{R} \to \mathbb{R}$ with the property that for each $\sigma \in \Sigma$, there is $m_\sigma \in M_\mathbb{R}$ with the property that $\psi(\rho) = \langle m_\sigma, \rho \rangle = -a_\rho$. This is similar to the support functions considered in §2 except that $\psi$ is only well defined up to a linear function on $N_{\mathbb{R}}$ (this follows from (2.2)) and the $m_\sigma$ need not be integral. Then we say that **a** is *convex* if $\psi$ is a convex function on $N_{\mathbb{R}}$. The convex classes form a cone $\operatorname{cpl}(\Sigma) \subset A_{n-1}(X) \otimes \mathbb{R}$ which has the following nice structure.

PROPOSITION 3.1. *If $X$ is a simplicial projective toric variety, then* $\operatorname{cpl}(\Sigma) \subset A_{n-1}(X) \otimes \mathbb{R} = H^{1,1}(X, \mathbb{R})$ *is a strongly convex polyhedral cone with nonempty interior in $H^{1,1}(X,\mathbb{R})$. Furthermore, the interior of this cone is precisely the Kähler cone of $X$.*

As observed in [**Bat3**], the first part of the proposition follows from [**OP**] (see also [**Rei2**]), and the second part follows easily in the smooth case. When $X$ is simplicial, [**AGM2**] gives a careful definition the Kähler cone and shows that Proposition 3.1 continues to hold in this case. We will see later that this proposition has implications for both symplectic geometry and mirror symmetry.

Support functions $\psi$ corresponding to Kähler classes are *strictly convex* and are described in [**Bat3**] using the primitive collections from §2 as follows .

PROPOSITION 3.2. *If $X$ is a simplicial projective toric variety, then a support function $\psi$ coming from $[\sum_\rho a_\rho D_\rho] \in A_{n-1}(X) \otimes \mathbb{R}$ is strictly convex if and only if for every primitive collection $\mathcal{P} = \{\rho_1, \ldots, \rho_k\}$, we have*

$$\psi(\rho_1 + \cdots + \rho_k) > \psi(\rho_1) + \cdots + \psi(\rho_k).$$

The dual of the Kähler cone is the *Mori cone* of effective 1-cycles modulo numerical equivalence. Then Proposition 3.2 can be interpreted as describing generators for the Mori cone (see [**Bat2**], [**OP**] and [**Rei2**] for more on the Mori cone).

EXAMPLE. Consider fans $\Sigma$ in $\mathbb{R}^3$ whose 1-dimensional cone generators are

$$e_0 = (0,0,-2),\ e_1 = (1,1,1),\ e_2 = (1,-1,1),\ e_3 = (-1,-1,1),\ e_4 = (-1,1,1).$$

Think of $e_1, e_2, e_3, e_4$ as the upper vertices of a cube and $e_0$ as lying on the negative $z$-axis. We will use the integer lattice generated by $e_1, e_2, e_3$. Note that $e_0 = -e_1 - e_3$ and $e_4 = e_1 - e_2 + e_3$.

There are several ways to get a complete fan from these generators. For example, the cones $\sigma_{1234}$, $\sigma_{012}$, $\sigma_{023}$, $\sigma_{034}$ and $\sigma_{041}$ (where $\sigma_{1234}$ is the cone with generators $e_1, e_2, e_3, e_4$, etc.) and their faces determine a singular fan $\Sigma$. But if we subdivide $\sigma_{1234}$ into $\sigma_{123}$ and $\sigma_{341}$, then we get a smooth fan $\Sigma_1$. Similarly, we can subdivide $\sigma_{1234}$ into $\sigma_{124}$ and $\sigma_{234}$ to get another smooth fan $\Sigma_2$. The toric varieties corresponding to $\Sigma$, $\Sigma_1$ and $\Sigma_2$ will be denoted $X$, $X_1$ and $X_2$ respectively.

The primitive collections for $\Sigma_1$ are $\{e_2, e_4\}$ and $\{e_0, e_1, e_3\}$. Hence, using Proposition 3.2, we see that a support function $\psi$ is strictly convex if and only if

$$\begin{aligned}\psi(e_2 + e_4) &> \psi(e_2) + \psi(e_4)\\ \psi(e_0 + e_1 + e_3) &> \psi(e_0) + \psi(e_1) + \psi(e_3).\end{aligned}$$

If we let $\psi(e_i) = -a_i$ and use the relations $e_2 + e_4 = e_1 + e_3$ and $e_0 + e_1 + e_3 = 0$, these inequalities are equivalent to

$$\begin{aligned}a_2 + a_4 &> a_1 + a_3\\ a_0 + a_1 + a_3 &> 0.\end{aligned} \tag{3.1}$$

To determine the Kähler cone of $X_1$, we have to interpret (3.1) in terms of the Chow group $A_2(X_1)\otimes\mathbb{R}$. However, for $X_1$, the exact sequence (2.2) can be written

$$0 \longrightarrow \mathbb{Z}^3 \xrightarrow{\alpha} \mathbb{Z}^5 \xrightarrow{\beta} \mathbb{Z}^2 \longrightarrow 0, \tag{3.2}$$

where $\beta$ maps $(a_0, a_1, a_2, a_3, a_4)$ to $(s,t) = (a_0 + a_1 + a_3, a_0 + a_2 + a_4)$. Using $s, t$ as coordinates on $A_2(X_1) \otimes \mathbb{R} \simeq \mathbb{R}^2$, the inequalities (3.1) can be written $t > s > 0$. By Proposition 3.1, this is the Kähler cone of $X_1$. Thus $\mathrm{cpl}(\Sigma_1)$ is $t \geq s \geq 0$.

We now turn our attention to $X_2$. The primitive collections for $\Sigma_2$ are $\{e_1, e_3\}$ and $\{e_0, e_2, e_4\}$, which gives inequalites similar to (3.1) (just interchange $a_1, a_3$ with $a_2, a_4$). Since (3.2) depends only on the 1-dimensional cones of a fan, we see that $A_2(X_1) \otimes \mathbb{R}$ is the same $\mathbb{R}^2$ with the same coordinates $s, t$. The only difference is that the Kähler cone of $X_2$ is given by $s > t > 0$ and $\mathrm{cpl}(\Sigma_2)$ is $s \geq t \geq 0$

Hence the first quadrant in $\mathbb{R}^2$ is divided into cones $t \geq s \geq 0$ and $s \geq t \geq 0$ whose interiors are the Kähler cones of the smooth toric varieties $X_1$ and $X_2$. Also, the ray $s = t > 0$ corresponds to ample divisors on the singular toric variety $X$.

As we will see in §7, this is an example of the *secondary fan* or *GKZ decomposition*. We should also mention that $X_1$ and $X_2$ are related by a *flop* (see [**Rei2**]).

## §4. Symplectic geometry

Besides the quotient $\mathbb{P}^n = (\mathbb{C}^{n+1} - \{0\})/\mathbb{C}^*$ considered in §2, there is the related quotient

$$\mathbb{P}^n = S^{2n+1}/S^1$$

where $S^1 \subset \mathbb{C}^*$ acts on the unit sphere $S^{2n+1} \subset \mathbb{C}^{n+1}$ in the usual way. In this section, we will use *symplectic reduction* to generalize this construction to simplicial projective toric varieties.

**The construction.** Let $X = X_\Sigma$ be the toric variety determined by a fan $\Sigma$ in $N_\mathbb{R} \simeq \mathbb{R}^n$. We will assume that $X$ is simplicial and projective. To simplify notation, let $r = |\Sigma(1)|$. As in §2, we have the group $G = \mathrm{Hom}_\mathbb{Z}(A_{n-1}(X), \mathbb{C}^*)$. The maximal compact subgroup of $G$ is

$$G_\mathbb{R} = \mathrm{Hom}_\mathbb{Z}(A_{n-1}(X), S^1). \tag{4.1}$$

The inclusion $G \subset (\mathbb{C}^*)^r$ gives an action of $G_\mathbb{R}$ on $\mathbb{C}^r$.

Now consider the map (called the *moment map*)

$$\mu_\Sigma : \mathbb{C}^r \xrightarrow{\mu} \mathbb{R}^r \xrightarrow{\beta_\mathbb{R}} A_{n-1}(X) \otimes \mathbb{R}, \tag{4.2}$$

where $\mu$ defined by $\mu(z_1, \ldots, z_r) = \frac{1}{2}(|z_1|^2, \ldots, |z_r|^2)$ and $\beta_\mathbb{R}$ comes from the exact sequence

$$0 \longrightarrow M_\mathbb{R} \longrightarrow \mathbb{R}^r \xrightarrow{\beta_\mathbb{R}} A_{n-1}(X) \otimes \mathbb{R} \longrightarrow 0$$

obtained by tensoring (2.2) with $\mathbb{R}$. Note that $\mu_\Sigma$ is constant on $G_\mathbb{R}$-orbits.

Since $X$ is projective and simplicial, $A_{n-1}(X) \otimes \mathbb{R} \simeq H^2(X, \mathbb{R}) \simeq H^{1,1}(X, \mathbb{R})$. Recall from §3 that the *Kähler cone* in $H^{1,1}(X, \mathbb{R})$ consists of all possible Kähler classes on $X$. Under these isomorphisms, we get a cone in $A_{n-1}(X) \otimes \mathbb{R}$, also called the Kähler cone. Then we modify the construction $X = \big(\mathbb{C}^r - Z(\Sigma)\big)/G$ from §2 as follows.

THEOREM 4.1. *Let $X = X_\Sigma$ be a projective simplicial toric variety, and assume that $\mathbf{a} \in A_{n-1}(X) \otimes \mathbb{R}$ is in the Kähler cone. Then $\mu_\Sigma^{-1}(\mathbf{a}) \subset \mathbb{C}^r - Z(\Sigma)$, and the natural map*

$$\mu_\Sigma^{-1}(\mathbf{a})/G_\mathbb{R} \longrightarrow \big(\mathbb{C}^r - Z(\Sigma)\big)/G = X$$

*is a diffeomorphism which preserves the class of the symplectic form (to be explained below).*

PROOF. When $X$ is smooth, a proof that we have a diffeomorphism can be found in Guillemin's recent book [**Gui**, Appendix 1], and the statement about the class of the symplectic from follows from equation (1.6) of [**Gui**, Appendix 2] (the $\lambda_i$ in [**Gui**] are $-a_i$ in our notation). This proof can be modified to work in the simplicial case. One can also use results in [**Kir**] to prove the theorem. The version in [**Aud**] is somewhat incomplete since the Kähler cone is not mentioned. □

EXAMPLES. 1. When $X = \mathbb{P}^n$, the map $\beta_\mathbb{R}$ in (4.2) is the map $\mathbb{R}^{n+1} \to \mathbb{R}$ defined by $(a_0, \ldots, a_n) \mapsto \sum_{i=0}^n a_i$. Thus $\mu_\Sigma(z_0, \ldots, z_n) = \frac{1}{2}\sum_{i=0}^n |z_i|^2$. Since the Kähler cone is $\mathbb{R}^+$, we see that $\mu_\Sigma^{-1}(a)$ is a sphere for any $a > 0$. Hence we recover the usual description of $\mathbb{P}^n$ as a quotient of the $(2n+1)$-sphere.

2. Consider the toric varieties $X_1$ and $X_2$ from the example in §3. These have the same group $G_\mathbb{R}$ acting on $\mathbb{C}^5$ and the same moment map $\mu_{\Sigma_1} = \mu_{\Sigma_2}$, which we write as $\mu : \mathbb{C}^5 \to \mathbb{R}^2$. Then $\mu$ is given by

$$\mu(z_0, z_1, z_2, z_3, z_4) = \tfrac{1}{2}\big(|z_0|^2 + |z_1|^2 + |z_3|^2, |z_0|^2 + |z_2|^2 + |z_4|^2\big)$$

(this follows from the description of $\beta$ in (3.2)). Since we determined the Kähler cones of $X_1$ and $X_2$ in §3, it follows that

$$\mu^{-1}(s,t)/G_\mathbb{R} \simeq \begin{cases} X_1, & \text{if } t > s > 0 \\ X_2, & \text{if } s > t > 0. \end{cases}$$

**Symplectic manifolds and Hamiltonian actions.** To understand the construction given in Theorem 4.1, one needs to discuss symplectic geometry. A symplectic structure on a real manifold $M$ is a closed, nondegenerate 2-form $\omega$. The symplectic form $\omega$ converts functions into vector fields as follows: if $f$ is a $C^\infty$ function on $M$, then there is a unique vector field $X_f$ on $M$ with the property that $\omega(X, X_f) = X(f)$ for any vector field $X$. We call $X_f$ the *Hamiltonian* of $f$. Basic references on symplectic geometry are [**Aud**] and [**Kir**], and the reader might also want to consult [**Ati1**], [**Ati2**] and [**GS**].

Now suppose that a compact connected Lie group $G_\mathbb{R}$ acts on $M$. This action induces an infinitesimal action of the Lie algebra $\mathfrak{g}_\mathbb{R}$ where every $\lambda \in \mathfrak{g}_\mathbb{R}$ gives a vector field $X_\lambda$ on $M$. Then the action is *Hamiltonian* if:

1. The symplectic form $\omega$ is invariant under the group action.
2. For each $\lambda \in \mathfrak{g}_\mathbb{R}$, the vector field $X_\lambda$ is Hamiltonian (i.e., is the Hamiltonian vector field of some $C^\infty$ function on $M$).

A basic property of a Hamiltonian action is that it has a *moment map*

$$\mu : M \longrightarrow \mathfrak{g}_\mathbb{R}^*,$$

which has the property that for every $\lambda \in \mathfrak{g}_\mathbb{R}$, the vector field $X_\lambda$ is the Hamiltonian of the function $\lambda \circ \mu : M \to \mathbb{R}$.

EXAMPLES. 1. The most basic example is $\mathbb{C}^r$ endowed with the symplectic form $\omega = \sum_{j=1}^r dx_j \wedge dy_j$, where $z_j = x_j + iy_j$. It is easy to check that the natural action of $(S^1)^r$ on $\mathbb{C}^r$ is Hamiltonian and the moment map

$$\mu : \mathbb{C}^r \longrightarrow (\mathbb{R}^r)^*$$

is defined by $\mu(z_1, \ldots, z_r) = \frac{1}{2}(|z_1|^2, \ldots, |z_r|^2)$ (this uses the basis of $(\mathbb{R}^r)^*$ dual to the standard basis of the Lie algebra $\mathbb{R}^r$ of $(S^1)^r$).

2. When $G_{\mathbb{R}}$ is the group coming from a toric variety $X$ as described in (4.1), then the action of $G_{\mathbb{R}}$ on $\mathbb{C}^r$ is Hamiltonian and the moment map is *exactly* as described in (4.2) provided we identify $\mathbb{R}^r$ with it dual $(\mathbb{R}^r)^*$. Thus Theorem 4.1 tells us how to construct a toric variety using the moment map of a Hamiltonian action.

**Symplectic reduction.** Given a symplectic manifold $M$ with a Hamiltonian action of $G_{\mathbb{R}}$ and moment map $\mu : M \to \mathfrak{g}_{\mathbb{R}}^*$, we can use this data to construct other symplectic manifolds by the process of *symplectic reduction*. If $\mathbf{a} \in \mathfrak{g}_{\mathbb{R}}$ is a regular value of $\mu$, then $\mu^{-1}(\mathbf{a})$ is a manifold, but the restriction of the symplectic form $\omega$ to $\mu^{-1}(\mathbf{a})$ will fail to be symplectic (it won't be nondegenerate). However, if $G_{\mathbb{R}}$ acts freely on $\mu^{-1}(\mathbf{a})$, then the restriction of $\omega$ descends to the quotient $\mu^{-1}(\mathbf{a})/G_{\mathbb{R}}$ as a symplectic form. This is what we mean by symplectic reduction.

If we look back at Theorem 4.1, we see that the basic assertion of the theorem is that we can construct smooth projective toric varieties by symplectic reduction. Furthermore, we can now explain what it means for the diffeomorphism

$$\mu_\Sigma^{-1}(\mathbf{a})/G_{\mathbb{R}} \longrightarrow \big(\mathbb{C}^r - Z(\Sigma)\big)/G = X$$

to preserve the class of the symplectic form: the symplectic reduction $\mu_\Sigma^{-1}(\mathbf{a})/G_{\mathbb{R}}$ has a natural symplectic structure coming from $\omega$ on $\mathbb{C}^r$, and $X$ has a symplectic structure coming from any Kähler form whose cohomology class is $\mathbf{a} \in A_{n-1}(X)\otimes\mathbb{R}$. The above diffeomorphism need not map one symplectic form to the other, but it does preserve their cohomology classes.

**Delzant polytopes.** In addition to the Hamiltonian action used to construct a smooth projective toric variety $X$, we also have an action of the real torus $T_{\mathbb{R}} = (S^1)^n$ on $X$. It is well-known that if we give $X$ the symplectic structure coming from an ample divisor $D = \sum_\rho a_\rho D_\rho$ (so $a_\rho \in \mathbb{Z}$), then the action of $T_{\mathbb{R}}$ is Hamiltonian, and the moment map

$$\mu_X : X \longrightarrow M_{\mathbb{R}}$$

(note that $M_{\mathbb{R}} = N_{\mathbb{R}}^*$ and $N_{\mathbb{R}}$ is the Lie algebra of $T_{\mathbb{R}}$) can be described explicitly (see [**Ful**, Sect. 4.2] or [**Oda1**, Sect. 2.4]). Furthermore, the image of the moment map is, up to translation, *precisely* the polytope $\Delta_D$ defined in (1.1), and the induced map

$$X/T_{\mathbb{R}} \longrightarrow \Delta_D$$

is a homeomorphism.

With the publication of Delzant's thesis [**Del**], it became possible to view the above results in a broader context. Namely, given a smooth projective toric $X$ and a class $[D] = [\sum_\rho a_\rho D_\rho]$ in its Kähler cone (so $a_\rho \in \mathbb{R}$), we get the polytope

$$\Delta_D = \{m \in M_{\mathbb{R}} : \langle m, \rho\rangle \geq -a_\rho\} \subset M_{\mathbb{R}}.$$

This polytope is no longer integral, but the normals to its facets are integral, and since $X$ is smooth, the normals to the facets meeting at each vertex of $\Delta_D$ form a basis of the lattice $N$ (this is just another way of saying that the normal fan of $\Delta_D$ is the fan of $X$). Such a polytope is called a *Delzant polytope*.

EXAMPLE. Consider the toric varieties $X_1$ and $X_2$ from the example in §3. we leave it for the reader to show that, up to translation, the Delzant polytope in $\mathbb{R}^3$ corresponding to $D = a_0D_0 + \cdots + a_4D_4$ is defined by the inequalities

$$x, y, z \geq 0, \quad x + z \leq s, \quad x - y + z \geq s - t,$$

where as usual $s = a_0 + a_1 + a_3$ and $t = a_0 + a_2 + a_4$. For $t > s > 0$, this gives the Delzant polytope for $X_1$, and for $s > t > 0$ we get the polytope for $X_2$. It is a good exercise to draw these polytopes so you can see what happens when $s = t$. (For the best picture, have the first octant face away from you and let the $xz$-plane be horizontal.)

If we give the toric variety $X$ a symplectic structure whose class lies in $[D]$, then the action of $T_{\mathbb{R}}$ is still Hamiltonian and the Delzant polytope $\Delta_D$ is again the image of the moment map. Conversely, given *any* Delzant polyotope $\Delta$, one can construct a smooth projective toric variety $X_\Delta$ with $\Delta$ as the image of the moment map (see [**Gui**]).

What is more remarkable is Delzant's purely symplectic characterization of smooth projective toric varieties (see [**Del**]).

THEOREM 4.2. *Let $X$ be a real $2n$-dimensional compact connected symplectic manifold with an effective Hamiltonian action of $(S^1)^n$. Then the image of the moment map is a Delzant polytope $\Delta$ and $X$ is diffeomorphic (as a Hamiltonian $(S^1)^n$-space) to the smooth projective toric variety $X_\Delta$ determined by $\Delta$.*

One way to understand this theorem is to observe that for an effective Hamiltonian action of $(S^1)^m$ on a connected symplectic manifold $M$, we always have $2m \leq \dim M$. Thus Theorem 4.2 characterizes what happens when $M$ is compact and the dimension of the torus $(S^1)^m$ is as large as possible.

## §5. Torus coordinates and toric ideals

Besides the homogeneous coordinates of §2, toric varieties have intrinsic coordinates living on the torus. This is because a basis $e_1, \ldots, e_n$ of $M$ induces an isomorphism $T = N \otimes_{\mathbb{Z}} \mathbb{C}^* \simeq (\mathbb{C}^*)^n$, giving coordinates $t_1, \ldots, t_n$ on $T$. Then, for $m = \sum_{i=1}^n a_i e_i \in M$, the character $\chi^m$ from §1 is the Laurent monomial $\mathbf{t}^m = \Pi_{i=1}^n t_i^{a_i}$, and the coordinate ring of $T$ is $\mathbb{C}[t_1^{\pm 1}, \ldots, t_n^{\pm 1}]$. These coordinates don't extend to the whole toric variety, but they are still very useful.

**The toric variety of a polytope.** Given an $n$-dimensional integral convex polytope $\Delta \subset M_{\mathbb{R}}$, we get the toric variety $X_\Delta$ and an ample divisor $D$ as mentioned in §1. Then (1.2) can be written as

$$H^0(X_\Delta, \mathcal{O}_X(D)) = \bigoplus_{m \in \Delta \cap M} \mathbb{C} \cdot \mathbf{t}^m = L(\Delta), \tag{5.1}$$

so that we can think of the global sections of $D$ in terms of Laurent polynomials.

The most concrete way to get $D$ from $\Delta$ is to represent $\Delta \subset M_{\mathbb{R}}$ by inequalities $\langle m, \rho \rangle \geq -a_\rho$ (so the $\rho$'s are normals to the facets of $\Delta$), and then $D = \sum_\rho a_\rho D_\rho$.

From a more sophisticated point of view, $D$ is the divisor associated to the support function $\psi : N_{\mathbb{R}} \to \mathbb{R}$ defined by

$$\psi(u) = \min\{\langle m, u\rangle : m \in \Delta\}.$$

Since $D$ is ample, the global sections of some multiple $kD$ give a projective embedding of $X_\Delta$. We can use this to construct $X_\Delta$ as follows. Let $l(k\Delta) = \dim L(k\Delta)$ be the number of integer points of $k\Delta$, and consider the map

$$\Psi : (\mathbb{C}^*)^n \longrightarrow \mathbb{P}^{l(k\Delta)-1}$$

defined by

$$\Psi(t_1, \dots, t_n) = (\mathbf{t}^{m_1}, \dots, \mathbf{t}^{m_{l(k\Delta)}}) \tag{5.2}$$

where $k\Delta \cap M = \{m_1, \dots, m_{l(k\Delta)}\}$. Then, since $\Psi$ extends to an embedding of $X_\Delta$, it is clear that $X_\Delta$ is the closure of $\Psi((\mathbb{C}^*)^n)$ in $\mathbb{P}^{l(k\Delta)-1}$. Later in this section we will use this approach to define non-normal toric varieties.

A more algebraic method of constructing $X_\Delta$ is due to Batyrev [**Bat4**]. Given $\Delta$, consider the cone over $\Delta \times \{1\} \subset M_{\mathbb{R}} \oplus \mathbb{R}$ (in the terminology of [**BB1**], this is a *Gorenstein cone*). The integer points of the cone give a semigroup algebra $S_\Delta$. Since $(m, k) \in M \oplus \mathbb{Z}$ is in the cone if and only if $m \in k\Delta$, $S_\Delta$ is the subring of $\mathbb{C}[t_0, t_1^{\pm 1}, \dots, t_n^{\pm 1}]$ spanned by Laurent monomials $t_0^k \mathbf{t}^m$ with $k \geq 0$ and $m \in k\Delta$. This ring can be graded by setting $\deg(t_0^k \mathbf{t}^m) = k$, and one can show that

$$X_\Delta = \operatorname{Proj}(S_\Delta).$$

Since $S_\Delta$ is the coordinate ring of an affine toric variety, it is Cohen-Macaulay and hence $X_\Delta$ is arithmetically Cohen-Macaulay. This ring will play an important role in the next section.

We can also relate $S_\Delta$ to the coordinate ring $S$ of §2. If $\alpha = [D] \in A_{n-1}(X_\Delta)$ is the class of $D = \sum_\rho a_\rho D_\rho$, then by [**BC**], we can define a ring isomorphism

$$S_\Delta \simeq \bigoplus_{k=0}^{\infty} S_{k\alpha} \subset S \tag{5.3}$$

by sending the Laurent monomial $t_0^k \mathbf{t}^m$ to the monomial $\Pi_\rho x_\rho^{k a_\rho + \langle m, \rho\rangle}$. This is a special case of the isomorphism (2.5).

**Newton polytopes.** There are many situations where instead of a polytope, the initial data is a Laurent polynomial corresponding to a finite set $\mathcal{A} \subset M = \mathbb{Z}^n$ of exponents, which we write as

$$f = \sum_{m \in \mathcal{A}} c_m \mathbf{t}^m, \quad c_m \neq 0.$$

The convex hull $\Delta = \operatorname{Conv}(\mathcal{A})$ is called the *Newton polytope* of $f$ and is used in many contexts (see, for example, [**AVG**], [**Dan3**], [**DL**], [**GKZ1**], [**Kho**], [**Kus**], [**McD**], [**Var**]). This polytope might not be $n$-dimensional, but if we use the lattice $\mathbb{Z}\mathcal{A}$ generated by $\mathcal{A}$, then we get a toric variety $X_\Delta$ of dimension $= \operatorname{rank}(\mathbb{Z}\mathcal{A})$ and, as in (5.1), an ample line bundle on $X_\Delta$ whose global sections are $L(\Delta) = \oplus_{m \in \Delta \cap \mathbb{Z}\mathcal{A}} \mathbb{C} \cdot \mathbf{t}^m$. In particular, our Laurent polynomial $f$ is a global section.

A variant of this is that sometimes one is given Laurent polynomials $f_1, \ldots, f_s$ corresponding to possibly different sets of exponents $\mathcal{A}_1, \ldots, \mathcal{A}_s \subset \mathbb{Z}^n$. In this situation, we get polytopes $\Delta_i = \mathrm{Conv}(\mathcal{A}_i)$, and we let

$$\Delta = \Delta_1 + \cdots + \Delta_s \tag{5.4}$$

be their *Minkowski sum*. Also let $L(\Delta_i) = \oplus_{m \in \Delta_i \cap \mathbb{Z}\mathcal{A}} \mathbb{C}\cdot\mathbf{t}^m$, where $\mathcal{A} = \mathcal{A}_1 \cup \cdots \cup \mathcal{A}_s$. From Proposition 2.4 of [**BB3**], we get the following result which is useful when studying complete intersections in toric varieties.

PROPOSITION 5.1. *Given $f_1, \ldots, f_s$ and $\Delta$ as above, the toric variety $X_\Delta$ has divisors $D_i$ whose global sections are $L(\Delta_i)$, and $\mathcal{O}_{X_\Delta}(D_i)$ is generated by these sections. In particular, each $f_i$ is a global section of $\mathcal{O}_{X_\Delta}(D_i)$.*

**Non-normal toric varieties.** We can generalize the construction (5.2) as follows. Given a finite set of exponents $\mathcal{A} = \{m_1, \ldots, m_\ell\} \subset \mathbb{Z}^n$, we get a map

$$\Psi : (\mathbb{C}^*)^n \longrightarrow \mathbb{C}^\ell$$

defined by

$$\Psi(t_1, \ldots, t_n) = (\mathbf{t}^{m_1}, \ldots, \mathbf{t}^{m_\ell}). \tag{5.5}$$

The Zariski closure of $\Psi((\mathbb{C}^*)^n) \subset \mathbb{C}^\ell$ is an affine variety denoted ${}^aX_{\mathcal{A}}$ (the superscript $^a$ refers to affine).

If $d = \mathrm{rank}(\mathbb{Z}\mathcal{A})$, then one can show that $T = {}^aX_{\mathcal{A}} \cap (\mathbb{C}^*)^\ell$ is isomorphic to $(\mathbb{C}^*)^d$ and is Zariski open in ${}^aX_{\mathcal{A}}$. Furthermore, the natural action of $T$ on itself extends to an action on ${}^aX_{\mathcal{A}}$ (see [**Stu2**, Lemma 13.4]). This sounds like the definition of toric variety, but *normality* is missing. Nevertheless, we will refer to ${}^aX_{\mathcal{A}}$ as a toric variety, and we will see in this section and in §9 that these toric varieties are very useful. Basic references for non-normal toric varieties are [**GKZ1**] and [**Stu2**].

EXAMPLE. When $\mathcal{A} = \{2, 3\} \subset \mathbb{Z}$, we get the map $\Psi(t) = (t^2, t^3)$, which leads to the cuspidal cubic $y^2 = x^3$ in $\mathbb{C}$. This is clearly a non-normal toric variety.

The following proposition explains how ${}^aX_{\mathcal{A}}$ relates to the usual kind of affine toric variety. See [**Stu2**] for the proof.

PROPOSITION 5.2. *The normalization of ${}^aX_{\mathcal{A}}$ is the affine toric variety $X_\sigma$, where $M = \mathbb{Z}\mathcal{A}$ and $\sigma \subset N_{\mathbb{R}}$ is dual to the convex polyhedral cone* Cone$(\mathcal{A})$ *generated by $\mathcal{A}$. Hence ${}^aX_{\mathcal{A}}$ is normal if and only if $\mathbb{N}\mathcal{A} = \mathbb{Z}\mathcal{A} \cap \mathrm{Cone}(\mathcal{A})$, where $\mathbb{N}\mathcal{A}$ is the set of all non-negative integer linear combinations of $\mathcal{A}$.*

We next consider projective toric varieties defined using $\mathcal{A} \subset \mathbb{Z}^n$. In practice, there are two ways of doing this. The first method assumes that $\mathcal{A}$ lies in an affine hyperplane in $\mathbb{Z}^n$ not passing through the origin. In this case, it is easy to see that ${}^aX_{\mathcal{A}} \subset \mathbb{C}^\ell$ is defined by homogeneous polynomials. Then ${}^aX_{\mathcal{A}}$ is the affine cone of a projective variety in $\mathbb{P}^{\ell-1}$ denoted $X_{\mathcal{A}}$. One can prove that $X_{\mathcal{A}}$ is a toric variety (possibly non-normal) of dimension equal to the dimension of the affine span of $\mathcal{A}$ (= the dimension of the convex hull Conv$(\mathcal{A})$).

A second method for creating projective toric varieties starts with an *arbitrary* subset $\mathcal{A} = \{m_1, \ldots, m_\ell\} \subset \mathbb{Z}^n$ and considers the map

$$\Psi(t_1, \ldots, t_n) = (\mathbf{t}^{m_1}, \ldots, \mathbf{t}^{m_\ell}) \in \mathbb{P}^{\ell-1}. \tag{5.6}$$

Then, as in [**GKZ1**], we define $X_{\mathcal{A}}$ to be the closure in $\mathbb{P}^{\ell-1}$ of the image of $\Psi$. The dimension of $X_{\mathcal{A}}$ again equals the dimension of $\mathrm{Conv}(\mathcal{A})$.

These two approaches are related as follows. Given $\mathcal{A}$ as in the second method, $\mathcal{A} \times \{1\} \subset \mathbb{Z}^{n+1}$ lies in an affine hyperplane not passing through the origin. Then the affine cone of $X_{\mathcal{A}}$ is easily seen to be ${}^aX_{\mathcal{A}\times\{1\}}$. Thus $X_{\mathcal{A}}$, as defined by the second method, equals $X_{\mathcal{A}\times\{1\}}$, as defined by the first.

The normalization of the projective toric variety $X_{\mathcal{A}}$ can be computed using Proposition 5.2. This is a bit delicate because of the distinction between normality and projective normality—it is possible for $X_{\mathcal{A}}$ to be normal without ${}^aX_{\mathcal{A}}$ being so (see [**Har**, Ex. 3.18 on p. 23] for an example).

PROPOSITION 5.3. *The normalization of $X_{\mathcal{A}}$ is $X_\Delta$, where $\Delta = \mathrm{Conv}(\mathcal{A})$.*

PROOF. We regard $X_{\mathcal{A}}$ as arising from (5.6). As noted above, its affine cone is ${}^aX_{\mathcal{A}\times\{1\}}$, so by Proposition 5.2, the normalization of ${}^aX_{\mathcal{A}\times\{1\}}$ comes from the cone over $\mathcal{A} \times \{1\}$. This equals the cone over $\Delta \times \{1\}$, which is *exactly* the cone used in Batyrev's construction of $X_\Delta$ earlier in this section. Thus, the normalization of ${}^aX_{\mathcal{A}}$ is the affine cone of $X_\Delta$, which implies that the map $X_\Delta \to X_{\mathcal{A}}$ is finite and birational. The proposition now follows from Zariski's Main Theorem. □

EXAMPLE. To see how non-normal projective toric varieties can occur, consider a complete toric variety $X$ (in the usual sense) with an ample divisor $D$. If we let $\mathcal{A} = \Delta_D \cap M$ and use the basis of $H^0(X, \mathcal{O}_X(D))$ given by Laurent monomials, we get a map $\Psi : X \to \mathbb{P}^{\ell-1}$ as above. Since $D$ need not be very ample, $\Psi$ need not be an embedding. But (5.6) shows that the image $\Psi(X)$ is precisely the toric variety $X_{\mathcal{A}}$. It follows from Proposition 5.3 that $X$ *is the normalization of the image of the map to projective space given by an ample line bundle on* $X$.

There is a nice criterion for normality which involves the Hilbert polynomial of $X_{\mathcal{A}}$ and the Ehrhart polynomial of the polytope $\Delta = \mathrm{Conv}(\mathcal{A})$. By [**Stu2**, Ch. 13], the Hilbert polynomial of $X_{\mathcal{A}}$ is given by

$$H_{\mathcal{A}}(k) = |\{m_{i_1} + \cdots + m_{i_k} : m_{i_1}, \ldots, m_{i_k} \in \mathcal{A}\}|, \quad k \gg 0,$$

and, as usual, the Ehrhart polynomial of $\Delta$ is

$$E_\Delta(k) = |\mathbb{Z}\mathcal{A} \cap k\Delta|, \quad k \geq 0$$

These polynomials have the same leading term, which implies that the degree of $X_{\mathcal{A}} \subset \mathbb{P}^{\ell-1}$ is the normalized volume of $\Delta$ (see [**Stu2**, Thm. 4.16]). Then we have the following result of Sturmfels [**Stu2**, Thm. 13.11].

THEOREM 5.4. *The toric variety $X_{\mathcal{A}} \subset \mathbb{P}^{\ell-1}$ is normal if and only if the Hilbert polynomial $H_{\mathcal{A}}$ equals the Ehrhart polynomial $E_{\mathcal{A}}$.*

**Toric ideals.** A familiar example from algebraic geometry is the twisted cubic $(x, y, z) = (t, t^2, t^3)$ in $\mathbb{C}^3$. This is a special case of the construction (5.5). The ideal of the twisted cubic is $\langle y - x^2, z - x^3\rangle \subset \mathbb{C}[x, y, z]$ and is our first example of a toric ideal. As we will soon see, the simple form of its generators is no accident.

Given $\mathcal{A} = \{m_1, \ldots, m_\ell\} \subset \mathbb{Z}^d$, the affine toric variety ${}^aX_{\mathcal{A}}$ is defined by an ideal $I_{\mathcal{A}} \subset \mathbb{C}[x_1, \ldots, x_\ell]$, which we call a *toric ideal*. In terms of elimination theory, the ideal $I_{\mathcal{A}}$ arises from the equations $x_i - \mathbf{t}^{m_i} = 0$, $1 - yt_1 \cdots t_n = 0$ by eliminating $y, t_1, \ldots, t_n$ (the equation $1 - yt_1 \cdots t_n = 0$ guarantees that $(t_1, \ldots, t_n) \in (\mathbb{C}^*)^n$).

A toric ideal $I_{\mathcal{A}}$ is homogeneous when $\mathcal{A}$ lies in an affine hyperplane missing the origin, in which case $I_{\mathcal{A}}$ defines the projective toric variety $X_{\mathcal{A}}$. Also, if $X_{\mathcal{A}}$ is projectively normal, then $\mathbb{C}[x_1,\dots,x_\ell]/I_{\mathcal{A}}$ is isomorphic to the ring $S_\Delta$ of (5.3). So facts about toric ideals give useful information about the coordinate rings of projective toric varieties.

As with the twisted cubic, toric ideals are generated by *binomials*, which are differences of monomials. To state the result, note that a vector $\mathbf{a} \in \mathbb{Z}^\ell$ can be written $\mathbf{a}^+ - \mathbf{a}^-$, where $\mathbf{a}^+$ and $\mathbf{a}^-$ have non-negative entries and disjoint support.

LEMMA 5.5. *If $\mathcal{A} = \{m_1,\dots,m_\ell\} \subset \mathbb{Z}^n$, then the toric ideal $I_{\mathcal{A}}$ can be written*

$$I_{\mathcal{A}} = \langle \mathbf{x}^{\mathbf{a}^+} - \mathbf{x}^{\mathbf{a}^-} : \mathbf{a} = (a_1,\dots,a_\ell) \in \mathbb{Z}^\ell,\ \textstyle\sum_{i=1}^{\ell} a_i m_i = 0\rangle.$$

In [**Stu1**] and [**Stu2**], one can find a wealth of results about toric ideals, including facts about Gröbner bases and relations to secondary fans. We will discuss these ideas briefly in §7. Applications to enumeration, sampling, integer programming, and primitive partition identities are given in [**Stu2**, Ch. 5 and 6].

We close this section with an unsolved problem about toric ideals.

CONJECTURE 5.6. *If $X_{\mathcal{A}}$ is a smooth projectively normal toric variety, then the toric ideal $I_{\mathcal{A}}$ is generated by quadratic binomials.*

In this situation, it is known that $I_{\mathcal{A}}$ is generated by binomials of degree at most $d = \operatorname{rank}(\mathbb{Z}\mathcal{A})$ (see [**Stu2**, Thm. 13.14]).

## §6. Cohomology of toric hypersurfaces

A complete toric variety $X$, being rational, is a very special kind of variety. But as an ambient space, $X$ can be home to some interesting subvarieties (we will see some Calabi-Yau examples in §8). In particular, using the homogeneous coordinates of §2 or the torus coordinates of §5, it is easy to describe hypersurfaces and complete interesections in $X$. In this section, we will study hypersurfaces $Y \subset X$ and the associated affine hypersurfaces $Y \cap T \subset T$, where $T$ is the torus of $X$. We will end the section with some remarks about complete intersections.

**Affine and projective hypersurfaces.** Given a Laurent polynomial $f \in \mathbb{C}[t_1^{\pm 1},\dots,t_n^{\pm 1}]$, the equation $f = 0$ defines an affine hypersurface

$$Z_f \subset T = (\mathbb{C}^*)^n.$$

We can assume that $f \in L(\Delta)$ for some convex integral polytope $\Delta \subset \mathbb{R}^n$ (for example, let $\Delta$ be the Newton polytope of $f$). We will also assume that $\Delta$ has dimension $n$ (it is easy to reduce to this case). Then we have a toric variety $X_\Delta$ containing $T$, and $f \in L(\Delta)$ defines a projective hypersurface

$$Y_f \subset X_\Delta$$

since $f$ can be regarded as a global section of an ample line bundle on $X_\Delta$. Note that $Z_f = Y_f \cap T$, though $Y_f$ need not be the Zariski closure of $Z_f$. The latter happens, for instance, if $f$ is a single monomial $\mathbf{t}^m$.

When $\Delta$ is the Newton polytope of $f$, there is a nice relation between the topology of $Z_f$ and the vertices of $\Delta$. Using the map

$$\text{(6.1)} \qquad \log : (\mathbb{C}^*)^n \longrightarrow \mathbb{R}^n, \quad (t_1,\dots,t_n) \mapsto (\log|t_1|,\dots,\log|t_n|)$$

one can show that unbounded connected components of $\mathbb{R}^n - \log(Z_f)$ correspond to vertices of $\Delta$ and in each such component contains a translate of the corresponding cone in the normal fan of $\Delta$. There is also a version of this (using the moment map) for $Y_f$—see [**GKZ**, Sect. 1.B and 1.C of Ch. 6].

Since $Z_f$ or $Y_f$ could be very singular for an arbitrary $f \in L(\Delta)$, we need some sort of genericity condition on $f$. There are two conditions which are used in practice. First, $f$ is *nondegenerate* or *$\Delta$-regular* if $Y_f \subset X_\Delta$ meets every torus orbit transversely, i.e., if for every orbit $O$, $Y_f \cap O$ is smooth of codimension 1 in $O$. In particular, a nondegenerate $Y_f$ misses all fixed points of $T$, and since fixed points correspond to vertices of $\Delta$, it follows that $\Delta$ is the Newton polytope of $f$. By Bertini's Theorem, a generic $f \in L(\Delta)$ is nondegenerate (see [**Kho**]).

The second condition applies when $X_\Delta$ is simplicial, which for $\Delta$ means that every vertex is contained in precisely $n$ faces (such a polytope is called *simple*). In this case, $X_\Delta$ is a $V$-manifold, and $f \in L(\Delta)$ is *quasi-smooth* if $Y_f \subset X_\Delta$ is a $V$-submanifold, as defined in [**BC**, Sect. 3]. When $X_\Delta$ is smooth, this means that $Y_f$ is smooth. One can also show that nondegenerate implies quasi-smooth, though the converse is not true (the smooth cubic $y^2z = x^3 - z^3$ in $\mathbb{P}^2$ is an example). The notion of quasi-smooth was introduced by Danilov (see [**Dan2**, Sect. 14]).

In the simplicial case, we can use the homogeneous coordinate ring $S = \mathbb{C}[x_\rho]$ of §2 to describe hypersurfaces in $X_\Delta$. If $\beta = [D] \in A_{n-1}(X_\Delta)$ is the class of the line bundle $D$ determined by $\Delta$, then there is a natural isomorphism

$$L(\Delta) \simeq S_\beta \tag{6.2}$$

described in (5.3). Thus, $f \in L(\Delta)$ corresponds to a homogeneous polynomial $F \in S_\beta$. The equation $F = 0$ gives a hypersurface in $\mathbb{C}^{\Sigma(1)} - Z(\Sigma)$ (where $\Sigma$ is the fan of $X_\Delta$). By Theorem 2.1, this descends to a hypersurface $Y_F \subset X_\Delta$, and one can check that $Y_F = Y_f$. Furthermore, $Y_F$ is quasi-smooth if and only if the partial derivatives $\partial F/\partial x_\rho$ have no common zeros in $\mathbb{C}^{\Sigma(1)} - Z(\Sigma)$ (see [**BC**, Sect. 3]).

When the defining polynomial $f$ or $F$ is clear from context, we will write $Z \subset T$ in the affine case and $Y \subset X_\Delta$ in the projective case.

**Cohomology of affine hypersurfaces.** For a nondegenerate affine hypersurface $Z = Z_f \subset T = (\mathbb{C}^*)^n$, the cohomology groups $H^i(Z)$ (always with complex coefficients) carry natural mixed Hodge structures. In [**DK**], the dimension $h^{p,q}(H^i(Z))$ of the $(p,q)$ Hodge component of $\mathrm{Gr}^{\mathcal{W}}_{p+q} H^i(Z)$ (where $\mathcal{W}$ is the weight filtration) is computed using the combinatorics of the Newton polytope $\Delta$ of $f$.

It suffices to compute $h^{p,q}(H^i_c(Z))$ since $H^i_c(Z)$ is dual (as a mixed Hodge structure) to $H^{2n-2-i}(Z)$. In fact, it is sufficient to compute

$$e^{p,q}(Z) = \sum_i (-1)^i h^{p,q}(H^i_c(Z))$$

since the Gysin map

$$H^i_c(Z) \longrightarrow H^{i+2}_c(T)$$

is an isomorphism of mixed Hodge structures (suitably shifted) for $i > n-1$ and $H^i_c(Z) = 0$ for $i < n-1$ (recall that $Z$ is smooth and affine).

In the special case when $\Delta$ is simplicial, computing $e^{p,q}(Z)$ is fairly easy. First, for a face $\Gamma \subset \Delta$, define $\phi_i(\Gamma)$ for $0 < i \le \dim \Gamma$ by the formulas

$$
\begin{aligned}
\phi_i(\Gamma) &= \sum_{k=1}^{i} (-1)^{i-k} \binom{\dim \Gamma + 1}{i-k} l^*(k\Gamma) \\
&= (-1)^{\dim \Gamma + 1} \sum_{k=0}^{\dim \Gamma + 1 - i} (-1)^{i+k} \binom{\dim \Gamma + 1}{i+k} l(k\Gamma)
\end{aligned} \tag{6.3}
$$

where $l(k\Gamma)$ (resp. $l^*(k\Gamma)$) is the number of integer points in $k\Gamma$ (resp. in the relative interior of $k\Gamma$). The two representations of $\phi_i(\Gamma)$ are related to the remarkable properties of the Ehrhart polynomial from §5 (see [**Bat4**, Sect. 2] for more details).

When $p > q$, [**DK**, 5.7] gives the formula

$$
e^{p,q}(Z) = (-1)^{n+p+q} \sum_{\dim \Gamma > p} (-1)^{\dim \Gamma} \binom{n - \dim \Gamma}{n-1-p-q} \phi_{\dim \Gamma - p}(\Gamma),
$$

where the sum is over all faces $\Gamma$ of $\Delta$ of dimension $> p$. Since $e^{p,q}(Z) = e^{q,p}(Z)$, this gives us everything except $e^{p,p}(Z)$. However, we also have the identity

$$
(-1)^{n-1} \sum_q e^{p,q}(Z) = (-1)^p \binom{n}{p+1} + \phi_{n-p}(\Delta), \tag{6.4}
$$

from [**DK**, 4.4], which now enables us to compute $e^{p,p}(Z)$.

The paper [**DK**] also describes an algorithm for computing the full mixed Hodge structure of $Z_f \subset X_\Delta$ for *any* polytope $\Delta$. There are tables giving explicit formulas when $1 \le \dim(\Delta) \le 4$ (see [**DK**, 5.11]).

One way to represent the numbers $e^{p,q}(Z)$ is via the $E$-polynomial

$$
E(Z; u, v) = \sum_{p,q} e^{p,q}(Z) u^p v^q.
$$

Then (6.4) describes $E(Z; u, 1)$, which tells us about the Hodge filtration. The weight filtration, on the other hand, concerns $E(Z; u, u)$. This polynomial is studied in [**BB2**] and [**DL**]. We should also mention that explicit formulas for $E(Z; u, v)$ (for $\Delta$ arbitary, not just simplicial) can be found in [**BB2**].

We next describe some work of Batyrev [**Bat4**] on representing cohomology classes of $Z \subset T$ as residues of forms. Here, we will focus on the mixed Hodge structure of $H^i(Z)$. The dual of the above Gysin map is the natural map $H^k(T) \to H^k(Z)$, which is an isomorphism for $k < n-1$ and injective for $i = n-1$. Because of this, we define the *primitive cohomology* of $Z$ by the exact sequence

$$
0 \longrightarrow H^{n-1}(T) \longrightarrow H^{n-1}(Z) \longrightarrow H_0^{n-1}(Z) \longrightarrow 0. \tag{6.5}
$$

It follows that $H_0^{n-1}(Z)$ has a mixed Hodge structure. The graded pieces of the Hodge filtration will be denoted $\mathrm{Gr}_{\mathcal{F}}^p H_0^{n-1}(Z)$. In [**Bat4**, Cor. 3.14], Batyrev observes that (6.4) can be reformulated as

$$
\dim \mathrm{Gr}_{\mathcal{F}}^p H_0^{n-1}(Z) = \phi_{p+1}(\Delta). \tag{6.6}
$$

EXAMPLE. Let $\Delta$ be a convex integer polygon in $\mathbb{R}^2$. Then $Z \subset (\mathbb{C}^*)^2$ is a curve which can be obtained by removing $m$ points from a smooth complete curve of genus $g$. As is well-known, this determines the Hodge numbers of $H^1(Z)$, and using (6.5), we obtain $h^{1,0}(H_0^1(Z)) = h^{0,1}(H_0^1(Z)) = g$ and $h^{1,1}(H_0^1(Z)) = m - 3$. However, (6.6) and (6.3) imply that

$$h^{0,1}(H_0^1(Z)) = \phi_1(\Delta) = l^*(\Delta)$$
$$h^{1,0}(H_0^1(Z)) + h^{1,1}(H_0^1(Z)) = \phi_2(\Delta) = l(\Delta) - 3.$$

It follows that $g = l^*(\Delta)$ and $m = l(\Delta) - l^*(\Delta)$ = the number of integer points on the boundary of $\Delta$. Thus we can see the mixed Hodge structure of $Z$ geometrically. This example can also be done using the formulas of [**DK**].

The next step is to represent cohomology classes in $H_0^{n-1}(Z)$ algebraically. For this purpose, let $g = t_0 f(t_1, \ldots, t_n) - 1$, which is in the ring $S_\Delta$ introduced in §5. We also set

$$g_i = t_i \frac{\partial g}{\partial t_i}, \quad 0 \le i \le n.$$

Then $g_0 = t_0 f$ and $g_i \in S_\Delta$ has degree 1 in $S_\Delta$ because $t_0$ appears to the first power. These polynomials generate the graded ideal $J_{f,\Delta} = \langle g_0, \ldots, g_n \rangle \subset S_\Delta$. The following result is proved in [**Bat4**, Sect. 6].

THEOREM 6.1. *There is a natural isomorphism*

$$\mathrm{Gr}_{\mathcal{F}}^p H_0^{n-1}(Z) \simeq (S_\Delta / J_{f,\Delta})_{n-p}.$$

The idea behind this isomorphism is that a polynomial $t_0^{n-p} g(t_1, \ldots, t_n) \in (S_\Delta)_{n-p}$ gives a $n$-form

$$\frac{g}{f^{n-p}} \frac{dt_1}{t_1} \wedge \cdots \wedge \frac{dt_n}{t_n}$$

on $T - Z$ whose residue lies in $\mathrm{Gr}_{\mathcal{F}}^p H_0^{n-1}(Z)$.

**Cohomology of projective hypersurfaces.** We now turn our attention to the projective hypersurface $Y = Y_f \subset X = X_\Delta$, where as usual $f \in L(\Delta)$ is nondegenerate. In addition, we will assume that $X$ is simplicial. Then the natural map

$$H^i(X) \longrightarrow H^i(Y)$$

is an isomorphism for $i < n - 1$ and is injective for $i = n - 1$. In analogy with the affine case, we define the *primitive cohomology* by the exact sequence

$$0 \longrightarrow H^{n-1}(X) \longrightarrow H^{n-1}(Y) \longrightarrow H_0^{n-1}(Y) \longrightarrow 0.$$

Since $X$ is simplicial, $Y$ is quasi-smooth and hence $H_0^{n-1}(Y)$ has a pure Hodge structure. Letting $h_0^{p,n-1-p}$ denote the dimension of the appropriate Hodge component, [**DK**, 5.5] can be restated as

$$h_0^{p,n-1-p} = (-1)^n \sum_{\dim \Gamma > p} (-1)^{\dim \Gamma} \phi_{\dim \Gamma - p}(\Gamma),$$

where $\phi_i(\Gamma)$ is as in (6.3). This formula implies that $h_0^{n-1,0} = \phi_1(\Delta) = l^*(\Delta)$.

We have already seen that $Y \subset X$ can be defined in two ways, either using $f \in L(\Delta)$ or $F \in S_\beta$, where $f$ and $F$ are related via (6.2). Then we have the following results from [**Bat4**] and [**BC**].

THEOREM 6.2. *Let $X = X_\Delta$ be simplicial and $Y = Y_f$ be nondegenerate. If $I^{(1)} \subset S_\Delta$ be the ideal generated by those $t_0^k\mathbf{t}^m \in S_\Delta$ for which $m$ is an interior point of $k\Delta$, and $H_f$ is its image in $S_\Delta/J_{f,\Delta}$, then there are natural isomorphisms*

$$H_0^{p,n-1-p}(Y) \simeq \mathrm{Gr}_{\mathcal{F}}^p \mathcal{W}_{n-1} H_0^{n-1}(Z) \simeq (H_f)_{n-p},$$

*where $Z = Z_f$ is the corresponding affine hypersurface.*

THEOREM 6.3. *Let $X = X_\Delta$ be simplicial and $Y = Y_F$ be quasi-smooth. If $J(F) = \langle \frac{\partial F}{\partial x_\rho} \rangle \subset S$ be the Jacobian ideal of $F$, and $\beta_0 = \left[\sum_\rho D_\rho\right]$ be the anticanonical class, then there is a natural isomorphism*

$$H_0^{p,n-1-p}(Y) \simeq (S/J(F))_{(n-p)\beta-\beta_0}$$

*when $p \neq n/2-1$, and for $p = n/2-1$, we have an exact sequence*

$$0 \to H^{n-2}(X) \xrightarrow{\cup[Y]} H^n(X) \to (S/J(F))_{(n/2+1)\beta-\beta_0} \to H_0^{n/2-1,n/2}(Y) \to 0.$$

Theorem 6.3 generalizes a classic result of Griffiths on the cohomology of projective hypersurfaces (see [**Gri**] and [**PS**]). The basic idea is similar to Theorem 6.1: a homogeneous polynomial $G \in S_{(n-p)\beta-\beta_0}$ gives a $n$-form on $X - Y$ whose residue lies in $H_0^{n-1-p}(Y)$ (see [**BC**] for a precise description). The complication in the case $p = n/2-1$ comes from the exact sequence

$$0 \to H^{n-2}(X) \xrightarrow{\cup[Y]} H^n(X) \to H^n(X-Y) \to H_0^{n-1}(Y) \to 0.$$

When $X$ is projective space (or even a weighted projective space), the sequence implies $H^n(X-Y) \simeq H_0^{n-1}(Y)$, but this can fail in the toric case. Fortunately, this only affects $H_0^{n/2-1,n/2}(Y)$.

REMARKS. 1. While Theorem 6.1 uses the vanishing of $H^i(X, \mathcal{O}_X(Y))$ for $Y$ ample and $i > 0$, Theorem 6.3 uses the Bott-Steenbrink-Danilov vanishing theorem:

$$H^i(X, \widehat{\Omega}_X^p(Y)) = 0, \quad Y \text{ ample}, \ i > 0. \tag{6.7}$$

This result is stated in [**Dan2**] and [**Oda2**] without proof. In the simplicial case, a proof appeared in [**BC**, Sect. 7], and a general proof (using characteristic $p > 0$) can be found in [**BFLM**]. A generalization of (6.7) is the vanishing theorem:

$$H^i(X, \mathcal{W}_k\widehat{\Omega}_X^p(\log(-K))(Y)) = 0, \quad Y \text{ ample}, \ i > 0. \tag{6.8}$$

Here, $\mathcal{W}_k\widehat{\Omega}_X^p(\log(-K)) = \widehat{\Omega}_X^{p-k} \wedge \widehat{\Omega}_X^k(\log(-K))$ is the usual weight filtration and $-K = \sum_\rho D_\rho$ is the anticanonical divisor. When $X$ is simplicial, (6.8) was proved in [**BC**], and (as noted by Lauritzen) the general case follows from (6.7).

2. Using the isomorphism (5.3), one can relate the ideal $H_f \subset S_\Delta/J_{f,\Delta}$ to the ideal generated by $\Pi_\rho x_\rho$ in $S/\langle x_\rho \partial F/\partial x_\rho \rangle$. This leads to a natural isomorphism

$$H_0^{p,n-1-p}(Y) \simeq (S/J_1(F))_{(n-p)\beta-\beta_0},$$

where $J_1(F)$ is the ideal quotient

$$J_1(F) = \langle x_\rho \partial F/\partial x_\rho \rangle : \Pi_\rho x_\rho$$

(see [**BC**, Sect. 11]). For a weighted projective space, $J(F)$ equals $J_1(F)$, but in general the relation between these ideals is not well understood. The ideal $J_1(F)$ arises naturally in certain mirror symmetry contexts (see [**MP**, 5.36]).

It can also happen that one is interested in a hypersurface $Y \subset X = X_\Delta$ which is big and nef but not ample. In the toric context, big and nef mean that $Y$ corresponds to a $n$-dimensional integer polytope $\Delta'$ which is a Minkowski summand of $\Delta$ (i.e., $\Delta' + \Delta'' = \mu\Delta$ for some integer polyope $\Delta''$ and $\mu \in \mathbb{Z}$—see [**BB3**, Sect. 2]). In this case, the map $H^i(X) \to H^i(Y)$ need not be an isomorphism for $i < n-1$, and Theorems 6.2 and 6.3 can also fail. An example of the latter is given by the proper transform of a degree 8 hypersurface in a resolution of $\mathbb{P}(1,1,2,2,2)$. We will say more about this when we study Calabi-Yau hypersurfaces in §8.

Besides the cohomology of toric hypersurfaces, one can also study their moduli (see [**AGM2**], [**Bat4**] and [**BC**]). The resulting variations of Hodge structure are closely connected with hypergeometric functions, which will be discussed in §9.

**Complete intersections.** Complete intersections in toric varieties can be studied from several points of view. In the affine case, suppose we have $f_i \in L(\Delta_i)$ as in Proposition 5.1. Then we get the affine complete intersection

$$Z_{f_1} \cap \cdots \cap Z_{f_s} \subset T.$$

To compute the cohomology of this variety, we use the *Cayley trick*. Consider the toric variety $T \times \mathbb{C}^s$ with variables $t_1, \ldots, t_n, \lambda_1, \ldots, \lambda_s$, and let

$$\mathcal{F} = \lambda_1 f_1 + \cdots \lambda_s f_s - 1.$$

This gives the affine hypersurface $Z_{\mathcal{F}} \subset T \times \mathbb{C}^s$, and one obtains

$$E(Z_{f_1} \cap \cdots \cap Z_{f_s}; u, v) = (uv-1)^s - (uv)^{1-s} E(Z_{\mathcal{F}}; u, v)$$

under suitable nondegeneracy hypotheses. This follows by considering the projection $Z_{\mathcal{F}} \to T$ (see [**DK**, 6.2]). Hence we can reduce to the hypersurface case (note, however, that $Z_{\mathcal{F}}$ is a hypersurface in $T \times \mathbb{C}^s$ instead of a torus).

Turning to the projective case, let $\Delta = \Delta_1 + \cdots + \Delta_s$ be as in (5.4), and consider the complete intersection

$$Y_{f_1} \cap \cdots \cap Y_{f_s} \subset X = X_\Delta,$$

which we assume to be nondegenerate. Now consider the projective bundle

$$\mathbb{P}(\mathcal{E}) = \mathbb{P}(\mathcal{O}_X(D_1) \otimes \cdots \otimes \mathcal{O}_X(D_s)),$$

where $D_i$ is the divisor corresponding to $\Delta_i$ in Proposition 5.1. If $\pi : \mathbb{P}(\mathcal{E}) \to X$ is the natural projection, then $\pi_*(\mathcal{O}_{\mathbb{P}(\mathcal{E})}(1)) = \mathcal{O}_X(D_1) \otimes \cdots \otimes \mathcal{O}_X(D_s)$. Hence there is a unique section of $\mathcal{O}_{\mathbb{P}(\mathcal{E})}(1)$ corresponding to $(f_1, \ldots, f_s)$. This section defines a hypersurface $\mathcal{Y} \subset \mathbb{P}(\mathcal{E})$, and one can show that the natural map

$$\mathbb{P}(\mathcal{E}) - \mathcal{Y} \longrightarrow X - Y_{f_1} \cap \cdots \cap Y_{f_s}$$

is a $\mathbb{C}^{s-1}$ bundle in the Zariski topology. Hence

$$H^i_c(X - Y_{f_1} \cap \cdots \cap Y_{f_s}) \simeq H_c^{i+2(s-1)}(\mathbb{P}(\mathcal{E}) - \mathcal{Y}),$$

which is an isomorphism of mixed Hodge structures of degree $(s-1, s-1)$. This is discussed in more detail in [**BB1**].

From the point of view of the homogeneous coordinate ring $S$ of $X$, each $f_i$ corresponds to a homogeneous polynomial $F_i \in S$. If we assume that each $\mathcal{O}_X(D_i)$

is ample, then it follows from [**CCD**] that the homogeneous coordinate ring of $\mathbb{P}(\mathcal{E})$ is $S \otimes \mathbb{C}[y_1, \ldots, y_s]$, where the variables $y_i$ have the property that

$$F = y_1 F_1 + \cdots + y_s F_s \tag{6.9}$$

is homogeneous and defines the hypersurface $\mathcal{Y} \subset \mathbb{P}(\mathcal{E})$. This explains why the above construction can be regarded as the projective version of the Cayley trick.

It should also possible to represent primitive cohomology classes of $Y_{f_1} \cap \cdots \cap Y_{f_s}$ using the Jacobian ideal of $F \in S \otimes \mathbb{C}[y_1, \ldots, y_s]$. This has not been done in general, but the case $X = \mathbb{P}^n$ has been studied in [**ENV**], [**Konn**], [**LT**] and [**Ter**]. Furthermore, [**Dim**] and [**Nag**] treat this case from the toric point of view, which we now explain. Let $F_1, \ldots, F_s$ be homogeneous polynomials (in the usual sense) in $\mathbb{C}[x_0, \ldots, x_n]$ of degrees $d_1, \ldots, d_s$ at least 2. This gives the complete intersection $Y_{F_1} \cap \cdots \cap Y_{F_s} \subset \mathbb{P}^n$ of dimension $n - s$. The homogeneous coordinate ring of $\mathbb{P}(\mathcal{E})$ is $R = \mathbb{C}[x_0, \ldots, x_n, y_1, \ldots, y_s]$ and is graded by $\mathbb{Z}^2$, where

$$\deg(x_i) = (1, 0), \quad \deg(y_j) = (-d_j, 1).$$

The polynomial $F$ of (6.9) has degree $\beta = (0,1)$ in $R$ and, defining primitive cohomology in the usual way, one obtains a natural isomorphism

$$H_0^{p,n-s-p}(Y_{F_1} \cap \cdots \cap Y_{F_s}) \simeq (R/J(F))_{(n-p)\beta - \beta_0}, \tag{6.10}$$

where $\beta = \deg(F) = (0,1)$, $\beta_0 = \deg(x_0 \cdots x_n y_1 \cdots y_s) = (n + 1 - \sum_{j=1}^s d_j, s)$, and $J(F)$ is the Jacobian ideal. Note the similarity with the second part of Theorem 6.3. It should be possible to prove a version of (6.10) for complete intersections of ample hypersurfaces in an arbitrary complete simplicial toric variety.

## §7. Secondary fans and polytopes

The *secondary polytope* was discovered in 1988 by Gelfand, Kapranov and Zelevinsky [**GKZ1**], with further developments by Billera, Filliman and Sturmfels [**BFS**] and Oda and Park [**OP**]. We will discuss the secondary polytope and its normal fan, which is called the *secondary polytope* or *GKZ decomposition*. At the end of the section we will also mention the *Gröbner fan* of a toric ideal.

**The secondary fan.** We begin with the secondary fan, following [**OP**]. As noted in §3, different toric varieties can have closely related Kähler cones. This happens because different fans in $N_\mathbb{R}$ can share the same 1-dimensional cones. To study the general case, fix a finite set of primitive vectors $\mathcal{B} \subset N$, where as usual $N \simeq \mathbb{Z}^n$ and $M$ is its dual, and assume that $\mathrm{Cone}(\mathcal{B}) = N_\mathbb{R}$. We will consider *all* complete fans $\Sigma$ in $N_\mathbb{R}$ whose 1-dimensional cone generators $\Sigma(1)$ lie in $\mathcal{B}$.

Given $\mathcal{B}$, we define $A_\mathcal{B} \simeq \mathbb{R}^{|\mathcal{B}|-n}$ by the exact sequence

$$0 \longrightarrow M_\mathbb{R} \xrightarrow{\alpha} \bigoplus_{\rho \in \mathcal{B}} \mathbb{R} \cdot e_\rho \xrightarrow{\beta} A_\mathcal{B} \longrightarrow 0, \tag{7.1}$$

where $\alpha(m) = \sum_{\rho \in \mathcal{B}} \langle m, \rho \rangle e_\rho$. Also let $\mathcal{B}^0 = \{\beta(e_\rho) : \rho \in \mathcal{B}\} \subset A_\mathcal{B}$. Then the pair $(\mathcal{B}^0, A_\mathcal{B})$ is the *linear Gale transform* of $(\mathcal{B}, N_\mathbb{R})$. The cone

$$A_\mathcal{B}^+ = \mathrm{Cone}(\mathcal{B}^0) = \{\textstyle\sum_{\rho \in \mathcal{B}} a_\rho \beta(e_\rho) : a_\rho \geq 0\} \subset A_\mathcal{B},$$

is strongly convex since $\mathrm{Cone}(\mathcal{B}) = N_\mathbb{R}$. From (2.2), we see that if $\Sigma$ is a fan with $\Sigma(1) = \mathcal{B}$, then $A_\mathcal{B} = A_{n-1}(X_\Sigma) \otimes \mathbb{R}$ and $A_\mathcal{B}^+$ is the cone generated by effective divisors on $X_\Sigma$.

Now suppose that $\Sigma$ is a simplicial projective fan with $\Sigma(1) = \mathcal{B}$. From §3, we have the cone $\mathrm{cpl}(\Sigma) \subset A_{\mathcal{B}}$ of convex classes, and it is easy to see that $\mathrm{cpl}(\Sigma) \subset A_{\mathcal{B}}^{+}$. Furthermore, the interior of $\mathrm{cpl}(\Sigma)$ is the Kähler cone of $X_\Sigma$. In the example from §3, these cones filled up $A_{\mathcal{B}}^{+}$ as we varied $\Sigma$. In general, this doesn't always happen, which is why we must allow fans $\Sigma$ where $\Sigma(1)$ is a proper subset of $\mathcal{B}$. Here, $A_{\mathcal{B}}$ no longer equals $A_{n-1}(X_\Sigma) \otimes \mathbb{R}$, so that the definition of convex needs to be modified. We say that $\sum_{\rho \in \mathcal{B}} a_\rho \beta(e_\rho)$ is *convex* if for every $\sigma \in \Sigma$, there is $m_\sigma \in M_{\mathbb{R}}$ such that $\langle m_\sigma, \rho \rangle \geq -a_\rho$ for all $\rho \in \mathcal{B}$, with equality holding for $\rho \in \sigma$. We again get a strongly convex polyhedral cone $\mathrm{cpl}(\Sigma) \subset A_{\mathcal{B}}^{+}$ which has nonempty interior when $\Sigma$ is simplicial and projective. According to [**OP**, Cor. 3.6], the cones $\mathrm{cpl}(\Sigma)$ fit together as follows.

THEOREM 7.1. *If* $\mathrm{Cone}(\mathcal{B}) = N_{\mathbb{R}}$, *then, as* $\Sigma$ *ranges over all simplicial projective fans in* $N_{\mathbb{R}}$ *with* $\Sigma(1) \subset \mathcal{B}$, *the cones* $\mathrm{cpl}(\Sigma)$ *and their faces form a fan in* $A_{\mathcal{B}}$ *whose support is* $A_{\mathcal{B}}^{+}$.

The fan of this theorem is the *secondary fan* or *GKZ decomposition* of $\mathcal{B}$. An example with two maximal cones was given in §3. In general, the structure of the secondary fan is quite interesting. For example, two cones $\mathrm{cpl}(\Sigma)$ and $\mathrm{cpl}(\Sigma')$ with a common codimension 1 face are related by a "flop" (as in the example from §3) or by adding or subtracting a single 1-dimensional cone to $\Sigma$. Also, faces on the boundary of $A_{\mathbb{R}}^{+}$ correspond to certain "degenerate" fans. This is all explained in [**OP**, Sect. 3].

The secondary fan has a nice relation to the moment map from §4. If you look back at its construction, you'll see that the group $G$ and the moment map $\mu_\Sigma$ of $X_\Sigma$ depend only on $\Sigma(1)$. Thus, given $\mathcal{B} \subset N$ as above, we get the moment map

$$\mu_{\mathcal{B}} : \mathbb{C}^{\mathcal{B}} \longrightarrow A_{\mathcal{B}}^{+}.$$

By Theorem 4.1, if $\mathbf{a}$ is in the interior of $\mathrm{cpl}(\Sigma)$, then

$$\mu_{\mathcal{B}}^{-1}(\mathbf{a})/G_{\mathbb{R}} \simeq X_\Sigma$$

when $\Sigma$ is projective, simplicial and satisifies $\Sigma(1) = \mathcal{B}$. In [**CK**], it is shown that this holds under the weaker hypothesis $\Sigma(1) \subset \mathcal{B}$ (with $\Sigma$ still projective and simplicial). Hence, the secondary fan gives a complete picture of the toric varieties we can build via symplectic reduction from a given moment map.

**The secondary polytope.** Given a finite subset $\mathcal{A} \subset N_{\mathbb{R}} \simeq \mathbb{R}^n$, we can describe a very interesting polytope using certain triangulations of its convex hull $\Delta^* = \mathrm{Conv}(\mathcal{A})$. We will assume that $\Delta^*$ is $n$-dimensional. Then a *triangulation* of $\mathcal{A}$ is a triangulation $T$ of $\Delta^*$ such that the vertices of each simplex in $T$ lie in $\mathcal{A}$ (though not every element of $\mathcal{A}$ need be used). Furthermore, the triangulation $T$ is *regular* or *coherent* if there is a function $\psi : \Delta^* \to \mathbb{R}$ which is affine on each simplex of $T$ and strictly convex. See [**GKZ1**] or [**BFS**] for precise definitions and proofs of the results stated below.

If $T$ is a triangulation of $\mathcal{A}$, we get the point

$$\phi_T = \sum_{\sigma \in T} \sum_{u \in \sigma} \mathrm{Vol}(\sigma)\, e_u \in \bigoplus\nolimits_{u \in \mathcal{A}} \mathbb{R} \cdot e_u. \tag{7.2}$$

The convex hull of these points is the *secondary polytope* of $\mathcal{A}$, denoted $\Sigma(\mathcal{A})$. One can show that the dimension of $\Sigma(\mathcal{A})$ is $|\mathcal{A}| - n - 1$ and that its vertices are precisely the points $\phi_T$ for which $T$ is a regular triangulation of $\mathcal{A}$. This polytope has some very interesting combinatorial properties [**GKZ1**], [**Stu2**].

The secondary polytope also has a normal fan $\mathcal{N}(\Sigma(\mathcal{A}))$, which is described using an affine version of the Gale transform. From $\mathcal{A}$, we get the set $\mathcal{A} \times \{1\} \subset N_{\mathbb{R}} \otimes \mathbb{R}$, and as in (7.1), we can construct an exact sequence

$$0 \longrightarrow M_{\mathbb{R}} \oplus \mathbb{R} \xrightarrow{\alpha} \bigoplus_{u \in \mathcal{A}} \mathbb{R} \cdot e_u \xrightarrow{\beta} A_{\mathcal{A}} \longrightarrow 0, \tag{7.3}$$

where $\alpha(m, \lambda) = \sum_{u \in \mathcal{A}} (\langle m, u \rangle + \lambda) e_u$. Also set $\mathcal{A}^0 = \{\beta(e_u) : u \in \mathcal{A}\}$. Then the pair $(\mathcal{A}^0, A_{\mathcal{A}})$ is the *affine Gale transform* of $(\mathcal{A}, N_{\mathbb{R}})$. Note that $\sum_{u \in \mathcal{A}} \beta(e_u) = 0$ and that $A_{\mathcal{A}}$ has dimension $|\mathcal{A}| - n - 1$.

Using the dual map $\beta^* : A_{\mathcal{A}}^* \to \bigoplus_{u \in \mathcal{A}} \mathbb{R} \cdot e_u$, one can show that the image of $\beta^*$ is parallel to the affine span of the secondary polytope $\Sigma(\mathcal{A})$. It follows that the normal fan of $\Sigma(\mathcal{A})$ lives naturally in $A_{\mathcal{A}}$. This is the *secondary fan of* $\mathcal{A}$, denoted $\mathcal{N}(\Sigma(\mathcal{A}))$. The maximal cones of the secondary fan correspond to vertices of $\Sigma(\mathcal{A})$ and hence to regular triangulations of $\mathcal{A}$. Note also that $\mathcal{N}(\Sigma(\mathcal{A}))$ is a complete fan in $A_{\mathcal{A}}$ since it comes from a polytope.

**The enlarged secondary fan.** The secondary fan $\mathcal{N}(\Sigma(\mathcal{A}))$ might seem rather different from the secondary fan defined earlier. To see the relation, let $\mathcal{B} = \Sigma(1)$, where $\Sigma$ is a projective fan, not necessarily simplicial. Then, as before, we get the secondary fan of $\mathcal{B}$, whose support is the strongly convex cone $A_{\mathcal{B}}^+ \subset A_{\mathcal{B}}$. Following [**AGM2**], we can enlarge this to a complete fan as follows. Fix an ample divisor $D = \sum_\rho a_\rho D_\rho$ with $a_\rho > 0$ for all $\rho$. This implies that 0 is an interior point of the corresponding polytope $\Delta \subset M_{\mathbb{R}}$. Then the *dual polytope* $\Delta^\circ \subset N_{\mathbb{R}}$ is defined by

$$\Delta^\circ = \{u \in N_{\mathbb{R}} : \langle m, u \rangle \geq -1 \text{ for all } m \in \Delta\}.$$

Note that 0 is an interior point of $\Delta^\circ$, though $\Delta^\circ$ need not be integral. Also, the normal fan $\Sigma$ of $\Delta$ is obtained by taking cones over proper faces of $\Delta^\circ$. In particular, the vertices of $\Delta^\circ$ are $(1/a_\rho)\rho$ for $\rho \in \mathcal{B} = \Sigma(1)$. Now let

$$\mathcal{A} = \mathrm{Vert}(\Delta^\circ) \cup \{0\} = \{(1/a_\rho)\rho : \rho \in \Sigma(1)\} \cup \{0\}.$$

We leave it to the reader to show that there is a natural isomorphism $A_{\mathcal{A}} \simeq A_{\mathcal{B}}$ which carries $\mathcal{A}^0$ to $\mathcal{B}^0 \cup \{-\sum_{\rho \in \mathcal{B}} a_\rho \beta(e_\rho)\}$. Under this isomorphism, we can compare the secondary fans of $\mathcal{A}$ and $\mathcal{B}$ as follows.

PROPOSITION 7.2. *Let $\mathcal{A}$ and $\mathcal{B}$ be as above. Then:*

1. *Under the isomorphism $A_{\mathcal{A}} \simeq A_{\mathcal{B}}$, the cone of $\mathcal{N}(\Sigma(\mathcal{A}))$ given by a regular triangulation $T$ of $\mathcal{A}$ lies in $A_{\mathcal{B}}^+$ if and only if 0 is contained in every maximal simplex of $T$.*
2. *For such a triangulation $T$, let $\Sigma'$ is the fan obtained by taking cones over simplices of $T$. Then $\Sigma'$ is a projective simplicial fan with $\Sigma'(1) \subset \mathcal{B}$, and the cone* $\mathrm{cpl}(\Sigma')$ *is the cone of $\mathcal{N}(\Sigma(\mathcal{A}))$ given by to $T$.*
3. *Conversely, every maximal cone* $\mathrm{cpl}(\Sigma') \subset A_{\mathcal{B}}^+$ *comes from such a triangulation $T$. We get $T$ by intersecting $\Delta^\circ$ with the cones in $\Delta^\circ$.*

In this situation, we call $\mathcal{N}(\Sigma(\mathcal{A}))$ the *enlarged secondary fan* of $\mathcal{B}$. A consequence of this proposition is that there is a bijective correspondence between maximal cones of the secondary fan of $\mathcal{B}$ and regular triangulations of $\mathcal{A}$ whose maximal cones all contain 0.

EXAMPLE. The toric variety $X = X_\Sigma$ from §3 has cone generators

$$e_0 = (0,0,-2),\ e_1 = (1,1,1),\ e_2 = (1,-1,1),\ e_3 = (-1,-1,1),\ e_4 = (-1,1,1)$$

and maximal cones $\sigma_{1234}$, $\sigma_{012}$, $\sigma_{023}$, $\sigma_{034}$ and $\sigma_{041}$. Let $\Delta$ be polytope associated to the anticanonical divisor $D_0 + D_1 + D_2 + D_3 + D_4$, which is ample. One can compute that the dual polytope of $\Delta$ is

$$\Delta^\circ = \text{Conv}(e_0, e_1, e_2, e_3, e_4). \tag{7.4}$$

In §8, we will see that $\Delta$ and $\Delta^\circ$ are examples of *reflexive polytopes*.

Using $\mathcal{A} = \text{Vert}(\Delta^\circ) \cup \{0\} = \{e_0, e_1, e_2, e_3, e_4, 0\}$, one sees that there are exactly four triangulations of $\mathcal{A}$, which means that the enlarged secondary fan has four maximal cones. Two of the triangulations have 0 as a vertex, and the corresponding cones were described in §3. We leave it to the reader to determine the other two cones in the fan and the corresponding triangulations of $\mathcal{A}$.

One can also describe the enlarged secondary fan using the total space of the line bundle over $X_\Delta$ given by $-\sum_\rho a_\rho D_\rho$ (see [**AGM2**]). This is closely related to an alternate approach to the secondary fan $\mathcal{N}(\Sigma(\mathcal{A}))$ which appears in [**OP**]. We will see in §10 that the enlarged secondary fan and the secondary polytope are used in mirror symmetry.

Other references for secondary polytopes and the GKZ-decomposition are [**Loe**] and [**Par**]. A significant generalization of the secondary polytope, called the *fiber polytope*, is described in [**BS**] and [**Zie**].

**The Gröbner fan.** If our finite set $\mathcal{A}$ is integral, i.e., $\mathcal{A} = \{u_1, \dots, u_\ell\} \subset N$, then we can use term orders on a toric ideal to create a fan closely related to the secondary fan of $\mathcal{A}$. The basic idea is that $\mathcal{A} \times \{1\} \subset N \oplus \mathbb{Z}$ determines the toric ideal $I_{\mathcal{A}\times\{1\}} \subset \mathbb{C}[x_1, \dots, x_\ell]$. This ideal defines the affine toric variety ${}^aX_{\mathcal{A}\times\{1\}}$, which by §5 is the affine cone over $X_\mathcal{A} \subset \mathbb{P}^{\ell-1}$. For this reason, we write the toric ideal as $I_\mathcal{A}$ instead of $I_{\mathcal{A}\times\{1\}}$. By Lemma 5.5, we can express this ideal as

$$I_\mathcal{A} = \langle \mathbf{x}^{\mathbf{a}^+} - \mathbf{x}^{\mathbf{a}^-} : \mathbf{a} = (a_1, \dots, a_\ell) \in \mathbb{Z}^\ell,\ \textstyle\sum_{i=1}^\ell a_i m_i = 0,\ \sum_{i=1}^\ell a_i = 0\rangle. \tag{7.5}$$

We can use elements $\omega \in \mathbb{R}^\ell$ to create initial ideals of $I_\mathcal{A}$ (in the sense of Gröbner theory) as follows: if $f = \sum_\mathbf{a} c_\mathbf{a} \mathbf{x}^\mathbf{a}$, define $\text{LT}_\omega(f) = \sum_{\omega\cdot\mathbf{a}\ \text{maximal}} c_\mathbf{a}\mathbf{x}^\mathbf{a}$, and then set

$$\text{LT}_\omega(I_\mathcal{A}) = \langle \text{LT}_\omega(f) : f \in I_\mathcal{A}\rangle.$$

For generic $\omega$, this is the initial ideal of $I_\mathcal{A}$ for some term order, and one can show that every initial ideal arises in this way [**Stu2**]. The interesting aspect is that different $\omega$'s can give the same initial ideal. We define

$$\omega \sim \omega' \iff \text{LT}_\omega(I_\mathcal{A}) = \text{LT}_{\omega'}(I_\mathcal{A}).$$

One can prove that each equivalence class is a relatively open convex polyhedral cone. Furthermore, if $\alpha$ is the map from (7.3), then $\omega + \alpha(m,\lambda) \sim \omega$ for all $(m,\lambda) \in M_\mathbb{R} \oplus \mathbb{R}$. Hence it makes sense to take the quotient by the image of $\alpha$. This leads to the following result.

PROPOSITION 7.3. *Under the map $\beta : \mathbb{R}^\ell \to A_{\mathcal{A}}$ from* (7.3), *we have:*

1. *The closures of the images of the equivalence classes form a complete fan in $A_{\mathcal{A}}$ called the* Gröbner fan.
2. *The Gröbner fan of $\mathcal{A}$ refines the secondary fan $\mathcal{N}(\Sigma(\mathcal{A}))$.*

This is proved in [**Stu2**]. One can also show that the Gröbner fan is the normal fan of a polytope called the *state polytope*. Algorithms for computing the state polytope and Gröbner fan are given in [**Stu2**], and the reader may also wish to consult [**MR**]. Although the Gröbner fan of $I_{\mathcal{A}}$ may be strictly finer than the secondary fan of $\mathcal{A}$, there are situations where the Gröbner fan is very useful. An example from mirror symmetry can be found in [**HLY1**].

## §8. Reflexive polytopes and Calabi-Yau hypersurfaces

In this section, we will create some interesting families of Calabi-Yau varieties using toric geometry. The key idea will be Batyrev's notion of *reflexive polytopes*. At the end of the section we will also consider complete intersections and nef-partitions.

**Singular Calabi-Yau varieties.** The quintic threefold in $\mathbb{P}^4$ is one of the best-known examples of a Calabi-Yau manifold. However, there are many contexts (including toric geometry) where *singular* Calabi-Yau varieties arise naturally. Hence we define Calabi-Yau as follows. Let $Y$ be a $d$-dimensional normal projective complex variety. Assume $\mathcal{O}_Y(K_Y) = \widehat{\Omega}^d_Y \simeq \mathcal{O}_Y$ and

$$H^1(Y, \mathcal{O}_Y) \simeq \cdots \simeq H^{d-1}(Y, \mathcal{O}_Y) \simeq \{0\}.$$

Then $Y$ is *canonical Calabi-Yau* if it has at worst canonical singularities. Furthermore, $Y$ is *minimal Calabi-Yau* if it has at worst $\mathbb{Q}$-factorial terminal singularities.

Since canonical singularities are Cohen-Macaulay and a Calabi-Yau has trivial canonical bundle, a singular Calabi-Yau is always Gorenstein. Thus its singularities are either Gorenstein canonical or $\mathbb{Q}$-factorial Gorenstein terminal.

To understand why these singularities are appropriate for Calabi-Yau varieties, recall that a singularity is *canonical* if there is a local resolution of singularities $f : \widehat{Y} \to Y$ such that

$$K_{\widehat{Y}} = f^*(K_Y) + \sum_i a_i E_i,$$

where the sum is over the exceptional divisors $E_i$ of $f$ and $a_i \geq 0$. If in addition we have $a_i > 0$ for all $i$, then the singularity is *terminal*. Thus having terminal singularities means that any resolution must change the canonical class. In particular, if $Y$ is a singular minimial Calabi-Yau, then its resolutions cease to be Calabi-Yau since the canonical class is no longer trivial.

As for canonical singularities, there are many situations (including threefolds and the Calabi-Yaus to be constructed below) where having canonical singularities implies the existence of a partial resolution $f : \widehat{Y} \to Y$ such that $K_{\widehat{Y}} = f^*(K_Y)$ (so the canonical class doesn't change) and $\widehat{Y}$ has $\mathbb{Q}$-factorial terminal singularities (so any further resolution changes the canonical class). Thus, if we start from a singular canonical Calabi-Yau $Y$, then $\widehat{Y}$ (if it exists) is a minimal Calabi-Yau which is as close as possible to being smooth while remaining Calabi-Yau. In Batyrev's terminology, $\widehat{Y}$ is called a *maximal projective crepant partial desingularization* of $Y$ (a MPCP-desingularization for short).

For more background on canonical and terminal singularities (including some nice examples), see [**Rei5**].

**Reflexive polytopes and Fano toric varieties.** Following Batyrev [**Bat1**], we say that a $n$-dimensional integral convex polytope $\Delta \subset M_{\mathbb{R}}$ is *reflexive* if it contains the origin as an interior point and if its dual polytope

$$\Delta^\circ = \{u \in N_{\mathbb{R}} : \langle m, u\rangle \geq -1 \text{ for all } m \in \Delta\} \subset N_{\mathbb{R}}$$

is also integral. Since $(\Delta^\circ)^\circ = \Delta$, reflexive polytopes always come in pairs. It also follows that each facet $F$ of a reflexive polytope $\Delta$ is defined by an equation $\langle u_F, m\rangle = -1$ for some $u_F \in N$. This easily implies that 0 is the only point of $M$ in the interior of $\Delta$. The polytopes $\Delta$ and $\Delta^\circ$ from (7.4) are an example of a reflexive polytope and its dual.

It is known (see [**Bat1**]) that in each dimension, there are only finitely many reflexive polytopes up to unimodular equivalence, and work is underway to classify *all* reflexive polytopes in dimension 4 (see [**Ska**]).

Given a reflexive polytope $\Delta$, we get a toric variety $X_\Delta$. Since the facets of $\Delta$ are given by $\langle u_F, m\rangle = -1$, it follows easily that the divisor on $X_\Delta$ determined by $\Delta$ is precisely the anticanonical divisor $-K = \sum_\rho D_\rho$. Also, the anticanonical divisor is ample, which tells us that $X_\Delta$ is a Fano variety, and it is Gorenstein since $K$ is obviously Cartier. Conversely, one can show that every Gorenstein Fano toric variety arises from a reflexive polytope (see [**Bat1**]).

**Calabi-Yau hypersurfaces.** Now suppose that $Y \subset X_\Delta$ is a general anticanonical hypersurface in a Gorenstein Fano toric variety coming from a reflexive polytope $\Delta$. Then the adjunction formula

$$\widehat{\Omega}_Y^{n-1} \simeq \widehat{\Omega}_{X_\Delta}^n(-K) \otimes \mathcal{O}_Y$$

shows that $\widehat{\Omega}_Y^{n-1} \simeq \mathcal{O}_Y$. Since $\mathcal{O}_{X_\Delta}(-Y) \simeq \mathcal{O}_{X_\Delta}(K) = \widehat{\Omega}_{X_\Delta}^n$, we also get an exact sequence

$$0 \longrightarrow \widehat{\Omega}_{X_\Delta}^n \longrightarrow \mathcal{O}_X \longrightarrow \mathcal{O}_Y \longrightarrow 0,$$

which makes it easy to show that $H^i(Y, \mathcal{O}_Y) = 0$ for $0 < i < n-1$ (this uses Serre-Grothendieck duality and the vanishing of $H^i(X_\Delta, \mathcal{O}_{X_\Delta})$ for $i > 0$). Finally, Bertini theorems show that $Y$ has at most Gorenstein toric singularities, which are known to be canonical. It follows that $Y$ is a canonical Calabi-Yau variety.

When $Y$ is singular, we would like to desingularize it as much as possible while remaining Calabi-Yau. As one might expect, toric geometry tells us exactly what to do. We've seen that the cone generators of the normal fan are the vertices of $\Delta^\circ$. We will consider projective simiplicial fans $\Sigma$ which refine the normal fan of $\Delta$ and whose cone generators satisfy

$$\Sigma(1) = N \cap \Delta^\circ - \{0\}. \tag{8.1}$$

To see why this condition is relevant, remember that one desingularizes $X_\Delta$ by subdividing cones into simplicial ones and then adding new cone generators until we get smooth cones. For example, adding a new cone generator $\rho$ gives a birational map of toric varieties $f : X \to X_\Delta$, and if $\rho$ lies in the cone over the facet $\langle u, m_F\rangle = -1$ of $\Delta^\circ$, then using the techniques of [**Rei5**], one obtains

$$K_X = f^*(K_{X_\Delta}) - (\langle \rho, m_F\rangle + 1)D, \tag{8.2}$$

where $D$ is the exceptional divisior (and is the divisor of $X$ corresponding to $\rho$). Thus, as long as we add new cone generators coming from $N \cap \Delta^\circ$, we don't change the canonical class. But once we've used up all of $N \cap \Delta^\circ - \{0\}$, as $\Sigma$ does, then any further $\rho$'s must lie outside of $\Delta^\circ$ (since reflexive implies 0 is the only integer interior point). Hence $\langle \rho, m_F \rangle < -1$, and from (8.2), it follows that we have terminal singularities. Since $\Sigma$ is simplicial, we see that $X_\Sigma$ is $\mathbb{Q}$-factorial. Thus we have the following theorem (see [**Bat1**] for details).

THEOREM 8.1. *Let $\Delta \subset M_\mathbb{R}$ be a reflexive polytope, and let $\Sigma$ be a fan in $N_\mathbb{R}$ which refines the normal fan of $\Delta$ and satisfies* (8.1). *Then:*

1. *The general anticanonical hypersurface $Y \subset X_\Delta$ is a canonical Calabi-Yau variety.*
2. *The general anticanonical hypersurface $\widehat{Y} \subset X_\Sigma$ is a minimal Calabi-Yau variety. Furthermore, $\widehat{Y}$ is a MPCP-desingularization of its image $Y \subset X_\Delta$ under the map $X_\Sigma \to X_\Delta$.*

When $\Delta$ is a 4-dimensional reflexive polytope, one can show that the MPCP-desingularization $\widehat{Y} \subset X_\Sigma$ is smooth Calabi-Yau threefold (see [**Bat1**]).

It follows that for each fan $\Sigma$ as in the statement of the theorem, we get a family of minimal Calabi-Yau varieties. In terms of what we considered in §7, these fans correspond to certain maximal cones in the secondary fan of the set $\mathcal{B} = N \cap \Delta^\circ - \{0\} \subset N$.

Of course, since $\Delta$ gives the family $\widehat{Y} \subset X_\Sigma$ of minimal Calabi-Yaus, we can use the reflexive polytope $\Delta^\circ$ to construct a "dual" family $\widehat{Y}^\circ \subset X_{\Sigma^\circ}$ of minimal Calabi-Yau varieties, where $\Sigma^\circ(1) = M \cap \Delta - \{0\}$. We will see in §10 that $\widehat{Y}^\circ \subset X_{\Sigma^\circ}$ is conjectured to be the mirror family of $\widehat{Y} \subset X_\Sigma$. Because of the previous paragraph, the situation is complicated by the multiple choices for $\Sigma$ and $\Sigma^\circ$. This is related to the idea of *multiple mirrors*, also to be studied in §10.

There are also some nice formulas for Hodge numbers. The MPCP-desingularization $\widehat{Y} \subset X_\Sigma$ is simplicial, so that $H^*(\widehat{Y})$ has a pure Hodge structure. But we can't apply the results of §6 directly since $\widehat{Y}$ may fail to be ample (this happens, for example, when $X_\Delta = \mathbb{P}(1,1,2,2,2)$). However, we have the following results from [**Bat1**].

THEOREM 8.2. *Let $\Delta$ be a reflexive polytope of dimension $n \geq 4$ and let $\Sigma$ be a fan as in Theorem 8.1. If $\widehat{Y} \subset X_\Sigma$ is a general anticanonical hypersurface, then*

$$h^{n-2,1}(\widehat{Y}) = l(\Delta) - n - 1 - \sum_{\operatorname{codim}\Gamma=1} l^*(\Gamma) + \sum_{\operatorname{codim}\Gamma=2} l^*(\Gamma)l^*(\Gamma^\circ),$$

*where $\Gamma$ is a face of $\Delta$, $\Gamma^\circ$ is the corresponding dual face of $\Delta^\circ$ and, as in §6, $l(\Gamma)$ (resp. $l^*(\Gamma)$) is the number of integer (resp. interior integer) points in $\Gamma$. Also,*

$$h^{1,1}(\widehat{Y}) = l(\Delta^\circ) - n - 1 - \sum_{\operatorname{codim}\Gamma^\circ=1} l^*(\Gamma^\circ) + \sum_{\operatorname{codim}\Gamma^\circ=2} l^*(\Gamma^\circ)l^*(\Gamma),$$

*where now $\Gamma^\circ$ is a face of $\Delta^\circ$ and $\Gamma$ is the dual face.*

The remarkable symmetry evident between the formulas for $h^{1,1}$ and $h^{n-2,1}$ will be important when we discuss mirror symmetry in §10. A different approach to the study of $h^{1,1}$ can be found in [**Roa1**].

Although the above construction seems completely natural, it can take some thought to find the reflexive polytope. For example, a degree 7 hypersurface in $\mathbb{P}(1,1,1,2,2)$ is Calabi-Yau, yet the simplex giving $\mathbb{P}(1,1,1,2,2)$ in not reflexive, nor is the hypersurface Cartier. Here, the reflexive polytope $\Delta$ is (up to translation) the Newton polytope of all monomials of degree 7, and the toric variety $X_\Delta$ is a blow-up of $\mathbb{P}(1,1,1,2,2)$. Some early evidence for mirror symmetry came from a list of 7555 Calabi-Yau hypersurfaces in weighted projective spaces. The list exhibited an incomplete duality which was only fully understood after the definition of reflexive polytope (see [**CdK**]).

**Calabi-Yau complete intersections.** We next generalize the above construction to create families of Calabi-Yau complete intersections. In §6, we discussed some basic facts about complete intersections in toric varieties, though the special features of the Calabi-Yau case are due to [**LBor**], with subsequent work by [**BB1–3**]. Details of what follows can be found in these references.

The basic way to describe a Calabi-Yau complete intersection in a toric variety is through the idea of a nef-partition. Suppose that $\Delta$ is a $n$-dimensional reflexive polytope, so that $\mathrm{Vert}(\Delta^\circ)$ is the set of cone generators of the fan of $X = X_\Delta$. Then a *nef-partition* is a disjoint union

$$\mathrm{Vert}(\Delta^\circ) = E_1 \cup \cdots \cup E_r \tag{8.3}$$

such that the divisors $D_j = \sum_{\rho\in E_j} D_\rho$ are Cartier and nef (i.e., generated by global sections). Equivalently, a nef-partition is a partition of the anticanonical divisor $-K = \sum_\rho D_\rho$ into a sum of nef Cartier divisors. Furthermore, if $\Delta_j$ is the polytope associated to the divisor $D_j$, then it follows that

$$\Delta = \Delta_1 + \cdots + \Delta_r. \tag{8.4}$$

We will always assume that $n - r \geq 1$.

In this situation, let $Y_j \in |D_j|$ be generic. Then the complete intersection

$$V = Y_1 \cap \cdots \cap Y_r \subset X_\Delta$$

is a canonical Calabi-Yau. Furthermore, as in Theorem 8.1, suppose that $\Sigma$ is a fan refining the normal fan of $\Delta$ with the property that $\Sigma(1) = N \cap \Delta^\circ - \{0\}$. We can regard $D_j$ as divisors on $X_\Sigma$, and if $\widehat{Y}_j$ is a general member of $|D_j|$, then, as shown in [**BB3**], the complete intersection

$$\widehat{V} = \widehat{Y}_1 \cap \cdots \cap \widehat{Y}_r \subset X_\Sigma \tag{8.5}$$

is a minimal Calabi-Yau variety which is a MPCP-desingularization of the corresponding $V \subset X_\Delta$. There are formulas similar to those of Theorem 8.2 for the Hodge numbers $h^{n-r-1,1}(\widehat{V})$ and $h^{1,1}(\widehat{V})$.

In §10, we will see that the nef-partition giving the family of Calabi-Yau complete intersections $\widehat{V} \subset X_\Sigma$ naturally determines a "dual" family of Calabi-Yau complete intersections. This construction is used in mirror symmetry.

## §9. Resultants, discriminants and hypergeometric functions

This section will discuss briefly those aspects of resultants, discriminants and hypergeometric functions which pertain to toric varieties. Basic references are the book [**GKZ1**] and, for hypergeometric functions, the papers [**GKZ2**] and [**GKZ3**].

**$\mathcal{A}$-Resultants.** Given a finite set of exponents $\mathcal{A} \subset \mathbb{Z}^n$, we get the vector space $L(\mathcal{A})$ of Laurent polynomials

$$f = \sum_{m \in \mathcal{A}} c_m \mathbf{t}^m.$$

For simplicity, we will assume that $\mathcal{A}$ is affinely independent in $\mathbb{Z}^n$. Then let $\nabla_{\mathcal{A}} \subset L(\mathcal{A})^{n+1}$ be the Zariski closure of those $(n+1)$-tuples of polynomials $(f_0, \ldots, f_n) \in L(\mathcal{A})^{n+1}$ where there is $\mathbf{t} \in (\mathbb{C}^*)^n$ such that $f_0(\mathbf{t}) = \cdots = f_n(\mathbf{t}) = 0$. One can show that $\nabla_{\mathcal{A}}$ is a hypersurface in $L(\mathcal{A})^{n+1}$ and hence is defined by a polynomial $R_{\mathcal{A}} = 0$. Thus $R_{\mathcal{A}}$ is a polynomial in the coefficients $c_{i,m}$ of $f_i$, and in fact $R_{\mathcal{A}} \in \mathbb{Z}[c_{i,m}]$. This polynomial is the *$\mathcal{A}$-resultant*.

For example, if $\mathcal{A} = \{0, 1, \ldots, d\} \subset \mathbb{Z}$, then $R_{\mathcal{A}}$ is the usual resultant of two polynomials $f_0, f_1$ of degree $d$ in one variable. Many other examples can be found in [**GKZ3**] and [**Stu4**]. In general, the $\mathcal{A}$-resultant is sometimes called a *sparse resultant* since only the exponents in $\mathcal{A}$ are used. It is also possible to allow the polynomials $f_0, \ldots, f_n$ to have different exponents—this leads to what is known as the *$(\mathcal{A}_0, \ldots, \mathcal{A}_n)$-resultant* or *mixed sparse resultant*. Algorithms for computing sparse resultants are described in [**CE**] and [**Stu4**].

**Chow forms.** We can think of the $\mathcal{A}$-resultant in toric terms as follows. If $\mathcal{A}$ has $\ell$ elements, then we get the (possibly non-normal) toric variety $X_{\mathcal{A}} \subset \mathbb{P}^{\ell-1}$ as in (5.6). This projective variety has a *Chow form* $R_{X_{\mathcal{A}}}$, which is a polynomial in the Plücker coordinates $[m_0, \ldots, m_n]$ that vanishes precisely on those subspaces $L \in G(\ell, n-\ell-1)$ where $\mathbb{P}(L) \cap X_{\mathcal{A}} \neq \emptyset$.

It is customary to rewrite the Chow form as follows. If we regard $L$ as being defined by $n+1$ linear forms, then the coefficients of these forms give a $(n+1) \times \ell$ matrix $(c_{i,m})$, where we use elements $m \in \mathcal{A}$ to label the coordinates of $\mathbb{P}^{\ell-1}$. Then, replacing the Plücker coordinate $[m_0, \ldots, m_n]$ by the bracket polynomial $[m_0, \ldots, m_n] = \det(c_{i,m_j})$, we get a polynomial in the $c_{i,m}$ which is easily seen to coincide with the $\mathcal{A}$-resultant $R_{\mathcal{A}}$.

**Chow polytopes and secondary polytopes.** Using the Chow form, we can create the *Chow polytope* as follows. If we express $R_{X_{\mathcal{A}}}$ as a polynomial in the brackets $[\sigma] = [m_0, \ldots, m_n]$, then a term $T$ in $R_{X_{\mathcal{A}}}$ is of the form $T = c\Pi_\sigma [\sigma]^{a_\sigma}$, $c \neq 0$, and we assign $T$ the weight

$$\phi_T = \sum_{\sigma} \sum_{m \in [\sigma]} a_\sigma e_m \in \bigoplus_{m \in \mathcal{A}} \mathbb{R} \cdot e_m.$$

The Chow polytope of $X_{\mathcal{A}}$ is defined to be the convex hull of these points $\phi_T$.

The remarkable fact is that the Chow polytope of $X_{\mathcal{A}}$ is *exactly* the secondary polyotope $\Sigma(\mathcal{A})$ from §7 (which is the convex hull of the points $\phi_T$ defined in (7.2)). This is proved in [**GKZ1**, Ch. 9]. We know that the vertices of $\Sigma(\mathcal{A})$ correspond to regular triangulations of $\mathcal{A}$, and given such a triangulation, there is a precise formula for the corresponding term of $R_{X_{\mathcal{A}}}$. One can also formulate this in terms of certain toric degenerations of $X_{\mathcal{A}}$ (see [**KSZ1**]). Some explicit examples can be found in [**Stu4**], and [**Stu3**] has similar results concerning the $(\mathcal{A}_0, \ldots, \mathcal{A}_n)$-resultant.

**$\mathcal{A}$-Discriminants.** Given $X_{\mathcal{A}} \subset \mathbb{P}^{\ell-1}$ as above, let $\nabla_{\mathcal{A}} \subset L(\mathcal{A})$ be the Zariski closure of those Laurent polynomials $f$ for which the affine hypersurface $Z_f \subset (\mathbb{C}^*)^n$ is singular. Then $\nabla_{\mathcal{A}}$ is the affine cone over the dual variety $X_{\mathcal{A}}^\vee$. If $\nabla_{\mathcal{A}}$ has codimension 1, we define the *$\mathcal{A}$-discriminant* $\Delta_{\mathcal{A}}$ to be the defining equation of $\nabla_{\mathcal{A}}$ (or, equivalently, of $X_{\mathcal{A}}^{\vee}$). We set $\Delta_{\mathcal{A}} = 1$ in all other cases. Note that $\Delta_{\mathcal{A}}$ is a polynomial in the coefficients $c_m$ of $f = \sum_{m \in \mathcal{A}} c_m \mathbf{t}^m$, and, in fact, $\Delta_{\mathcal{A}} \in \mathbb{Z}[c_m]$.

EXAMPLES. 1. When $\mathcal{A}$ consists of all non-negative exponent vectors of degree at most $d$, $\Delta_{\mathcal{A}}$ is the usual discriminant of a homogeneous polynomial of degree $d$. More precisely, if $F$ is the homogenization of $f \in L(\mathcal{A})$, then $\Delta_{\mathcal{A}}(f) = \mathrm{Disc}(F)$.

2. Consider all linear combinations of the monomials $1, x, \ldots, x^p, y, yx, \ldots, yx^q$. Then $\Delta_{\mathcal{A}}$ is the usual resultant $R(f,g)$ of polynomials of degrees $p$ and $q$ respectively. This is an example of the Cayley trick and can be used to express any $(\mathcal{A}_0, \ldots, \mathcal{A}_n)$-resultant as a discriminant (see [**GKZ**, Prop. 1.7 of Ch. 9]).

3. An example where $\Delta_{\mathcal{A}} = 1$ is given by the Segre embedding $\mathbb{P}^1 \times \mathbb{P}^2 \to \mathbb{P}^3$. More generally, same is true for the Segre embedding of $\mathbb{P}^n \times \mathbb{P}^m$ whenever $n \neq m$.

**Principal $\mathcal{A}$-determinants.** Given $f \in L(\mathcal{A})$, the *principal $\mathcal{A}$-determinant* $E_{\mathcal{A}}(f)$ is defined to be the resultant

$$E_{\mathcal{A}}(f) = R_{\mathcal{A}}(f, t_1 \partial f/\partial t_1, \ldots, t_n \partial f/\partial t_n).$$

Note that the resultant makes sense since $t_i \partial f/\partial t_i \in L(\mathcal{A})$.

The nicest fact about $E_{\mathcal{A}}$ is that its Newton polytope is *precisely* the secondary polytope $\Sigma(\mathcal{A})$ (see [**GKZ**, Thm. 1.4 of Ch. 10]). One way to understand this result is to observe that $E_{\mathcal{A}}$ is the polynomial obtained from the Chow form $R_{X_{\mathcal{A}}}$ under the specialization which sends the bracket $[\sigma]$ to $\pm\mathrm{Vol}(\sigma)\Pi_{m\in\sigma} c_m$ (where $\pm\mathrm{Vol}(\sigma)$ is the signed volume of the simplex spanned by $m \in \sigma$).

When $X_{\mathcal{A}}$ is smooth, there is an especially nice relation between $\Delta_{\mathcal{A}}$ and $E_{\mathcal{A}}$.

THEOREM 9.1. *If $X_{\mathcal{A}} \subset \mathbb{P}^{\ell-1}$ is smooth and $Q = \mathrm{Conv}(\mathcal{A})$, then*

$$E_{\mathcal{A}}(f) = \pm \prod\nolimits_{\Gamma \subset Q} \Delta_{\mathcal{A} \cap \Gamma}(f_{|\Gamma}),$$

*where $\Gamma \subset Q$ are the nonempty faces and $f_{|\Gamma} = \sum_{m \in \mathcal{A} \cap \Gamma} c_m \mathbf{t}^m$.*

When $X_{\mathcal{A}}$ is singular, there is a more complicated but very elegant formula relating $\Delta_{\mathcal{A}}$ and $E_{\mathcal{A}}$—see [**GKZ**, Ch. 10] for the details.

Another way to see the relation between $\Delta_{\mathcal{A}}$ and $E_{\mathcal{A}}$ is to study the projective hypersurface $Y_f \subset X_{\mathcal{A}}$ defined by $f \in L(\mathcal{A})$. Then we have the following result.

PROPOSITION 9.2. *Given $X_{\mathcal{A}} \subset \mathbb{P}^{\ell-1}$, we have:*

1. *If $X_{\mathcal{A}}$ is smooth and $\Delta_{\mathcal{A}}$ is nonconstant, then $Y_f \subset X_{\mathcal{A}}$ is smooth if and only if $\Delta_{\mathcal{A}}(f) \neq 0$.*
2. *If $E_{\mathcal{A}}$ is nonconstant, then $Y_f \subset X_{\mathcal{A}}$ is nondegenerate (as defined in §6) if and only if $E_{\mathcal{A}}(f) \neq 0$.*

It is likely that the first part of this proposition remains true in the simplicial case, so that $Y_f$ should be quasi-smooth in the sense of §6 if and only if $\Delta_{\mathcal{A}}(f) \neq 0$. However, no proof has appeared in print.

As a final comment, note that resultants can be computed in terms of discriminants (using the Cayley trick) and vice versa (using $E_{\mathcal{A}}$ and Theorem 9.1 in the smooth case). Historically, $\mathcal{A}$-discriminants were first discovered in the context of $\mathcal{A}$-hypergeometric functions and led to the definition of $\mathcal{A}$-resultants. In practice, resultants are easier to compute (see, for example, [**Stu4**]).

**$\mathcal{A}$-Hypergeometric functions.** To define the $\mathcal{A}$-hypergeometric equations, consider $\mathcal{A} = \{m_1, \ldots, m_\ell\} \subset \mathbb{Z}^n$. For simplicity, we will assume that $\mathbb{Z}^n$ is affinely generated by $\mathcal{A}$, which means that $\mathbb{Z}^{n+1}$ is generated by $\mathcal{A} \times \{1\}$. Then, for each $\mathbf{a} = (a_1, \ldots, a_n) \in \mathbb{Z}^\ell$ satisfying the conditions

$$\sum_{i=1}^{\ell} a_i m_i = 0, \quad \sum_{i=1}^{\ell} a_i = 0, \tag{9.1}$$

consider the differential operator

$$\Box_{\mathbf{a}} = \prod_{a_j>0} \left(\frac{\partial}{\partial c_j}\right)^{a_j} - \prod_{a_j<0} \left(\frac{\partial}{\partial c_j}\right)^{-a_j} \tag{9.2}$$

where we regard $c_1, \ldots, c_\ell$ as variables on $L(\mathcal{A}) = \mathbb{C}^\ell$. We will also consider the differential operators

$$\mathcal{Z}_0 = \sum_{j=1}^{\ell} c_j \frac{\partial}{\partial c_j}, \quad \mathcal{Z}_i = \sum_{j=1}^{\ell} m_{ji} c_j \frac{\partial}{\partial c_j}, \quad i = 1, \ldots, n,$$

where $m_j = (m_{j1}, \ldots, m_{jn}) \in \mathbb{Z}^n$ are the elements of $\mathcal{A}$. Then, for a function $\Phi(c_1, \ldots, c_\ell)$ on $\mathbb{C}^\ell$, the *$\mathcal{A}$-hypergeometric system with exponents* $\beta_0, \ldots, \beta_n \in \mathbb{C}$ is the system of differential equations

$$\begin{aligned} \Box_{\mathbf{a}} \Phi &= 0, \quad \text{for all } \mathbf{a} \text{ satisfying (9.1)} \\ \mathcal{Z}_i \Phi &= \beta_i \Phi, \quad \text{for all } i = 0, \ldots, n. \end{aligned} \tag{9.3}$$

This system is holonomic, so that its solutions form a locally constant sheaf outside a hypersurface in $L(\mathcal{A}) = \mathbb{C}^\ell$. Two especially nice facts are first, that the generic number of linearly independent solutions of (9.3) is the normalized volume of the polytope $\mathrm{Conv}(\mathcal{A})$, and second, that $(c_1, \ldots, c_\ell)$ is generic if and only if the principal $\mathcal{A}$-determinant is nonvanishing, i.e., if and only if $E_{\mathcal{A}}(f) \neq 0$ for $f = \sum_{i=0} c_i \mathbf{t}^{m_i}$. Proofs can be found in [**GKZ3**].

A more direct connection with toric varieties can be seen by considering the symbols of the operators $\Box_{\mathbf{a}}$. Using $x_1, \ldots, x_\ell$ as variables on the dual of $L(\mathcal{A}) = \mathbb{C}^\ell$, the symbol of $\Box_{\mathbf{a}}$ is its Fourier transform $\check{\Box}_{\mathbf{a}}$, which is obtained by replacing $\partial/\partial c_j$ in (9.2) with $x_j$. If we write $\mathbf{a} = \mathbf{a}^+ - \mathbf{a}^-$ as in Lemma 5.5, then $\check{\Box}_{\mathbf{a}}$ becomes $x^{\mathbf{a}^+} - x^{\mathbf{a}^-}$. Since we do this for all $\mathbf{a}$ satisfying (9.1), it follows from (7.5) that we get the toric ideal $I_{\mathcal{A}}$. Hence the Fourier transform of (9.3) is supported on the toric variety $X_{\mathcal{A}}$. This toric variety plays an important role in the proofs of many results about $\mathcal{A}$-hypergeometric functions. Also, although (9.3) involves infinitely many equations $\Box_{\mathbf{a}}\Phi = 0$, one can always reduce to a finite number using a Gröbner basis for the toric ideal $I_{\mathcal{A}}$ (see [**HLY1**]).

There are several ways to write down solutions to the $\mathcal{A}$-hypergeometric system. For us, the most interesting method involves the periods of the affine hypersurface $Z_f \subset (\mathbb{C}^*)^n$ defined by $f = \sum_{i=0} c_i \mathbf{t}^{m_i}$. We will assume that $\mathcal{A} = \Delta \cap \mathbb{Z}^n$, where $\Delta \subset \mathbb{R}^n$ is an integer polytope. As in §5, we have the ring $S_\Delta \subset \mathbb{C}[t_0, t_1, \ldots, t_n]$. In [**Bat4**, Thm. 14.2], the following result is proved.

PROPOSITION 9.3. *Let $\mathcal{A} = \Delta \cap \mathbb{Z}^n$ and $t_0^k \mathbf{t}^m \in S_\Delta$ (so that $m \in k\Delta \cap \mathbb{Z}^n$). Then, for a $n$-cycle $\gamma \in H_n((\mathbb{C}^*)^n - Z_f)$ and $f = \sum_{i=0}^{\ell} c_i \mathbf{t}^{m_i} \in L(\mathcal{A})$ nondegenerate, the period integral*

$$\Pi(c_1, \ldots, c_\ell) = \Pi(f) = \int_\gamma \frac{\mathbf{t}^m}{f^k} \frac{dt_1}{t_1} \wedge \cdots \wedge \frac{dt_n}{t_n}$$

*satisfies the $\mathcal{A}$-hypergeometric system* (9.3) *for exponents* $(\beta_0, \ldots, \beta_n) = (-k, -m)$.

By Proposition 9.2, $Z_f$ is nondegenerate if and only if $E_\mathcal{A}(f) \neq 0$. This helps explain why the singular set of the $\mathcal{A}$-hypergeometric system is defined by $E_\mathcal{A} = 0$. The integral in Proposition 9.3 is an example of an *Euler integral*, which is the main object of study in [**GKZ2**].

It is also possible to write down series solutions of the $\mathcal{A}$-hypergeometric system. Formally, these solutions look like

$$\Phi_\gamma(c_1, \ldots, c_\ell) = \sum_{\mathbf{a}} \frac{\mathbf{c}^{\gamma+\mathbf{a}}}{\Pi_{j=1}^{\ell} \Gamma(\gamma_j + a_j + 1)}, \tag{9.4}$$

where $\gamma \in \mathbb{C}^\ell$, $\Gamma$ is the usual $\Gamma$-function, and the sum is over all $\mathbf{a}$ satisfying (9.1). By restricting to certain choices of $\gamma$ and $\beta_0, \ldots, \beta_n$, one can show that the above series converge locally when $(c_1, \ldots, c_\ell)$ lies in certain regions of $\mathbb{C}^\ell$ determined by a regular triangulation $T$ of $\mathrm{Conv}(\mathcal{A})$. This is described as follows. We saw in §7 that a regular triangulation gives a maximal cone of $A_\mathcal{A}$ in the secondary fan of $\mathcal{A}$. Using the exact sequence (7.3), we get a cone $C(T) \subset \bigoplus_{m \in \mathcal{A}} \mathbb{R} \cdot e_m \simeq \mathbb{R}^\ell$. Then one can prove that the series (9.4) converges locally for those $(c_1, \ldots, c_\ell)$ such that $-\log(c_1, \ldots, c_\ell)$ (as defined in (6.1)) lies in a suitable translate of $C(T)$. Furthermore, one gets all solutions of (9.3) in this way. See [**GKZ3**] for details.

As mentioned earlier, the Newton polytope of the principal $\mathcal{A}$-determinant $E_\mathcal{A}$ is the secondary polytope (with vertices corresponding to regular triangulations). In terms of (6.1), the regions of nice convergence of the series (9.4) correspond (up to sign) to unbounded components of $\mathbb{R}^\ell - \log(\{E_\mathcal{A} = 0\})$. This again illustrates why the singular set of $\mathcal{A}$-hypergeometric system is given by $E_\mathcal{A} = 0$.

Finally, Proposition 9.3 shows that the system (9.3) is closely tied to the torus. Roughly speaking, the equations $\mathcal{Z}_i \Pi = \beta_i \Pi$ express the invariance of the period integral $\Pi$ for $Z_f \subset (\mathbb{C}^*)^n$ under the infinitesimal automorphisms of $(\mathbb{C}^*)^n$. If we were to formulate a similar period integral for $Y_f \subset X_\Delta$, we would need to add further equations to (9.3) to account for automorphisms of $X_\Delta$. This *extended $\mathcal{A}$-hypergeometric system* is described in [**HLY1**].

## §10. Mirror symmetry

In 1991, a group of physicists made some startling enumerative predictions concerning rational curves on a quintic threefold (see [**CdGP**]). The basic idea was that a hard computation on the quintic threefold became easier by working on its "mirror". The first mirror constructions involved finite quotients of weighted projective spaces which, as Roan pointed out in 1992, can be described naturally using toric methods (see [**Roa2**]). After Batyrev's 1993 introduction of reflexive polytopes (see [**Bat1**]), toric geometry has become a basic tool of mirror symmetry.

We will not discuss mirror symmetry in general—the reader should consult [**Yau**], [**Mor3**] or [**CK**] for an introduction to this fascinating topic. Rather, we will concentrate on describing how toric geometry is used in mirror symmetry. This is a very active field, and the references given below are far from complete.

**Complex and Kähler moduli.** The complex moduli of varieties have been studied for many years. The infinitesimal deformations of a compact complex manifold $Y$ of dimension $d$ are given by $H^1(Y, \Theta_Y)$, which in nice cases is the tangent space to the complex moduli space. When $Y$ is Calabi-Yau, the isomorphism $\Omega_Y^d \simeq \mathcal{O}_Y$ implies $\Theta_Y \simeq \Omega_Y^{d-1}$, so that $H^1(Y, \Theta_Y) \simeq H^{d-1,1}(Y)$.

The intense study of Kähler moduli is more recent. The basic definition is as follows: given a Kähler manifold $Y$, the *complexified Kähler cone* is the cone

$$K_{\mathbb{C}}(Y) = \{B + iJ \in H^2(Y, \mathbb{C}) : J \text{ is Kähler}\}/\operatorname{Im} H^2(Y, \mathbb{Z}), \tag{10.1}$$

and the *complexified Kähler moduli space* is the quotient $K_{\mathbb{C}}(Y)/\operatorname{Aut}(Y)$. The exact structure of this quotient space depends on how $\operatorname{Aut}(Y)$ acts on the ordinary Kähler cone of $Y$ (as described in §3). In the Calabi-Yau case, it is conjectured that the Kähler cone is polyhedral modulo the action of $\operatorname{Aut}(Y)$ (see [**Mor1**] for a precise statement). This would imply the existence of a semi-toric compactification of the complexified Kähler moduli space. Assuming that $\operatorname{Aut}(Y)$ acts discretely, the tangent space to the Kähler moduli space (we will usually drop the adjective "complexified") is $H^{1,1}(Y)$.

So far, we've assumed that $Y$ is smooth, and in dimension 3, this is sufficient. Higher dimensional generalizations of mirror symmetry lead naturally to singular varieties, and one can define both complex and Kähler moduli when $Y$ is a minimal Calabi-Yau variety as in §8.

**The naive idea of mirror symmetry.** A Calabi-Yau threefold $Y$ together with a complexified Kähler class $\omega = B + iJ$ determine a $N = 2$ superconformal field theory (SCFT), and mirror symmetry suggests that there should be another Calabi-Yau $Y^\circ$ with class $\omega^\circ$ which in some sense interchanges the complex and Kähler structures of $Y$ and $Y^\circ$ but still gives the same $N = 2$ SCFT.

This interchange of complex and Kähler moduli in particular gives isomorphisms of the corresponding moduli spaces (the "mirror map") and hence induces isomorphisms

$$H^{d-1,1}(Y) \simeq H^{1,1}(Y^\circ), \quad H^{1,1}(Y) \simeq H^{d-1,1}(Y^\circ) \tag{10.2}$$

between their tangent spaces. But mirror symmetry is much more than these isomorphisms, for an isomorphism of $N = 2$ SCFTs also implies that certain trilinear functions on $H^{1,1}(Y)$ and $H^{d-1,1}(Y^\circ)$ should agree after a change of variables given by the mirror map. These trilinear functions are called *3-point functions* or *correlation functions* in the physics literature and are related to enumerative geometry and quantum cohomology (for $H^{1,1}(Y)$) and Hodge theory (for $H^{d-1,1}(Y^\circ)$). Unfortunately, these topics are beyond the scope of this survey.

The physical theories involved in mirror symmetry have yet to be defined rigorously, but there are "mathematical mirror symmetry conjectures" which capture the mathematically interesting consequences. This can be done in all dimensions, and versions of these conjectures are in [**Mor1–3**], [**Giv2**], [**Kont**] and [**Ver**]. Mirror symmetry for holomorphically symplectic manifolds is proved in [**Ver**], and for complete intersections in projective space, a version of mirror symmetry involving certain "virtual" numbers of rational curves of degree $d$ is proved in [**Giv1**]. Other than these exceptions, mathematical mirror symmetry is still largely conjectural. Hence, in what follows, our term "mirror" really means "conjectural mirror".

**Mirror symmetry for toric hypersurfaces.** In [**Bat1**], Batyrev used reflexive polytopes to construct mirrors for Calabi-Yau toric hypersurfaces. As in §8, the process starts with an $n$-dimensional reflexive polytope $\Delta \subset M_{\mathbb{R}}$, which by definition determines the anticanonical divisor on $X_\Delta$. This gives canonical Calabi-Yau hypersurfaces $Y \subset X_\Delta$. Then, if $\Sigma$ is a projective simplicial fan refining the normal fan of $\Delta$ and satisfying $\Sigma(1) = N \cap \Delta^\circ - \{0\}$, we obtain a family of minimal Calabi-Yau hypersurfaces $\widehat{Y} \subset X_\Sigma$.

To construct the mirror of this family, we repeat the above procedure using the dual polytope $\Delta^\circ$. Thus, for a suitable fan $\Sigma^\circ$ in $M_{\mathbb{R}}$ with $\Sigma^\circ(1) = M \cap \Delta - \{0\}$, the anticanonical divisor determines a family $\widehat{Y}^\circ \subset X_{\Sigma^\circ}$ of minimal Calabi-Yaus. This is conjectured to be the mirror of $\widehat{Y} \subset X_\Sigma$.

Evidence for the mirror relation between $\widehat{Y}$ and $\widehat{Y}^\circ$ comes from Theorem 8.2, which shows that

$$H^{n-2,1}(\widehat{Y}) \simeq H^{1,1}(\widehat{Y}^\circ), \quad H^{1,1}(\widehat{Y}) \simeq H^{n-2,1}(\widehat{Y}^\circ),$$

as predicted by (10.2) (since $\widehat{Y}$ and $\widehat{Y}^\circ$ have dimension $n-1$). Also, $H^{n-2,1}(\widehat{Y})$ has a subspace $H^{n-2,1}_{\text{poly}}(\widehat{Y})$ consisting of deformations of $\widehat{Y}$ obtained by varying its defining equation in $X_\Sigma$, and similarly, $H^{1,1}(\widehat{Y})$ has a subspace $H^{1,1}_{\text{toric}}(\widehat{Y})$ consisting of restrictions of $(1,1)$-classes on $X_\Sigma$. The *monomial-divisor mirror map* of [**AGM2**] gives natural isomorphisms

$$H^{n-2,1}_{\text{poly}}(\widehat{Y}) \simeq H^{1,1}_{\text{toric}}(\widehat{Y}^\circ), \quad H^{1,1}_{\text{toric}}(\widehat{Y}) \simeq H^{n-2,1}_{\text{poly}}(\widehat{Y}^\circ).$$

EXAMPLE. To compute the mirror of the quintic threefold $Y \subset \mathbb{P}^4$, note that $Y$ is an anticanonical hypersurface, and the corresponding polytope $\Delta \subset M_{\mathbb{R}} \simeq \mathbb{Z}^4$ is reflexive. Hence the mirror should be determined by the dual polytope $\Delta^\circ \subset N_{\mathbb{R}}$.

The toric variety $X_{\Delta^\circ}$ comes from the normal fan of $\Delta^\circ$. The cone generators of this fan are the vertices of $\Delta$, which are

$$(-1,-1,-1,-1),\ (4,-1,-1,-1),\ (-1,4,-1,-1),\ (-1,-1,4,-1),\ (-1,-1,-1,4).$$

These generate a sublattice $M' \subset M$ of index 125, and the quotient $M/M'$ is

$$H = \big\{(a_0,a_1,a_2,a_3,a_4) \in (\mathbb{Z}_5)^5 : \textstyle\sum_{i=0}^4 a_i \equiv 0 \bmod 5\big\}/\mathbb{Z}_5$$

where $\mathbb{Z}_5 \subset (\mathbb{Z}_5)^5$ is the diagonal subgroup. Using the lattice $M'$, the normal fan gives $\mathbb{P}^4$, so by [**Oda1**, Cor. 1.16], $X_{\Delta^\circ}$ is the quotient of $\mathbb{P}^4$ by $M/M' \simeq H$, where $[a_0,a_1,a_2,a_3,a_4] \in H$ acts on $\mathbb{P}^4$ by the map $(x_0,x_1,x_2,x_3,x_4) \mapsto (\zeta^{a_0}x_0, \zeta^{a_1}x_1, \zeta^{a_2}x_2, \zeta^{a_3}x_3, \zeta^{a_4}x_4)$ for $\zeta = \exp(2\pi i/5)$.

The homogeneous coordinate ring $S$ of $X_{\Delta^\circ}$ from §2 is the polynomial ring $S = \mathbb{C}[x_0,x_1,x_2,x_3,x_4]$, which is graded by the Chow group $A_3(X_{\Delta^\circ})$. From (2.2), we get

$$0 \longrightarrow \mathbb{Z}^4 \xrightarrow{\ \alpha\ } \mathbb{Z}^5 \xrightarrow{\ \beta\ } \mathbb{Z} \oplus H \longrightarrow 0,$$

where $\alpha$ is as usual and $\beta$ is given by

$$\beta(a_0,a_1,a_2,a_3,a_4) = \big(\textstyle\sum_{i=0}^4 a_i, [-a_1-a_2-a_3-a_4, a_1, a_2, a_3, a_4]\big) \in \mathbb{Z} \oplus H.$$

Thus $A_3(X_{\Delta^\circ}) \simeq \mathbb{Z} \oplus H$ and the grading on $S$ is obtained by letting a monomial $x_0^{a_0}x_1^{a_1}x_2^{a_2}x_3^{a_3}x_4^{a_4}$ have degree $\beta(a_0,a_1,a_2,a_3,a_4) \in \mathbb{Z} \oplus H$.

It follows that the anticanonical class has degree $\beta(1,1,1,1,1) = (5,0)$, and the only monomials in $S$ of this degree are $x_i^5$ and $x_0x_1x_2x_3x_4$ (this can be seen directly or using the isomorphism $S_{(5,0)} \simeq L(\Delta^\circ)$ from (6.2)). Furthermore, the automorphisms of $X_{\Delta^\circ}$ given by its torus show that any anticanonical hypersurface is isomorphic to one defined by an equation of the form

$$x_0^5 + x_1^5 + x_2^5 + x_3^5 + x_4^5 + 5\psi\, x_0x_1x_2x_3x_4 = 0.$$

This gives a 1-dimensional family of hypersurfaces $Y^\circ \subset X_{\Delta^\circ}$. In concrete terms, $Y^\circ$ is the quotient by $H$ of the hypersurface in $\mathbb{P}^4$ defined by the above equation.

Finally, to get the mirror, we pick a projective simplicial fan $\Sigma^\circ$ in $M_\mathbb{R}$ refining the normal fan of $\Delta^\circ$ such that $\Sigma^\circ(1) = M \cap \Delta - \{0\}$. Then the mirror family of the quintic threefold $Y$ is given by the hypersurfaces $\widehat{Y}^\circ \subset X_{\Sigma^\circ}$ which are the proper transforms of $Y^\circ \subset X_{\Delta^\circ}$. Since $Y \subset \mathbb{P}^4$ has 1-dimensional Kähler moduli (i.e., $h^{1,1} = 1$), we expect its mirror $\widehat{Y}^\circ$ to have 1-dimensional complex moduli. Note that $M \cap \Delta - \{0\}$ has *lots* of points besides the vertices, and hence there are *many* choices for $\Sigma^\circ$. A specific choice for $\Sigma^\circ$ is described in the appendix to [**Mor3**].

In general, once the mirror family has been found, one gets enumerative predictions on $\widehat{Y}$ by computing the *mirror map* and the *Yukawa coupling* on $\widehat{Y}^\circ$. Without getting into precise definitions, these objects are closely related to period integrals and Picard-Fuchs equations on a toric hypersurface, and the computations involve series expansions about certain "special" boundary points in the complex moduli space. These "special" points have *maximal unipotent monodromy* and their existence is suggested by the structure of the Kähler moduli space of the mirror. See [**Mor2-3**] for further details.

In many cases, the period integrals and Picard-Fuchs equations can be computed directly (see [**Mor4**] for some examples), though Proposition 9.3 suggests a connection with $\mathcal{A}$-hypergeometric equations. This has been studied carefully in [**HLY1-2**]. Further references for both of these methods can be found in [**Mor2**].

**Multiple mirrors and global moduli.** The theory described so far is only local. Given a family of minimal Calabi-Yau hypersurfaces $\widehat{Y} \subset X_\Sigma$ and its mirror $\widehat{Y}^\circ \subset X_{\Sigma^\circ}$, the complex moduli of $\widehat{Y}$ should correspond to the Kähler moduli of $\widehat{Y}^\circ$. The problem is that complex moduli typically form a quasi-projective variety, while Kähler moduli are often the quotient of a bounded domain by a finite group. These are very different types of mathematical objects.

Hence, the symmetry $\widehat{Y} \leftrightarrow \widehat{Y}^\circ$ only gives a local isomorphism between complex and Kähler moduli. To get a global version of mirror symmetry for toric hypersurfaces, we need to examine the role of the many fans $\Sigma$ and $\Sigma^\circ$ which can occur. This is the *multiple mirror* phenomenon. For complex moduli, we expect the $\widehat{Y}$'s for different $\Sigma$'s to be related by a series of flops, and hence they should have the same complex moduli (although the varieties themselves are not isomorphic).

The picture on the Kähler side is more interesting, for here, we've seen in §7 that the various Kähler cones for $\Sigma^\circ(1) = M \cap \Delta - \{0\}$ fit together to form the secondary fan of $\mathcal{B} = M \cap \Delta - \{0\}$. The support of this fan is a strongly convex cone in $A_{n-1}(X_\Delta) \otimes \mathbb{R}$. Gluing together the corresponding complexified Kähler cones (10.1) gives the *partially enlarged Kähler moduli space* from [**AGM1**]. (Strictly speaking, we are only dealing with the toric part of the moduli space, but we will ignore this detail.)

The partially enlarged moduli space is still too small to be quasi-projective. To get something bigger, we use the enlarged secondary fan from §7. This is determined by an ample divisor on a toric variety, which here is the anticanonical divisor. The corresponding fan is the secondary fan $\mathcal{N}(\Sigma(\mathcal{A}))$ for $\mathcal{A} = M \cap \Delta$. When we complexify the cones in this fan and glue them together, we get the *enlarged Kähler moduli space* from [**AGM1**], which corresponds to the whole complex moduli of the mirror.

This isomorphism between moduli spaces extends to certain natural compactifications. For simplicity, let's restrict to polynomial moduli, which come from those $f \in L(\Delta)$ which are nondegenerate (i.e., $E_{\mathcal{A}}(f) = 0$ for $\mathcal{A}$ as above) modulo automorphisms of the toric variety. If we instead use only automorphisms coming from the torus, we can compactify using the Newton polytope of the principal $\mathcal{A}$-determinant $E_{\mathcal{A}}$. The resulting compactification has a natural toric structure where the fixed points correspond to the vertices of the Newton polytope. But the Newton polytope of $E_{\mathcal{A}}$ is the secondary polytope $\Sigma(\mathcal{A})$, whose vertices correspond to cones in the enlarged Kähler moduli space. Furthermore, the fixed points for vertices corresponding to cones in the partially enlarged Kähler moduli space are precisely the "special" points mentioned earlier. For more of the mathematics behind this picture, see [**AGM2**].

The physics interpretation is also quite interesting: the cones coming from the partially enlarged Kähler moduli space correspond to different "Calabi-Yau phases" of the same SCFT, while the other cones in the enlarged moduli space correspond to "non-geometric phases". See [**AGM1**] and [**MP**] for more details.

**Nef-partitions and Gorenstein cones.** In §8, we described how a nef-partition $\mathrm{Vert}(\Delta^\circ) = E_1 \cup \cdots \cup E_r$ of a reflexive polytope $\Delta$ gave a Minkowski sum $\Delta = \Delta_1 + \cdots + \Delta_r$ in $M_{\mathbb{R}}$ and a canonical Calabi-Yau complete intersection $V = Y_1 \cap \cdots \cap Y_r \subset X_\Delta$. Furthermore, a projective simplicial fan $\Sigma$ refining the normal fan of $\Delta$ and satisfying $\Sigma(1) = N \cap \Delta^\circ - \{0\}$ gives the MPCP-desingularization $\widehat{V} = \widehat{Y}_1 \cap \cdots \cap \widehat{Y}_r \subset X_\Sigma$ as in (8.5). This is a family of minimal Calabi-Yau complete intersections.

To get the mirror family, we follow [**LBor**] and consider the polytopes

$$\nabla_i = \mathrm{Conv}(\{0\} \cup E_i) \subset N_{\mathbb{R}}, \quad i = 1, \dots, r. \tag{10.3}$$

Then $\nabla = \nabla_1 + \cdots + \nabla_r$ is a reflexive polytope in $N_{\mathbb{R}}$ with a natural nef-partition coming from (10.3). It is interesting to note that $\nabla^\circ \subset M_{\mathbb{R}}$ is different from our original $\Delta$. In fact, [**LBor**] shows that

$$\nabla^\circ = \mathrm{Conv}(\Delta_1, \dots, \Delta_r) \subset \Delta_1 + \cdots + \Delta_r = \Delta \subset M_{\mathbb{R}},$$

with a similar relation between $\Delta^\circ$ and $\nabla$ in $N_{\mathbb{R}}$.

Then the nef-partition for $\nabla$ gives a family of canonical Calabi-Yau complete intersections $V^\circ \subset X_\nabla$, and picking a projective simplicial fan $\Sigma^\circ$ refining the normal fan of $\nabla$ and satisfying $\Sigma^\circ(1) = M \cap \nabla^\circ - \{0\}$ gives a family $\widehat{V}^\circ \subset X_{\Sigma^\circ}$ of minimal Calabi-Yau complete intersections which is conjectured in [**LBor**] to be the mirror of $\widehat{V} \subset X_\Delta$. Evidence for this is presented in [**BB2-3**], [**Bv**] and [**LT**] (see also the references in [**BB3**]).

A significant generalization of this construction, which uses *Gorenstein cones*, appeared in [**BB1**]. We will not go into the details, but Gorenstein cones can explain all of the mirror constructions we've given so far, as well as describing mirrors for certain rigid Calabi-Yau manifolds (where the mirror may have a different dimension). Details can be found in [**BB1**].

**Reid's fantasy.** If a Calabi-Yau threefold $Y_1$ degenerates to a variety with only nodes as singularities, then one can resolve the singularities to obtain another Calabi-Yau $Y_2$ (which could fail to be Kähler). Locally, a vanishing cycle $S^3 \subset Y_1$ collapses to a node and is then resolved to give $\mathbb{P}^1 \simeq S^2 \subset Y_2$. Hence $Y_1$ and $Y_2$ can have quite different Betti numbers. In [**Rei3**], Reid speculates that the moduli of *all* Calabi-Yau threefolds many be connected in this (possibly non-Kähler) way.

In the physics literature, the singular Calabi-Yau between $Y_1$ and $Y_2$ is a *conifold* and going from $Y_1$ to $Y_2$ is a *conifold transition*. Such transitions were long thought to produce unacceptable singularities in the physical theories, but recently (see [**GMS**]), these difficulties were resolved by allowing certain non-perturbative string states (electrically charged black holes) on $Y_1$ to become massless on the conifold and to be intrepreted on $Y_2$ as elementary perturbative states (elementary particles). It follows that a Kähler version of Reid's fantasy would enable any two Calabi-Yau threefolds to be connected by a single physical theory. This implies that when we try to model the vacuum state of the universe using $\mathbb{R}^4 \times$ (Calabi-Yau threefold), we don't need to worry about which Calabi-Yau to use, since all possibilities can occur.

Although Reid's fantasy is still conjectural, it has been verified that for the 7555 Calabi-Yau threefolds mentioned in §8, their moduli are connected through Kähler varieties, though the singularities may be more complicated than nodes (see [**ACJM**] and [**CGGK**]). The basic idea is as follows. Suppose we have reflexive polytopes $\Delta_2 \subset \Delta_1$, which implies $\mathrm{Vert}(\Delta_2) \subset \mathrm{Vert}(\Delta_1)$. Then, for $Y_1 \subset X_{\Delta_1}$ defined by $f \in L(\Delta_1)$, we can degenerate $Y_1$ by letting the coefficients of $f$ corresponding to vertices not in $\Delta_2$ become zero. This gives a singular variety $\widetilde{Y} \subset X_{\Delta_1}$ which, when resolved, corresponds to a Calabi-Yau hypersurface $Y_2 \subset X_{\Delta_2}$. Once the MPCP-desingularizations are taken into account, we can link the corresponding moduli spaces. As already mentioned, the singularities of $\widetilde{Y}$ may be more complicated than just simple nodes, and at present there is no physical explanation of the transition from $Y_1$ to $Y_2$. But this method is sufficient to link up the 7555 Calabi-Yaus on the list, and as noted in [**Ska**], it may be sufficient to connect the moduli of all 3-dimensional Calabi-Yau toric hypersurfaces.

**Further remarks.** Although our discussion of mirror symmetry has been rather superficial (and has omitted some important ideas), it should be clear that toric geometry has a prominent role to play, if for no other reason than providing a rich supply of examples. Notice also that virtually everything in the earlier sections of the paper has been used. Symplectic geometry seems to be an exception, but this is only because we didn't describe Witten's *linear sigma models* [**Wit**], which are physical theories where toric varieties enter by means of symplectic reduction. See [**MP**] for more details and for some other interesting uses of toric geometry in mirror symmetry.

We should also mention that it is possible to compute the quantum cohomology of toric varieties (see [**Bat3**]). In addition, certain mirror symmetry calculations suggest that in some situations, the usual Hodge numbers $h^{p,q}$ need to be replaced by *string-theoretic Hodge numbers* $h^{p,q}_{\mathrm{st}}$. For Calabi-Yau complete intersections in a toric varieties, these numbers are computed in [**BB2**].

## §11. Other developments

Besides the topics reported on so far, the last few years have seen a lot of interesting work on other aspects of toric geometry. Here is a selection, with apologies for the many fine papers not mentioned.

**Very ample divisors.** As is well-known, an ample divisor D on a complete toric variety $X$ is very ample if $X$ is smooth or has dimension $\leq 2$. In general, $D$ may fail to be very ample, but [**EW**] proves that $(n-1)D$ is always very ample, where $n \geq 2$ is the dimension of $X$. Examples show that this result is sharp.

There is also a notion of *$k$-very ampleness* which measures the behavior of $D$ relative to 0-dimensional subschemes $Z \subset X$ with $h^0(\mathcal{O}_Z) = k+1$ (so that 0-ample means spanned by global sections and 1-ample means very ample). For smooth toric surfaces, [**DiR**] shows that $k$-very ampleness can be interpreted in terms of convexity properties of the support function of $D$ relative to the integer lattice.

**Embeddings into toric varieties.** Another nice fact about toric varieties concerns embeddings of a complete variety $Y$. The Chevalley criterion (proved by Kleiman) states that $Y$ can be embedded into a projective space if and only if every finite subset of $Y$ is contained in an affine open. When $Y$ is normal, [**Wło**] proves that we can embed $Y$ into a complete toric variety if and only if every *two element* subset of $Y$ lies in an affine open.

**Classifying toric varieties.** There are several ways one can try to classify toric varieties. For example, one can work one dimension at a time, which is the approach taken in §8 when we discussed reflexive polytopes. Another strategy is to classify smooth complete toric varieties according to their Picard number $\rho$. For $\rho = 2$, this is done in [**Kle**], which such varieties are shown to be projective (and [**ESch**] finds projective embeddings for which Conjecture 5.6 is satisfied.) Similarly, [**KS**] shows that smooth toric varieties with Picard number 3 are projective. Using this and the primitive collections defined in §2, a classification for smooth complete toric varieties with $\rho = 3$ is given in [**Bat2**]. The papers mentioned here also contain references to earlier work on classification.

**Invariants of toric varieties.** When a toric variety $X$ is smooth (resp. simplicial), its cohomology over $\mathbb{Z}$ (resp. $\mathbb{Q}$) is well-known. However, it is also possible to compute the rational intersection cohomology of a compact toric variety (see [**Fie**]), and since a toric variety has a natural torus action, there is also equivariant cohomology to consider, which is computed in [**Bif**]. Another topic is homology with closed supports (for both regular and intersection homology), which in degree $2n-2$ is studied in [**BBFK1**] and [**BBFK2**].

Turning to less topological invariants, one can study the K-theory (Grothendieck groups of vector bundles or coherent sheaves) of a toric variety, which coincide in this case and are computed in [**Morel**]. The Brauer group of a toric variety is discussed in [**DFM**] in the case of an algebraically closed field, and the split case over arbitrary fields is studied in [**For**]. It would be interesting to see what results could be obtained for non-split tori (to be defined below). The paper [**For**] also considers certain invariants of $X_\Sigma$ which depend only on the combinatorial type of the fan $\Sigma$. Similar questions about $\mathrm{Pic}(X_\Sigma)$ are studied in [**Eik**].

Another invariant known in the smooth (resp. simplicial) toric case is the Chow ring $A^*(X)$ (resp. $A^*(X) \otimes \mathbb{Q}$). For a general toric variety, the Chow groups $A_k(X)$ are computed in [**Dan2**, Sect. 10], and when $X$ is a complete toric variety, one can relate $A_k(X) \otimes \mathbb{Q}$ to a certain weight filtration of the Borel-Moore homology of $X$ (see [**Tot**]). Besides these classical Chow groups, there are also the *operational Chow groups* $A^k(X)$ of Fulton and MacPherson, which give a Chow cohomology ring $A^*(X)$. When $X$ is a complete toric variety, this ring is computed in [**FS**].

**Intersection theory on toric varieties.** As mentioned in the introduction, Hard Lefschetz for simplicial toric varieties was used in Stanley's proof of McMullen's conjectures about convex simplicial polytopes. A nice discussion of this may be found in [**Ful**, Sect. 5.6]. Since Hard Lefschetz for intersection cohomology is a very deep result, it is reasonable to ask if a simpler proof exists in the toric case. This led to the papers [**Oda1**] and [**Oda5**] on the de Rham cohomology of toric varieties, and Ishida [**Ish2**] finally succeeded in giving a self-contained toric proof (without reference to Hard Lefschetz) of McMullen's conjectures. (We should also mention the paper [**Dan1**], which studies the de Rham cohomology of toriodal varieties.)

McMullen eventually found a proof of his conjectures which avoids both toric varieties and Hard Lefschetz [**McM**]. His proof uses a certain *polytope algebra*. In [**FS**], it is shown that in dimension $n$, the polytope algebra is the inverse limit of the Chow cohomology rings $A^*(X) \otimes \mathbb{Q}$ over the directed system of *all* toric compactifications of the torus $(\mathbb{C}^*)^n$.

**Counting lattice points.** If $X_\Delta$ is the toric variety determined by an $n$-dimensional integer polytope $\Delta \subset M_\mathbb{R} \simeq \mathbb{R}^n$, then its Todd class can be written

$$Td(X_\Delta) = \sum_{\sigma\in\Sigma} r_\sigma[V(\sigma)] \in \textstyle\bigoplus_{k=0}^n A_k(X_\Delta),$$

where $\Sigma$ is the normal fan of $\Delta$, $[V(\sigma)]$ is the class of the orbit closure corresponding to $\sigma$, and $r_\sigma \in \mathbb{Q}$. By [**Dan2**], the number of integer points in $\Delta$ is given by

$$l(\Delta) = \sum_{\sigma\in\Sigma} r_\sigma \mathrm{Vol}(F(\sigma)),$$

where $\mathrm{Vol}(F(\sigma))$ is the normalized volume of the face of $\Delta$ corresponding to $\sigma$. There is a similar formula for the Ehrhart polynomial $E_\Delta(k) = l(k\Delta)$ for $k \geq 1$. See [**Ful**, Sect. 5.3] or [**More2**, Sect. 1.1] for a nice introduction to this topic.

These formulas reduce the problem of counting lattice points to finding an explicit expression for the Todd class. While this can be done for any given toric variety (see [**Ful**]), general formulas weren't available until recently, when several different solutions were found. In [**More2**], a formula for $r_\sigma$ is given which depends *only* on the cone $\sigma$ and not the fan in which it sits. For simplicial toric varieties, [**Pom**] introduces the idea of a "mock" Todd class, denoted $TD(X_\Delta)$, which is built using formulas from the smooth case. The difference $Td(X_\Delta) - TD(X_\Delta)$ is then described using functions which involve Dedekind sums and lead to explicit formulas for the component $Td^2(X_\Delta)$. (For the weighted projective space $\mathbb{P}(q_0, q_1, q_2, q_3)$, another approach to this computation can be found in [**Lat**].) A slighlty different definition of "mock" Todd class is given in [**CS**] and is used to obtain a third expression for $Td(X_\Delta)$. A corollary is an explicit formula for the number of lattice points in an integer simplex.

We should also mention that the Todd class is related to the total Chern class. For a singular toric variety, the formula $ch(X_\Delta) = \sum_{\sigma\in\Sigma}[V(\sigma)]$ is shown to hold in homology with closed supports (see [**BBF**]).

The quite different approach to the study of lattice points appears in [**Bri1**]. Here, the object of interest is the Laurent polynomial

$$l_\Delta(\mathbf{t}) = \sum_{m\in\Delta\cap M} \mathbf{t}^m,$$

where we still assume that $X_\Delta$ is simplicial. Using the Lefschetz-Riemann-Roch theorem from equivariant K-theory, $l_\Delta(\mathbf{t})$ can be written as a sum of rational functions in $\mathbf{t}$ determined by the cones at the vertices of $\Delta$. Hence we get a formula which not only counts lattice points (by setting $\mathbf{t} = (1,\ldots,1)$) but also describes the lattice points themselves. Simpler proofs of this result can be found in [**Ish1**] and [**SI1**], and a weighted version is in [**Bri2**]. We should also mention that these questions can be studied from a purely "polytope" point of view, which uses a *combinatorial Riemann-Roch theorem* and avoids toric methods. See [**KK**] for details.

**Rational points on toric varieties over number fields.** To define a toric variety over a number field $K$, first observe that a torus $T$ over $K$ is determined by a lattice $M \simeq \mathbb{Z}^n$ with an action of $\mathrm{Gal}(E/K)$, where $K \subset E$ is a finite Galois extension over which the torus splits, i.e., becomes isomorphic to $(\mathbb{G}_m)^n$. Then a toric variety $X$ over $K$ containing $T$ is determined by a fan in $M_\mathbb{R}$ which is invariant under $\mathrm{Gal}(E/K)$. See [**BT**] for details and references.

In this situation, one can study $K$-rational points on $X$ using the height function coming from a metrized ample line bundle $\mathcal{L}$ on $X$. The basic question concerns the asymptotics of $N(T,\mathcal{L},B)$, which is the number of $K$-rational points in the torus $T \subset X$ with height bounded by $B$. When $X$ is a smooth projective Fano toric variety over $K$, the natural line bundle to use is the anticanonical line bundle $\mathcal{L} = \mathcal{O}(-K_X)$. In [**BT**], the asymptotic formula

$$N(T,\mathcal{L},B) = c\,B\,(\log B)^{r-1}(1+o(1)), \quad B \gg 0$$

is proved, where $r$ is the rank of the Picard group $\mathrm{Pic}(X)$ over $K$ and $c$ is a nonzero constant. This verifies the toric case of a conjecture of Manin for Fano varieties over number fields (see [**FMT**]). The constant $c$ can be explicitly computed in terms of the geometry of the cone of effective divisors in $\mathrm{Pic}(X)_\mathbb{R}$, the order of the non-trivial part of the Brauer group of $X$, and a certain Tamagawa number associated with the metrized bundle $\mathcal{L}$ on $X$.

**Quotients of toric varieties.** In §2, we constructed a toric variety $X$ as the quotient $(\mathbb{C}^{\Delta(1)} - Z(\Sigma))/G$, which is a geometric quotient when $X$ is simplicial. Oda observed that $\mathbb{C}^{\Delta(1)} - Z(\Sigma)$ has a natural structure of a toric variety, so that we are taking the quotient of a toric variety by a subgroup of its torus. One can study this problem in general, and various types of quotients are possible, including *combinatorial quotients*, *GIT quotients* and *Chow quotients*. For GIT quotients, there can be several quotients in any given situation, and the different quotients are related by a chamber structure where quotients for chambers sharing a common wall are (in good cases) related by a flip. (This is similar to the secondary fan described in §7.) Furthermore, the Chow quotient can be described using a certain fiber polytope and is the inverse limit of the GIT quotients. See [**B-BS**], [**Hu**], [**KSZ2**] and [**Tha**] for details.

**Residues on toric varieties.** There is a huge literature on residues. For multidimensional residues, toric methods can be used in defining local residues (see

[**PT**]), and Gröbner methods, including the Gröbner fan from §7, can be used in computing global residues (see [**CDS**]). A version of global residues specific to toric varieties was introduced in [**Cox3**], and various properties of these *toric residues* have been studied in [**CCD**] and [**CD**]. Toric residues have also been used in mirror symmetry (see [**MP**]).

**Singularities of toric varieties.** Singularities of affine toric varieties have been the subject of several recent papers. For example, basic tools in deformation theory are the spaces $T^1_X$ and $T^2_X$ which describe infinitesimal deformations and obstructions, and for an affine toric variety $X$, these are computed in [**Alt1**]. One can also study deformations which are themselves toric, and in the case of isolated 3-dimensional toric Gorenstein singularities, the versal deformation can be constructed entirely by toric means (see [**Alt3**]). Furthermore, the irreducible components of the deformation space correspond to certain decompositions of a lattice polytope into Minkowski sums of other lattice polytopes. The role of Minkowski sums in the deformation theory of affine toric varieties is also explored in [**Alt2**]. Another reference for Gorenstein toric singularities is [**Nom**].

In studying the resolutions of singularities of an affine toric variety $X$, a divisor in a resolution of $X$ is *essential* if a birational copy of the divisor appears in every resolution of the variety, and it is *equivariant essential* if it appears (again up to birational equivalence) in every toric resolution. In [**BG-S**], it is shown that if $X$ comes from a cone $\sigma \subset N_{\mathbb{R}}$, then equivariant essential divisors correspond bijectively to minimal generators of the semigroup $\sigma \cap N - \{0\}$. Furthermore, in dimension 3, the same is true for essential divisors. Other papers dealing with the resolution of 3-dimensional toric singularities are [**ESpa**] and [**Pou**].

In another context, a conjecture of Shokurov on the minimal discrepancies of log-terminal singularities was verified for toric singularities in [**ABor**]. Hypersurface sections of toric singularities are considered in [**Tsu**]. It is also possible to discuss toric singularities without reference to a base field (or even a base scheme). This topic is studied in [**Kat**] and may have applications to arithmetic algebraic geometry.

For the special case of orbifold singularities, toric methods have been used to construct resolutions and to study the relation to moduli spaces of representations of the associated McKay quiver [**SI2**].

**Resolution of singularities.** A recent development is the discovery in [**Ad**] and [**BP**] of a simple proof of a weak form of resolution of singularities. The precise result is that if $X$ is a normal projective variety in characteristic 0 and $Y \subset X$ is a proper subvariety, then there is a birational morphism $f : \widehat{X} \to X$ such that $\widehat{X}$ is a smooth projective variety and $f^{-1}(Y)$ is a strict normal-crossings divisor. However, $f : \widehat{X} - f^{-1}(Y) \to X - Y$ might not be the identity, so we don't get a resolution of singularities in the usual sense. The proofs in [**Ad**] and [**BP**] are slightly different, but both make essential use of toric methods. One can also use these methods to study equivariant resolutions [**AW**].

## Conclusion

In this survey, we have attempted to convey the richness of the recent work in toric geometry. An unexpected consequence of all this activity is that it is less clear where to learn about the subject. One could start with the standard approach to toric varieties, as in [**Dan2**], [**Ful**], [**Oda2**]. (Other introductions include [**Ewa1**],

[**Oda3**], [**Oda4**] and [**Rei4**].) Alternatively, one could begin with the quotient construction of §2, where [**Cox2**] is one of many references. This is closely tied to the symplectic approach, as described in [**Aud**] and [**Gui**]. Yet another starting point would be the theory of non-normal toric varieties, as in [**GKZ1**] or [**Stu2**]. Finally, the books [**Ewa2**] and [**Stu2**] highlight the deep connection between toric varieties and polyhedral geometry.

Of course, there is no "best" approach to toric varieties. The multiplicity of entry points is actually a virtue, for it enables people from different areas of mathematics to learn about and contribute to this fascinating and accessible part of algebraic geometry.

The author is grateful to Gottfried Barthel, Victor Batyrev, Sheldon Katz, Niels Lauritzen, David Morrison and Bernd Sturmfels for numerous useful comments. The referee made many thoughtful suggestions which are greatly appreciated.

## References

[Ad] D. Abramovich and J. de Jong, *Smoothness, semistability and toroidal geometry*, preprint 1996.

[AW] D. Abramovich and and J. Wang, *Equivariant resolution of singularities in characteristic* 0, preprint, 1996.

[Alt1] K. Altmann, *Infinitesimal deformations and obstructions for toric singularities*, preprint, 1994.

[Alt2] ———, *Minkowski sums and homogenous deformations of toric varieties*, Tôhoku Math. J. **47** (1995), 151–184.

[Alt3] ———, *The versal deformation of an isolated toric Gorenstein singularity*, preprint 1994.

[AVG] V. Arnold, A. Varchenko and A. Gusein-Zade, *Singularities of Differentiable Maps, Volumes I–II*, Birkhäuser, Boston Basel Berlin, 1985–1988.

[AMRT] A. Ash, D. Mumford, M. Rapaport and Y. Tai, *Smooth Compactifications of Locally Symmetry Varieties*, Lie Groups: History, Frontiers and Applications IV, Math. Sci. Press, Brookline, MA, 1975.

[AGM1] P. Aspinwall, B. Greene and D. Morrison, *Calabi-Yau moduli space, mirror manifolds and spacetime topology change in string theory*, Nucl. Phys. **B416** (1994), 414–480.

[AGM2] ———, *The monomial-divisor mirror map*, Int. Math. Res. Notices (1993), 319–337.

[Ati1] M. Atiyah, *Angular momentum, convex polyhedra and algebraic geometry*, Proc. Edin. Math. Soc. **23** (1983), 121–138.

[Ati2] ———, *Convexity and commuting Hamiltonians*, Bull. London Math. Soc. **14** (1982), 1–15.

[Aud] M. Audin, *The Topology of Torus Actions on Symplectic Manifolds*, Progress in Math. **93**, Birkhäuser, Boston Basel Berlin, 1991.

[ACJM] A. Avram, P. Candelas, D. Jančić and M. Mandelberg, *On the connectedness of the moduli space of Calabi-Yau manifolds*, preprint, 1995.

[BBF] G. Barthel, J.-P. Brasselet and K.-H. Fieseler, *Classes de Chern des variétés toriques singulières*, C. R. Acad. Sci. Paris Sér I Math. **315** (1992), 187–192.

[BBFK1] G. Barthel, J.-P. Brasselet, K.-H. Fieseler and L. Kaup, *Diviseurs invariants et homomorphisme de Poincaré de variétés complexes*, to appear, Tôhoku Math. J. (1996).

[BBFK2] ———, *Invariante Divisoren und Schnitthomologie von torischen Varietäten*, Paramter Spaces, Banach Center Publications, volume 36, Polish Academy of Sciences, Warsaw, 1996, pp. 9–23.

[Bat1] V. Batyrev, *Dual polyhedra and mirror symmetry for Calabi-Yau hypersurfaces in toric varieties*, J. Algebraic Geometry **3** (1994), 493–535.

[Bat2] ———, *On the classification of smooth projective toric varieties*, Tôhoku Math. J. **43** (1991), 569–585.

[Bat3] ———, *Quantum cohomology rings of toric manifolds*, Journées de Géométrie Algébrique d'Orsay (Juillet 1992), Astérisque **218**, Sociéte Mathématique de France, Paris, 1993, pp. 9–34.

[Bat4] ———, *Variations of the mixed Hodge structure of affine hypersurfaces in algebraic tori*, Duke Math. J. **69** (1993), 349–409.

[BB1] V. Batyrev and L. Borisov, *Dual cones and mirror symmetry for generalized Calabi-Yau manifolds*, Essays in Mirror Symmetry II, S.-T. Yau, editor, to appear.

[BB2] ———, *Mirror duality and string-theoretic Hodge numbers*, preprint, 1995.

[BB3] ———, *On Calabi-Yau complete intersections in toric varieties*, Proceedings of Trento Conference (1994), to appear.

[BC] V. Batyrev and D. Cox, *On the Hodge structure of projective hypersurfaces in toric varieties*, Duke Math. J. **75** (1994), 293–338.

[BT] V. Batyrev and Y. Tschinkel, *Manin's conjecture for toric varieties*, preprint, 1995.

[Bv] V. Batyrev and D. van Straten, *Generalized hypergeometric functions and rational curves on Calabi-Yau complete intersections in toric varieties*, preprint, 1993.

[B-BS] A. Bialynicki-Birula and A. Sommese, *A conjecture about compact quotients by tori*, Adv. Studies in Pure Math. **8** (1986), 59–68.

[Bif] E. Bifet, *Cohomology, symmetry and perfection*, Publ. Mat. **36** (1992), 407–420.

[BFS] L. Billera, P. Filliman and B. Sturmfels, *Constructions and complexity of secondary polytopes*, Adv. in Math. **83** (1990), 155–179.

[BS] L. Billera and B. Sturmfels, *Fiber polytopes*, Ann. of Math. **135** (1992), 527–549.

[BP] F. Bogomolov and T. Pantev, *Weak Hironaka theorem*, preprint, 1996.

[ABor] A. Borisov, *Minimal discrepancies of toric singularities*, preprint, 1994.

[LBor] L. Borisov, *Towards the mirror symmetry for Calabi-Yau complete intersections in Gorenstein toric Fano varieties*, preprint, 1993.

[BG-S] C. Bouvier and G. Gonzalez-Springberg, *Système générateur minimal, diviseurs essentiels et G-désingularisations de variétiés toriques*, Tôhoku Math. J. **47** (1995), 125–149.

[Bri1] M. Brion, *Points entiers dans les polyèdres convexes*, Ann. Sci. École Norm. Sup. **21** (1988), 653–663.

[Bri2] ———, *Polyèdres et réseaux*, Enseign. Math. **38** (1992), 71–88.

[BFLM] A. Buch, J. Funch Thomsen, N. Lauritzen and V. Mehta, *Frobenius morphisms over $\mathbb{Z}/p^2$ and Bott vanishing*, preprint, 1995.

[CdGP] P. Candelas, X. de la Ossa, P. Green and L. Parkes, *A pair of Calabi-Yau manifolds as an exactly soluble superconformal field theory*, Essays on Mirror Manifolds, S.-T. Yau, editor, International Press, Hong Kong, 1992, pp. 31–95.

[CdK] P. Candelas, X. de la Ossa and S. Katz, *Mirror Symmetry for Calabi-Yau hypersurfaces in weighted $\mathbb{P}_4$ and extensions of Landau-Ginzburg theory*, Nucl. Phys. **B450** (1995), 267–292.

[CE] J. Canny and I. Emiris, *An efficient algorithm for the sparse mixed resultant*, Proc. 10th Inter. Symp. on Applied Algebra, Algebraic Algorithms and Error-Correcting Codes, G. Cohen, T. More and O. Moreno, editors, Lecture Notes in Computer Science **263**, Springer-Verlag, Berlin Heidelberg New York, 1993, pp. 89–104.

[CS] S. Cappell and J. Shaneson, *Genera of algebraic varieties and counting of lattice points*, Bull. AMS **30** (1994), 60–69.

[CCD] E. Cattani, D. Cox and A. Dickenstein, *Residues in toric varieties*, preprint, 1995.

[CD] E. Cattani and A. Dickenstein, *A global view of residues in the torus*, preprint, 1996.

[CDS] E. Cattani, A. Dickenstein and B. Sturmfels, *Computing multidimensional residues*, Effective Methods in Algebraic Geometry, MEGA 94, Progress in Math., Birkhäuser, Boston Basel Berlin, to appear.

[CGGK] T.-M. Chiang, B. Greene, M. Gross and Y. Kanter, *Black hole condensation and the web of Calabi-Yau manifolds*, preprint, 1995.

[Cox1] D. Cox, *The functor of a smooth toric variety*, Tôhoku Math. J. **47** (1995), 251–262.

[Cox2] ———, *The homogeneous coordinate ring of a toric variety*, J. Algebraic Geom. **4** (1995), 17–50.

[Cox3] ———, *Toric residues*, Ark. Mat, to appear.

[CK] D. Cox and S. Katz, *Mirror Symmetry and Algebraic Geometry*, in preparation.

[Dan1] V. Danilov, *De Rham complex on toroidal variety*, Algebraic Geometry, Proceedings of the US-USSR Symposium, 1989, S. Bloch, I. Dolgachev and W. Fulton, editors, Springer-Verlag, Berlin Heidelberg New York, 1991, pp. 26–38.

[Dan2] ———, *The geometry of toric varieties*, Russian Math. Surveys **33** (1978), 97–154.

[Dan3] ———, *Newton polyhedra and vanishing cohomology*, Funct. Anal. Appl. **13** (1979), 103–115.

[DK] V. Danilov and A. Khovanskii, *Newton polyhedra and an algorithm for computing Hodge-Deligne numbers*, Math. USSR Izvestiya **29** (1987), 279–298.

[Del] T. Delzant, *Hamiltoniens périodiques et image convex de l'application moment*, Bull. Soc. Math. France **116** (1988), 315–339.

[Dem] M. Demazure, *Sous-groupes algébriques de rang maximum du groupe de Cremona*, Ann. Sci. École Norm. Sup. **3** (1970), 507–588.

[DFM] F. DeMeyer, T. Ford and R. Miranda, *The cohomological Brauer group of a toric variety*, J. Algebraic Geom. **2** (1993), 137–154.

[DL] J. Denef and F. Loeser, *Weights of exponential sums, intersection cohomology, and Newton polyhedra*, Invent. Math. **106** (1991), 275–294.

[Dim] A. Dimca, *Residues and cohomology of complete intersections*, Duke Math. J. **78** (1995), 89–100.

[DiR] S. Di Rocco, *On higher embeddings of toric surfaces*, preprint, 1995.

[Eik] M. Eikelberg, *Picard groups of compact toric varieties and combinatorial classes of fans*, Results Math. **23** (1993), 251–293.

[ENV] H. Esnault, M. Nori and V. Srinivas, *Hodge type of projective varieties of low degree*, Math. Ann. **293** (1992), 1–6.

[Ewa1] G. Ewald, *Algebraic geometry and convexity*, Handbook of Convex Geometry, Vol. A, B, North-Holland, Amsterdam, 1993, pp. 603–626.

[Ewa2] ———, *Combinatorial Convexity and Algebraic Geometry*, Springer-Verlag, Berlin Heidelberg New York, 1996.

[ESch] G. Ewald and A. Schmeinck, *Representation of the Hirzebruch-Kleinschmidt varieties by quadrics*, Beiträge Algebra Geom. **34** (1993), 151–156.

[ESpa] G. Ewald and B. Spazier, *On the resolution of singularities in affine toric 3-varieties*, Results Math. **25** (1994), 234–241.

[EW] G. Ewald and U. Wessels, *On the ampleness of invertible sheaves in complete toric varieties*, Results Math. **19** (1991), 275–278.

[Fie] K.-H. Fieseler, *Rational intersection cohomology of projective toric varieties*, J. Reine Angew. Math. **413** (1991), 88-98.

[Fin] J. Fine, *Geometric progressions, convex polytopes and toric varieties*, in preparation.

[For] T. Ford, *Topological invariants of a fan associated to a toric variety*, Comm. Algebra **23** (1995), 4031–4045.

[FMT] J. Franke, Y. Manin and Y. Tschinkel, *Rational points of bounded height on Fano varieties*, Invent. Math. **95** (1989), 425–435.

[Ful] W. Fulton, *Introduction to Toric Varieties*, Princeton U. Press, Princeton, NJ, 1993.

[FS] W. Fulton and B. Sturmfels, *Intersection theory on toric varieties*, preprint, 1994.

[GKZ1] I. Gelfand, M. Kapranov and A. Zelevinsky, *Discriminants, Resultants, and Multidimensional Determinants*, Birkhäuser, Boston Basel Berlin, 1994.

[GKZ2] ———, *Generalized Euler integerals and $\mathcal{A}$-hypergeometric functions*, Adv. Math. **84** (1990), 255–271.

[GKZ3] ———, *Hypergeometric functions and toral manifolds*, Funct. Anal. Appl. **23** (1989), 94–106.

[Giv1] A. Givental, *Homological geometry and mirror symmetry*, Proceedings of the International Congress of Mathematicians, 1994, Zürich, Birkhäuser, Boston Basel Berlin, 1995, pp. 472–480.

[Giv2] ———, *Equivariant Gromov-Witten invariants*, preprint, 1996.

[Gri] P. Griffiths, *On the periods of certain rational integrals I*, Ann. of Math. **90** (1969), 460–495.

[Gue] M. Guest, *The topology of the space of rational curves on a toric variety*, Acta Math. **174** (1995), 119–145.

[Gui] V. Guillemin, *Moment Maps and Combinatorial Invariants of Hamiltonian $T^n$-spaces*, Progress in Math. **122**, Birkhäuser, Boston Basel Berlin, 1994.

[GS] V. Guillemin and S. Sternberg, *Convexity properties of the moment mapping*, Invent. Math. **67** (1982), 491–513.

[Har] R. Hartshorne, *Algebraic Geometry*, Graduate Texts in Math. **52**, Springer-Verlag, Berlin Heidelberg New York, 1977.

[Hoc] M. Hochster, *Rings of invariants of tori, Cohen–Macaulay rings generated by monomials, and polytopes*, Ann. of Math. **96** (1972), 318–337.

[HLY1] S. Hosono, B. Lian and S.-T. Yau, *GKZ-Generalized hypergeometric systems in mirror symmetry of Calabi-Yau hypersurfaces*, preprint, 1995.

[HLY2] ———, *Maximal degeneracy points of GKZ systems*, preprint, 1996.

[Hu] Y. Hu, Duke Math. J. **68** (1992), 151–184; Erratum **68** (1992), 609.

[Ish1] M.-N. Ishida, *Polyhedral Laurent series and Brion's equalities*, International J. Math. **1** (1990), 251–265.

[Ish2] ———, *Torus embeddings and algebraic intersection complexes, I and II*, preprint.

[Jac] K. Jaczewski, *Generalized Euler sequence and toric varieties*, Classification of Algebraic Varieties, C. Cilberto, E. Livorni and A. Sommese, editors, Contemporary Mathematics **162**, AMS, Providence, RI, 1994, pp. 227–247.

[Kaj] T. Kajiwara, *The functor of a toric variety with enough invariant effective Cartier divisors*, preprint, 1996.

[KK] J.-M. Kantor and A. Khovanskii, *Une application du thèoréme de Riemann-Roch combinatoire au polynôme d'Ehrhart des polytopes entiers de* $\mathbb{R}^d$, C. R. Acad. Sci. Paris Sér I Math. **317** (1993), 501–507.

[KSZ1] M. Kapranov, B. Sturmfels and A. Zelevinsky, *Chow polytopes and general resultants*, Duke Math. J. **67** (1992), 189–218.

[KSZ2] ———, *Quotients of toric varieties*, Math. Ann. **290** (1991), 643–655.

[Kat] K. Kato, *Toric singularities*, Am. J. Math. **116** (1994), 1073–1099.

[KKMS] G. Kempf, F. Knudsen, D. Mumford and B. Saint-Donat, *Torodial Embeddings I*, Lecture Notes in Math. **339**, Springer-Verlag, Berlin Heidelberg New York, 1973.

[Kho] A. Khovanskii, *Newton polyhedra and toroidal varieties*, Funct. Anal. Appl. **11** (1977), 289–296.

[Kir] F. Kirwan, *Cohomology of Quotients in Symplectic and Algebraic Geometry*, Princeton U. Press, Princeton, NJ, 1984.

[Kle] P. Kleinschmidt, *A classification of toric varieties with few generators*, Aequationes Math. **3** (1988), 254–266.

[KS] P. Kleinschmidt and B. Sturmfels, *Smooth toric varieties with small Picard number are projective*, Topology **30** (1991), 289–299.

[Konn] K. Konno, *On the variational Torelli problem for complete intersections*, Compositio Math. **78** (1991), 271–296.

[Kont] M. Kontsevich, *Homological algebra of mirror symmetry*, Proceedings of the International Congress of Mathematicians, 1994, Zürich, Birkhäuser, Boston Basel Berlin, 1995, pp. 120–139.

[Kus] A. Kushnirenko, *Polyèdres de Newton et nombres de Milnor*, Invent. Math. **32** (1976), 1–31.

[Lat] R. Laterveer, *Weighted complete intersections and lattice points*, Math. Z. **218** (1995), 213–218.

[LT] A. Libgober and J. Teitelbaum, *Lines on Calabi-Yau complete intersections, mirror symmetry, and Picard-Fuchs equations*, Int. Math. Res. Notices (1993), 29–39.

[Loe] F. Loeser, *Polytopes secondaires et discriminants*, Séminaire Bourbaki, volume 1990/91, exposés 730–744, Astérisque **201–203**, Sociéte Mathématique de France, Paris, 1991, pp. 387–420.

[McD] J. McDonald, *Fiber polytopes and fractional power series*, preprint, 1995.

[McM] P. McMullen, *On simple polytopes*, Invent. Math. **113** (1993), 419–444.

[MR] T. Mora and L. Robbiano, *The Gröbner fan of an ideal*, Computational Aspects of Commutative Algebra, L. Robbiano, editor, Academic Press, London, 1989, pp. 49–74.

[More1] R. Morelli, *The K-theory of a toric variety*, Adv. Math. **100** (1993), 154–182.

[More2] ———, *Pick's theorem and the Todd class of a toric variety*, Adv. Math. **100** (1993), 183–231.

[Mor1] D. Morrison, *Compactifications of moduli spaces inspired by mirror symmetry*, Journées de Géométrie Algébrique d'Orsay (Juillet 1992), Astérisque **218**, Sociéte Mathématique de France, Paris, 1993, pp. 243–271.

[Mor2] ———, *Making enumerative predictions by means of mirror symmetry*, preprint, 1995.

[Mor3] ———, *Mirror symmetry and rational curves on quintic threefolds: A guide for mathematicians*, J. Amer. Math. Soc. **6** (1993), 223–247.

[Mor4] ———, *Picard-Fuchs equations and mirror maps for hypersurfaces*, Essays on Mirror Manifolds, S.-T. Yau, editor, International Press, Hong Kong, 1992, pp. 241–264.

[MP] D. Morrison and M. R. Plesser, *Summing the instantons: Quantum cohomology and mirror symmetry in toric varieties*, Nucl. Phys. **B440** (1995), 279–354.

[Mus] I. Musson, *Differential operators on toric varieties*, J. Pure and Appl. Alg. **95** (1994), 303–315.

[Nag] J. Nagel, *The Abel-Jacobi map for complete intersections*, preprint, 1995.

[Nom] A. Noma, *Gorenstein toric singularities and convex polytopes*, Tôhoku Math. J. **43** (1991), 529–535.

[Oda1] T. Oda, *The algebraic de Rham theorem for toric varieties*, Tôhoku Math. J. **45** (1993), 231–247.

[Oda2] ———, *Convex Bodies and Algebraic Geometry*, Springer-Verlag, Berlin Heidelberg New York, 1988.

[Oda3] ———, *Geometry of Toric Varieties*, Proceedings of the Hyderabad Conference on Algebraic Groups, 1989, S. Ramanan, editor, Manoj Prakashan, Madras, 1991, pp. 407–440.

[Oda4] ———, *Recent topics on toric varieties*, Amer. Math. Soc. Transl. **172** (1996), 77–91.

[Oda5] ———, *Simple convex polytopes and the strong Lefschetz theorem*, J. Pure Appl. Algebra **71** (1991), 265–286.

[OP] T. Oda and H. Park, *Linear Gale transforms and Gelfand-Kapranov-Zelevinskij decompositions*, Tôhoku Math. J. **43** (1991), 375–399.

[Par] H. Park, *The Chow ring and GKZ-decomposition for $\mathbb{Q}$-factorial toric varieties*, Tôhoku Math. J. **45** (1993), 109–145.

[PT] M. Passare and A. Tsikh, *On connections between the local structure of holomorphic mappings, multidimensional residues, and generalized Mellin transforms*, Russian Acad. Sci. Dokl. Math. **46** (1993), 88–91.

[Per1] D. Perkinson, *Inflections of toric varieties*, preprint, 1995.

[Per2] ———, *Principal parts of line bundles on toric varieties*, Compositio Math., to appear.

[PS] C. Peters and J. Steenbrink, *Infinitesimal variation of Hodge structure and the generic Torelli theorem for projective hypersurfaces*, Classification of Algebraic and Analytic Manifolds, K. Ueno, editor, Progress in Math. **39**, Birkhäuser, Boston Basel Berlin, 1983, pp. 399–463.

[Pom] J. Pommerscheim, *Toric varieties, lattice points and Dedekind sums*, Math. Ann. **295** (1993), 1–24.

[Pou] N. Pouyanne, *Une résolutions de singularitiés toriques simpliciales des singularitiés quotient dimension trois*, Ann. Fac. Sci. Toulouse Math. **6** (1992), 363–398.

[Rei1] M. Reid, *Canonical 3-folds*, Journées de Géometrié Algébrique d'Angers, A. Beauville, editor, Sijthoff & Noordhoff, Alphen ann den Rijn, 1980, pp. 273–310.

[Rei2] ———, *Decomposition of toric morphisms*, Arithmetic and Geometry, M. Artin and J. Tate, editors, Progress in Math. **39**, Birkhäuser, Boston and Basel, 1983, pp. 395–418.

[Rei3] ———, *The moduli space of 3-folds with $K = 0$ may nevertheless by irreducible*, Math. Ann. **278** (1987), 329–334.

[Rei4] ———, *Review of "Convex Bodies and Algebraic Geometry: An Introduction to Toric Varieties" by T. Oda*, Bull. AMS **21** (1989), 360–364.

[Rei5] ———, *Young person's guide to canonical singularities*, Algebraic Geometry, Bowdoin 1985, S. Bloch, editor, Proc. Sym. Pure Math. **46**, Part 1, AMS, Providence, RI, 1987, pp. 345–414.

[Roa1] S.-S. Roan, *Picard groups of hypersurfaces in toric varieties*, preprint, 1995.

[Roa2] ———, *Topological properties of Calabi-Yau mirror manifolds*, Volume honoring Professor C. N. Yang, Bookcrafters, Inc., 1994, pp. 251–264.

[SI1] A. Sardo Infirri, *Lefschetz fixed point theorem and lattice points in convex polytopes*, Adv. Math. **116** (1995), 55–81.

[SI2] ———, *Resolutions of orbifold singularities and flows on the McKay quiver*, preprint, 1996.

[Sat] I. Satake, *On the arithmetic of tube domains*, Bull. Amer. Math. Soc. **79** (1973), 1076–1094.

[Ska] H. Skarke, *Weight systems for toric Calabi-Yau varieties and reflexivity of Newton polyhedra*, preprint, 1996.

[Sta] R. Stanley, *The number of faces of a convex simplicial polytope*, Adv. in Math. **35** (1980), 236–238.

[Stu1] B. Sturmfels, *Equations defining toric varieties*, this volume.

[Stu2] ———, *Gröbner Bases and Convex Polytopes*, University Lecture Series **8**, AMS, Providence, RI, 1996.

[Stu3] ———, *On the Newton polytope of the resultant*, J. Algebraic Combin. **3** (1994), 207–236.

[Stu4] ———, *Sparse eliminaton theory*, Computational Algebraic Geometry and Commutative Algebra, D. Eisenbud and L. Robbiano, editors, Cambridge Univ. Press, Cambridge, 1993, pp. 264–298.

[Ter] T. Terasoma, *Infinitesimal variation of Hodge structures and the weak global Torelli theorem for complete intersections*, Ann. of Math. **132** (1990), 213–235.

[Tha] M. Thaddeus, *Toric quotients and flips*, Topology, Geometry and Field Theory, World Sci. Publishing, River Edge, NJ, 1994, pp. 193–213.

[Tot] B. Totaro, *Chow groups of linear varieties*, preprint, 1994.

[Tsu] H. Tsuchihashi, *Hypersurface singularities of toric singularities*, Kodai Math. J. **14** (1991), 210–221.

[Var] A. Varchenko, *Zeta function of monodromy and Newton's diagram*, Invent. Math. **37** (1976), 253–262.

[Ver] M. Verbitsky, *Mirror symmetry for hyperkaehler manifolds*, preprint, 1995.

[Wit] E. Witten, *Phases of $N = 2$ theories in two dimensions*, Nucl. Phys. **B403** (1993), 159–222.

[Wło] J. Włodarczyk, *Embeddings in toric varieties and prevarieties*, J. Algebraic Geom. **2** (1993), 705–726.

[Yau] S.-T. Yau, editor, *Essays on Mirror Manifolds*, International Press, Hong Kong, 1992.

[Zie] G. Ziegler, *Lectures on Polytopes*, Graduate Texts in Math. **152**, Springer-Verlag, Berlin Heidelberg New York, 1995.

Department of Mathematics and Computer Science, Amherst College, Amherst, Massachusetts 01002

*E-mail address*: `dac@cs.amherst.edu`

Proceedings of Symposia in Pure Mathematics
Volume **62.2**, 1997

# Equations Defining Toric Varieties

Bernd Sturmfels

This article is based on my lecture at the AMS Summer Institute in Algebraic Geometry at Santa Cruz, July 1995. The topic is toric ideals, by which I mean the defining ideals of subvarieties of affine or projective space which are parametrized by monomials. With one exception there will be no proofs given in this paper. For most assertions which are stated without a reference, the reader can find a proof and combinatorial details in my monograph "Gröbner Bases and Convex Polytopes" [**Stu**]. Connections to recent developments in algebraic geometry and theoretical physics are discussed in the beautiful survey by David Cox [**Cox**].

## 1. Toric ideals and their polytopes

In many branches of mathematics and its applications one encounters algebraic varieties which are parametrized by monomials. Such varieties are called *toric varieties* in this article. This stands in contrast to common practise in algebraic geometry (see [**Cox**]), where toric varieties are assumed to be normal. From the point of view taken by the author it is more natural to start out with the following definition. Let $\mathcal{A} = (\mathbf{a}_1, \dots, \mathbf{a}_n)$ be any integer $d \times n$-matrix. Each column vector $\mathbf{a}_i = (a_{1i}, \dots, a_{di})^T$ is identified with a Laurent monomial $\mathbf{t}^{\mathbf{a}_i} = t_1^{a_{1i}} \cdots t_d^{a_{di}}$. The *toric ideal* $I_\mathcal{A}$ associated with $\mathcal{A}$ is the kernel of the $k$-algebra homomorphism

$$\mathbb{C}[x_1, x_2, \dots, x_n] \to \mathbb{C}[t_1, \dots, t_d, t_1^{-1}, \dots, t_d^{-1}],\ x_i \mapsto \mathbf{t}^{\mathbf{a}_i}.$$

Every vector $\mathbf{u}$ in $\mathbb{Z}^n$ can be written uniquely as $\mathbf{u} = \mathbf{u}_+ - \mathbf{u}_-$, where $\mathbf{u}_+$ and $\mathbf{u}_-$ are non-negative and have disjoint support. The difference of two monomials is called a *binomial.* An ideal generated by binomials is called a *binomial ideal.*

LEMMA 1.1.
(a) *The toric ideal $I_\mathcal{A}$ is generated by the binomials $\mathbf{x}^{\mathbf{u}_+} - \mathbf{x}^{\mathbf{u}_-}$, where $\mathbf{u}$ runs over all integer vectors in the kernel of the matrix $\mathcal{A}$.*
(b) *An ideal in $\mathbb{C}[x_1, \dots, x_n]$ is toric if and only if it is prime and binomial.*

---

1991 *Mathematics Subject Classification.* Primary 14M25, 52B20; Secondary 13P10.
Supported in part by the NSF and the David and Lucile Packard Foundation.

Here is an easy method for computing generators of $I_\mathcal{A}$. Assume for simplicity that $\mathcal{A} \subset \mathbb{N}^d$. Then our toric ideal equals the elimination ideal

$$(1.1) \qquad I_\mathcal{A} \quad = \quad \langle x_1 - \mathbf{t}^{\mathbf{a}_1}, x_2 - \mathbf{t}^{\mathbf{a}_2}, \cdots x_n - \mathbf{t}^{\mathbf{a}_n} \rangle \, \cap \, \mathbb{C}[x_1, x_2, \ldots, x_n],$$

which can be computed by lexicographic Gröbner bases in $\mathbb{C}[x_1, \ldots, x_n, t_1, \ldots, t_d]$. More efficient algorithms for the same task are described in Section 12.1 of [**Stu**].

The zero set of $I_\mathcal{A}$ in affine $n$-space is denoted $X_\mathcal{A}$ and called the *affine toric variety* defined by $\mathcal{A}$. The dimension of $X_\mathcal{A}$ equals the rank of the matrix $\mathcal{A}$. If all columns of $\mathcal{A}$ have the same coordinate sum, then the ideal $I_\mathcal{A}$ is homogeneous and defines a *projective toric variety* $Y_\mathcal{A}$ in $\mathbb{P}^{n-1}$. In what follows we identify $\mathcal{A}$ with the point configuration given by its columns.

EXAMPLES 1.2. Here are some familiar examples of projective toric varieties.

(a) The *twisted cubic curve* in $\mathbb{P}^3$ is defined by four equidistant points on a line:

$$\mathcal{A} \quad = \quad \begin{pmatrix} 3 & 2 & 1 & 0 \\ 0 & 1 & 2 & 3 \end{pmatrix}$$

The corresponding toric ideal is generated by three quadratic binomials:

$$I_\mathcal{A} \quad = \quad \langle x_1x_3 - x_2^2, \, x_1x_4 - x_2x_3, \, x_2x_4 - x_3^2 \rangle.$$

More generally, the *$r$-th Veronese embedding of* $\mathbb{P}^{n-1}$ equals $Y_\mathcal{A}$ for

$$\mathcal{A} \quad = \quad \big\{ (i_1, i_2, \ldots, i_n) \in \mathbb{N}^n \, : \, i_1 + i_2 + \cdots + i_n = r \big\}.$$

(b) All *rational normal scrolls* are toric. For instance, the *cubic scroll* $S_{2,1}$ in $\mathbb{P}^4$ is defined by the matrix

$$\mathcal{A} \quad = \quad \begin{pmatrix} 2 & 1 & 1 & 0 & 0 \\ 0 & 1 & 0 & 2 & 1 \\ 0 & 0 & 1 & 0 & 1 \end{pmatrix}$$

If we were to add a column vector $(0, 0, 2)$ to this matrix then we would get the quadratic Veronese embedding of $\mathbb{P}^2$ into $\mathbb{P}^5$.

(c) The Segre embedding of $\mathbb{P}^r \times \mathbb{P}^s$ into $\mathbb{P}^{rs+r+s}$ is toric. Here $I_\mathcal{A}$ is the ideal of $2 \times 2$-minors of an $(r+1) \times (s+1)$-matrix of indeterminates, and the configuration $\mathcal{A}$ consists of the $rs$ vertices of the product of two regular simplices $\Delta_r \times \Delta_s$.

(d) Consider a generic point in the Grassmannian of lines in $\mathbb{P}^3$. The closure of its orbit under the natural $(\mathbb{C}^*)^4$-action is the toric variety $Y_\mathcal{A}$, where

$$\mathcal{A} \quad = \quad \begin{pmatrix} 1 & 1 & 1 & 0 & 0 & 0 \\ 1 & 0 & 0 & 1 & 1 & 0 \\ 0 & 1 & 0 & 1 & 0 & 1 \\ 0 & 0 & 1 & 0 & 1 & 1 \end{pmatrix}$$

The closure of any torus orbit in a flag variety arises from a configuration $\mathcal{A}$ of weights in a $GL_n(\mathbb{C})$-module. That module is $\wedge^2\mathbb{C}^4$ in the above case. □

The $d$-dimensional algebraic torus $(\mathbb{C}^*)^d$ acts on affine $n$-space $\mathbb{C}^n$ via

$$(x_1, \ldots, x_n) \;\mapsto\; (x_1\mathbf{t}^{\mathbf{a}_1}, \ldots, x_n\mathbf{t}^{\mathbf{a}_n}).$$

The affine toric variety $X_{\mathcal{A}}$ is the closure of the $(\mathbb{C}^*)^d$-orbit of the point $(1,1,\ldots,1)$. A basic invariant of $X_{\mathcal{A}}$ is the *convex polyhedral cone* $pos(\mathcal{A})$ consisting of all non-negative linear combinations of column vectors in $\mathcal{A}$. For a projective toric variety $Y_{\mathcal{A}}$ we also consider the convex hull $conv(\mathcal{A})$ of the points in $\mathcal{A}$. This is a *convex polytope* of dimension $rank(\mathcal{A})-1$. Note that $pos(\mathcal{A})$ equals the cone over $conv(\mathcal{A})$.

PROPOSITION 1.3.
(a) *The $(\mathbb{C}^*)^d$-orbits on the affine toric variety $X_{\mathcal{A}}$ are in order-preserving bijection with the faces of the cone $pos(\mathcal{A})$.*
(b) *The $(\mathbb{C}^*)^d$-orbits on the projective toric variety $Y_{\mathcal{A}}$ are in order-preserving bijection with the faces of the polytope $conv(\mathcal{A})$.*

We next determine the degree of a projective toric variety $Y_{\mathcal{A}}$ in $\mathbb{P}^{n-1}$. Let $\mathcal{L}$ be the sublattice of $\mathbb{Z}^d$ affinely generated by $\mathcal{A}$. We normalize the volume form on $\mathcal{L}\otimes_{\mathbb{Z}}\mathbb{R}$ in such a way that each primitive lattice simplex has unit volume. The normalized volume of the polytope $conv(\mathcal{A})$ is a positive integer denoted $Vol(\mathcal{A})$.

THEOREM 1.4. *The degree of $Y_{\mathcal{A}}$ in $\mathbb{P}^{n-1}$ equals the normalized volume $Vol(\mathcal{A})$.*

It is instructive to verify Theorem 1.4 for the varieties in Examples 1.2.
(a) For the twisted cubic, $conv(\mathcal{A})$ is a line segment of normalized length 3.
(b) For the cubic scroll $S_{2,1}$, $conv(\mathcal{A})$ is a quadrangle with three edges of unit length and one edge of length two. Its normalized area equals $Vol(\mathcal{A}) = 3$.
(c) The product of simplices $\Delta_r \times \Delta_s$ can be triangulated into $\binom{r+s}{r}$ simplices of unit volume. Hence the degree of the Segre variety equals $\binom{r+s}{r}$.
(d) For the generic torus orbit on the Grassmannian of lines in $\mathbb{P}^3$, the polytope $conv(\mathcal{A})$ is a regular octahedron. It can be triangulated into four tetrahedra of unit volume. Therefore $Y_{\mathcal{A}}$ is a toric threefold of degree four in $\mathbb{P}^5$.

The toric ideals $I_{\mathcal{A}}$ in Example 1.2 are generated by quadratic binomials, and their varieties $Y_{\mathcal{A}}$ are projectively normal (see Section 2). The quadratic generators are easy to find, in view of the special structure of the matrices $\mathcal{A}$. The simplicity of these geometric examples is misleading: for a general matrix $\mathcal{A}$ it is difficult to identify generators for $I_{\mathcal{A}}$. One objective of this article is to describe methods for studying and solving this problem. We illustrate this issue for two families of toric ideals which arise from an application in computational statistics.

EXAMPLE 1.5. (The toric variety of the Birkhoff polytope)
Let $\mathcal{A}$ be the set of $p\times p$-permutation matrices. Here $d=p^2$, $n=p!$, and $conv(\mathcal{A})$ is the *Birkhoff polytope* consisting of all doubly-stochastic matrices. The dimension of $conv(\mathcal{A})$ and hence of the toric variety $Y_{\mathcal{A}}$ is $(p-1)^2$. The variables in the toric ideal $I_{\mathcal{A}}$ are indexed by permutations. For instance, for $p=3$ we have

$$I_{\mathcal{A}} \quad = \quad \langle\, x_{123}x_{231}x_{312} \;-\; x_{132}x_{213}x_{321}\,\rangle.$$

In general, the toric ideal $I_{\mathcal{A}}$ is generated by forms of degree $p$. It is a challenging combinatorial problem to determine $Vol(\mathcal{A}) = degree(Y_{\mathcal{A}})$. The known values are:

| $p$ | 3 | 4 | 5 | 6 | 7 |
|---|---|---|---|---|---|
| $dim(Y_{\mathcal{A}})$ | 4 | 9 | 16 | 25 | 36 |
| $degree(Y_{\mathcal{A}})$ | 3 | 352 | 4718075 | 14666561365176 | 17832560768358341943028 |

Another open problem is to bound the degree of the universal Gröbner basis of $I_{\mathcal{A}}$.

EXAMPLE 1.6. (Looks like Segre but isn't)
Fix integers $r \leq s \leq t$. Let $n = rst$, $d = rs + rt + st$ and identify $\mathbb{Z}^d$ with the direct sum $\mathbb{Z}^{r\times s} \oplus \mathbb{Z}^{r\times t} \oplus \mathbb{Z}^{s\times t}$. We denote the standard basis vectors in the three components as $\mathbf{e}_{ij}$, $\mathbf{e}'_{ik}$ and $\mathbf{e}''_{jk}$ respectively. Our configuration in this example is

$$\mathcal{A} \quad = \quad \{ \mathbf{e}_{ij} \oplus \mathbf{e}'_{ik} \oplus \mathbf{e}''_{jk} \; : \; i = 1, \ldots, r, \; j = 1, \ldots, s, \; k = 1, \ldots, t \}.$$

The toric ideal $I_\mathcal{A}$ is the kernel of the ring map

$$\mathbb{C}[x_{ijk}] \; \rightarrow \; \mathbb{C}[u_{ij}, v_{ik}, w_{jk}], \quad x_{ijk} \; \mapsto \; u_{ij} \cdot v_{ik} \cdot w_{jk},$$

where the indices $i, j, k$ run as in $\mathcal{A}$. The smallest example is $r = s = t = 2$, where

$$I_\mathcal{A} \quad = \quad \langle x_{111}x_{122}x_{212}x_{221} \; - \; x_{112}x_{121}x_{211}x_{222} \rangle.$$

At first glance the projective toric variety $Y_\mathcal{A}$ looks similar to a Segre variety. But this is a deception. The following questions are widely open for general $r, s, t \geq 3$:

- Characterize the faces of $conv(\mathcal{A})$, i.e. the torus orbits on $Y_\mathcal{A}$.
- Determine the normalized volume of $conv(\mathcal{A})$, i.e. the degree of $Y_\mathcal{A}$.
- Find minimal generators for $I_\mathcal{A}$, or at least bound their degree.

## 2. Normal toric varieties

In algebraic geometry one often assumes that $X_\mathcal{A}$ is normal and that $Y_\mathcal{A}$ is projectively normal. This imposes strong combinatorial restrictions on the configuration $\mathcal{A}$. We first discuss these restrictions and then we present known results and open problems concerning the generators of the corresponding toric ideals $I_\mathcal{A}$.

Let $\mathbb{N}\mathcal{A}$ denote the semigroup spanned by $\mathcal{A}$. Throughout this paper we assume that $\mathbb{N}\mathcal{A} \cap -\mathbb{N}\mathcal{A} = \{0\}$. This condition means that the semigroup algebra $\mathbb{C}[\mathcal{A}] := \mathbb{C}[x_1, \ldots, x_n]/I_\mathcal{A}$ has no non-trivial units. The semigroup $\mathbb{N}\mathcal{A}$ lies in the intersection of the abelian group $\mathbb{Z}\mathcal{A}$ and the convex polyhedral cone $pos(\mathcal{A})$:

$$\mathbb{N}\mathcal{A} \quad \subseteq \quad pos(\mathcal{A}) \cap \mathbb{Z}\mathcal{A}. \tag{2.1}$$

We say that the configuration $\mathcal{A}$ is *normal* if equality holds in (2.1).

LEMMA 2.1. *The affine variety $X_\mathcal{A}$ is normal if and only if $\mathcal{A}$ is normal.*

If $\mathcal{A}$ is not normal, then one can replace it (e.g. using Algorithm 13.2 in [**Stu**]) by the unique minimal finite subset $\mathcal{B}$ of $\mathbb{Z}^d$ such that

$$\mathbb{N}\mathcal{B} \quad = \quad pos(\mathcal{A}) \cap \mathbb{Z}\mathcal{A}.$$

In this case $X_\mathcal{B}$ is the *normalization* of $X_\mathcal{A}$.

Singularities of toric varieties are characterized as follows:

LEMMA 2.2. *The affine toric variety $X_\mathcal{A}$ is smooth if and only if the semigroup $\mathbb{N}\mathcal{A}$ is isomorphic to the free semigroup $\mathbb{N}^r$ for some $r$.*

EXAMPLE 2.3. The affine toric surface $X_{\mathcal{A}}$ defined by $\mathcal{A} = \{(2,0),(1,1),(0,2)\}$ is normal but not smooth. It is the cone over a smooth quadric curve in $\mathbb{P}^2$.

From now on we assume that the ideal $I_{\mathcal{A}}$ is homogeneous. It defines a projective toric variety $Y_{\mathcal{A}}$ in $\mathbb{P}^{n-1}$. The intersection of $Y_{\mathcal{A}}$ with the affine chart $\{x_i \neq 0\} \simeq \mathbb{A}^{n-1}$ equals the affine toric variety $X_{\mathcal{A}-\mathbf{a}_i}$ defined by the configuration

$$\mathcal{A} - \mathbf{a}_i \quad = \quad \{\mathbf{a}_1 - \mathbf{a}_i, \ldots, \mathbf{a}_{i-1} - \mathbf{a}_i, \mathbf{a}_{i+1} - \mathbf{a}_i, \ldots, \mathbf{a}_n - \mathbf{a}_i\}.$$

Thus $Y_{\mathcal{A}}$ has an open cover consisting of $n$ affine toric varieties. In general the number $n$ can be lowered, by the following proposition.

PROPOSITION 2.4. *The projective toric variety $Y_{\mathcal{A}}$ is covered irredundently by the affine varieties $X_{\mathcal{A}-\mathbf{a}_i}$ where $\mathbf{a}_i$ runs over the vertices of the polytope $conv(\mathcal{A})$.*

Thus $Y_{\mathcal{A}}$ is normal (resp. smooth) if and only if each affine chart $X_{\mathcal{A}-\mathbf{a}_i}$ in the above cover is normal (resp. smooth). In particular, $Y_{\mathcal{A}}$ is normal if and only if

$$\mathbb{N}(\mathcal{A} - \mathbf{a}_i) \quad = \quad pos(\mathcal{A} - \mathbf{a}_i) \cap \mathbb{Z}(\mathcal{A} - \mathbf{a}_i) \qquad \text{for} \quad i = 1, \ldots, n.$$

Let $H_{\mathcal{A}}(s)$ denote the *Hilbert polynomial* of $Y_{\mathcal{A}}$, that is, $H_{\mathcal{A}}(s) = dim_k(\mathbb{C}[\mathcal{A}]_s)$ for $s \gg 0$. The *Ehrhart polynomial* $E_{\mathcal{A}}$ of the lattice polytope $conv(\mathcal{A})$ is defined by

$$E_{\mathcal{A}}(s) \quad = \quad \#\big( s \cdot conv(\mathcal{A}) \ \cap \ \mathbb{Z}^d \big) \qquad \text{for all } s \in \mathbb{N}.$$

The relationship between the Hilbert polynomial and the Ehrhart polynomial was studied by A. Khovanskii in [**Kho**].

PROPOSITION 2.5. *A projective toric variety $Y_{\mathcal{A}}$ is normal if and only if its Hilbert polynomial $H_{\mathcal{A}}$ is equal to its Ehrhart polynomial $E_{\mathcal{A}}$.*

A much stronger requirement is to ask that $Y_{\mathcal{A}}$ be *projectively normal*, which means that the affine cone $X_{\mathcal{A}}$ over $Y_{\mathcal{A}}$ is normal, i.e., $\mathbb{N}\mathcal{A} = pos(\mathcal{A}) \cap \mathbb{Z}\mathcal{A}$.

EXAMPLE 2.6. (Normal versus projectively normal)
Let $d = 2, n = 4, r \geq 4$ and $\mathcal{A} = \{(r,0),(r-1,1),(1,r-1),(0,r)\}$. Here $Y_{\mathcal{A}}$ equals the projective line $\mathbb{P}^1$. It is smooth (hence normal) but not projectively normal. The toric ideal $I_{\mathcal{A}} \subset \mathbb{C}[x_1,x_2,x_3,x_4]$ is minimally generated by one quadric and $r-1$ binomials of degree $r-1$. In this example $X_{\mathcal{A}}$ is not Cohen-Macaulay. □

A well-known result due to M. Hochster states that normal affine toric varieties are Cohen-Macaulay. A consequence of this fact is the following degree bound.

THEOREM 2.7. *If $Y_{\mathcal{A}}$ is a projectively normal $r$-dimensional toric variety then the homogeneous toric ideal $I_{\mathcal{A}}$ is generated by binomials of degree at most $r$.*

This degree bound is sharp since the toric hypersurface $x_0^r = x_1 x_2 \cdots x_r$ is projectively normal. It is unknown whether Theorem 2.7 extends to Gröbner bases. In the following conjecture we do not allow any linear changes of coordinates.

CONJECTURE 2.8. *If $Y_{\mathcal{A}}$ is projectively normal $r$-dimensional toric variety, then $I_{\mathcal{A}}$ has a Gröbner basis consisting of binomials of degree at most $r$.*

To appreciate the distinction between generators and Gröbner bases consider the following example: If $d = 3, n = 8$ and $\mathcal{A} = \{(3,0,0),(0,3,0),(0,0,3),(2,1,0),$ $(1,2,0),(2,0,1),(1,0,2),(0,2,1),(0,1,2)\}$, then $I_{\mathcal{A}}$ is generated by quadrics but has no quadratic Gröbner basis. It is unknown whether $\mathbb{C}[\mathcal{A}]$ is a Koszul algebra.

Another problem is to better understand the effect of smoothness on the degrees of the defining equations.

CONJECTURE 2.9. *If $Y_{\mathcal{A}}$ is a smooth and projectively normal toric variety, then the toric ideal $I_{\mathcal{A}}$ is generated by quadratic binomials.*

It may even be conjectured that in this case $I_{\mathcal{A}}$ possesses a quadratic Gröbner basis. It was briefly believed in January '95 that Conjecture 2.9 had been proven, but that proof was withdrawn. The answer is affirmative for scrolls by a result of Ewald and Schmeinck [**EwS**]. Conjecture 2.9 is also known to be true for toric surfaces. This is a consequence of the following theorem due to Bruns, Gubeladze and Trung [**BGT**]

THEOREM 2.10. *Let $Y_{\mathcal{A}}$ be a projectively normal toric surface and suppose that the polygon conv$(\mathcal{A})$ has at least four lattice points on its boundary. Then $I_{\mathcal{A}}$ possesses a quadratic lexicographic Gröbner basis.*

There is an important sufficient condition for $Y_{\mathcal{A}}$ to be projectively normal. (It is not necessary; see Example 13.17 in [**Stu**]). A *triangulation* $\Delta$ of the configuration $\mathcal{A}$ is a triangulation of the polytope *conv*$(\mathcal{A})$ whose vertices lie in $\mathcal{A}$. We call the triangulation $\Delta$ *unimodular* if each simplex in $\Delta$ is a primitive lattice simplex.

THEOREM 2.11. *There exists a unimodular triangulation of $\mathcal{A}$ if and only if some initial monomial ideal of $I_{\mathcal{A}}$ is radical. In this case $Y_{\mathcal{A}}$ is projectively normal.*

The following theorem was established by Knudsen and Mumford in the early days of toric geometry. It is a key ingredient in Mumford's proof of the semi-stable reduction theorem. See [**KKMS**] for details and the proof of Theorem 2.12.

THEOREM 2.12. *Let $Q$ be any lattice polytope in $\mathbb{Z}^d$. There exists an integer $m \gg 0$ such that the configuration $mQ \cap \mathbb{Z}^d$ possesses a unimodular triangulation.*

It would be interesting to find an effective version of this theorem.

PROBLEM 2.13. *Does there exist a bound $M(d)$ such that, for every $m \geq M(d)$ and every lattice polytope $Q$ in $\mathbb{Z}^d$, the set $mQ \cap \mathbb{Z}^d$ has a unimodular triangulation ?*

In our discussion so far $Y_{\mathcal{A}}$ was given as an explicit subvariety of some projective space $\mathbb{P}^{n-1}$. What can be said about $Y_{\mathcal{A}}$ as an abstract scheme, independently of any choice of a very ample line bundle ? This is where *polyhedral fans* enter the picture. Let $Q$ be any polytope in a real vector space $V$. For a face $F$ of $Q$ we consider the set of linear functionals on $Q$ which attain their maximum at $F$. This is a convex polyhedral cone in the dual space $V^*$. The collection of these cones, as $F$ runs over all faces of $Q$, is a polyhedral fan. It is called the *normal fan* of $Q$.

PROPOSITION 2.14. *Two projective toric varieties $Y_{\mathcal{A}}$ and $Y_{\mathcal{B}}$ have isomorphic normalizations if and only if their polytopes conv$(\mathcal{A})$ and conv$(\mathcal{B})$ have the same normal fan.*

Here "isomorphic" refers to an isomorphism of equivariant torus embeddings. If $Y_{\mathcal{A}}$ is normal, then the normal fan of *conv*$(\mathcal{A})$ retains just enough information to remember $Y_{\mathcal{A}}$ as an abstract torus embedding. But it forgets the specific line bundle which was used to map $Y_{\mathcal{A}}$ into $\mathbb{P}^{n-1}$. In the synthetic approach to toric varieties, as presented in the books [**Ful**] and [**Oda**], the normal fan comes before the polytope. Starting with any complete fan, one constructs an abstract complete normal toric variety by gluing affine pieces as in Proposition 2.4. See also [**Cox**].

EXAMPLE 2.15. Fix positive integers $i < j < k$ and consider the configuration

$$\mathcal{A} \quad := \quad conv(\{(i,j,k),(i,k,j),(j,i,k),(j,k,i),(k,i,j),(k,j,i)\}) \cap \mathbb{Z}^3.$$

The normal fan of the hexagon $conv(\mathcal{A})$ is independent of the choice of $i,j,k$. All toric surfaces $Y_\mathcal{A}$ arising for different choices of $i,j,k$ are normal and isomorphic to one another. The abstract scheme $Y_\mathcal{A}$ equals $\mathbb{P}^2$ blown up at three points. All ideals $I_\mathcal{A}$ possess quadratic lexicographic Gröbner bases, by Theorem 2.10.

In many applications of toric geometry one encounters toric varieties $Y_\mathcal{A}$ which are not normal. Or sometimes they are normal but this is difficult to verify. We describe two instances of the latter kind arising from representation theory.

Let $G$ be a connected semi-simple algebraic group over $\mathbb{C}$. Fix a maximal torus $(\mathbb{C}^*)^d$ in $G$, let $P$ be a parabolic subgroup containing $(\mathbb{C}^*)^d$ and consider the flag variety $G/P$. The following result is due to R. Dabrowski [**Dab**].

THEOREM 2.16. *The closure in $G/P$ of a generic $(\mathbb{C}^*)^d$-orbit is normal.*

It is open whether the generic torus orbit closures are *projectively normal* for all very ample line bundles on $G/P$. It is also open whether the closures of all (non-generic) $(\mathbb{C}^*)^d$-orbits are normal, or even projectively normal. This latter conjecture is known to be true in the case when $G/P$ is the classical Grassmannian $Gr_r(\mathbb{C}^d)$ in its Plücker embedding. In this case $\mathcal{A}$ is a subset of the *hypersimplex*

$$\{(i_1,\ldots,i_d) \in \{0,1\}^d \,:\, i_1 + \cdots + i_d = r\}$$

such that $\mathcal{A}$ consists of the incidence vectors of the bases of a realizable *matroid.* For instance, in Example 1.2 (d) that matroid is the uniform rank 2 matroid on 4 elements. We refer to [**GGMS**] for an introduction to matroids from an algebro-geometric point of view. The following result is due to Neil White [**Wh1**].

THEOREM 2.17. *Let $\mathcal{A}$ be the set of incidence vectors of the bases of a matroid. Then $Y_\mathcal{A}$ is projectively normal.*

We close with a reformulation of a combinatorial conjecture in [**Wh2**].

CONJECTURE 2.18. *Let $\mathcal{A}$ be the set of incidence vectors of the bases of a matroid. Then the homogeneous toric ideal $I_\mathcal{A}$ is generated by quadratic binomials.*

## 3. Binomial Zoo

Our objective is to understand the minimal generators and Gröbner bases of the toric ideal $I_\mathcal{A}$. To this end we introduce the following three definitions. A binomial $\mathbf{x}^{\mathbf{u}_+} - \mathbf{x}^{\mathbf{u}_-}$ in $I_\mathcal{A}$ is called a *circuit* if its support (i.e. the set of variables appearing in that binomial) is minimal with respect to inclusion. We write $\mathcal{C}_\mathcal{A}$ for the set of all circuits in $I_\mathcal{A}$. Geometrically speaking, we consider all images of $X_\mathcal{A}$ under projection into coordinate subspaces of $\mathbb{P}^{n-1}$; such an image is called a circuit of $X_\mathcal{A}$ if it has codimension 1 in its coordinate subspace. We define the *universal Gröbner basis* $\mathcal{U}_\mathcal{A}$ to be the union of all reduced Gröbner bases of $I_\mathcal{A}$. We say that a binomial $\mathbf{x}^{\mathbf{u}_+} - \mathbf{x}^{\mathbf{u}_-}$ in $I_\mathcal{A}$ lies in the *Graver basis* $Gr_\mathcal{A}$ if there exists no other binomial $\mathbf{x}^{\mathbf{v}_+} - \mathbf{x}^{\mathbf{v}_-} \in I_\mathcal{A}$ such that $\mathbf{x}^{\mathbf{v}_+}$ divides $\mathbf{x}^{\mathbf{u}_+}$ and $\mathbf{x}^{\mathbf{v}_-}$ divides $\mathbf{x}^{\mathbf{u}_-}$.

PROPOSITION 3.1. *For any toric ideal $I_{\mathcal{A}}$ we have the inclusions*

$$\mathcal{C}_{\mathcal{A}} \subseteq \mathcal{U}_{\mathcal{A}} \subseteq Gr_{\mathcal{A}}. \tag{3.1}$$

*Each of the four combinations of strict or non-strict inclusions is possible.*

EXAMPLES 3.2.

1. For the twisted cubic curve in Example 1.2 (a) we have

$$\begin{aligned} \mathcal{C}_{\mathcal{A}} &= \{x_1x_3 - x_2^2,\, x_2x_4 - x_3^2,\, x_1^2x_4 - x_2^3,\, x_1x_4^2 - x_3^3\} \quad \text{and} \\ \mathcal{U}_{\mathcal{A}} &= Gr_{\mathcal{A}} = \mathcal{C}_{\mathcal{A}} \cup \{x_1x_4 - x_2x_3\}. \end{aligned}$$

   In this example the circuits do not generate the ideal $I_{\mathcal{A}}$ although they do define the twisted cubic as a subscheme of $\mathbb{P}^3$.
2. For the Veronese surface in $\mathbb{P}^5$, the set of circuits $\mathcal{C}_{\mathcal{A}}$ equals the universal Gröbner basis $\mathcal{U}_{\mathcal{A}}$, but the Graver basis $Gr_{\mathcal{A}}$ properly contains $\mathcal{U}_{\mathcal{A}}$.

In general, the circuits of a projective toric variety $Y_{\mathcal{A}}$ do not define $Y_{\mathcal{A}}$ scheme-theoretically. But they do set-theoretically. We state this result in the affine case.

THEOREM 3.3. *The toric variety $X_{\mathcal{A}}$ is cut out by its circuits: $Rad\langle\mathcal{C}_{\mathcal{A}}\rangle = I_{\mathcal{A}}$. The embedded components of $\langle\mathcal{C}_{\mathcal{A}}\rangle$ are supported on torus orbits of $X_{\mathcal{A}}$, i.e. every associated prime of $\langle\mathcal{C}_{\mathcal{A}}\rangle$ has the form $I_{F\cap\mathcal{A}}$ for some face $F$ of the cone $pos(\mathcal{A})$.*

Theorem 3.3 was proved jointly with David Eisenbud in [**EiSt**]. The main result in [**EiSt**] concerns the decomposition of arbitrary binomial schemes. It underlines the importance of toric ideals as combinatorial building blocks in algebraic geometry:

THEOREM 3.4. *Let $I$ be any binomial ideal in $\mathbb{C}[x_1,\dots,x_n]$. Then $I$ possesses a primary decomposition into primary binomial ideals. In particular, every associated prime of $I$ has the form $I_{\mathcal{A}} + \langle x_{i_1},\dots,x_{i_r}\rangle$ for some configuration $\mathcal{A}$.*

There is an important class of toric varieties for which both inclusions in (3.1) are equalities. They are called *unimodular* and defined by the following theorem.

THEOREM 3.5. *For a configuration $\mathcal{A}$ the following conditions are equivalent:*

1. *Every triangulation of $\mathcal{A}$ is unimodular.*
2. *For every subset $\mathcal{B}$ of $\mathcal{A}$, the quotient group $\mathbb{Z}\mathcal{A}/\mathbb{Z}\mathcal{B}$ is free abelian.*
3. *Every initial monomial ideal of $I_{\mathcal{A}}$ is a radical ideal.*
4. *In every circuit of $I_{\mathcal{A}}$ both monomials are square-free.*

If these four equivalent conditions hold, then we call $\mathcal{A}$ *unimodular.*

PROPOSITION 3.6. *If $\mathcal{A}$ is unimodular then $X_{\mathcal{A}}$ is normal.*

EXAMPLE 3.7. The term "circuits" originates from the following class of unimodular toric varieties. Let $G$ be a finite directed graph. We label the vertices of $G$ by $1, 2, \dots, d$ and we encode the edges of $G$ as differences of unit vectors in $\mathbb{Z}^d$:

$$\mathcal{A}_G = \big\{\mathbf{e}_i - \mathbf{e}_j \in \mathbb{Z}^d \,:\, (i,j) \text{ is a directed edge of } G\big\}.$$

The unimodularity of $\mathcal{A}_G$ is a basic result in matroid theory. The circuits in the toric ideal $I_{\mathcal{A}_G}$ correspond to the directed circuits in $G$. The following examples of circuits of length five illustrate this correspondence:

| Directed circuit in $G$ | $\leftrightarrow$ | Circuit in $I_{\mathcal{A}_G}$ |
|---|---|---|
| $1 \to 2 \to 3 \to 4 \to 5 \to 1$ | $\leftrightarrow$ | $x_{12}x_{23}x_{34}x_{45}x_{51} - 1$ |
| $1 \to 2 \to 3 \leftarrow 4 \to 5 \to 1$ | $\leftrightarrow$ | $x_{12}x_{23}x_{45}x_{51} - x_{43}$ |
| $1 \to 2 \to 3 \leftarrow 4 \to 5 \leftarrow 1$ | $\leftrightarrow$ | $x_{12}x_{23}x_{45} - x_{15}x_{43}$ |

If $G$ is the complete directed bipartite graph $K_{r,s}$ then $I_{\mathcal{A}_G}$ is the ideal of $2 \times 2$-minors of an $r \times s$-matrix of indeterminates and $Y_{\mathcal{A}_G}$ is the Segre variety $\mathbb{P}^{r-1} \times \mathbb{P}^{s-1}$. The degrees of the circuits in $I_{\mathcal{A}_G}$ range from 2 to $max\{r, s\}$.

If $G$ is a complete graph, then $\mathcal{A}_G$ equals the *root system* of type $A_{n-1}$. The toric variety $X_{\mathcal{A}_G}$ is the closure of a generic $(\mathbb{C}^*)^n$-orbit in the *adjoint representation* of $GL_n(\mathbb{C})$. Passing to subgraphs corresponds to passing to non-generic orbit closures in the adjoint representation. Thus the affine toric varieties defined by directed graphs are precisely the closures of $(\mathbb{C}^*)^n$-orbit in the adjoint representation of $GL_n(\mathbb{C})$. These toric varieties are all normal by Proposition 3.6. □

A configuration $X_{\mathcal{A}}$ is called *hereditarily normal* if all closures of torus orbits on $X_{\mathcal{A}}$ are normal. This is equivalent to the requirement that $X_{\mathcal{B}}$ is normal for every subset $\mathcal{B}$ of $\mathcal{A}$. Unimodular configurations are hereditarily normal, but not conversely. Normality of torus orbits for abitrary semisimple algebraic groups was studied by J. Morand [**Mor**]. She showed that the root systems of types $A_n$, $B_2$, $C_2$, $D_4$ are hereditarily normal and that all other root systems are not hereditarily normal.

The following criterion underlines the geometric significance of the circuits.

THEOREM 3.8. *An affine toric variety $X_{\mathcal{A}}$ is hereditarily normal if and only if every circuit in $I_{\mathcal{A}}$ has at least one square-free monomial.*

The Graver basis $Gr_{\mathcal{A}}$ has the following geometric interpretation. Consider the closure of the affine variety $X_{\mathcal{A}} \subset \mathbb{C}^n$ in the $n$-fold product of projective lines $\mathbb{P}^1 \times \mathbb{P}^1 \times \cdots \times \mathbb{P}^1$. It is defined by a toric ideal $I_{\Lambda(\mathcal{A})}$ in $\mathbb{C}[x_1, y_1, x_2, y_2, \cdots, x_n, y_n]$. which is homogeneous with respect to each pair of variables $(x_i, y_i)$. Here $\Lambda(\mathcal{A})$ is a certain configuration of $2n$ vectors in $\mathbb{Z}^{n+d}$, which is called the *Lawrence lifting* of $\mathcal{A}$. A key property of such configurations $\Lambda(\mathcal{A})$ is the following.

THEOREM 3.9. *The defining ideal $I_{\Lambda(\mathcal{A})}$ of a toric subvariety of $\mathbb{P}^1 \times \cdots \times \mathbb{P}^1$ is minimally generated by its Graver basis $Gr_{\Lambda(\mathcal{A})}$.*

The original ideal $I_{\mathcal{A}}$ is recovered from $I_{\Lambda(\mathcal{A})}$ by dehomogenizing, that is, by replacing the variables $y_1, \ldots, y_n$ by 1. This induces a bijection between the Graver basis of $\Lambda(\mathcal{A})$ and the Graver basis of $\mathcal{A}$. Under this bijection we have the following geometric interpretation of the Graver basis.

COROLLARY 3.10. *The Graver basis of an affine toric variety consists of the minimal generators of the ideal of its closure in $\mathbb{P}^1 \times \cdots \times \mathbb{P}^1$.*

For instance, if $\mathcal{A}$ is the configuration in Example 1.2 (a), then

$$\begin{aligned} I_{\Lambda(\mathcal{A})} \quad = \quad & \langle\, x_1x_3y_2^2 - x_2^2y_1y_3,\ x_1x_4y_2y_3 - x_2x_3y_1y_4,\ x_2x_4y_3^2 - x_3^2y_2y_4, \\ & x_1^2x_4y_2^3 - x_2^3y_1^2y_4,\ x_1x_4^2y_3^3 - x_3^3y_1y_4^2 \,\rangle. \end{aligned}$$

This ideal defines a toric surface in $\mathbb{P}^1 \times \mathbb{P}^1 \times \mathbb{P}^1 \times \mathbb{P}^1$. It is the closure of the affine cone over the twisted cubic curve. We get the Graver basis for $I_{\mathcal{A}}$, the ideal of the twisted cubic, by setting $y_1, y_2, y_3, y_4$ to 1 in these five binomials.

## 4. Degree Bounds

The following problem is of great interest in computational algebraic geometry.

CONJECTURE 4.1. *Let $Y$ be an irreducible projective variety of degree $d$. Then the homogeneous prime ideal of $Y$ is generated by forms of degree at most $d$.*

Many authors have concentrated on proving stronger inequalities involving the *(Castelnuovo-Mumford) regularity* $reg(Y)$ of $Y$. Let $I_Y$ denote the ideal of $Y$ in $\mathbb{C}[x_1, \ldots, x_n]$. Then $reg(Y)$ is the maximum of the numbers $deg(\sigma) - i$ where $\sigma$ is a minimal $i$-th syzygy of $I_Y$. Here $i = 0$ is for ideal generators, $i = 1$ for syzygies among them, etc... Following [**BS**], two equivalent definitions of *regularity* are:

1. $reg(Y)$ is the maximum degree of an element in the reduced Gröbner basis of $I_{\phi(Y)}$ with respect to the reverse lexicographic term order, where $I_{\phi(Y)}$ is the image of $I_Y$ under a generic linear automorphism $\phi \in GL_n(\mathbb{C})$;
2. $reg(Y)$ is the smallest integer such that $H^i(\mathbb{P}^{n-1}, \mathcal{I}_Y(s)) = 0$ for all $i \geq 0$ and $s \geq reg(Y) - i$, where $\mathcal{I}_Y$ is the ideal sheaf of $Y$.

The following conjecture due to Eisenbud and Goto [**EG**] implies Conjecture 4.1.

CONJECTURE 4.2. *Let $Y$ be an irreducible projective variety, not contained in any hyperplane. Then $reg(Y) \leq degree(Y) - codim(Y) + 1$.*

The Eisenbud-Goto inequality holds when $Y$ is arithmetically Cohen-Macaulay. Hence, by Hochster's Theorem, it holds for projectively normal toric varieties. Conjecture 4.2 was proved for curves by Gruson-Lazarsfeld-Peskine [**GLP**], and for irreducible smooth surfaces and 3-folds by Lazarsfeld [**Laz**] and Ran [**R**].

Conjectures 4.1 and 4.2 are widely open in general, even for toric varieties $Y_\mathcal{A}$. Recall from Theorem 1.4 that the degree of $Y_\mathcal{A}$ equals the volume of the polytope $conv(\mathcal{A})$. Also the regularity of $Y_\mathcal{A}$ can be expressed combinatorially, using the simplicial representation of *Koszul homology* (see e.g. Theorem 12.12 in [**Stu**]). A class of toric varieties for which the Eisenbud-Goto inequality is valid was identified by Irena Peeva and the author in [**PS**]. For toric varieties $Y_\mathcal{A}$ in codimension 2 we explicitly construct the minimal free resolution of $I_\mathcal{A}$. Our construction implies:

THEOREM 4.3. *Let $Y_\mathcal{A}$ be a projective toric variety of codimension 2 in $\mathbb{P}^{n-1}$, not contained in any hyperplane. Then $reg(Y_\mathcal{A}) \leq degree(Y_\mathcal{A}) - 1$.*

EXAMPLE 4.4. The following family of toric surfaces in $\mathbb{P}^4$ shows that the inequality in Theorem 4.3 is tight. For any integer $d \geq 3$ let

$$\mathcal{A} \quad = \quad \begin{pmatrix} 1 & 1 & 1 & 1 & 1 \\ 0 & 1 & 1 & 0 & 0 \\ 0 & 0 & 1 & 1 & d \end{pmatrix}.$$

The surface $Y_\mathcal{A}$ is not arithmetically Cohen-Macaulay, its toric ideal $I_\mathcal{A}$ is generated by one quadric and $d$ forms of degree $d$, and $degree(Y_\mathcal{A}) = Vol(conv(\mathcal{A})) = d + 1$.

Given an arbitrary toric variety in higher codimension, the following is the currently best general bound for its regularity in terms of its degree.

THEOREM 4.5. *Let $Y_\mathcal{A}$ be a projective toric variety in $\mathbb{P}^{n-1}$. Then*

$$reg(Y_\mathcal{A}) \quad \leq \quad n \cdot degree(Y_\mathcal{A}) \cdot codim(Y_\mathcal{A}). \tag{4.1}$$

We shall outline the proof of Theorem 4.5. First, we make use of the fact that regularity is upper semi-continuous with respect to flat families. This implies

$$reg(I_{\mathcal{A}}) \quad \leq \quad reg(in_{\prec}(I_{\mathcal{A}})) \tag{4.2}$$

for any term order $\prec$ on $\mathbb{C}[x_1, \ldots, x_n]$. The *Taylor resolution* (see page 439 in [**Eis**]) for the monomial ideal $in_{\prec}(I_{\mathcal{A}})$ implies the inequality

$$reg(\, in_{\prec}(I_{\mathcal{A}})\,) \quad \leq \quad n \cdot maxgen(\, in_{\prec}(I_{\mathcal{A}})\,), \tag{4.3}$$

where $maxgen(\,\cdot\,)$ denotes the maximal degree of any minimal generator. In view of the inclusion $\mathcal{U}_{\mathcal{A}} \subseteq Gr_{\mathcal{A}}$ in Proposition 3.1, it suffices to prove the following:

LEMMA 4.6. *Let $Y_{\mathcal{A}}$ be a projective toric variety. Every binomial in the Graver basis $Gr_{\mathcal{A}}$ of the homogeneous ideal $I_{\mathcal{A}}$ has degree at most $degree(Y_{\mathcal{A}}) \cdot codim(Y_{\mathcal{A}})$.*

For a set $S$ of polynomials let $maxdeg(S)$ denote the maximum degree of any element in $S$. The proof of Lemma 4.6 is derived from the next two inequalities:

$$maxdeg(\mathcal{C}_{\mathcal{A}}) \quad \leq \quad degree(Y_{\mathcal{A}}) \tag{4.4}$$

This holds because the hypersurface defined by any circuit is a projection of $Y_{\mathcal{A}}$. The second inequality is

$$maxdeg(Gr_{\mathcal{A}}) \quad \leq \quad codim(Y_{\mathcal{A}}) \cdot maxdeg(\mathcal{C}_{\mathcal{A}}). \tag{4.5}$$

This follows from a standard inequality for Hilbert bases of integer monoids. See Chapter 4 in [**Stu**] for details.

While the inquality (4.4) is tight, there seems to be some room for improvement left in (4.5). In my lecture at Santa Cruz I asked whether even the equality $maxdeg(Gr_{\mathcal{A}}) = maxdeg(\mathcal{C}_{\mathcal{A}})$ might be true. This would have implied Conjecture 4.1 for all toric varieties. Unfortunately this equality was too good to be true. Serkan Hosten and Rekha Thomas found the following counterexample.

EXAMPLE 4.7. We shall present a projective toric variety in $\mathbb{P}^9$ which satisfies

$$16 \; = \; maxdeg(Gr_{\mathcal{A}}) \; = \; maxgen(I_{\mathcal{A}}) \quad > \quad maxdeg(\mathcal{C}_{\mathcal{A}}) = 15. \tag{4.6}$$

Our starting point is the affine toric surface $X_{\mathcal{B}} \subset \mathbb{C}^5$ defined by

$$\mathcal{B} \quad = \quad \begin{pmatrix} 1 & 3 & 4 & 6 & 0 \\ 0 & 0 & 0 & -5 & 1 \end{pmatrix}.$$

We then construct their Lawrence lifting $\mathcal{A} := \Lambda(\mathcal{B})$. In other words, $I_{\mathcal{A}}$ is the toric ideal defining the closure of $X_{\mathcal{B}}$ in $\mathbb{P}^1 \times \mathbb{P}^1 \times \mathbb{P}^1 \times \mathbb{P}^1 \times \mathbb{P}^1$. This ideal equals

$$\begin{aligned} I_{\mathcal{A}} = \langle\, & x_2 y_1^3 - x_1^3 y_2,\; x_3 y_1^4 - x_1^4 y_3,\; x_3^3 y_2^4 - x_2^4 y_3^3,\; x_4 x_5^5 y_2^2 - x_2^2 y_4 y_5^5,\; x_4 x_5^5 y_1^6 - x_1^6 y_4 y_5^5, \\ & \underline{x_4^2 x_5^{10} y_3^3 - x_3^3 y_4^2 y_5^{10}},\; x_4 x_5^5 y_1^2 y_3 - x_1^2 x_3 y_4 y_5^5,\; x_4 x_5^5 y_1^3 y_2 - x_1^3 x_2 y_4 y_5^5,\; x_3 y_1 y_2 - x_1 x_2 y_3, \\ & x_1 x_4 x_5^5 y_2 y_3 - x_2 x_3 y_1 y_4 y_5^5,\; x_2^2 x_4 x_5^5 y_3^3 - x_3^3 y_2^2 y_4 y_5^5,\; x_1 x_3^2 y_2^3 - x_2^3 y_1 y_3^2,\; x_1^2 x_3 y_2^2 - x_2^2 y_1^2 y_3, \\ & x_2 x_4 x_5^5 y_1 y_3^2 - x_1 x_3^2 y_2 y_4 y_5^5,\; x_1^2 x_4 x_5^5 y_3^2 - x_3^2 y_1^2 y_4 y_5^5,\; x_4^2 x_5^{10} y_1 y_2 y_3^2 - x_1 x_2 x_3^2 y_4^2 y_5^{10} \,\rangle. \end{aligned}$$

By Theorem 3.9, these 16 minimal generators of $I_{\mathcal{A}}$ coincide with the Graver basis $Gr_{\mathcal{A}}$. The first six binomials in this list are the circuits. Thus (4.6) holds for this example. We remark that the projective variety $Y_{\mathcal{A}} \subset \mathbb{P}^9$ defined by $I_{\mathcal{A}}$ satisfies $codim(Y_{\mathcal{A}}) = 3$, $degree(Y_{\mathcal{A}}) = 54$, and $reg(Y_{\mathcal{A}}) = 17$. □

In light of this counterexample, I wish to propose the following improvement of (4.5). Consider any circuit $C \in \mathcal{C}_{\mathcal{A}}$ and regard its support $supp(C)$ as a subset of $\mathcal{A}$. The lattice $\mathbb{Z}(supp(C))$ has finite index in the (possibly bigger) lattice $\mathbb{R}(supp(C)) \cap \mathbb{Z}\mathcal{A}$. We call this index the *index* of the circuit $C$. We define the *true degree* of the circuit $C$ to be the product $degree(C) \cdot index(C)$. It can be shown that the true degree of any circuit is bounded above by $degree(Y_{\mathcal{A}})$. The following conjecture would thus imply Conjecture 4.1 for projective toric varieties.

CONJECTURE 4.8. *The degree of any element in the Graver basis $Gr_{\mathcal{A}}$ of a toric ideal $I_{\mathcal{A}}$ is bounded above by the maximal true degree of any circuit $C \in \mathcal{C}_{\mathcal{A}}$.*

Note that Conjecture 4.8 is consistent with Example 4.7 because the underlined circuit $x_4^2 x_5^{10} y_3^3 - x_3^3 y_4^2 y_5^{10}$ has index two and hence its true degree is 30.

## References


[BS] D. Bayer and M. Stillman, *A criterion for detecting m-regularity*, Inventiones mathematicae **87** (1987), 1-11.

[BGT] W. Bruns, J. Gubeladze and N.V. Trung, *Normal polytopes, triangulations and Koszul algebras*, Journal für die Reine und Angewandte Mathematik, to appear.

[Cox] D. Cox, *Recent developments in toric geometry*, this volume.

[Dab] R. Dabrowski, *On normality of the closure of a generic torus orbit in $G/P$*, Pacific Journal of Mathematics **172** (1996), 321–330.

[EG] D. Eisenbud and S. Goto, *Linear free resolutions and minimal multiplicity*, Journal of Algebra **88** (1984), 89–133.

[Eis] D. Eisenbud, *Commutative Algebra with a View Toward Algebraic Geometry*, Springer Verlag, 1994.

[EiSt] D. Eisenbud and B. Sturmfels, *Binomial ideals*, Duke Mathematical Journal **84** (1996), 1–45.

[EwS] G. Ewald and A. Schmeinck, *Representation of the Hirzebruch-Kleinschmidt varieties by quadrics*, Beiträge zur Algebra und Geometrie **34** (1993), 151–156.

[Ful] W. Fulton, *Introduction to Toric Varieties*, Princeton University Press, 1993.

[GGMS] I.M. Gel'fand, R.M. Goresky, R.D. MacPherson and V. Serganova, *Combinatorial geometries, convex polyhedra and Schubert cells*, Advances in Math. **63** (1987), 301–316.

[GLP] L. Gruson, R. Lazarsfeld and C. Peskine, *On a theorem of Castelnuovo, and the equations defining space curves*, Inventiones mathematicae **72** (1983), 491–506.

[KKMS] G. Kempf, F. Knudsen, D. Mumford and B. Saint-Donat, *Toroidal Embeddings*, Springer Lecture Notes in Mathematics, Volume 339, 1973.

[Kho] A. Khovanskii, *The Newton polytope, the Hilbert polynomial and sums of finite sets*, Functional Analysis and its Applications **26** (1992), 276–281.

[Laz] R. Lazarsfeld, *A sharp Castelnuovo bound for smooth surfaces*, Duke Mathematical Journal **55** (1987), 423–429.

[Mor] J. Morand, *Closures of torus orbits in adjoint representations of semisimple groups*, Manuscript, Universität Basel (1995).

[Oda] T. Oda, *Convex Bodies and Algebraic Geometry: an Introduction to the Theory of Toric Varieties*, Springer Verlag, New York, 1988.

[PS] I. Peeva and B. Sturmfels, *Syzygies of codimension 2 lattice ideals*, Manuscript, UC Berkeley (1996).

[Ran] Z. Ran, *Local differential geometry and generic projections of threefolds*, Journal of Differential Geometry **32** (1990), 131–137.

[Stu] B. Sturmfels, *Gröbner Bases and Convex Polytopes*, American Mathematical Society, University Lecture Series, Vol. 8, Providence, RI, 1995.

[Wh1] N. White, *The basis monomial ring of a matroid*, Advances in Math. **24** (1977), 292–297.

[Wh2] N. White, *A unique exchange property for bases*, Linear Algebra and its Applications **31** (1980), 81-91.


DEPARTMENT OF MATHEMATICS, UNIVERSITY OF CALIFORNIA, BERKELEY, CA 94720
*E-mail address*: `bernd@math.berkeley.edu`

## Selected Titles in This Series

(*Continued from the front of this publication*)

38 **Richard V. Kadison, Editor,** Operator algebras and applications (Queens University, Kingston, Ontario, July/August 1980)

37 **Bruce Cooperstein and Geoffrey Mason, Editors,** The Santa Cruz conference on finite groups (University of California, Santa Cruz, June/July 1979)

36 **Robert Osserman and Alan Weinstein, Editors,** Geometry of the Laplace operator (University of Hawaii, Honolulu, March 1979)

35 **Guido Weiss and Stephen Wainger, Editors,** Harmonic analysis in Euclidean spaces (Williams College, Williamstown, Massachusetts, July 1978)

34 **D. K. Ray-Chaudhuri, Editor,** Relations between combinatorics and other parts of mathematics (Ohio State University, Columbus, March 1978)

33 **A. Borel and W. Casselman, Editors,** Automorphic forms, representations and $L$-functions (Oregon State University, Corvallis, July/August 1977)

32 **R. James Milgram, Editor,** Algebraic and geometric topology (Stanford University, Stanford, California, August 1976)

31 **Joseph L. Doob, Editor,** Probability (University of Illinois at Urbana-Champaign, Urbana, March 1976)

30 **R. O. Wells, Jr., Editor,** Several complex variables (Williams College, Williamstown, Massachusetts, July/August 1975)

29 **Robin Hartshorne, Editor,** Algebraic geometry – Arcata 1974 (Humboldt State University, Arcata, California, July/August 1974)

28 **Felix E. Browder, Editor,** Mathematical developments arising from Hilbert problems (Northern Illinois University, Dekalb, May 1974)

27 **S. S. Chern and R. Osserman, Editors,** Differential geometry (Stanford University, Stanford, California, July/August 1973)

26 **Calvin C. Moore, Editor,** Harmonic analysis on homogeneous spaces (Williams College, Williamstown, Massachusetts, July/August 1972)

25 **Leon Henkin, John Addison, C. C. Chang, William Craig, Dana Scott, and Robert Vaught, Editors,** Proceedings of the Tarski symposium (University of California, Berkeley, June 1971)

24 **Harold G. Diamond, Editor,** Analytic number theory (St. Louis University, St. Louis, Missouri, March 1972)

23 **D. C. Spencer, Editor,** Partial differential equations (University of California, Berkeley, August 1971)

22 **Arunas Liulevicius, Editor,** Algebraic topology (University of Wisconsin, Madison, June/July 1970)

21 **Irving Reiner, Editor,** Representation theory of finite groups and related topics (University of Wisconsin, Madison, April 1970)

20 **Donald J. Lewis, Editor,** 1969 Number theory institute (State University of New York at Stony Brook, Stony Brook, July 1969)

19 **Theodore S. Motzkin, Editor,** Combinatorics (University of California, Los Angeles, March 1968)

18 **Felix Browder, Editor,** Nonlinear operators and nonlinear equations of evolution in Banach spaces (Chicago, April 1968)

17 **Alex Heller, Editor,** Applications of categorical algebra (New York City, April 1968)

16 **Shiing-Shen Chern and Stephen Smale, Editors,** Global analysis, Part III (University of California, Berkeley, July 1968)

15 **Shiing-Shen Chern and Stephen Smale, Editors,** Global analysis, Part II (University of California, Berkeley, July 1968)

14 **Shiing-Shen Chern and Stephen Smale, Editors,** Global analysis, Part I (University of California, Berkeley, July 1968)

13 **Dana S. Scott (Part 1) and Thomas J. Jech (Part 2), Editors,** Axiomatic set theory (University of California, Los Angeles, July/August 1967)

(See the AMS catalog for earlier titles)

ISBN 0-8218-0895-8

9 780821 808955